Collins

White Rose Maths

AQA GCSE 9-1

Higher

Student Book 2

Caroline Hamilton and Ian Davies

William Collins' dream of knowledge for all began with the publication of his first book in 1819.
A self-educated mill worker, he not only enriched millions of lives, but also founded a flourishing publishing house. Today, staying true to this spirit, Collins books are packed with inspiration, innovation and practical expertise.
They place you at the centre of a world of possibility and give you exactly what you need to explore it.

Collins. Freedom to teach.

Published by Collins
An imprint of HarperCollins*Publishers*
The News Building
1 London Bridge Street
London
SE1 9GF

HarperCollins*Publishers*
Macken House
39/40 Mayor Street Upper
Dublin 1
D01 C9W8
Ireland

Browse the complete Collins catalogue at www.collins.co.uk

10 9 8 7 6 5 4 3 2 1

ISBN: 978-0-00-866960-7

British Library Cataloguing-in-Publication Data
A catalogue record for this publication is available from the British Library.

Series editors: Ian Davies and Caroline Hamilton
Authors: Matthew Ainscough, Rob Clasper, Rhiannon Davies and Sahar Shillabeer
Publisher: Katie Sergeant
Product manager: Richard Toms
Development editor: Karl Warsi
Editorial: Richard Toms, Amanda Dickson and Deborah Dobson
Proofreading and answer checking: Eric Pradel, Steven Matchett, Marie Taylor, Trevor Senior and Anna Cox
Cover designer: Sarah Duxbury
Typesetter: Jouve India Private Limited
Production controller: Alhady Ali
Printed and bound in India

This book contains FSC™ certified paper and other controlled sources to ensure responsible forest management.

For more information visit: www.harpercollins.co.uk/green

Text acknowledgements
The publishers gratefully acknowledge the permission granted to reproduce the copyright material in this book. Every effort has been made to trace copyright holders and to obtain their permission for the use of copyright material.

Contents

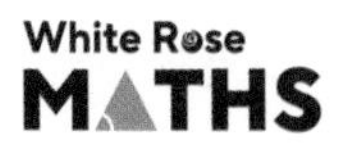

Contents

Introduction

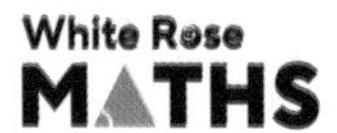

How to use this book

Welcome to the **Collins White Rose Maths AQA GCSE 9–1 Higher tier** course.

There are two Student Books in the series:

- **Student Book 1** covers Number, Algebra, and Ratio, proportion and rates of change.
- **Student Book 2** covers Geometry and measures, Probability, and Statistics.

Sometimes you will need some knowledge of a different area of mathematics within the topic you are studying. For example, you may need to set up and solve an algebraic equation when solving a geometry problem. You will often be able to use your earlier knowledge and skills from Key Stage 3 to help you do this.

Here is a short guide to how to get the most out of this book. We hope you enjoy continuing your learning journey.

Caroline Hamilton and Ian Davies, series editors

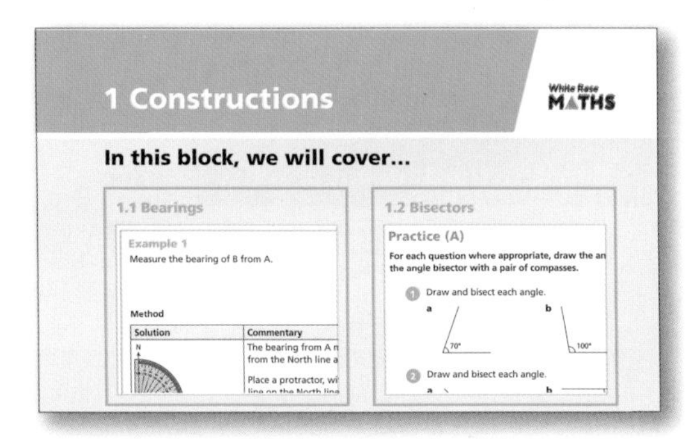
1 Constructions

In this block, we will cover...

1.1 Bearings

Example 1

Measure the bearing of B from A.

Method

Solution	Commentary
N	The bearing from A ... from the North line ... Place a protractor, ...

1.2 Bisectors

Practice (A)

For each question where appropriate, draw the angle bisector with a pair of compasses.

1 Draw and bisect each angle.

2 Draw and bisect each angle.

Block overviews Each block of related chapters starts with a visual introduction to the key concepts and learning you will encounter.

Are you ready? Before you start each part of a chapter, remind yourself of the maths you should already know with these questions. If you need more practice, refer to the *Collins White Rose Maths Key Stage 3* course or the series of *Collins White Rose Maths AQA GCSE 9–1 Foundation Student Books*.

Explanatory text Key words and concepts are explained before moving on to worked examples.

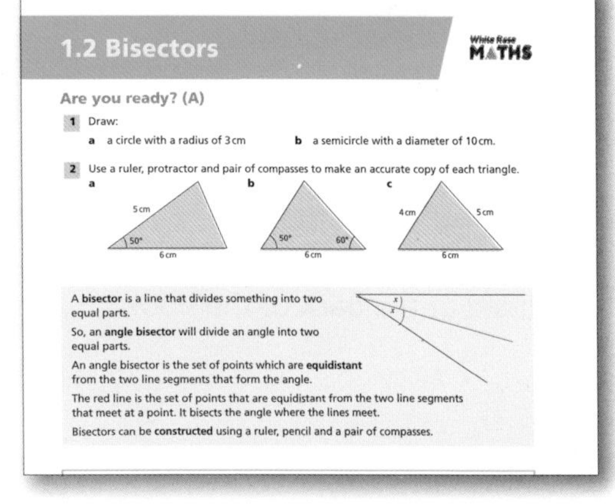
1.2 Bisectors

Are you ready? (A)

1 Draw:

a a circle with a radius of 3 cm   b a semicircle with a diameter of 10 cm.

2 Use a ruler, protractor and pair of compasses to make an accurate copy of each triangle.

A **bisector** is a line that divides something into two equal parts.

So, an **angle bisector** will divide an angle into two equal parts.

An angle bisector is the set of points which are **equidistant** from the two line segments that form the angle.

The red line is the set of points that are equidistant from the two line segments that meet at a point. It bisects the angle where the lines meet.

Bisectors can be **constructed** using a ruler, pencil and a pair of compasses.

Using your calculator

In this chapter, you will learn about trigonometric ratios, which are very useful in solving problems in geometry. The values of these ratios are stored on your scientific calculator, and can be accessed by using the keys 'sin', 'cos' and 'tan'.

As you work through the chapter, make sure you know how to use your calculator to calculate with these ratios.

Using your calculator Where appropriate, you are given advice on how to use the features of your calculator to find or check answers. Not all calculators work in the same way, so make sure you know how your model works.

Worked examples Learn how to approach different types of questions with worked examples that clearly walk you through the process of answering. Visual representations are provided to help when necessary.

Example

B is on a bearing of 070° from A.

The distance between A and B is 8 cm.

Draw a diagram to represent the information.

Method

Solution	Commentary
N	Mark a point labelled A and draw a North line.

Introduction

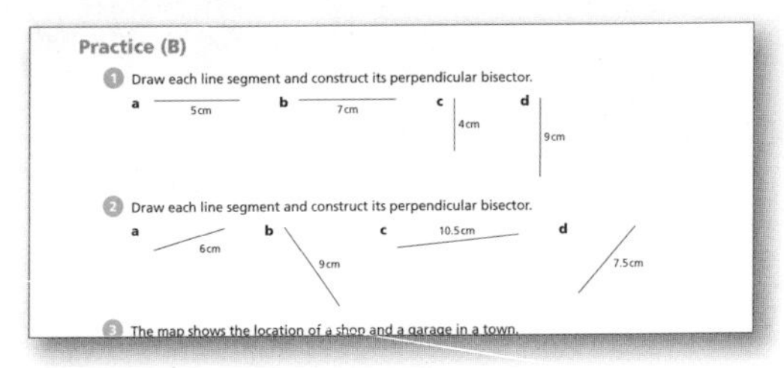

Practice (B)

1. Draw each line segment and construct its perpendicular bisector.
   a 5cm b 7cm c 4cm d 9cm
2. Draw each line segment and construct its perpendicular bisector.
   a 6cm b 9cm c 10.5cm d 7.5cm
3. The map shows the location of a shop and a garage in a town.

Practice Put what you have just learned into practice. Sometimes symbols are used in questions or whole sections to show when you should, or should not, use a calculator. If there is no symbol, the question or section can be approached in either way.

Many of the Practice sections conclude with a **What do you think?** exercise to encourage further exploration.

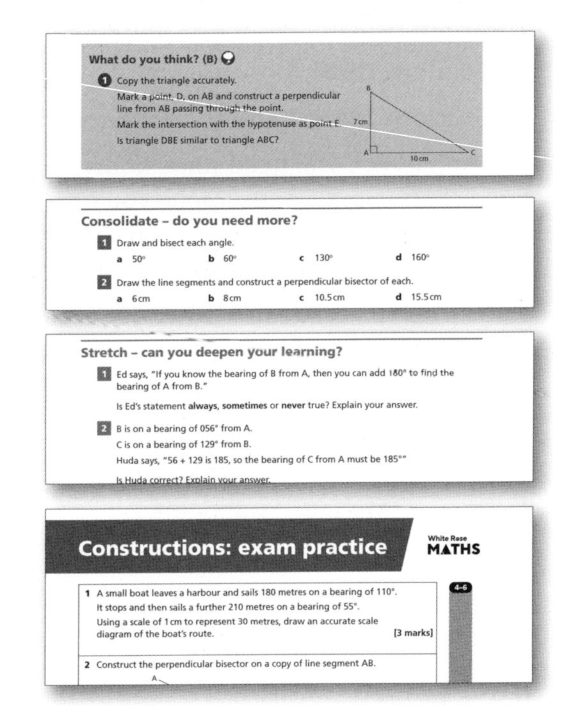

What do you think? (B)

1. Copy the triangle accurately.
   Mark a point, D, on AB and construct a perpendicular line from AB passing through the point.
   Mark the intersection with the hypotenuse as point E.
   Is triangle DBE similar to triangle ABC?

Consolidate – do you need more?

1. Draw and bisect each angle.
   a 50° b 60° c 130° d 160°
2. Draw the line segments and construct a perpendicular bisector of each.
   a 6cm b 8cm c 10.5cm d 15.5cm

Stretch – can you deepen your learning?

1. Ed says, "If you know the bearing of B from A, then you can add 180° to find the bearing of A from B."
   Is Ed's statement **always**, **sometimes** or **never** true? Explain your answer.
2. B is on a bearing of 056° from A.
   C is on a bearing of 129° from B.
   Huda says, "56 + 129 is 185, so the bearing of C from A must be 185°"
   Is Huda correct? Explain your answer.

Constructions: exam practice

White Rose MATHS

4-6

1. A small boat leaves a harbour and sails 180 metres on a bearing of 110°. It stops and then sails a further 210 metres on a bearing of 55°. Using a scale of 1cm to represent 30 metres, draw an accurate scale diagram of the boat's route. [3 marks]
2. Construct the perpendicular bisector on a copy of line segment AB.

Consolidate Reinforce what you have learned in the chapter with additional practice questions.

Stretch Take your learning further and challenge yourself to apply it in new ways or different areas of maths.

Exam practice At the end of each block, you will find exam-style questions to practise your learning. These are organised into grade bands of 4–6 and 7–9. There is extra practice at the end of the three main parts of the book.

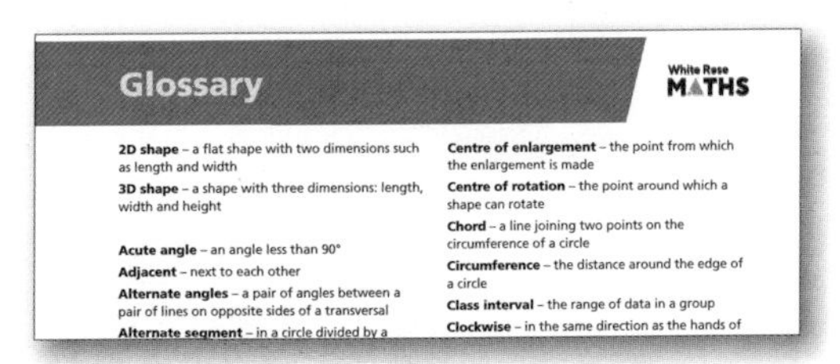

Glossary

White Rose MATHS

2D shape – a flat shape with two dimensions such as length and width
3D shape – a shape with three dimensions: length, width and height
Acute angle – an angle less than 90°
Adjacent – next to each other
Alternate angles – a pair of angles between a pair of lines on opposite sides of a transversal
Alternate segment – in a circle divided by a
Centre of enlargement – the point from which the enlargement is made
Centre of rotation – the point around which a shape can rotate
Chord – a line joining two points on the circumference of a circle
Circumference – the distance around the edge of a circle
Class interval – the range of data in a group
Clockwise – in the same direction as the hands of

Glossary Look up the meanings of any key words and phrases you are not sure about.

Answers Check your work using the answers provided at the back of the book.

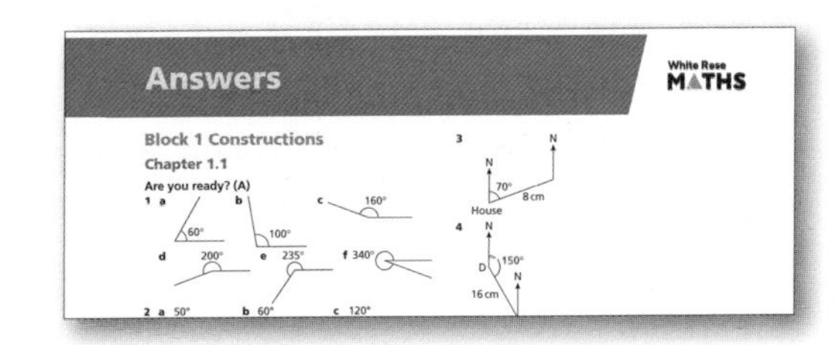

Answers

White Rose MATHS

Block 1 Constructions
Chapter 1.1
Are you ready? (A)

1 Constructions

In this block, we will cover...

1.1 Bearings

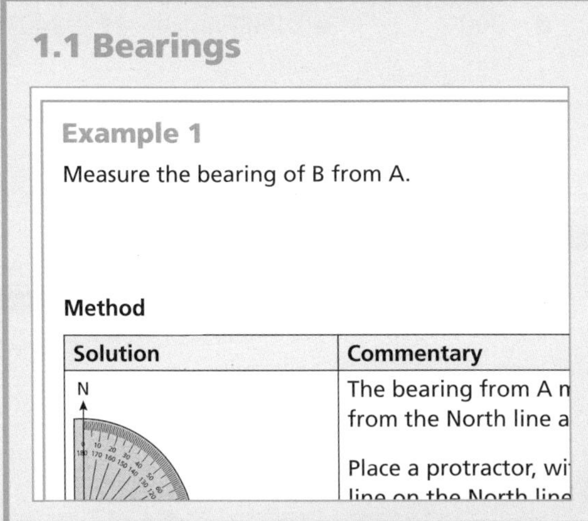

Example 1
Measure the bearing of B from A.
Method
Solution
Commentary
N
The bearing from A m
from the North line a
Place a protractor, wi

1.2 Bisectors

Practice (A)

For each question where appropriate, draw the an the angle bisector with a pair of compasses.

1 Draw and bisect each angle.

a 70°

b 100°

2 Draw and bisect each angle.

1.3 Perpendiculars to and from a line

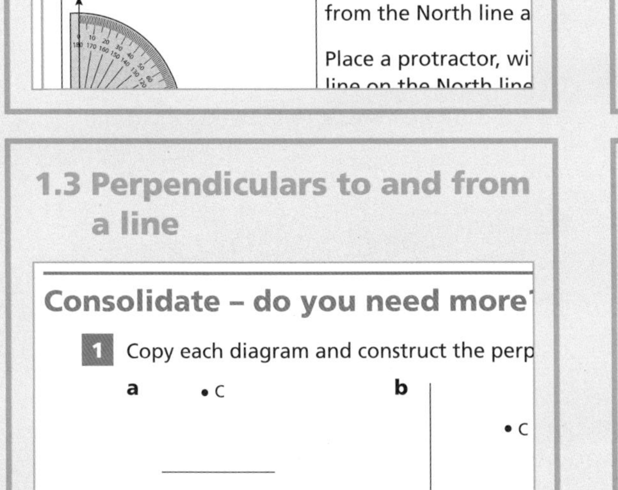

1.4 Loci

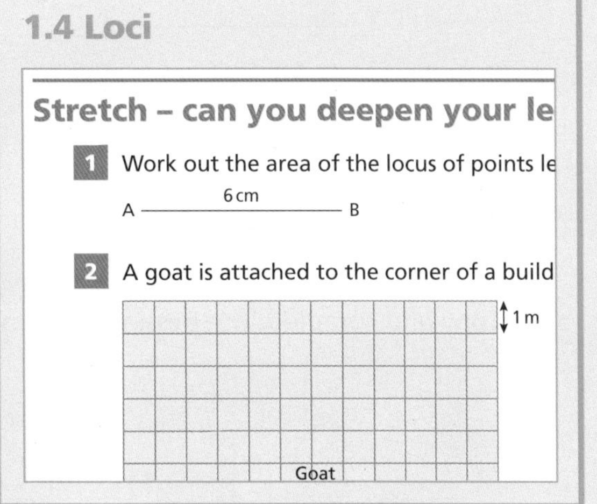

Stretch – can you deepen your le
1 Work out the area of the locus of points le
6 cm
A
B
2 A goat is attached to the corner of a build
1 m
Goat

1.1 Bearings

Are you ready? (A)

1 Use a ruler and protractor to draw each of these angles.

a 60° **b** 100° **c** 160° **d** 200° **e** 235° **f** 340°

2 Measure the angles.

a

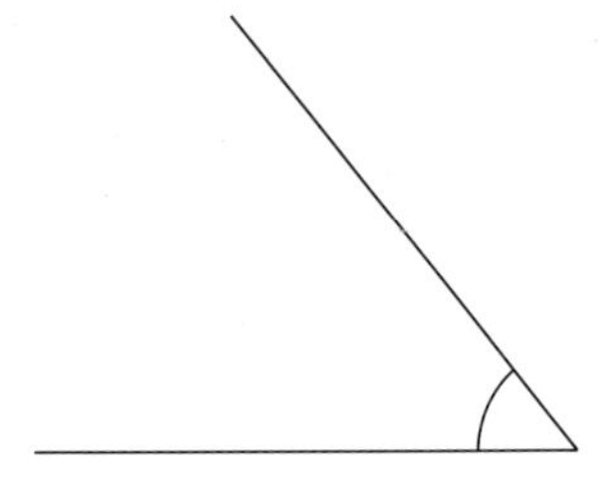

b

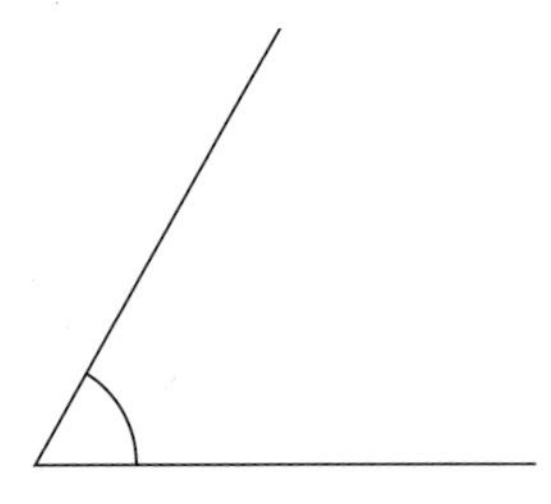

c

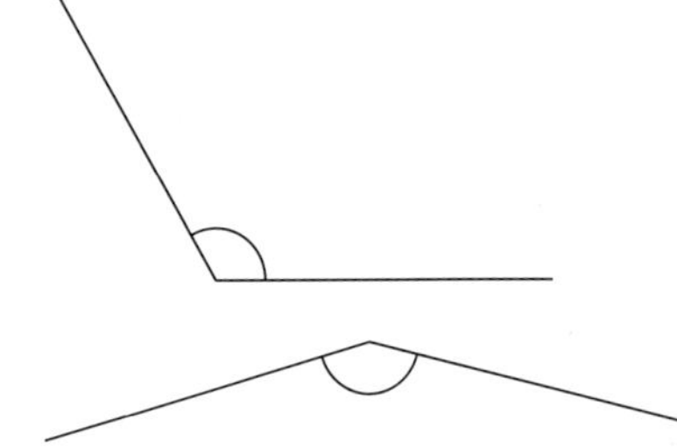

d

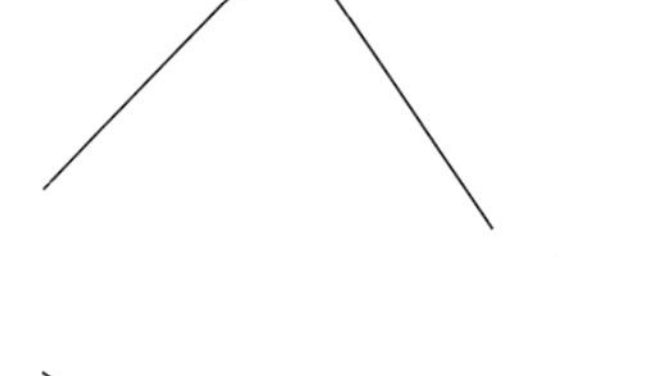

e

f

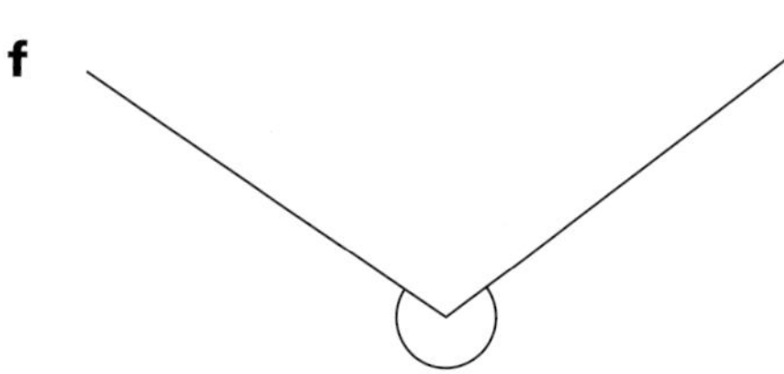

3 Copy and complete the missing compass points.

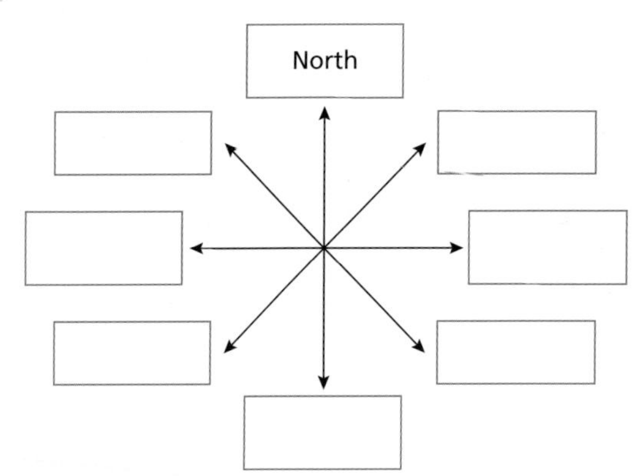

A **bearing** is an angle in degrees, measured clockwise from North. Bearings are written with three figures so you should include a leading zero when the bearing is less than 100°, e.g. 050°

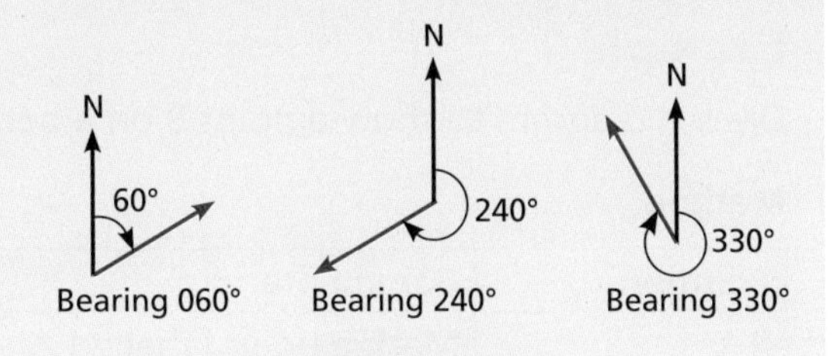

When asked to find the bearing of B from A, the 'from A' part tells you where to place your protractor.

Sometimes bearings questions are not drawn to scale, and you will need to use your knowledge of angles in parallel lines to solve problems.

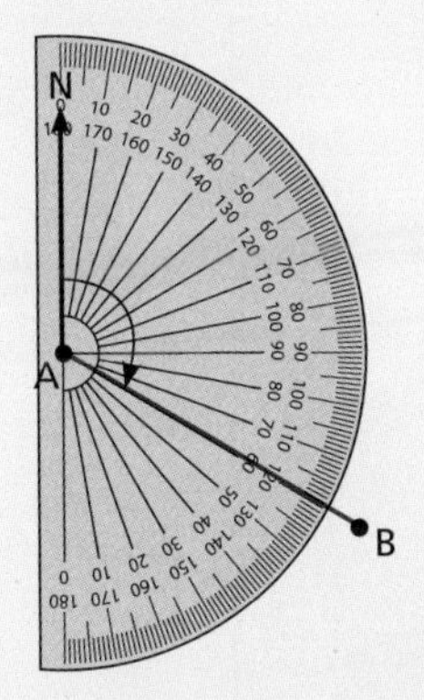

Angles in parallel lines are covered in Chapter 3.1

Example 1

Measure the bearing of B from A.

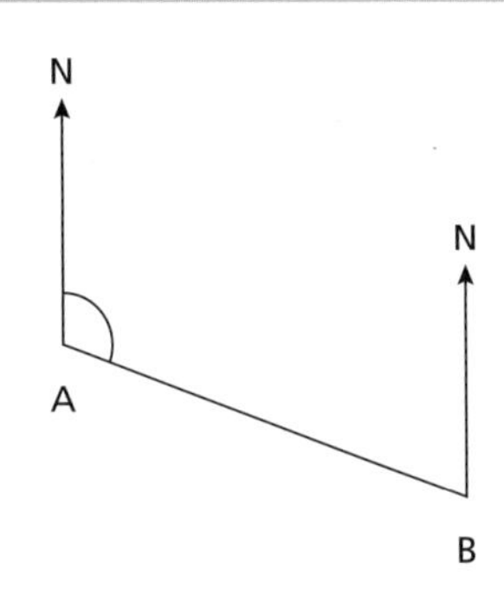

Method

Solution	**Commentary**
N N A B 110°	The bearing from A means you need to measure clockwise from the North line at A. Place a protractor, with the centre on the point A and the 0° line on the North line from A. Measure clockwise from North. The line segment from A to B meets the protractor at 70° and 110°. As the angle is obtuse, it must be 110°.

Example 2

Draw a diagram to show a point B on a bearing of 060° from A.

Method

Solution	Commentary
 N A	Mark a point labelled A and draw a North line.
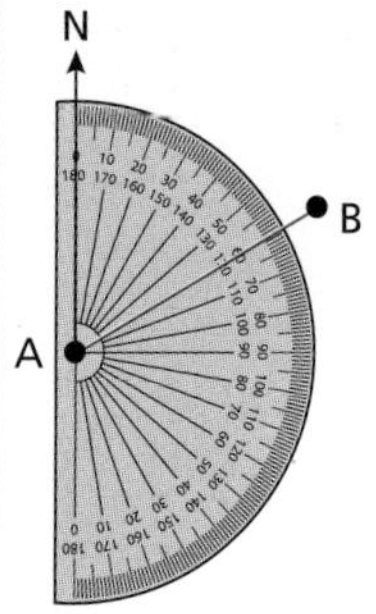 N B A	Place a protractor, using the point A as the centre. Measure and mark a point at 60°. Label the point B.
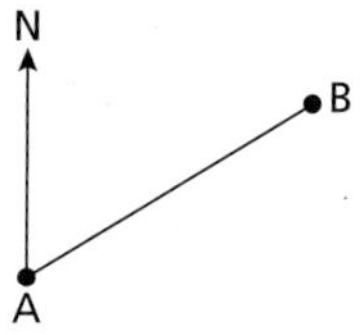 N B A	Draw a line from A to the point B.

Example 3

Draw a diagram to show point Q on a bearing of 240° from P.

Method

Solution	Commentary
 N P	Mark a point labelled P and draw a North line.
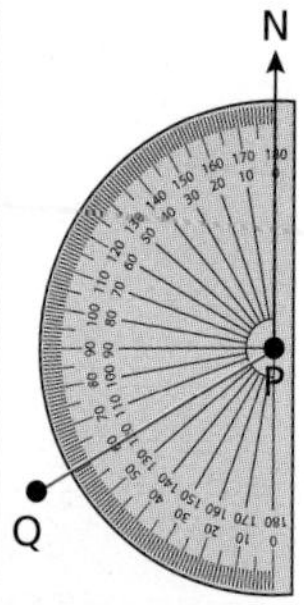 N P Q	Work out the obtuse angle from North by calculating. 360° – 240° = 120° Place a protractor, using the point P as the centre. Measure and mark a point at 120° anticlockwise from the North line drawn. Label the point Q.

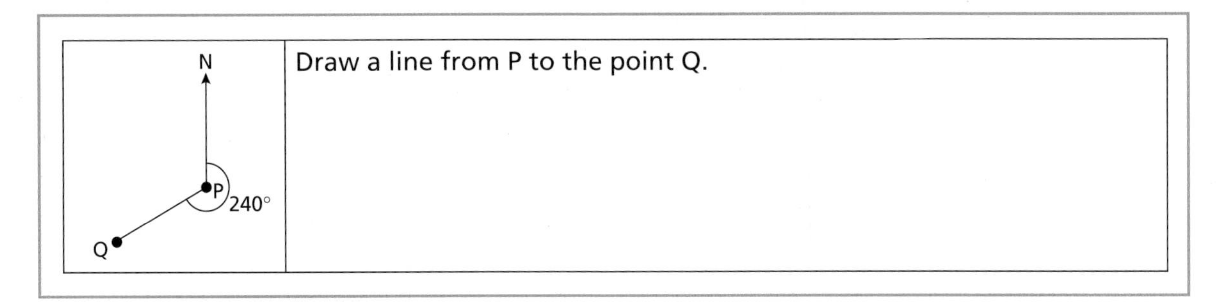

Draw a line from P to the point Q.

Practice (A)

1. Measure the bearing of B from A in each diagram.

a

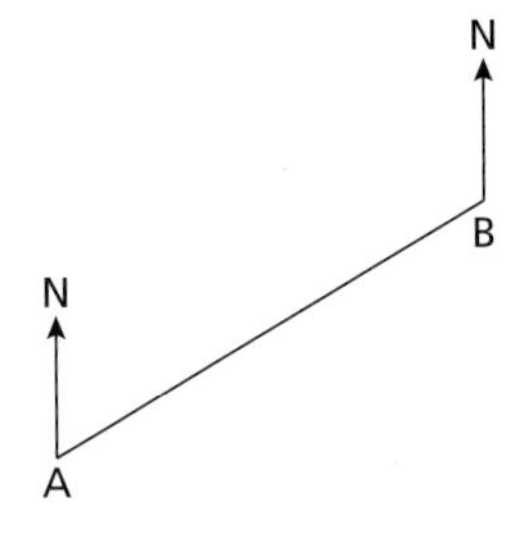

b

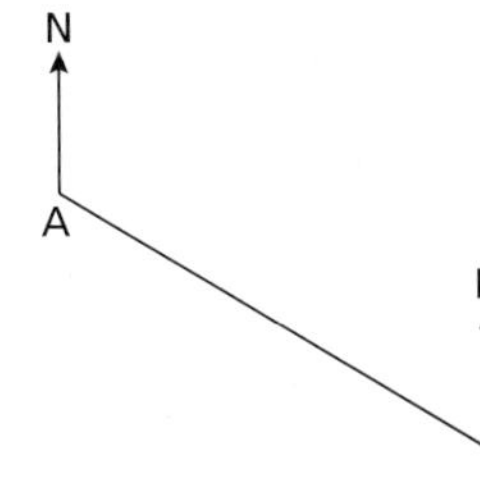

c

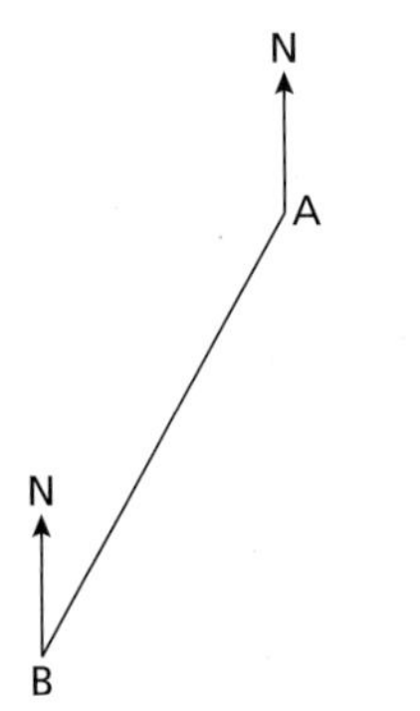

d

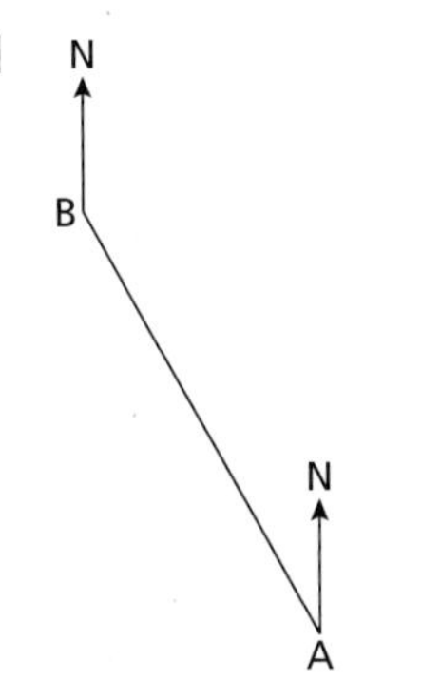

2. Measure the bearing of Q from P in each diagram.

a

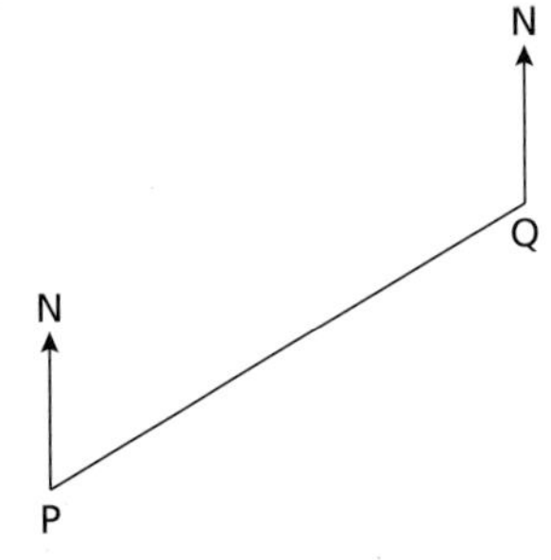

b

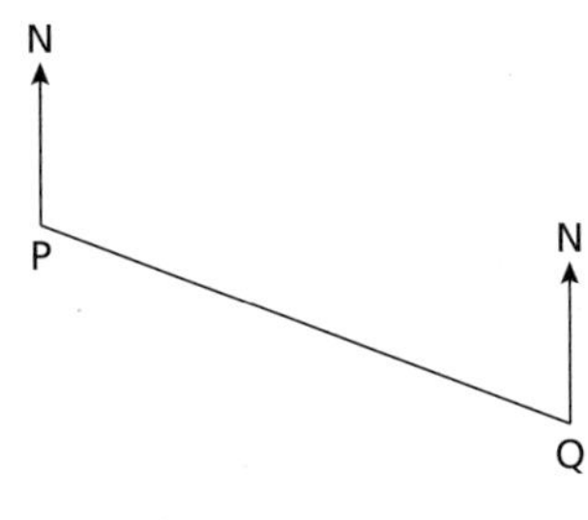

c

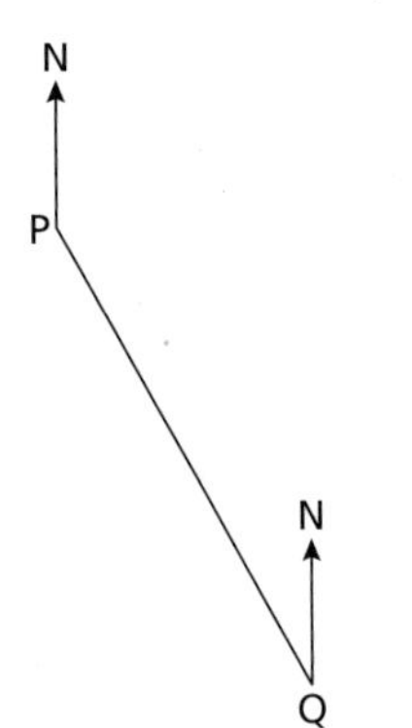

d

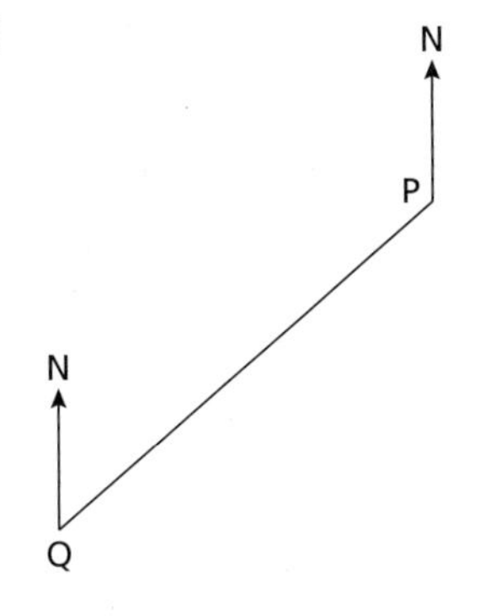

3. Draw diagrams to represent each description.

   a. The bearing of B from A is 070°
   b. The bearing of D from C is 100°
   c. The bearing of Q from P is 200°
   d. The bearing of M from L is 300°

4. Without measuring, estimate the bearing of B from A in each diagram.

a

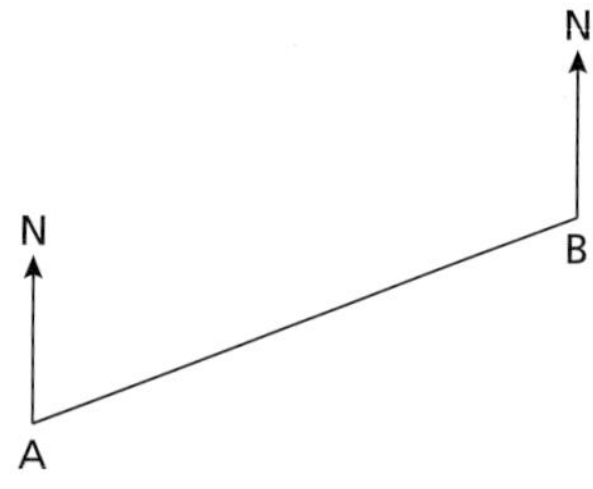

b

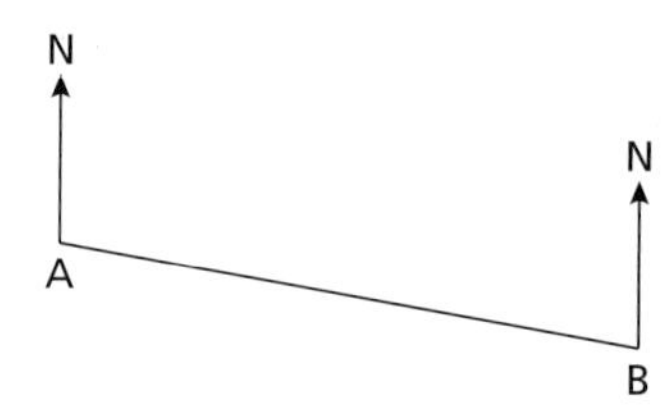

c

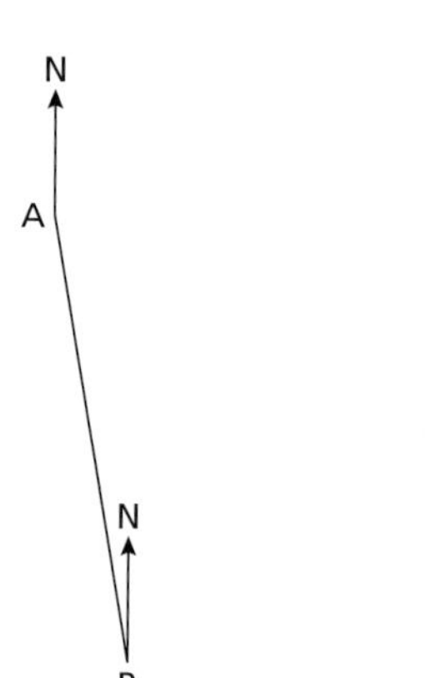

d

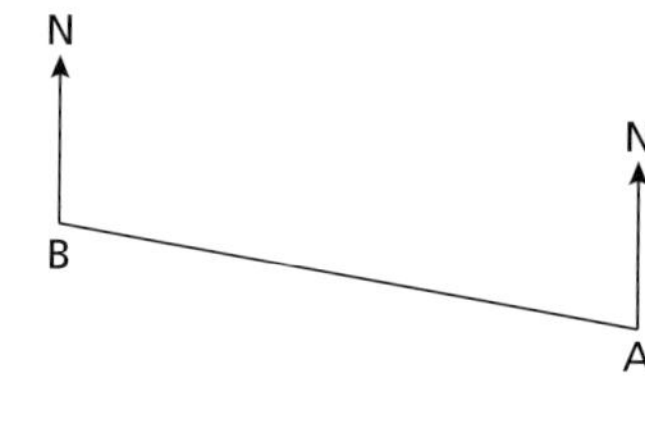

Check how close your answers are by measuring the bearings.

What do you think?

1. Here is a parallelogram.

   Which statements are **true**?

   A The bearing of B from A is the same as the bearing of C from D.

   B The bearing of C from B is the same as the bearing of D from A.

   C The bearing of C from A is the same as the bearing of D from B.

2. The diagram shows that the bearing of B from A is 070°.

   Work out the bearing of A from B.

   Draw a diagram to check your answer.

You will see bearings again in Block 6 (Right-angled triangles) and Block 7 (Non-right-angled triangles).

Are you ready? (B)

1 Use a ruler and protractor to draw line segments of the given lengths.

a 5 cm **b** 8 cm **c** 7.5 cm **d** 10.3 cm

2 Draw the bearings.

The bearing of B from A is: **a** 050° **b** 100° **c** 215°

3 The diagram shows points A and B on a map.

The scale of the map is 1 cm represents 2 km.

a Work out the actual distance between points A and B.

b The actual distance between two towns is 16 km.

Work out the distance between them on the map.

Example

B is on a bearing of 070° from A.

The distance between A and B is 8 cm.

Draw a diagram to represent the information.

Method

Solution	Commentary
N A	Mark a point labelled A and draw a North line.
N A	Place a protractor, using the point A as the centre. Measure and mark a point at 70°.
N B A	Line the point A and the new point up with a ruler and draw an 8 cm line. Mark a point at the end of the line and label it as B. N B A

Practice (B)

1. Draw diagrams to represent each statement.

   a B is on a bearing of 080° from A.

   The distance between A and B is 6 cm.

   b Q is on a bearing of 160° from P.

   The distance between P and Q is 10 cm.

   c M is on a bearing of 200° from L.

   The distance between L and M is 8 cm.

2. The map shows the locations of a church and a museum.

   a Measure the bearing of the museum from the church.

   b Measure the bearing of the church from the museum.

   On the map, 1 cm represents 5 metres.

   c What is the actual distance between the church and the museum in metres?

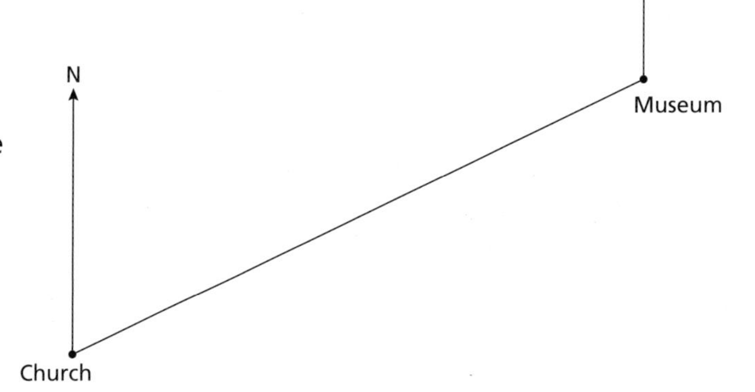

3. Mario cycles 16 miles on a bearing of 070° from his house along a straight road.

   Draw a scale diagram of Mario's route, using 1 cm to represent 2 miles.

4. Denholme and Queensbury are 8 km apart.

   Queensbury is on a bearing of 150° from Denholme.

   Draw a scale diagram of the location of Denholme and Queensbury, using 1 cm to represent 0.5 km.

5. Kath travels from home on a bearing of 100° for 10 km.

   She then travels due East for 8 km.

   Draw a scale diagram of Kath's journey, using 1 cm to represent 2 km.

6. The point B is on a bearing of 080° from A.

   The point C is on a bearing of 160° from B.

   Draw a diagram showing the points A, B and C.

7. Jackson travels 10 km on a bearing of 100° from his house.

   He then travels 8 km on a bearing of 040° to reach his office.

   a Draw a scale diagram of Jackson's journey, using 1 cm to represent 2 km.

   b Measure the bearing of Jackson's office from his house.

Consolidate – do you need more?

1. Draw the bearing of B from A, when the bearing is:

   a 100° **b** 120° **c** 150° **d** 200° **e** 250°

2. Draw diagrams to represent each statement.

   a B is on a bearing of 100° from A.
   The distance between A and B is 10 cm.

   b B is on a bearing of 120° from A.
   The distance between A and B is 10 cm.

   c B is on a bearing of 140° from A.
   The distance between A and B is 6 cm.

   d B is on a bearing of 170° from A.
   The distance between A and B is 6 cm.

   e B is on a bearing of 340° from A.
   The distance between A and B is 12 cm.

Stretch – can you deepen your learning?

1. Ed says, "If you know the bearing of B from A, then you can add 180° to find the bearing of A from B."

   Is Ed's statement **always**, **sometimes** or **never** true? Explain your answer.

2. B is on a bearing of 056° from A.

   C is on a bearing of 129° from B.

   Huda says, "56 + 129 is 185, so the bearing of C from A must be 185°"

   Is Huda correct? Explain your answer.

1.2 Bisectors

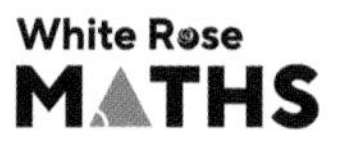

Are you ready? (A)

1 Draw:

a a circle with a radius of 3 cm

b a semicircle with a diameter of 10 cm.

2 Use a ruler, protractor and pair of compasses to make an accurate copy of each triangle.

a

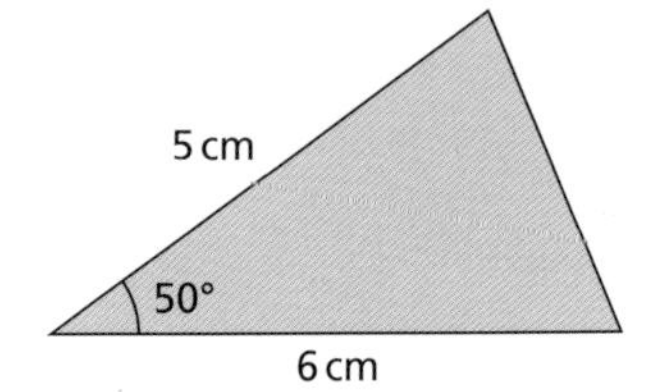

b

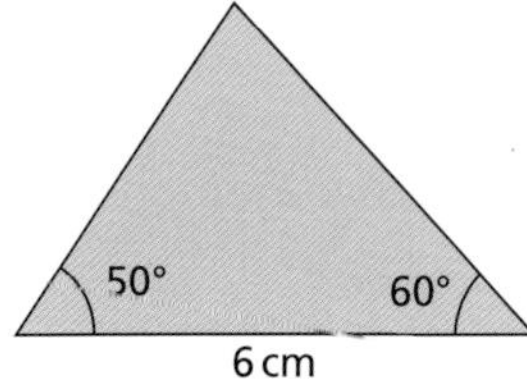

c

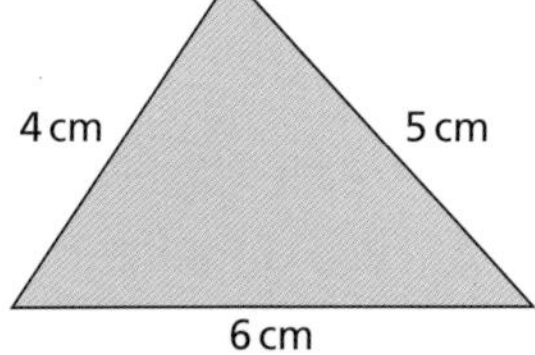

A **bisector** is a line that divides something into two equal parts.

So, an **angle bisector** will divide an angle into two equal parts.

An angle bisector is the set of points which are **equidistant** from the two line segments that form the angle.

The red line is the set of points that are equidistant from the two line segments that meet at a point. It bisects the angle where the lines meet.

Bisectors can be **constructed** using a ruler, pencil and a pair of compasses.

Example

Draw an angle of 50° and construct its angle bisector.

Method

Solution	Commentary
	Start by drawing an angle of 50°. Make the arms of the angle about 5 cm long.
	Use a pair of compasses to draw an arc of radius about 4 cm from the vertex of the angle that cuts both arms of the angle. Label these points of intersection A and B.

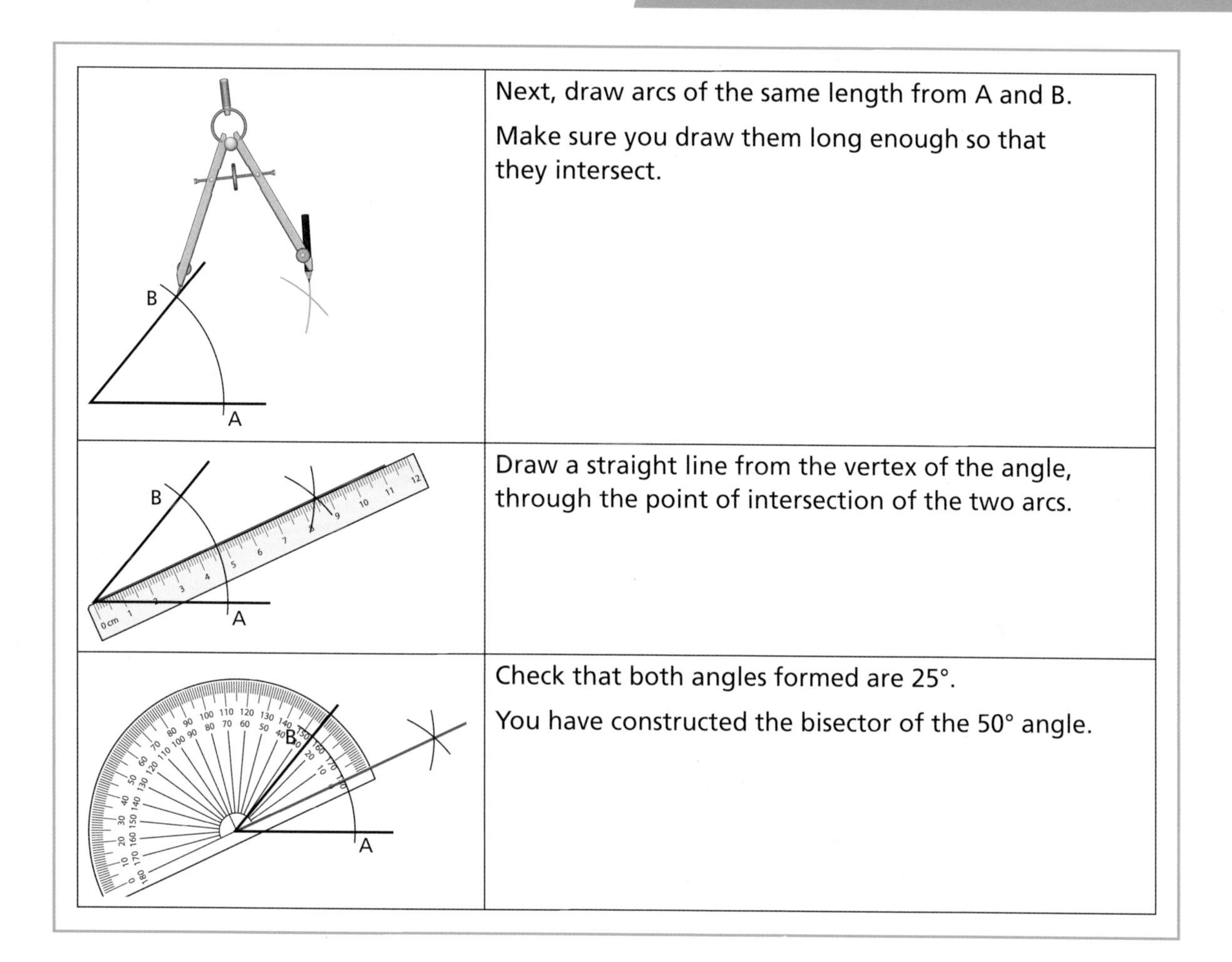

	Next, draw arcs of the same length from A and B. Make sure you draw them long enough so that they intersect.
	Draw a straight line from the vertex of the angle, through the point of intersection of the two arcs.
	Check that both angles formed are 25°. You have constructed the bisector of the 50° angle.

Practice (A)

For each question where appropriate, draw the angle with a protractor and then construct the angle bisector with a pair of compasses.

1. Draw and bisect each angle.

a 70°

b 100°

c 150°

2. Draw and bisect each angle.

a

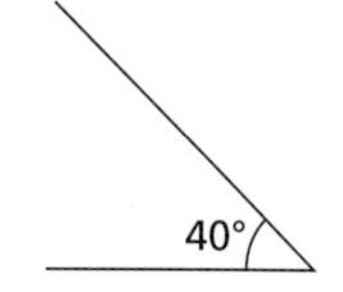

b

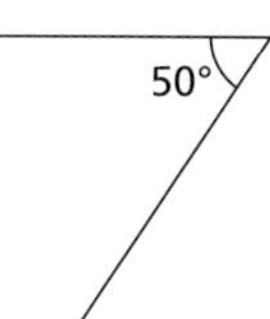

c

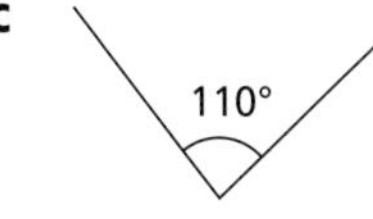

3 Copy the diagram and construct the angle bisector of AB and BC.

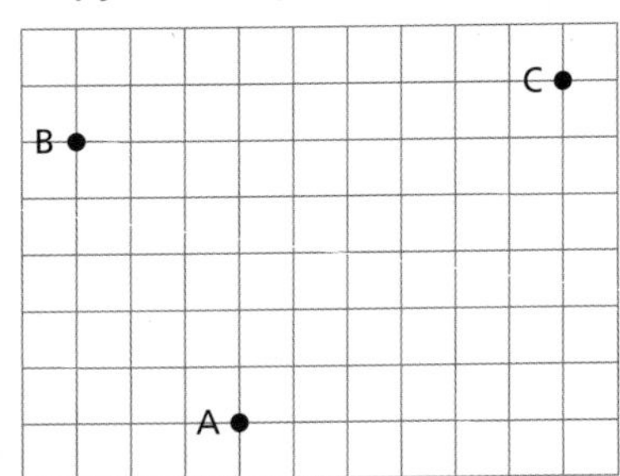

4 Draw each diagram accurately and bisect the angle ABC.

a

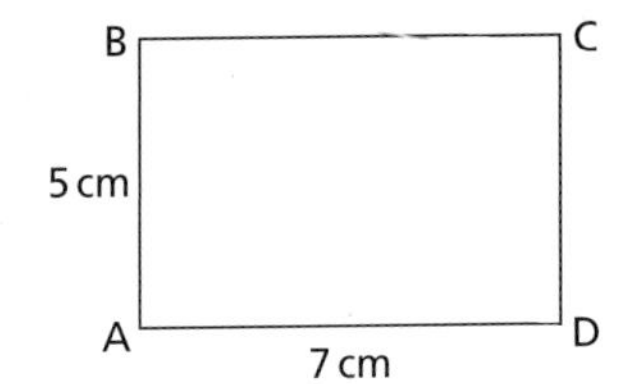

b

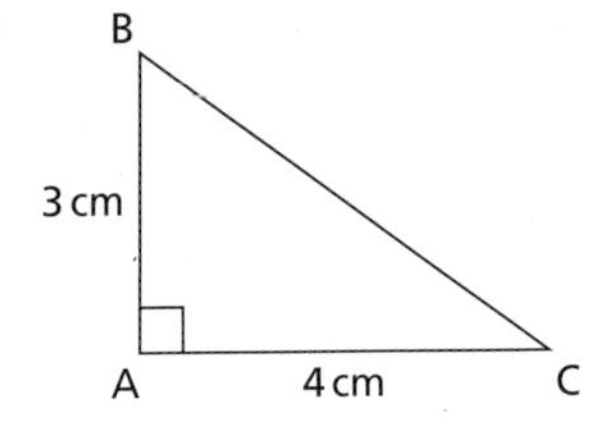

c

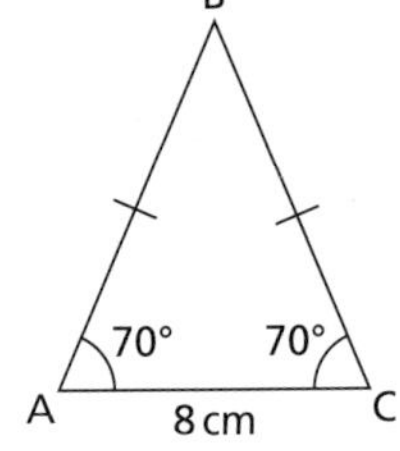

5 The plan view of a garden is shown.

a Draw a scale diagram of the garden using 1 cm to represent 1 m.

b Show all the points equidistant from the house and the fence.

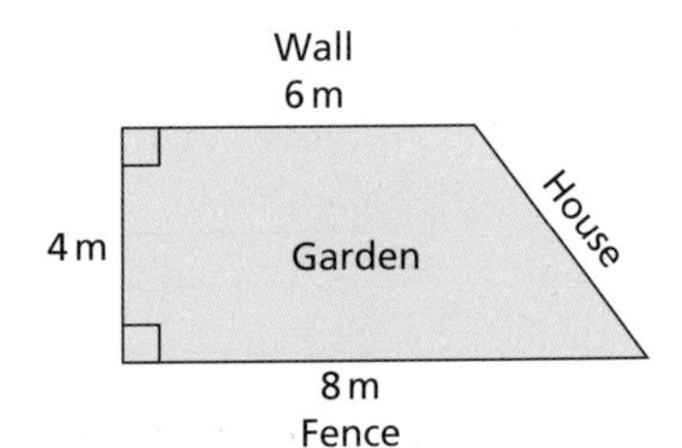

What do you think? (A)

1 Seb has drawn triangle ABC.

He thinks that the bisector of angle ABC will bisect the line AC.

Is he correct?

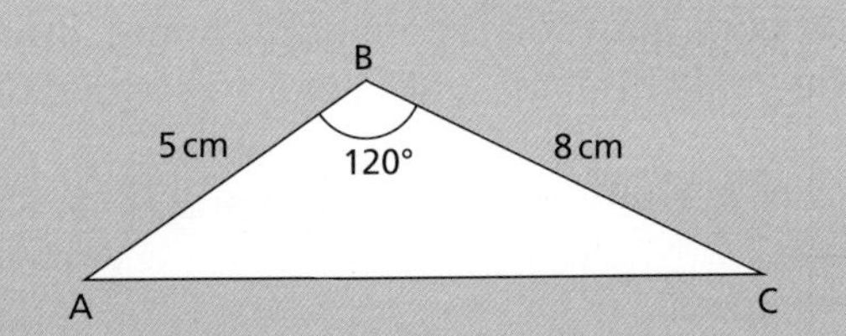

2 Use a ruler and a protractor to draw the triangle accurately. Then draw the angle bisector for each of the three angles.

Using the intersection of the three angle bisectors as the centre, draw the circle that touches all three sides of the triangle.

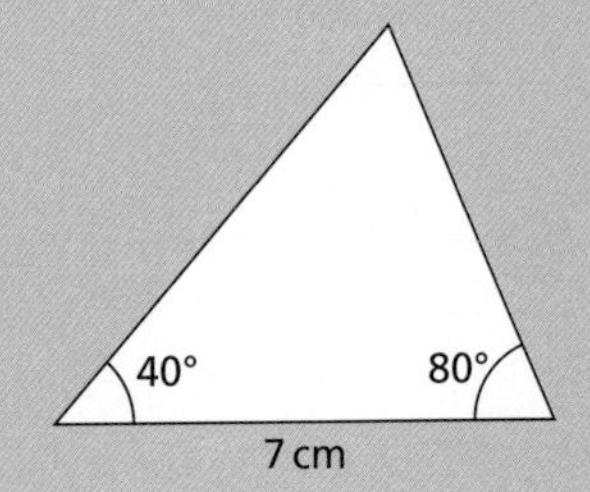

This is called an incircle.

Investigate with other triangles.

Are you ready? (B)

1 Copy each line, then draw another line perpendicular to each one.

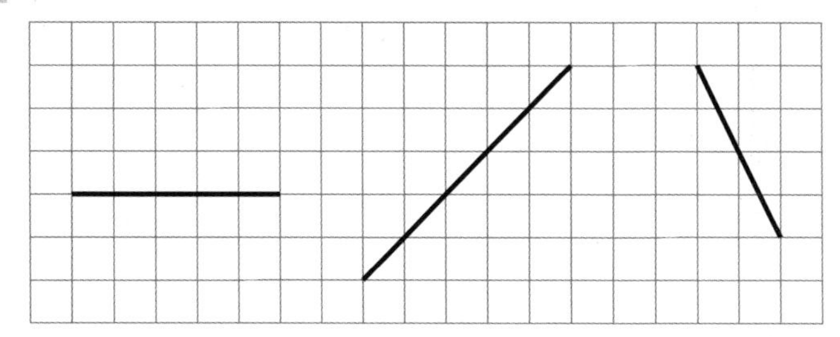

2 Draw a right-angled triangle with a base of 5 cm and a perpendicular height of 3 cm.

A **perpendicular bisector** is a line that divides a line segment into two equal parts and is perpendicular to the original line.

All of the points on a perpendicular bisector are equidistant from the end points of the line segment that it bisects.

Here, the perpendicular bisector of the line segment AB has been constructed.

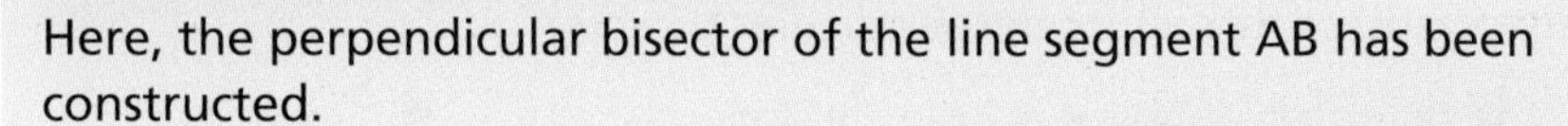

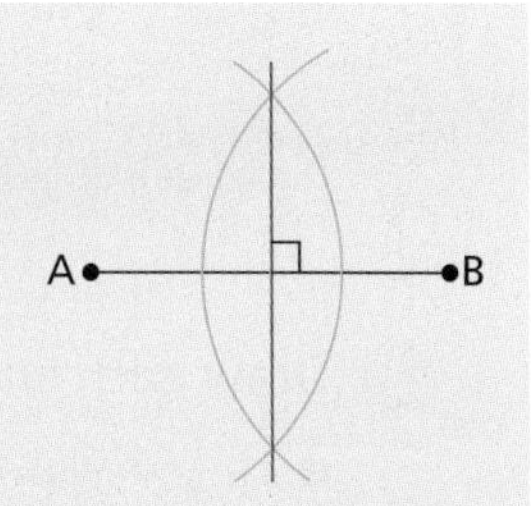

Example

Draw an 8 cm line segment and construct its perpendicular bisector.

Method

Solution	Commentary
A B	Draw an 8 cm line segment and label it as AB.
A B	Draw arcs of length about 6 cm from both end points of AB. The radius of the arcs must be more than half the length of the line segment.
A B	Join the points of intersection of the arcs with a straight line.

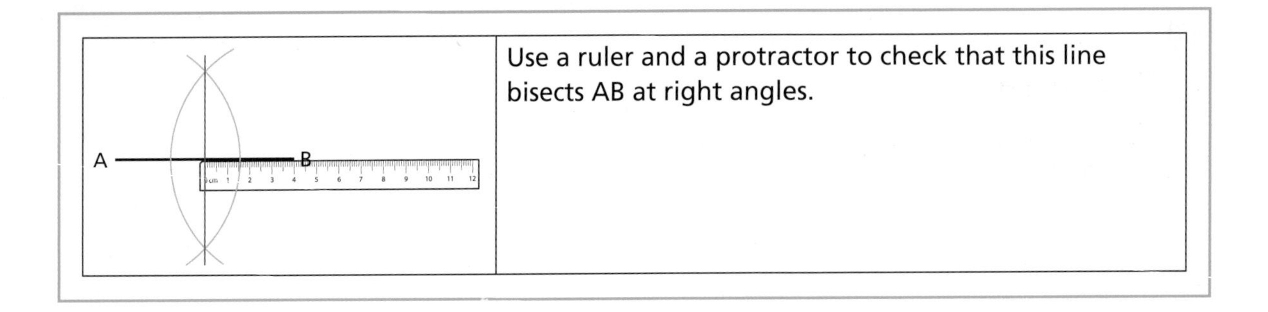

Use a ruler and a protractor to check that this line bisects AB at right angles.

Practice (B)

1. Draw each line segment and construct its perpendicular bisector.

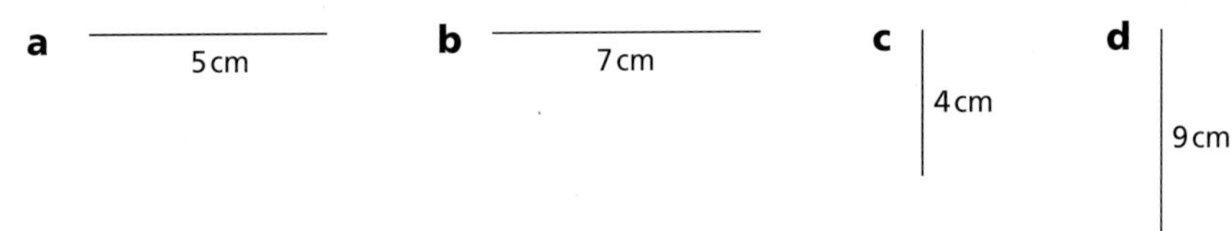

2. Draw each line segment and construct its perpendicular bisector.

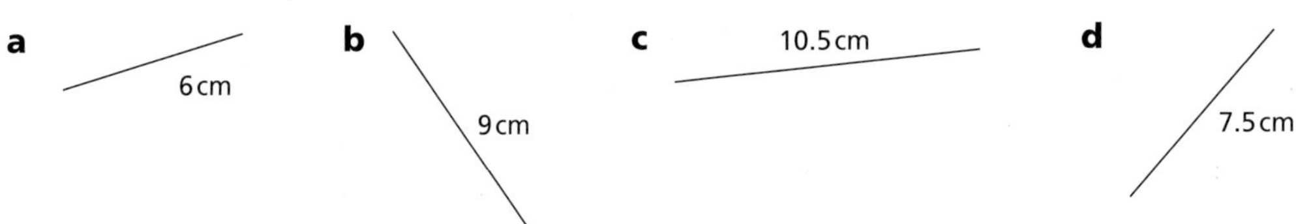

3. The map shows the location of a shop and a garage in a town.

   Copy the diagram and draw all of the points equidistant from the shop and the garage.

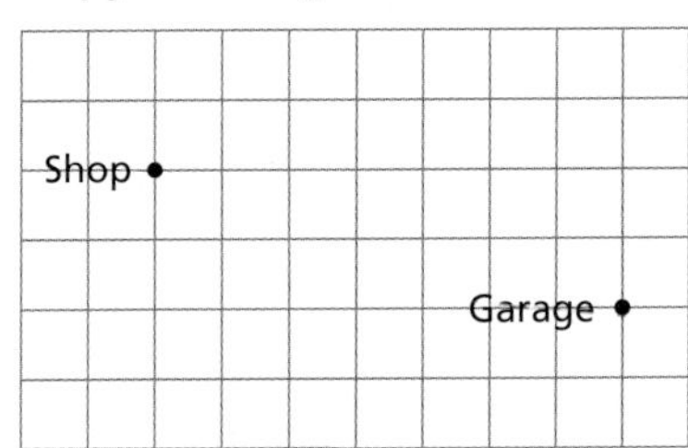

4. Copy the triangle to the given dimensions and bisect the side QR.

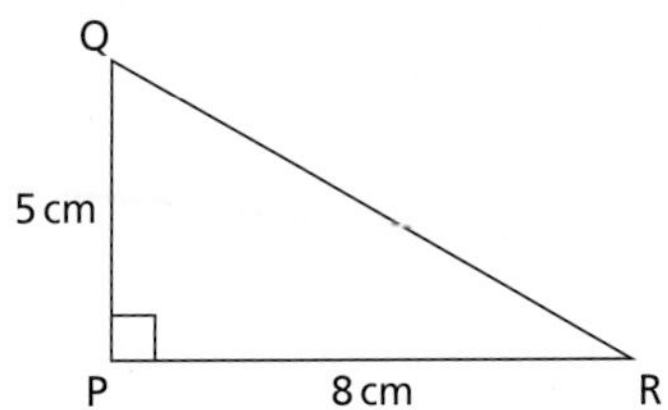

What do you think? (B)

1 Construct an equilateral triangle and bisect each of its sides.

Using the intersection of the three bisectors as the centre, draw the circle that touches all three vertices of the triangle.

This is called a circumcircle.

Repeat with different types of triangle.

2 Junaid says, "If you draw a square and bisect one of the angles, the bisector will also bisect the opposite angle."

Is Junaid correct?

Investigate with other quadrilaterals.

3 **a** Draw a right-angled triangle and label the vertices A, B and C, where BC is the hypotenuse.

Show that the perpendicular bisectors of AB and AC meet at the midpoint of BC.

Investigate with other right-angled triangles.

b Investigate where the perpendicular bisectors of the sides meet for:

i equilateral triangles **ii** isosceles triangles.

Consolidate – do you need more?

1 Draw and bisect each angle.

a 50° **b** 60° **c** 130° **d** 160°

2 Draw the line segments and construct a perpendicular bisector of each.

a 6 cm **b** 8 cm **c** 10.5 cm **d** 15.5 cm

3 Sketch the triangle ABC.

Show the point where the perpendicular bisector of BC meets the bisector of ∠BAC.

The symbol ∠ means 'angle'.

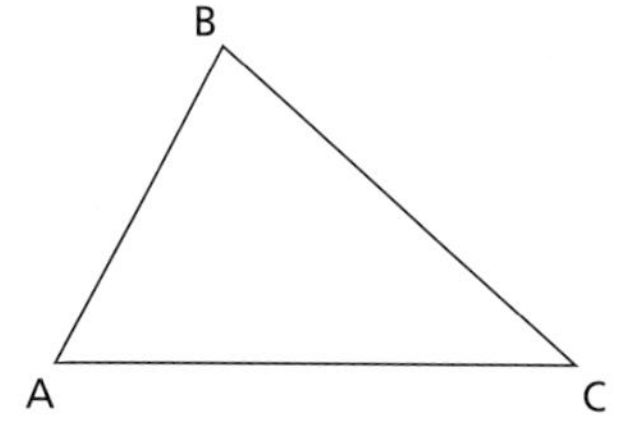

Stretch – can you deepen your learning?

1 Copy the rectangle and identify the point that is:

- equidistant from A and C
- and equidistant from AB and AD.

2 The angle bisector theorem states, "An angle bisector of a triangle divides the opposite side into two segments that are proportional to the other two sides of the triangle."

Investigate this theorem.

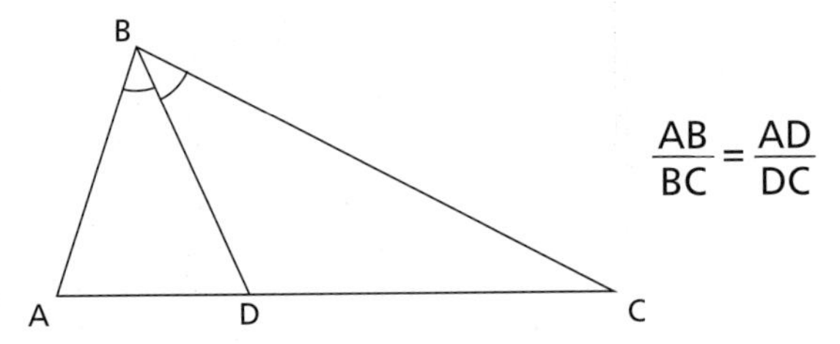

$$\frac{AB}{BC} = \frac{AD}{DC}$$

3 **a** Without using a protractor, construct a 30° angle.

b Using your construction from part **a**, what other angles can be constructed without using a protractor?

4 **a** Construct a regular hexagon using only a ruler and a pair of compasses.

b Investigate the construction of other regular polygons.

1.3 Perpendiculars to and from a line

Are you ready? (A)

1 Draw the line segment and construct its perpendicular bisector.

7 cm

2 Draw the two points and construct the perpendicular bisector between them.

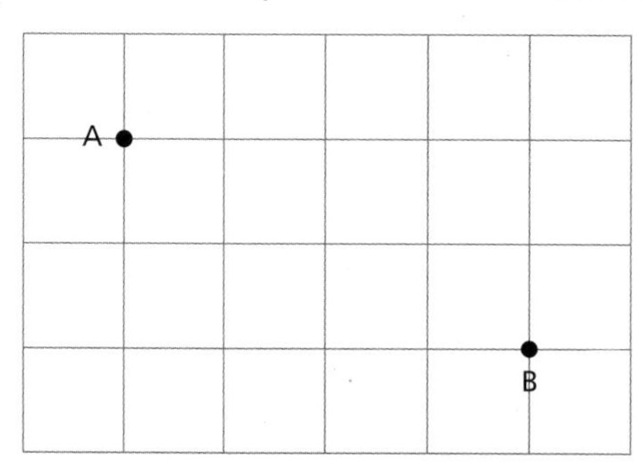

The method for constructing a perpendicular bisector of a line segment can be adapted to construct:

- a perpendicular *from* a given point *to* a line segment, and
- a perpendicular *to* a given point *on* a line segment

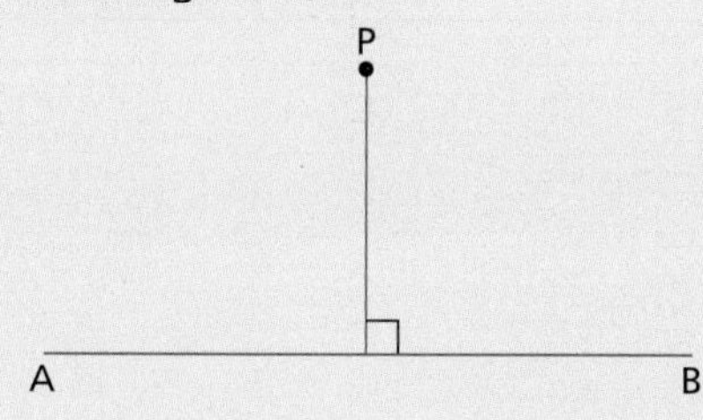

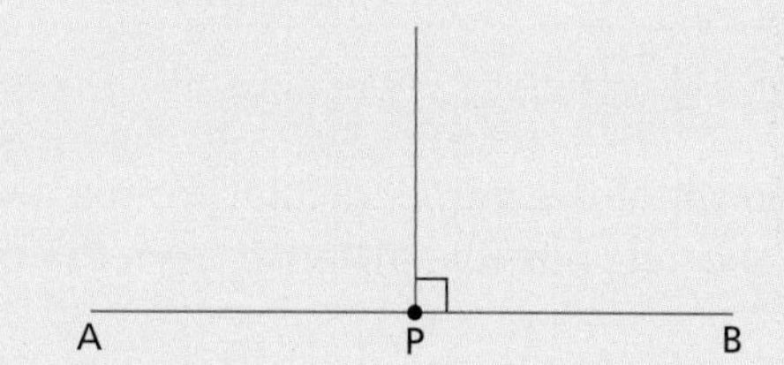

Example

Use a ruler and a pair of compasses to construct the perpendicular to AB passing through P.

P●

A ———————— B

Method

Solution	Commentary
P● A X Y B	Place the point of a pair of compasses on P and draw arcs that cut AB. Label the points of intersection X and Y.
P A X Y B	Now draw two more arcs from X and Y to construct the perpendicular bisector of XY. This will go through P and will be perpendicular to AB.

The length of the perpendicular from a given point to a line is the shortest possible distance from the point to the line.

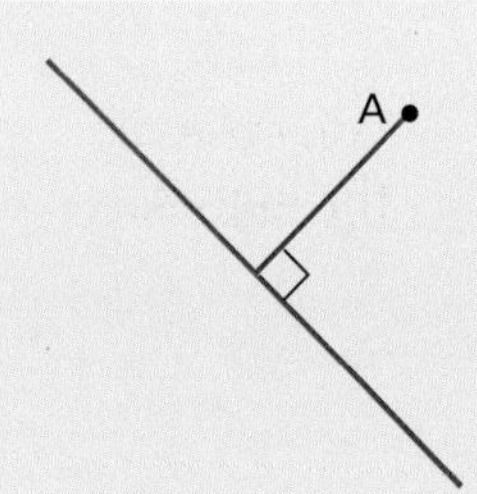

Practice (A)

1. Copy the diagrams and construct the perpendicular to each line segment passing through A.

   a •A

   b •A

   c

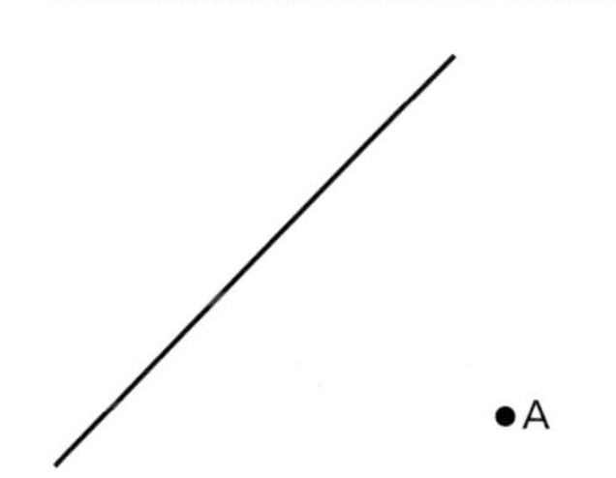

   d

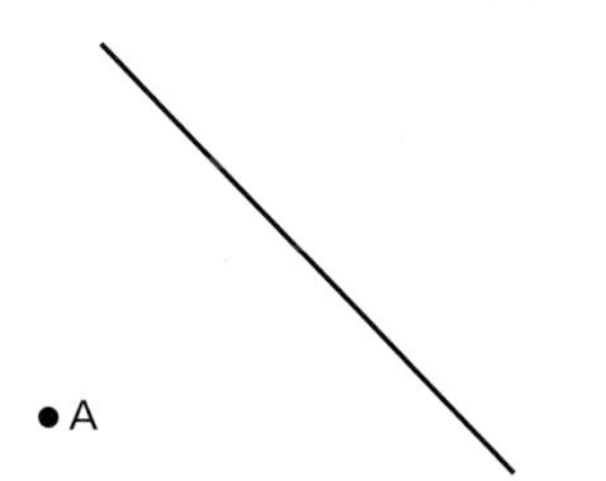

2. Copy the diagrams and construct the perpendicular to each line segment passing through A.

   a

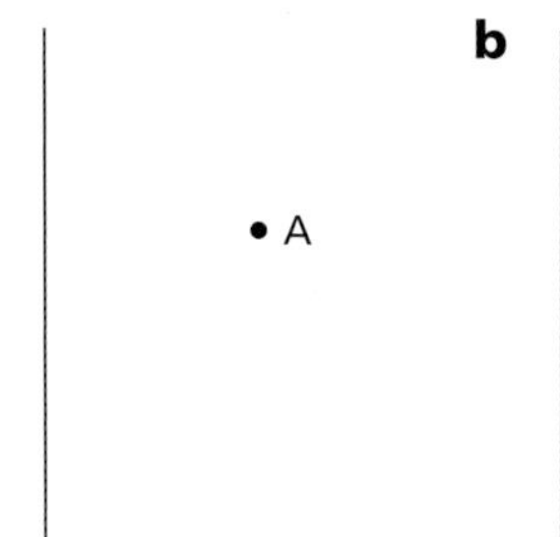

   b •A

   c

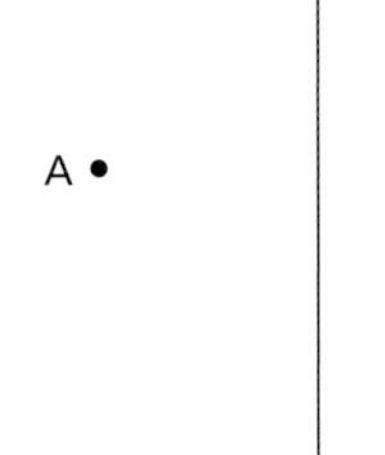

   d A •

3. Copy the diagrams and construct the perpendicular to each line segment passing through A.

   a **b**

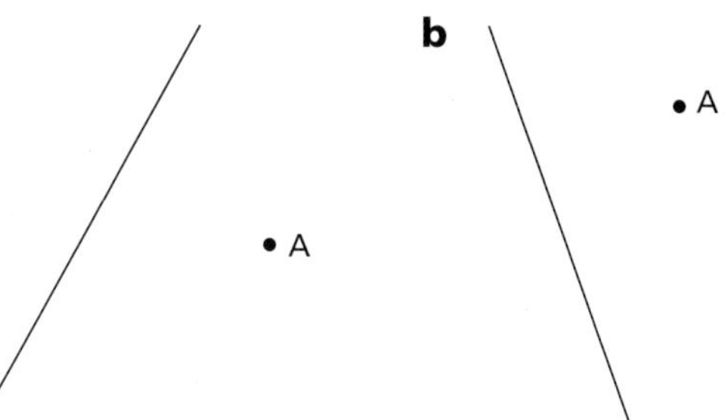

   c

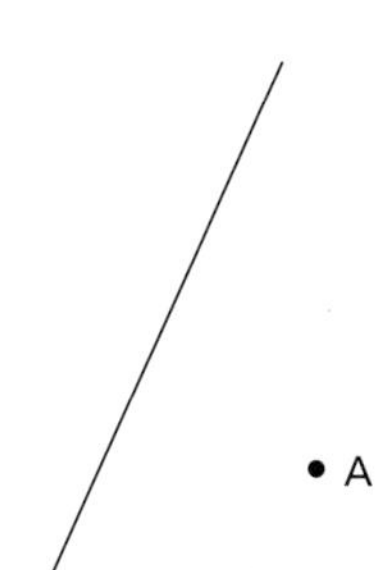

4. Copy the diagram and draw the shortest line segment from point C to the line AB.

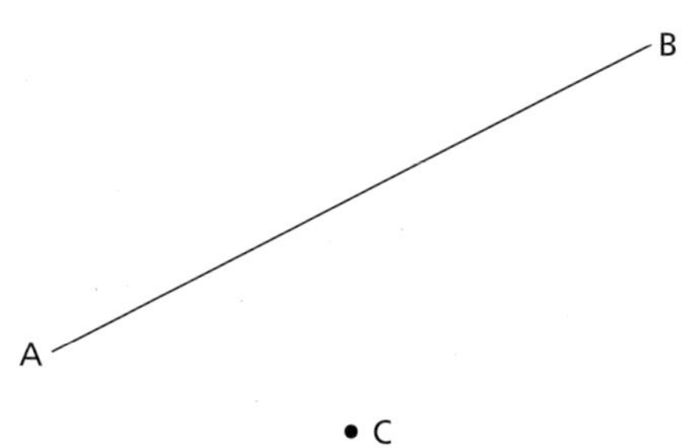

5 The scale diagram shows the location of a house (H), a barn (B) and a garage (G).

There is a straight path between the house and the barn.

Another straight path from the garage is perpendicular to the path between the house and the barn.

Copy the diagram and draw the two paths.

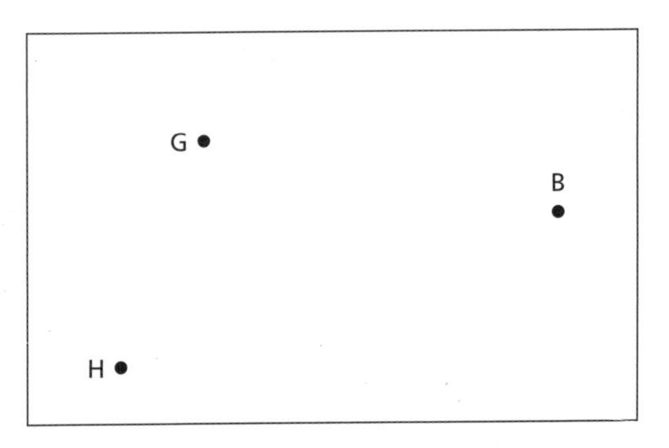

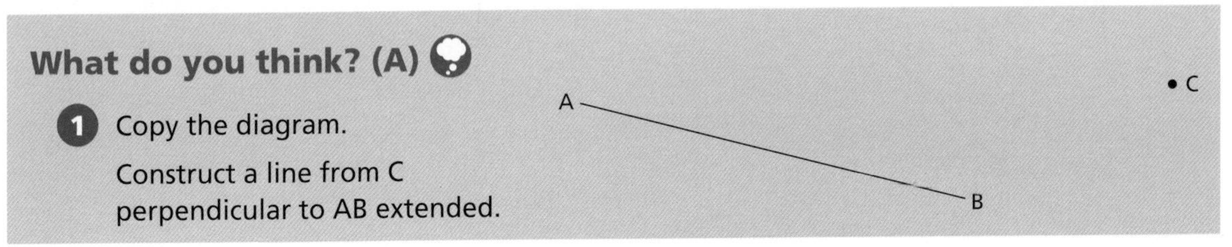

What do you think? (A)

1 Copy the diagram.

Construct a line from C perpendicular to AB extended.

Are you ready? (B)

1 Copy the line to the length shown and mark a point O on the line 6 cm from A.

Draw a circle with a radius of 3 cm, using O as the centre.

A ——— B (10 cm)

Example

Use a ruler and a pair of compasses to construct the perpendicular to PQ passing through point X.

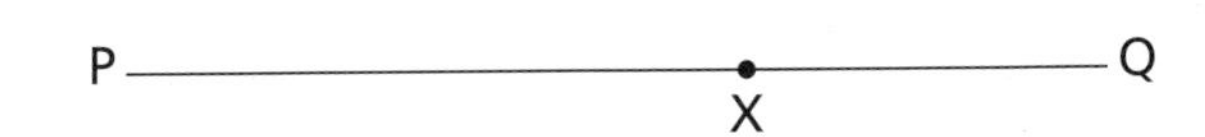

Method

Solution	Commentary
P, A, X, B, Q	Place the point of a pair of compasses on X and draw arcs on either side of X. Label these points A and B.
P, A, X, B, Q	Place the point of your compasses on A and draw arcs above and below PQ. Then do the same with the point of your compasses on B, such that the arcs intersect with the ones from A. Now draw the perpendicular bisector of AB.

Practice (B)

1. Copy each diagram and construct the perpendicular to AB passing through O.

   a

   b

   c

2. The triangle PQR is shown.

   The point O is 4 cm from the point Q.

   Copy the triangle and construct the locus of all points perpendicular to QR passing through O.

   Loci (the plural of 'locus') are covered in Chapter 1.4

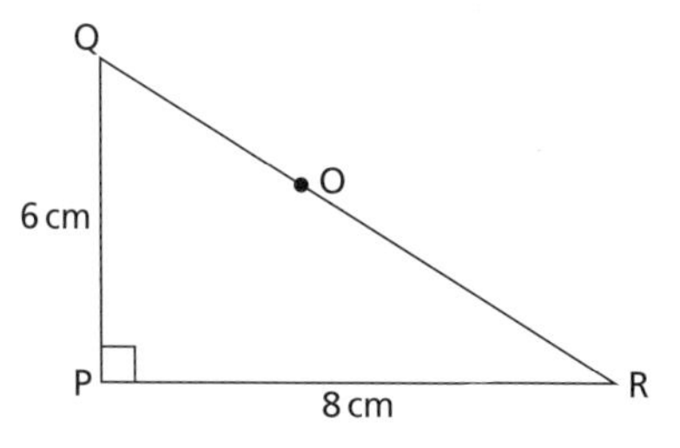

3. Draw a line AB that is 8 cm long.

   a Construct the perpendicular bisector of AB.

   b Construct the perpendicular to AB, passing through the point A.

   You might find it helpful to extend AB.

What do you think? (B)

1. Copy the triangle accurately.

   Mark a point, D, on AB and construct a perpendicular line from AB passing through the point.

   Mark the intersection with the hypotenuse as point E.

   Is triangle DBE similar to triangle ABC?

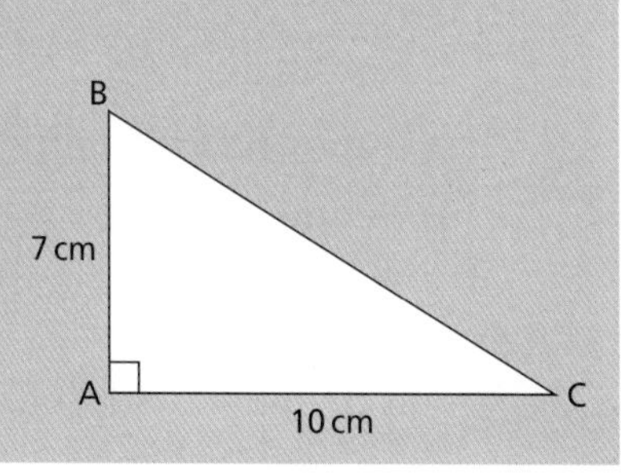

Consolidate – do you need more?

1 Copy each diagram and construct the perpendicular from point C to the line.

a 

b 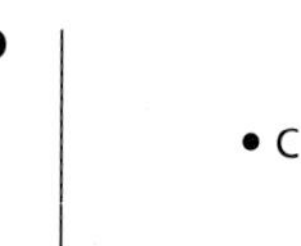

c 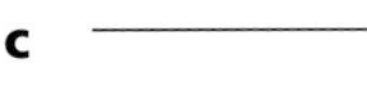

d 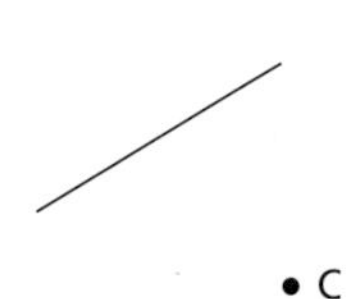

e 

f 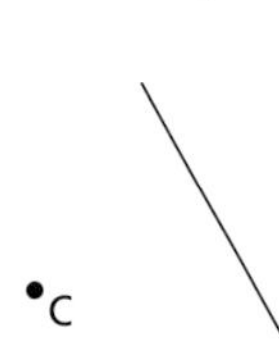

2 Copy each diagram and construct the perpendicular to the line passing through O.

a 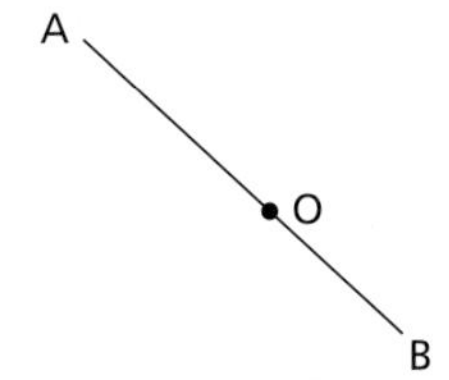

b 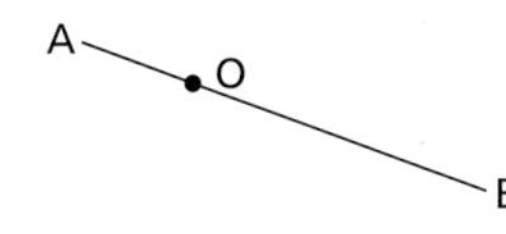

c 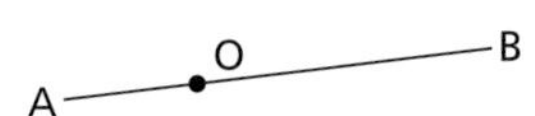

d 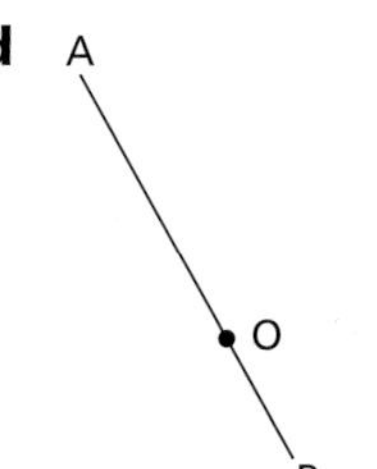

e 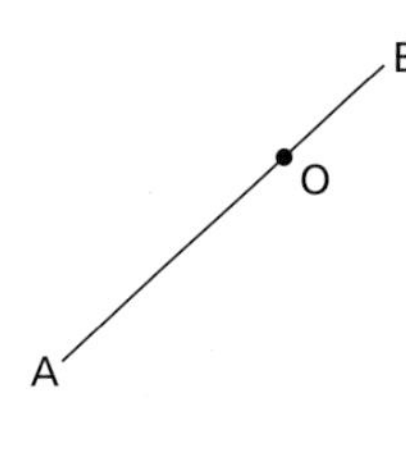

f 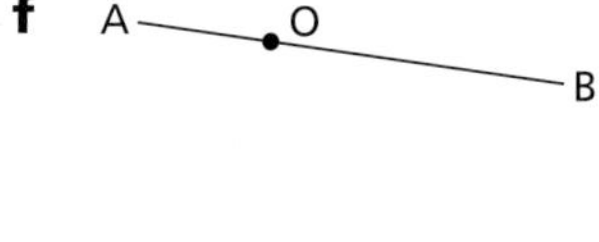

Stretch – can you deepen your learning?

1 Copy the diagram.

a Construct the perpendicular to AB from the point C.

b Compare the gradients of the two lines.

What do you notice? Will this always happen?

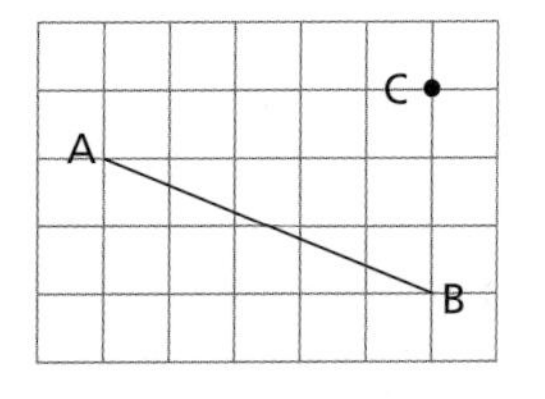

2 A is the point with coordinates (–2, –5) and B is the point with coordinates (2, 3).

a Draw x- and y-axes numbered from –5 to 5. Plot points A and B and join them to form the line segment AB.

b Work out the equation of the line that passes through A and B.

c Work out the coordinates of the midpoint of AB.

d Draw the perpendicular bisector of AB and work out its equation.

e Investigate the relationship between gradients of straight lines and lines that are perpendicular to them.

1.4 Loci

Are you ready? (A)

1 Mark a point, O, and construct the set of all points 4 cm from O.

2 Copy the diagram and construct the set of all points equidistant from AB and AC.

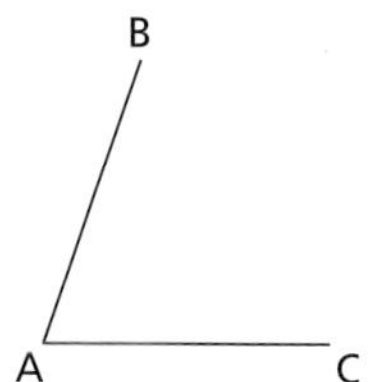

3 Copy the line.

Draw all the points equidistant from A and B.

A ——————————————————————— B

A **locus** (plural **loci**) is a set of points that all have a particular property.

In the circle, OA = OB = OC = 4 cm. The circle is the **locus** of the points that are 4 cm from O.

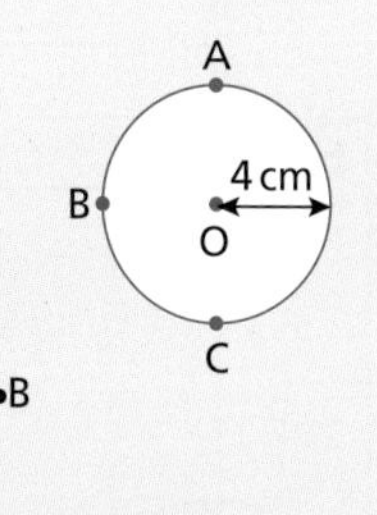

Another example is the locus of all points equidistant from two other points, A and B. This means all the points that are the same distance from A as they are from B. This would be the perpendicular bisector of the line segment joining A and B.

Example 1

ABCD is a square of side length 16 cm. O is at the centre of the square.

a Draw the locus of all points 7 cm from O.

b Shade the locus of all points inside the square that are more than 7 cm from O.

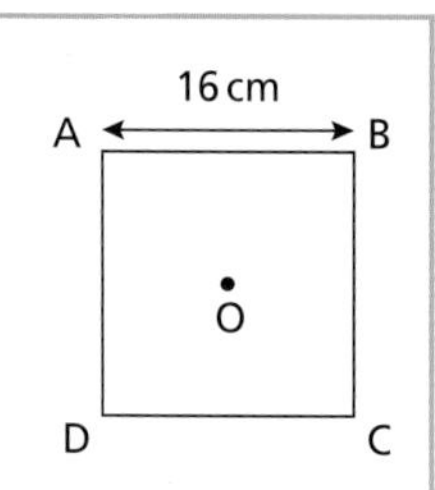

Method

Solution	Commentary
a	All the points that are exactly 7 cm from O form a circle, centre O with radius 7 cm.
b	The points inside the circle are less than 7 cm from O. So you need to shade the region outside the circle, but within the square.

Example 2

ABCD is a square of side length 10 cm.

a Draw the locus of all points within the square that are 3 cm from A.

b Shade the locus of all points within the square that are less than 3 cm from A.

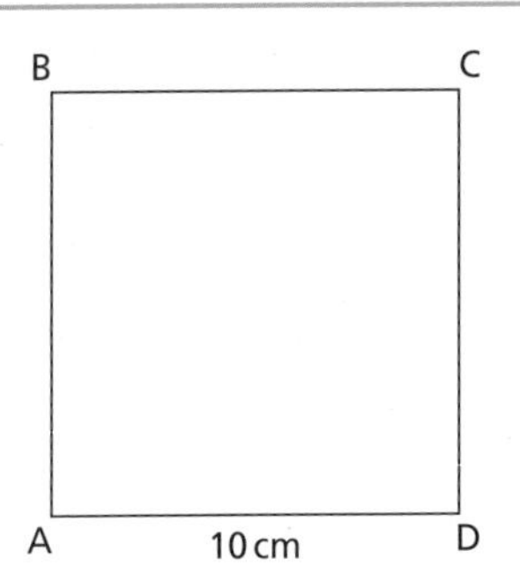

Method

Solution	Commentary
a [diagram: square ABCD with compasses drawing an arc from A]	Set a pair of compasses to 3 cm and draw an arc from the point A.
b [diagram: square ABCD with shaded quarter circle at A]	The points inside the quarter circle are less than 3 cm from A. Shade the region within the quarter circle and within the square.

Practice (A)

1. Mark a point O on a page. Draw the locus of all points exactly 3 cm from O.

2. Mark a point A on a page.
   - **a** Draw the locus of all points exactly 4 cm from A.
   - **b** Draw the locus of all points exactly 3 cm from A.
   - **c** Shade the region with points more than 3 cm from A but less than 4 cm from A.

3. Draw a line segment AB, 10 cm long.
   - **a** Construct the locus of all points 7 cm from A.
   - **b** Construct the locus of all points 5 cm from B.
   - **c** Shade the region with points that are less than 7 cm from A and less than 5 cm from B.

4. Draw a line AB 10 cm long. P is a point exactly 6 cm from A and 8 cm from B.
   Show that there are exactly two possible locations for P.

5. Draw the rectangle ABCD to the given dimensions.

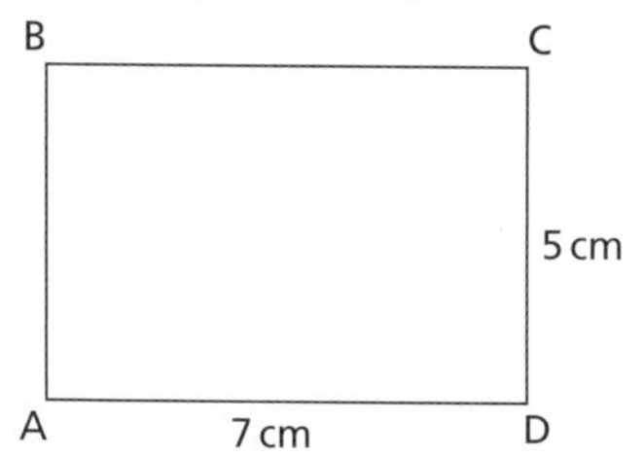

   a Draw the loci of the points within the rectangle that are:
      i exactly 4 cm from A   **ii** exactly 5 cm from C.

   b Shade the region within the rectangle with points that are less than 4 cm from A and less than 5 cm from C.

What do you think? (A)

1. Here is a straight line.

   A ——— 8 cm ——— B

   Is it possible to identify points that are less than 4 cm from A and less than 3 cm from B? Explain your answer.

2. The diagram represents the loci of points 5 cm from A, 3 cm from B and 4 cm from C.
   The letters **i** to **vii** represent the different regions within the diagram.

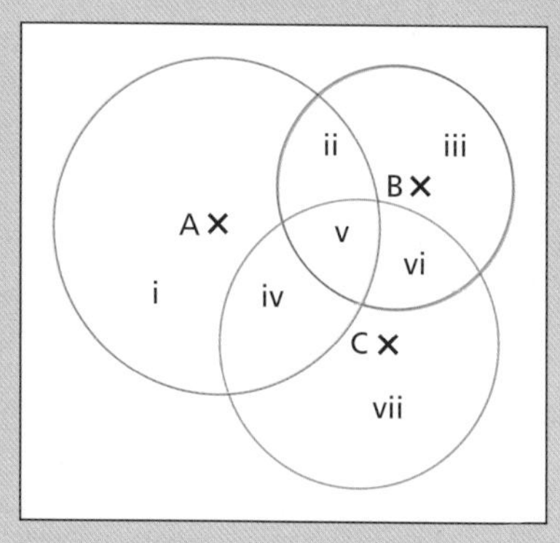

Not drawn to scale

   State the letter in a region which represents the sets of points that are:

   a less than 5 cm from A and less than 3 cm from B

   b less than 4 cm from C and less than 5 cm from A

   c less than 5 cm from A, less than 3 cm from B and less than 4 cm from C.

   Is there more than one answer in parts **a** to **c**? Explain your answer.

Are you ready? (B)

1 Make two copies of the rectangle using the given dimensions.

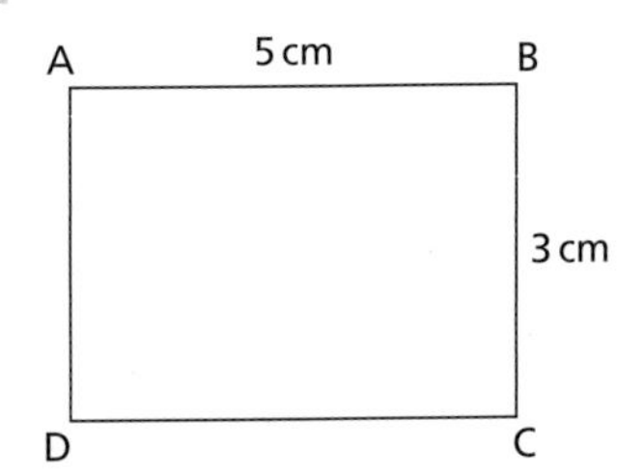

a Shade the region of all points within the rectangle that are less than 2 cm from D.

b Shade the region of all points within the rectangle that are more than 2 cm from B.

Example 1

ABCD is a square of side length 10 cm.

a Draw the locus of all points within the square that are exactly 2 cm from AD.

b Draw the locus of all points within the square that are exactly 6 cm from D.

c Shade the region of all points within the square that are less than 2 cm from AD **and** less than 6 cm from D.

Method

Solution	Commentary
a A B D C 2 cm	The locus is a line parallel to AD and 2 cm away from it.
b A B 6 cm D C 2 cm	The locus of points exactly 6 cm from D would be a circle with centre D with radius 6 cm. You only need to draw the part of the circle inside the square.
c A B 6 cm D C 2 cm	Less than 2 cm from AD means the region will be on the left-hand side of the locus drawn in part **a**. Less than 6 cm from D means the region will also need to be inside the sector drawn in part **b**. Shade the region that satisfies both these criteria.

Example 2

Draw the locus of all points 2 cm from the line AB.

A ——— 8 cm ——— B

Method

Solution	Commentary
A B	Set a pair of compasses to 2 cm and draw an arc from the point A. This arc shows the locus of all points 2 cm from A. Draw another arc from the point B. This arc shows the locus of all points 2 cm from B.
A B	Use a ruler to draw lines 2 cm above and below AB connecting the pair of arcs.

Practice (B)

1. Copy the line AB to the given length.

   Construct the locus of all points exactly 3 cm from line AB.

   A ——— 5 cm ——— B

2. Copy the square to the given dimensions.

   Construct the locus of all points that are 2 cm outside the square.

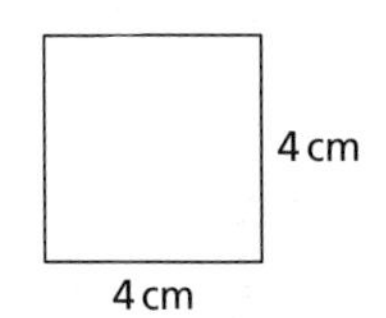

3. Copy the rectangle ABCD to the given dimensions.

   Draw the locus of points within the rectangle that are:

   a 4 cm from AB

   b 1 cm from BC.

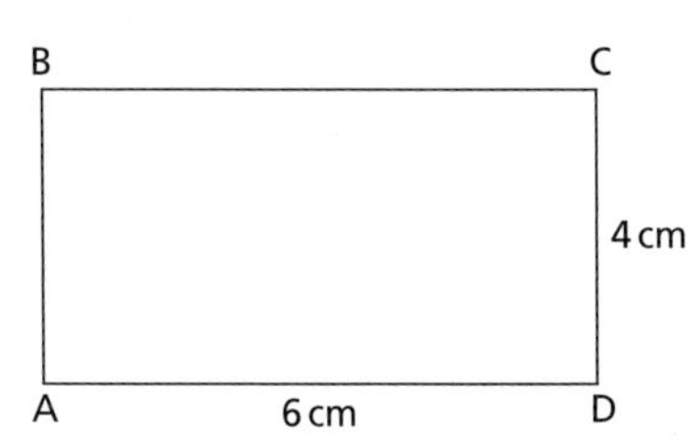

4. The diagram shows the plan of a room.

   a Draw a scale diagram of the room using 1 cm to represent 2 m.

   b Use your diagram to draw the locus of points within the room that are:

   i 8 m from the wall AD

   ii 10 m from the wall AB.

   c Shade the region within the room with points that are less than 8 m from AD and more than 10 m from AB.

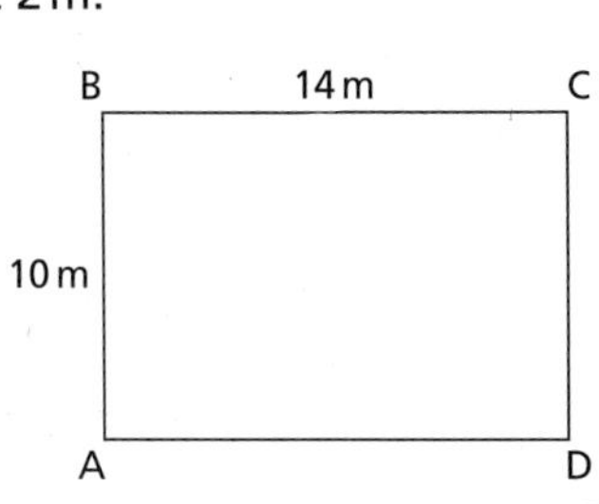

5 **a** Copy the rectangle PQRS to the given dimensions.

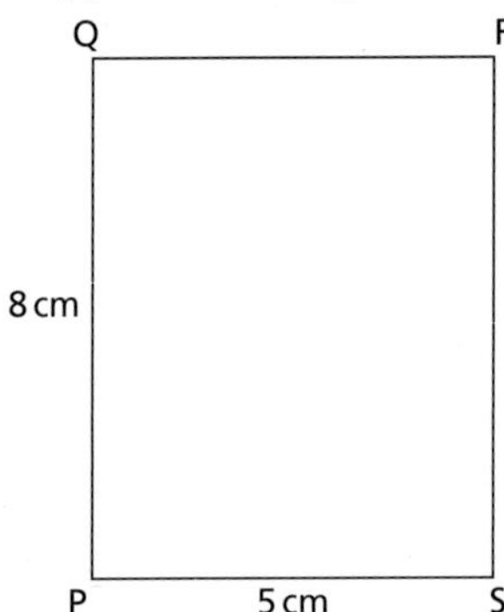

Draw the locus of points within the rectangle that are:

i 3 cm from QR   **ii** 4 cm from R.

b Shade the region within the rectangle with points more than 3 cm from QR and less than 4 cm from R.

6 The diagram shows a plan of Faith's garden.

Faith wants to plant a tree that is situated:

- more than 4 m from the house
- less than 5 m from B.

Draw a scale diagram of the garden using 1 cm to represent 2 m. Shade the region where Faith could plant the tree.

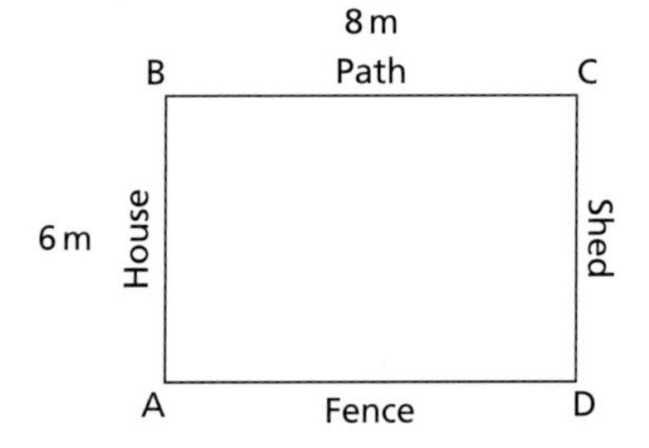

7 The rectangle ABCD is drawn on a centimetre grid.

Describe the shaded region.

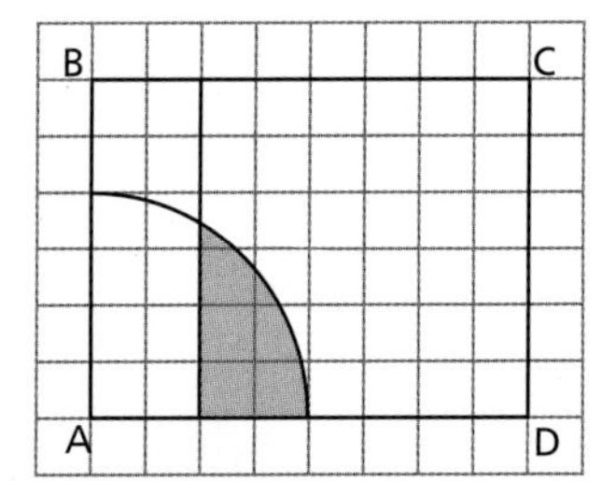

8 Make a scale drawing of the rectangle using 1 cm to represent 4 metres.

Shade the region with points that are:

- closer to PQ than PS
- more than 12 m from Q
- more than 6 m from QR.

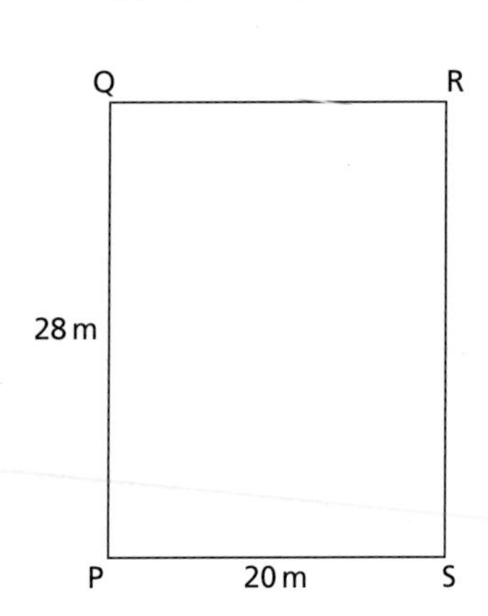

9 Copy the diagram onto a centimetre grid.

Shade the region with points that are:

- less than 5 cm from A
- more than 3 cm from M
- closer to AB than to BC.

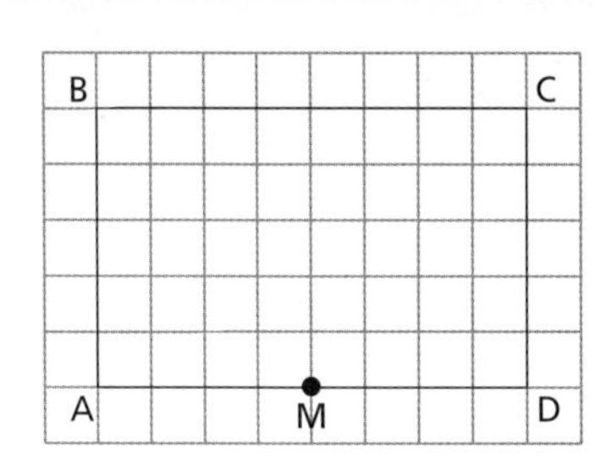

10 Copy the shape to the given dimensions.

Shade the region with points that are:

- more than 4 cm from AB
- less than 6 cm from F
- more than 2 cm from AF.

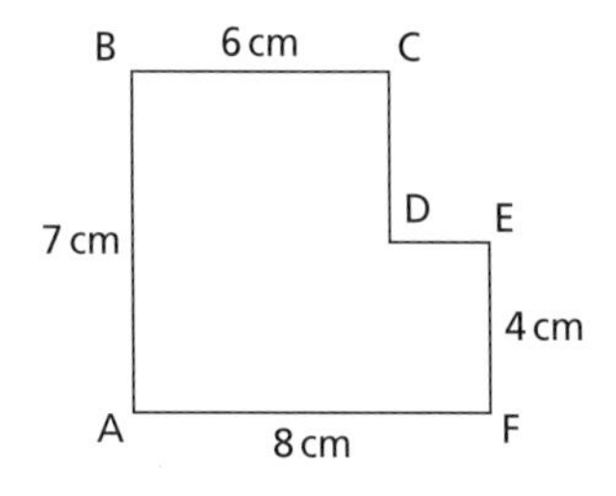

11 Copy the shape to the given dimensions.

Shade the region with points that are:

- less than 4 cm from AD
- less than 4 cm from BC
- more than 6 cm from D.

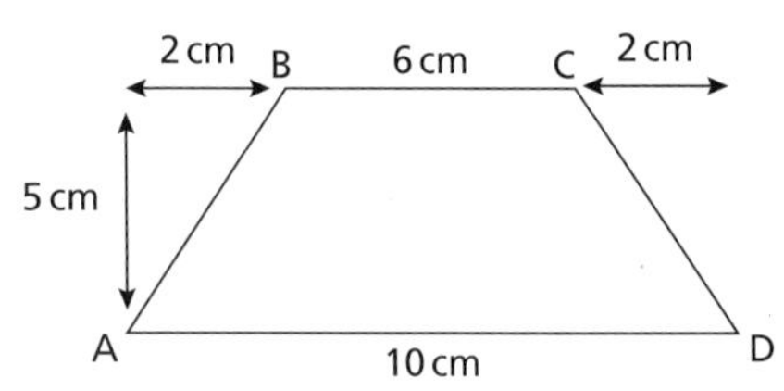

12 Here is a map showing the location of a radio transmitter and a power line. The map is drawn on a centimetre grid.

In order for a radio to work, it must be less than 12 km from the radio transmitter but more than 6 km from the power line.

Make a copy of the diagram using 1 cm to represent 3 km.

Shade the region where the radio will work.

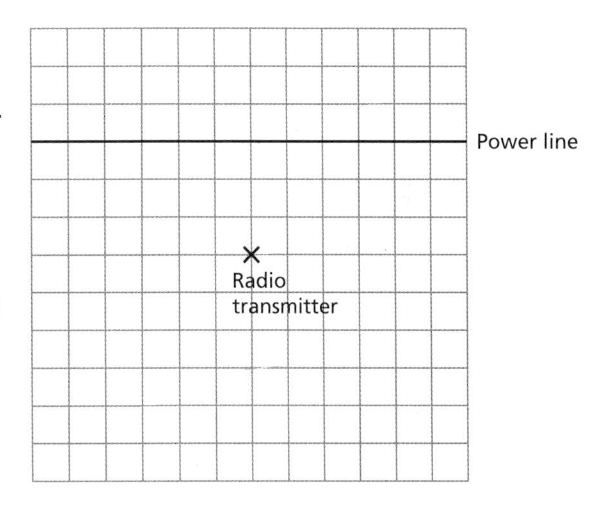

13 The diagram shows the plan of a garden drawn on a centimetre grid.

A gardener wants to lay a water pipe across a garden so that it is equidistant to the two hedges.

a Make a copy of the diagram and draw the location of the pipe.

The gardener also wants to connect a sprinkler system to the centre of the pipe.

b Mark the location of the sprinkler.

The sprinkler system can water the garden in all directions up to 6 metres.

c Shade the region of the garden that cannot be covered by the sprinkler system.

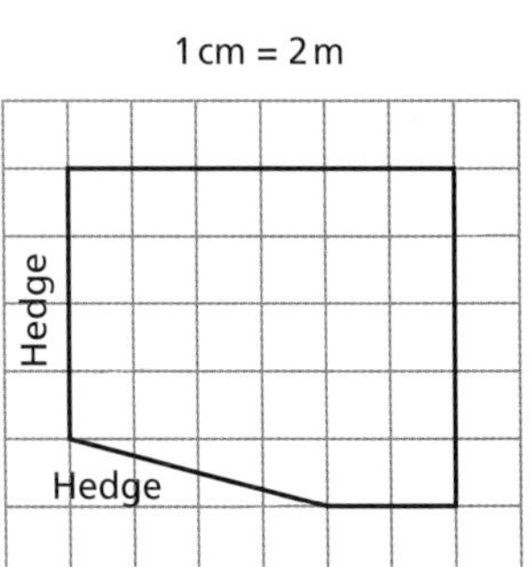

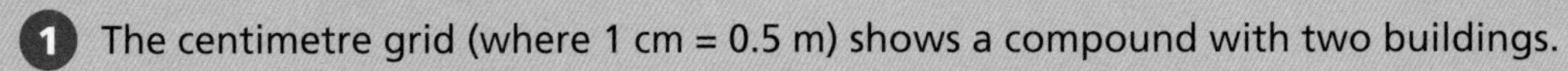

What do you think? (B)

1 The centimetre grid (where 1 cm = 0.5 m) shows a compound with two buildings.

Two security sensors are to be placed in the compound. Each security sensor can detect anything in a range of 6 m.

Security sensors must be placed on the side of a building or wall.

Which two locations would be best for the security sensors to cover as much of the compound as possible?

Which locations would be the least effective for installing the security sensors?

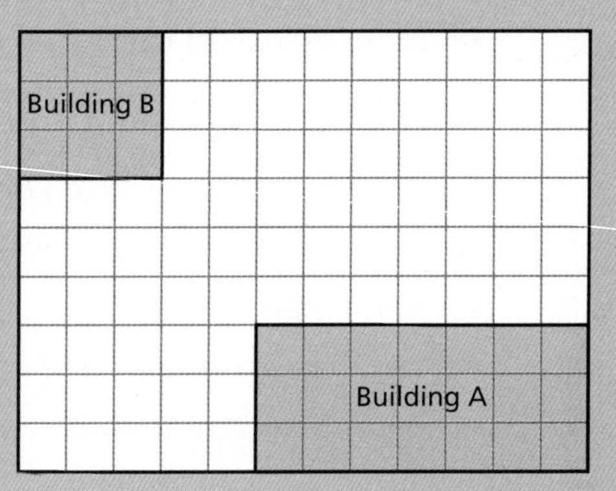

Consolidate – do you need more?

1 Copy the lines to the given lengths.

Construct the locus of all points 2 cm from each line.

a 7 cm

b 5 cm

c 6 cm

2 Copy the shapes to the given dimensions.

Draw the locus of all points 3 cm from each shape.

a 3 cm, 5 cm

b 4 cm, 6 cm

c 4 cm, 4 cm

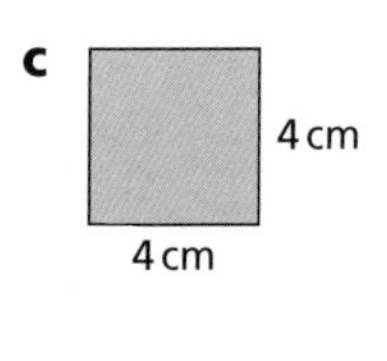

3 Make four copies of the rectangle to the given dimensions.

Shade the region within the rectangle with points that are:

a less than 4 cm from AB and more than 5 cm from A

b more than 4 cm from AB and more than 5 cm from A

c more than 4 cm from CD and more than 4 cm from B

d less than 4 cm from CD and more than 4 cm from D.

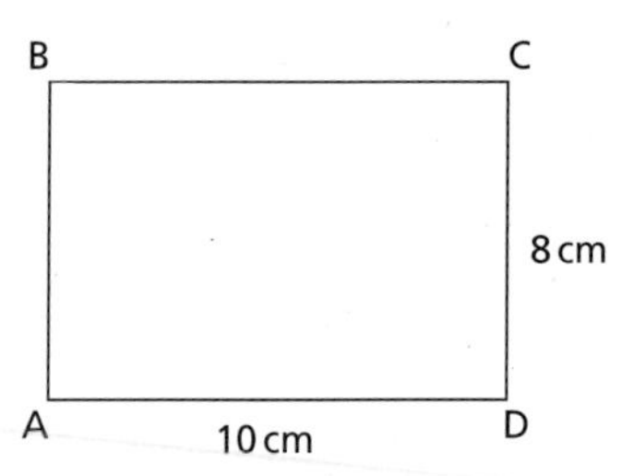

4 Copy the rectangle to the given dimensions.

Shade the region within the rectangle with points that are:

- closer to PQ than RS
- closer to QR than PS
- more than 2 cm from Q.

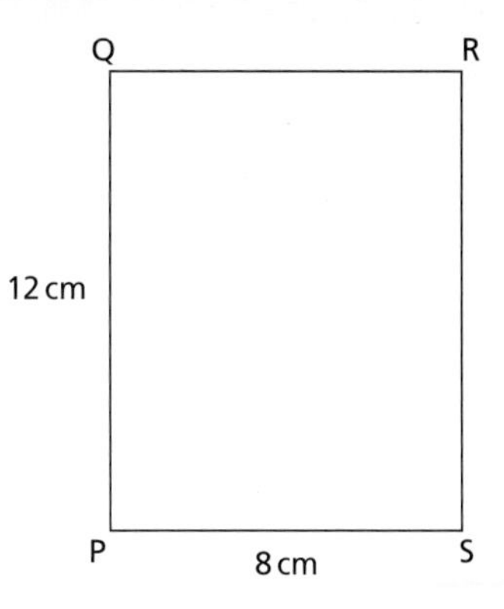

Stretch – can you deepen your learning?

1 Work out the area of the locus of points less than 3 cm from the line AB.

A —— 6 cm —— B

2 A goat is attached to the corner of a building with a 5 m rope in a grass field.

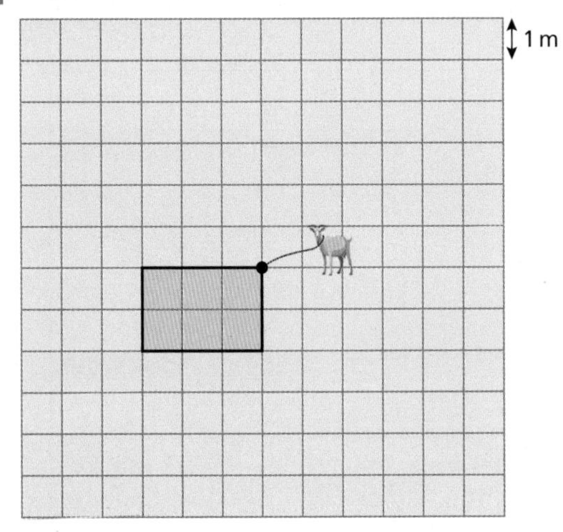

Work out the area of the grass that can be eaten by the goat.

3 A square is shown.

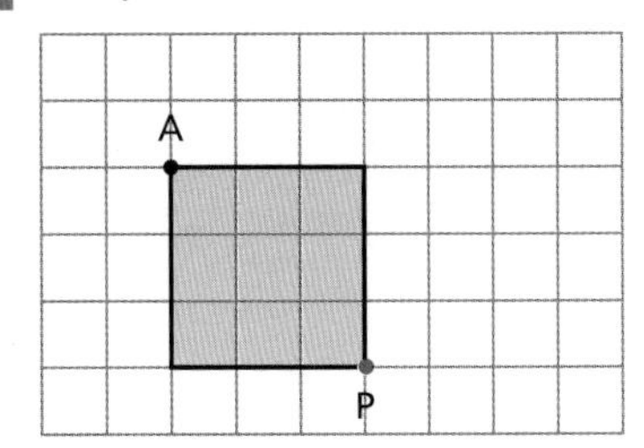

Draw the locus of the point A when the square is rotated 90° clockwise about the point P.

Constructions: exam practice

1 A small boat leaves a harbour and sails 180 metres on a bearing of 110°.

It stops and then sails a further 210 metres on a bearing of 55°.

Using a scale of 1 cm to represent 30 metres, draw an accurate scale diagram of the boat's route. **[3 marks]**

2 Construct the perpendicular bisector on a copy of line segment AB.

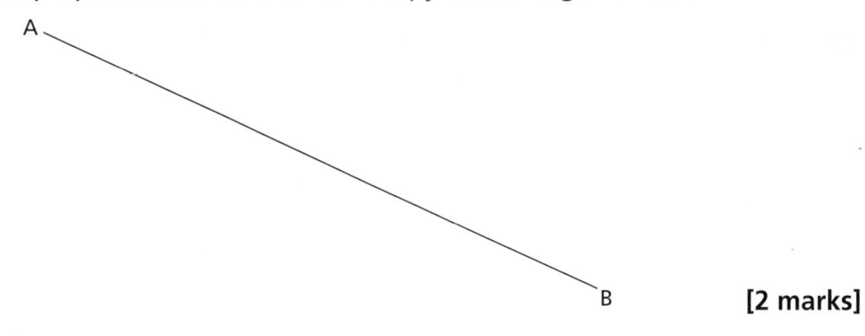

[2 marks]

3 Bisect a copy of the obtuse angle PQR.

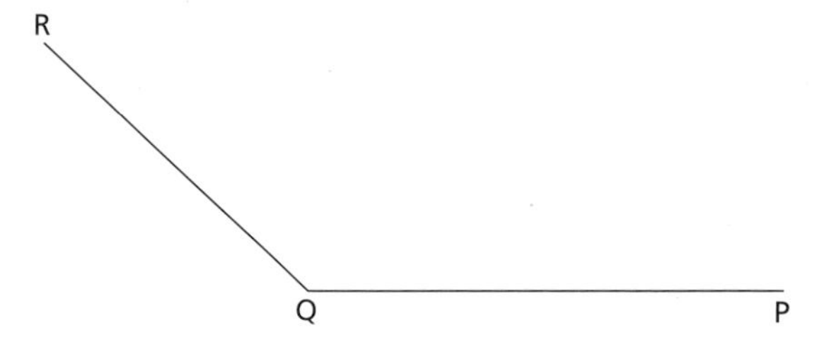

[2 marks]

4 Construct a triangle with sides of 6 cm, 7 cm and 9 cm. **[3 marks]**

5 The bearing of P from Q is 110°.

Ed thinks the bearing of Q from P will be 250°.

Show that Ed is wrong. **[2 marks]**

6 The diagram shows the base of a water tank.

A drain surrounding the base of the tank is to be constructed 2 metres away from the base of the tank.

On a scale diagram, draw accurately the position of the drain. **[3 marks]**

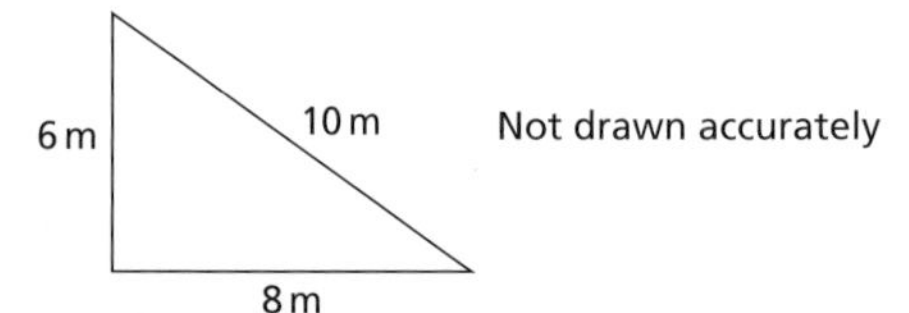

2 Congruence and similarity

White Rose
MATHS

In this block, we will cover...

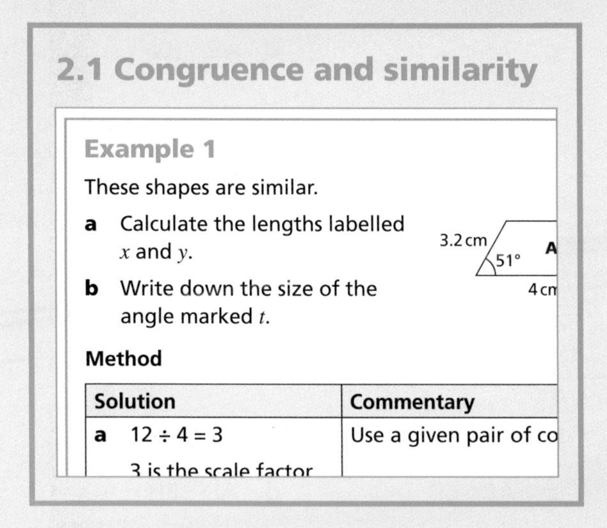
2.1 Congruence and similarity
Example 1
These shapes are similar.
a Calculate the lengths labelled x and y.
b Write down the size of the angle marked t.
3.2 cm
51°
4 cm
Method
Solution
Commentary
a 12 ÷ 4 = 3
3 is the scale factor
Use a given pair of co

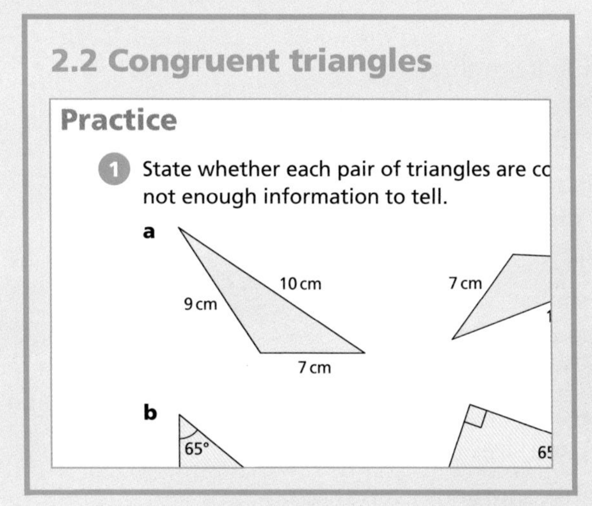
2.2 Congruent triangles
Practice
State whether each pair of triangles are co
not enough information to tell.
a
10 cm
9 cm
7 cm
7 cm
b
65°

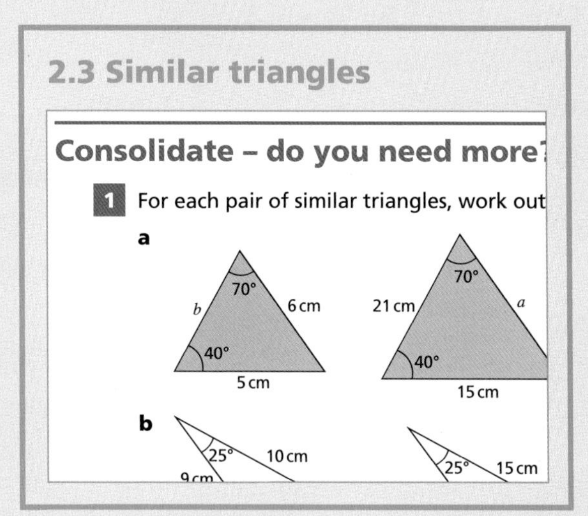
2.3 Similar triangles
Consolidate – do you need more
For each pair of similar triangles, work out
a
70°
b
6 cm
40°
5 cm
21 cm
70°
a
40°
15 cm
b
25°
10 cm
25°
15 cm

2.1 Congruence and similarity

White Rose MATHS

Are you ready?

1 Simplify the ratios.

a 10 : 12 **b** 35 : 100 **c** 7.5 : 3 **d** $4a : 10a$

2 Here are two similar rectangles.

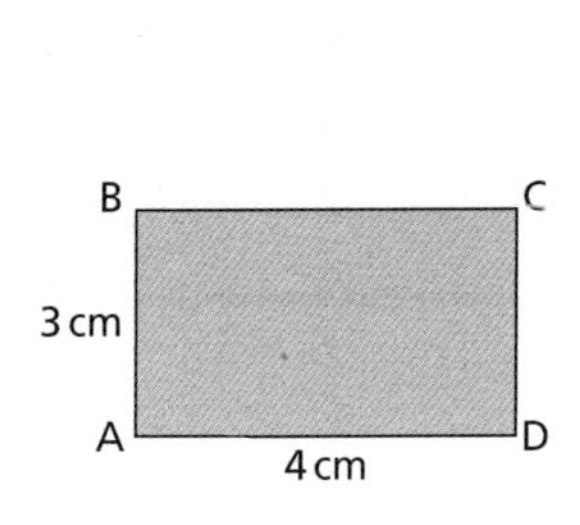

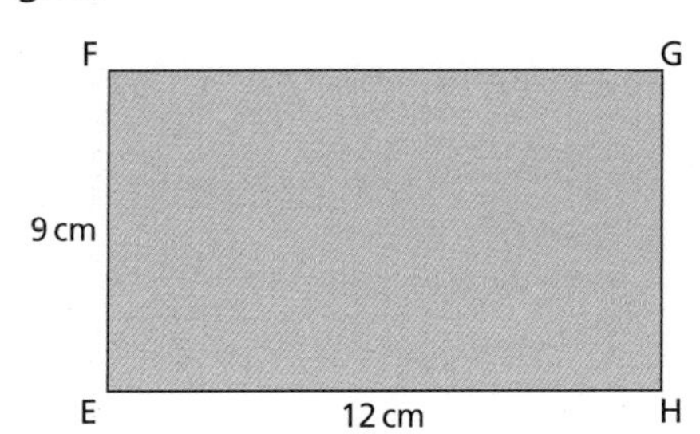

Write each ratio in its simplest form.

a AB : AD **b** EF : EH **c** AB : EF **d** AD : EH

The side lengths of shape B are three times the corresponding side lengths of shape A.

Notice that the ratio of any pair of corresponding sides in shape A and shape B is 1 : 3

4 cm A 6 cm

12 cm B 18 cm

When two figures are **similar**, their angles are the same and corresponding lengths are in the same ratio.

The angles of both rectangles are all 90°. The angles in a shape don't change when it is enlarged.

Two shapes are **congruent** if they are exactly the same shape and size.

The rectangles below are congruent even though they are in different orientations.

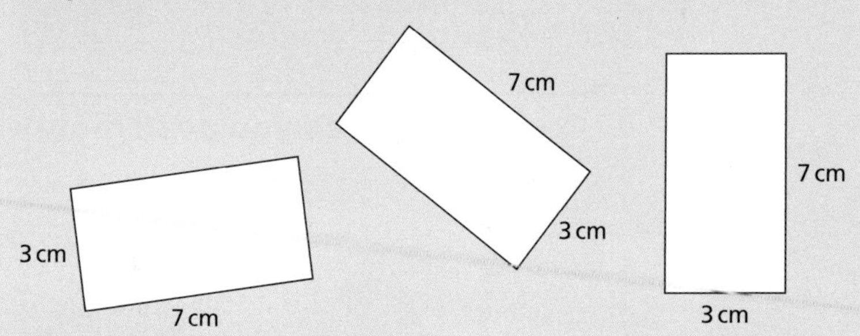

Any reflection, rotation or translation of a shape will produce an image that is congruent to the original shape.

These transformations are covered in Block 4.

Example 1

These shapes are similar.

a Calculate the lengths labelled x and y.

b Write down the size of the angle marked t.

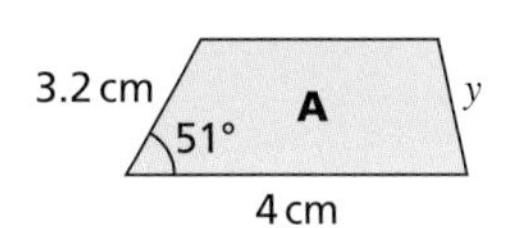

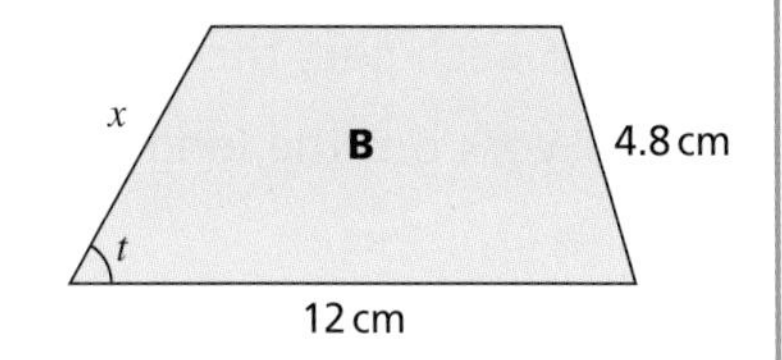

Method

Solution	Commentary
a $12 \div 4 = 3$ 3 is the scale factor.	Use a given pair of corresponding sides to find the **scale factor**.
$3.2 \times 3 = 9.6$ $x = 9.6$ cm	To get from shape A to shape B, multiply by 3
$4.8 \div 3 = 1.6$ $y = 1.6$ cm	To get from shape B to shape A, divide by 3
b $t = 51°$	The angles remain the same size when a shape is enlarged.

Example 2

Which triangles are congruent to A?

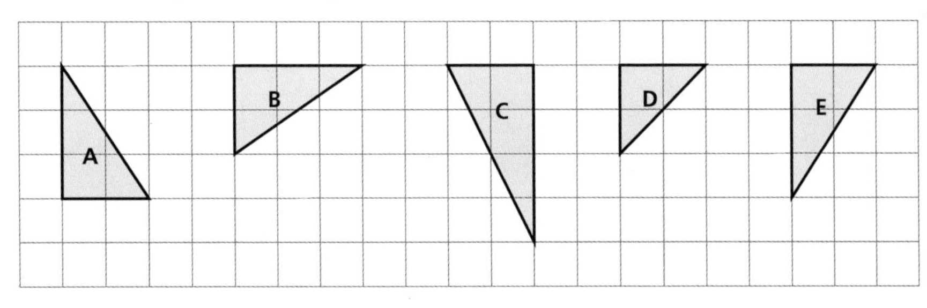

Method

Solution	Commentary
B and E	B and E are right-angled triangles with perpendicular sides of 2 units and 3 units, and so they are both congruent to A. Note that C and D do not have the same dimensions as A, therefore they are not congruent.

Practice

1 In each pair, the shapes are similar.

Work out the lengths of the sides labelled with letters.

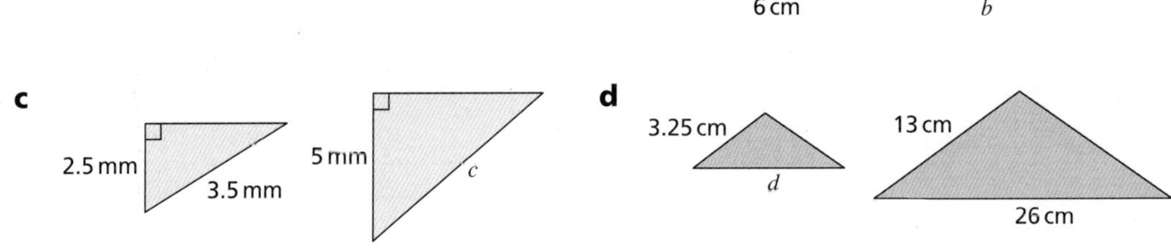

2 Here are two similar trapezia.

a Work out the lengths labelled a and b.

b Write down the values of c and d.

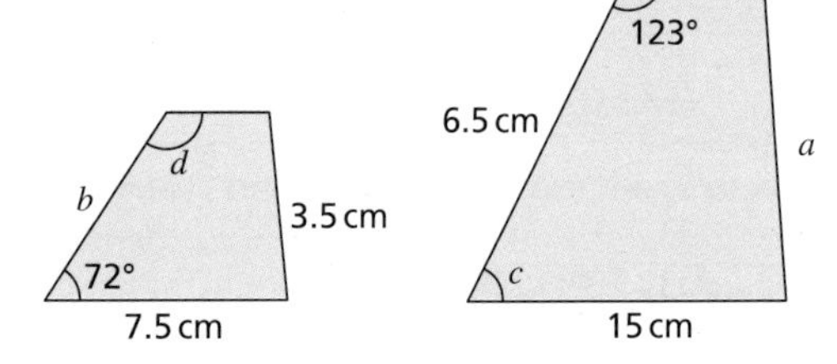

3 Here are some shapes:

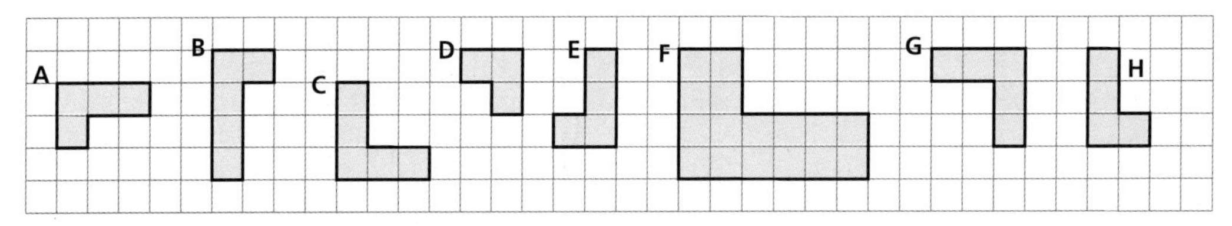

a Which two shapes are congruent to shape A?

b Write the letter of two more shapes that are congruent.

4 Copy shape X onto a grid and draw three more shapes that are congruent to X.

Vary the orientations of your shapes and compare your answers with a partner's.

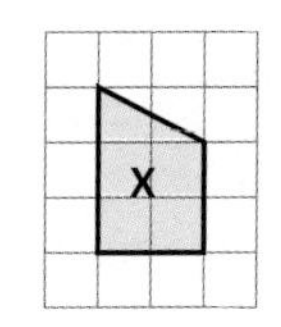

5 Here is rectangle A:

3 cm **A** 4 cm

Which of the rectangles below are similar to rectangle A?

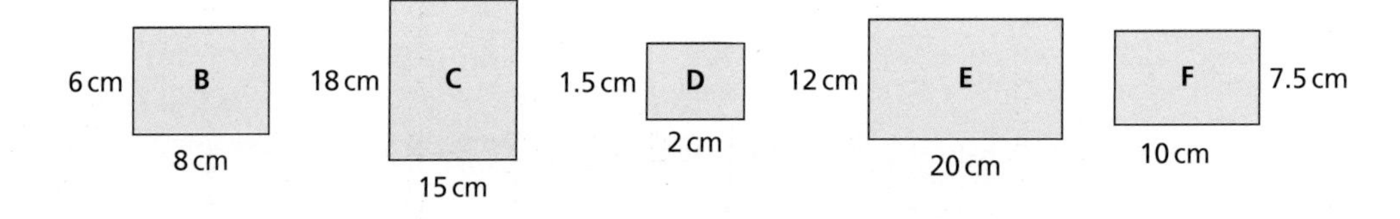

6. Copy the triangle onto a grid.

   Draw two more triangles that are similar to the triangle.

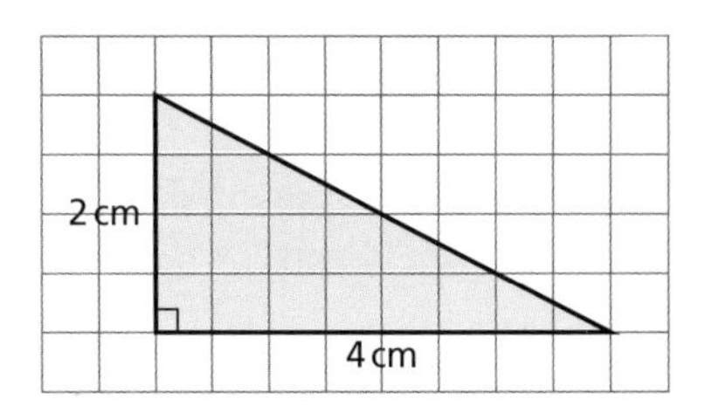

7. Rectangles P and Q are similar.

   P has an area of 34 cm².

   Work out the length of x.

P

8.5 cm

13 cm

Q

x

8. The three rectangles are similar.

   Work out the values of x and y.

3 cm

5 cm

12 cm

x

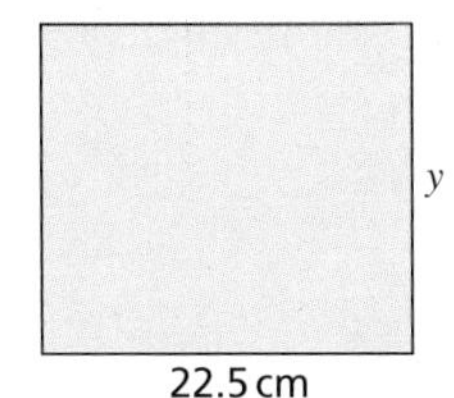

9. These quadrilaterals are similar.

   a Work out the lengths of a and b.

   b Write down the size of angle c.

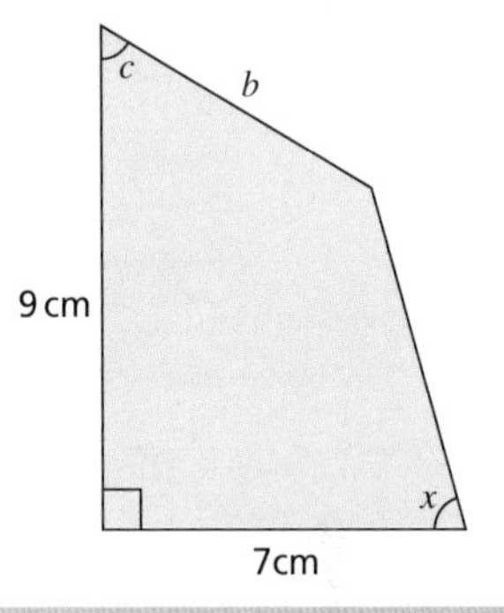

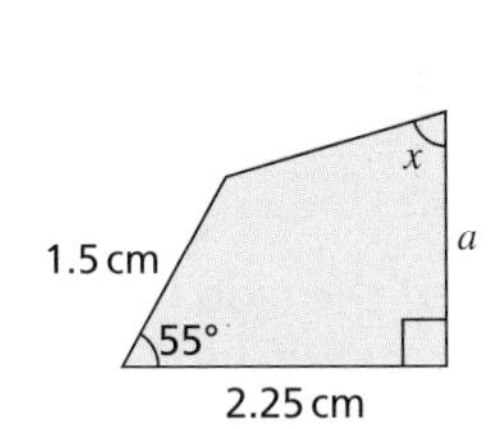

What do you think?

1. Huda says that a regular polygon is always similar to another regular polygon with the same number of sides.

   Is she correct?

2. Samira says, "If a pair of shapes are similar, they must be congruent."

   Ed says, "If a pair of shapes are congruent, they must be similar."

   a Explain why Samira is wrong.

   b Explain why Ed is correct.

3. Shown below is an example of a shape that tessellates. The shapes fit together without any gaps and can cover the whole of a surface.

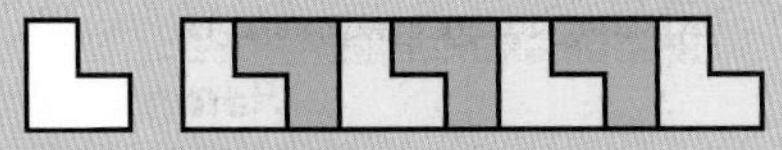

Show that shape B tessellates by drawing more congruent shapes.

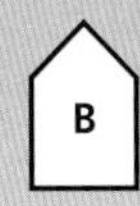

Explore other shapes and whether or not they tessellate.

Consolidate – do you need more?

1 For each pair of similar shapes, work out the lengths of sides x and y.

a

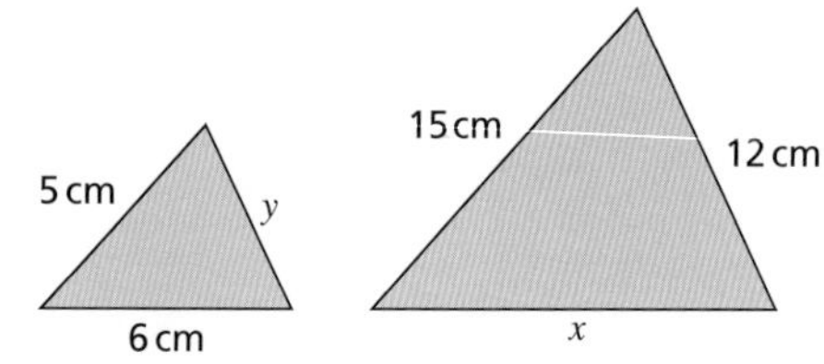

b

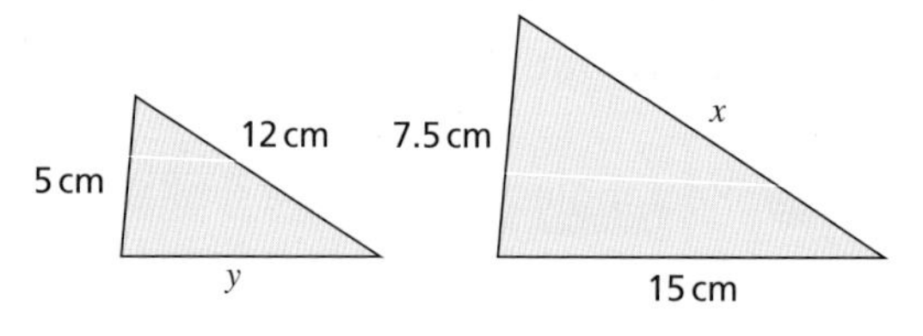

c

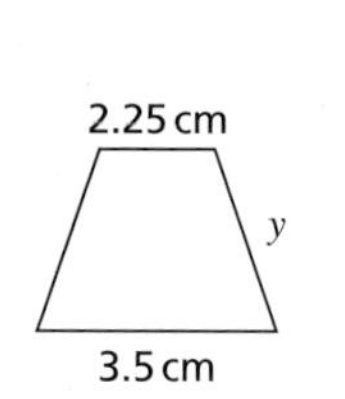

9 cm

7 cm

x

d

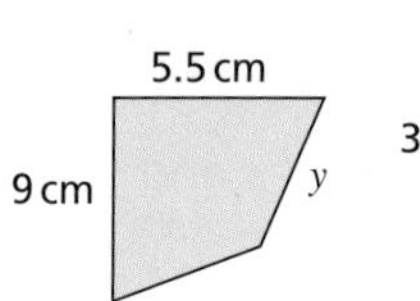

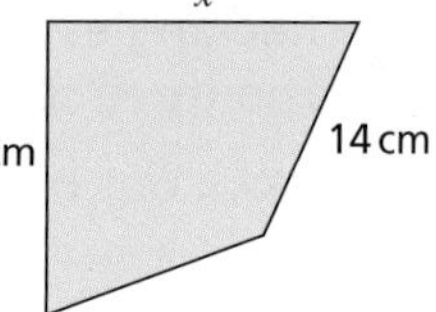

2 These two triangles are similar.

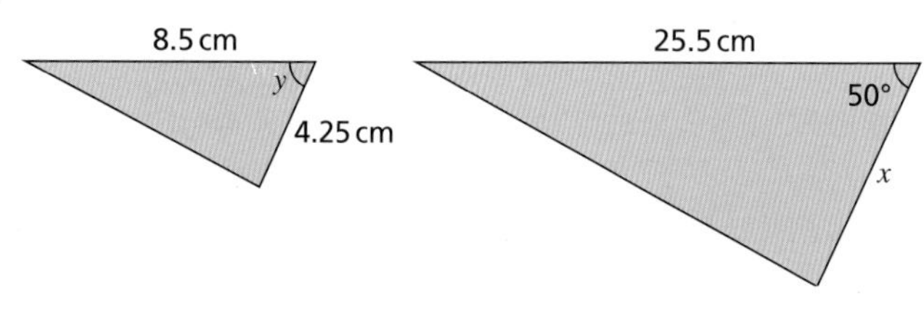

a Work out the length of the side labelled x.

b Write down the size of the angle y.

3 For each set of shapes, state the letter of any shape congruent to shape A.

a

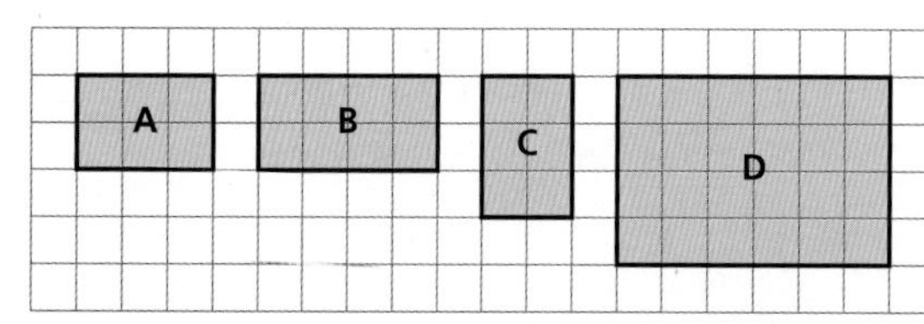

b

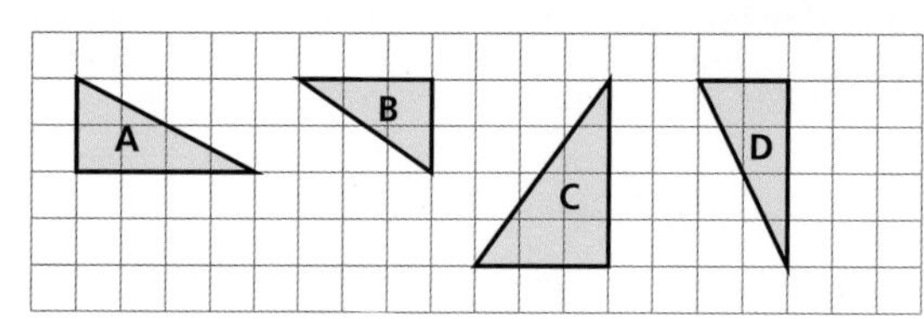

c

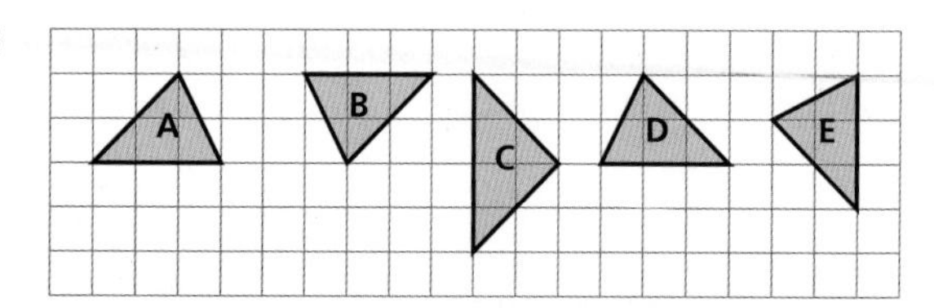

Stretch – can you deepen your learning?

1 Triangles ABC and DEF are similar.

Work out the value of x.

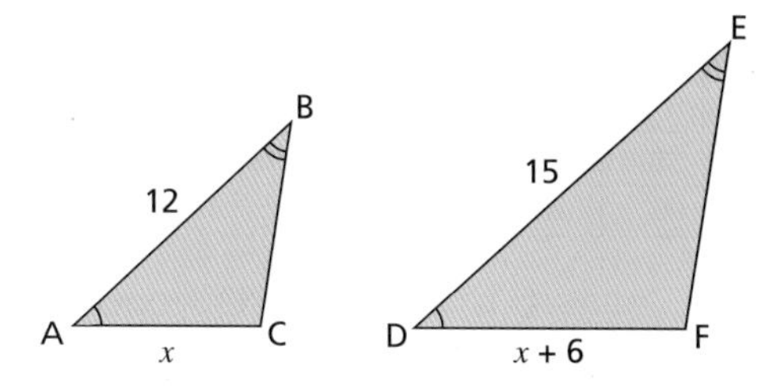

2 Rectangles ABCD and AEFG are similar.

Work out the shaded area.

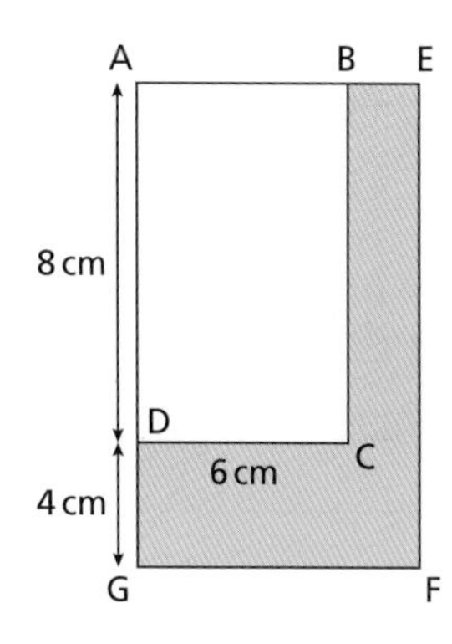

3 ABD and BDC are similar triangles.

Work out the perimeter of the quadrilateral ABCD.

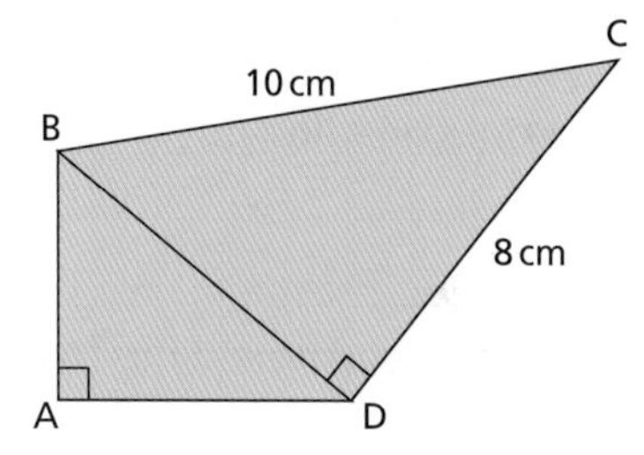

4 Here are two similar triangles.

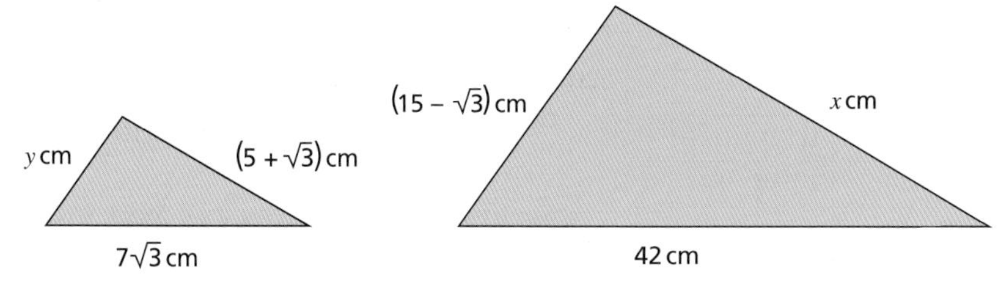

a Work out the value of x. Give your answer in the form $a\sqrt{b} + c$ where a, b and c are integers.

b Work out the value of y. Give your answer in the form $\frac{a\sqrt{b} + c}{d}$ where a, b, c and d are integers.

2.2 Congruent triangles

Are you ready?

1 State whether each triangle is **isosceles, equilateral** or **scalene**.

a 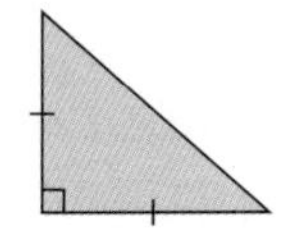

b 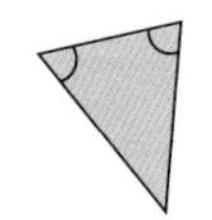

c 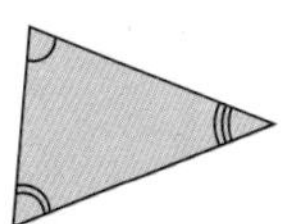

d 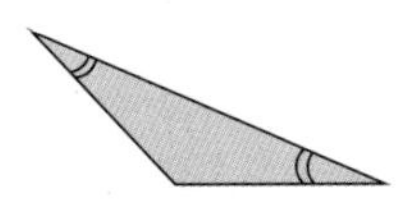

e 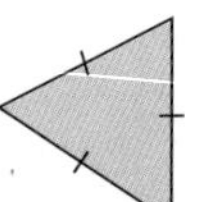

Congruent shapes are exactly the same shape and size. There are four sets of conditions which determine if two triangles are congruent.

Side, Side, Side (SSS)	**Side, Angle, Side (SAS)**
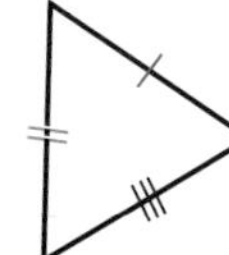 Three pairs of corresponding sides are equal.	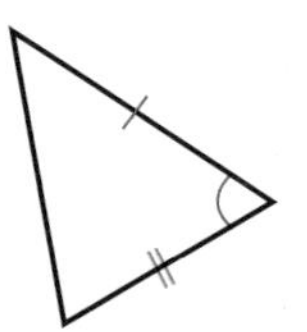 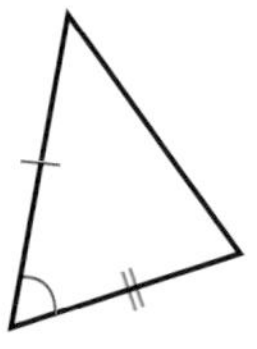 Two sides and the angle between them in one triangle are equal to the corresponding sides and angle of the other triangle.
Angle, Angle, Side (AAS) or Angle, Side, Angle (ASA)	**Right angle, Hypotenuse, Side (RHS)**
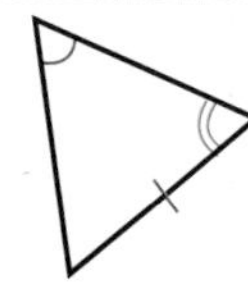 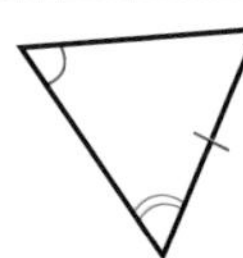 Two angles and one side of a triangle are equal to the corresponding angles and side of the other triangle.	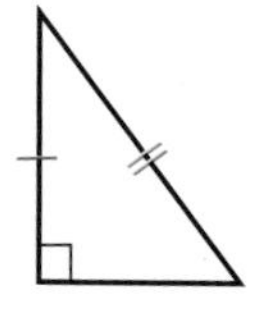 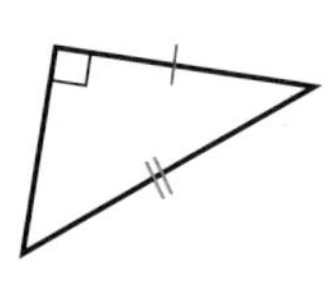 One side and the hypotenuse in a right-angled triangle are equal to one side and the hypotenuse in the other right-angled triangle.

To prove two triangles are congruent, you need to provide evidence that one of these conditions has been met and state which condition you have used.

Example 1

Show that only one of B, C and D is congruent to triangle A.

A

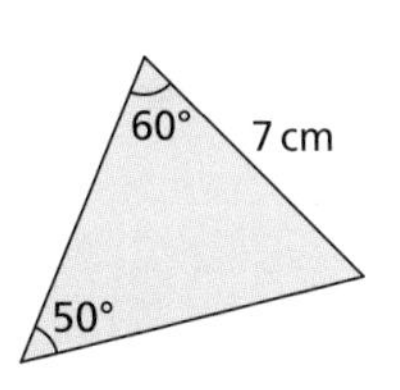

B

60°
7 cm
50°

C

50°
60°
7 cm

D

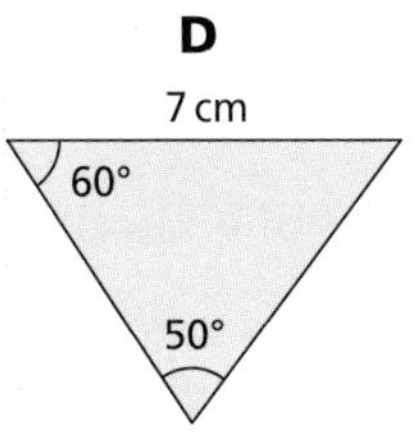

Method

Solution	Commentary
Triangle B is not congruent to A as the 7 cm length of B is adjacent to the two given angles. Triangle C is not congruent to A as the 7 cm length of C is adjacent to the given 50° angle, therefore does not correspond to the 7 cm length of A.	All the diagrams show two angles and one side, so check for the AAS condition. The 7 cm sides of B and C do not correspond with the 7 cm side of A, therefore B and C are not congruent to A.
Triangle D is congruent to A as the 7 cm length is adjacent to the given 60° angle. The triangles are congruent using the condition AAS.	D is congruent as it has two corresponding angles and the 7 cm side given corresponds with the 7 cm side of A.

Example 2

ABCD is a parallelogram.

Show that the triangles ABD and CBD are congruent.

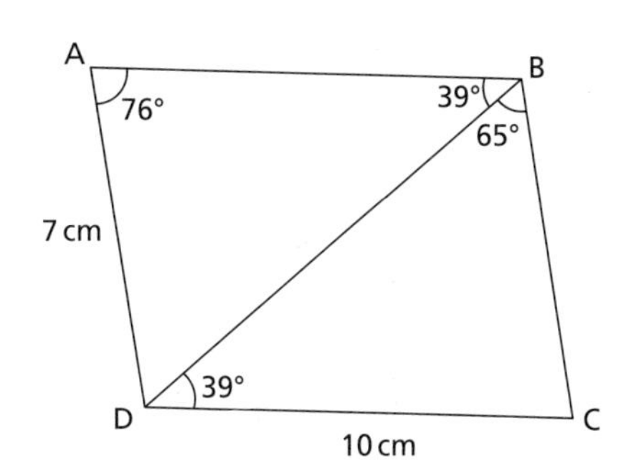

Method A

Solution	Commentary
AD = BC DC = AB Opposite sides of a parallelogram are equal in length.	Identify equal sides using properties of a parallelogram.
BD is a shared side.	BD is common to both triangles.
The triangles are congruent using the condition SSS.	State the condition you have used to show the triangles are congruent.

Method B

Solution	Commentary
AD = BC DC = AB Opposite sides of a parallelogram are equal in length.	Identify equal sides using properties of a parallelogram.
∠BCD = 180° – (39° + 65°) ∠BCD = 76° ∠BCD = ∠BAD	You can work out the angle BCD by subtracting the two known angles of the triangle BCD from 180°.
The triangles are congruent using the condition SAS.	State the condition you have used to show the triangles are congruent.

Practice

1. State whether each pair of triangles are congruent, not congruent or if there is not enough information to tell.

a

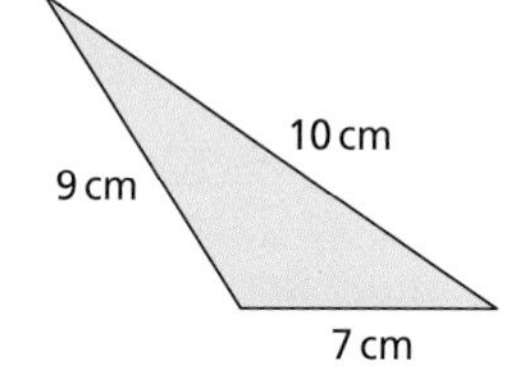

b

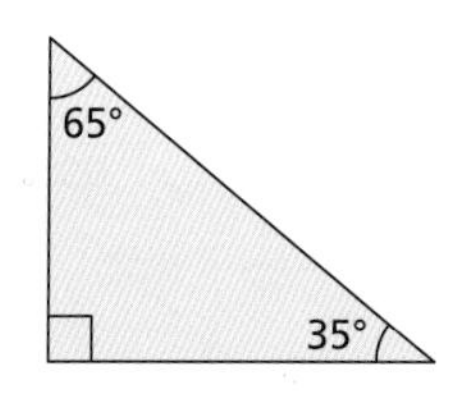

c

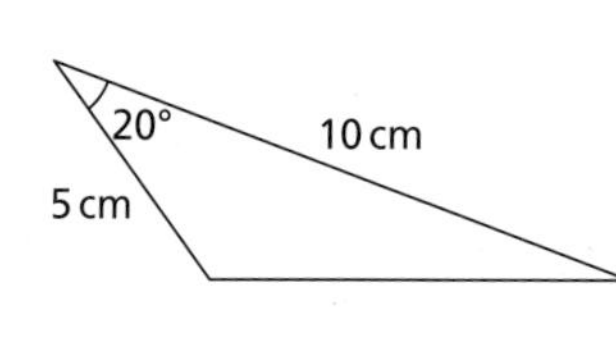

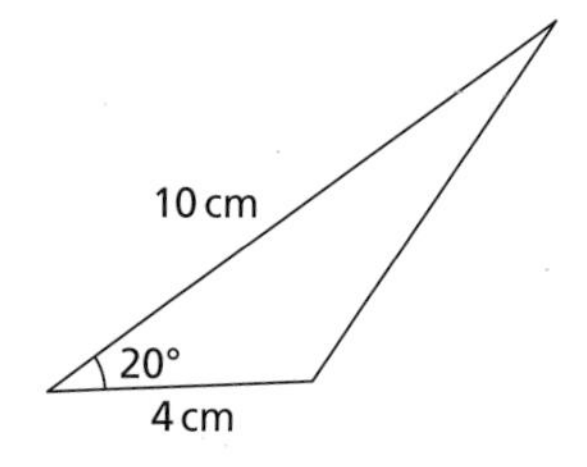

2. The two triangles are congruent.

   a Which side of triangle ABC corresponds to DE?

   b Which side is between angle ACB and angle ABC?

   c Write down the length of side DE.

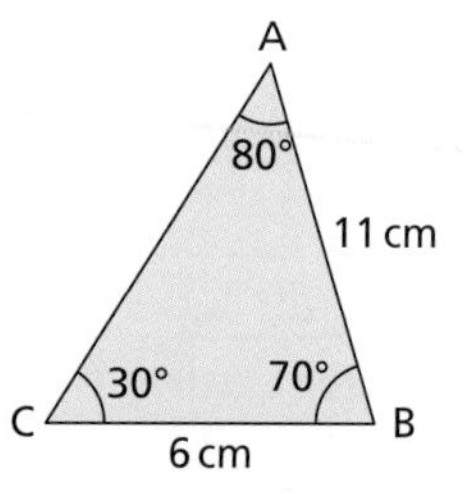

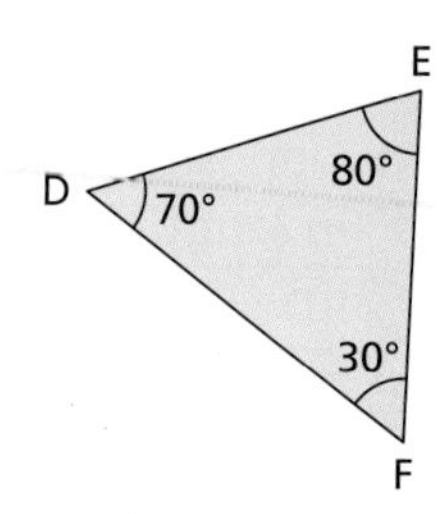

3. Here are three triangles, A, B and C.

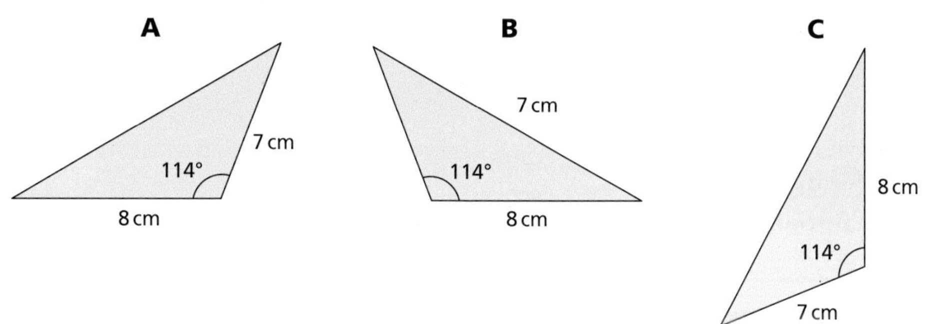

Which two triangles are congruent? Give a reason for your answer.

4. Four triangles A, B, C and D are shown.

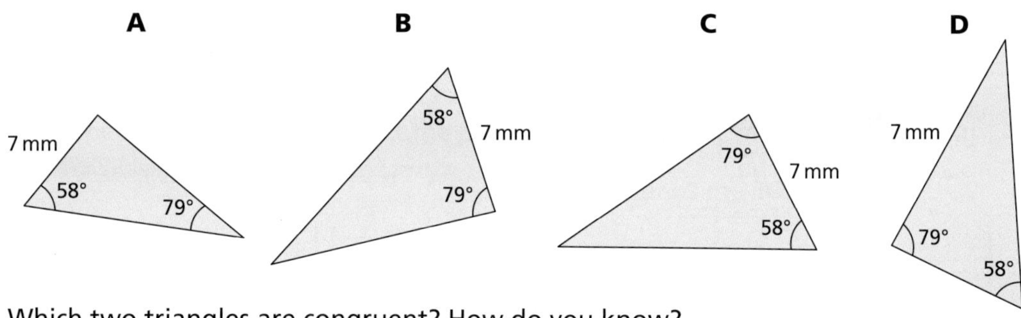

Which two triangles are congruent? How do you know?

5. Explain why each pair of triangles are congruent.

a 10 mm, 7 mm; 7 mm, 10 mm

b 11 cm, 3 cm, 9 cm; 3 cm, 11 cm, 9 cm

c 12 m, 65°, 7 m; 12 m, 65°, 7 m

6. Each diagram shows a shape split into two triangles.

For each diagram, prove that the pairs of triangles formed are congruent to each other. Give a reason for each answer.

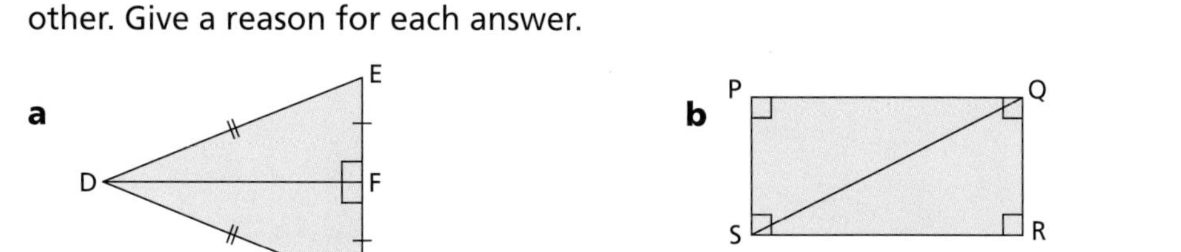

7 The two triangles are congruent.

Work out the values of x and y.

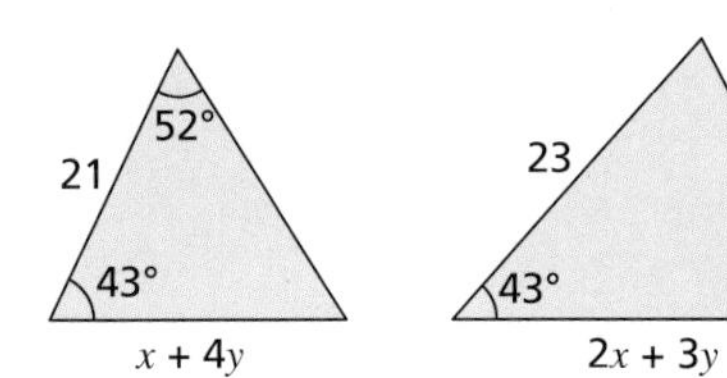

8 AE and CD are parallel.

Prove that triangles ABE and BCD are congruent.

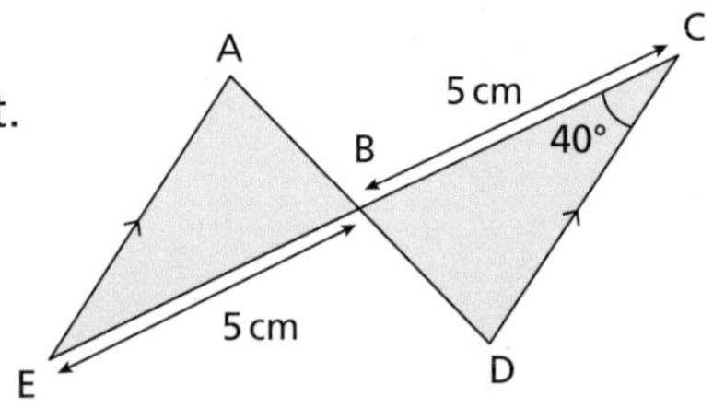

9 AB is parallel to DE.

a Show that triangles ABC and CDE are similar.

b What information would you need to prove that triangles ABC and CDE are congruent?

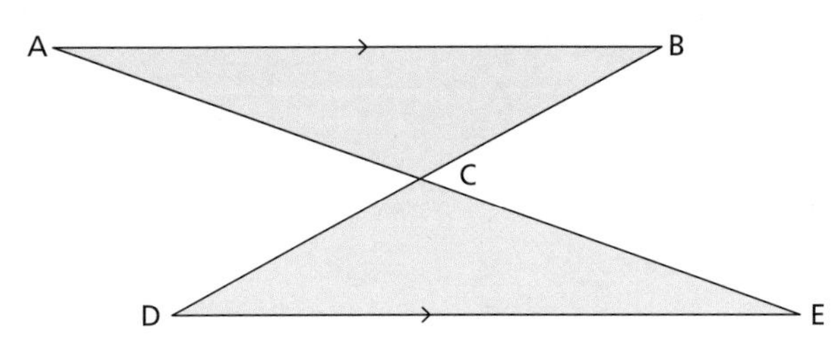

What do you think?

1 Two triangles are shown.

Chloe says, "I only need one more piece of information to prove that the triangles are congruent."

Which rule of congruence is Chloe going to use?

Is there more than one rule she could use?

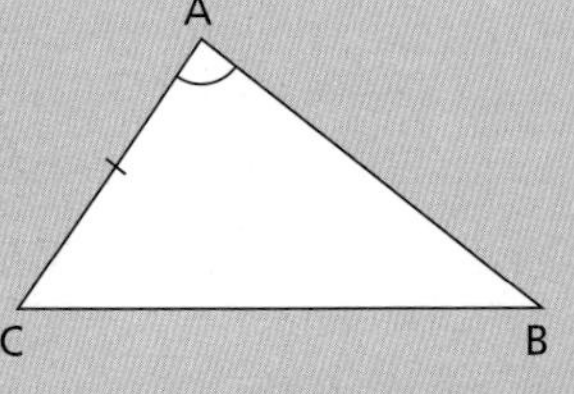

2 All quadrilaterals can be split into two congruent triangles.

Investigate the statement.

3 Here is a rhombus, ABCD.

In how many ways can you prove that triangles ABC and ADC are congruent?

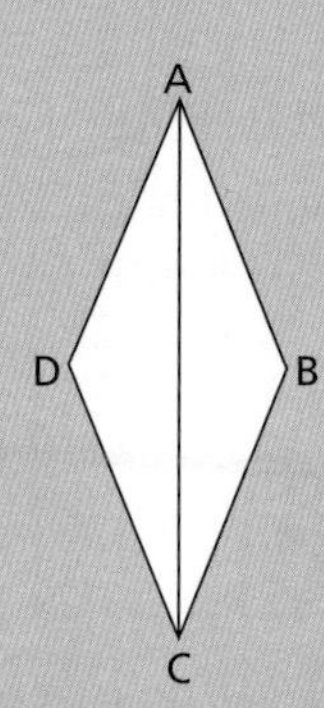

Consolidate – do you need more?

1 State the condition which proves that each pair of triangles are congruent.

a

11 cm
8 cm
9 cm
8 cm
11 cm
9 cm

b

5 cm
35°
8 cm
5 cm
35°
8 cm

c

54°
9 mm
61°

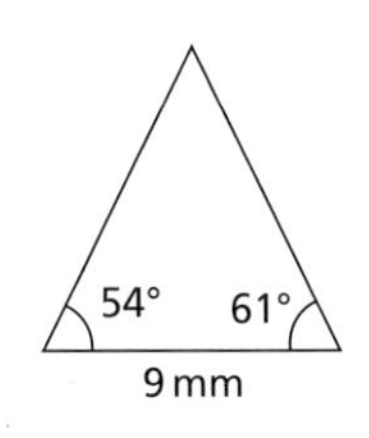

d

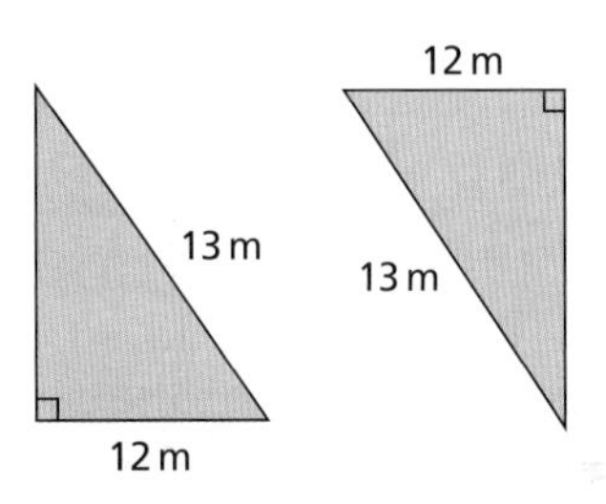

e

15 km
125°
24°

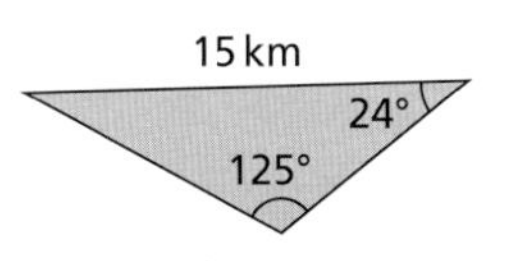

f

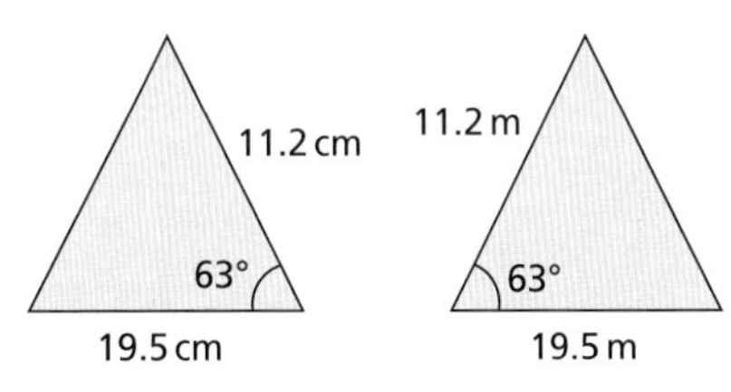

g

7.3 m
6.7 m
9.4 m

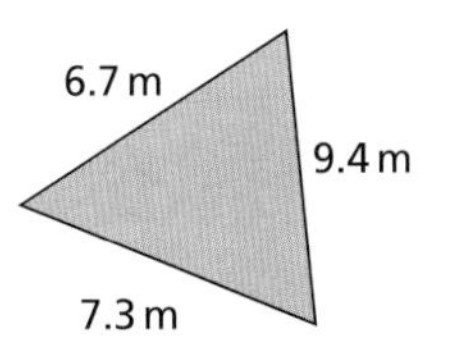

h

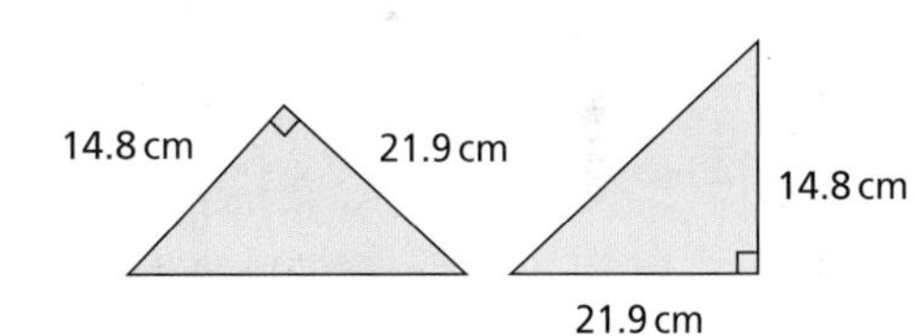

i

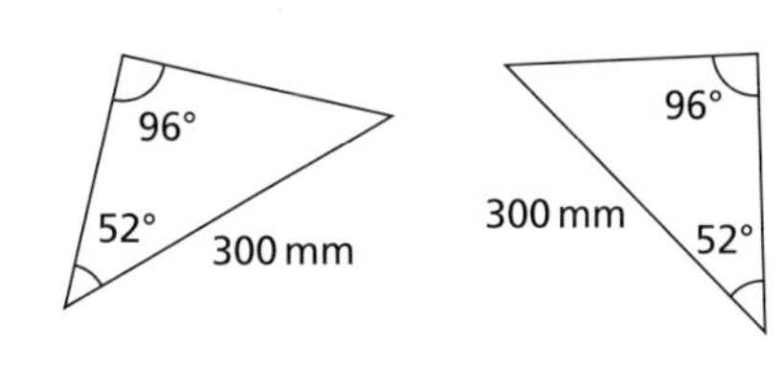

2 Here are three triangles, A, B and C.

A

5 cm
120°
11 cm

B

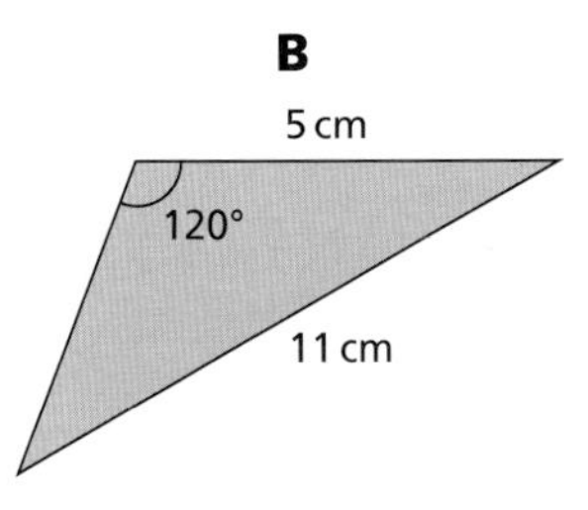

C

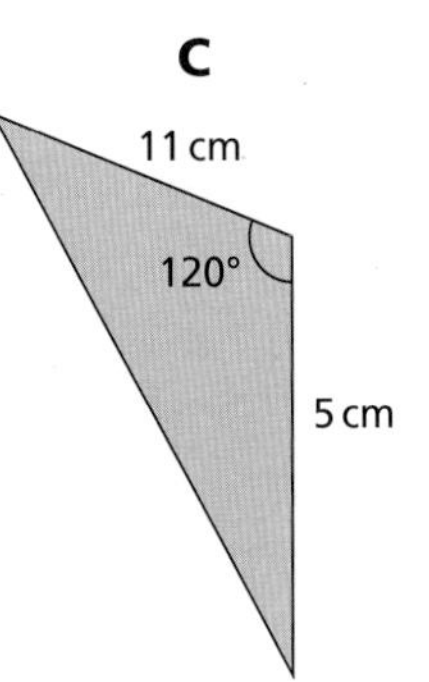

Which two triangles are congruent? State a reason for your answer.

3 ABCD is a parallelogram.

Prove that triangles ABD and BCD are congruent.

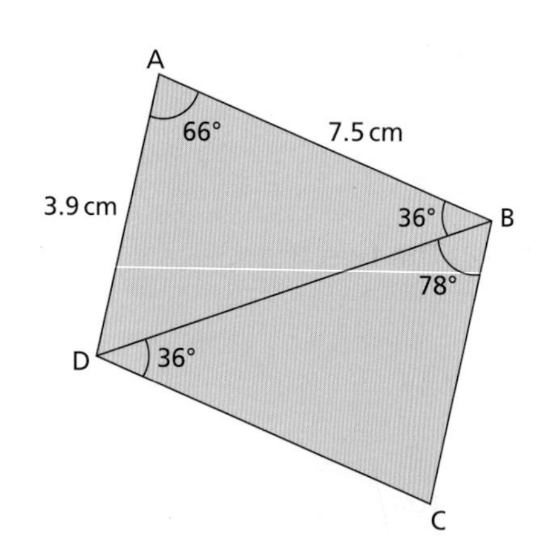

Stretch – can you deepen your learning?

1 ABCD and EFGH are both squares.

Each vertex of EFGH touches the perimeter of ABCD.

Prove that the triangles formed are congruent.

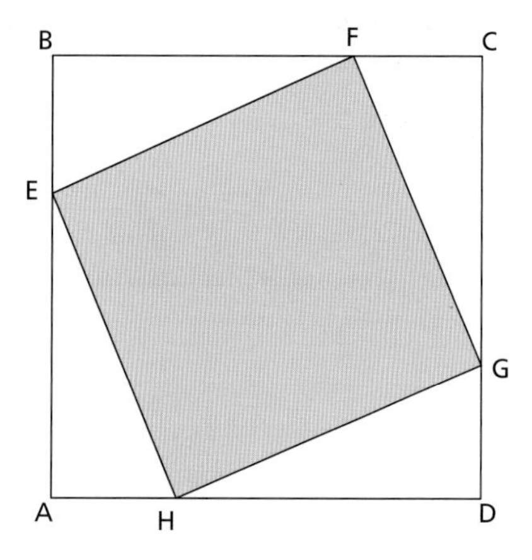

2 ABCD and CEFG are congruent rectangles.

Prove that the triangles BCG and CDE are congruent.

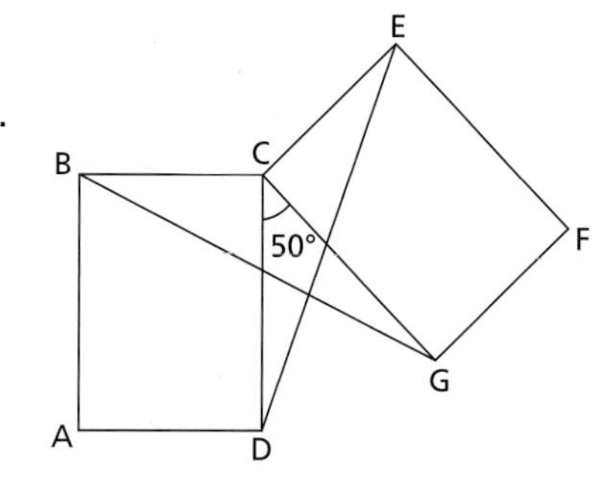

3 ABCD is an isosceles trapezium.

M is the midpoint of AB.

X is the point on BC such that BX : XC = 1 : 4

N is the point on CD such that CN : ND = 2 : 3

Y is the point on BC such that BY : YC = k : 1

The triangles BMX and CYN are congruent.

Work out the value of k.

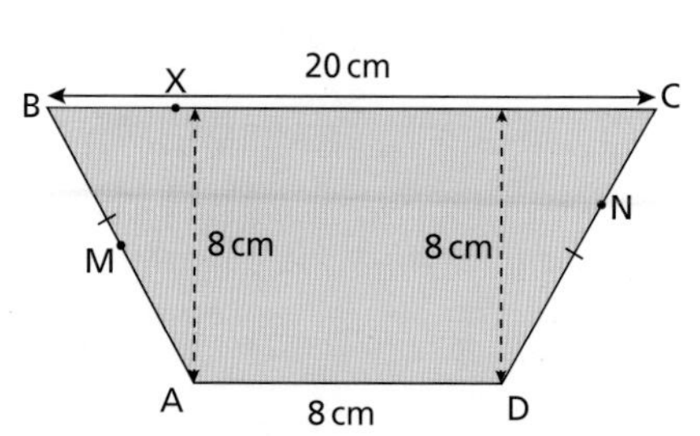

2.3 Similar triangles

Are you ready?

1 The two triangles are similar.

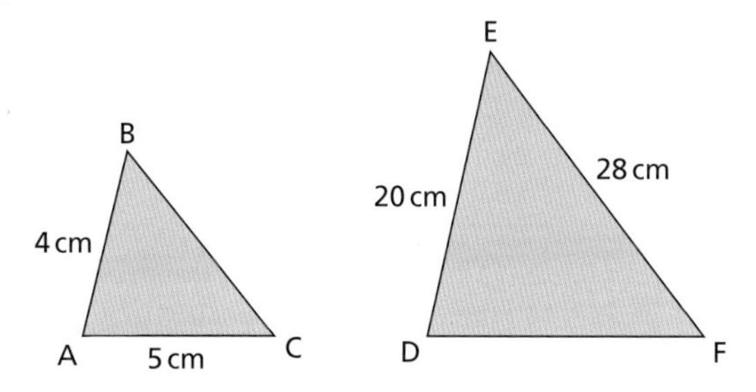

a Work out the length of DF.

b Work out the length of BC.

2 Here are two congruent triangles.

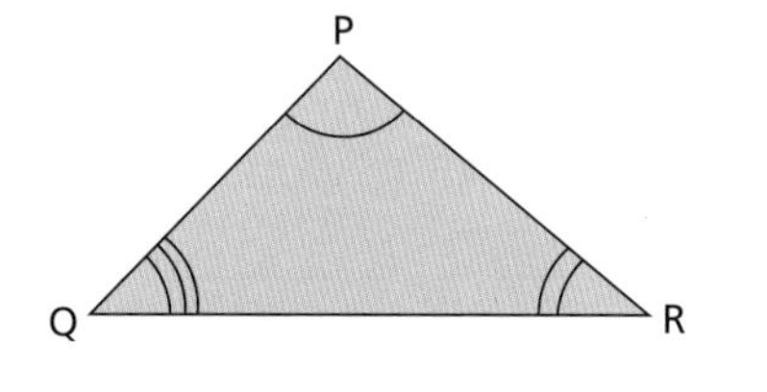

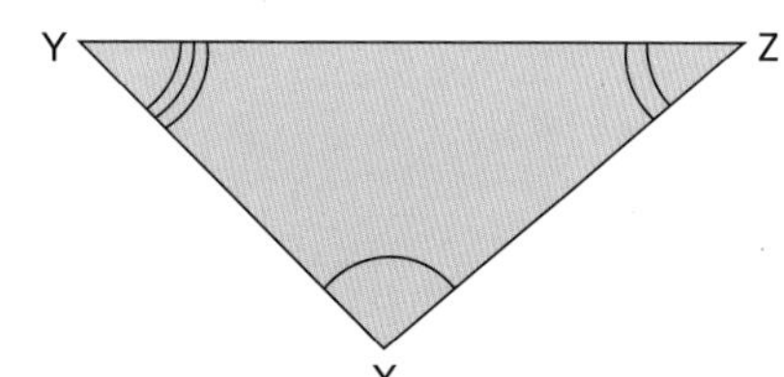

a Write the name of the side which corresponds to PQ.

b Write the name of the side which corresponds to YZ.

c Write the name of the angle which corresponds to ∠PRQ.

Two triangles are similar if all three of their angles are equal. One triangle will be an exact enlargement of the other.

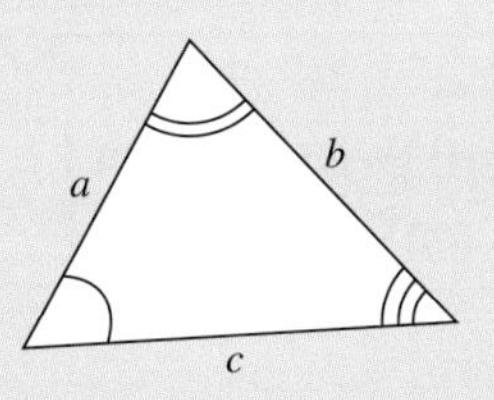

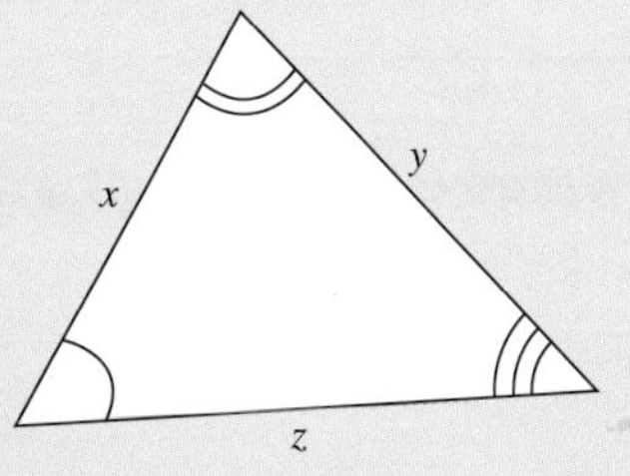

Here, the scale factor of enlargement is $\frac{x}{a}$ or $\frac{y}{b}$ or $\frac{z}{c}$. The ratios $x : a$, $y : b$ and $z : c$ are all equivalent.

To show that two triangles are similar, you have to show that **either** they have equal angles **or** that the corresponding sides are all in the same ratio.

Example 1

Triangles ABC and DEF are similar.

a Work out the length of EF.

b Work out the length of AC.

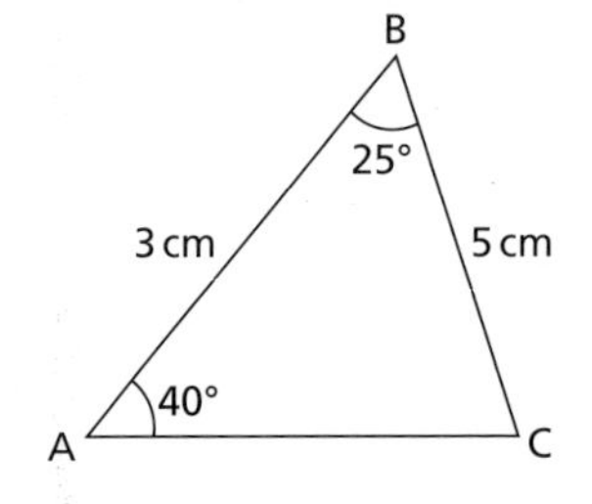

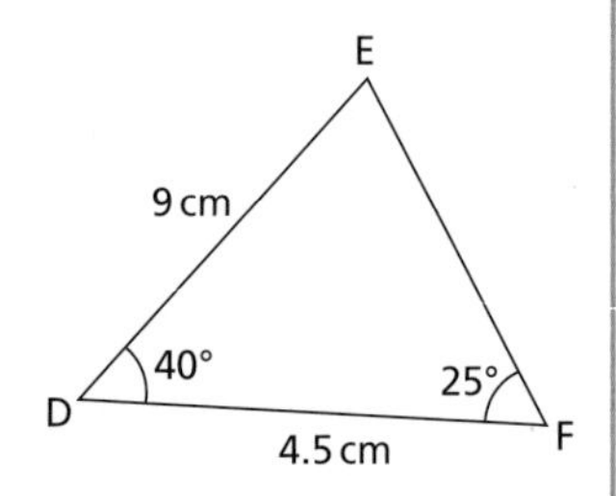

Method

Solution	Commentary
a	First you need to identify the sides that are corresponding. AB and DF are corresponding, as they both lie between the 40° and 25° angles.
Scale factor = 4.5 ÷ 3 = 1.5	Divide the length of DF by the length of AB to work out the scale factor.
EF = 5 × 1.5 = 7.5 EF = 7.5 cm	EF corresponds to BC as they are both opposite the 40° angle. Multiply the length of BC by 1.5
b 9 ÷ 1.5 = 6 AC = 6 cm	AC corresponds to DE as they are both opposite the 25° angle. Divide the length of DE by 1.5

Example 2

Triangles ABC and ADE are similar.

Work out the lengths of:

a BC **b** EC

Method

Solution	Commentary
a	It can help to draw the two triangles separately.
AB = 4 + 12 = 16 cm Scale factor = AB ÷ AD = 16 ÷ 4 Scale factor = 4	Side AB corresponds to side AD. You can use these to find the scale factor.
6 × 4 = 24 BC = 24 cm	Side BC corresponds to side ED.
b 3.8 × 4 = 15.2 AC = 15.2 cm	Work out the length of AC by multiplying AE by the scale factor.
15.2 – 3.8 = 11.4 EC = 11.4 cm	Subtract the length of AE from the length of AC to work out the length of EC.

Example 3

The diagram shows triangles ABD and ECD.

EC is parallel to AB.

Show that triangles ABD and ECD are similar.

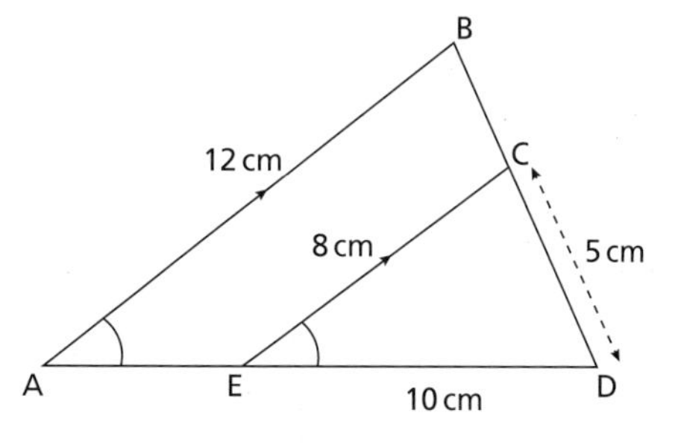

Method

Solution	Commentary
∠BAD = ∠CED as corresponding angles are equal. ∠ABD = ∠ECD as corresponding angles are equal. ∠BDA = ∠CDE as it is the same angle.	Use your knowledge of angle geometry to identify equal angles.
The triangles have the same angles so they are similar.	State your conclusion clearly, justifying why the triangles are similar.

Practice

1. In each part, the two triangles are similar.

   Work out the sides and angles labelled with letters in each pair of triangles.

   a

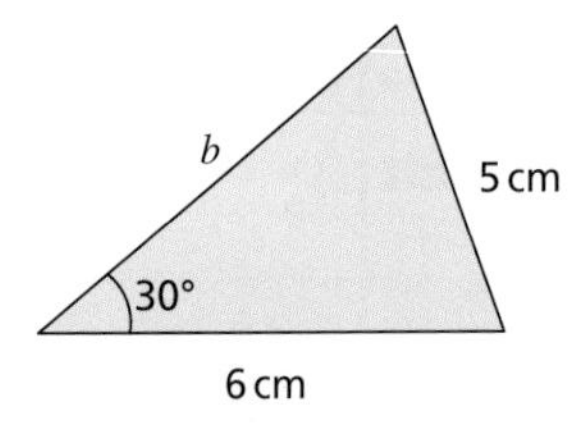

   b

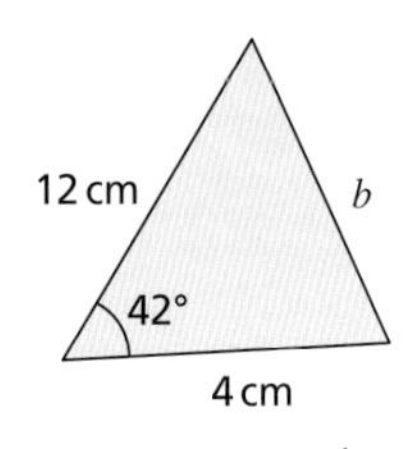

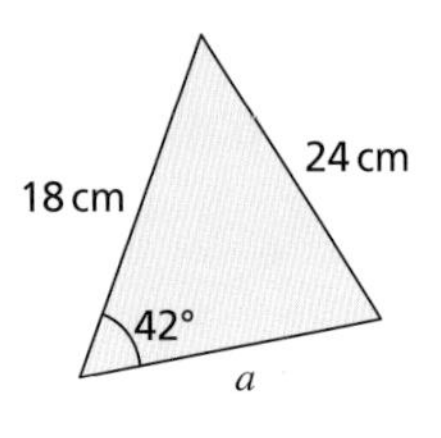

   c

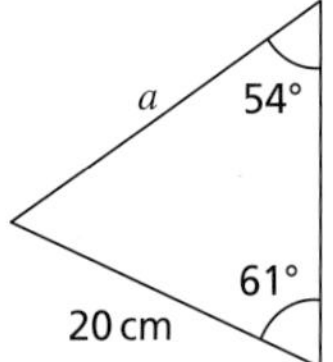

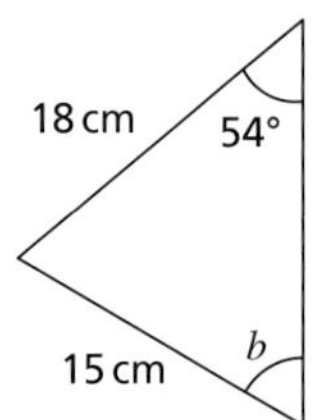

2. The triangles are similar.
   a Work out the length of a.
   b Work out the length of b.

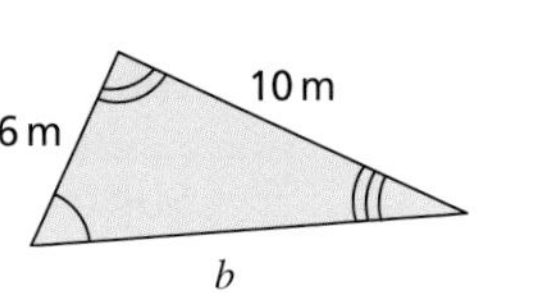

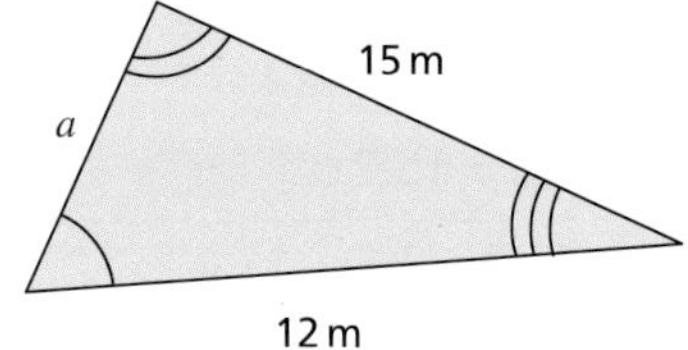

3. The triangles are similar.
   a Work out the length of a.
   b Work out the length of b.

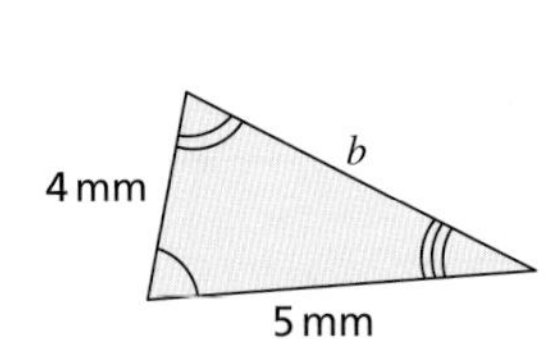

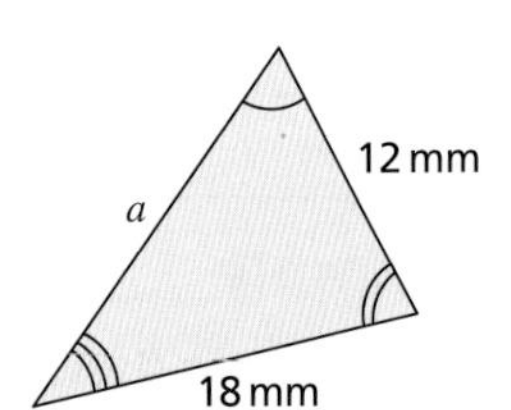

4. The triangles are similar.
   a Work out the length of x.
   b Work out the length of y.

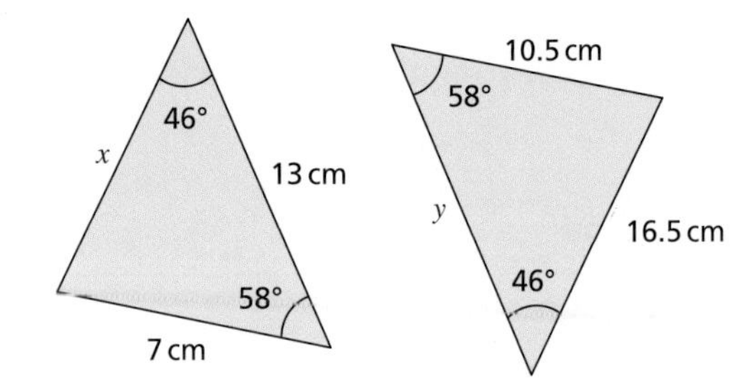

5. The triangles are similar.
   a Work out the length of a.
   b Work out the length of b.

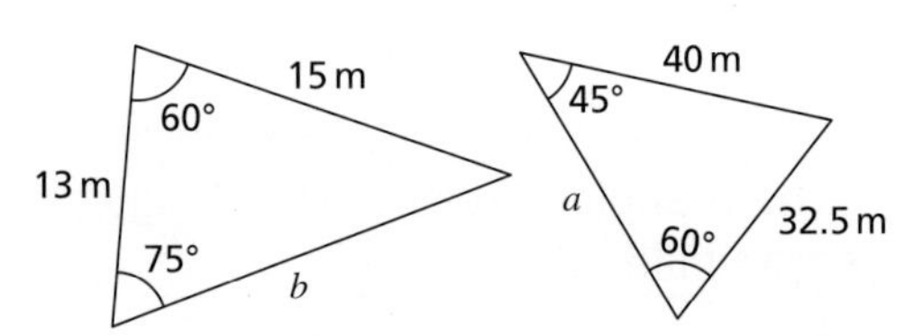

6 Here is the triangle ADE.

BC is parallel to DE.

DE = 18 cm, BC = 12 cm, AB = 8 cm

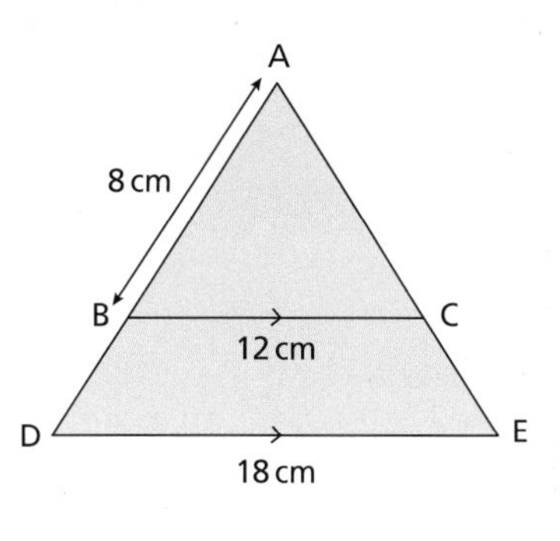

a Explain why triangles ABC and ADE are similar.

b Work out the length of side AD.

c Work out the length of side BD.

7 AB and DE are parallel line segments.

Work out the length of DC.

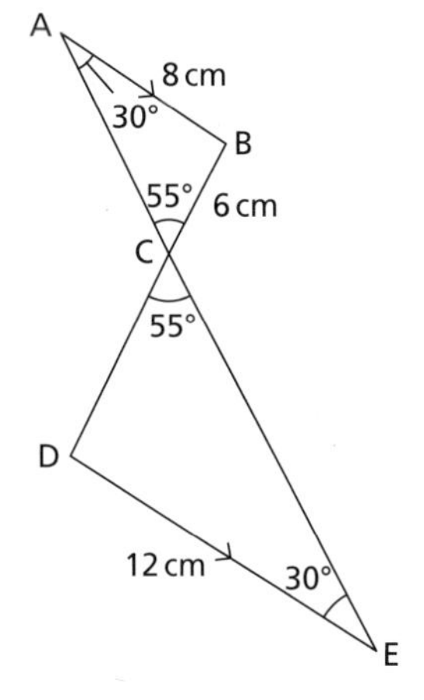

8 AB and DE are parallel line segments.

a Work out the length of DC.

b Work out the length of AC.

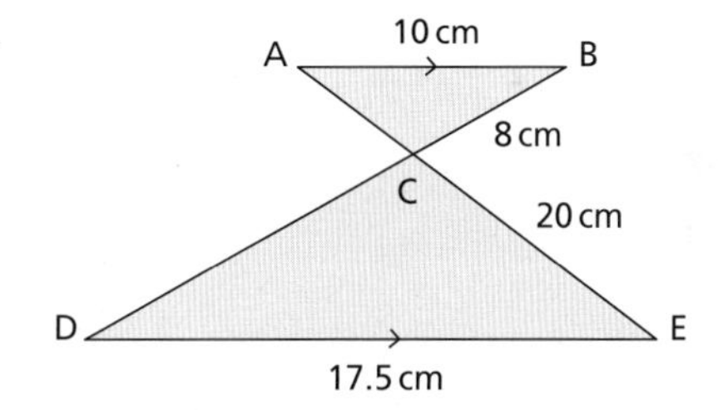

9 Show that these triangles are similar.

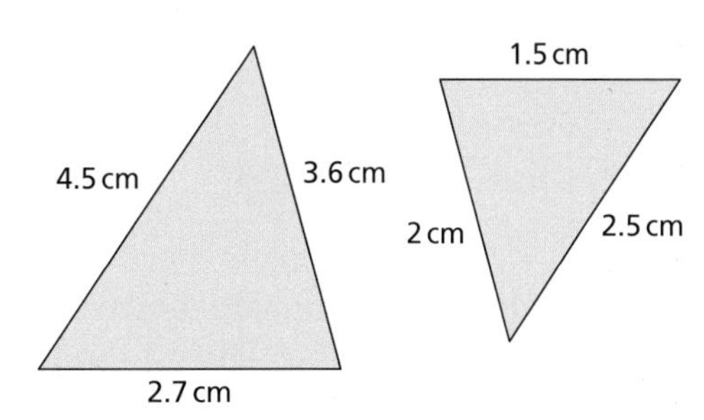

10 ABD and BCD are similar triangles.

a Work out the length of BD.

b Work out the perimeter of triangle ABC.

You may need to refer to Chapter 6.2 (Trigonometry).

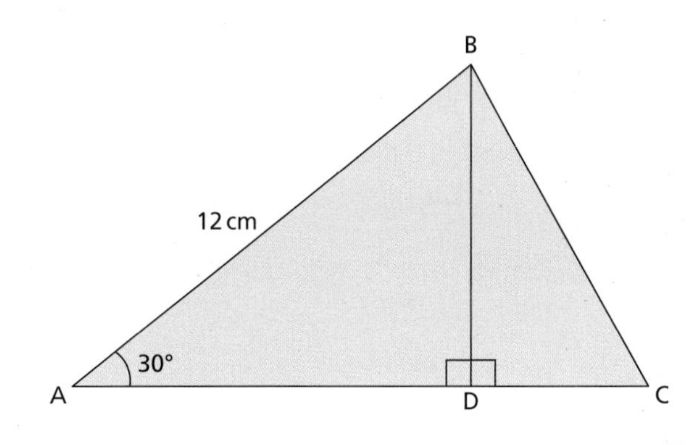

11 AB and DE are parallel line segments.

Work out the value of x.

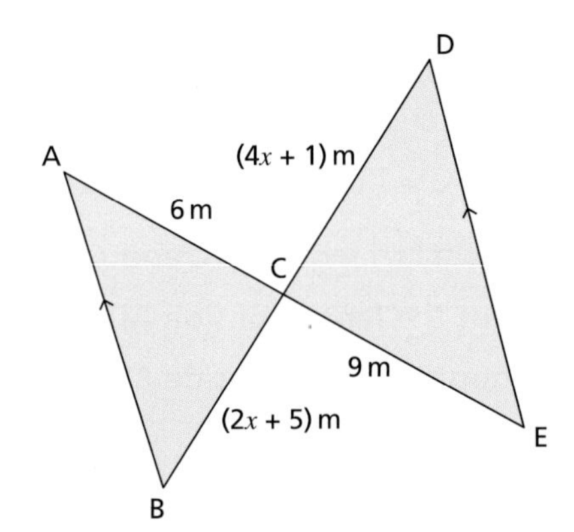

12 Here are two similar triangles.

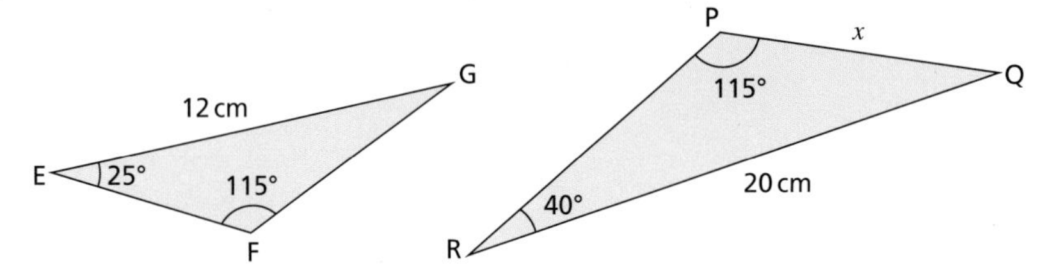

a Write an expression for the length of EF in terms of x.

b The perimeter of triangle EFG is 38.4 cm.

Write an expression for the length of FG in terms of x.

c The perimeter of PQR is 48 cm and PR = 10.35 cm.

Calculate the length of each unknown side of the triangles.

What do you think?

1 Here are three similar triangles:

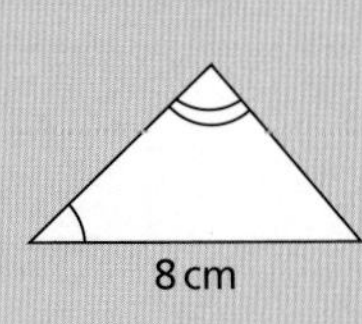

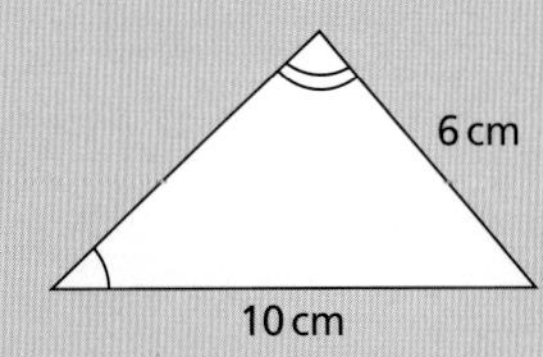

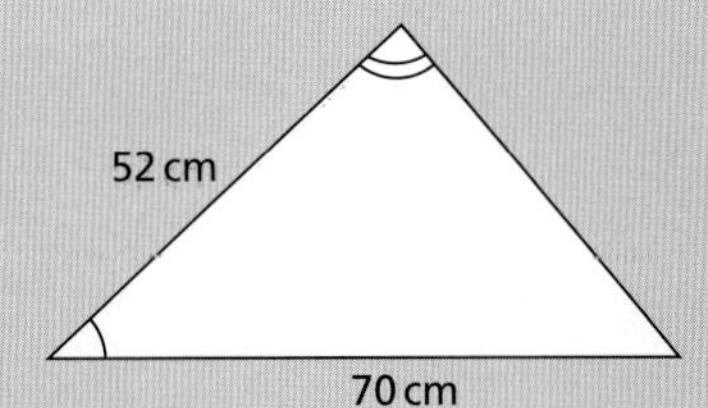

What information can you work out about the triangles?
Compare your findings and methods with a partner.

2 Here is a triangle.

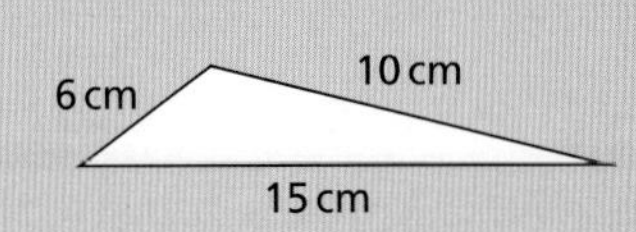

The triangle is enlarged. One of the side lengths of the new triangle measures 8 cm.

a Work out the possible scale factors for the enlargement.

b Sketch all the possible enlarged triangles.

3 Samira says, "If two angles in a pair of triangles are equal then the triangles are similar."

Explain why she is correct.

Consolidate – do you need more?

1 For each pair of similar triangles, work out the lengths of sides a and b.

a

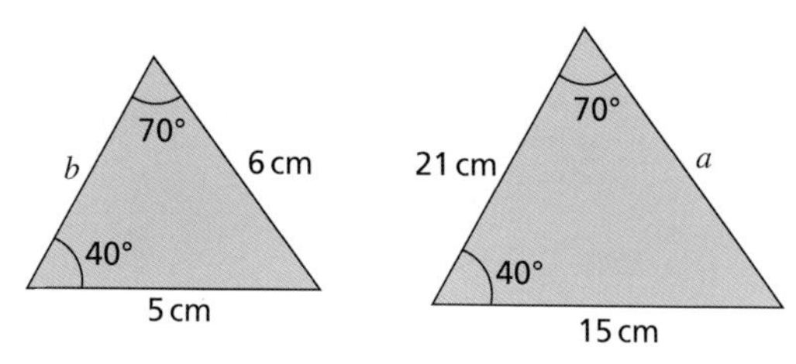

b

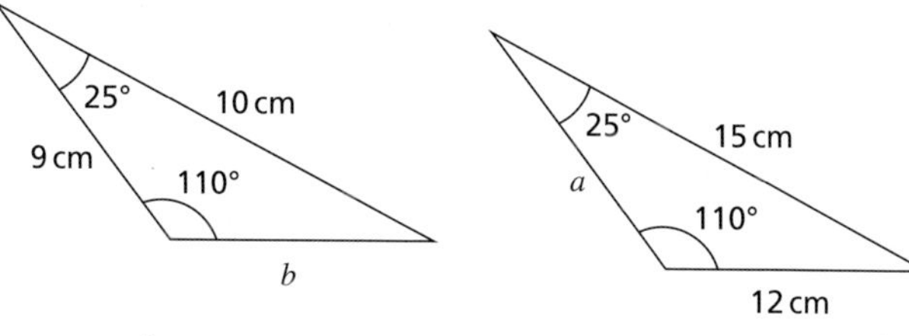

c

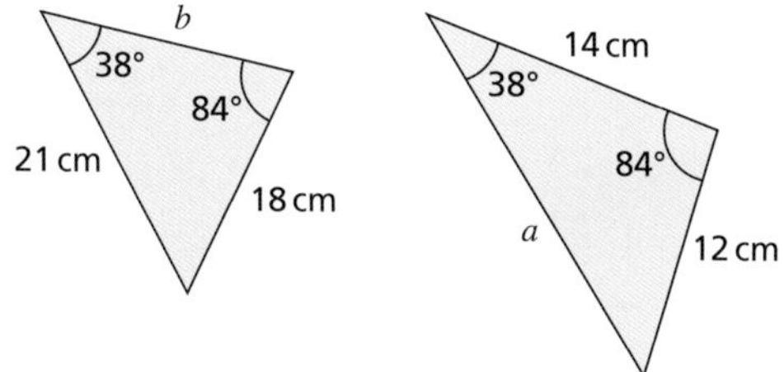

2 For each pair of similar triangles, work out the lengths of sides a and b.

a

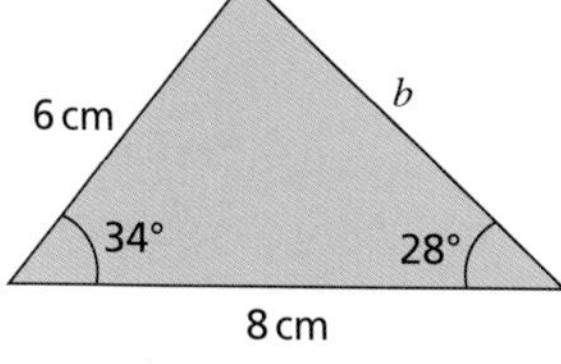

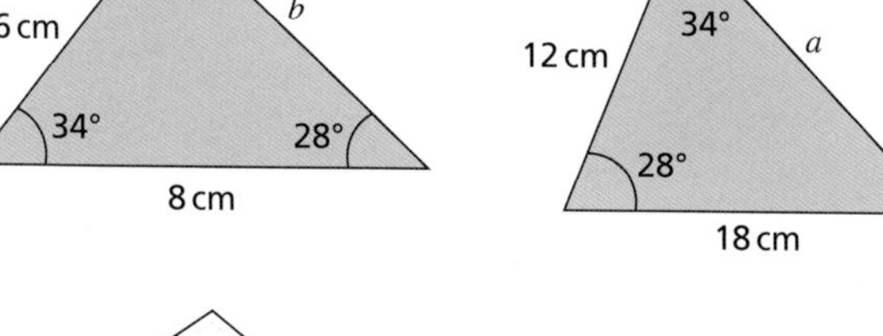

b

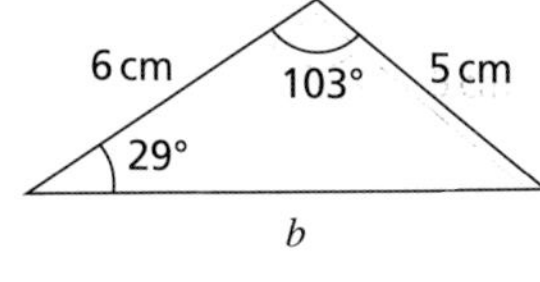

c

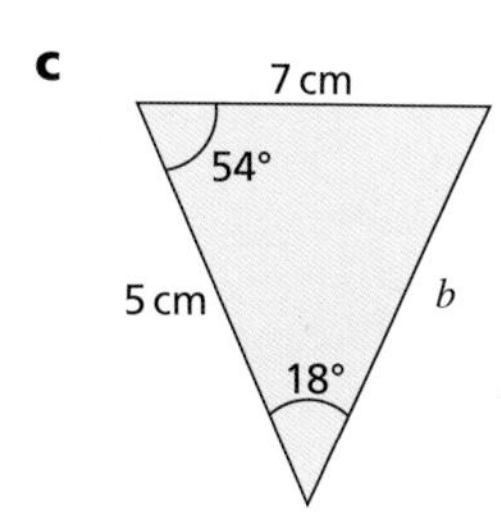

Stretch – can you deepen your learning?

1 These two triangles are similar.

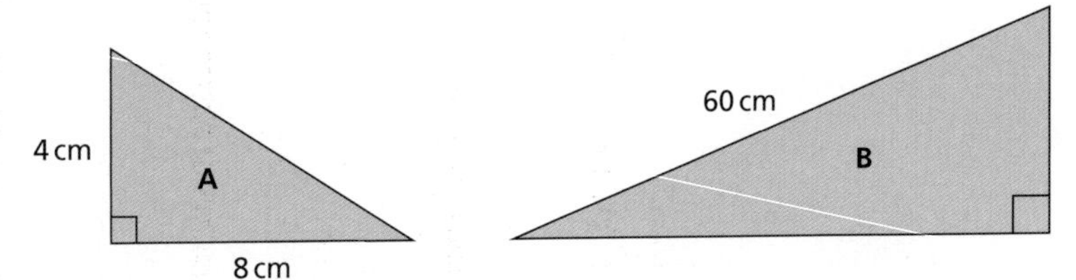

Work out the perimeter of triangle B. Give your answer in the form $(a + \sqrt{b})$ cm, where a and b are integers.

2 ABD and ECD are similar triangles.

There are two possible values for x.

Work out each of the values.

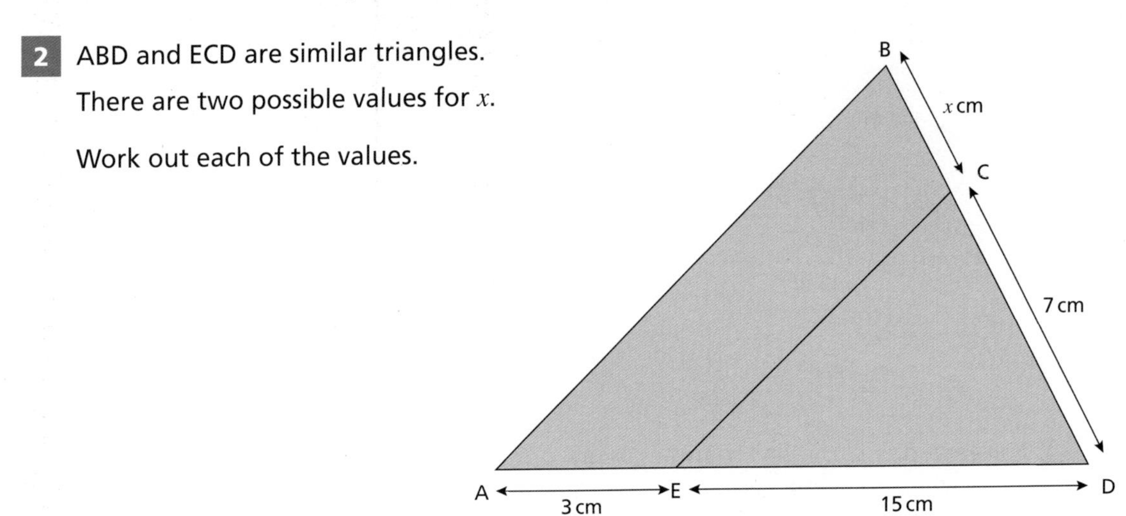

Congruence and similarity: exam practice

4–6

1 The diagram shows rectangle ABCD.

State three different pairs of congruent triangles shown in the diagram. **[3 marks]**

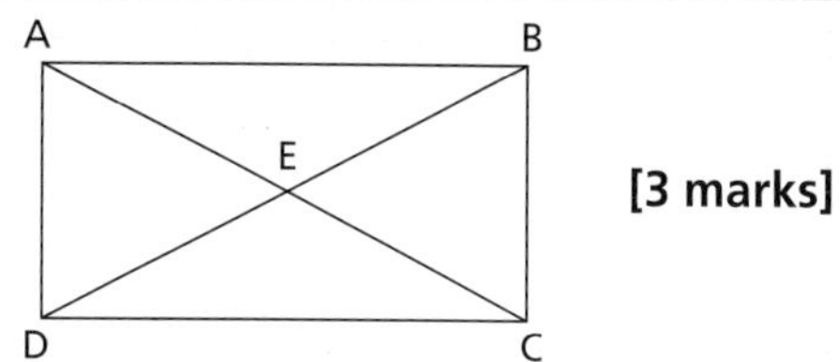

2 Here is rectangle A.

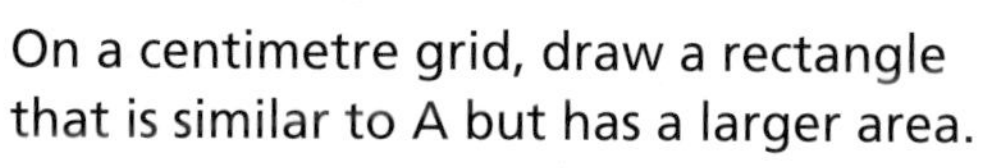

On a centimetre grid, draw a rectangle that is similar to A but has a larger area. **[1 mark]**

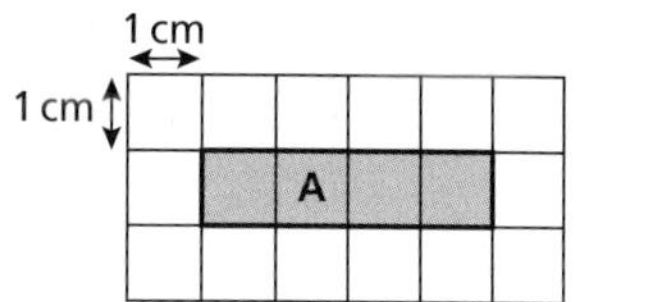

3 Quadrilaterals ABCD and PQRS are similar.

Calculate the length of CD. **[3 marks]**

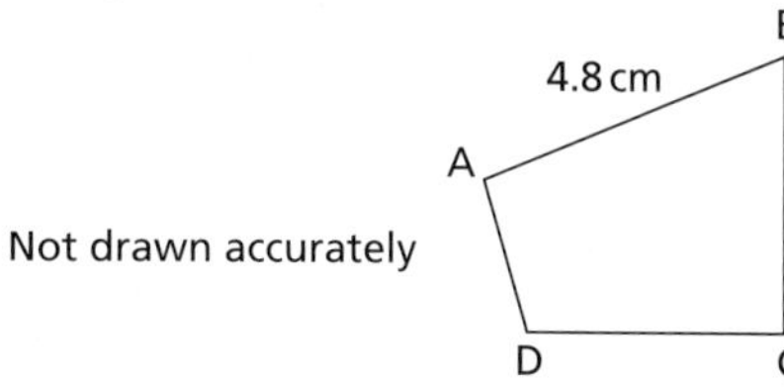

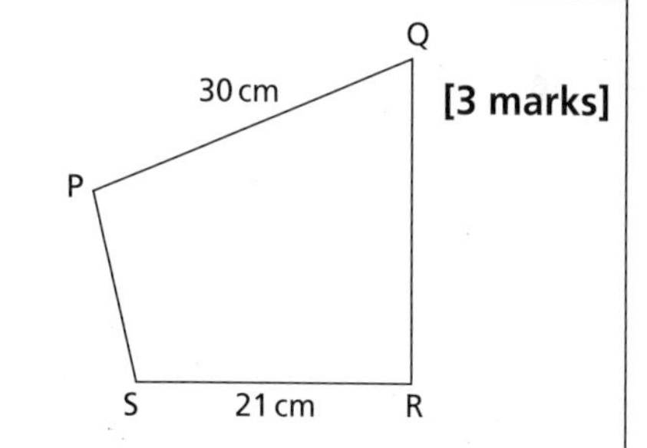

4 Triangle X has sides of length $5a$, $6a$ and $8a$ cm.

Triangle Y has sides of length $5b$, $6b$ and $8b$ cm.

Explain why triangle X is similar to triangle Y and state any further information needed to ensure that the two triangles would be congruent. **[2 marks]**

7–9

5 Work out the length of x. **[3 marks]**

Not drawn accurately

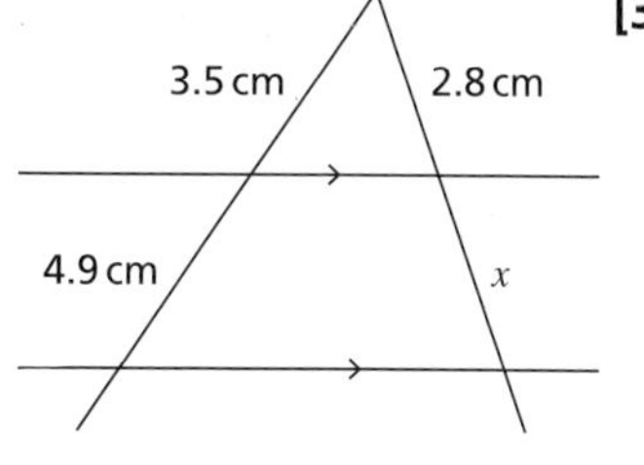

6 In triangle ABC, X is the midpoint of BC, and AX is perpendicular to BC.

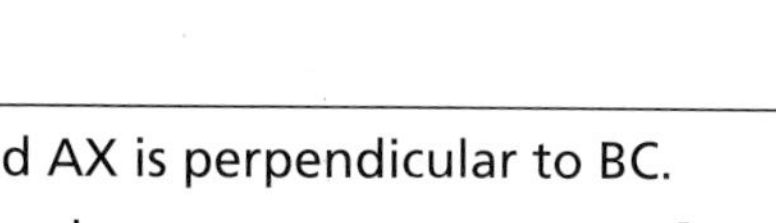

Prove that ABX and AXC are congruent triangles. **[4 marks]**

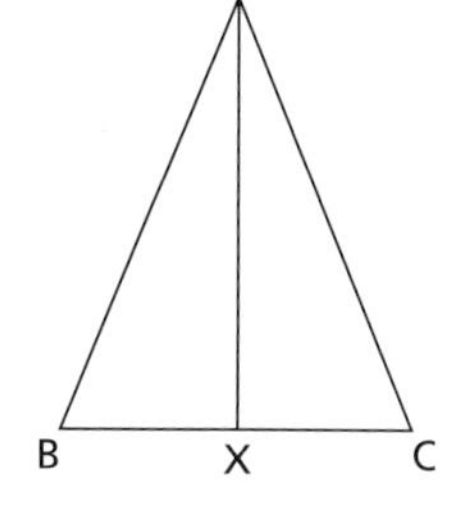

3 Angles and circles

In this block, we will cover...

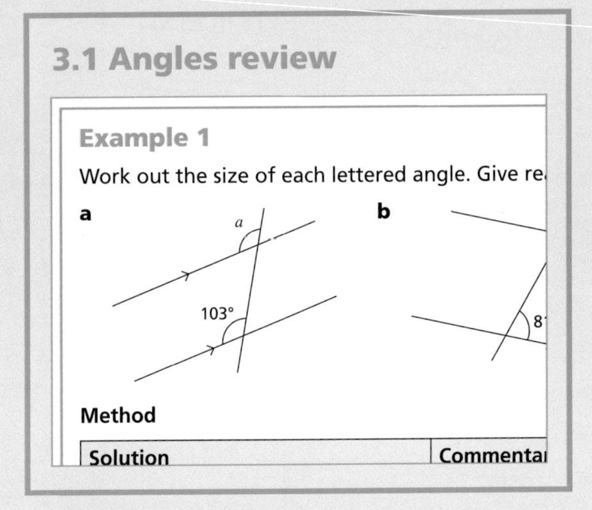

3.1 Angles review

Example 1

Work out the size of each lettered angle. Give re

a

b

Method

Solution	Commentar

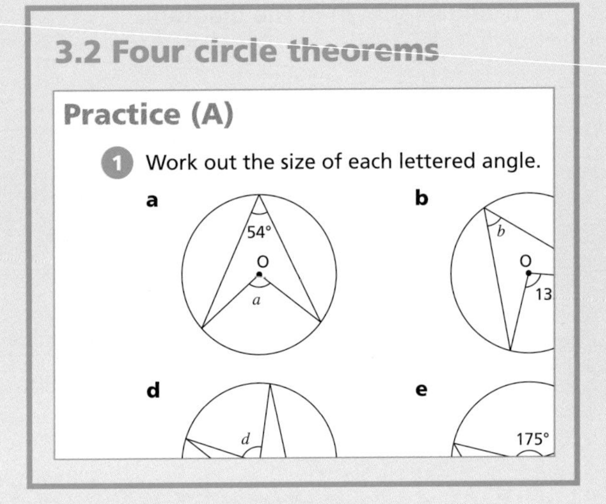

3.2 Four circle theorems

Practice (A)

1. Work out the size of each lettered angle.

a

b

d

e

3.3 Four more circle theorems

Consolidate – do you need more

1. M is the midpoint of the chord AB in the c

   $\angle BOM = 40°$

   Write down the size of angle OBM.

   Give a reason for your answer.

2. M is the midpoint of the chord AB in the c

   $\angle OAM = 35°$

3.4 Parts of circles

Stretch – can you deepen your le

1. The radius of the sector is the same length

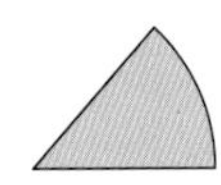

The square and the sector are put togethe

ABCDE has a perimeter of 55 cm.

AB = 12 cm

Work out the size of angle ABC.

3.1 Angles review

White Rose MATHS

Are you ready? (A)

1 Work out the size of each lettered angle.

a

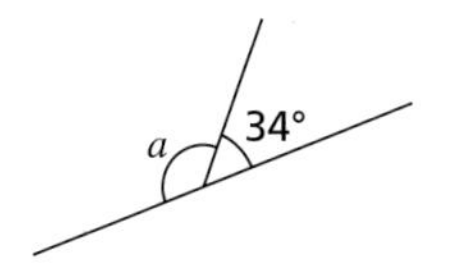

b

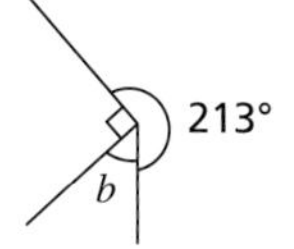

c

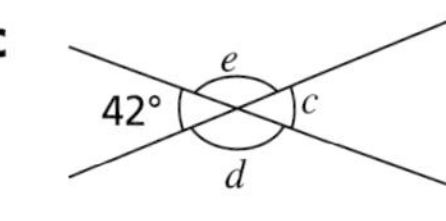

2 Work out the size of each lettered angle.

a

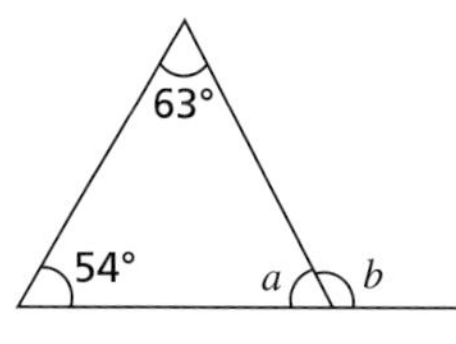

b

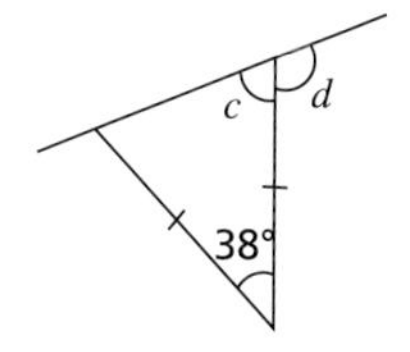

c

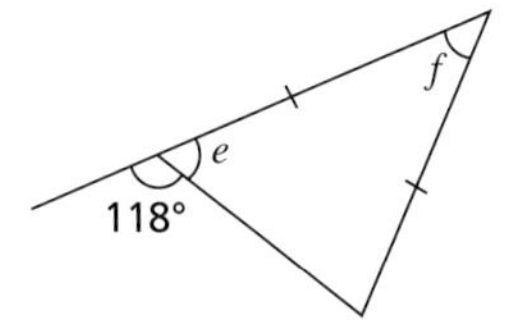

3 Work out the size of each lettered angle.

a

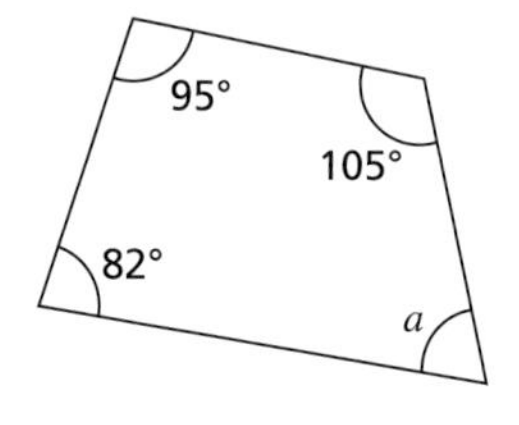

b

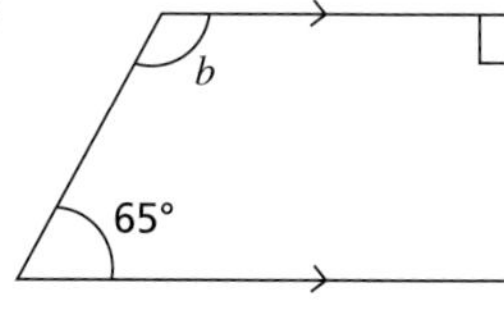

c

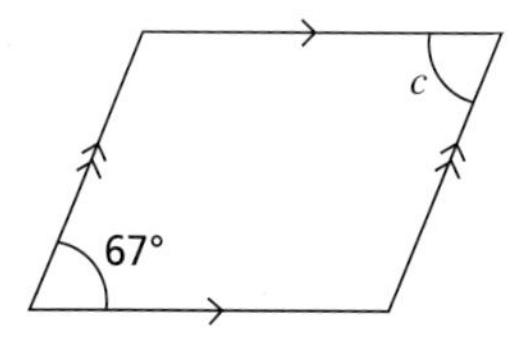

You will know many angle relationships from earlier study.

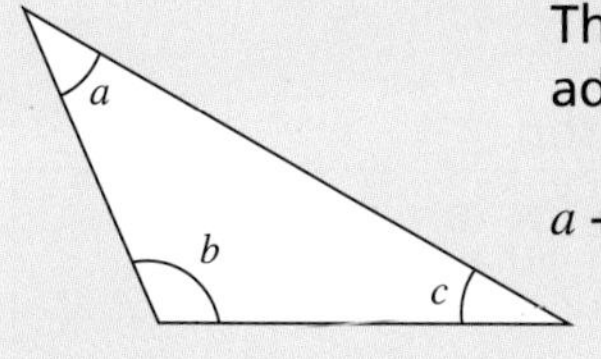

The angles in a triangle add up to 180°.

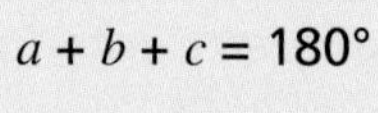

$a + b + c = 180°$

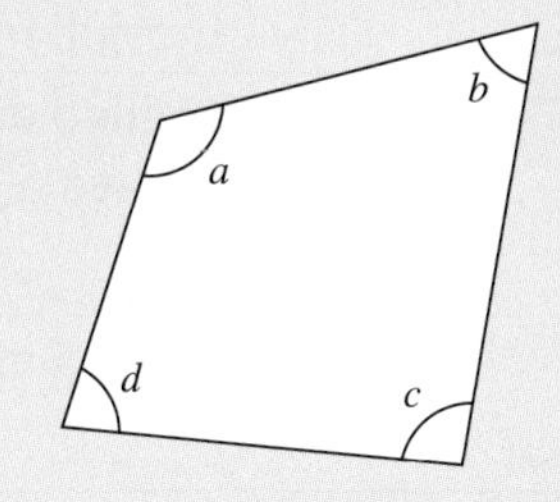

The angles in a quadrilateral add up to 360°.

$a + b + c + d = 360°$

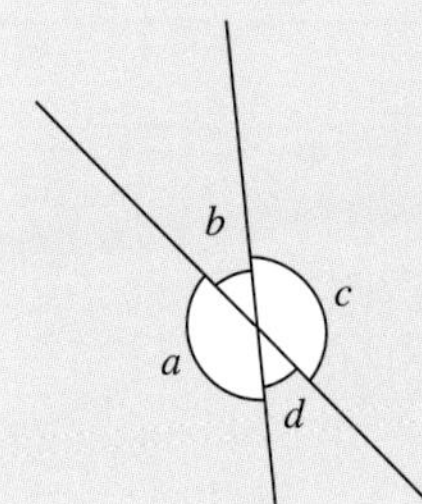

Vertically opposite angles are equal.

$a = c$ and $b = d$

Adjacent angles on a straight line add up to 180°.

$a + b = 180°$

These rules relate to angles formed when a **transversal** line cuts **parallel** lines.

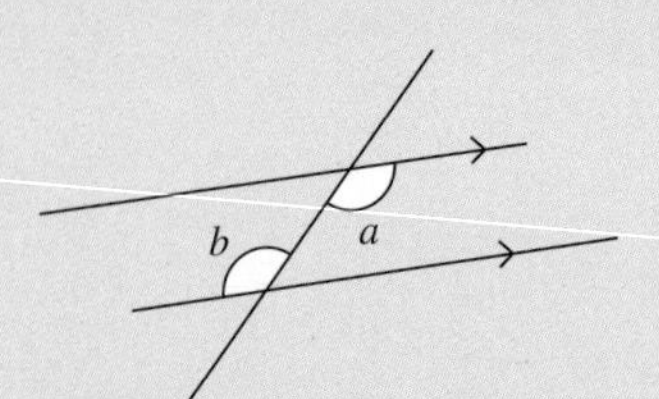

Alternate angles are equal.

$a = b$

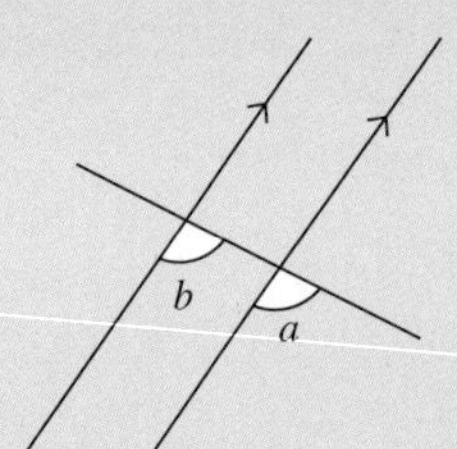

Corresponding angles are equal.

$a = b$

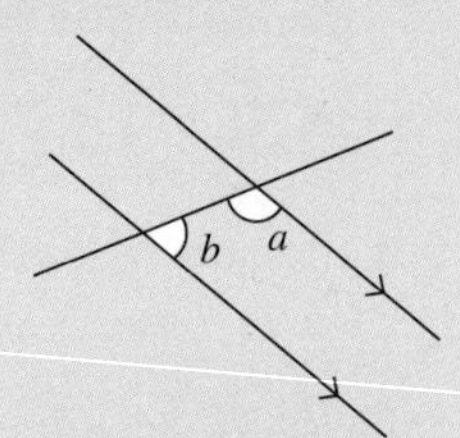

Co-interior angles add up to 180°.

$a + b = 180°$

Example 1

Work out the size of each lettered angle. Give reasons for your answers.

a

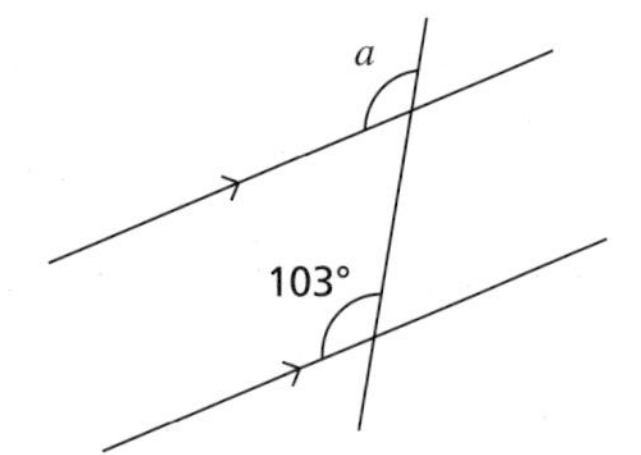

b

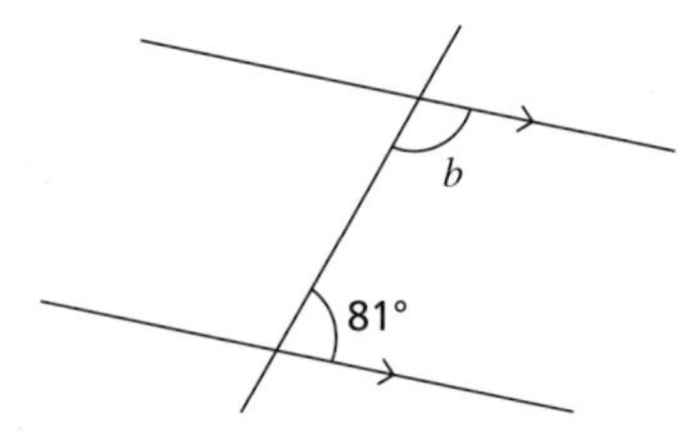

c

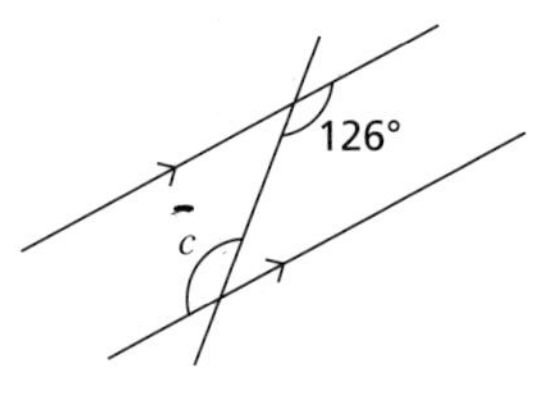

Method

Solution	Commentary
a $a = 103°$ Corresponding angles are equal.	The lettered angle is corresponding to the angle of 103°. Give the full mathematical reason.
b $b = 180° - 81° = 99°$ Co-interior angles have a sum of 180°.	The labelled angles are co-interior to each other and so have a sum of 180°.
c $c = 126°$ Alternate angles are equal.	The labelled angles are alternate to each other.

Example 2

AC and DF are parallel.

B lies on AC and E lies on DF.

EBF is a triangle.

∠DEB = 125° and ∠CBF = 42°

Work out the size of ∠EBF. Give reasons for your answer.

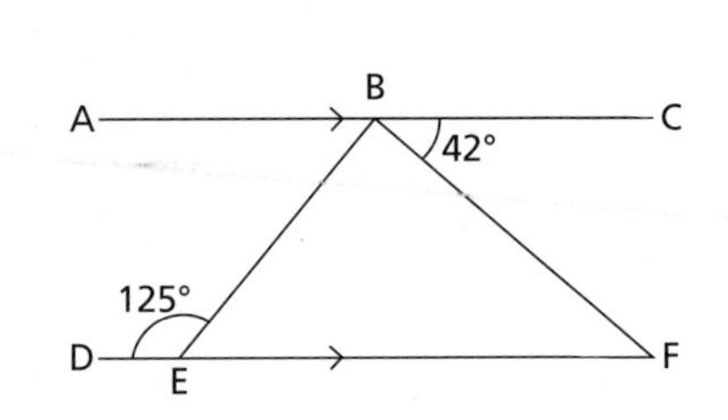

Method

Solution	Commentary
$\angle BEF = 180° - 125° = 55°$ Angles on a straight line have a sum of 180°.	Use the fact that DF is a straight line.
$\angle EFB = \angle CBF$ $\angle EFB = 42°$ Alternate angles are equal.	Use the fact that AC is parallel to DF.
$\angle EBF = 180° - 55° - 42°$ $\angle EBF = 83°$ Angles in a triangle have a sum of 180°.	You now know two of the angles in triangle EBF and so can work out the third.

Practice (A)

1. Work out the size of each lettered angle.
Give reasons for your answers.

a

b

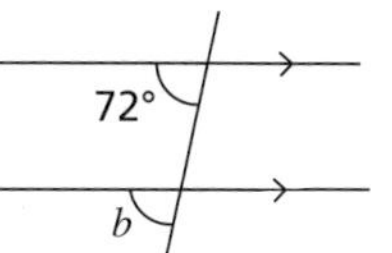

c

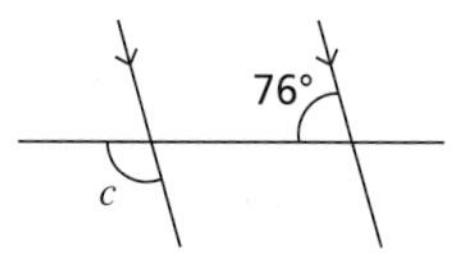

d

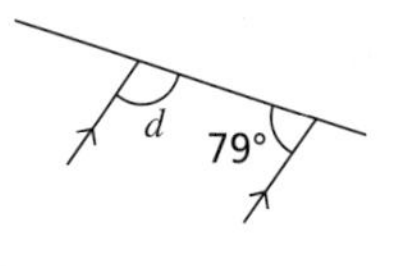

e

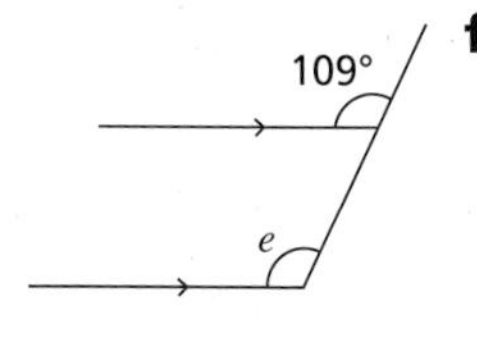

f

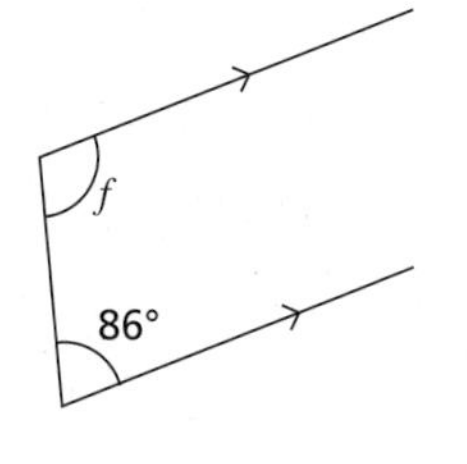

g

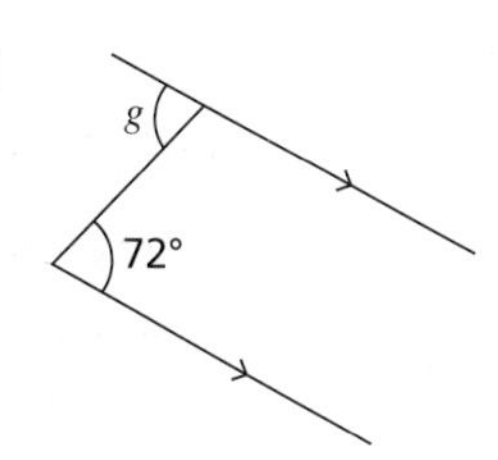

2 LM and NP are parallel, and O is a point on NP.

LN = LO

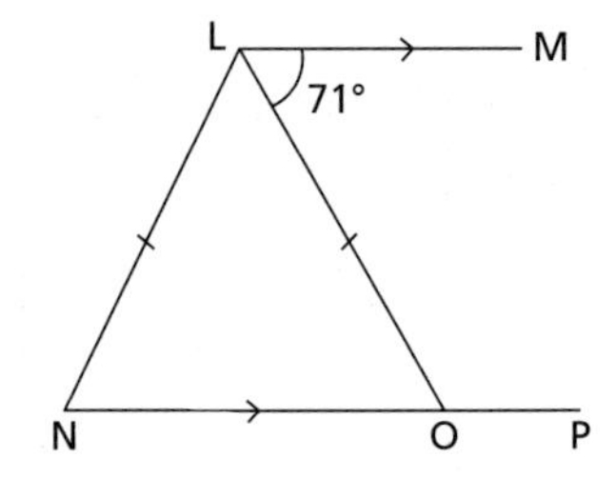

a Write the size of angle LON.
Give a reason for your answer.

b Work out the size of angle NLO.
Give a reason for your answer.

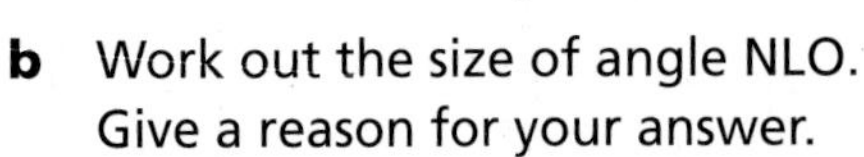

3 AB and CD are parallel.

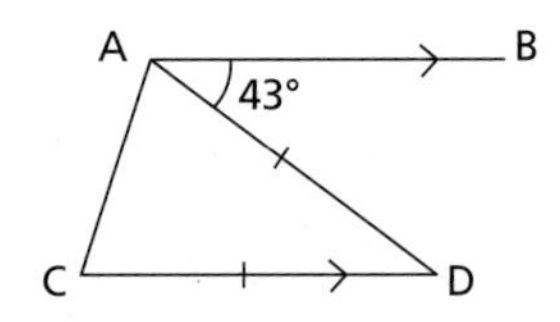

a Write the size of angle ADC.
Give a reason for your answer.

b Work out the size of angle DAC.
Give a reason for your answer.

4 Work out the size of each lettered angle. Give reasons for your answers.

a

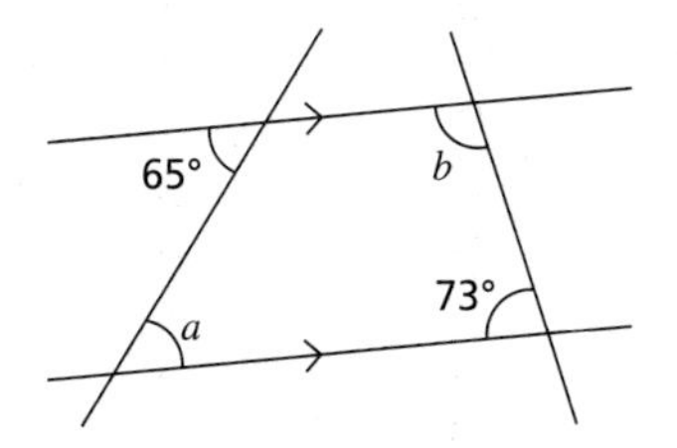

b

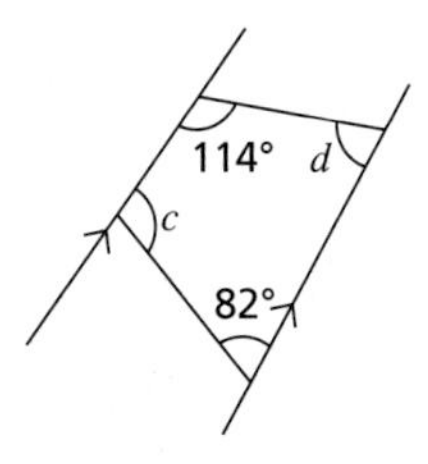

c

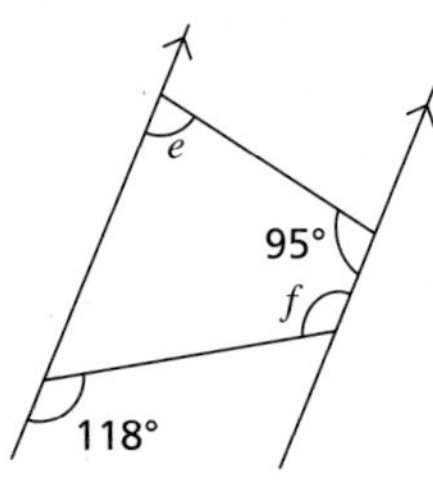

5 ACE and BCD are triangles.

AE and BD are parallel.

Work out the size of angle ACE.

Give reasons for your answer.

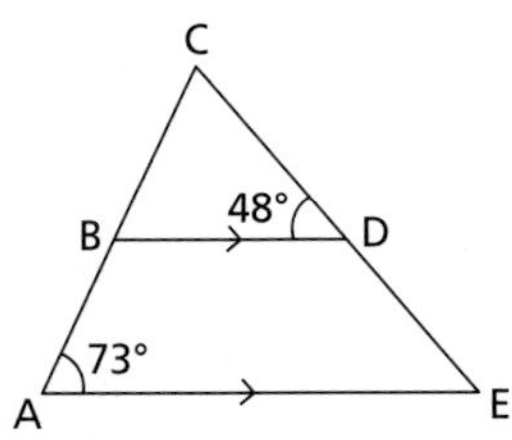

6 Use angle rules to work out the values of x, y and z.

a

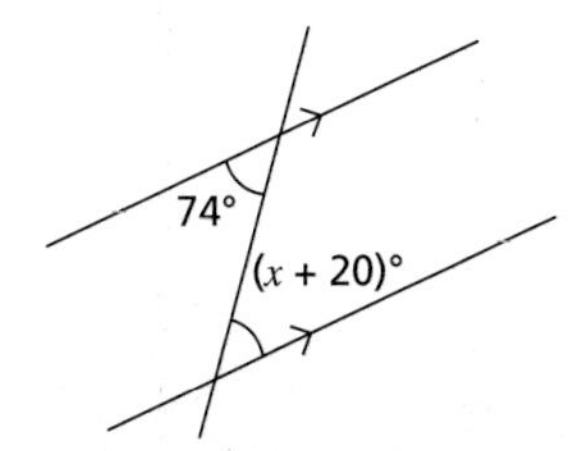

b

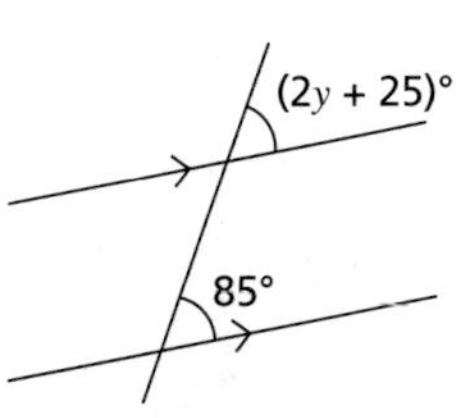

c

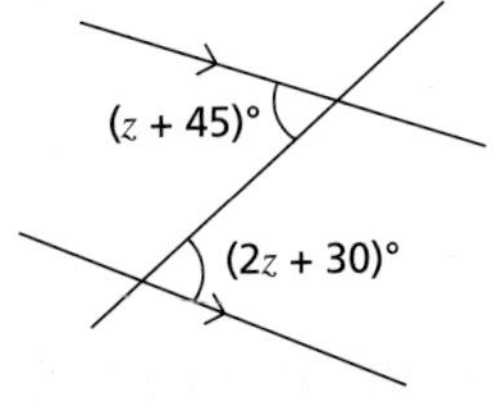

What do you think? (A)

1 Determine if any of the line segments in the diagrams are parallel to each other, justifying your answers.

The diagrams are not accurately drawn.

a

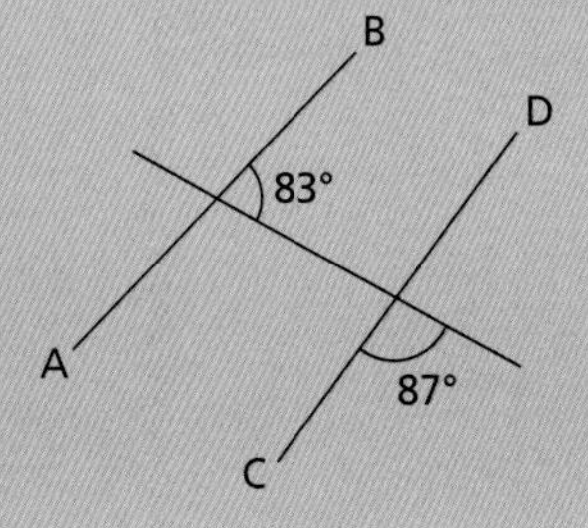

b

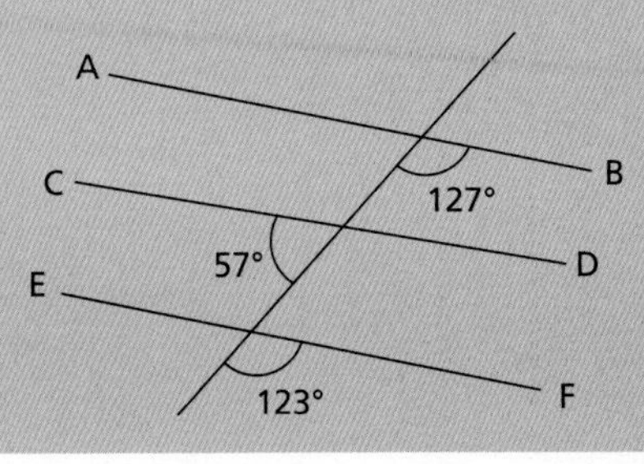

2 BK and CL are parallel.

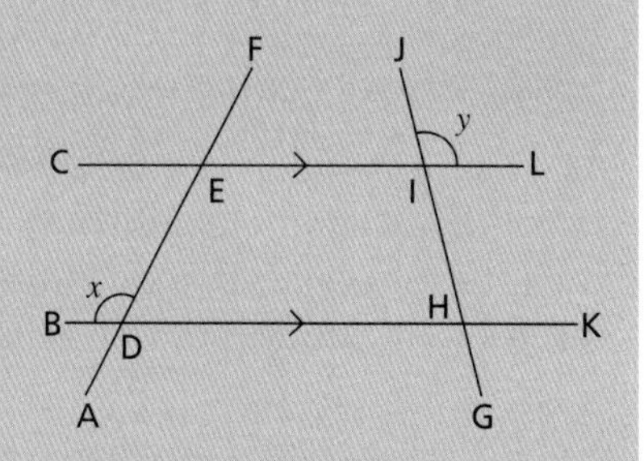

a Ed says, "If I know the values of x and y, I can work out every other angle."

Show that Ed is correct.

b Choose any two angles such that even if you knew their values, you wouldn't be able to work out the value of every other angle.

3 Work out the size of each lettered angle.

a

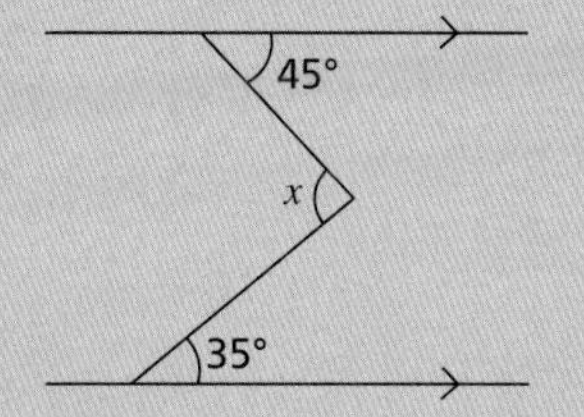

b

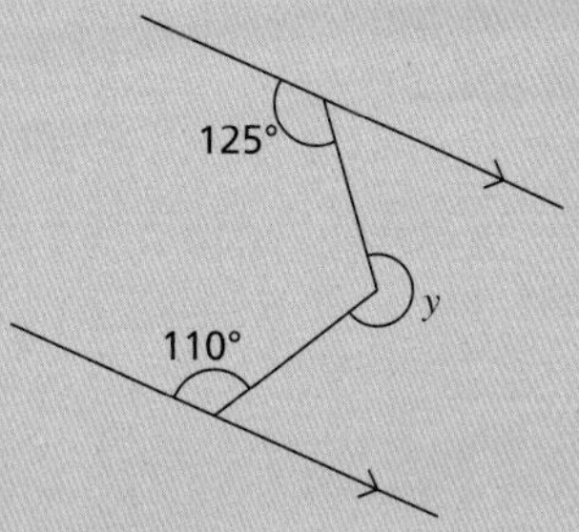

Compare your method with a partner's.

4 Prove that the triangles ABC and ECF are similar.

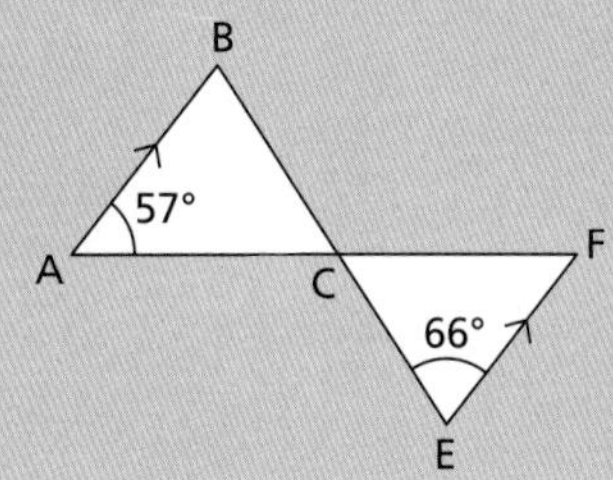

See Chapter 2.3 for a reminder of similar triangles.

Are you ready? (B)

1 Work out the size of each lettered angle.

a

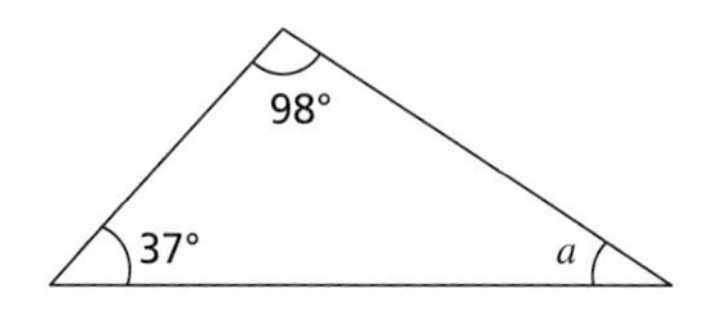

b

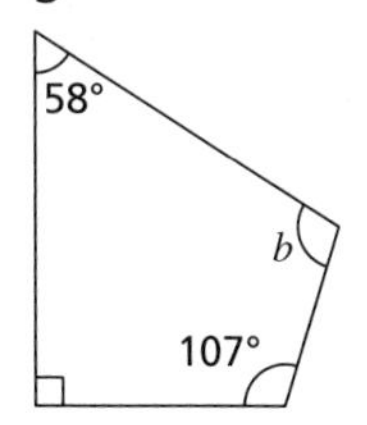

c

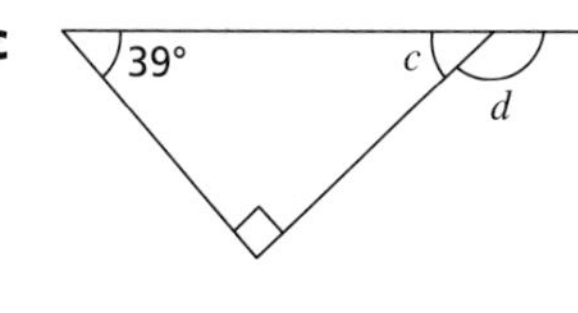

2 Write the mathematical name of a polygon with:

a five sides **b** six sides **c** eight sides.

An **interior** angle is an angle inside a polygon.

A polygon is **regular** if all its angles are equal in size and all its sides are equal in length.

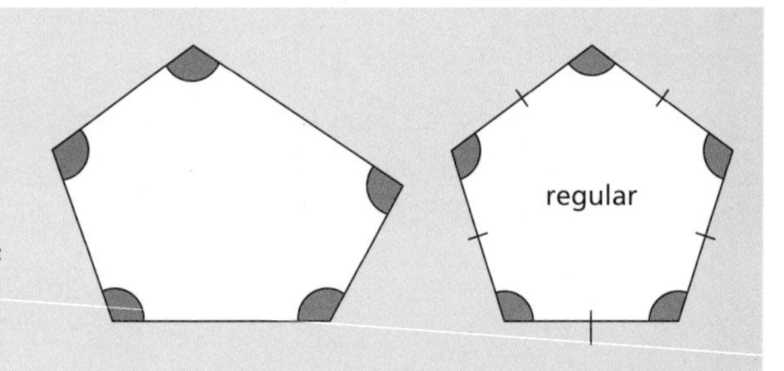

Splitting a polygon into triangles from the same vertex can help you work out the sum of the interior angles in a polygon. A pentagon can be split into three triangles.

As each triangle has an angle sum of 180°, the pentagon has an interior angle sum of 540° (3 × 180°).

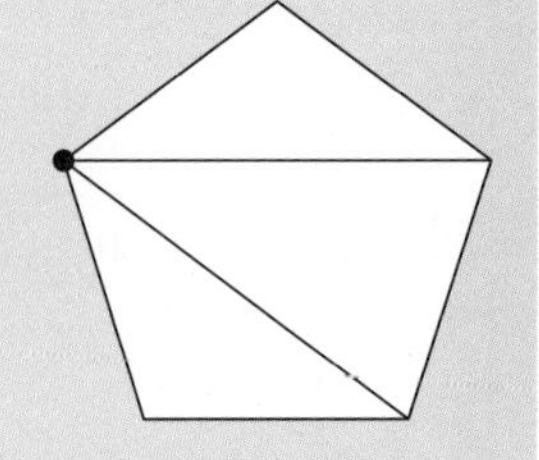

The number of triangles a polygon can be split into is always two fewer than the number of sides the polygon has, so the sum of the interior angles in a polygon with n sides is $(n - 2) \times 180°$.

An **exterior** angle is an angle between the side of a polygon and a line extended from the adjacent side.

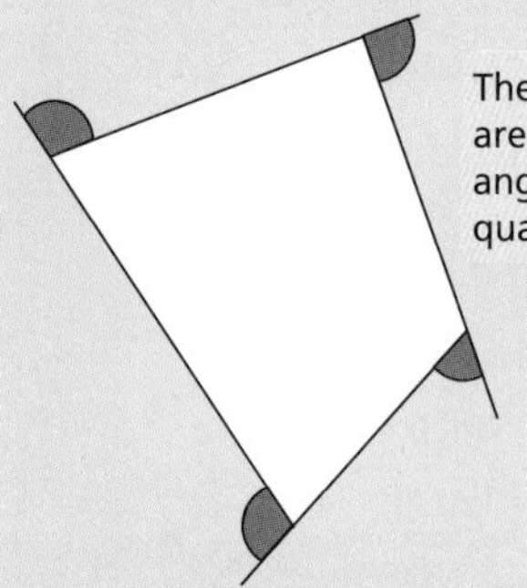

The shaded angles are the exterior angles of the quadrilateral.

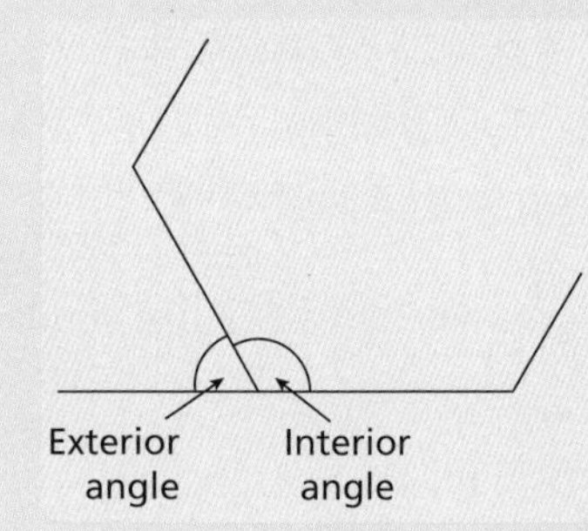

Interior and exterior angles of a polygon are adjacent on a straight line, so they sum to 180°.

The sum of the **exterior** angles of any polygon is 360°, as between them they form one complete turn.

Example

Work out the size of angle a.

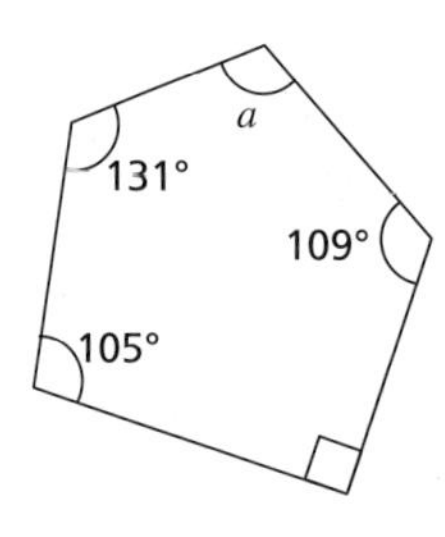

Method

Solution	Commentary
Angle sum = (5 – 2) × 180 = 3 × 180 = 540°	Substitute $n = 5$ into $(n - 2) \times 180°$
Total of known angles = 105° + 131° + 109° + 90° = 435° a = 540° – 435° = 105°	Subtract the sum of the given angles (including the right angle) from 540° to work out the size of angle a.

Practice (B)

1. For each regular polygon, work out:

   i the sum of the interior angles **ii** the size of each interior angle.

   a

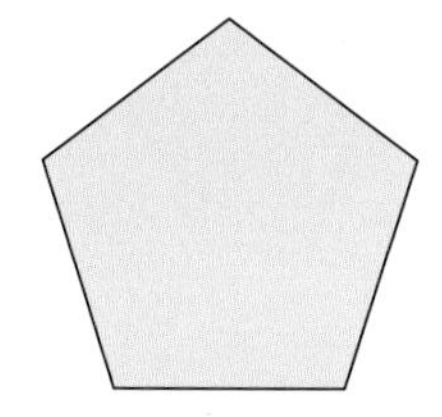

   b

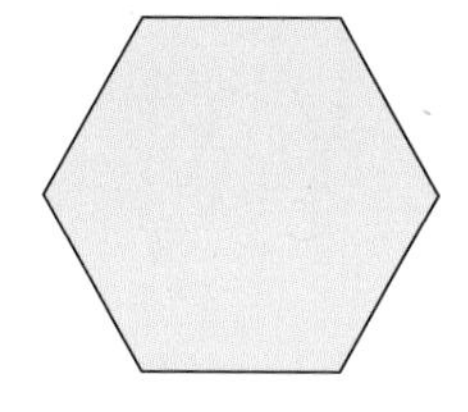

   c

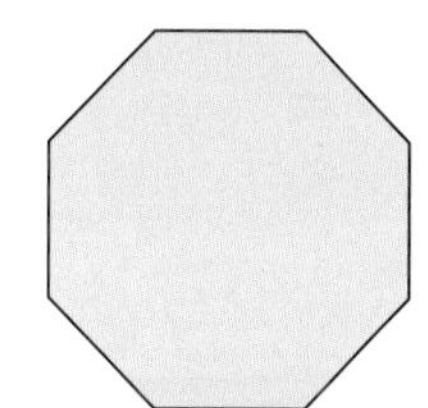

2. A pentagon is shown.

   a Write down the sum of the interior angles.

   b Work out the size of angle a.

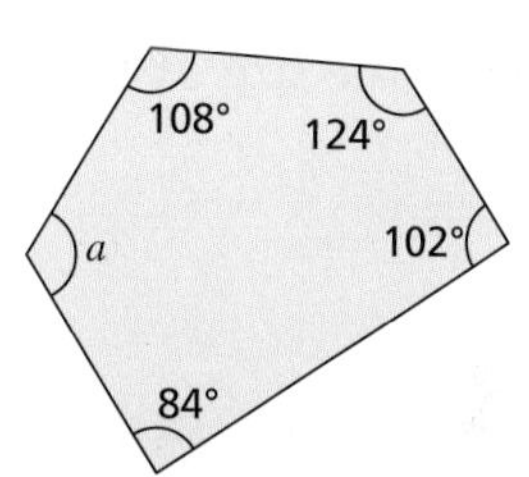

3. Here is a hexagon.

   Work out the size of angle b.

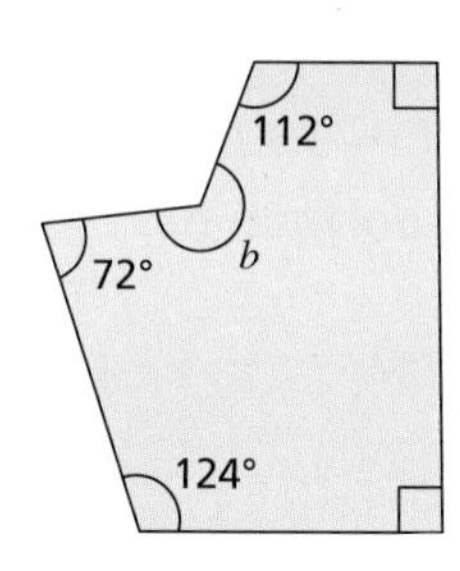

4. Work out the size of each lettered angle.

   a

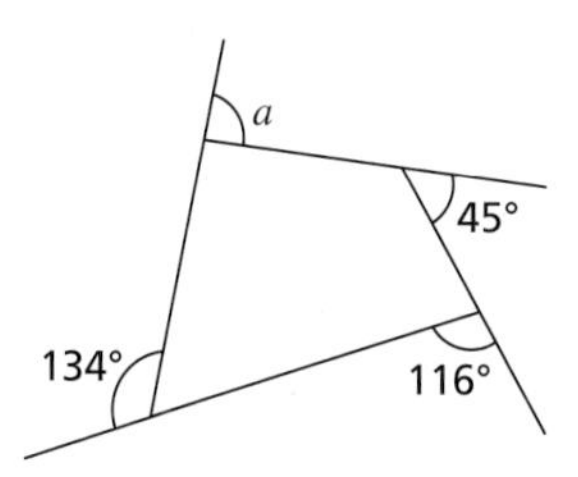

   b

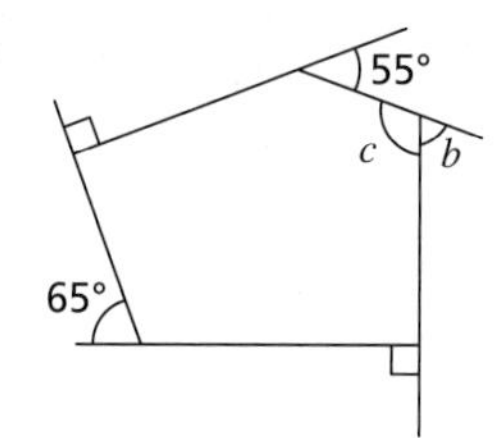

5. ABDEF is a pentagon and ABC is a straight line.

   Work out the size of angles f and g.

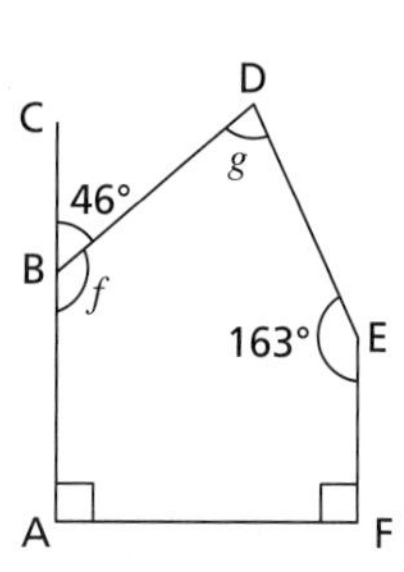

6. Here is a regular hexagon.

   Work out the size of angle x.

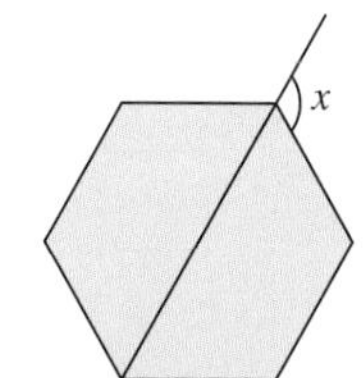

7. Here is a regular pentagon.

   Work out the size of angle y.

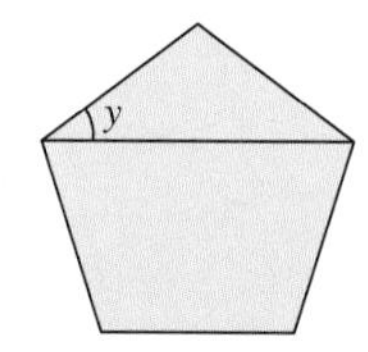

8 Each of the shapes are regular polygons.

Work out the size of angle y in each case.

a

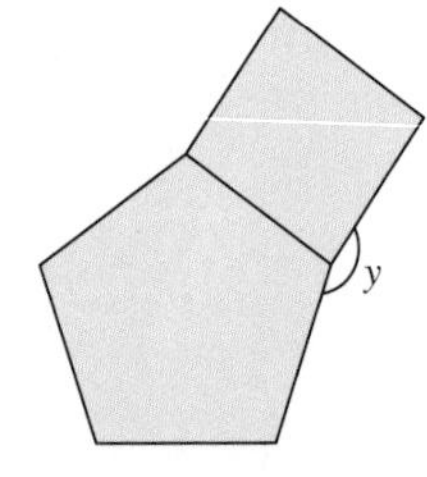

b

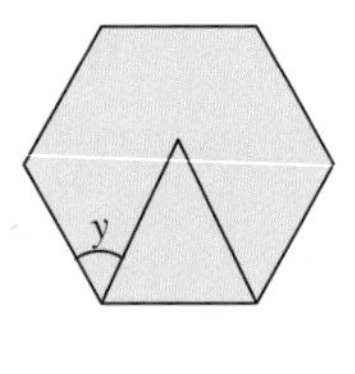

c

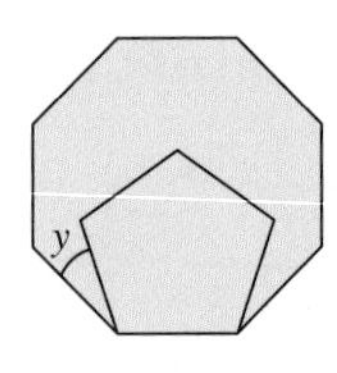

9 Here are two regular pentagons and a square.

Work out the size of angle x.

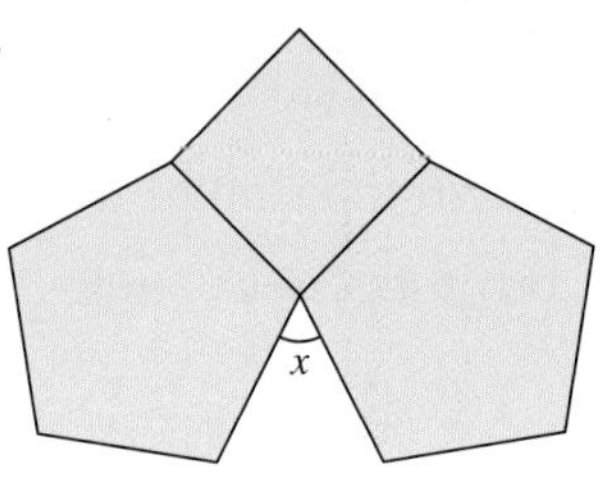

10 Here are two sides of a regular polygon.

Work out how many sides the polygon has.

150°

What do you think? (B)

1 Which of the statements are possible?

A regular polygon has an exterior angle of 14°	A regular polygon has an interior angle of 170°	A regular polygon has an exterior angle of 18°

For any statements which are possible, work out the number of sides the polygon has.

2 How many types of regular shape tessellate around a point? How do you know?

Tessellating shapes fit together without any gaps.

Consolidate – do you need more?

1 Write the size of each lettered angle. Give a reason for each answer.

a

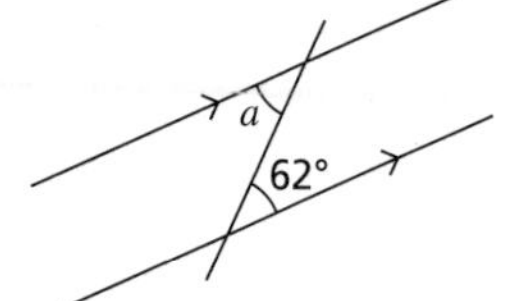

b

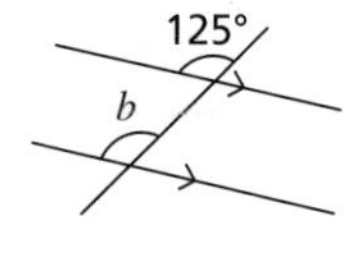

c

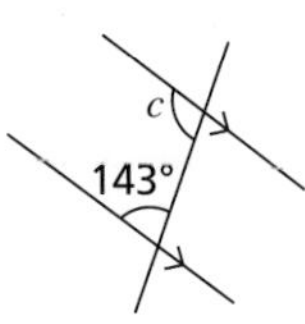

d

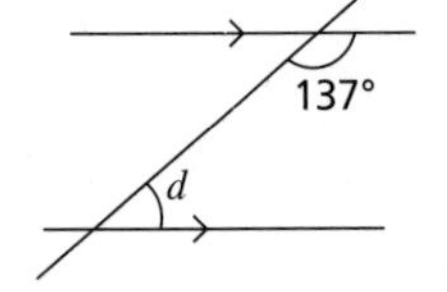

e

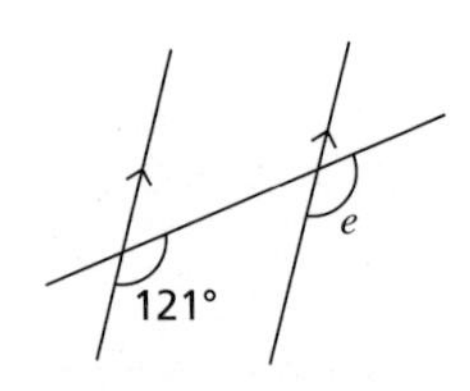

f

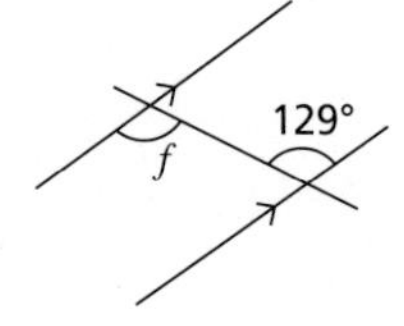

2 **a** Work out the size of angle x.

b Work out the size of angle y.

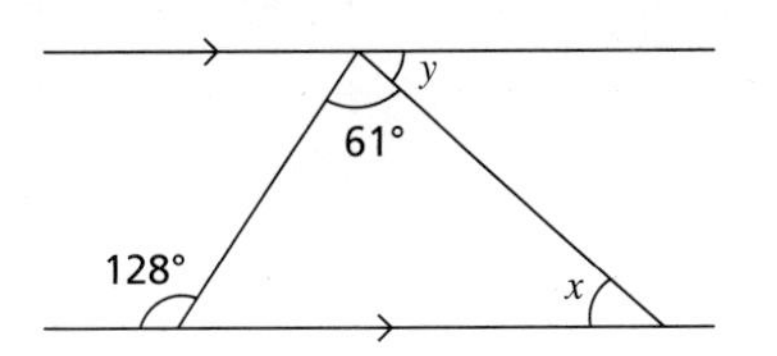

3 Work out the size of each angle labelled x.

a

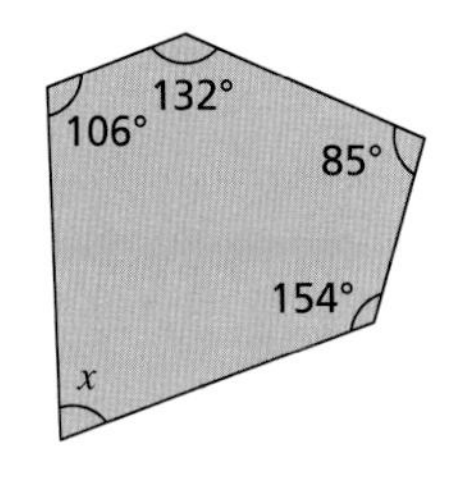

b

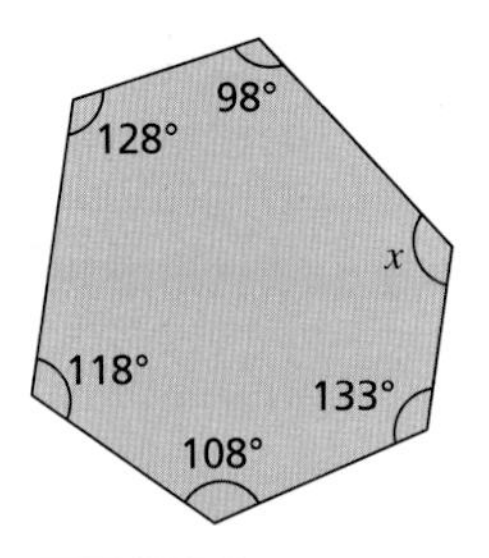

c

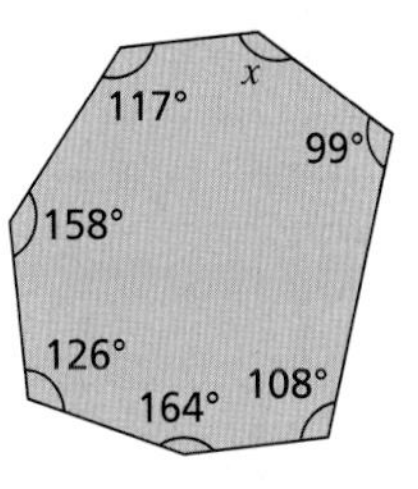

d

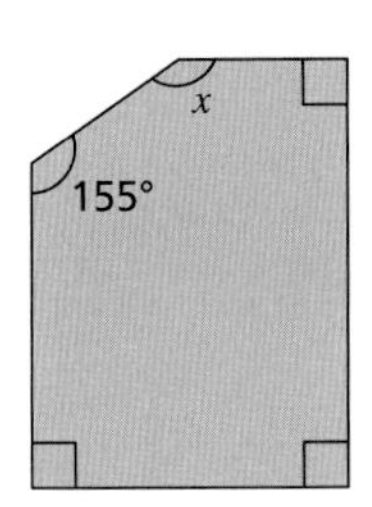

e

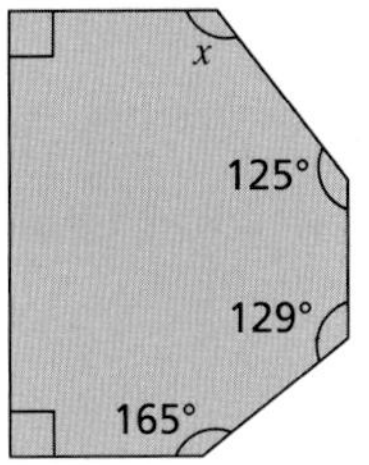

f

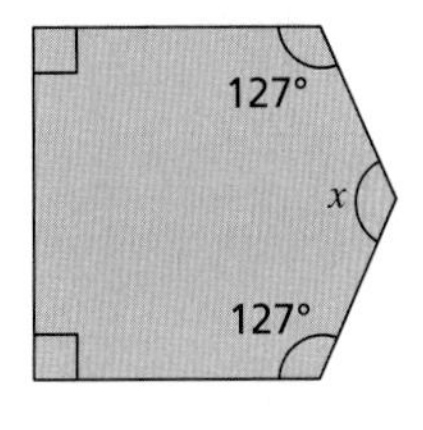

Stretch – can you deepen your learning?

1 Work out the values of x and y.

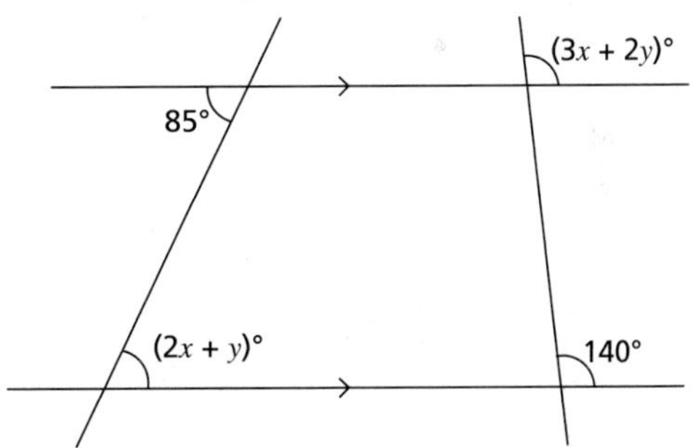

2 Here is a trapezium.

Work out the value of x.

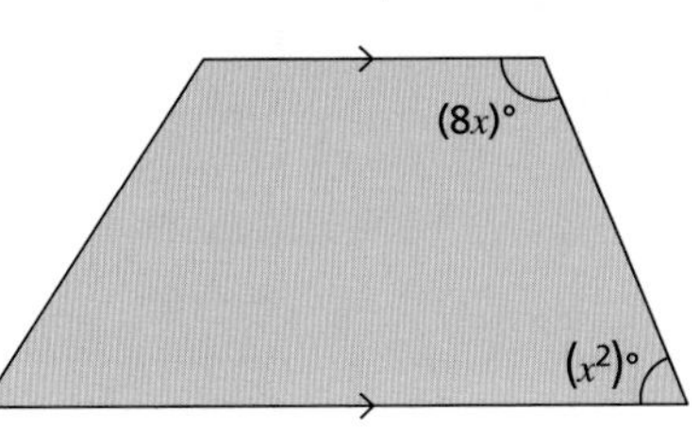

3 AE and BD are parallel and CD = BD.

Show that $y = 2x$.

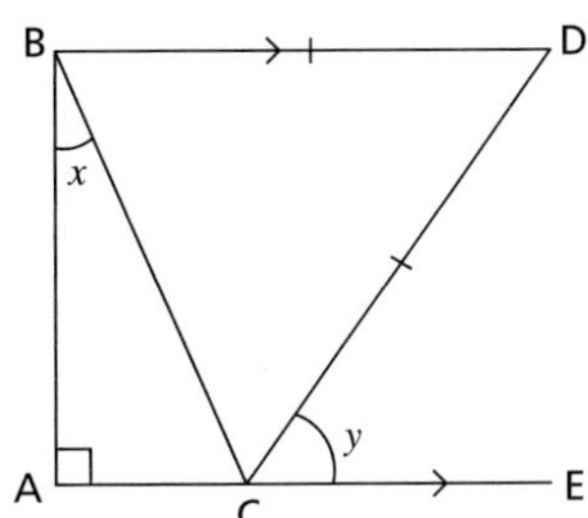

4 The diagram shows part of a regular polygon.

Work out the number of sides the polygon has.

(3x + 6)°

(14 – x)°

3.2 Four circle theorems

White Rose MATHS

Are you ready? (A)

1 The diagram shows a circle, centre O.

Match each word with one of the letters A to E.

Diameter	Radius	Circumference	Chord	Segment

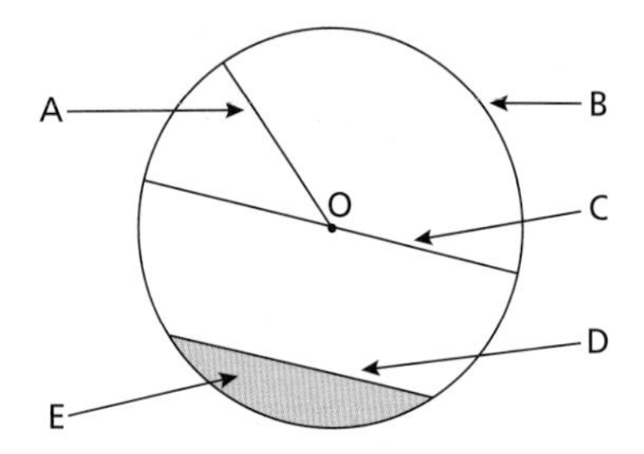

2 Work out the size of each lettered angle.

a

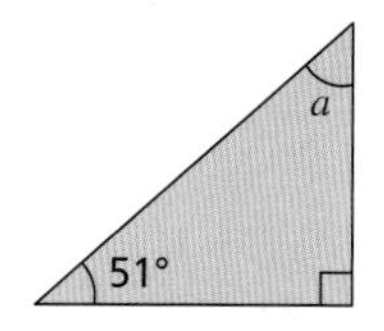

b

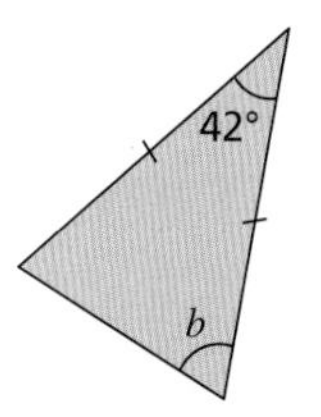

c

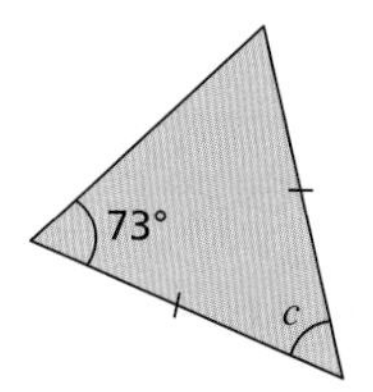

The **circle theorems** are a set of angle properties linked to geometric shapes in circles. You should be able to prove them and use them to work out other angles.

Angles at the centre and the circumference

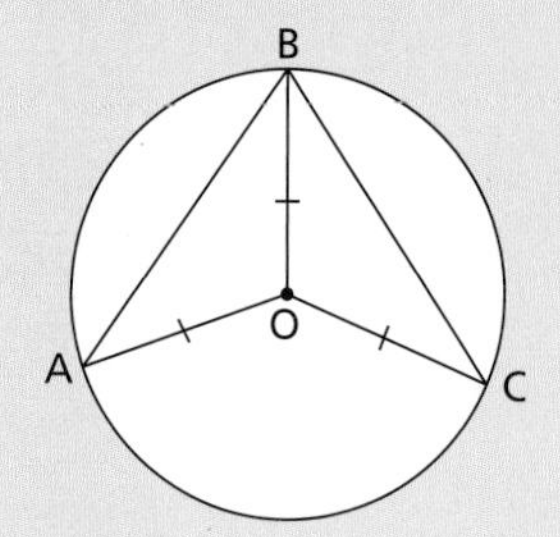

In the diagram, OA = OB = OC, as they are all radii.

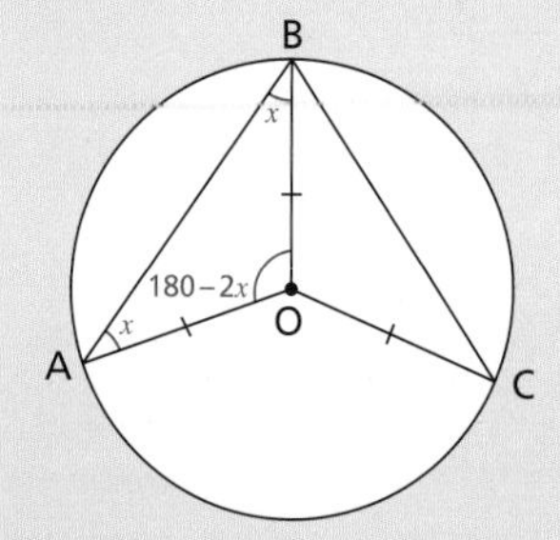

Let ∠ABO = x

Then ∠BAO = x as the triangle AOB is isosceles.

Because the sum of angles in a triangle is 180°, ∠AOB must be $(180 - 2x)°$.

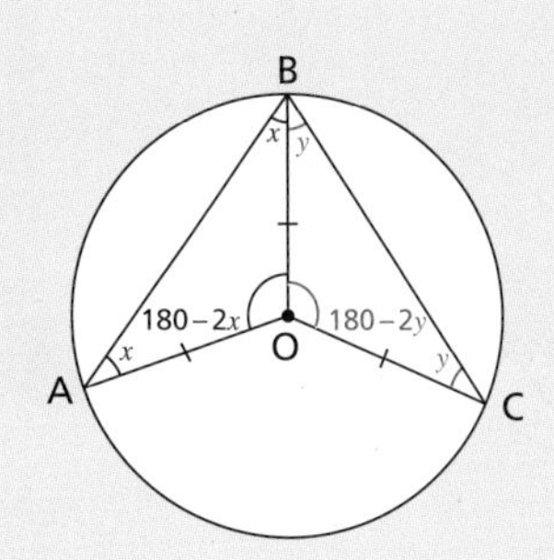

In the same way, let $\angle OBC = y$

Then $\angle BOC$ must be $(180 - 2y)°$.

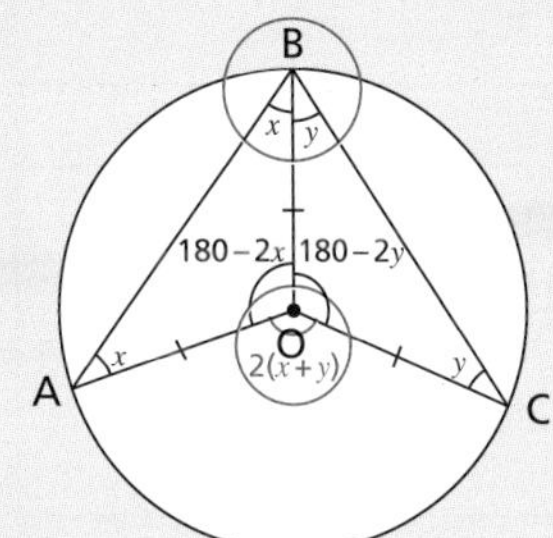

To work out $\angle AOC$,

subtract $(180 - 2x)°$ and $(180 - 2y)°$ from 360°,
giving $\angle AOC = (2x + 2y)°$

$= 2(x + y)°$

So $\angle AOC = 2 \times \angle ABC$

This rule is usually given as: The angle at the centre of a circle is twice the angle at the circumference.

Angles in a semicircle

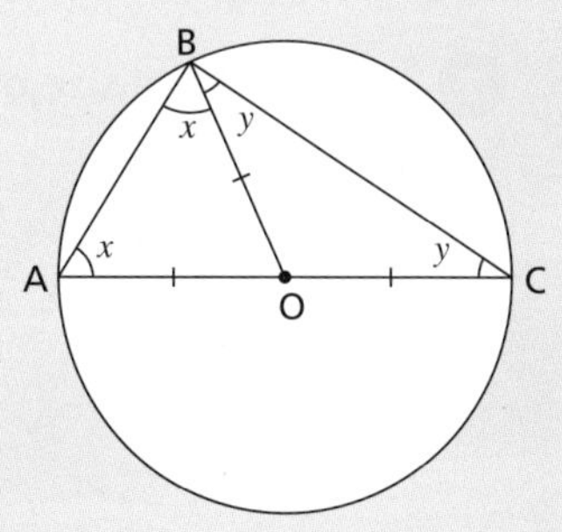

The line AOC is the diameter of the circle, centre O.

OA = OB = OC as they are all radii, and two isosceles triangles can be formed.

As the sum of the angles of triangle ABC is 180°, $2x + 2y = 180°$

Dividing both sides of this equation by 2 gives $x + y = 90°$

$\angle ABC = x + y = 90°$

This rule is usually given as: The angle in a semicircle is a right angle.

Example 1

Work out the size of angle a.
Give a reason for your answer.

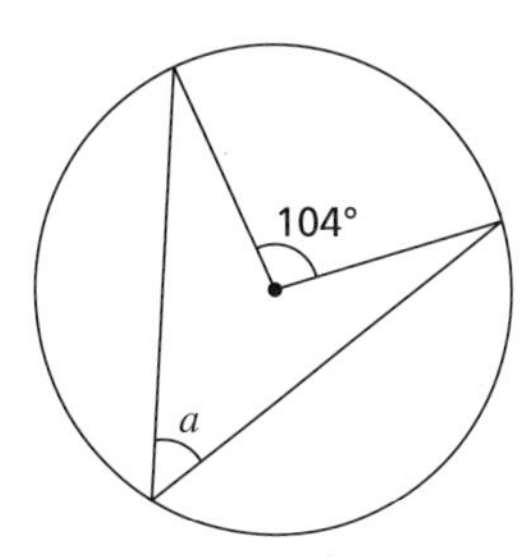

Method

Solution	Commentary
$a = 104° \div 2$ $= 52°$	Divide the given angle by 2 to work out the size of angle a.
The angle at the centre is twice the angle at the circumference.	State the rule you have used.

Example 2

AOC is the diameter of the circle, centre O.

Work out the size of angle BAC. Give a reason for your answer.

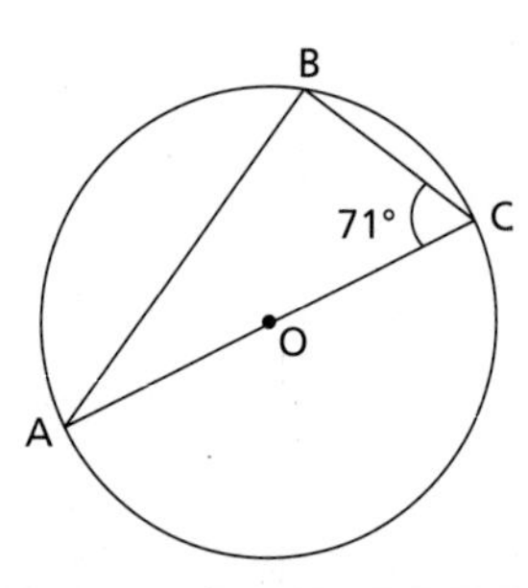

Method

Solution	Commentary
∠ABC = 90°, as the angle in a semicircle is 90° 90° + 71° = 161°	At each stage of your working, state the angle rule you have used.
∠BAC = 180° – 161° = 19°, as angles in a triangle add up to 180°	

Practice (A)

1 Work out the size of each lettered angle.

a

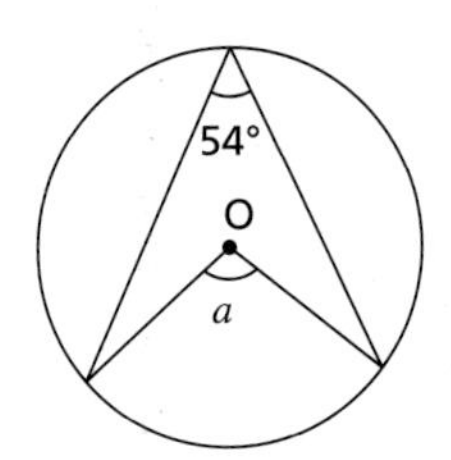

b

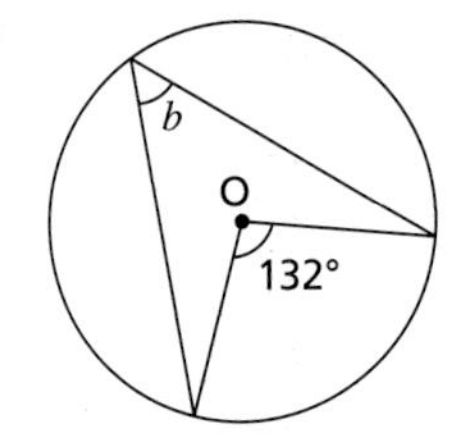

c

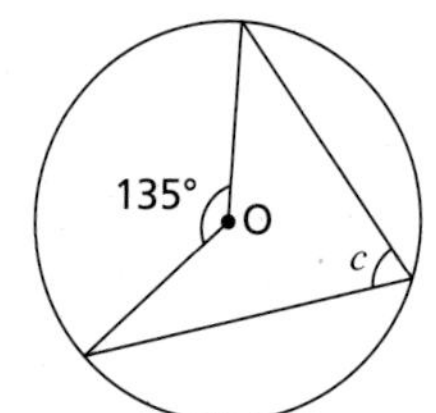

d

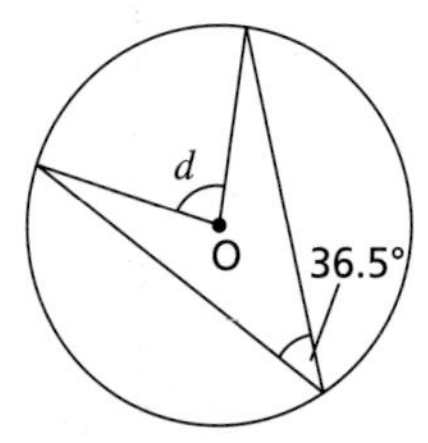

e

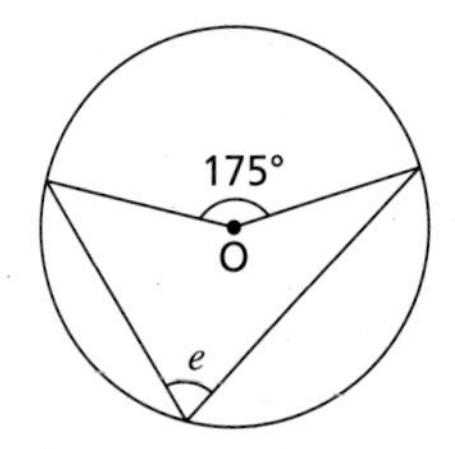

f

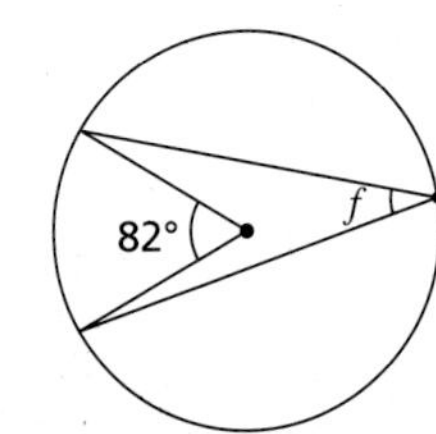

g

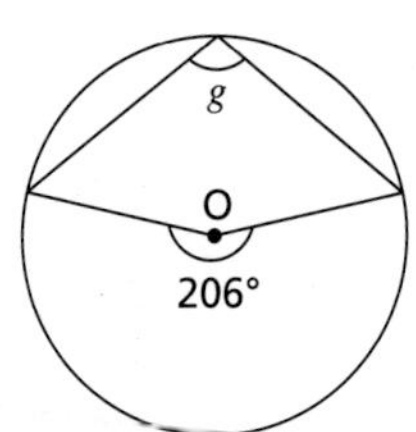

h

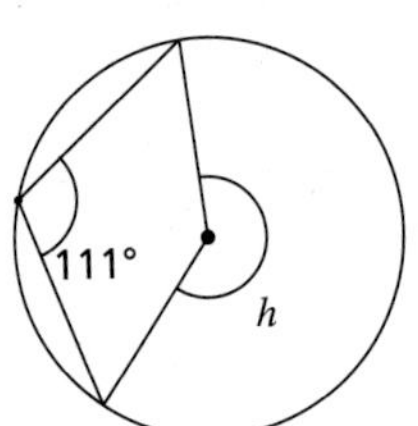

i

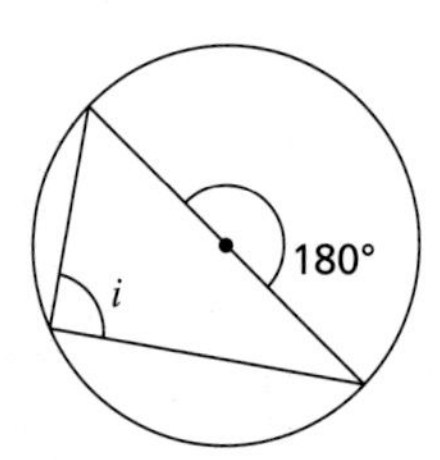

2 A, B and C are all points on the circle, centre O.

a Which of these angles touches the circumference?

b Work out the size of angle y.

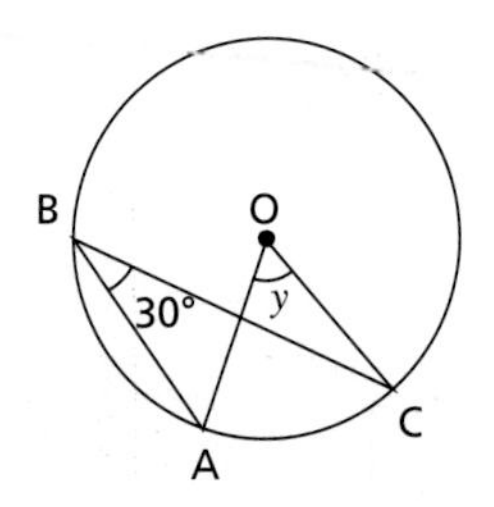

3 Work out the size of each lettered angle.

a

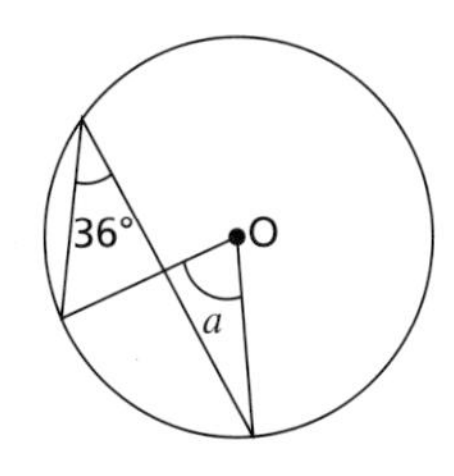

b

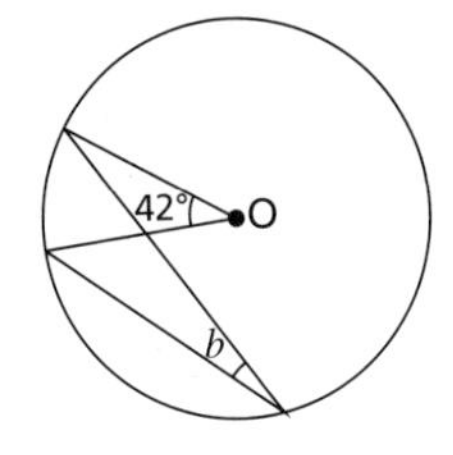

c

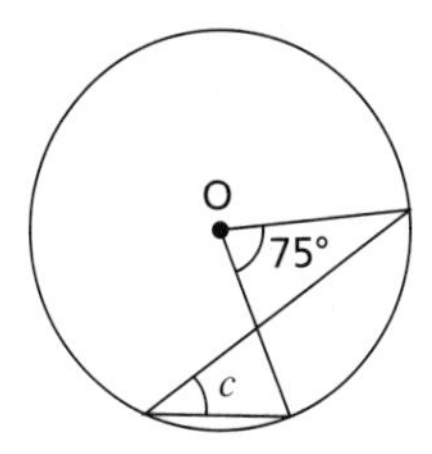

4 Work out the size of each angle labelled x.

a

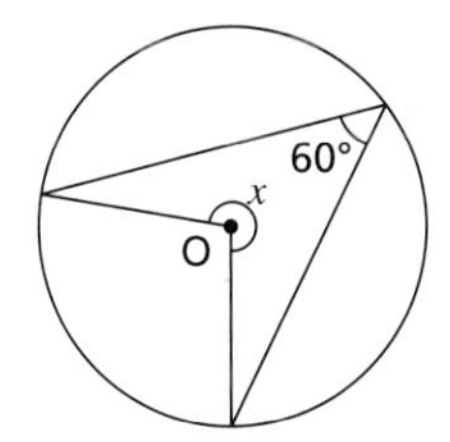

b

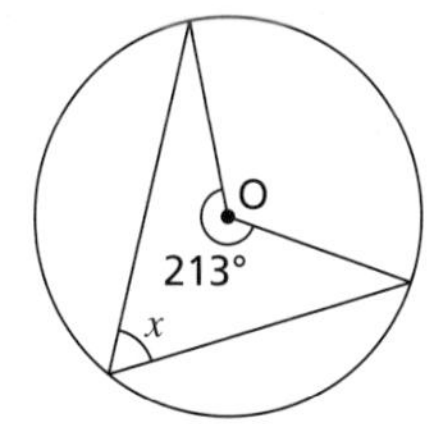

5 A, B and C are points on the circle, centre O.

Work out the size of angle ABC.
Give reasons for your answer.

6 Work out the size of each lettered angle.

a

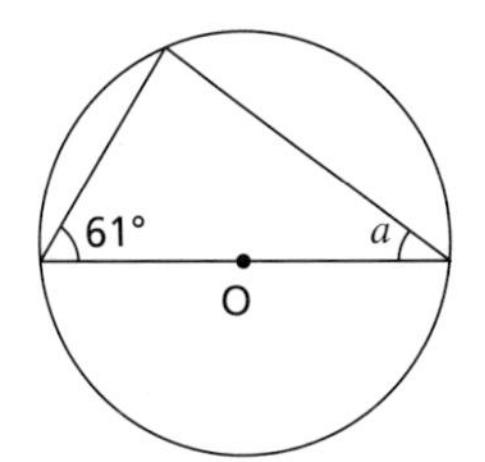

b

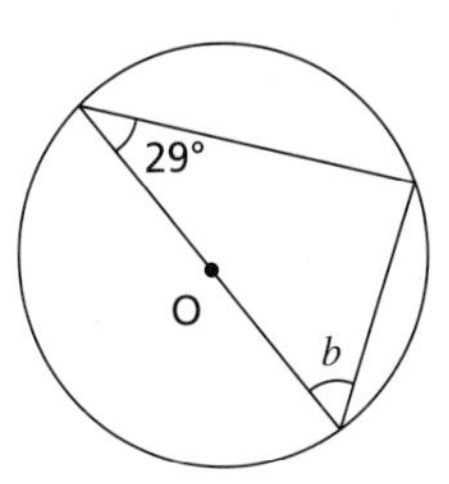

c

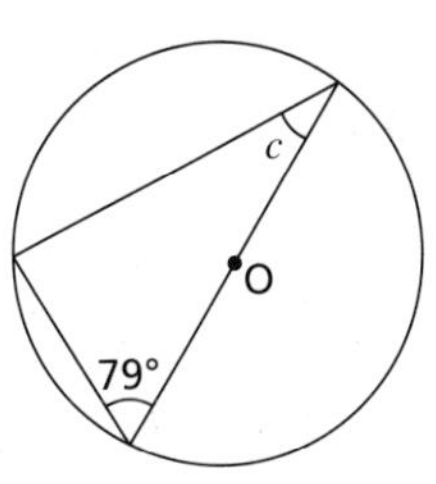

d

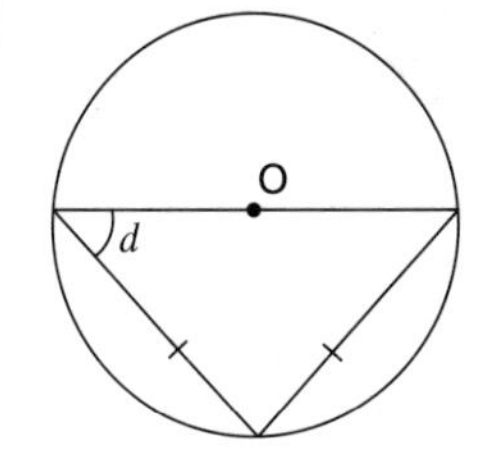

7 Work out the size of each lettered angle in these semicircles.

a

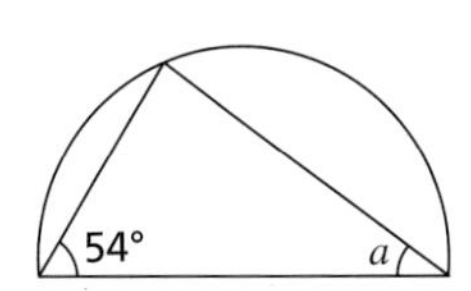

b

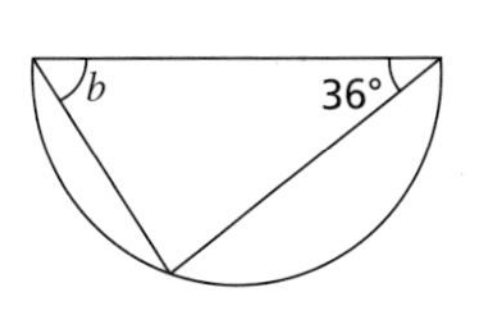

c

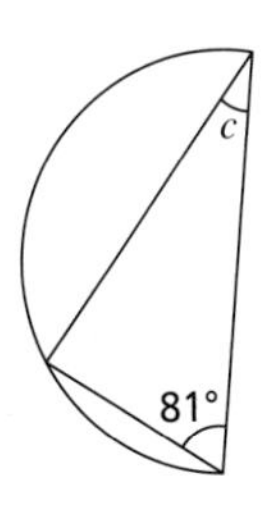

8 AOB is the diameter of a circle.

$\angle CBA = 28°$

Work out the size of angle CAB.

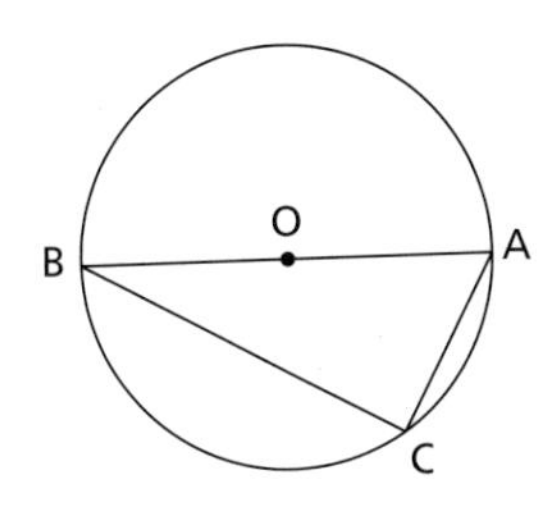

9. AOC is the diameter of a circle.

   AB = 9 cm and BC = 7 cm

   a Work out the length of the diameter of the circle.

   Give your answer to 3 significant figures.

   b Work out the size of angle BAC.

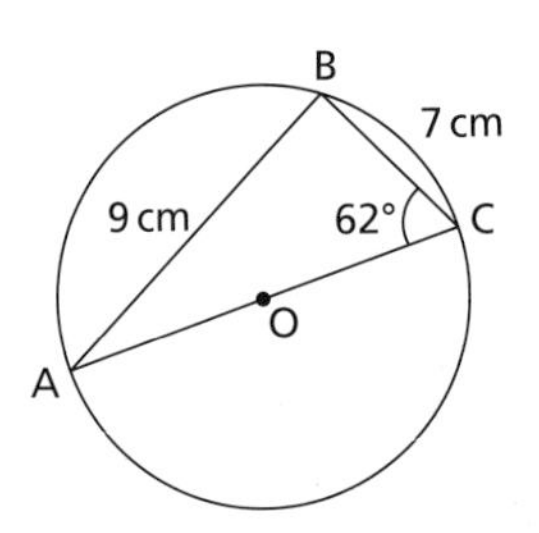

What do you think? (A)

1. A, B and C are points on the circumference of the circle, centre O.

   Ed says, "There isn't enough information to work out the angle OCB."

   Show that Ed is incorrect.

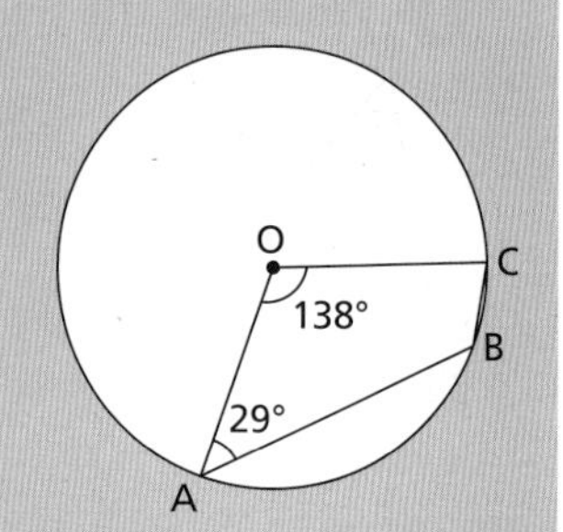

2. A, B and C are points on the circumference of a circle, centre O.

   Seb says, "ABC is a right angle, because AOC is the diameter of the circle."

   Flo says, "ABC is a right angle because the angle at the centre is twice the angle at the circumference."

   Who do you agree with? Explain your answer.

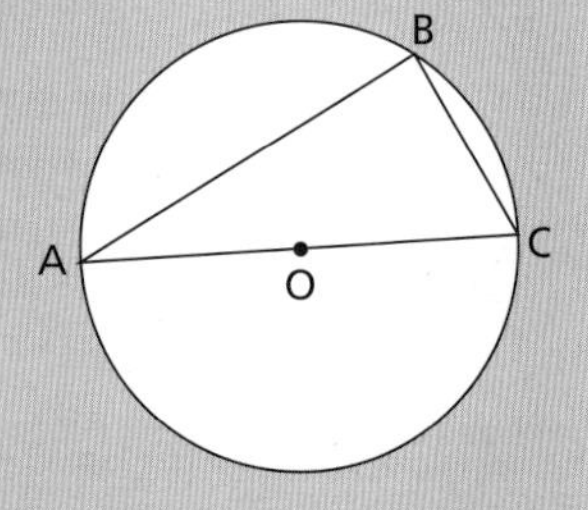

3. **a** Prove that $\angle XOZ = 2 \times \angle XYZ$

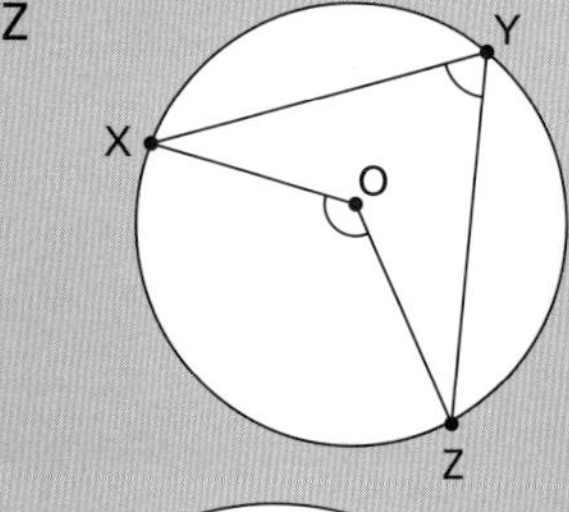

   b Prove that $\angle PQR = 90°$

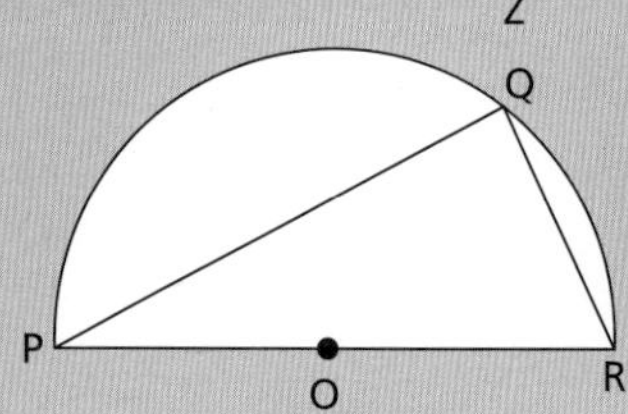

Angles in the same segment

A **chord** is a line segment that joins two points on the circumference of a circle, and splits the circle into two **segments** (a **major segment** and a **minor segment**).

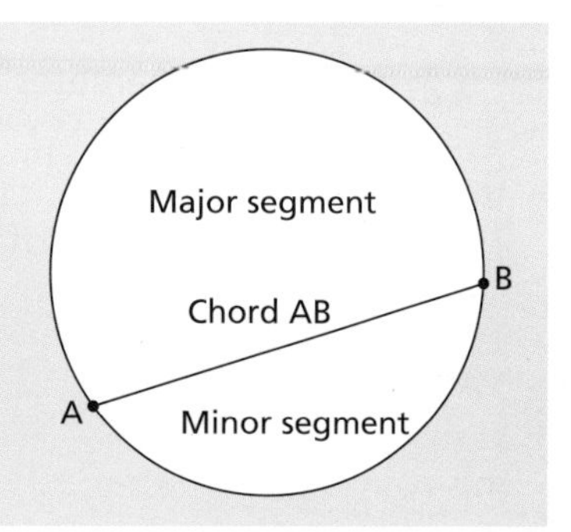

Here is an important theorem relating to chords and segments in a circle.

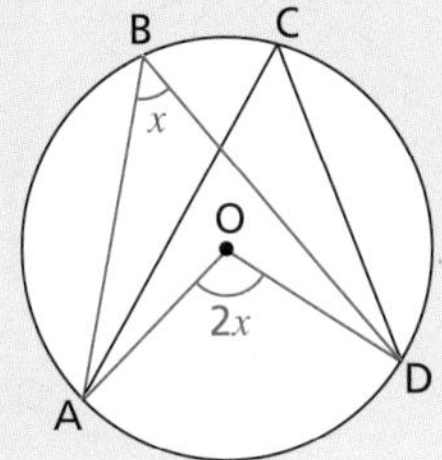
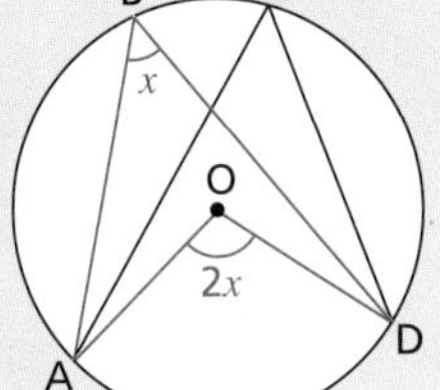

A, B, C and D are points on the circumference of a circle, centre O.

Let $\angle ABD = x$

Then $\angle AOD = 2x$ because the angle at the centre is twice the angle at the circumference.

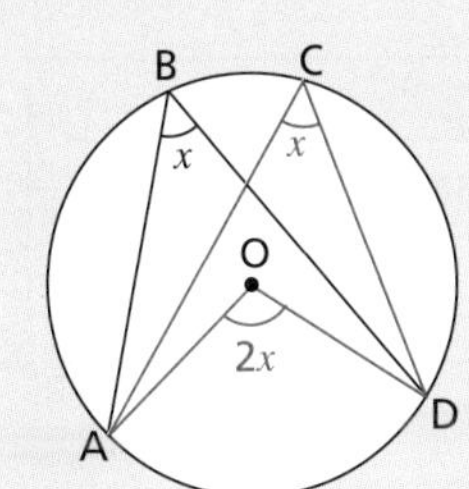

Using the same rule,
as $\angle AOD = 2x$, $\angle ACD = x$

So $\angle ABD = \angle ACD$

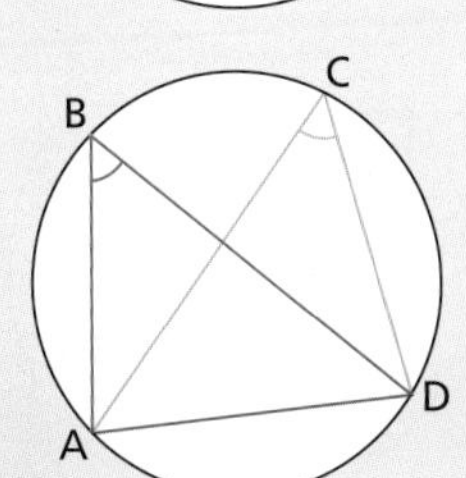

$\angle ABD$ and $\angle ACD$ are formed in the same segment of the circle.

They are both **subtended** by the chord AD.

The rule is usually given as: Angles in the same segment are equal.

Angles in a cyclic quadrilateral

A quadrilateral is **cyclic** if all four of its vertices touch the circumference of a circle.

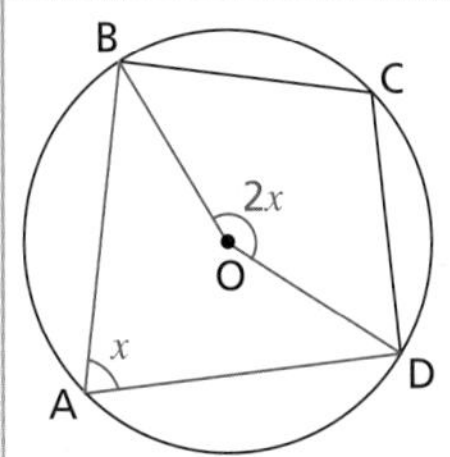	A, B, C and D are points on the circumference of the circle, centre O. Let $\angle BAD = x$, then $\angle BOD = 2x$ because the angle at the centre is twice the angle at the circumference.
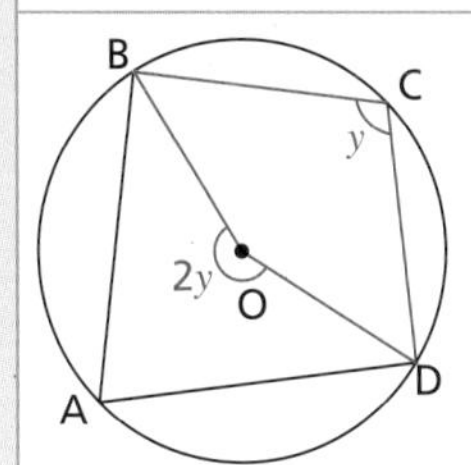	In the same way, let $\angle BCD = y$ Then the reflex angle $\angle BOD = 2y$
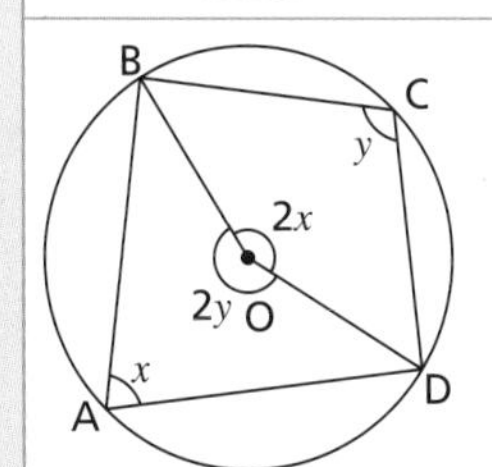	$2x + 2y = 360°$ because the angles around a point have a sum of 360°. Dividing both sides of the equation by 2 gives $x + y = 180°$

This rule is usually stated as: Opposite angles in a cyclic quadrilateral add up to 180°.

Example 1

A, B, C and D are points on the circumference of the circle, centre O.

Work out the size of angle b. Give reasons for your answer.

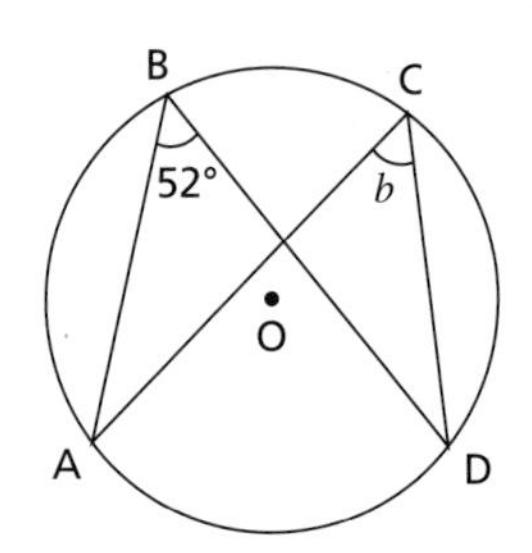

Method

Solution	Commentary
$b = 52°$ Angles in the same segment are equal.	Both angles are in the same segment and are therefore equal.

Example 2

Work out the sizes of angles a and b.
Give reasons for your answers.

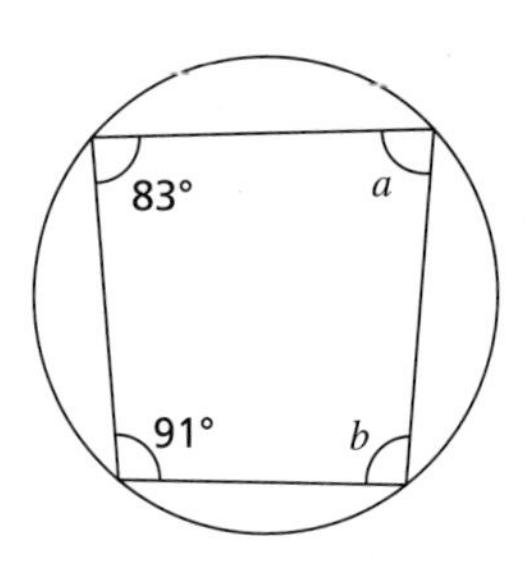

Method

Solution	Commentary
$a = 180° - 91°$ $= 89°$ Opposite angles in a cyclic quadrilateral have a sum of 180°.	Angle a is opposite the 91° angle. Subtract 91° from 180° to work out the size of angle a.
$b = 180° - 83°$ $= 97°$ Opposite angles in a cyclic quadrilateral have a sum of 180°.	Angle b is opposite the 83° angle. Subtract 83° from 180° to work out the size of angle b. You could also find angle b by using the angle sum of a quadrilateral is 360°.

Practice (B)

1. Write down the size of each lettered angle.

a

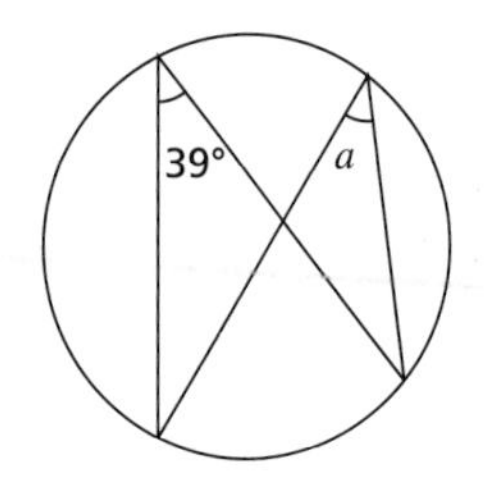

b

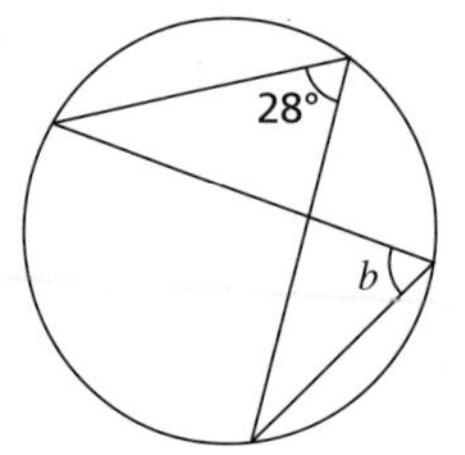

c

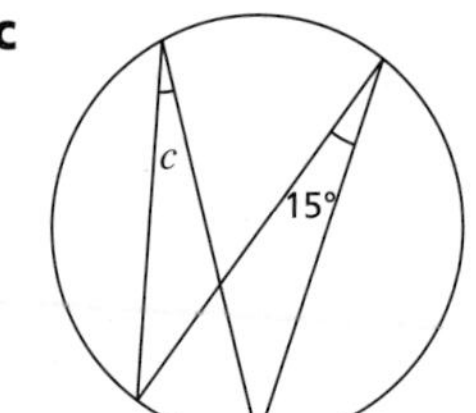

d

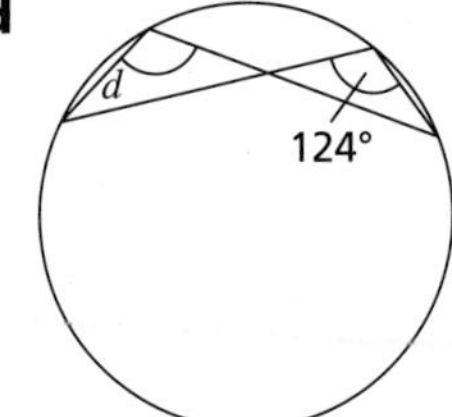

2 Write down the size of each lettered angle.

a

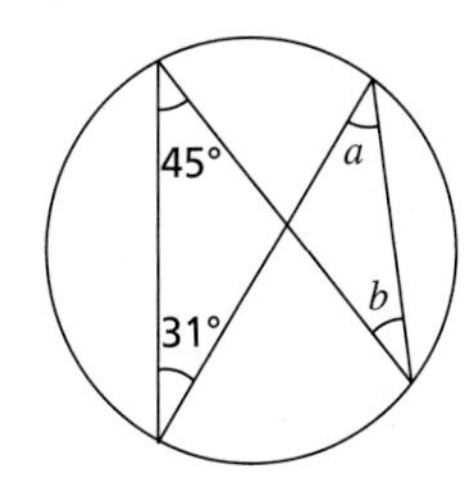

b

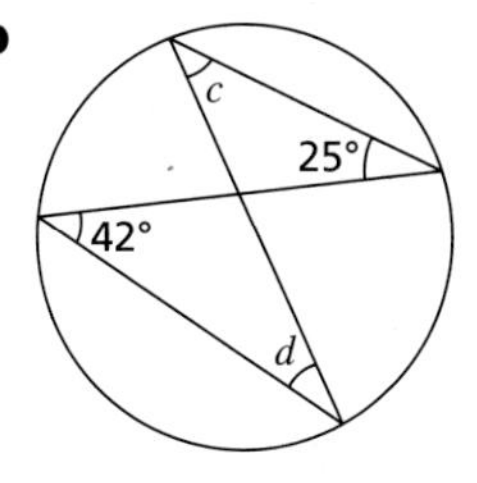

c

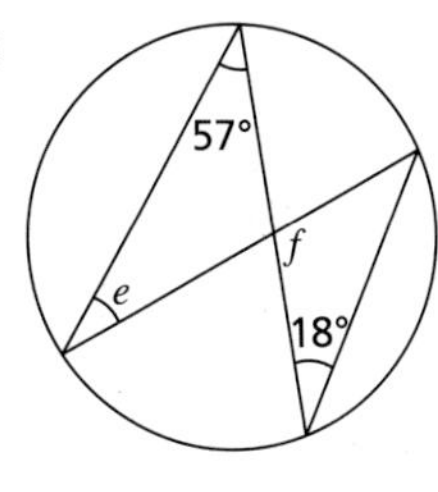

d

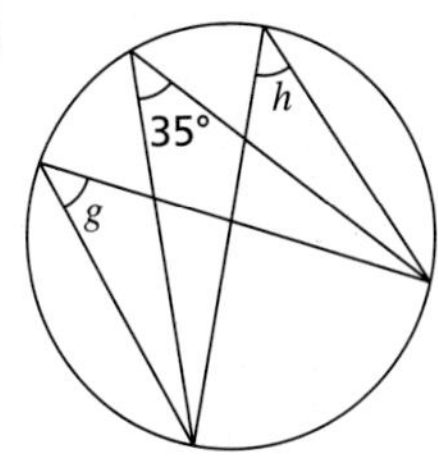

3 A, B, C and D are points on the circumference of a circle.

Work out the size of angle x.
Give reasons for your answer.

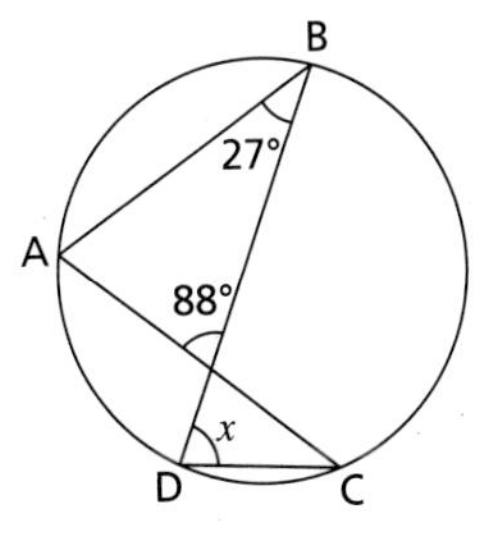

4 A, B, C and D are points on the circumference of a circle, centre O.

a Work out the size of angle ABD.
Give a reason for your answer.

b Work out the size of angle ACD.
Give a reason for your answer.

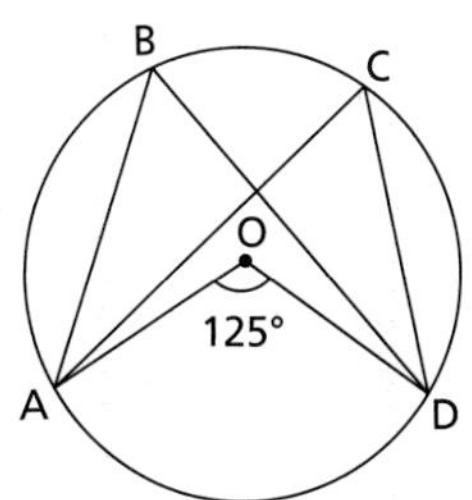

5 Work out the size of each lettered angle.

a

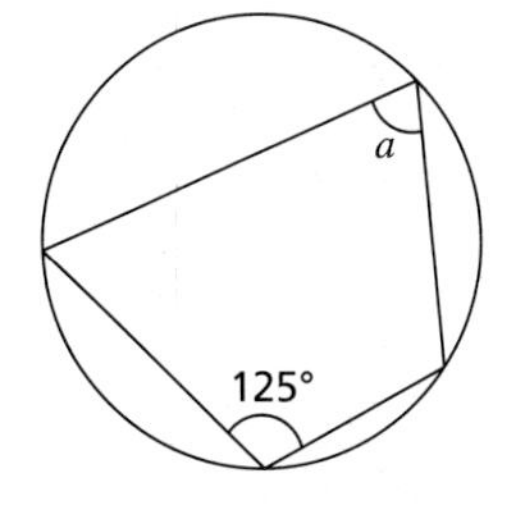

b

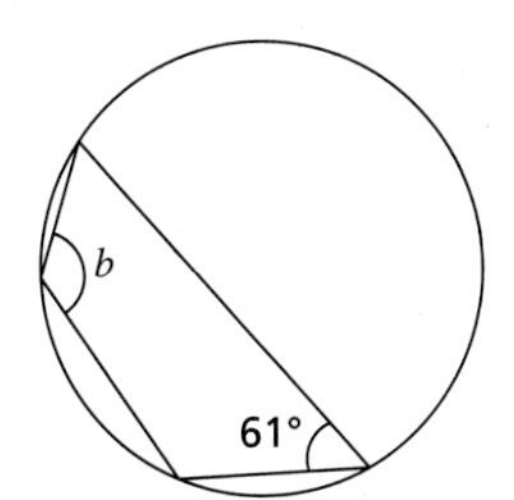

c

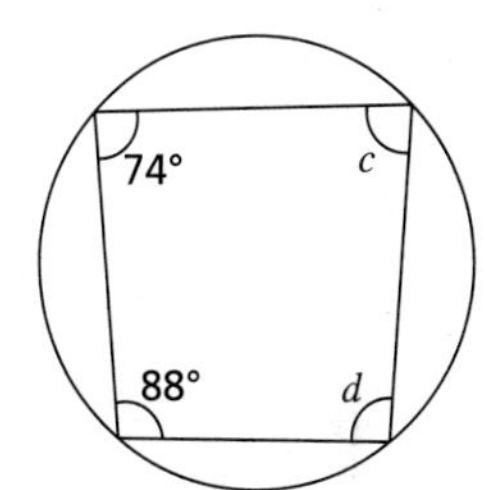

d

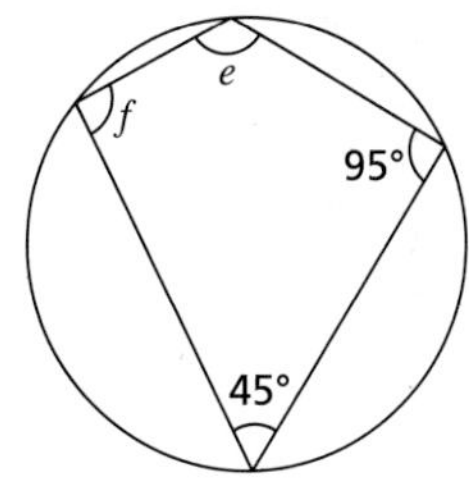

e

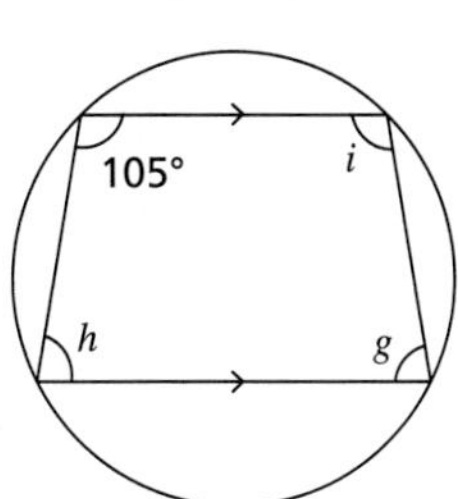

f

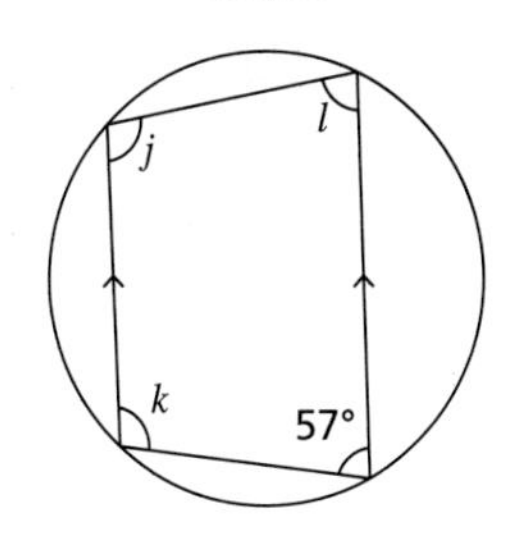

6 In each diagram, ADE is a straight line.

Work out the size of each lettered angle.

a

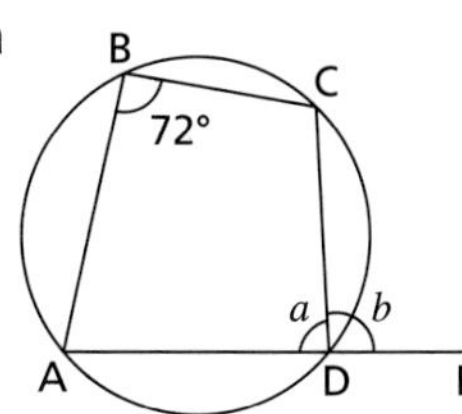

b

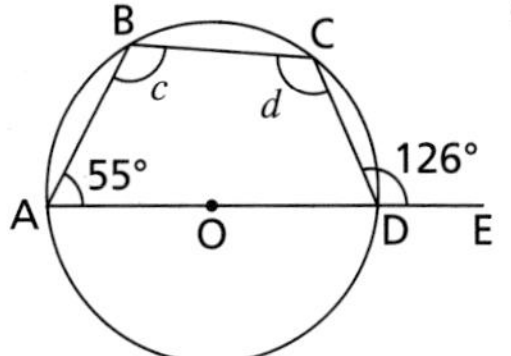

c

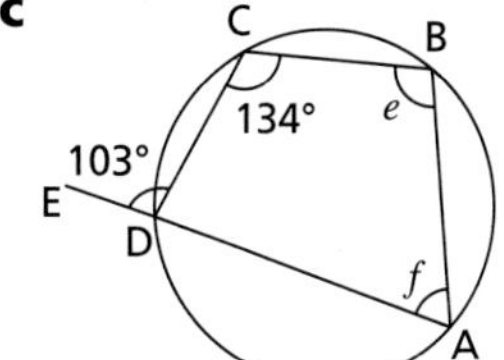

d

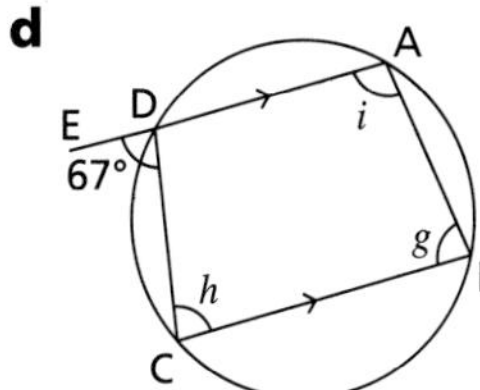

7 Work out the size of each lettered angle.

a

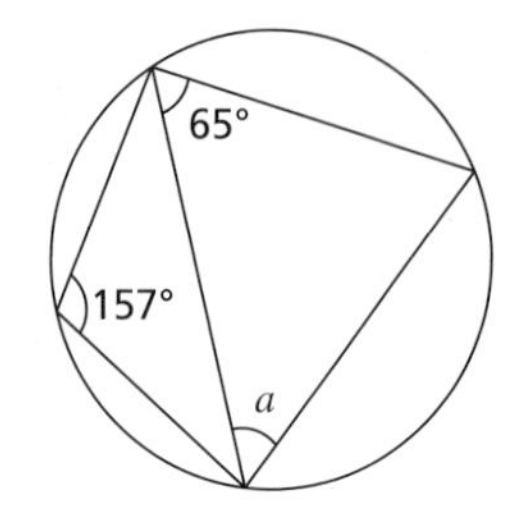

b

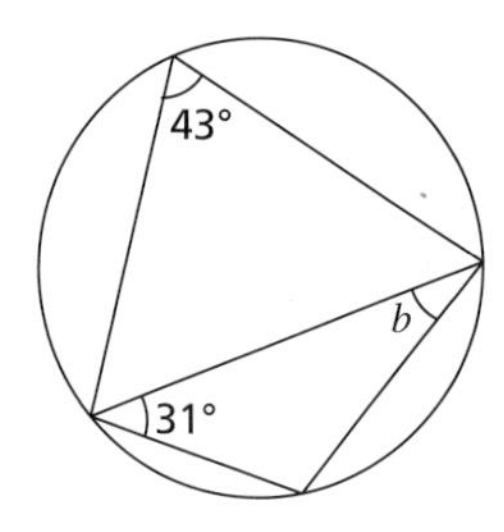

8 A, B, C and D are points on the circumference of a circle.

Work out the size of each lettered angle.
Give reasons for your answers.

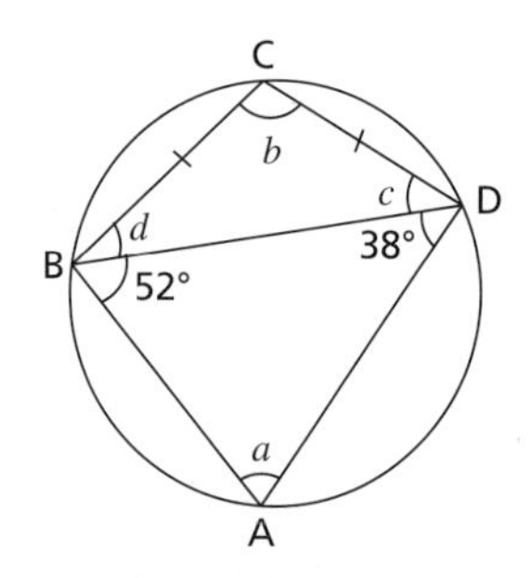

What do you think? (B)

1 "If at least one of the angles in a cyclic quadrilateral is a right angle, then it must be either a square or a rectangle."

Is the statement **true** or **false**? Explain your answer.

2 Draw a circle and mark two points on the circumference labelled A and B.

- Draw a line from A to a point on the circumference, then to B.
- Draw another line from A to a different point on the circumference, then to B, so that two triangles are formed.

Are the triangles **congruent**, **similar** or **neither**? Explain your answer.

3 **a** Prove that $\angle WXY + \angle WZY = 180°$

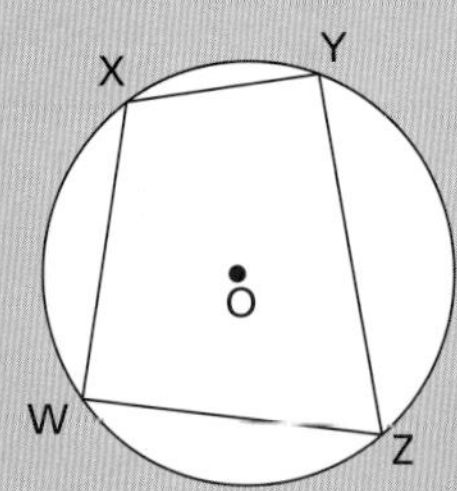

b Prove that $\angle WXZ = \angle WYZ$

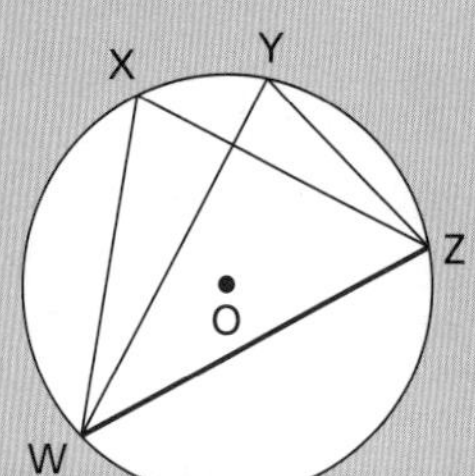

Consolidate – do you need more?

1 Work out the size of each lettered angle.

a

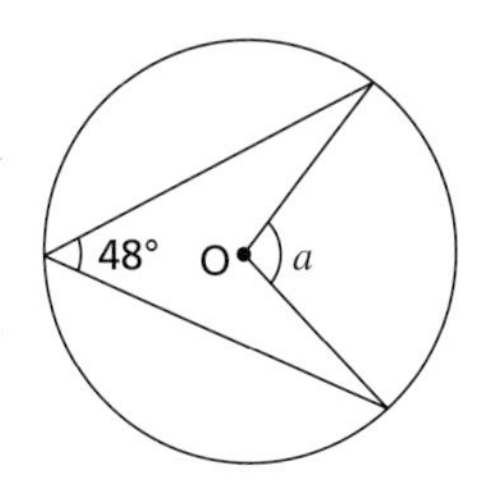

b

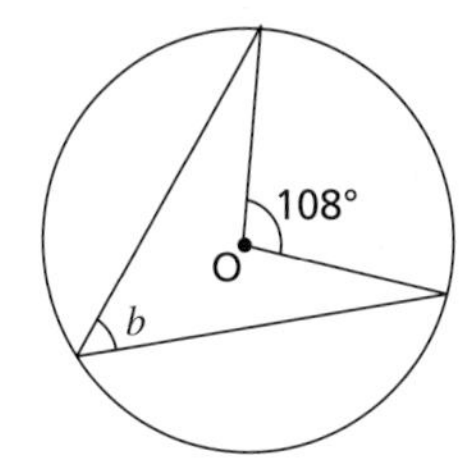

c

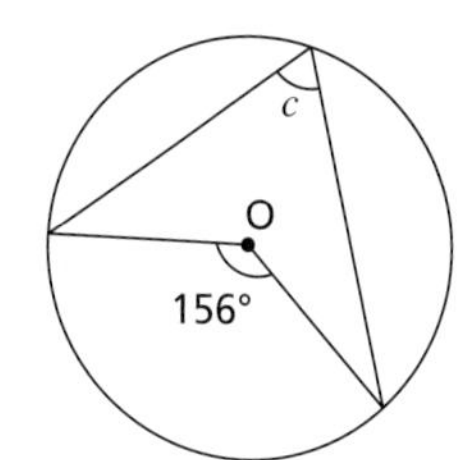

d

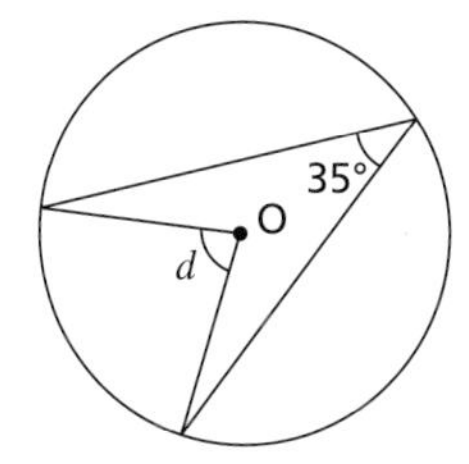

e

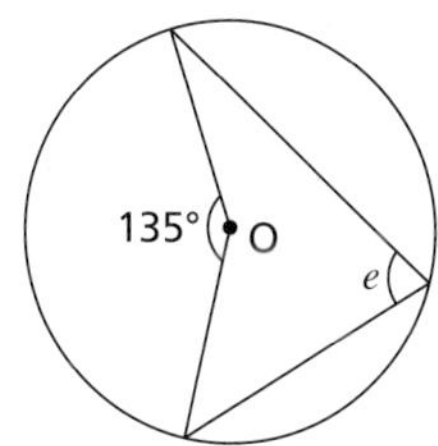

f

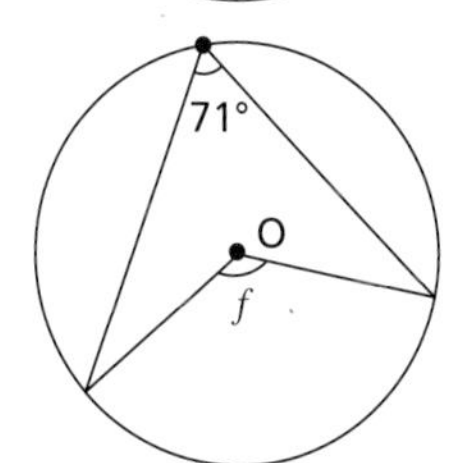

2 Work out the size of each lettered angle.

a

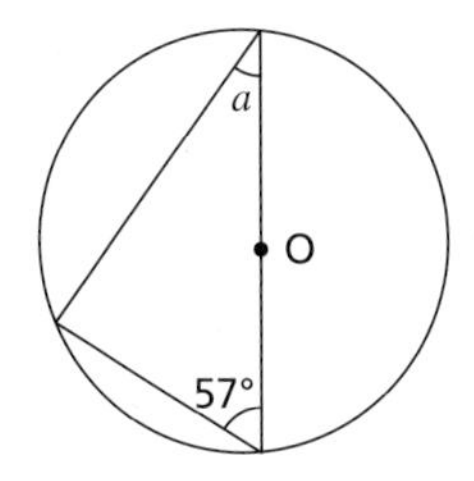

b

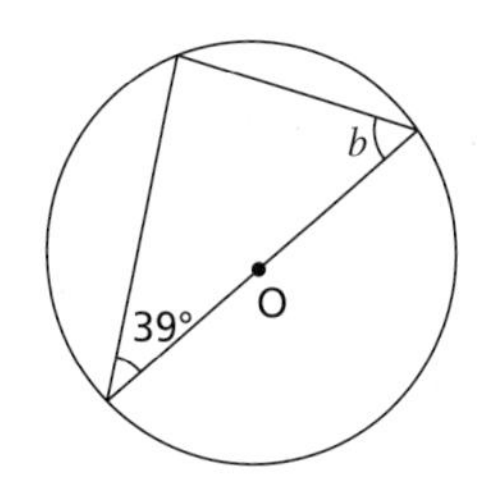

c

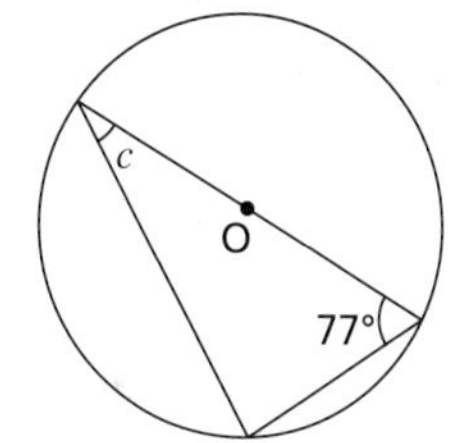

3 Write the size of each lettered angle.

a

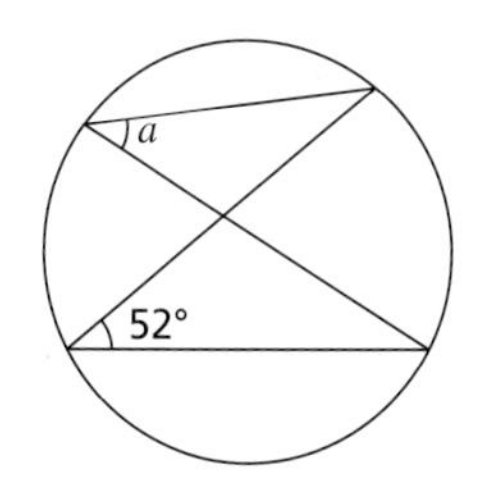

b

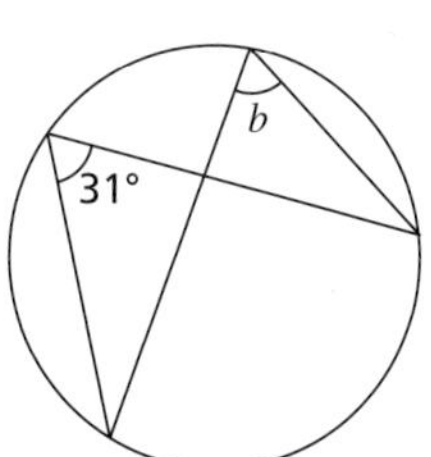

c

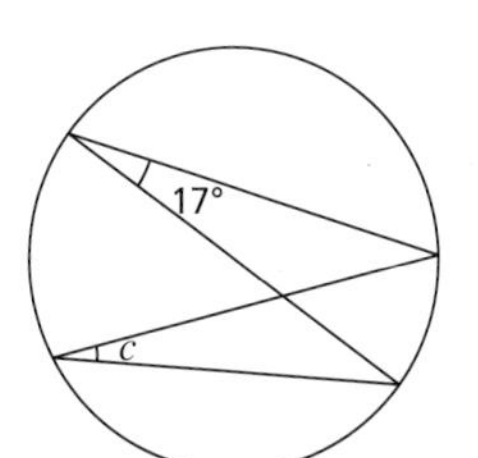

d

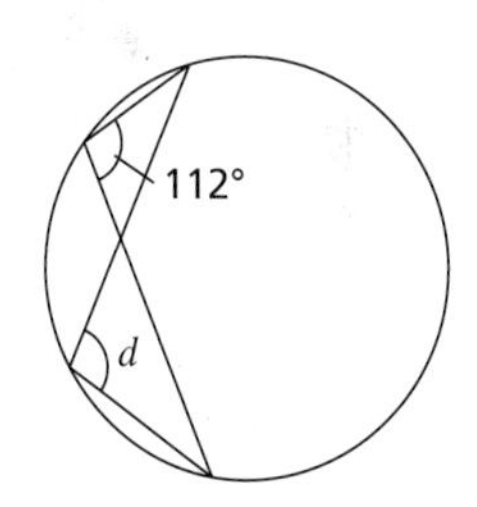

4 Work out the size of each lettered angle.

a

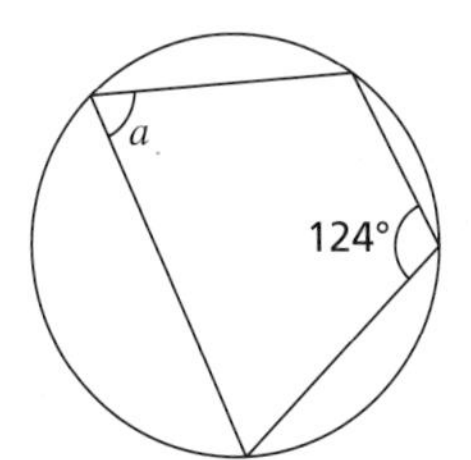

b

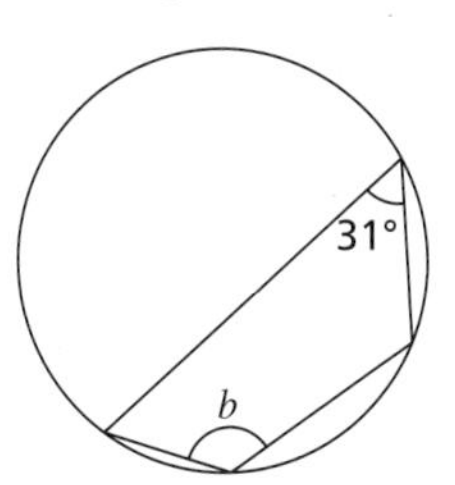

c

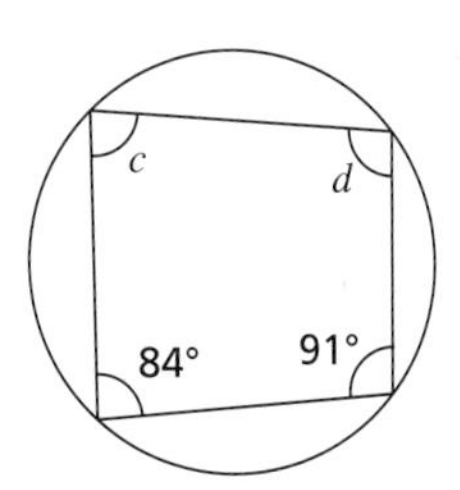

Stretch – can you deepen your learning?

1 A, B, C and D are points on the circumference of the circle, centre O.

Work out the size of angle DBC.
Give reasons for your answer.

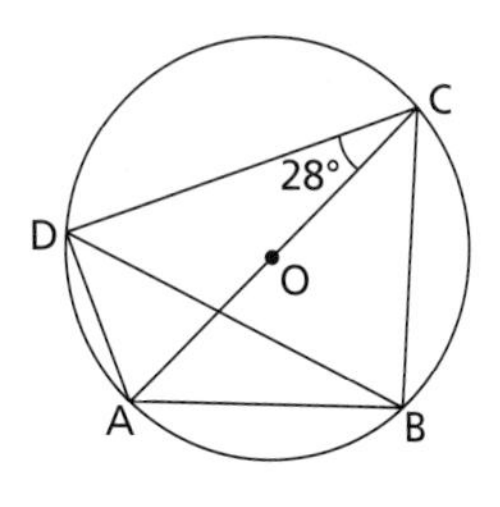

2 A, B, C and D are points on the circumference of the circle, centre O.

Work out the size of angle BAC.
Give reasons for your answer.

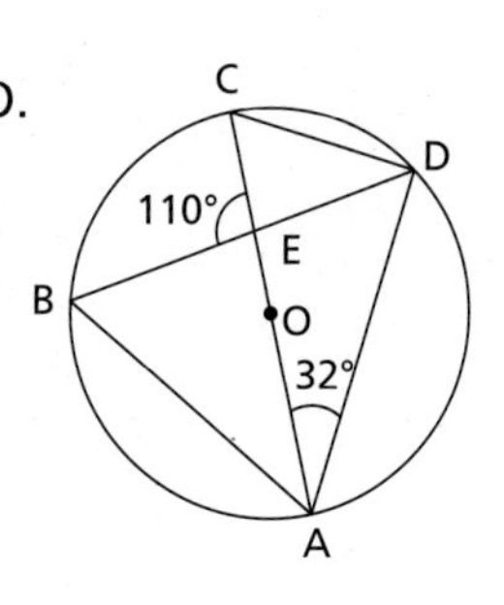

3 Here is a circle, centre O.

Show that $x = 90° - y$

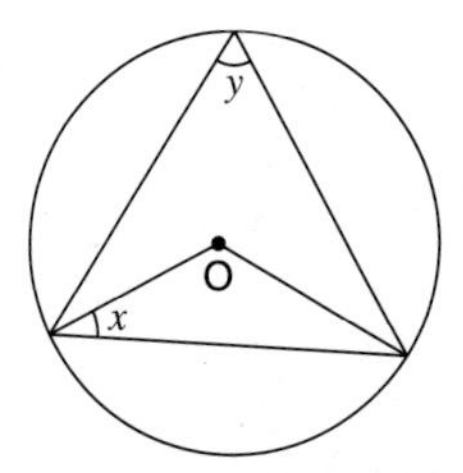

4 A, B and C are points on the circumference of the circle, centre O.

Work out the area of the triangle. Give your answer to 3 significant figures.

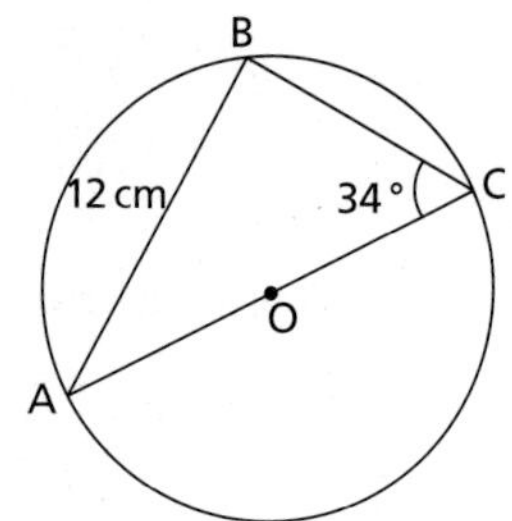

5 A, B, C, D and E are points on the circumference of the circle, centre O.

$\angle AED = (2x + 65)°$

$\angle ACD = (x + y)°$

Work out the values of x and y.

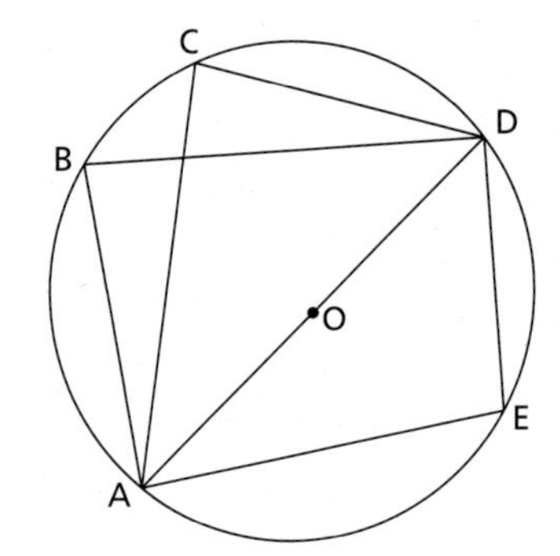

Here is a summary of the circle theorems covered in this chapter.

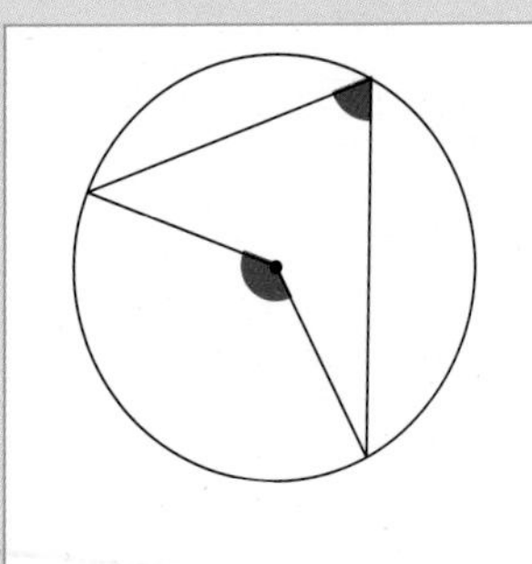	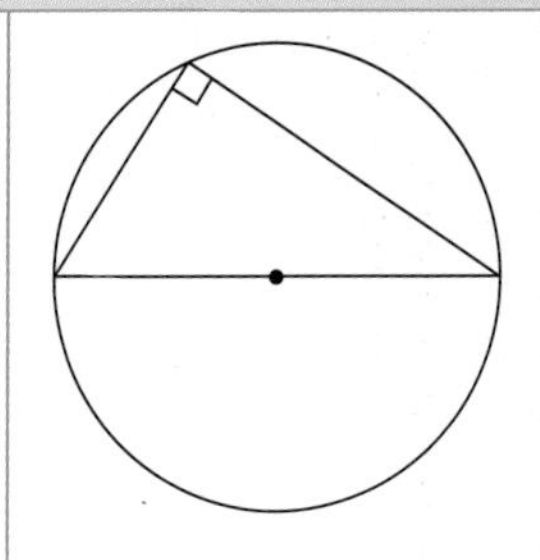	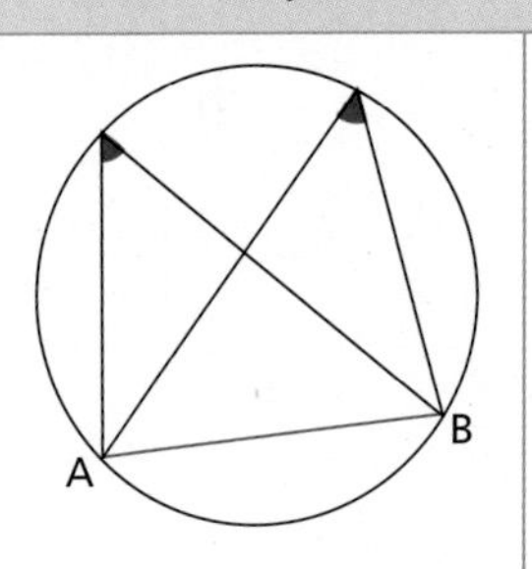	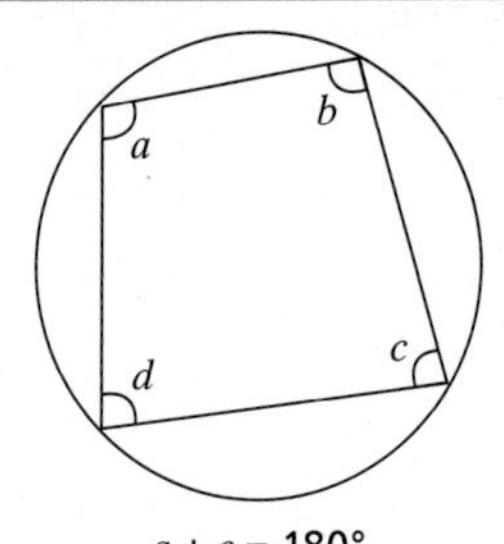 $a + c = 180°$ $b + d = 180°$
The angle at the centre is twice the angle at the circumference.	The angle in a semicircle is a right angle.	Angles in the same segment are equal.	Opposite angles in a cyclic quadrilateral have a sum of 180°.

3.3 Four more circle theorems

Are you ready? (A)

1 Write the mathematical name of each line segment shown.

a

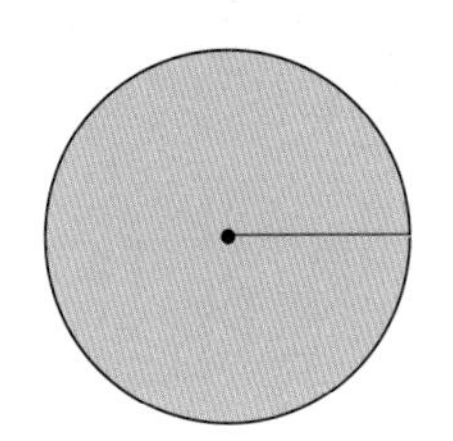

b

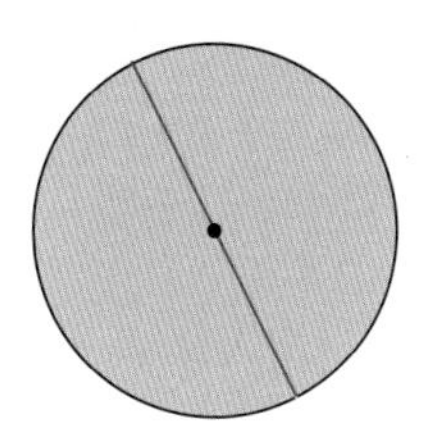

c

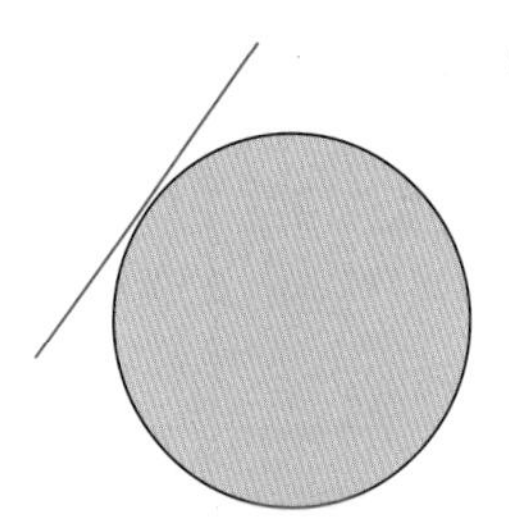

d

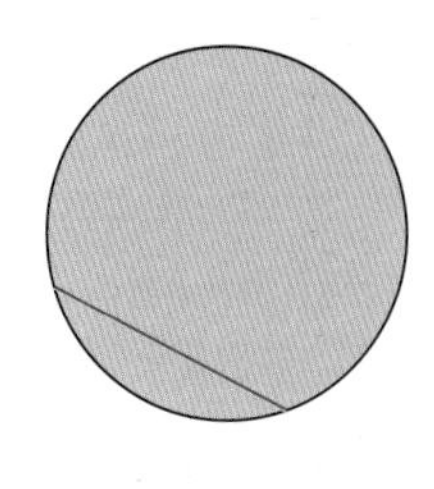

2 Calculate the size of each lettered angle.

a

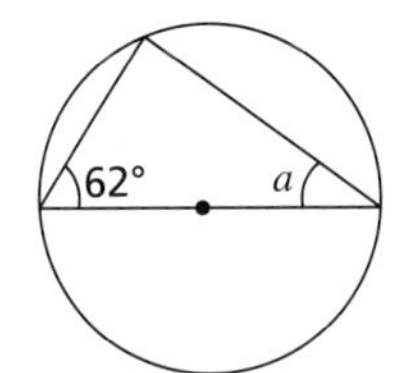

b

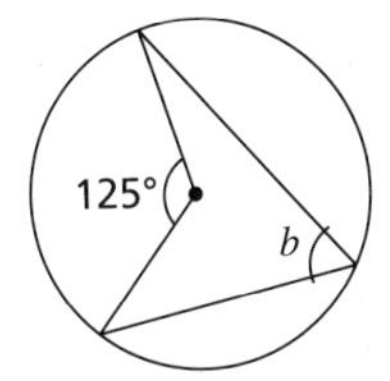

c

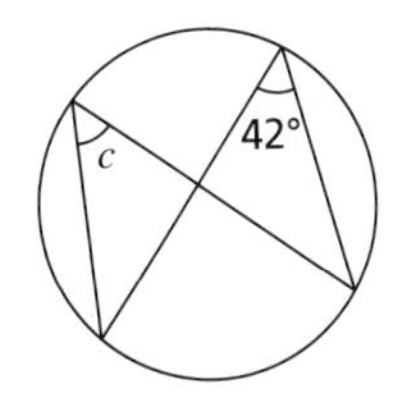

d

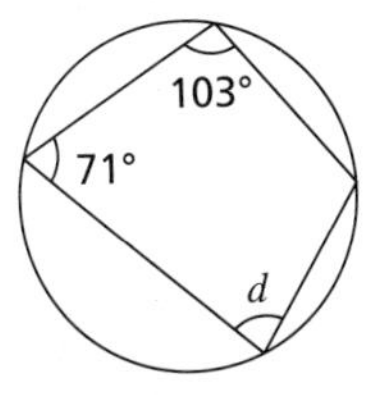

The perpendicular from the centre to a chord

Consider a perpendicular from the centre, O, of a circle to chord AB.

In the triangles AOC and BOC,
$\angle OCA = \angle OCB$ as both are 90°.

OA = OB as both are radii of the circle.

OC is common to both triangles.

So triangles AOC and BOC are congruent using the RHS condition.

This means AC = BC.

This is usually stated as:
The perpendicular from the centre of a circle to a chord bisects the chord.

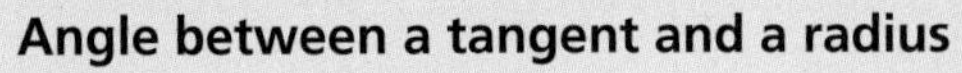

Angle between a tangent and a radius

A **tangent** is a straight line that touches the circumference of a circle at one point.

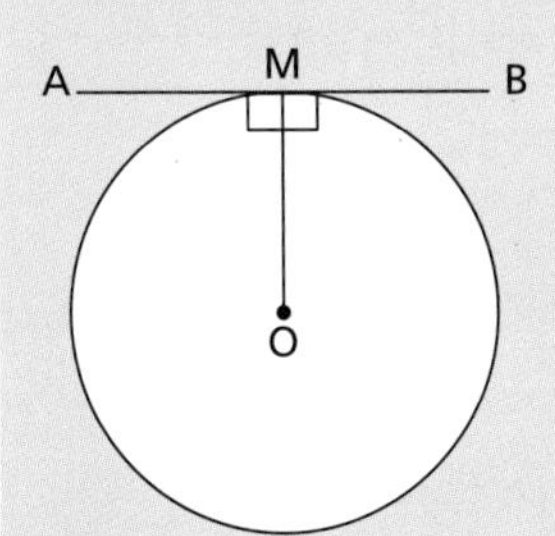

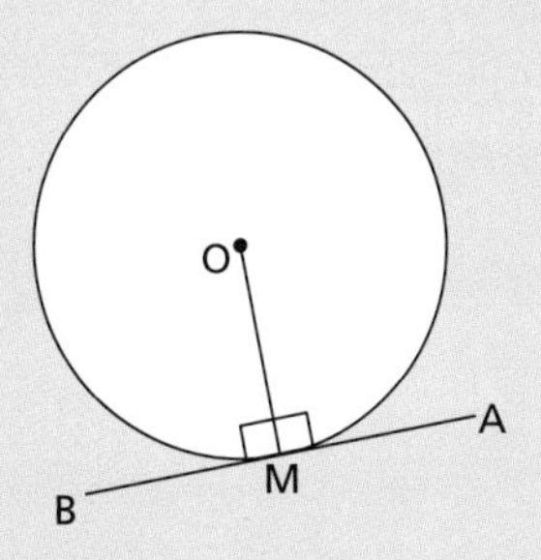

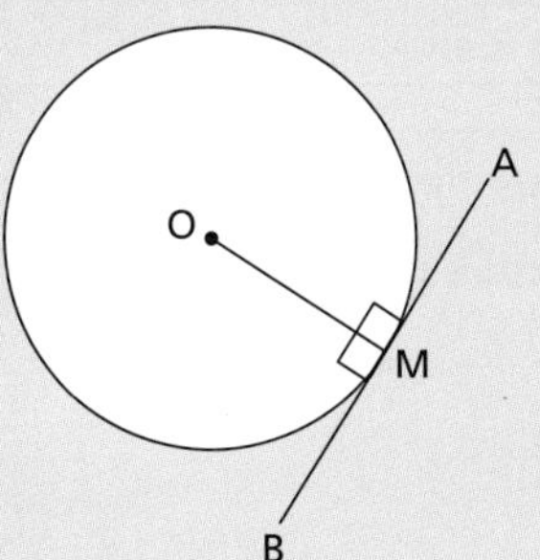

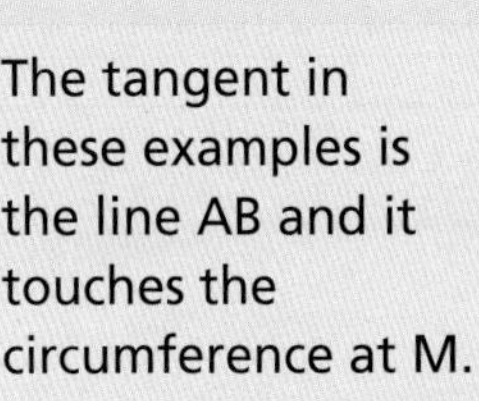
The tangent in these examples is the line AB and it touches the circumference at M.

The angle between a tangent and a radius is a right angle. This can be proved by a process called **proof by contradiction**, but you do not need to know this for GCSE.

This is usually stated as: The angle between a tangent and a radius is a right angle.

Two tangents from the same point

The diagram shows two tangents from point B to the circle, centre O, that meet the circle at points A and C.

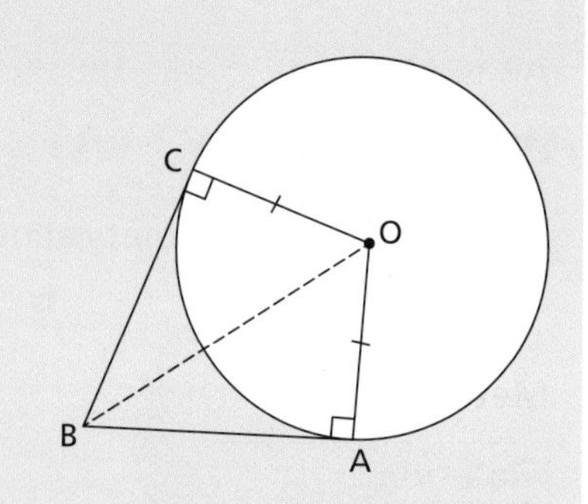

In triangles ABO and CBO,
OA = OC

OB is common to both triangles.

∠OAB = ∠OCB as both are 90°, using the fact that the angle between a tangent and a radius is a right angle.

So triangles AOC and BOC are congruent using the RHS condition.

As the triangles are congruent, then AB = BC.

This is usually stated as: Two tangents from the same point are equal in length.

Example 1

AB is a chord of the circle, centre O.

C is the midpoint of AB.

The radius of the circle is 6 cm.

AB = 8 cm.

Work out the length of OC, giving your answer to 3 significant figures.

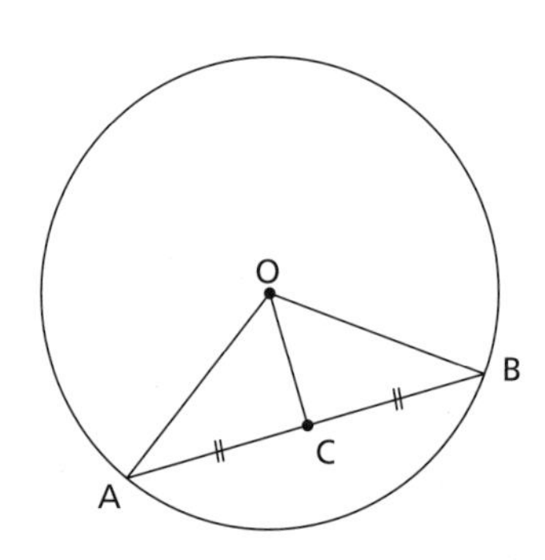

Method

Solution	Commentary
AC = 4 cm Angle AOC = 90° as the perpendicular from the centre of a circle bisects a chord.	The perpendicular from the centre of a circle bisects a chord, so as C is the midpoint of AB, OC must be perpendicular to AB.
Using Pythagoras' theorem: $OC^2 + 4^2 = 6^2$ $OC^2 = 36 - 16 = 20$ $OC = \sqrt{20} = 4.47$ cm (to 3 s.f.)	As triangle AOC is right-angled, you can use Pythagoras' theorem to work out the length of OC.

Example 2

AB is a tangent to the circle, centre O, at the point A.

Work out the size of angle AOB.
Give reasons for your answer.

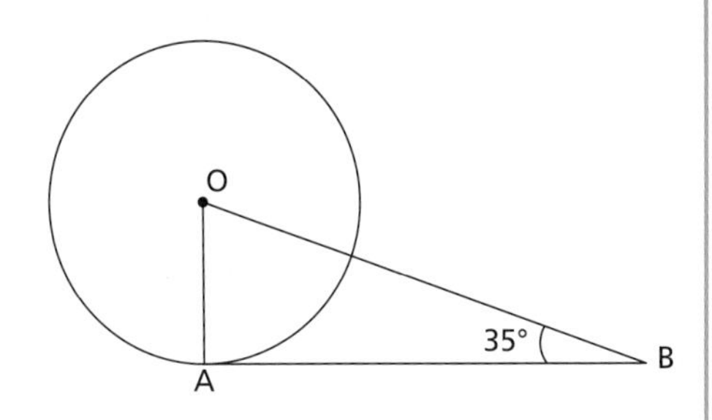

Method

Solution	Commentary
$\angle OAB = 90°$ The angle between a tangent and a radius is 90°.	OA is a radius and AB is a tangent. State the circle theorem you have used.
$\angle AOB = 180° - 90° - 35° = 55°$ The sum of the angles in a triangle is 180°.	Use the fact that the sum of the angles in a triangle is 180°.

Example 3

XY and ZY are tangents to the circle, centre O.

OZ = 5 cm

XY = 12 cm

Work out the perimeter of the quadrilateral OXYZ.
Give reasons for your answer.

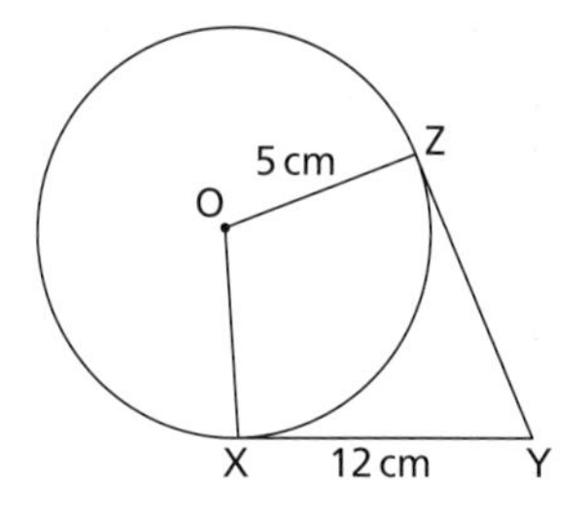

Method

Solution	Commentary
OX = 5 cm as OX and OZ must be equal since they are both radii of the circle. YZ = 12 cm as YZ = XY because two tangents from the same point are equal in length.	Work out the lengths of the sides you do not know already, giving clear reasons.
Perimeter = 5 + 5 + 12 + 12 = 34 cm	To work out the perimeter, find the sum of all four lengths.

Practice (A)

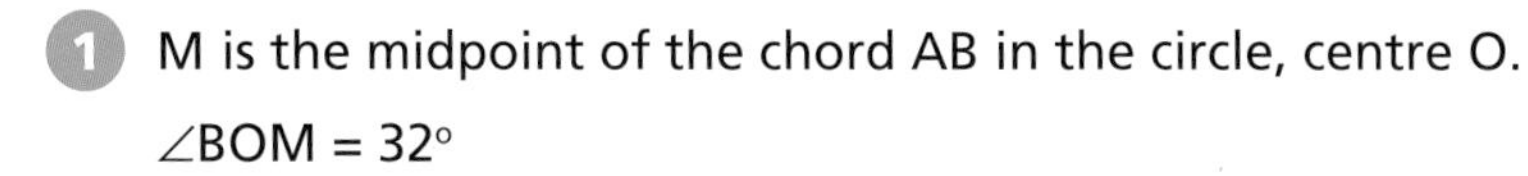

1. M is the midpoint of the chord AB in the circle, centre O.

   $\angle BOM = 32°$

   a State the size of angle OMB, giving a reason for your answer.

   b Work out the size of angle OBM.

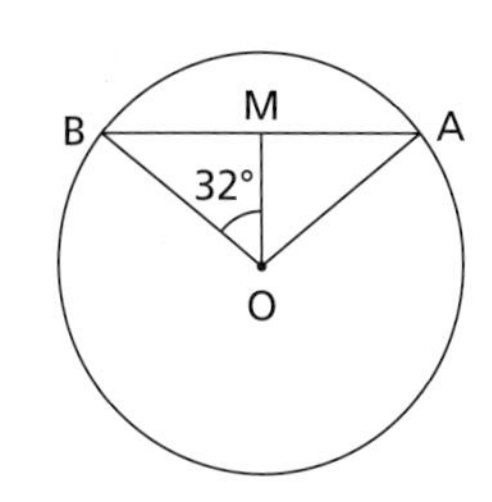

2 M is the midpoint of the chord AB in the circle, centre O.

$\angle OAM = 28°$

Work out the size of angle AOM.

Give reasons for your answer.

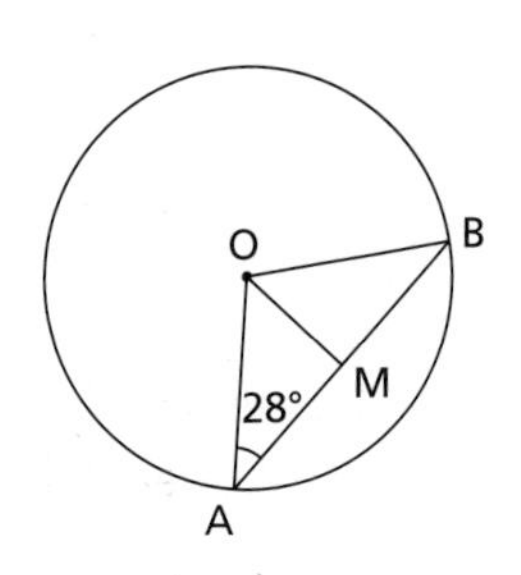

3 M is the midpoint of the chord DC in the circle, centre O.

$\angle OCM = 32°$

OM = 4 cm

a Work out the length of OC.

Give your answer to 1 decimal place.

b Work out the length of CM.

Give your answer to 1 decimal place.

c Work out the length of CD.

Give your answer to 1 decimal place.

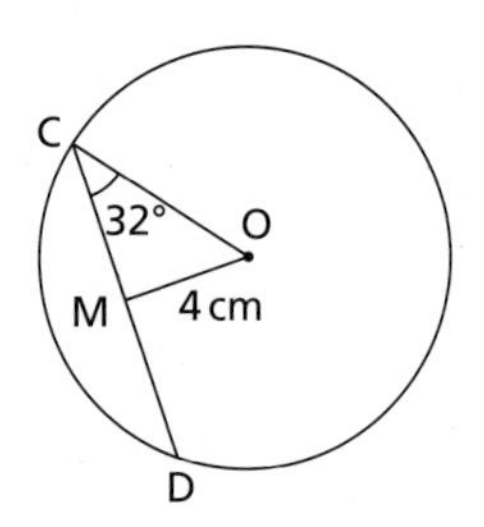

You will need to use trigonometry, which is reviewed in Block 6.

4 M is the midpoint of the chord AC in the circle, centre O.

$\angle OAM = 39°$

Work out the size of angle MOB.

Give reasons for your answer.

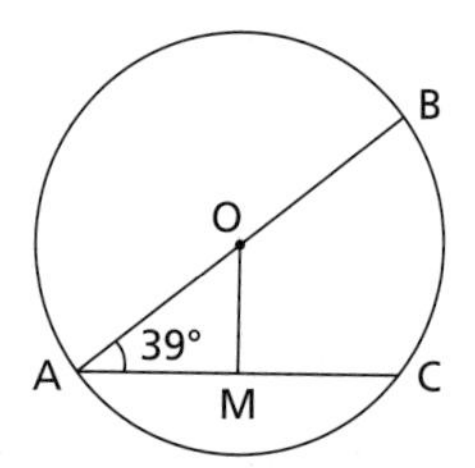

5 In each diagram, O is the centre of the circle and AB is a tangent.

Work out the size of each lettered angle.

a

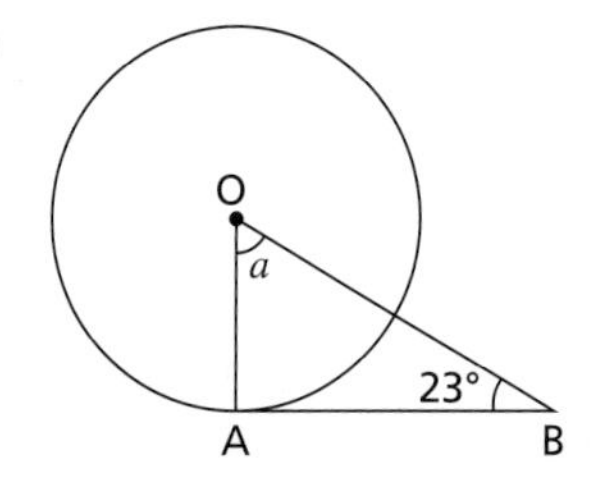

b

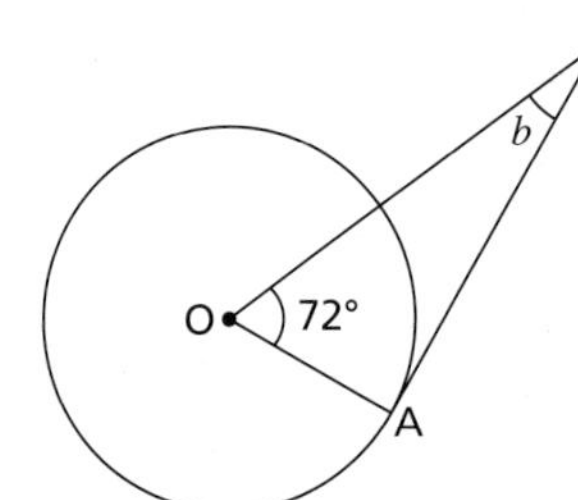

c

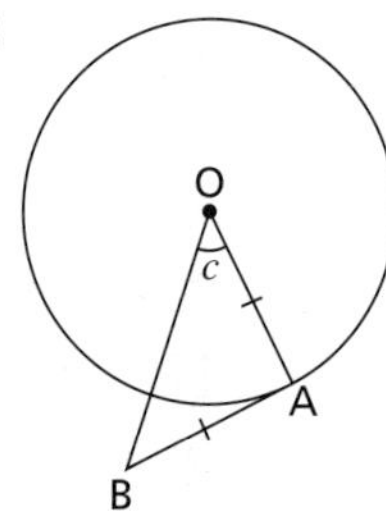

6 CD is a tangent to the circle, centre O.

$\angle BDA = 57°$

a Work out the size of angle CDB.

Give a reason for your answer.

b Work out the size of angle BAD.

Give a reason for your answer.

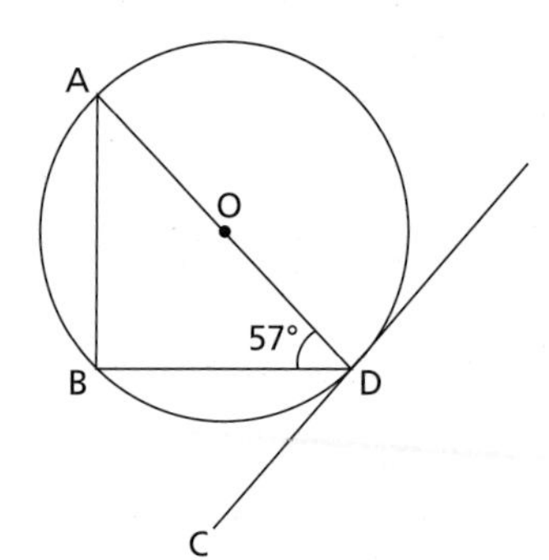

7 AC is a tangent to the circle at B.

AD is a tangent to the circle at E.

O is the centre of the circle.

Calculate the size of angle BAE.

Give reasons for each stage of your workings.

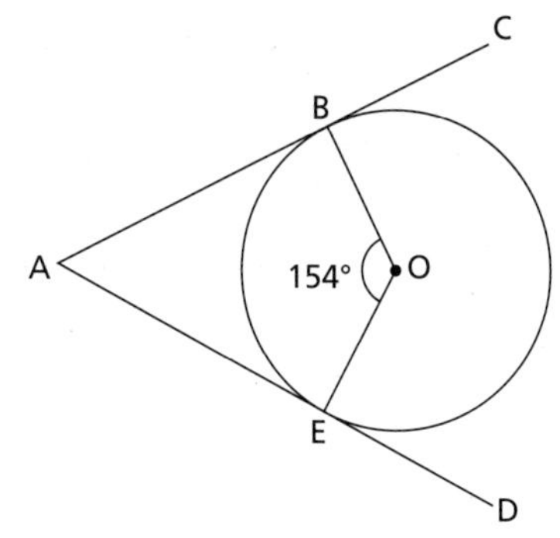

8 AB is a tangent to the circle, centre O.

Work out the value of x.

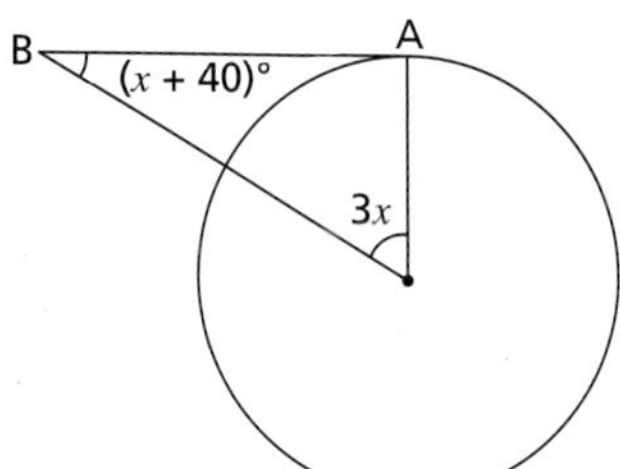

9 A, B, C and D are points on the circumference of the circle, centre O.

AE is a tangent to the circle.

$\angle AED = 36°$

Work out the size of angle ABD.

Give reasons for your answer.

10 AB and CB are tangents to the circle, centre O.

a Write down the length of AB.

Give a reason for your answer.

b Work out the size of angle x.

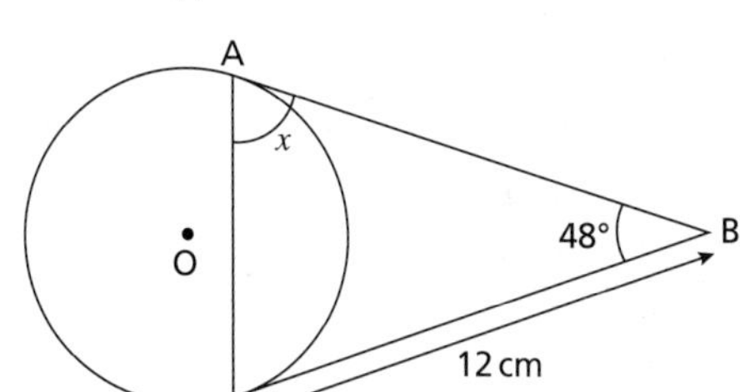

11 QS is a tangent to the circle at R.

QU is a tangent to the circle at T.

O is the centre of the circle.

The radius of the circle is 6 cm and QR = 8 cm.

a Work out the length of OQ.

b Work out the area of the quadrilateral QROT.

12 AB and BC are tangents to the circle at A and C.

O is the centre of the circle.

Calculate the length AB.

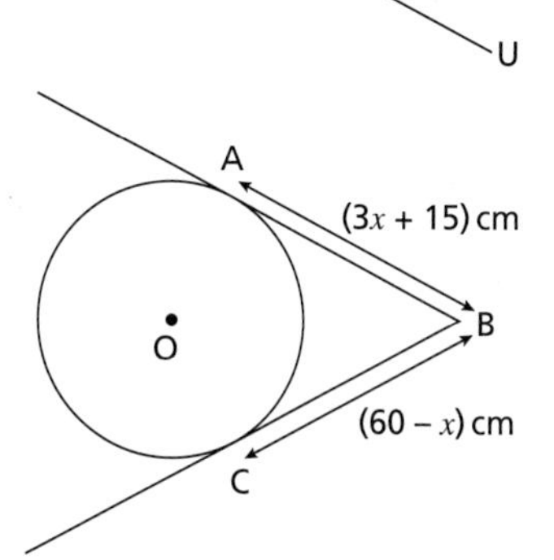

Are you ready? (B)

1 In which diagrams is the shaded region a segment of the circle?

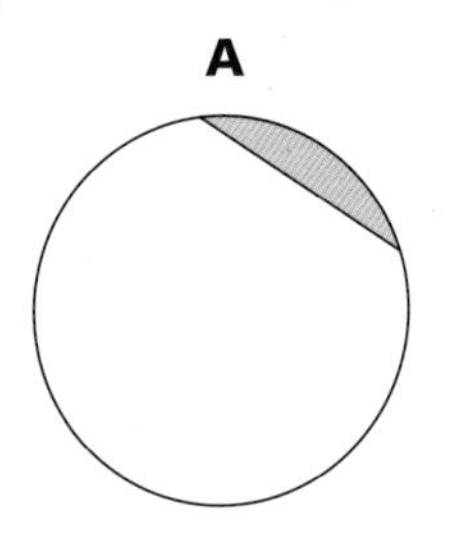

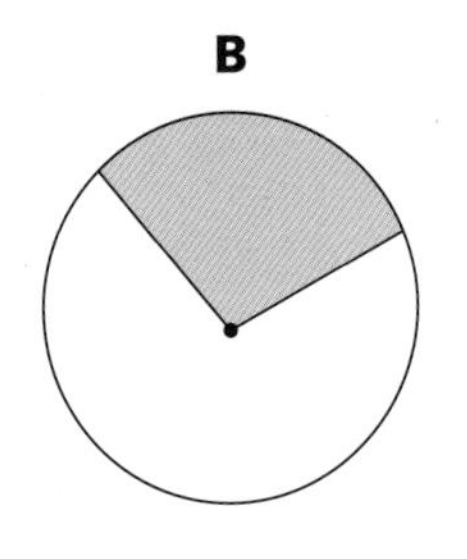

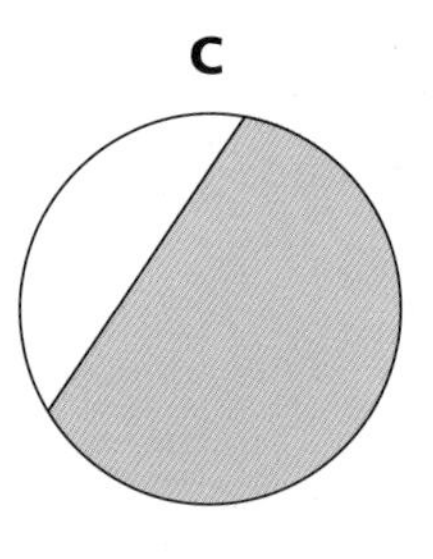

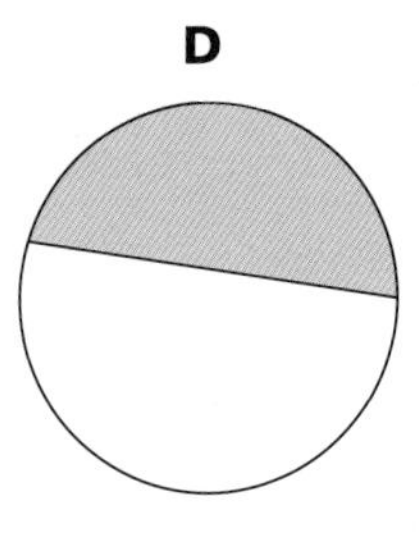

Alternate segment theorem

The diagrams show angles between a chord and a tangent, and angles in the other (or **alternate**) segment.

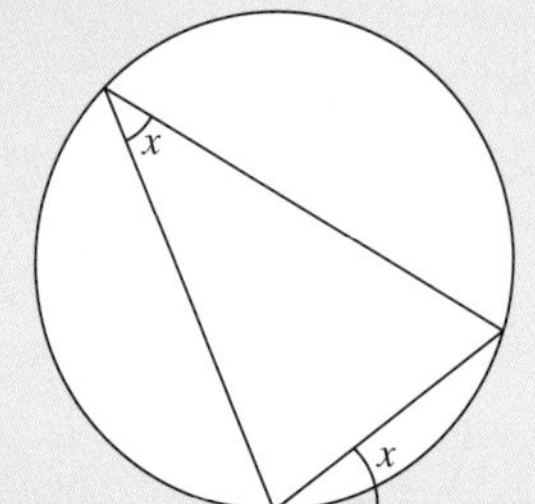

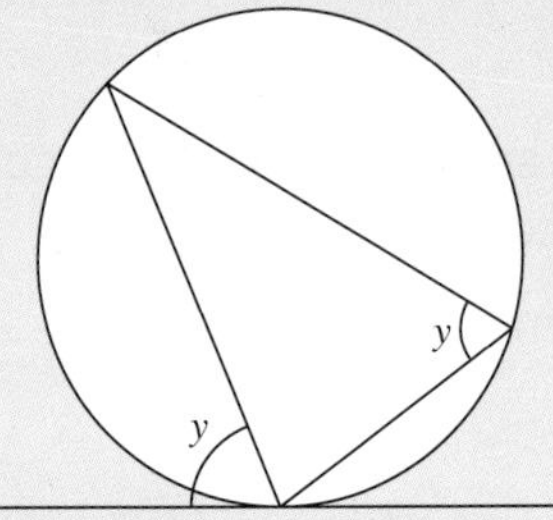

You can prove that the angle between a chord and a tangent is equal to the angle in the alternate segment.

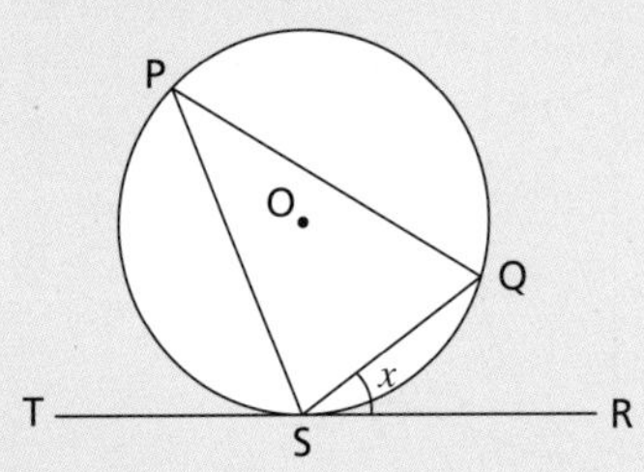

Consider triangle PQS in a circle centre O, where tangent RT meets the circle at S.

Let $\angle QSR = x$

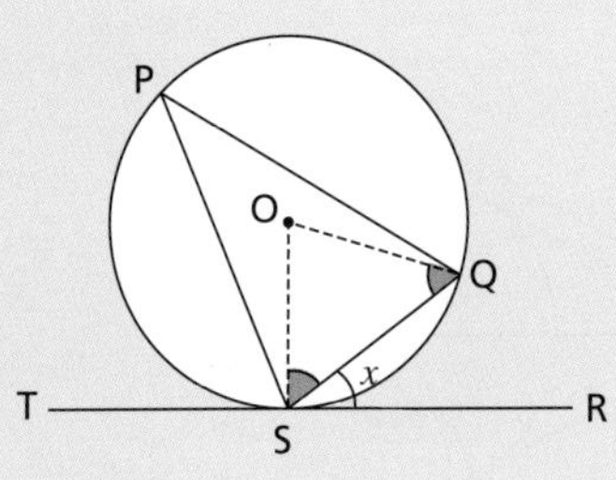

$\angle OSQ = (90 - x)°$ using the fact that the angle between a tangent and a chord is a right angle.

$\angle OQS = (90 - x)°$ as well, as triangle OSQ is isosceles (both OS and OQ are radii).

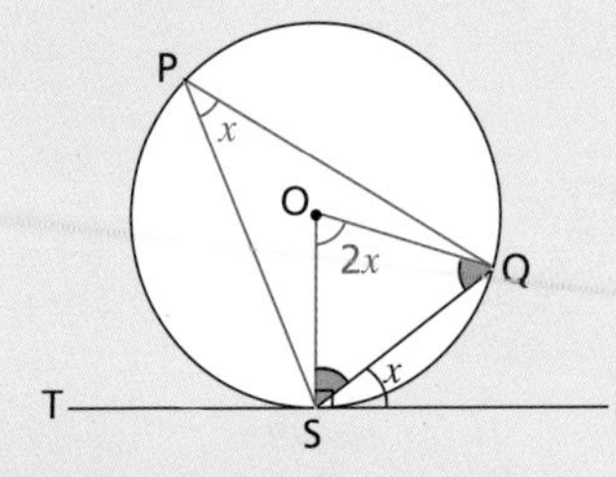

Using the fact that angles in a triangle add up to 180°, then $\angle SOQ = 2x$.

Hence $\angle SPQ = x$, because the angle at the centre of a circle is twice the angle at the circumference.

This is usually stated as:
The angle between a chord and a tangent is equal to the angle in the alternate segment.

Example

B, C and D are all points on the circumference of a circle.

AE is a tangent to the circle at the point B.

Work out the size of angle x, giving a reason for your answer.

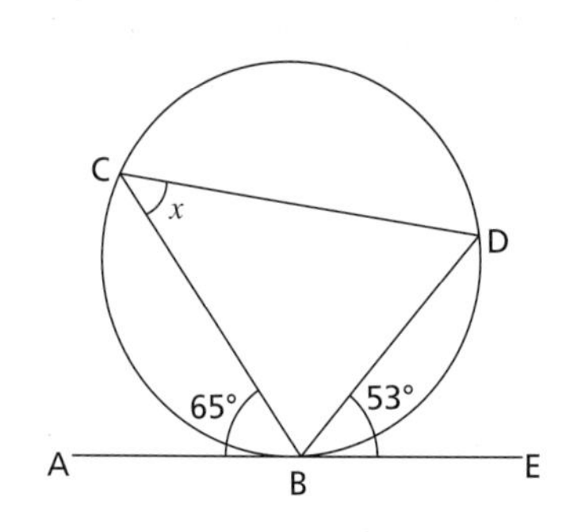

Method

Solution	Commentary
$x = 53°$ Angles in alternate segments are equal.	Look carefully to decide which is the required angle. In this case x is in the alternate segment to angle DBE. Angle CDB is alternate to the 65° angle.

Practice (B)

1. Copy the diagrams and mark the angle in the alternate segment to the angle shown.

a

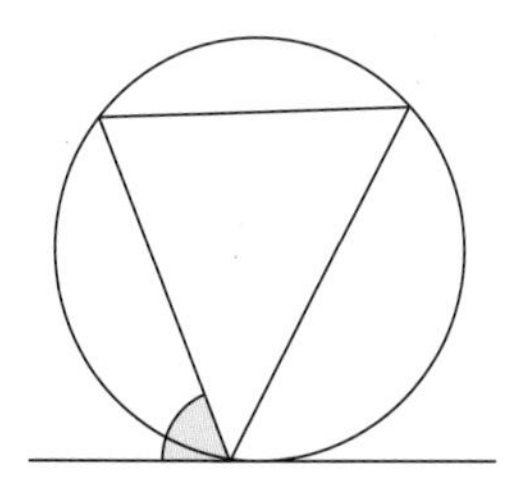

b

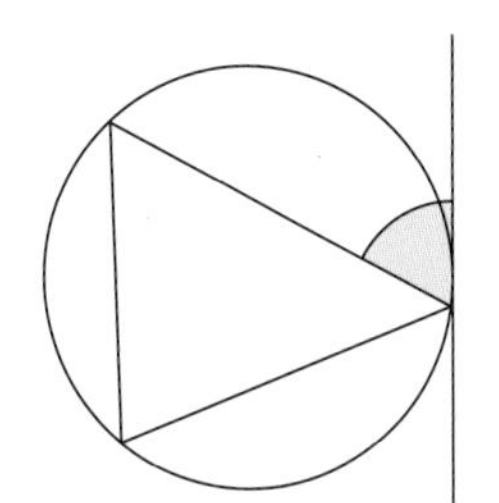

2. Work out the size of each lettered angle.

a

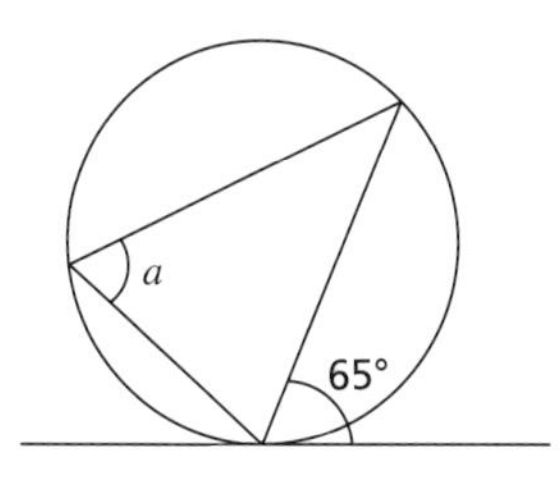

b

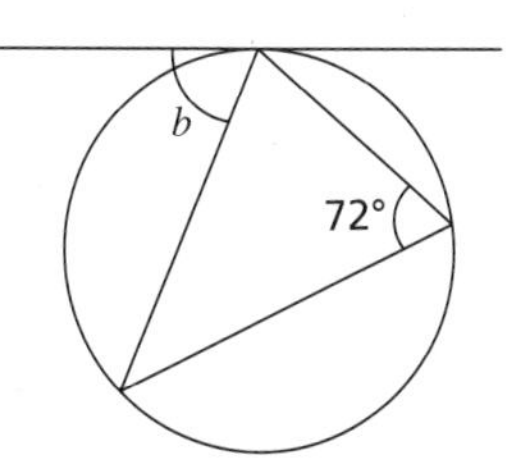

c

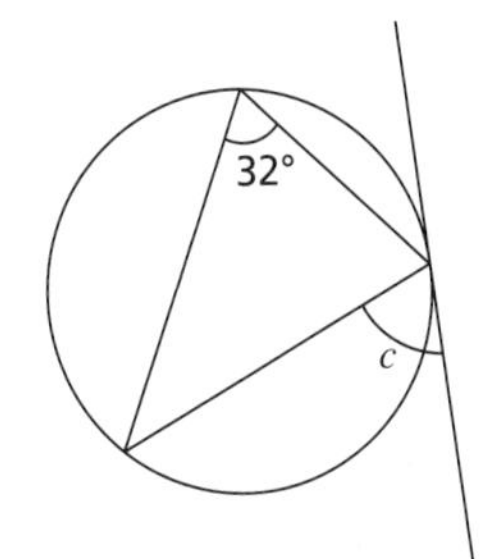

3 Work out the size of each lettered angle.

a

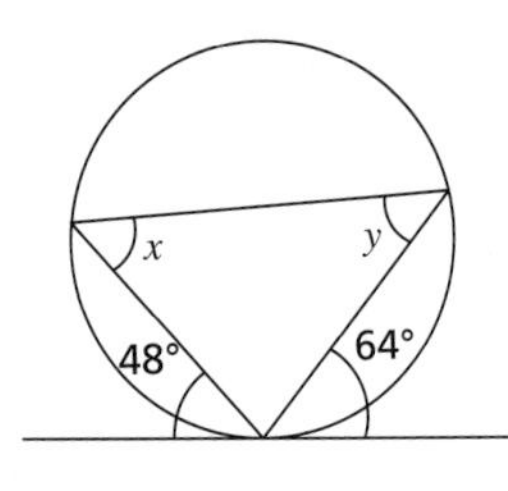

b

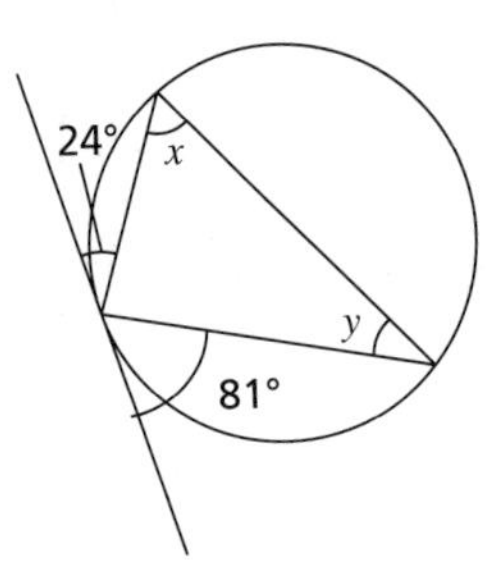

4 The points B, C and D are all points on the circumference of a circle.

AE is a tangent to the circle at point D.

Work out the size of:

a ∠BCD **b** ∠CDE **c** ∠BDC

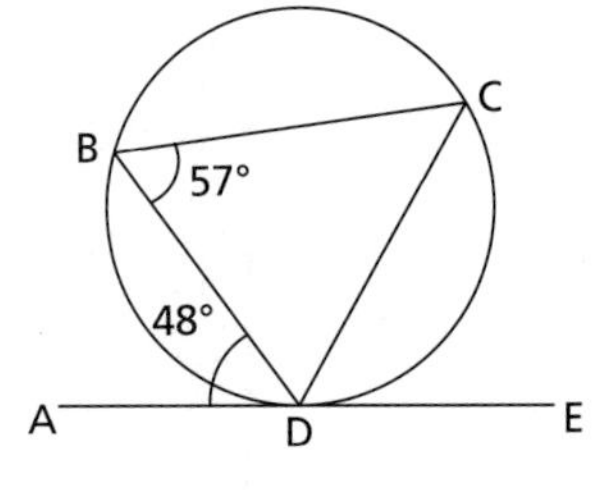

5 The points B, D and E are all points on the circumference of a circle.

ABC is a tangent to the circle.

a Work out the size of angle x.

Give reasons for your answer.

b Work out the size of angle y.

Give reasons for your answer.

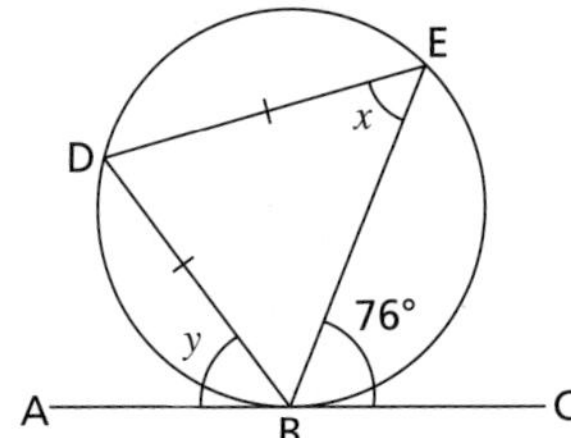

6 P, N and L are points on the circumference of a circle, centre O.

KM is a tangent to the circle at L.

Work out the size of angle MLN.

Give reasons for your answer.

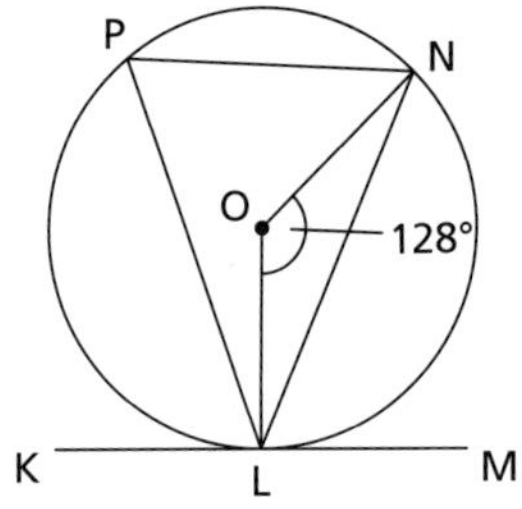

7 S, U and V are points on the circumference of a circle, centre O.

RT is a tangent to the circle at the point S.

Work out the size of angle OSV.

Give reasons for your answer.

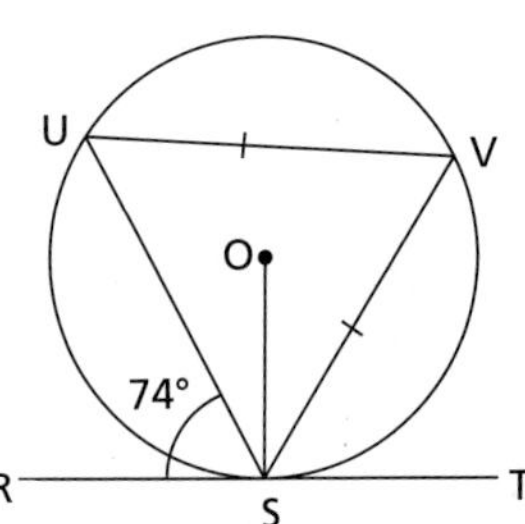

8 A, B and C are points on the circumference of a circle, centre O.

DE is a tangent to the circle at point A.

∠CAE = 32°

a Work out the size of angle OCA.

Give reasons for your answer.

b Work out the size of angle DAB.

Give reasons for your answer.

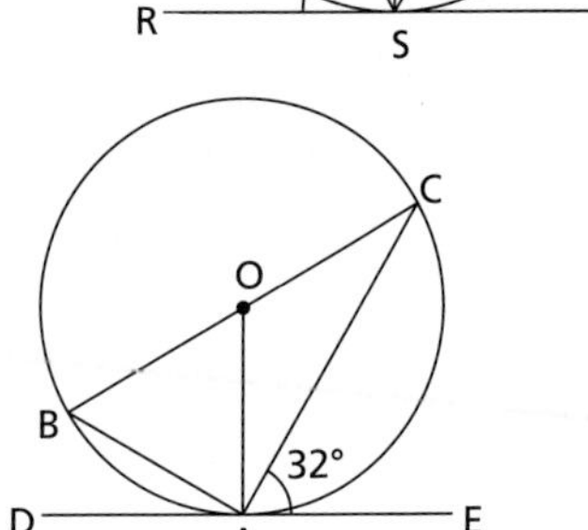

What do you think?

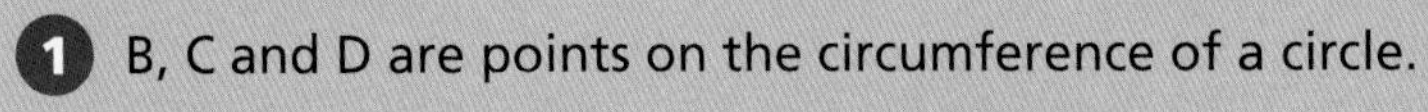

1. B, C and D are points on the circumference of a circle.
   AE is a tangent to the circle at the point B.
   Huda says, "If CD is parallel to AE, then triangle BCD is isosceles."
   Is she correct?

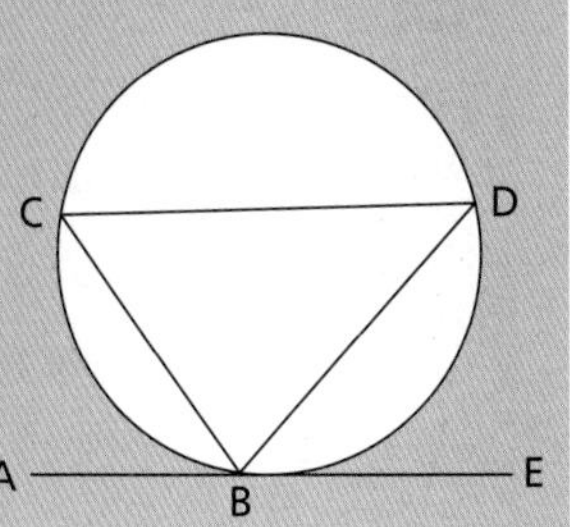

Consolidate – do you need more?

1 M is the midpoint of the chord AB in the circle, centre O.

$\angle BOM = 40°$

Write down the size of angle OBM.

Give a reason for your answer.

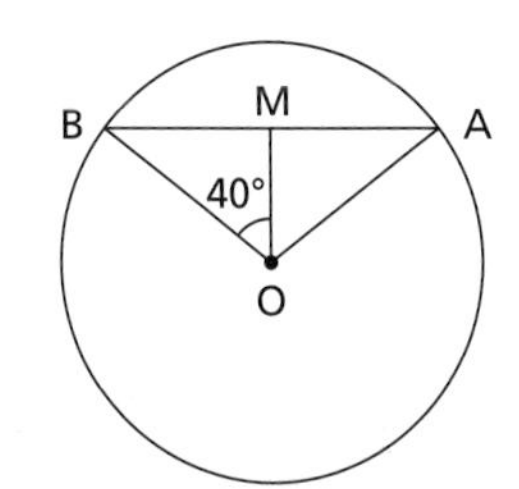

2 M is the midpoint of the chord AB in the circle, centre O.

$\angle OAM = 35°$

Work out the size of angle AOM.

Give reasons for your answer.

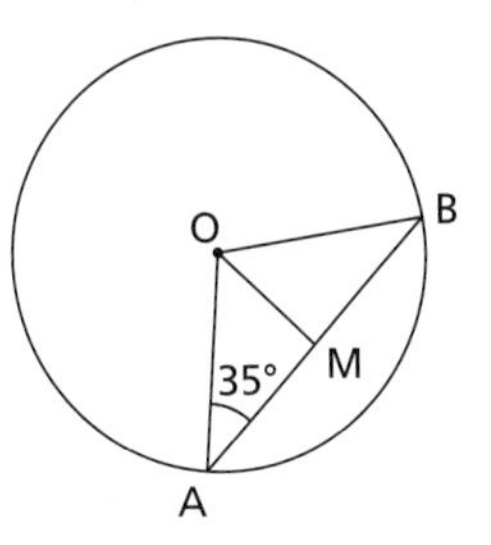

3 In each diagram, O is the centre of the circle and AB is a tangent.

Work out the size of each lettered angle.

a

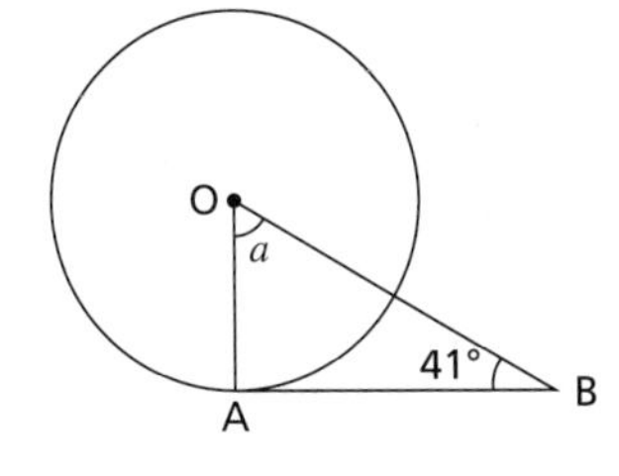

b

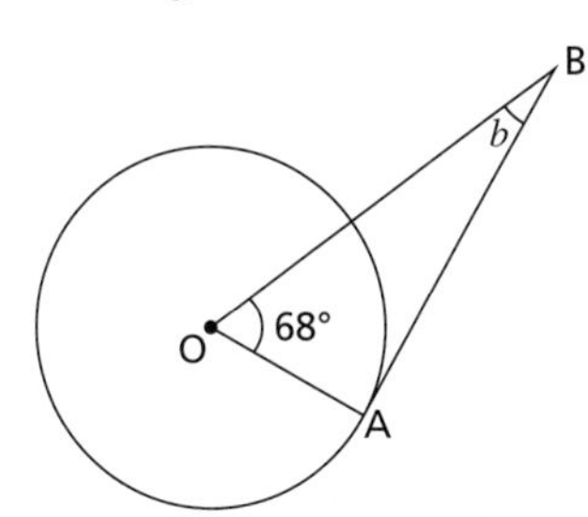

c

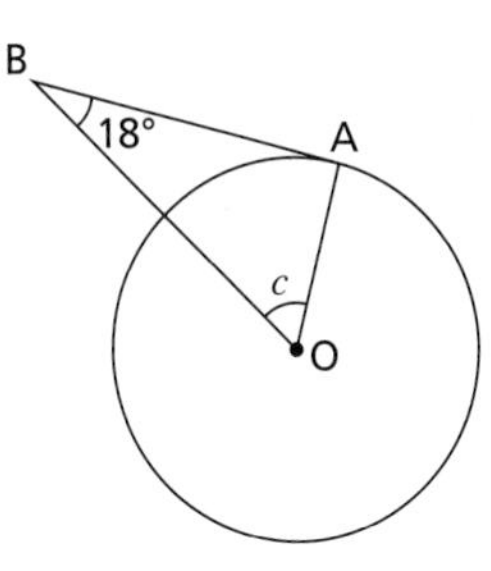

4 AB and CB are tangents to the circle, centre O.

Write down the length of AB.

Give a reason for your answer.

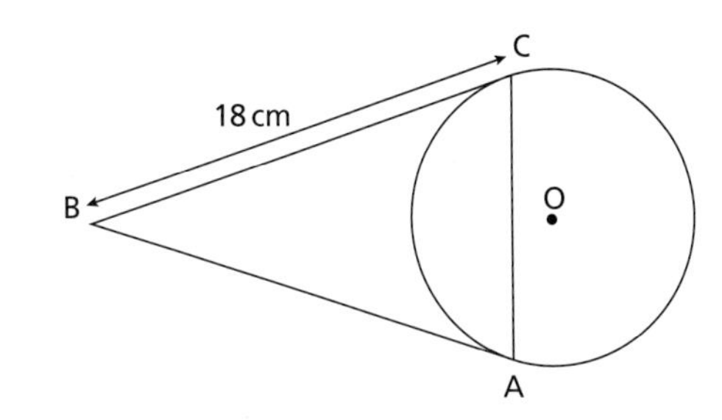

5 QS is a tangent to the circle at R.

QU is a tangent to the circle at T.

O is the centre of the circle.

The radius of the circle is 3 cm and QR = 4 cm.

Work out the length of OQ.

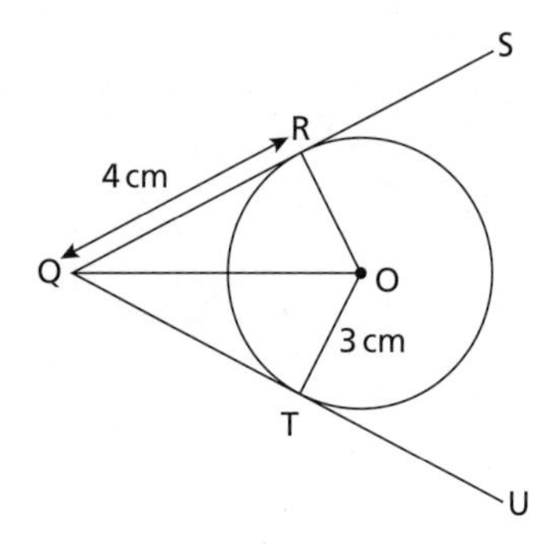

6 Work out the size of each lettered angle.

a

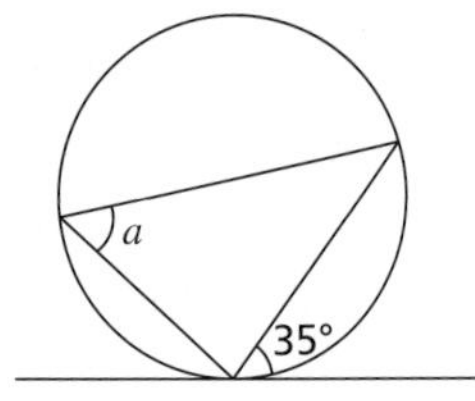

b

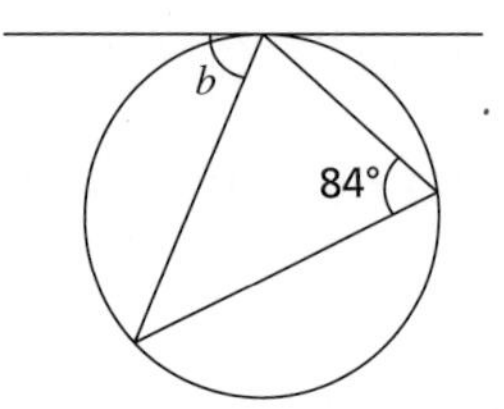

c

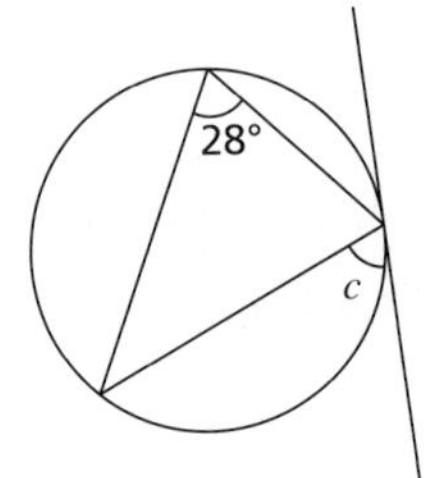

d

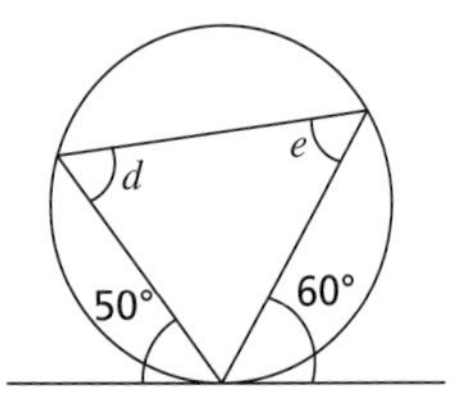

e

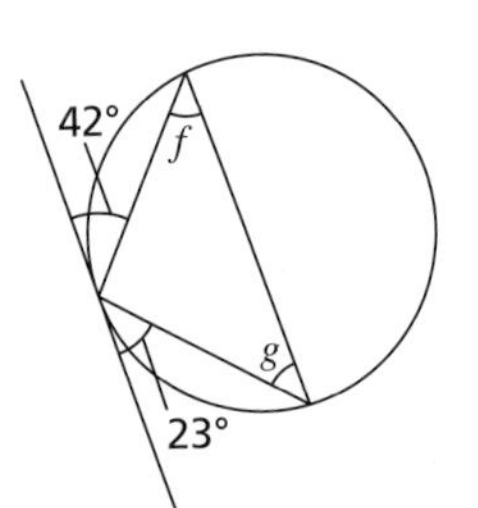

f

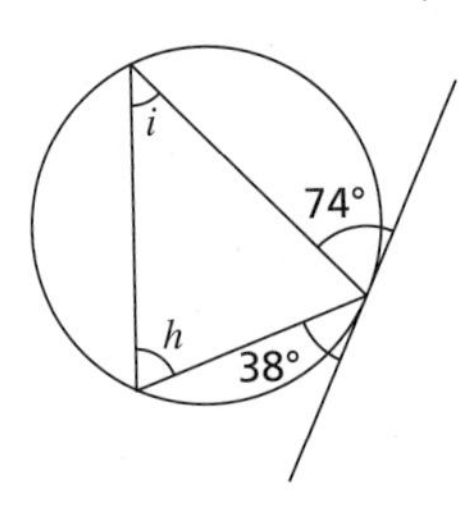

Stretch – can you deepen your learning?

1 B, F, E and D are points on the circumference of a circle.

AC is a tangent at the point B.

BF is parallel to DE.

∠FBD = 78°

∠EFD = 35°

Work out the size of angle ABF.

Give reasons for your answer.

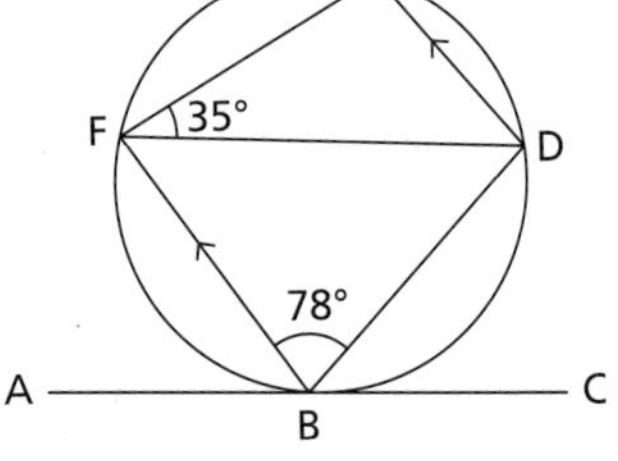

2 AC is a tangent to the circle at the point B.

B, D and E are points on the circumference.

∠BDE = $(2x + 5y)$°

∠ABD = $(3x + 2y)$°

Work out the values of x and y.

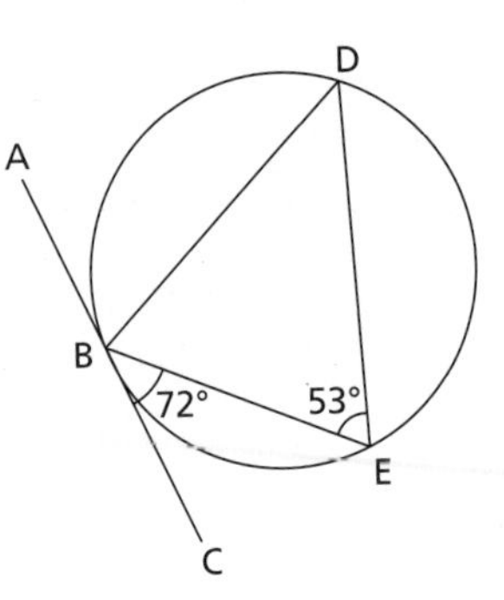

3 AC is a tangent to the circle at the point B.

B, D, E and F are all points on the circumference of the circle.

EB = EF

Work out the value of x in degrees.

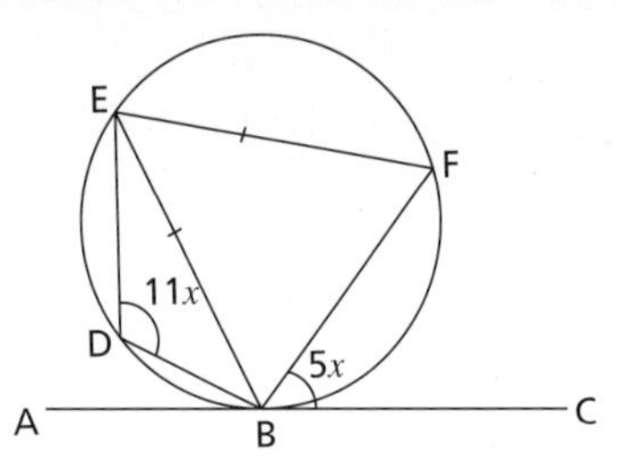

4 A, B and C are points on the circumference of the circle, centre O.

AB = BC

DE is a tangent to the circle at point C.

$\angle OCB = x$

Write an expression in terms of x for the size of angle ACD.

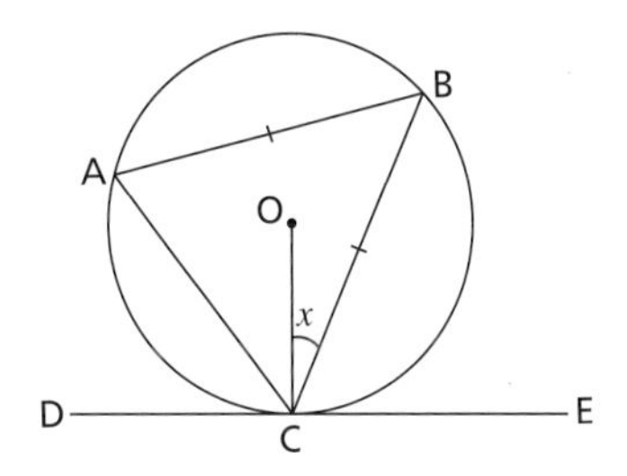

5 Q, S and T are points on the circumference of a circle, centre O.

PR is a tangent to the circle at point Q.

Prove that $\angle STQ = \angle PQS$

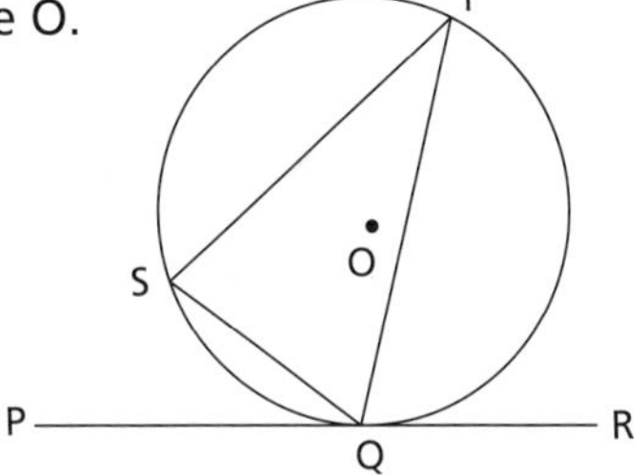

The circle theorems covered in this chapter are summarised below.

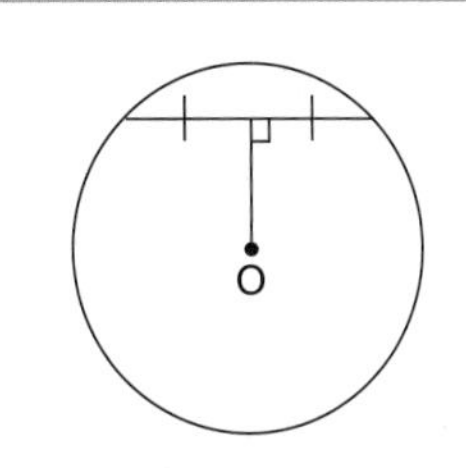	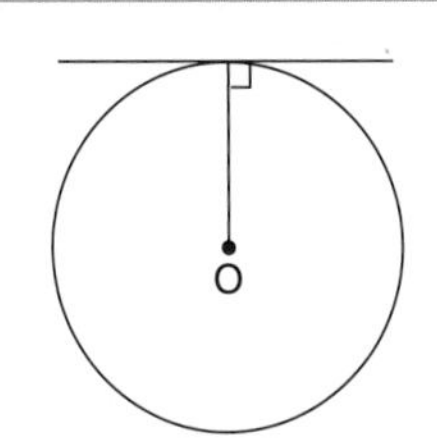	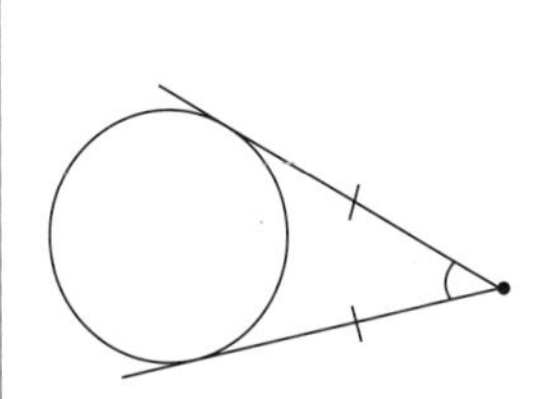	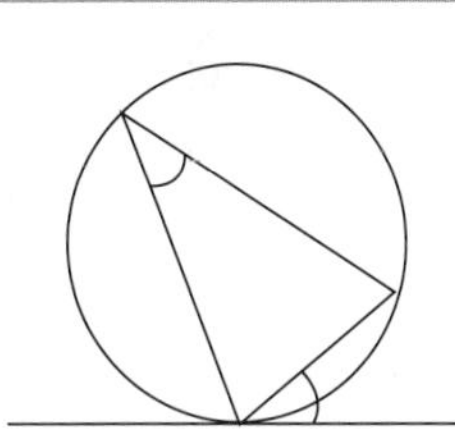
A perpendicular from the centre to a chord bisects the chord.	The angle between a tangent and a radius is a right angle.	Two tangents to a circle from the same point are equal in length.	The angle between a chord and a tangent is equal to the angle at the circumference in the alternate segment. This is known as the **alternate segment theorem**.

3.4 Parts of circles

Are you ready? (A)

1 Work out the circumference of each circle. Give your answers to 3 significant figures.

a

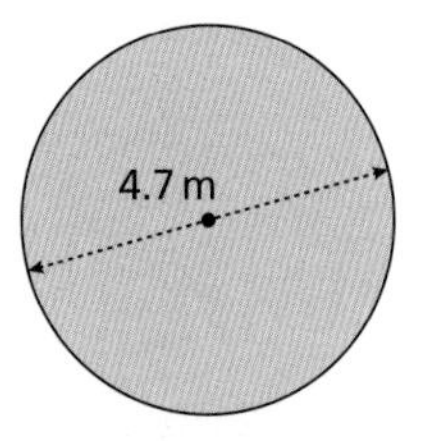

b

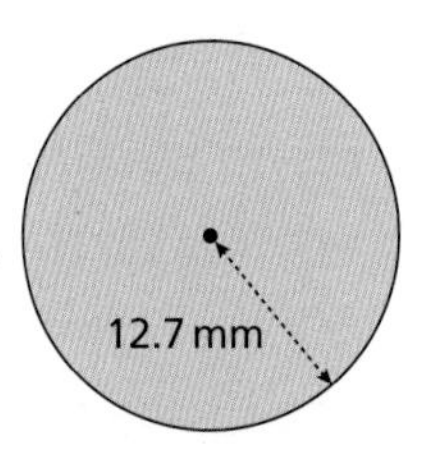

2 Work out the circumference of each circle. Give your answers in terms of π.

a

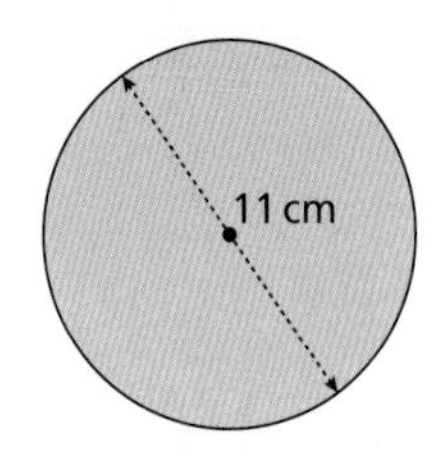

b

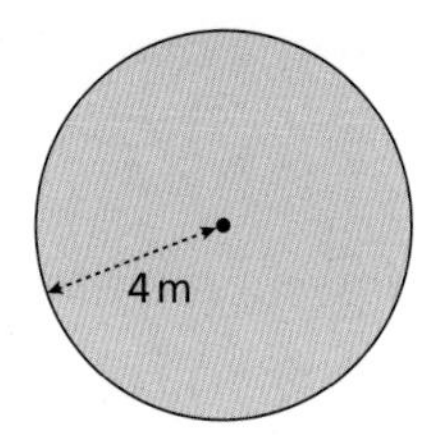

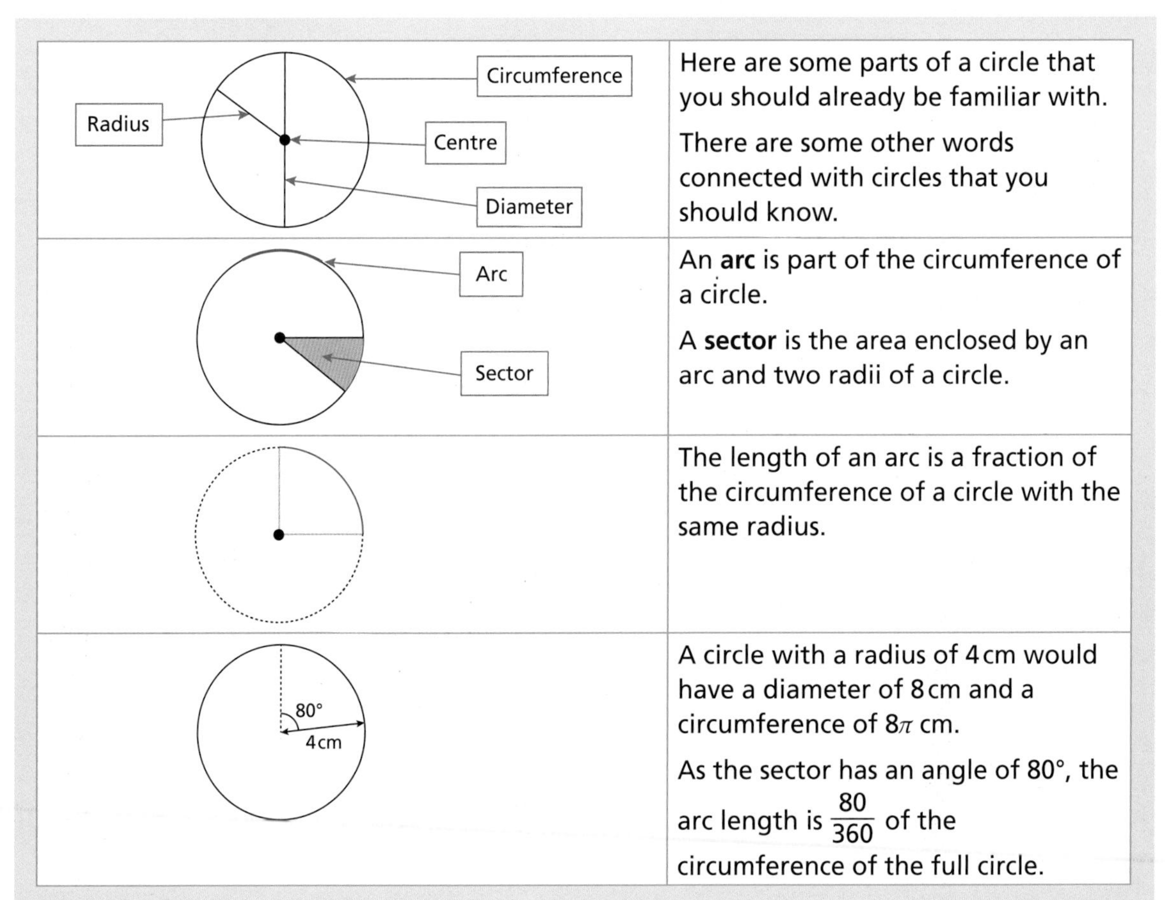

Here are some parts of a circle that you should already be familiar with.

There are some other words connected with circles that you should know.

An **arc** is part of the circumference of a circle.

A **sector** is the area enclosed by an arc and two radii of a circle.

The length of an arc is a fraction of the circumference of a circle with the same radius.

A circle with a radius of 4 cm would have a diameter of 8 cm and a circumference of 8π cm.

As the sector has an angle of 80°, the arc length is $\frac{80}{360}$ of the circumference of the full circle.

In general, for angle θ and diameter d, arc length = $\frac{\theta}{360} \times \pi \times d$

Example

Here is the sector AOB.

a Work out the length of the arc AB.

Give your answer to 3 significant figures.

b Work out the perimeter of the sector.

Give your answer to 3 significant figures.

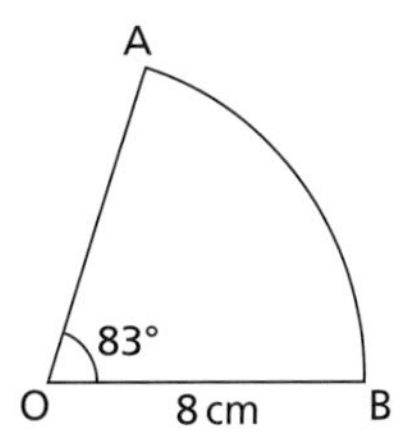

Method

Solution	Commentary
a Arc length = $\frac{\theta}{360} \times \pi \times d$	The length of the arc AB is a fraction of the total circumference.
Arc length = $\frac{83}{360} \times \pi \times 16$	The total circumference is $\pi \times 16$ The fraction is $\frac{83}{360}$
Arc length = 11.588 986...	
Arc length = 11.6 cm (to 3 s.f.)	Round your answer to 3 significant figures.
b Perimeter = AB + OA + OB	The perimeter of AOB is made up of the arc length AB plus the straight lengths OA and OB.
Perimeter = 11.6 + 8 + 8 Perimeter = 27.6 cm (to 3 s.f.)	To find the perimeter, you add all these lengths together.

Practice (A)

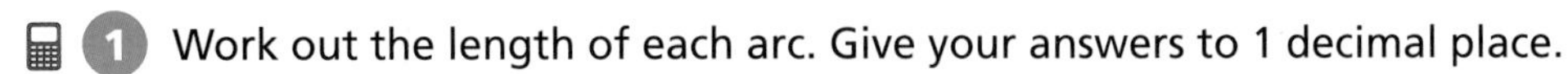

1 Work out the length of each arc. Give your answers to 1 decimal place.

a

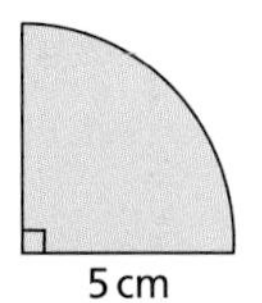

b

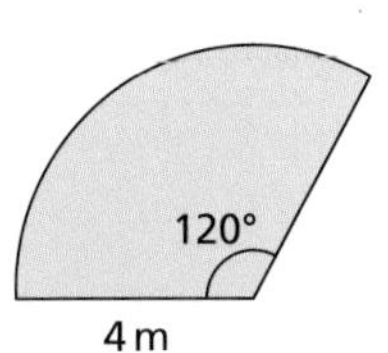

c

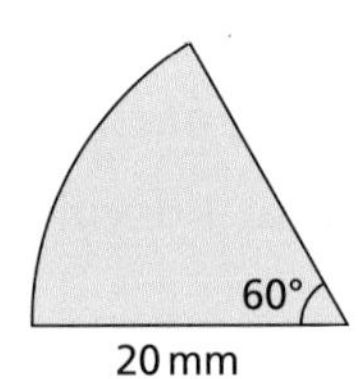

d

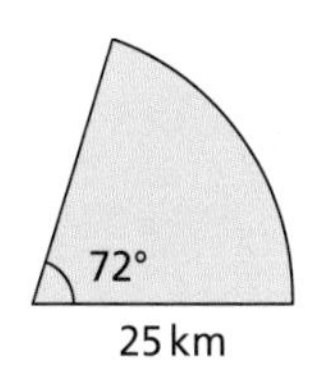

2 Work out the length of each arc. Give your answers to 3 significant figures.

a

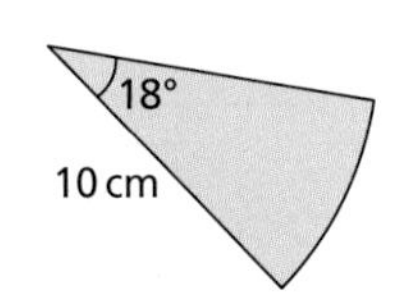

b

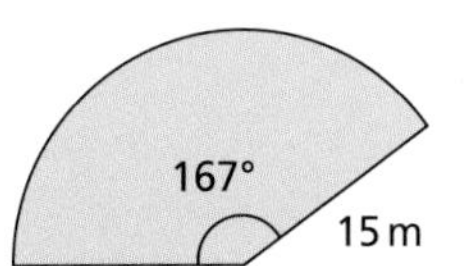

c

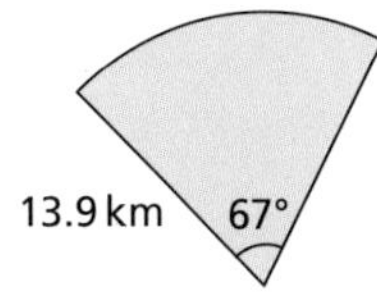

d

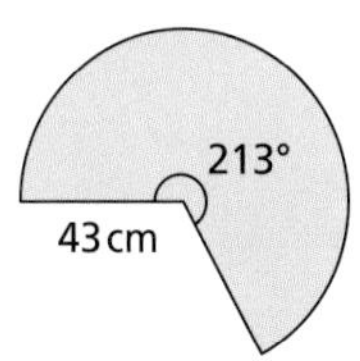

3 Work out the length of each arc. Give your answers in terms of π.

a

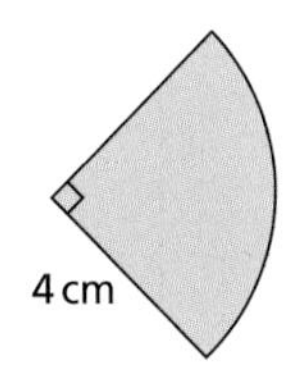

b

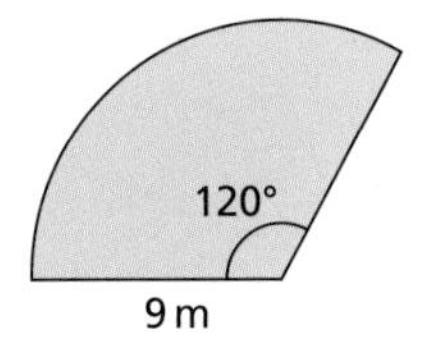

c

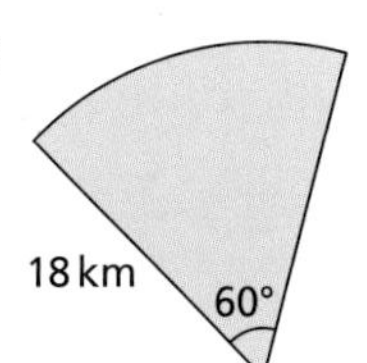

d

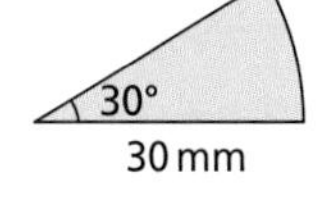

4 Work out the length of each arc. Give your answers to 3 significant figures.

a

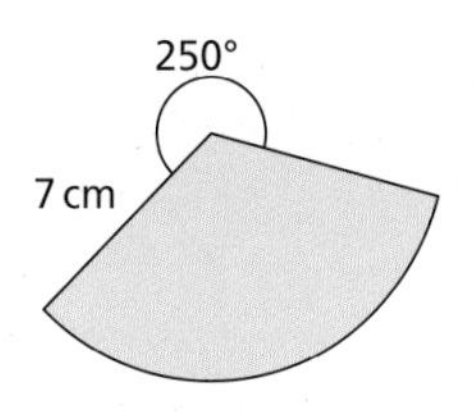

b

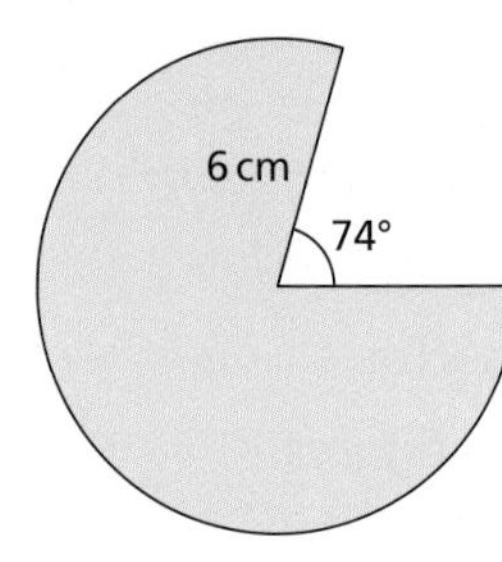

c

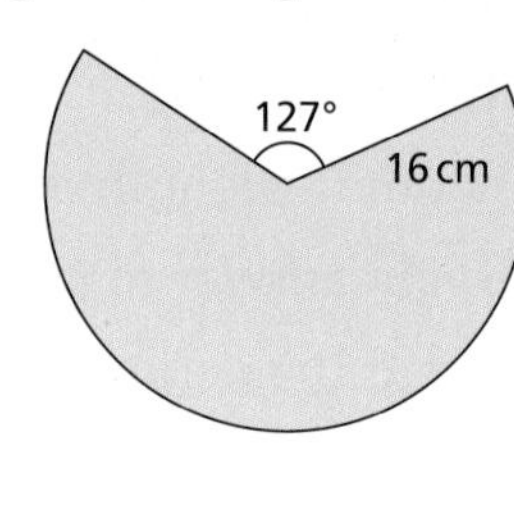

5 The sector AOB is shown.

a Work out the length of the arc AB.
Give your answer to 2 decimal places.

b Work out the perimeter of the sector AOB.
Give your answer to 2 decimal places.

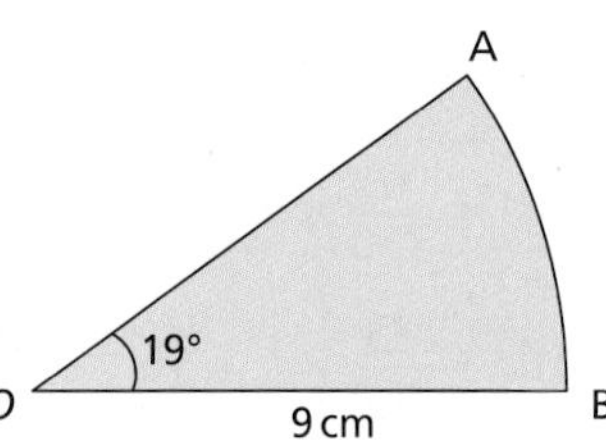

6 Work out the perimeter of this sector.
Give your answer to 3 significant figures.

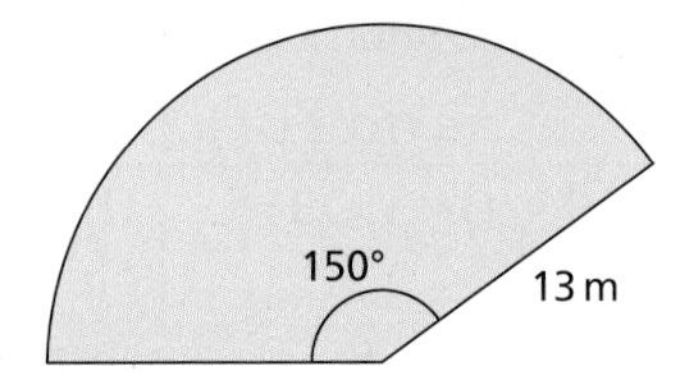

7 Work out the perimeter of this sector.
Give your answer in terms of π.

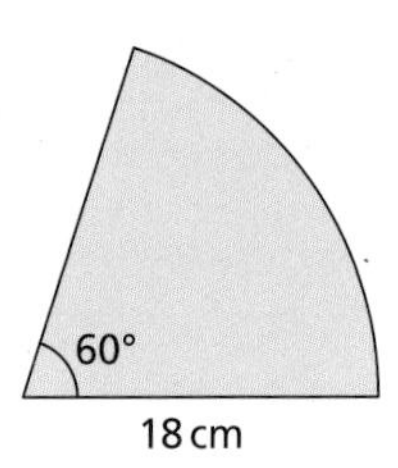

8 The arc length of the sector is 15 cm.
Work out the length of the radius.
Give your answer to 1 decimal place.

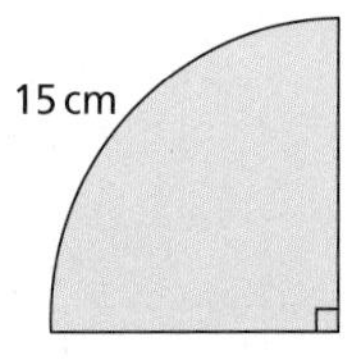

9 Work out the length of each radius. Give your answers to 1 decimal place.

a Arc length = 20 cm **b** Arc length = 35 m **c** Arc length = 14.7 km

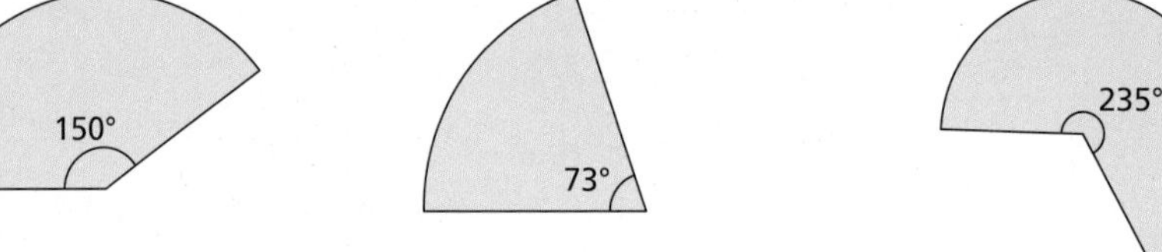

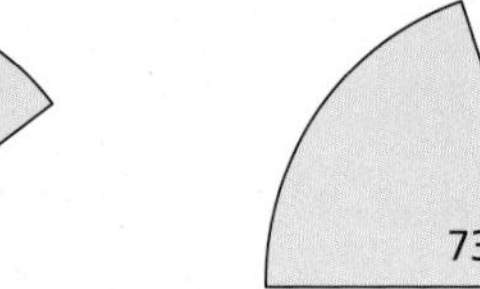

10 The arc length of the sector is 15π cm.
Work out the length of the radius.

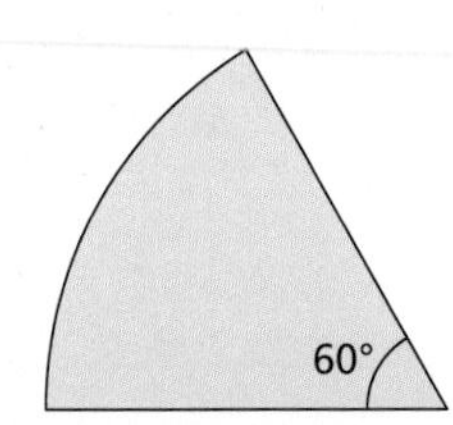

11 The arc length is 28 cm.

Work out the value of θ.

Give your answer to the nearest degree.

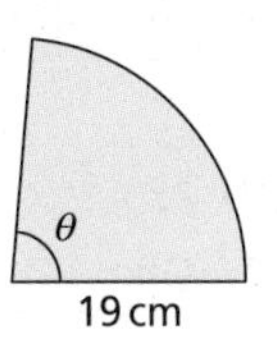

12 The perimeter of the sector is 35 m.

Work out the value of θ.

Give your answer to the nearest degree.

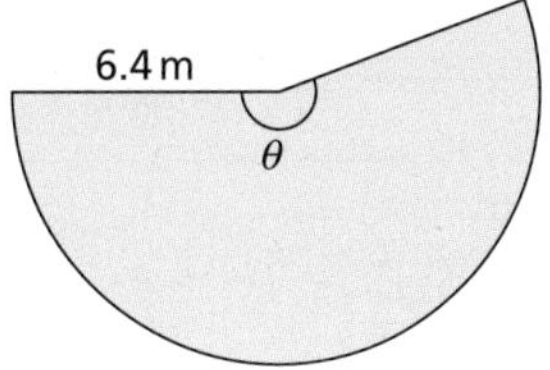

13 The perimeter of the sector is 68.4 cm.

Work out the length of the radius.

Give your answer to 3 significant figures.

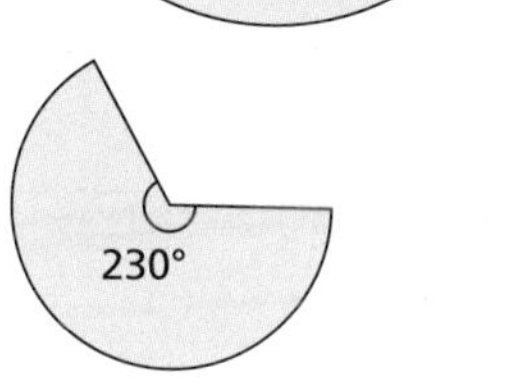

Are you ready? (B)

1 Work out the area of each circle. Give your answers to 3 significant figures.

a

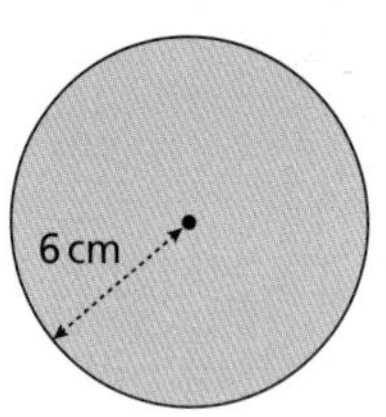

b

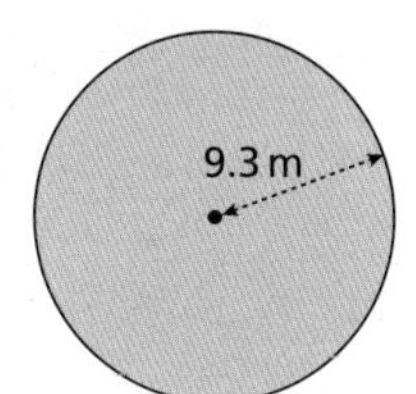

c

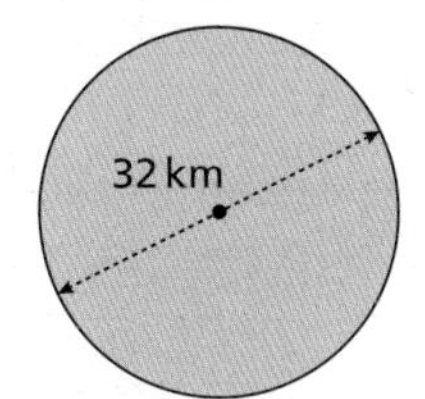

2 Work out the area of each circle. Give your answers in terms of π.

a

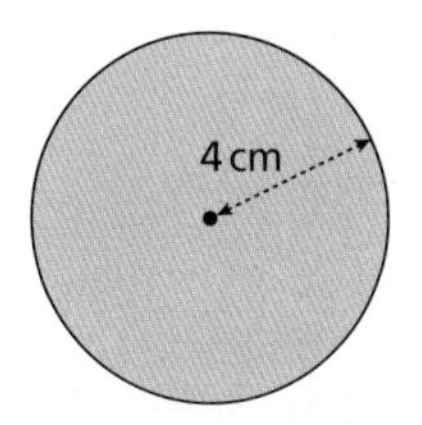

b

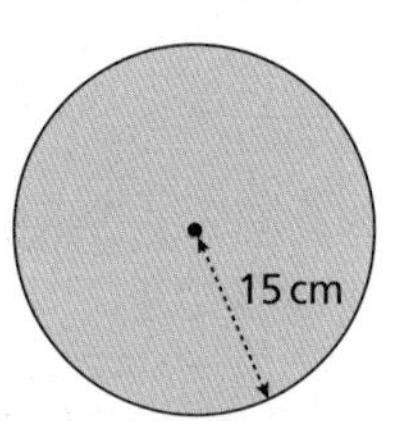

c

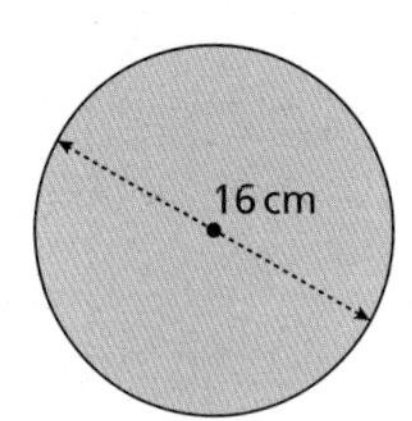

The area of a sector is a fraction of the area of a circle with the same radius.

This circle has an area of 36π cm^2.

The sector represents a quarter of the circle.

Therefore, the area of the sector must be one-quarter of the area of the circle, 9π cm^2.

In general, for angle θ and radius r,

area of sector $= \frac{\theta}{360} \times \pi \times r^2$

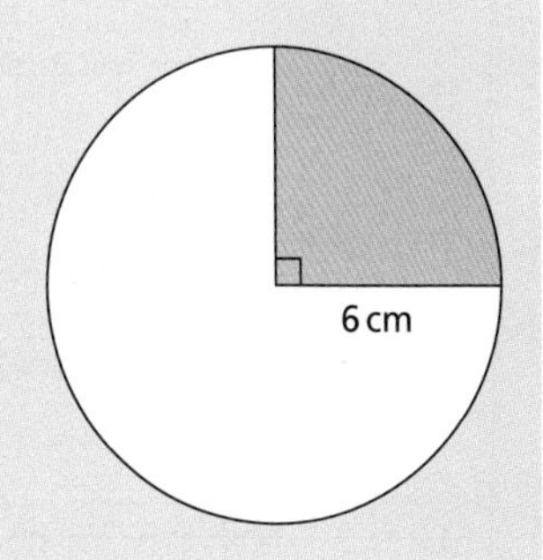

Example 1

Here is a sector.

Work out the area of the sector. Give your answer to 3 significant figures.

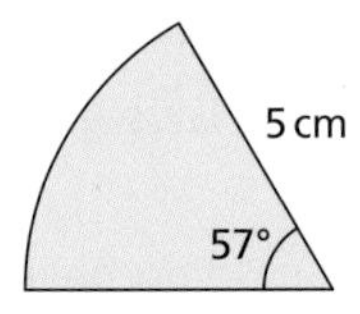

Method

Solution	Commentary
Area of sector = $\frac{57}{360} \times \pi \times 5^2$	The area of the sector is a fraction of the area of a circle with a radius of 5 cm. The area of the circle is $\pi \times 5^2$ and the sector represents $\frac{57}{360}$ of the circle.
Area of sector = 12.43547…. = 12.4 cm^2 (to 3 s.f.)	Calculate the area using the formula. Round your answer to 3 significant figures.

Example 2

The sector AOB has an area of 80 cm^2.

Work out the length of OA.

Give your answer to 3 significant figures.

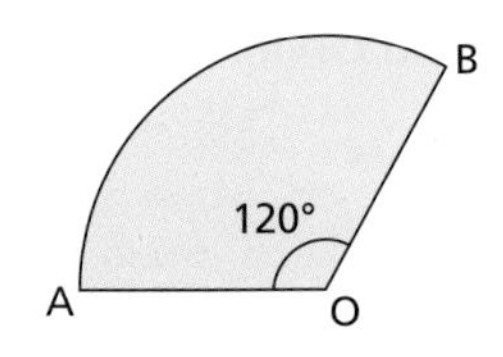

Method

Solution	Commentary
Area of sector = $\frac{\theta}{360} \times \pi \times r^2$ $80 = \frac{120}{360} \times \pi \times r^2$	Substitute the information into the formula.
$80 = \frac{1}{3} \times \pi \times r^2$ × 3 (both sides) $240 = \pi r^2$ ÷ π (both sides) $\frac{240}{\pi} = r^2$ $\sqrt{\ }$ (both sides) $\sqrt{\frac{240}{\pi}} = r$	$\frac{120}{360}$ can be simplified to $\frac{1}{3}$ Rearrange the equation to make r the subject. Rearranging equations is covered in *Collins White Rose Maths AQA GCSE 9–1 Higher Student Book 1*, Blocks 7 and 8.
OA = 8.740387….	Calculate $\sqrt{\frac{240}{\pi}}$ to work out the length of OA.
OA = 8.74 cm (to 3 s.f.)	Round your answer to 3 significant figures.

Practice (B)

1 Work out the area of the semicircle.

Give your answer to 1 decimal place.

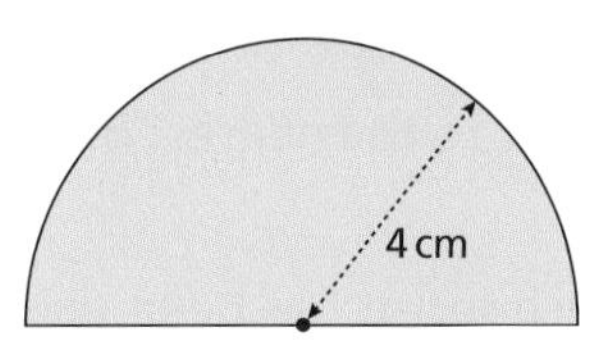

2 Work out the area of each sector. Give your answers to 3 significant figures.

a

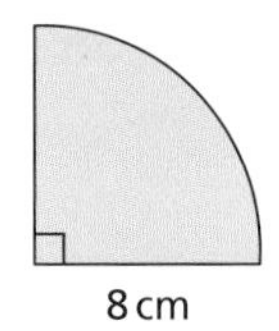

b

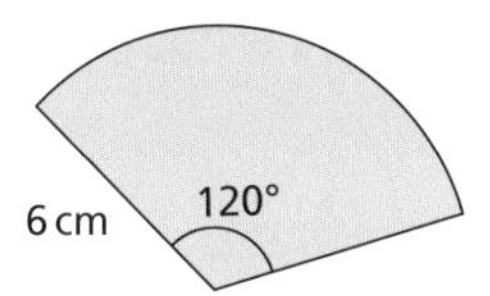

c

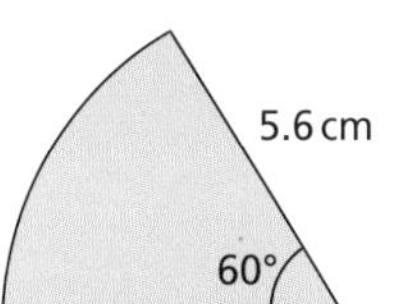

3 Work out the area of each sector. Give your answers to 3 significant figures.

a

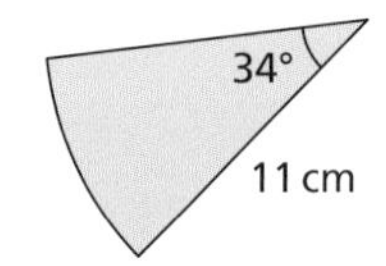

b

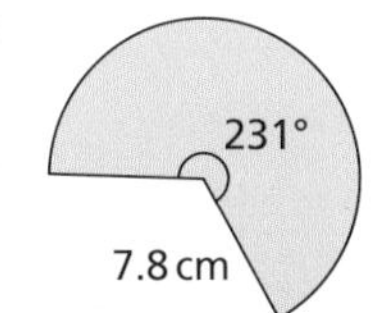

c

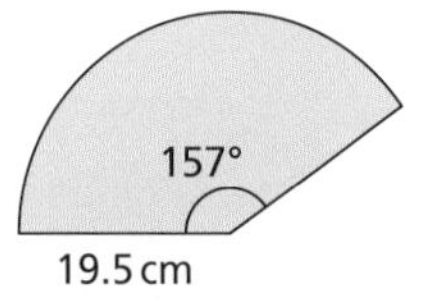

4 Work out the area of each sector. Give your answers in terms of π.

a

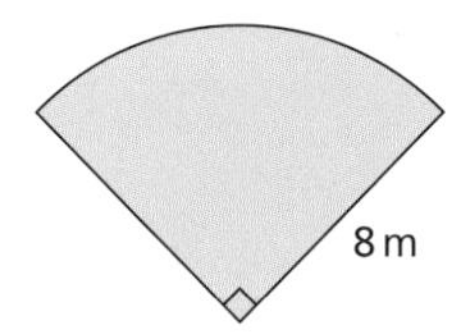

b

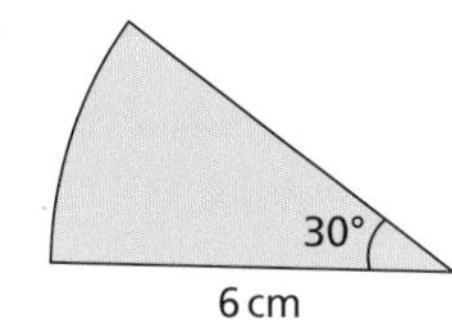

c

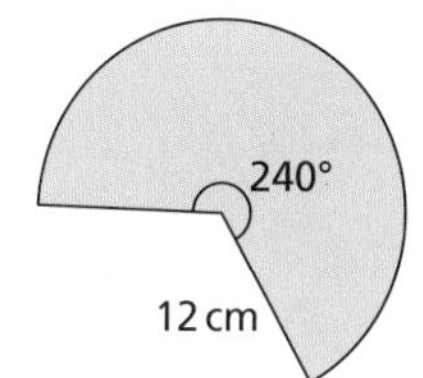

5 Work out the area of the sector.

Give your answer in terms of π.

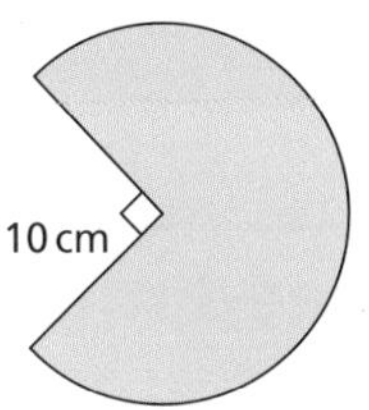

6 The sector has an area of 100 cm^2.

Work out the length of the radius.

Give your answer to 1 decimal place.

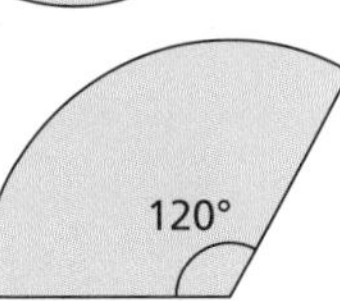

7 The area of each sector has been given.

Work out the length of the radius of each sector.

Give your answers to 3 significant figures.

a Area = 80 cm^2

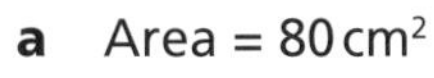

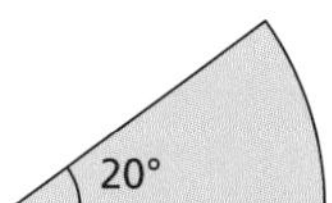

b Area = 90 cm^2

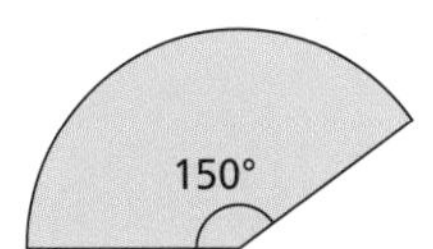

c Area = 132.5 mm^2

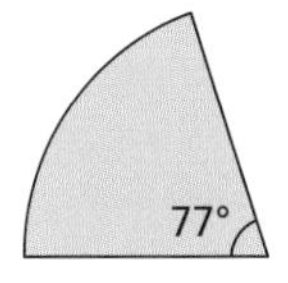

8 The area of the sector is $60\,cm^2$.

Work out the value of θ.

Give your answer to the nearest degree.

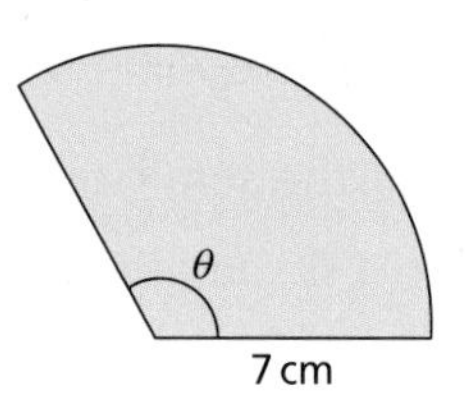

9 The area of each sector has been given.

Work out the value of θ in each sector. Give your answers to the nearest degree.

a Area = $80\,cm^2$

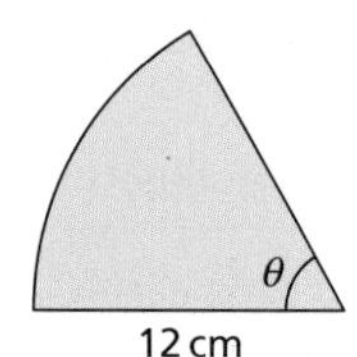

b Area = $200\,mm^2$

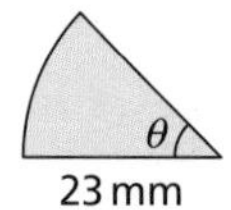

c Area = $68.5\,cm^2$

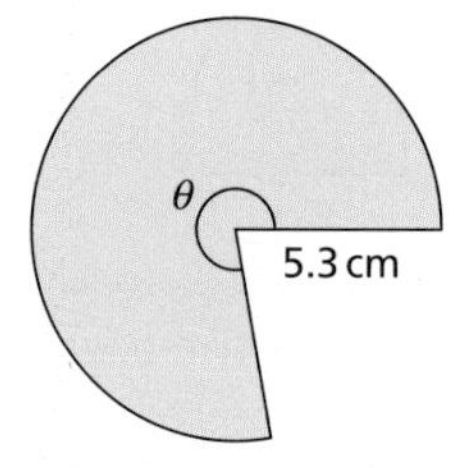

What do you think?

1 Huda says, "If you double the length of the radius of a sector, then the area will also double."

Show that Huda is wrong.

2 This sector has an area of $15\,cm^2$.

Work out the perimeter of the sector.

What would happen to the perimeter if the area of the sector is doubled?

Explain your answer.

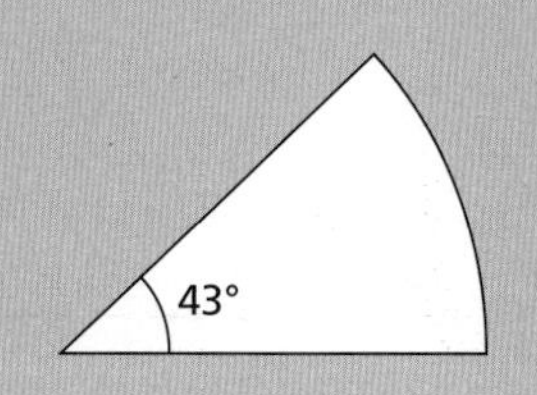

Consolidate – do you need more?

1 Work out the length of each arc. Give your answers to 3 significant figures.

a

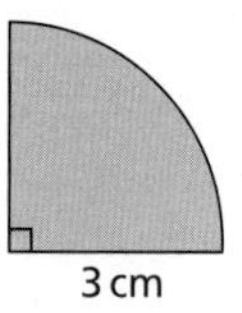

b

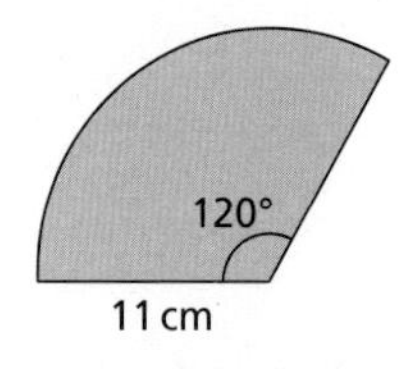

c

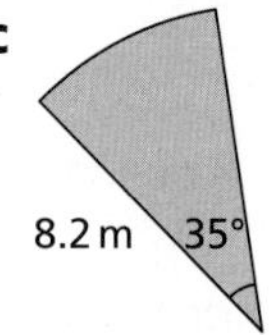

d

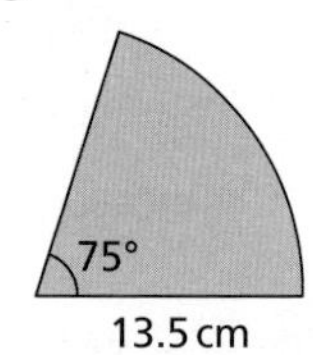

e

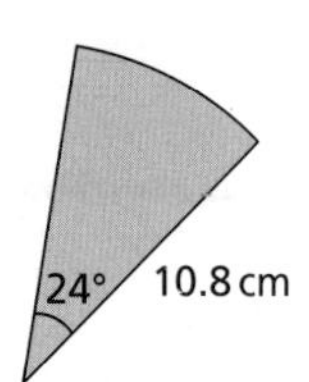

f

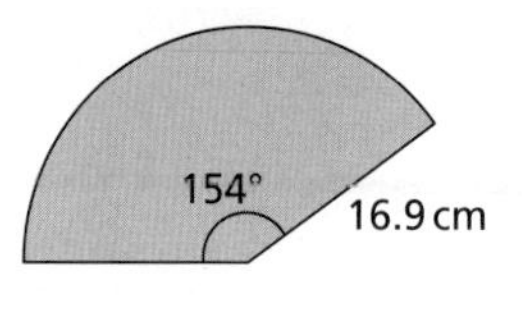

g

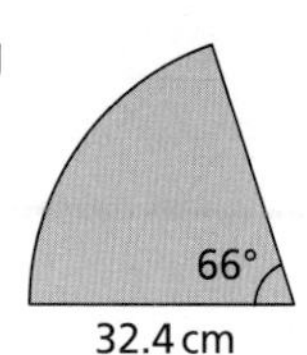

h

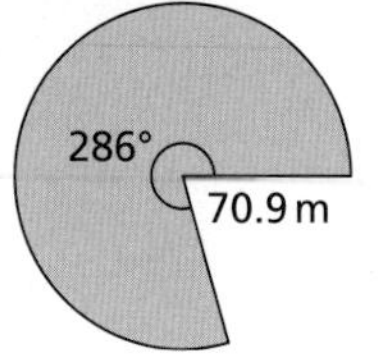

2 Work out the area of each sector. Give your answers to 3 significant figures.

a

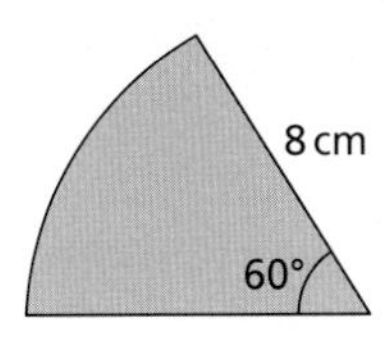

b

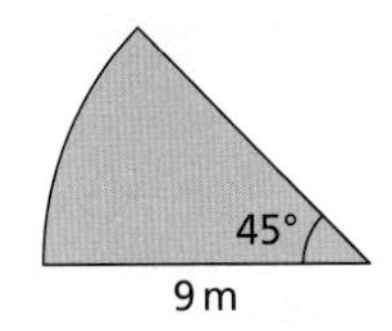

c

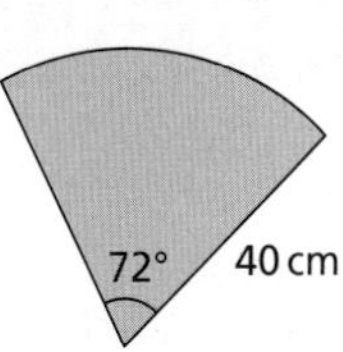

d

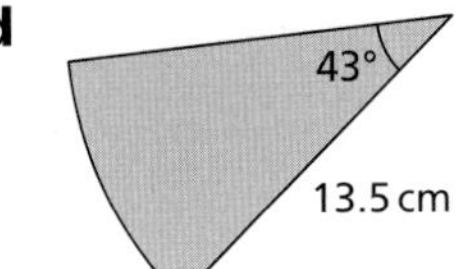

e

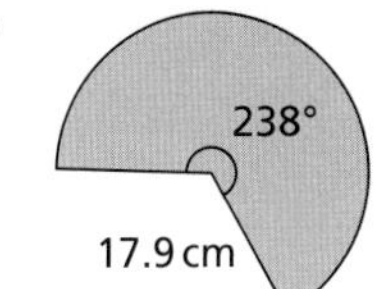

f

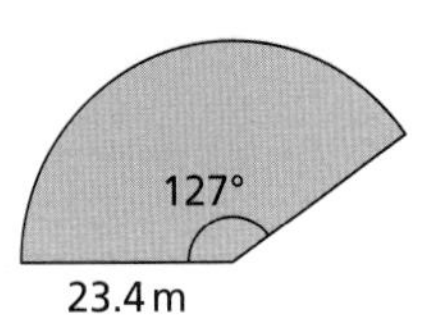

Stretch – can you deepen your learning?

1 The radius of the sector is the same length as a side of the square.

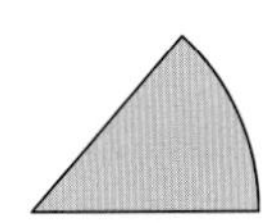

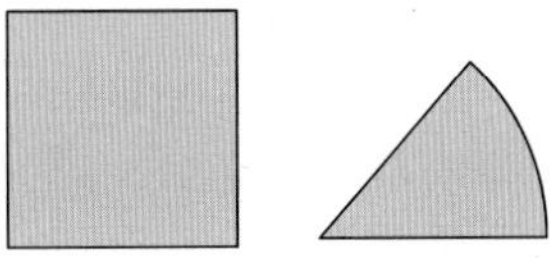

The square and the sector are put together to form the shape ABCDE.

ABCDE has a perimeter of 55 cm.

AB = 12 cm

Work out the size of angle ABC.

Give your answer to the nearest degree.

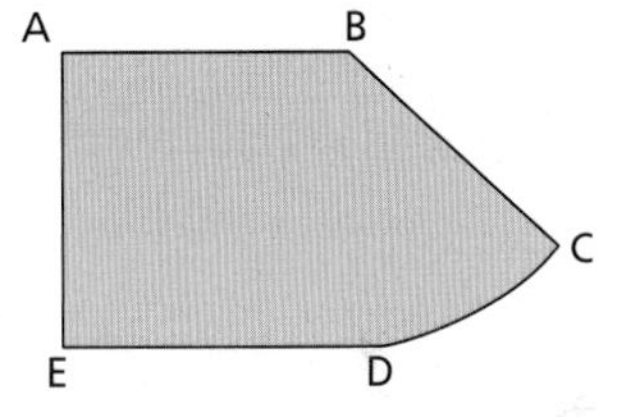

2 The shaded area is a segment of the circle, centre O, of which AOB is a sector.

a Work out the perimeter of the sector AOB.

Give your answer to 3 significant figures.

b Work out the area of the shaded segment.

Give your answer to 3 significant figures.

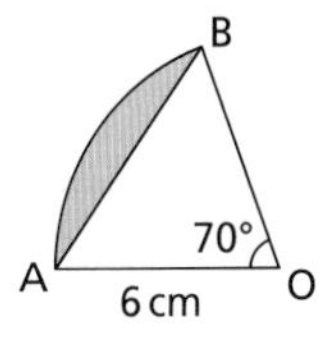

3 The area of the sector is 9 cm^2.

Work out the radius of the sector in terms of π.

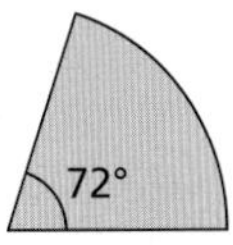

4 The diagram shows a semicircle in a quadrant of another circle.

Work out the area of the shaded region.

Give your answer in terms of π.

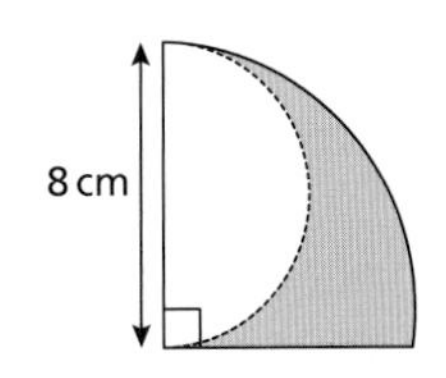

5 ABD and CBD are sectors, each with a radius of 20 cm.

Work out the shaded area.

Give your answer in terms of π.

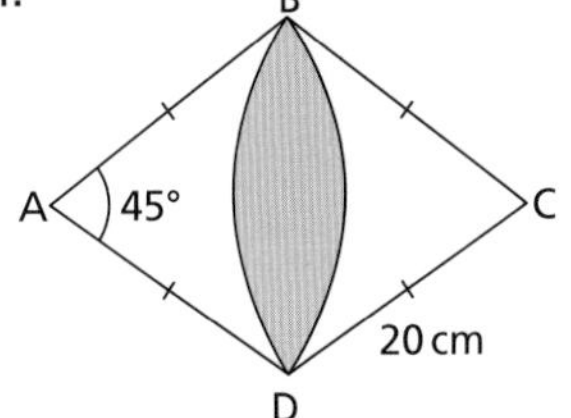

Angles and circles: exam practice

* *Diagrams are not accurately drawn.*

4–6

1 Work out the size of an interior angle of a regular decagon. **[2 marks]**

2 In the quadrilateral, angle b is twice the size of angle a.

Work out the size of angle c. **[4 marks]**

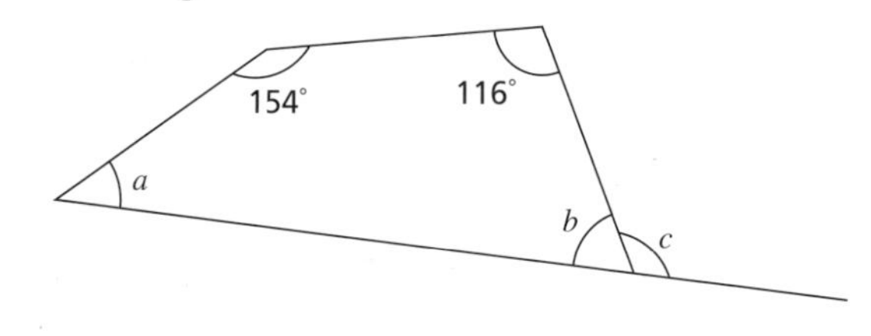

3 Each exterior angle of a regular polygon is 9°.

How many sides does the polygon have? **[2 marks]**

7–9

4 A cyclic quadrilateral is shown.

Work out the value of x. **[2 marks]**

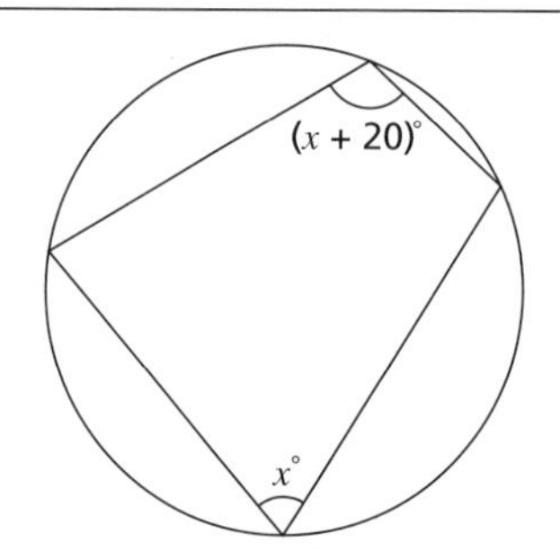

5 A, B and C are points on a circle.

DCB is a straight line and XAY is a tangent to the circle.

Work out the size of angle ACD, giving reasons at each stage of your working. **[3 marks]**

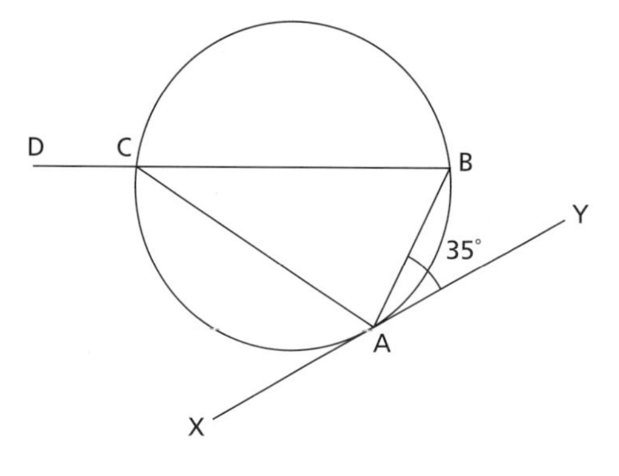

4 Transformations

In this block, we will cover...

4.1 Reflection and rotation

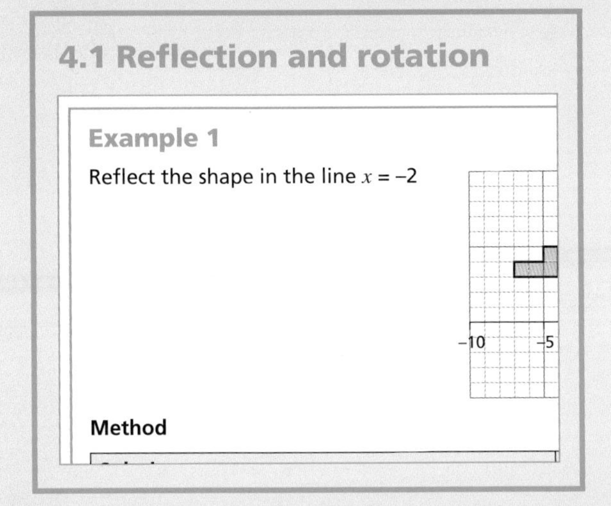

Example 1
Reflect the shape in the line x = −2
−10
−5
Method

4.2 Translation

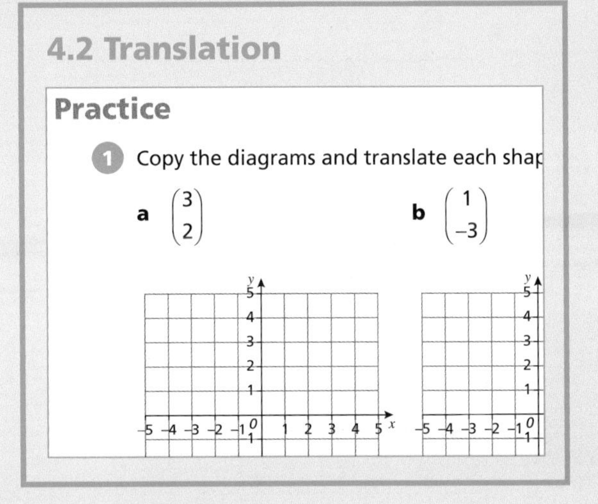

Practice
1 Copy the diagrams and translate each shap
a
b

4.3 Enlargement

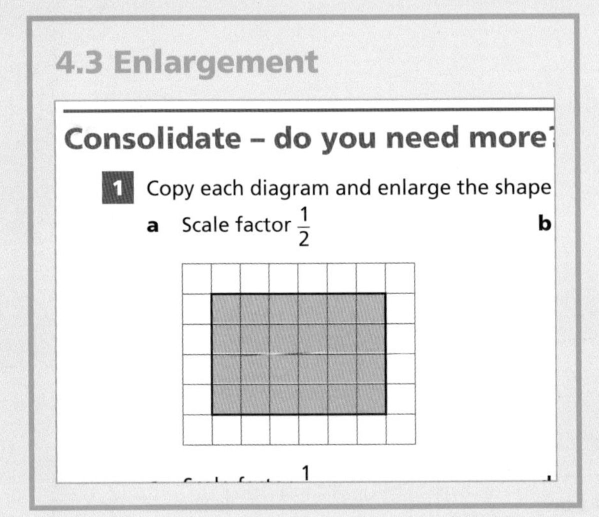

Consolidate – do you need more
1 Copy each diagram and enlarge the shape
a Scale factor 1/2
b

4.4 Combinations of transformations

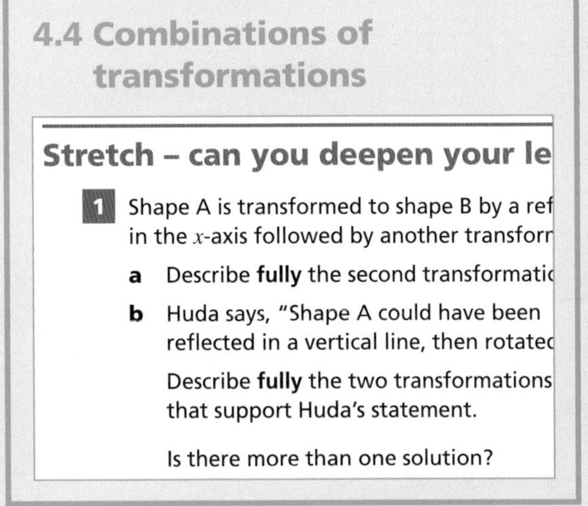

Stretch – can you deepen your le
1 Shape A is transformed to shape B by a ref
in the x-axis followed by another transforr
a Describe fully the second transformati
b Huda says, "Shape A could have been
reflected in a vertical line, then rotate
Describe fully the two transformations
that support Huda's statement.
Is there more than one solution?

Are you ready? (A)

1 Copy each shape and reflect it in the dashed line.

a

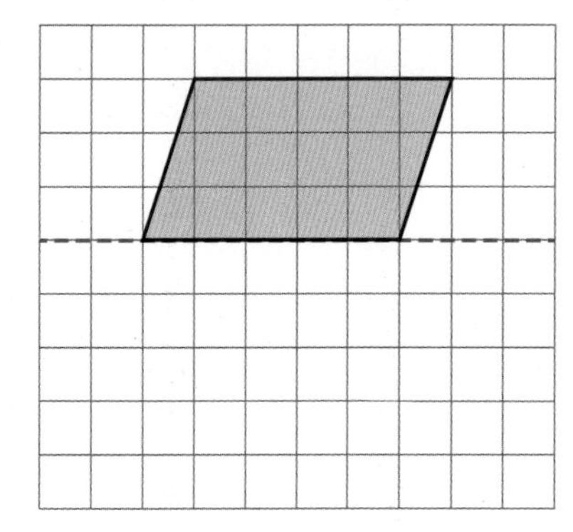

b

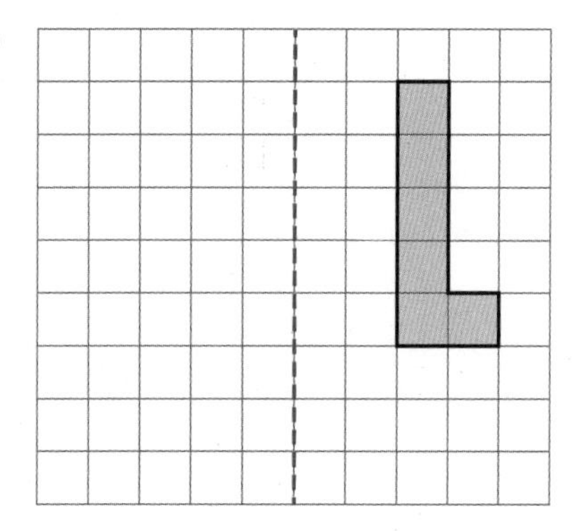

2 Write the equation of each line.

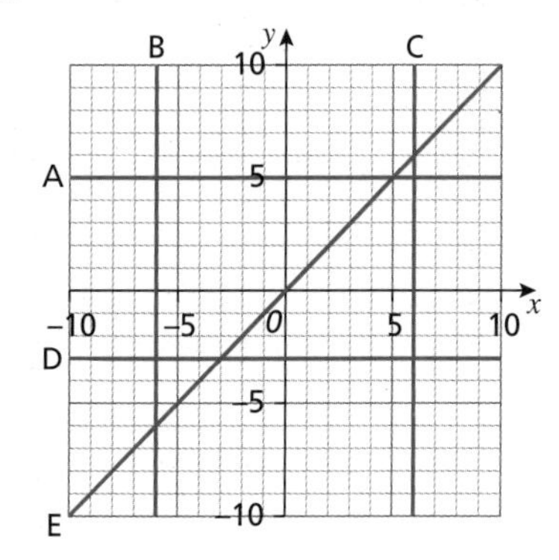

A **reflection** is a transformation that produces the image of a shape in a mirror line.

The points on the mirror line do not change position. These are called **invariant** points.

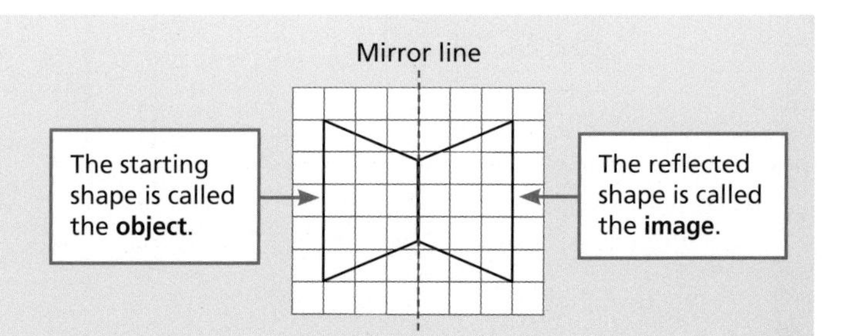

A reflected image is always congruent to the original object. This means that it is exactly the same size and shape as the object. Each point of the image is the same perpendicular distance from the mirror line as the corresponding point on the object.

On a coordinate grid, shapes are generally reflected in vertical or horizontal lines (with equations of the form $x = a$ or $y = b$) or in the diagonal lines $y = x$ or $y = -x$.

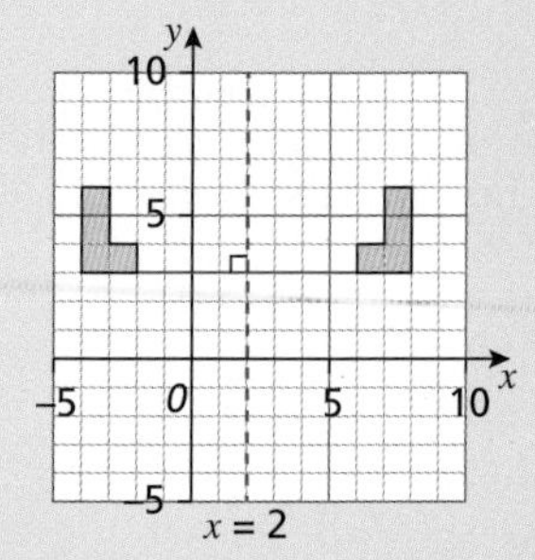

Example 1

Reflect the shape in the line $x = -2$

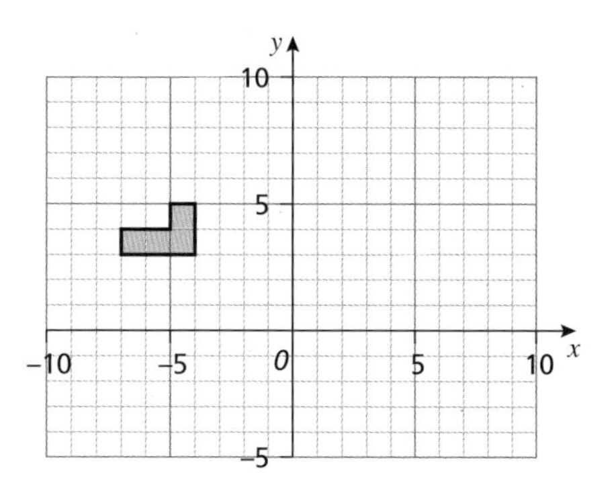

Method

Solution	Commentary
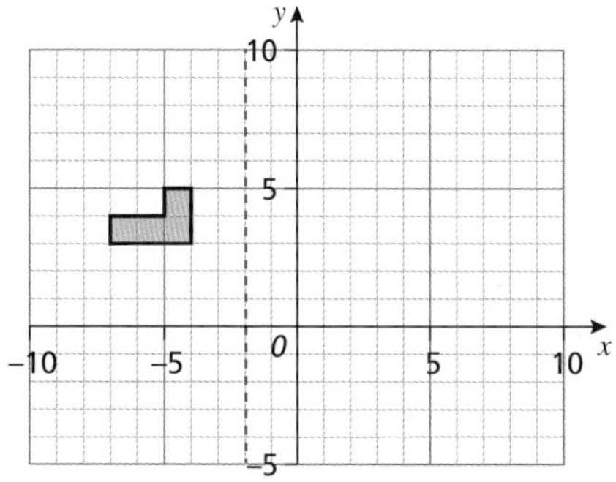	Draw the line $x = -2$, a vertical line which intersects the x-axis at the point (0, −2).
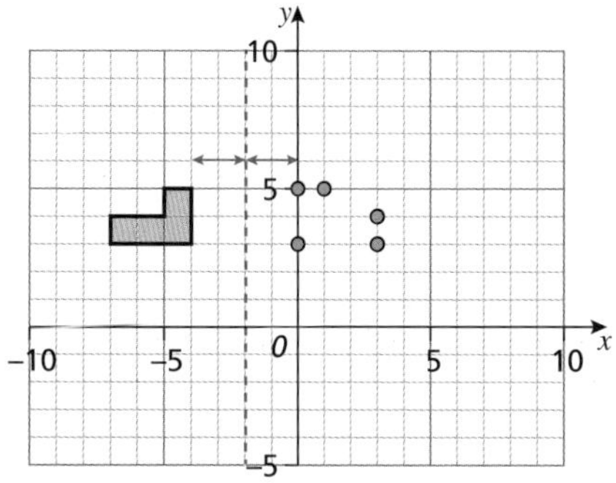	Plot each vertex of the image the same distance from the mirror line as the object, but on the other side.
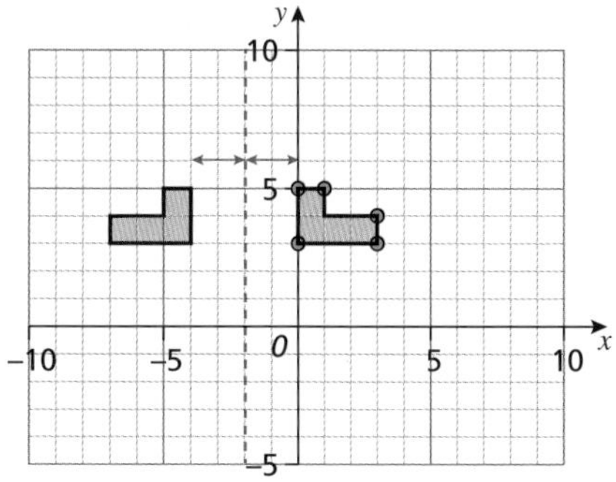	Join the points to complete the reflection.

Example 2

Reflect the shape in the line $y = x$.

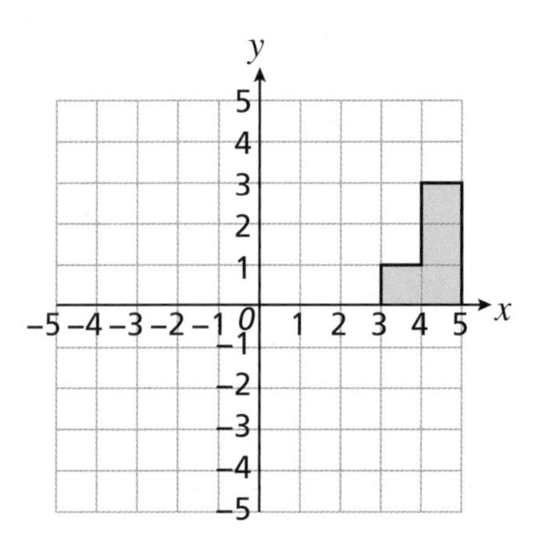

Method

Solution	Commentary
	The line $y = x$ is a diagonal line that goes through (0, 0), (1, 1), (2, 2), and so on. To reflect the shape, count from each vertex to the mirror line and then count the same distance on the other side of the mirror line. Because the mirror line is diagonal, you should count diagonally through the squares. For example, from (3, 0) to the mirror line is one-and-a-half diagonals of the squares. The diagonal lines must be perpendicular to the mirror line.

Practice (A)

1 Copy each shape onto squared paper and then draw the reflection in each of the given lines.

a Reflect the shape in $x = -2$

b Reflect the shape in $y = 1$

c Reflect the shape in the x-axis.

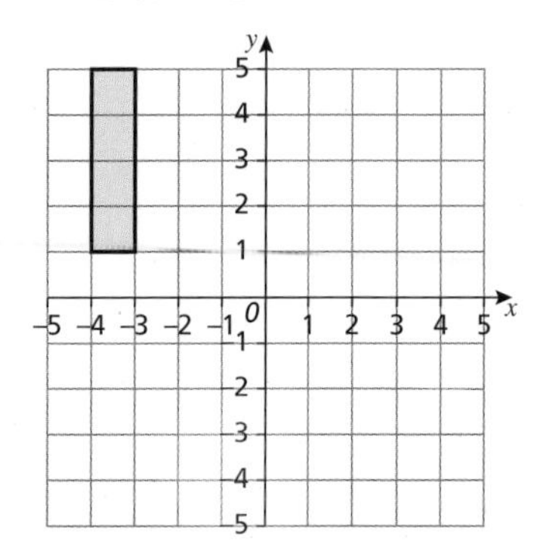

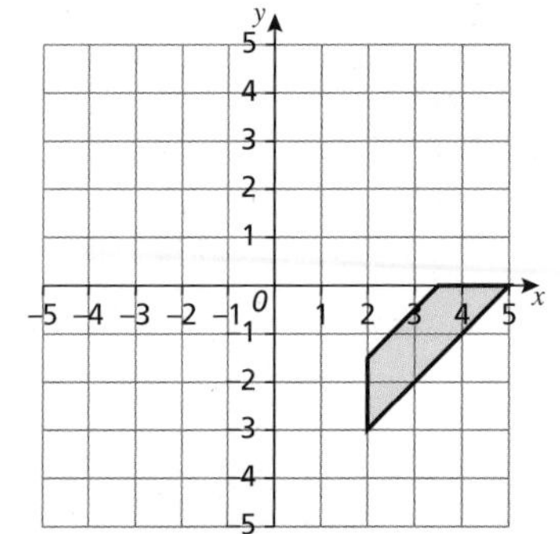

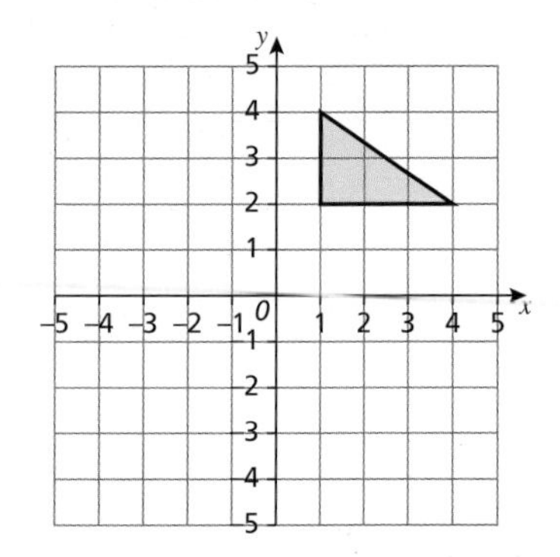

2 Write the equation of the line of reflection for each pair of shapes.

a

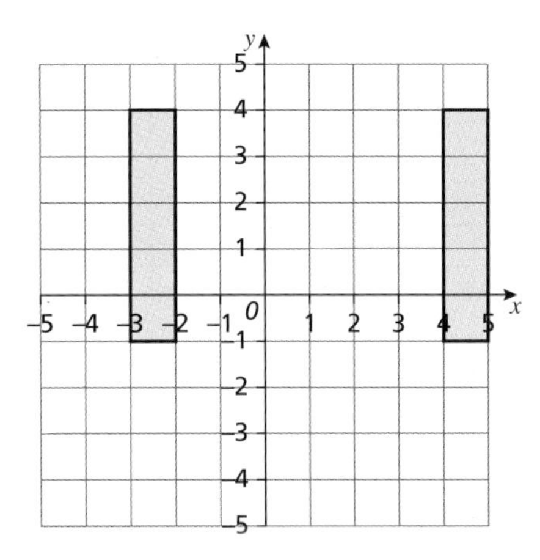

b

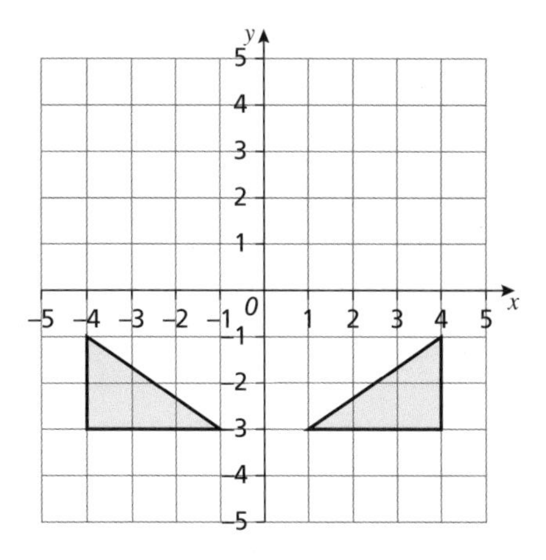

c

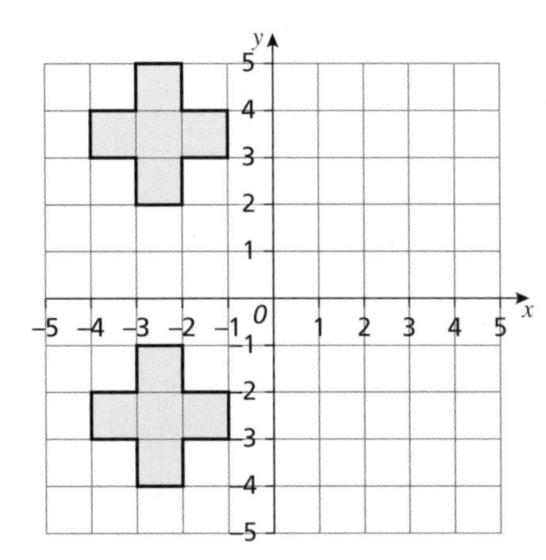

3 Copy the shape onto squared paper, and then draw the reflection in each of the given lines.

a Reflect the shape in the line $y = 3$

b Reflect the shape in the line $x = 3$

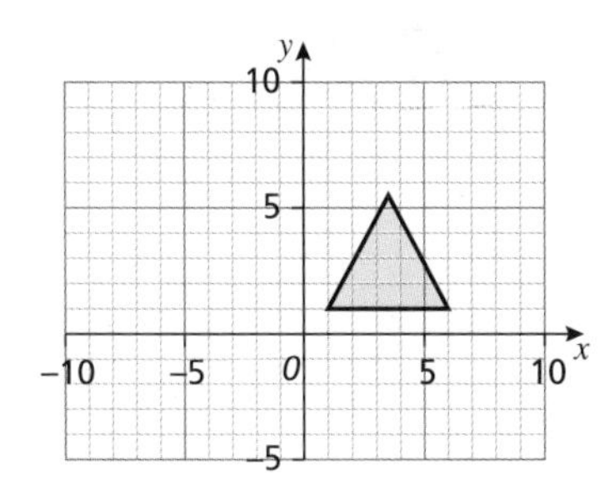

4 Which of these reflections are correct for the given mirror line?

A

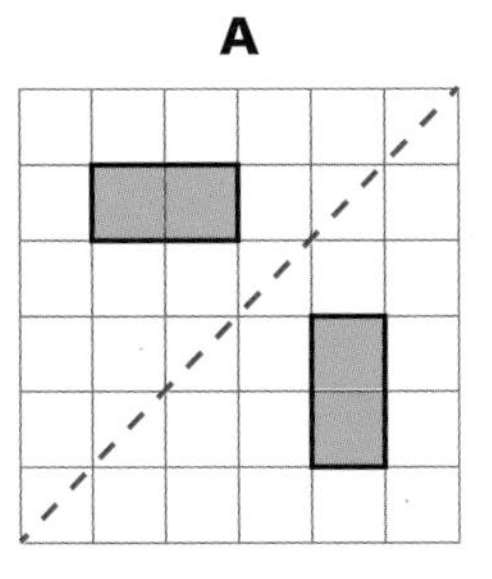

B

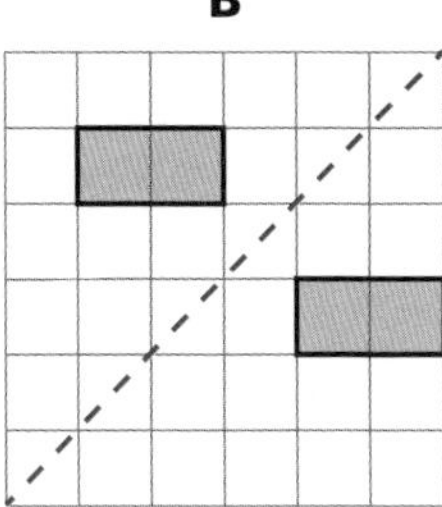

C

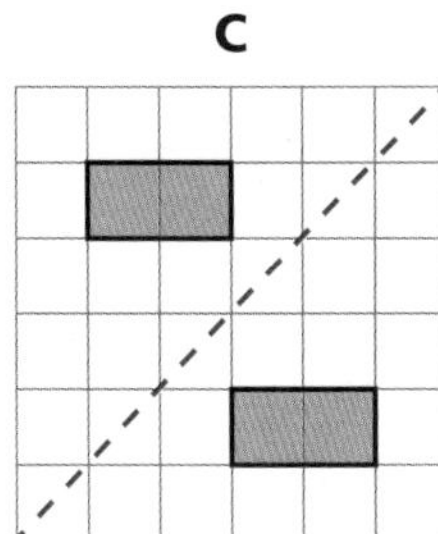

5 Copy each shape and reflect it in the line $y = x$.

a

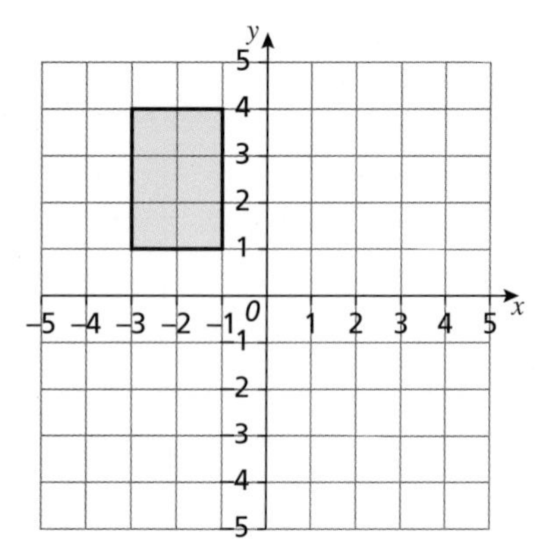

b

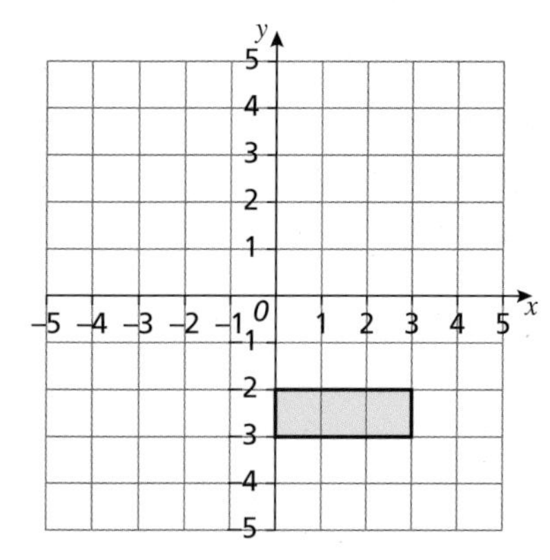

c

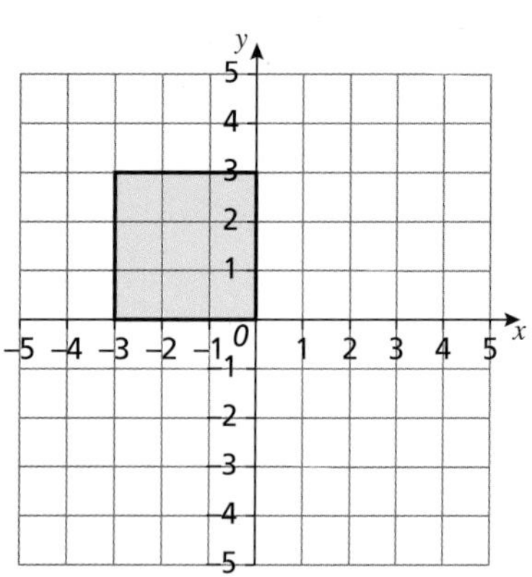

d

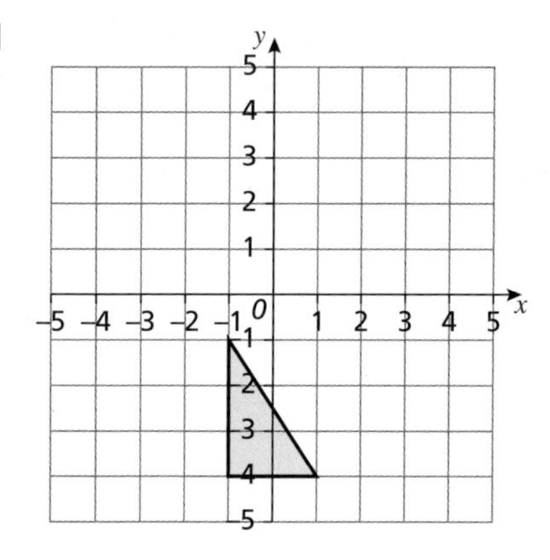

6 Copy each shape and reflect it in the line $y = -x$.

a

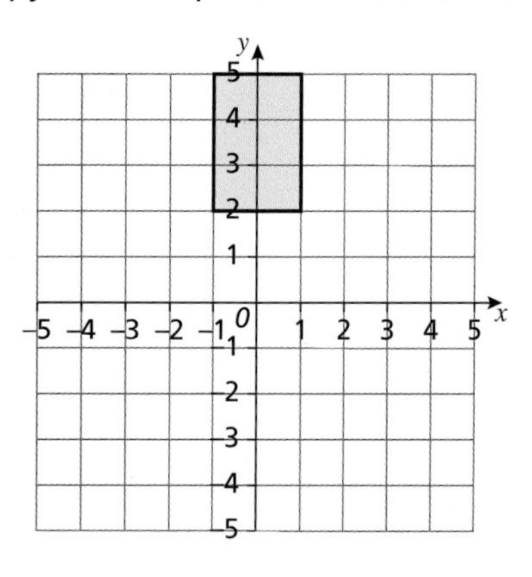

b

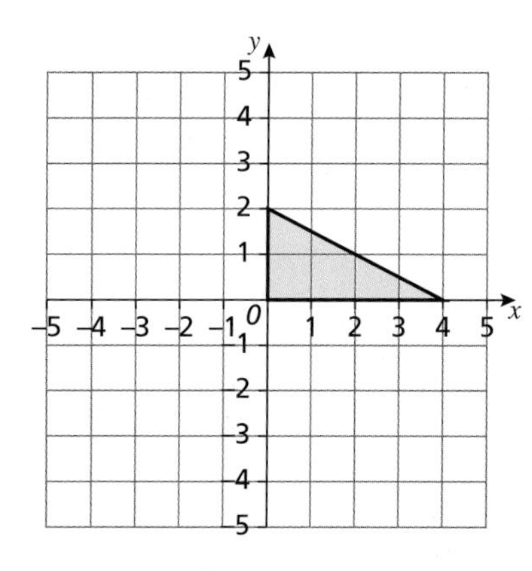

c

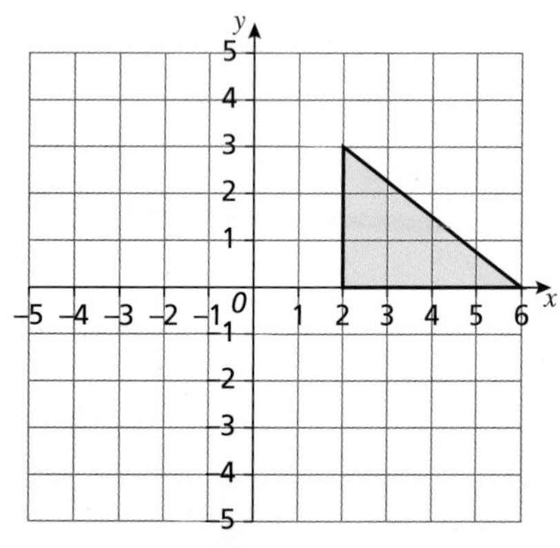

d

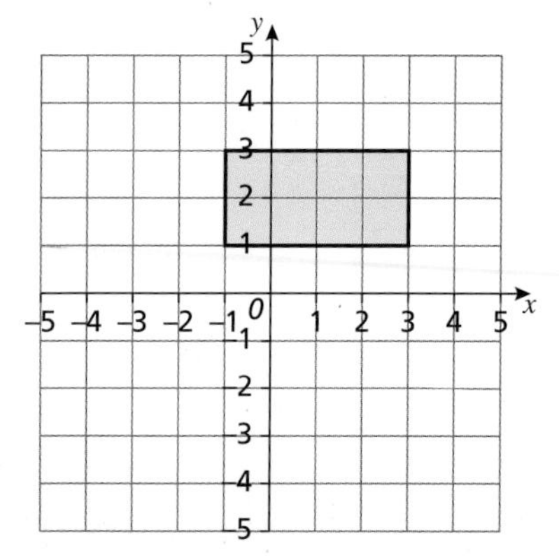

7 Copy the shape onto a grid.

a Reflect the shape in the line $x = -1$

Label this shape P.

b Reflect the shape P in the line $y = x$.

Label this shape Q.

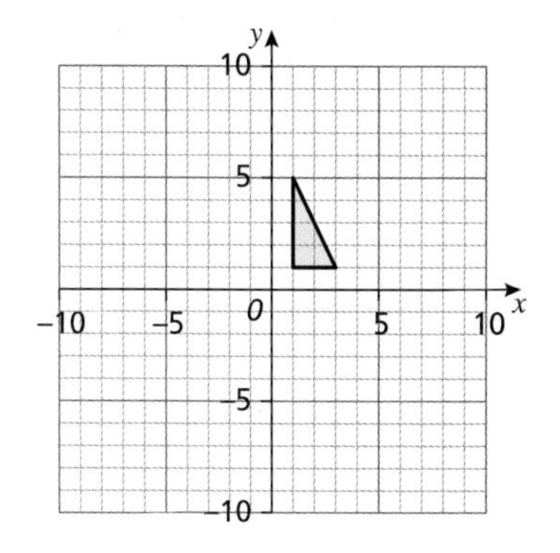

What do you think? (A)

1 Here is a shape on a grid.

Ed says, "When I reflect the shape, the vertices of the image have the same coordinates as the vertices of the object."

Write the equations of all of the lines Ed could have reflected the shape in.

Think of any other shapes that could achieve the same result.

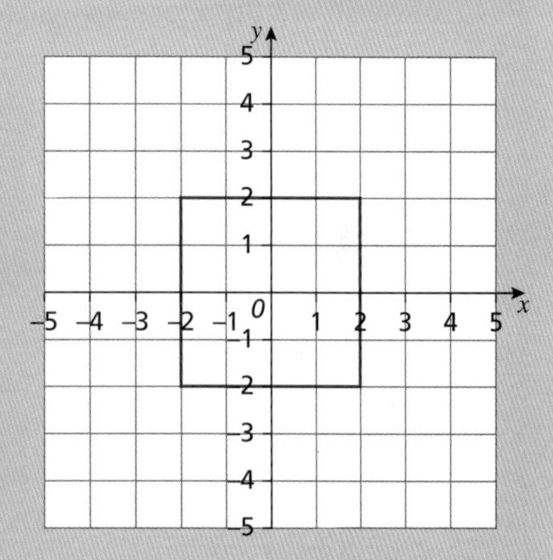

2 Draw a shape on a grid and reflect it in the line $y = x$.

What do you notice about the coordinates of the midpoints of the line segments joining corresponding vertices?

Are you ready? (B)

1 State the order of rotational symmetry for each shape.

a

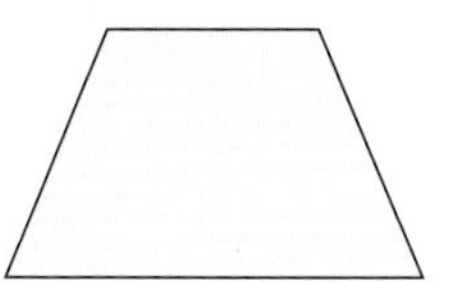

b

c

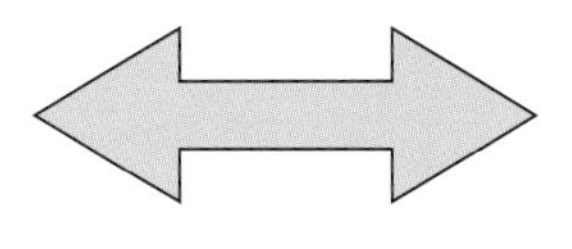

2 Draw each shape after these rotations.

i 180° ii 90° clockwise

a

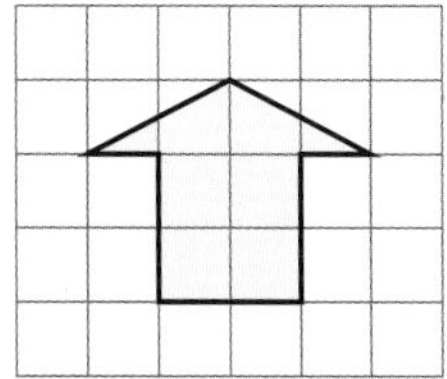

b

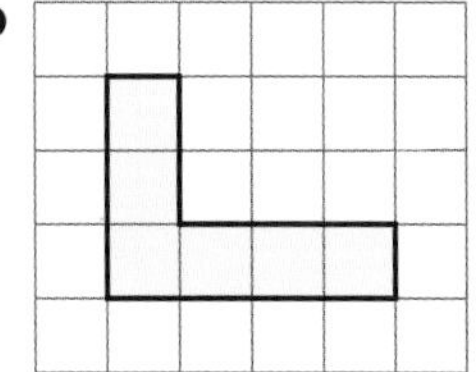

c

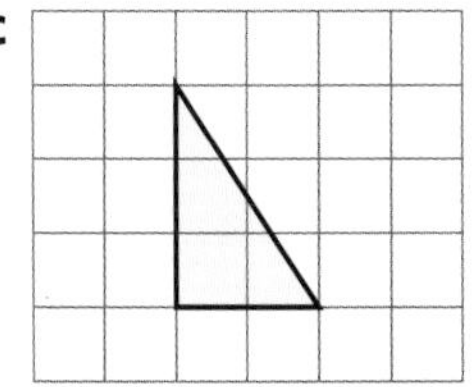

d

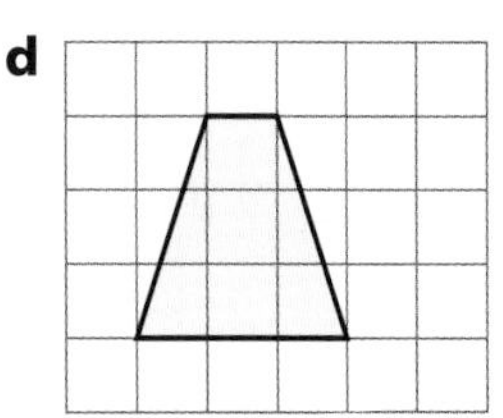

A **rotation** turns a shape around a fixed point, called the centre of rotation. Rotations can be clockwise or anticlockwise and are often through 90°, 180° or 270°.

You can use tracing paper to help with rotations.

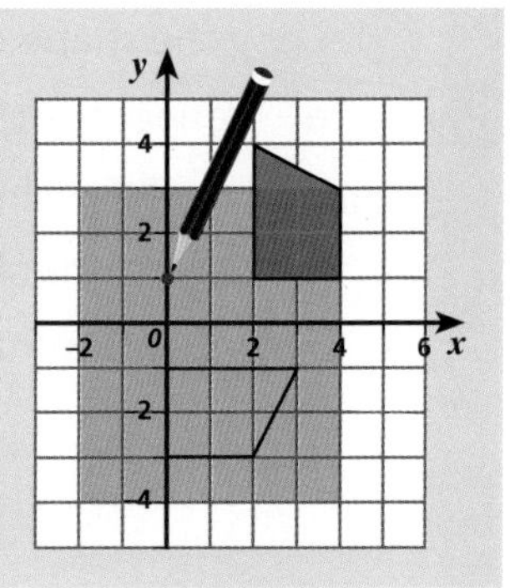

Example 1

Rotate the shape 90° clockwise about the point (1, 2).

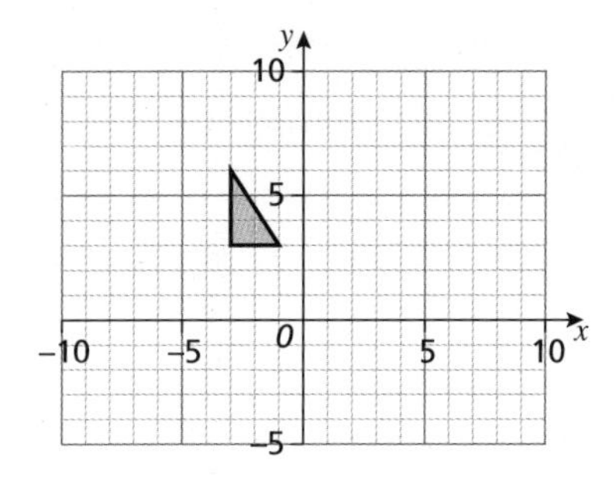

Method

Solution	Commentary
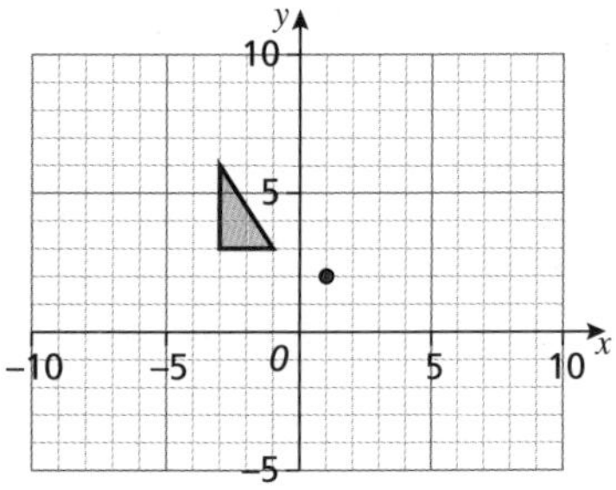	Mark the point (1, 2) on the grid.
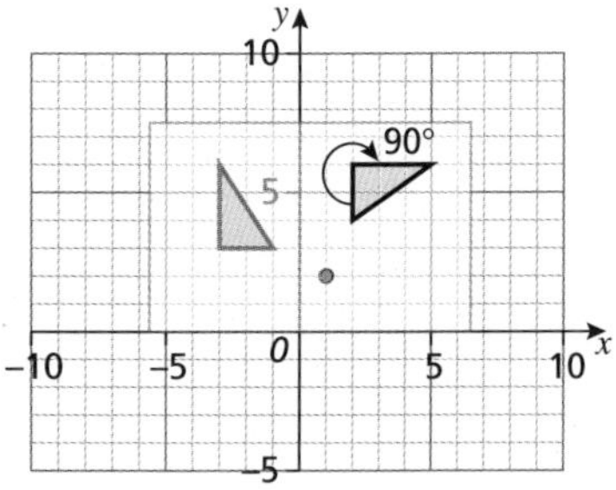	Place tracing paper over the grid and trace the triangle and the point. Place a pencil on the point and turn the tracing paper 90° clockwise.
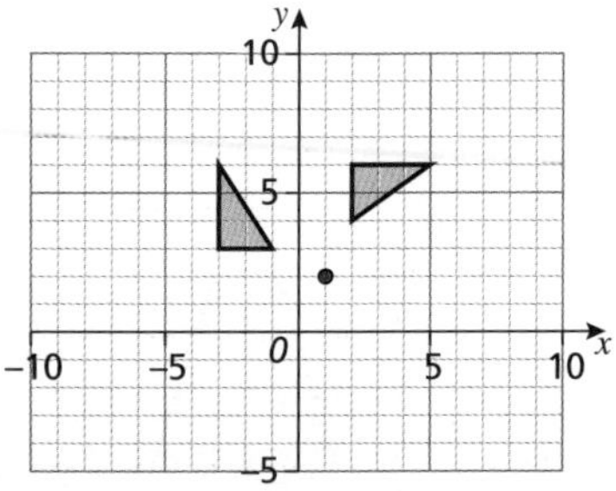	Draw the image in the same place indicated by the shape on the tracing paper to complete the rotation.

Example 2

Describe fully the transformation from shape A to shape B.

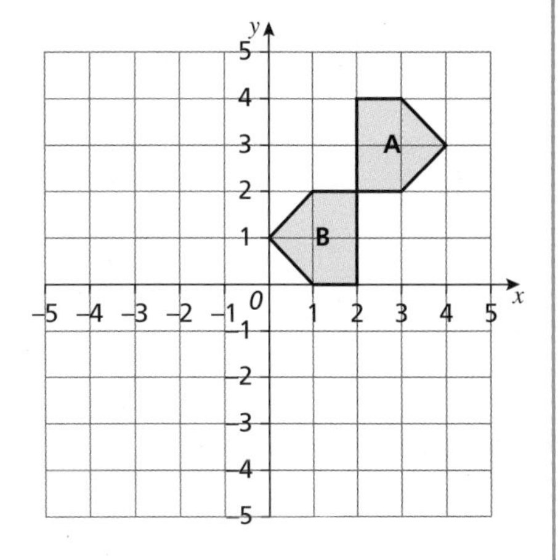

Method

Solution	Commentary
	You can identify the transformation as a rotation as the shape has been turned. You may be able to 'spot' the centre of rotation or use tracing paper to help you find it. You can then use tracing paper to check the angle of rotation, in this case 180°.
A rotation of 180° about the point (2, 2).	Usually, you would state the direction of the rotation, but for 180° this isn't required because clockwise and anticlockwise would give the same result.

Example 3

Describe fully the transformation from shape X to Y.

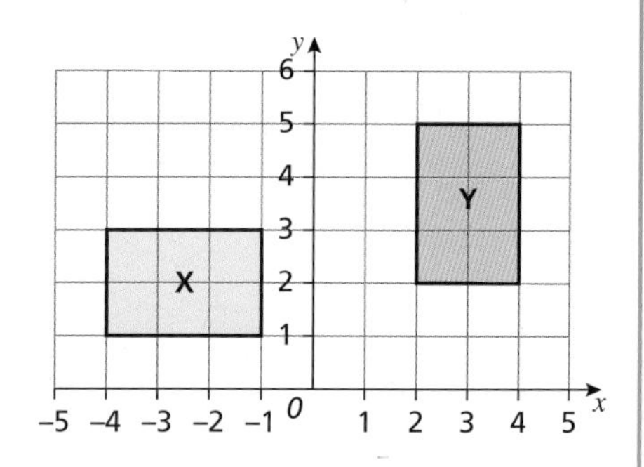

Method

Solution	Commentary
	You can identify the transformation as a rotation of 90° by eye. You can then use trial and error to try to find the centre of rotation, or you can use bisectors. Construct the perpendicular bisector of the line segment joining any pair of corresponding points on the object and the image.
	Repeat with another pair of corresponding points. The centre of rotation is where the bisectors meet.
A rotation of 90° clockwise about the point (1, 0)	Fully describe the rotation.

Practice (B)

1 Copy each shape onto a grid and rotate it through the given angle about the centre of rotation.

a 90° clockwise about the point (0, 0)

b 90° anticlockwise about the point (2, 1)

c 180° about the point (1, 0)

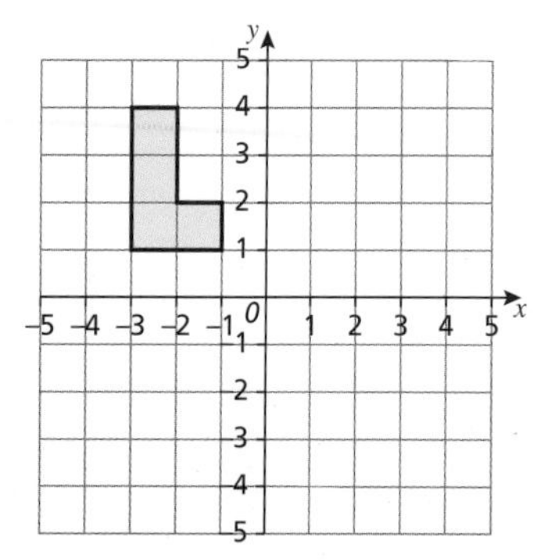

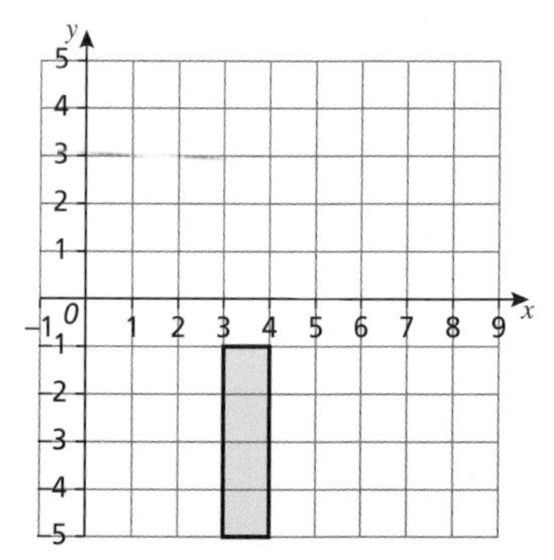

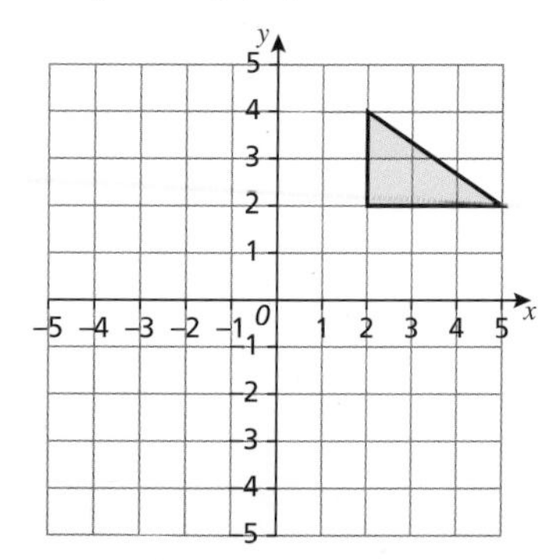

2 Copy the shape onto a grid and rotate it 270° anticlockwise about the origin.

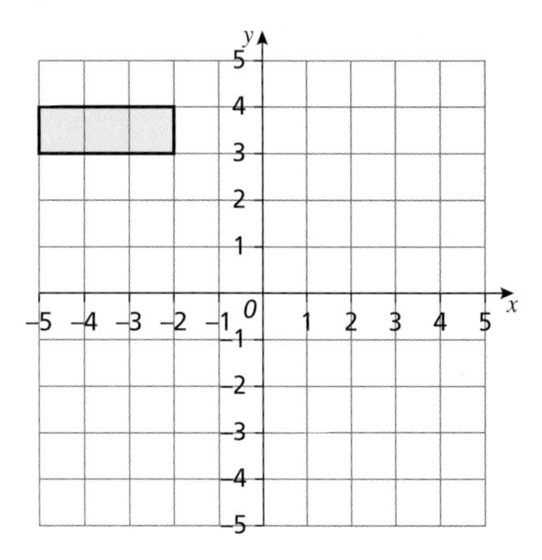

3 Copy each shape onto a grid and rotate it through the given angle about the centre of rotation.

a 90° clockwise about (–3, 2) **b** 90° clockwise about (1, 2)

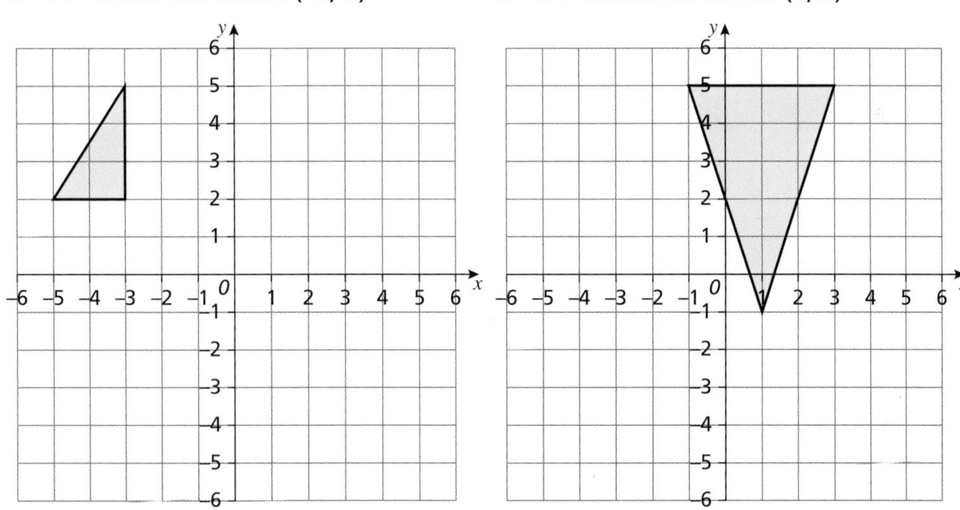

c 180° about (5, –6)

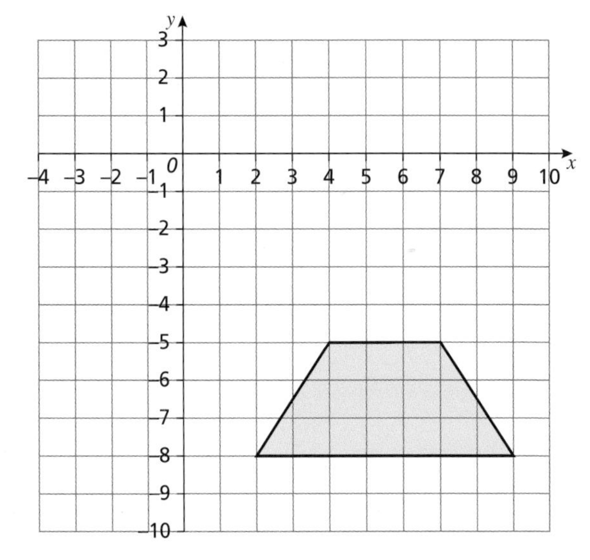

4 Which option, A, B or C, shows the correct rotation of rectangle Z by 90° clockwise about the point P?

A **B** **C**

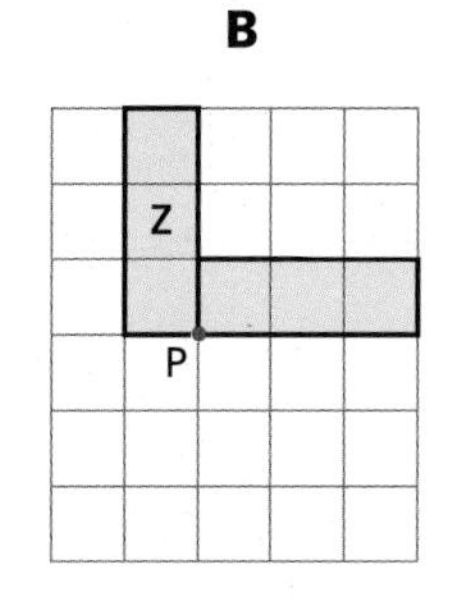

5 Describe fully the transformation that maps:

a shape A onto shape A′

b shape B onto shape B′.

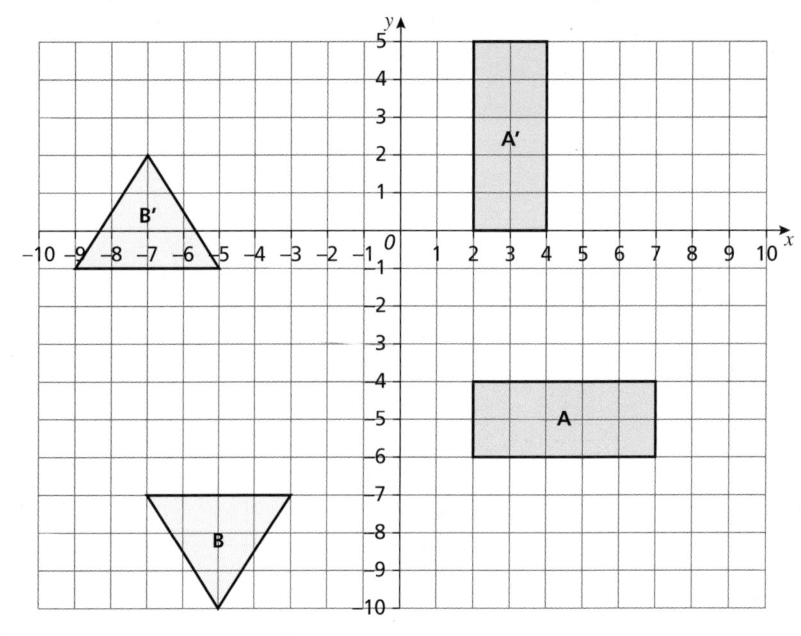

6 Fully describe the transformation which maps the red rectangle onto the blue rectangle.

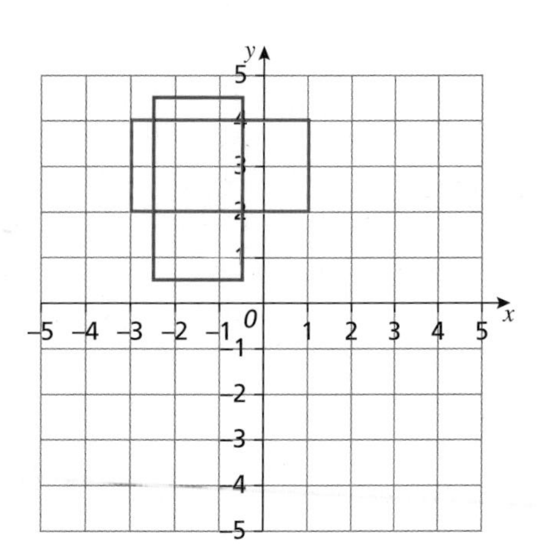

What do you think? (B)

1. Ed says, "If you join the corresponding vertices of an object and an image, the lines will intersect at the centre of rotation."

   Is this **always, sometimes** or **never** true?

2. Is it possible to accurately rotate a shape 45° clockwise?

   Investigate with different shapes.

   Investigate rotating a shape through other angles.

3. Huda has rotated a shape through 540° clockwise.

   State two rotations which would give the same result as Huda's.

4. Investigate when rotations and reflections result in some points being invariant.

Consolidate – do you need more?

1. Copy each shape onto a grid and reflect it in the given line of reflection.

   a y-axis

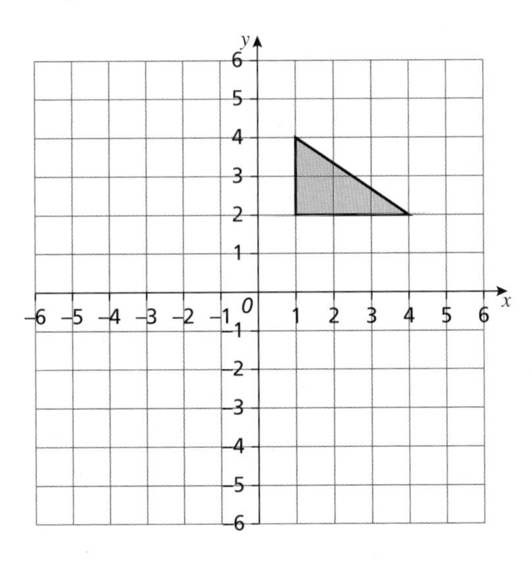

   b $x = 1$

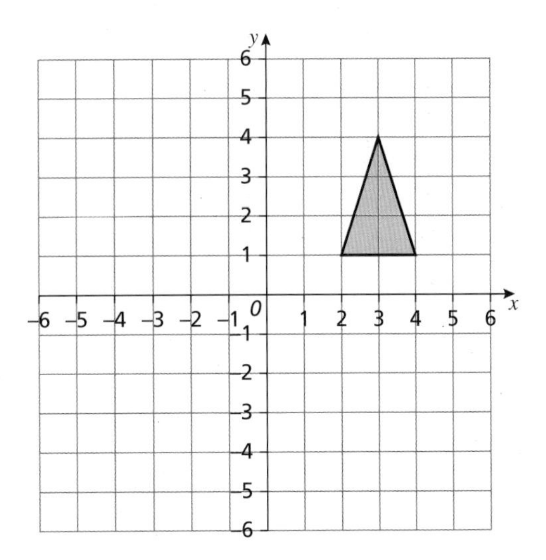

c $y = -1$

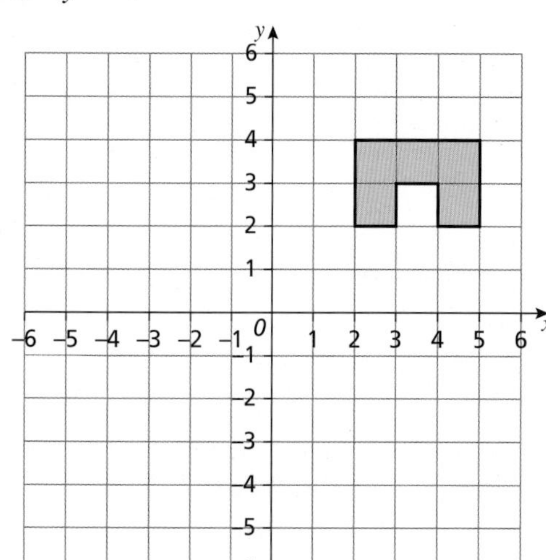

d $x = 3$

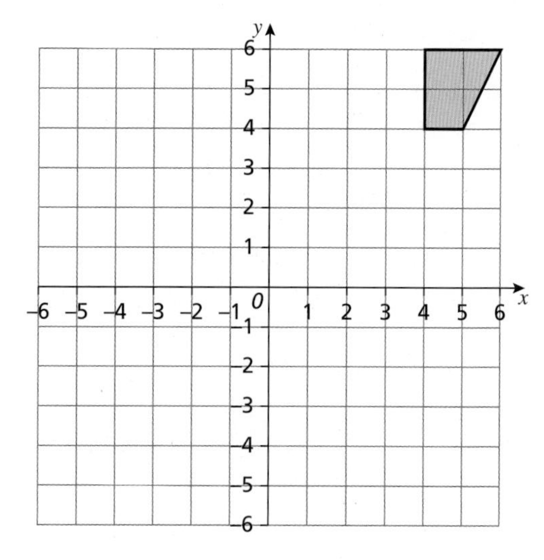

2 Copy each shape onto a grid and reflect it in the line $y = x$.

a

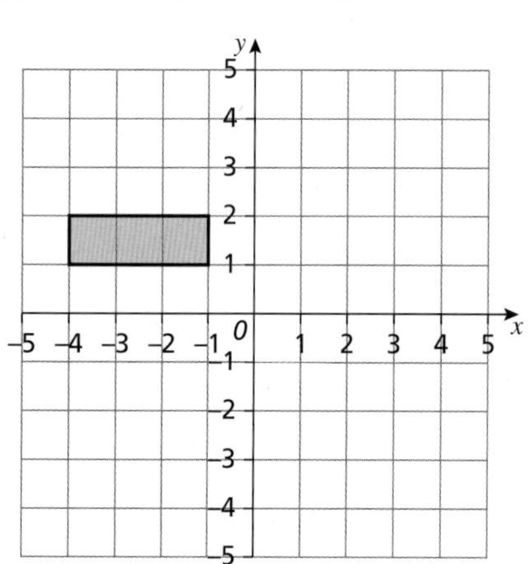

b

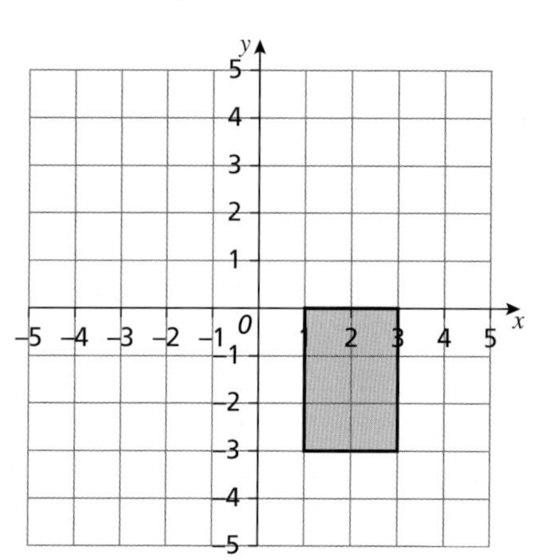

c

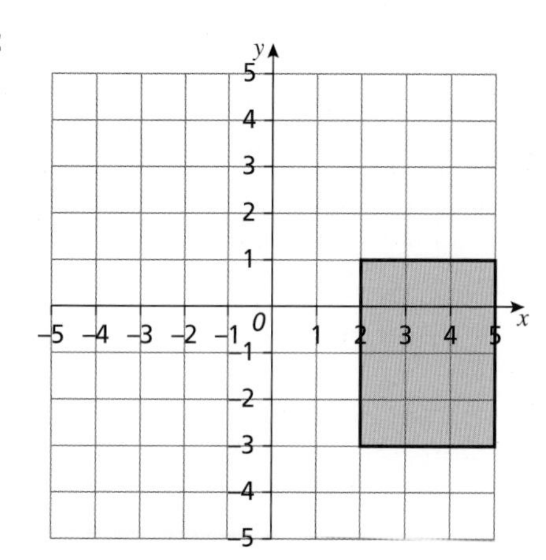

3 Describe fully the transformation that maps shape A to shape B.

a

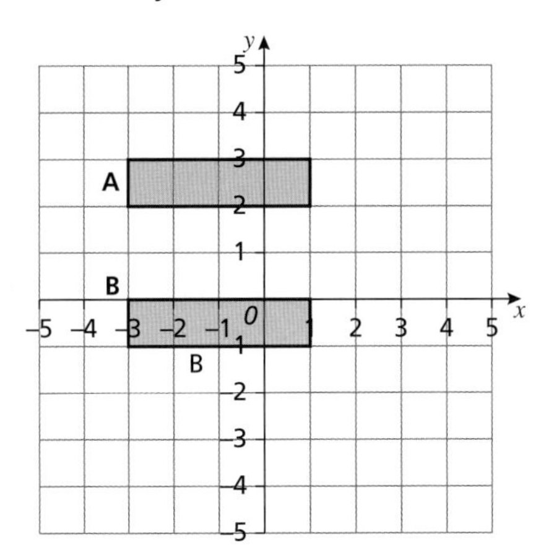

b

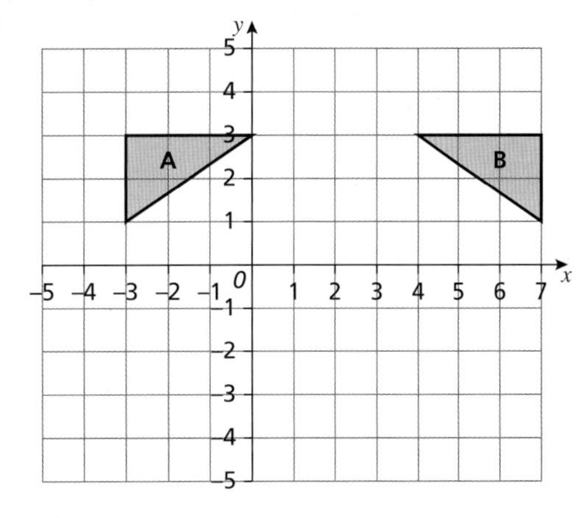

c

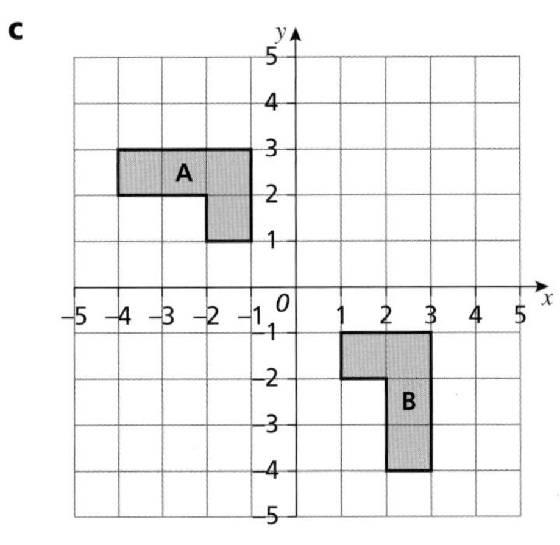

4 Copy each shape onto a grid and rotate it through the given angle about the given centre of rotation.

a 90° clockwise about (0, 0)

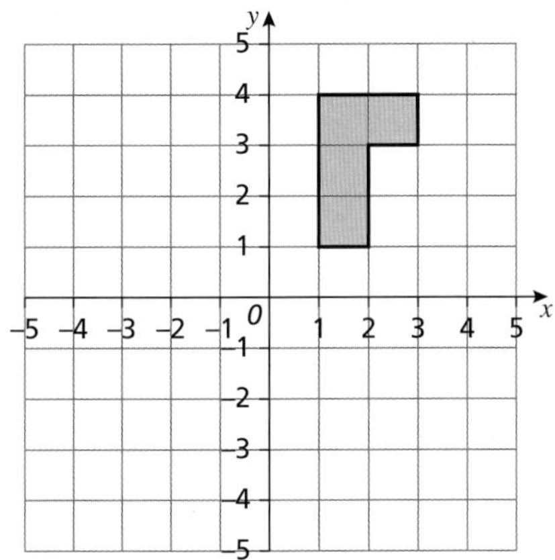

b 90° anticlockwise about (1, 0)

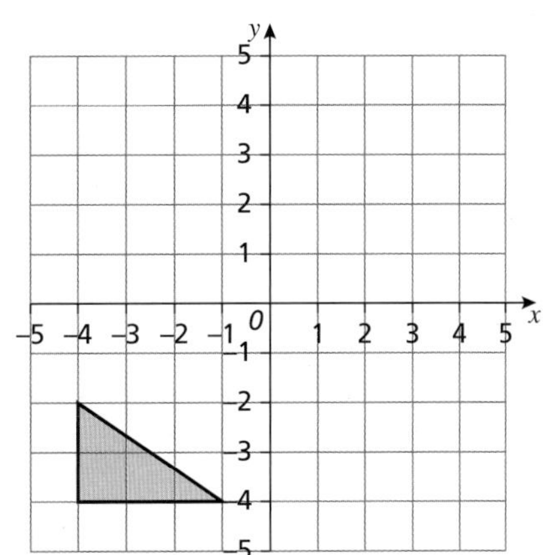

c 180° about (0, –2)

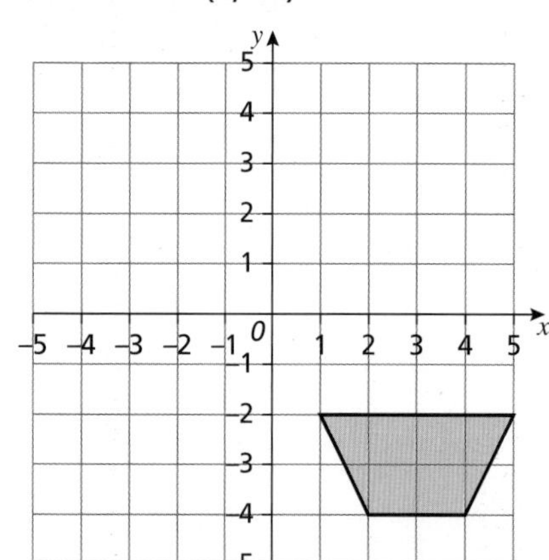

d 90° clockwise about (–2, –2)

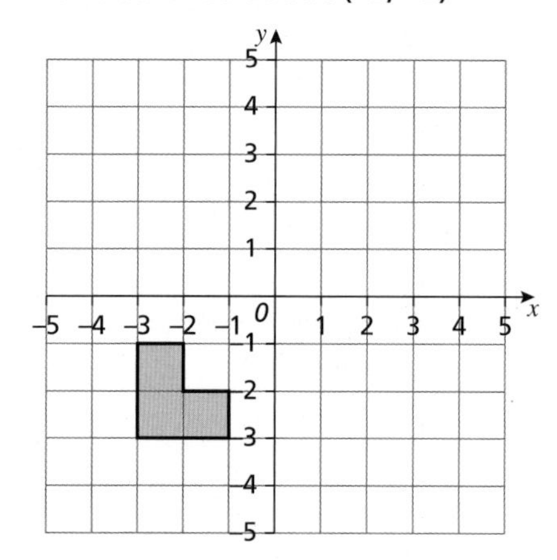

Stretch – can you deepen your learning?

1 The vertices of a trapezium have the coordinates (–1, 1), (–1, 3), (2, 3) and (3, 1).

The trapezium is reflected in the line $y = 2$

Without drawing a diagram, work out the coordinates of the vertices of the new shape.

2 Triangle P has been drawn on the grid.

The triangle is reflected in the line $x = -1$ to triangle Q.

Triangle P is reflected in the line $y = 2$ to triangle R.

Triangle Q is reflected in the line $y = 2$ to triangle S.

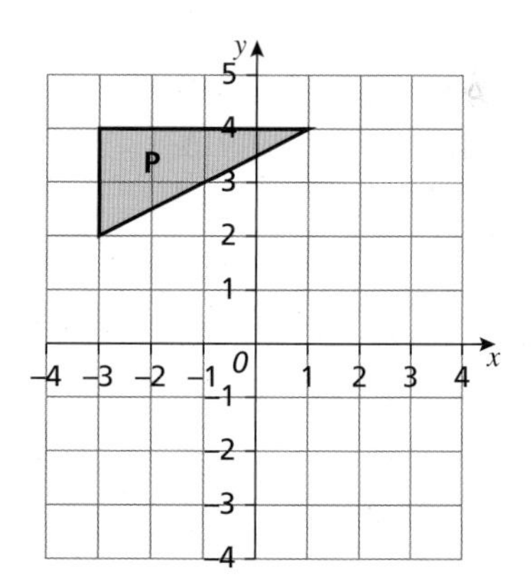

a What is the mathematical name for the shape enclosed by P, Q, R and S?

b Work out the area of the shape.

3 The rectangle shown is rotated about the point (m, n).

One of the vertices of the new rectangle has the coordinates (2, 5).

How many pairs of values for m and n can you find?

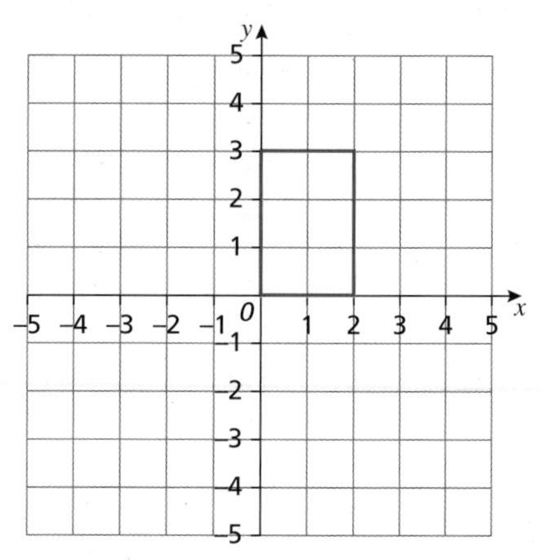

4 Point A has coordinates (2, 3). It is reflected in the line $x + y = 7$ to give point A′.

Work out an equation of the straight line that passes through A and A′.

4.2 Translation

White Rose MATHS

Are you ready?

1 Match each column vector to its description.

$\begin{pmatrix} -6 \\ 1 \end{pmatrix}$	1 right and 6 up
$\begin{pmatrix} 6 \\ 1 \end{pmatrix}$	6 left and 1 down
$\begin{pmatrix} -6 \\ -1 \end{pmatrix}$	1 left and 6 up
$\begin{pmatrix} 6 \\ -1 \end{pmatrix}$	6 right and 1 up
$\begin{pmatrix} -1 \\ 6 \end{pmatrix}$	6 right and 1 down
$\begin{pmatrix} 1 \\ 6 \end{pmatrix}$	6 left and 1 up

A **translation** is an example of a transformation. It moves a shape vertically, horizontally, or both. Translations are described using vector notation.

The number at the top of the vector represents a move to the right (if positive) or to the left (if negative).

The number on the bottom of the vector represents a move up (if positive) or down (if negative).

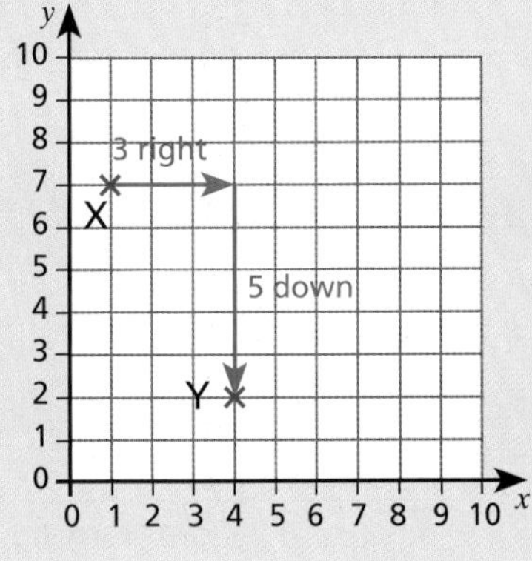

The vector describing the translation from X to Y is written

$\begin{pmatrix} 3 \\ -5 \end{pmatrix}$ → 3 right, → 5 down

Example 1

Translate shape A by the vector $\begin{pmatrix} -2 \\ 3 \end{pmatrix}$.

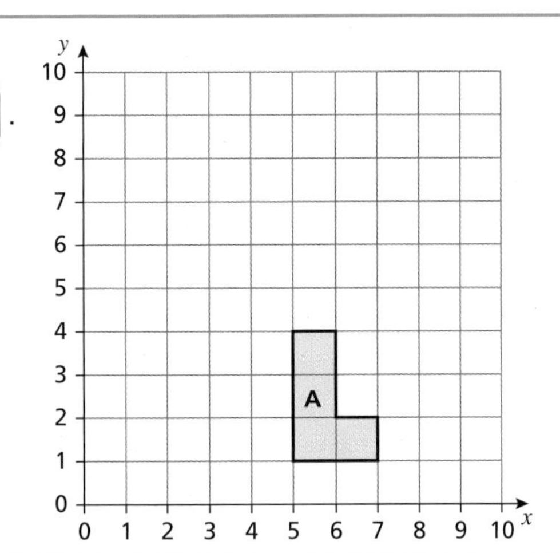

Method

Solution	Commentary
	-2 in the vector $\begin{pmatrix}-2\\3\end{pmatrix}$ means move 2 squares to the left. Choose a vertex of the shape and count 2 squares to the left. 3 in the vector $\begin{pmatrix}-2\\3\end{pmatrix}$ means move 3 squares up. From the point you reached after counting to the left, move 3 squares up. Repeat for all the vertices of the shape. The translated shape should be congruent to the original shape. Remember that 'congruent' means it is exactly the same shape and size.

Example 2

Describe fully the translation from shape A to B.

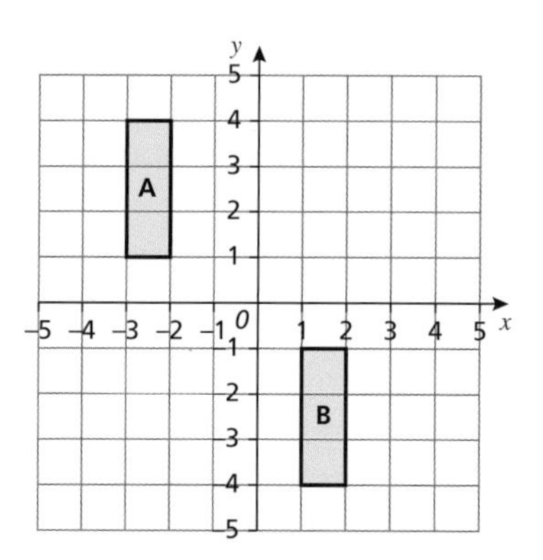

Method

Solution	Commentary
	Start with one vertex on shape A and find the corresponding vertex on shape B. Count the squares for the horizontal movement. For this question, it is 4 squares to the right so the top component in the column vector will be 4 Then count the squares for the vertical movement. For this question, it is 5 squares downwards so the bottom component in the column vector will be -5
A translation by the vector $\begin{pmatrix}4\\-5\end{pmatrix}$	

Practice

1. Copy the diagrams and translate each shape by the given vector.

a $\begin{pmatrix} 3 \\ 2 \end{pmatrix}$ **b** $\begin{pmatrix} 1 \\ -3 \end{pmatrix}$ **c** $\begin{pmatrix} -3 \\ 2 \end{pmatrix}$

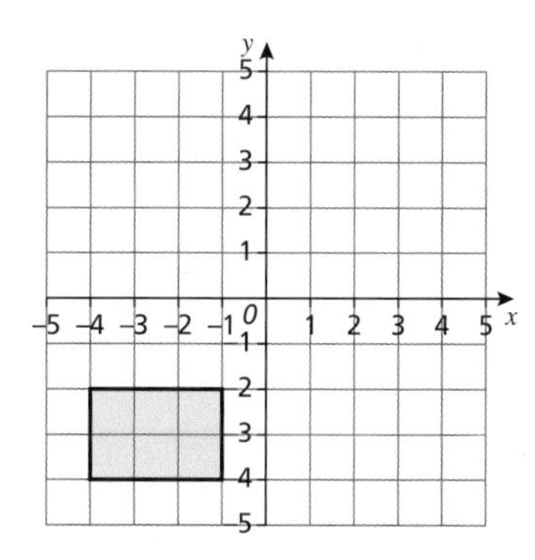
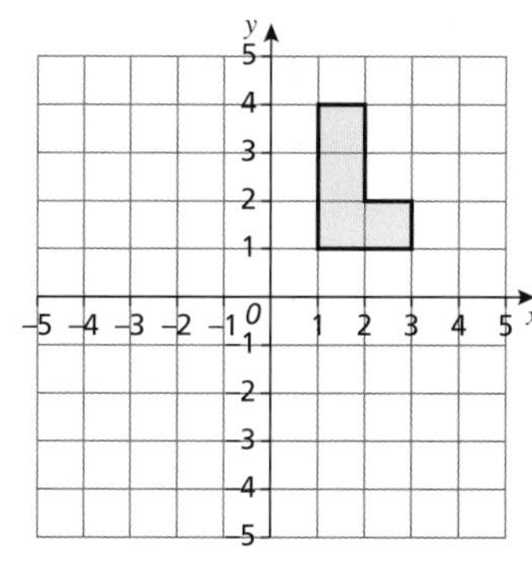
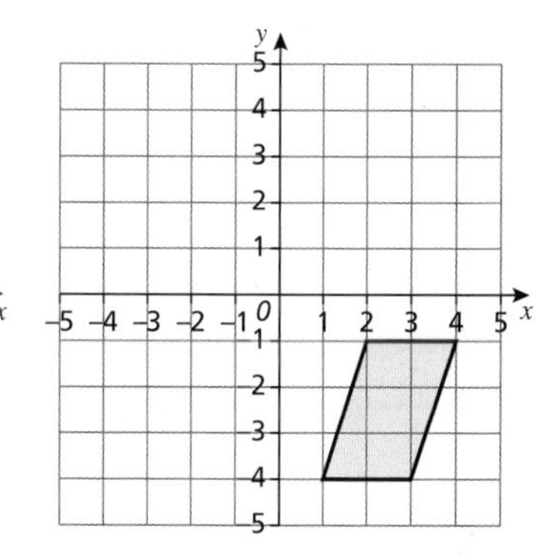

2. Describe fully the translation from shape A to shape B in each diagram.

a

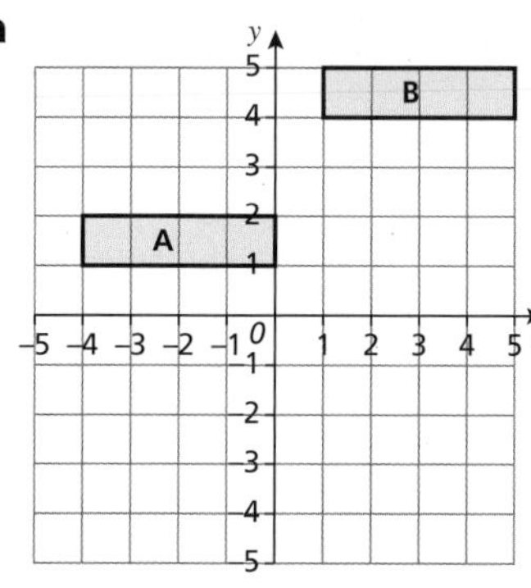

b

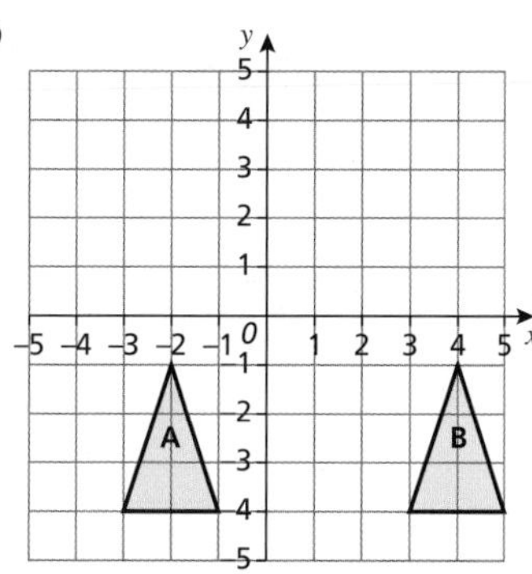

c

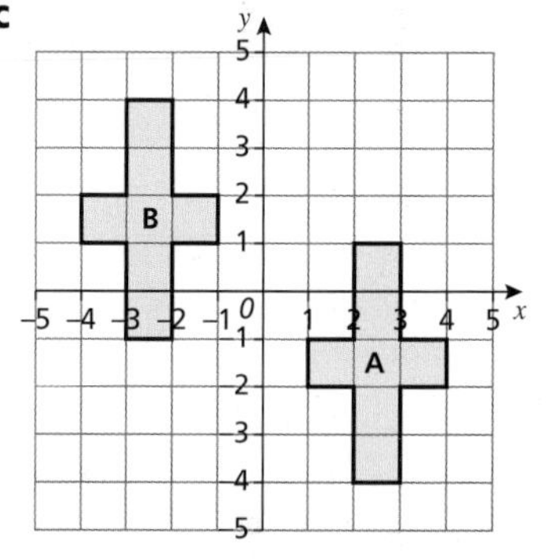

3. Shape A is translated by the vector $\begin{pmatrix} 4 \\ -3 \end{pmatrix}$ to shape B.

Shape B is translated by the vector $\begin{pmatrix} -2 \\ 1 \end{pmatrix}$ to shape C.

Describe the transformation from shape A to shape C.

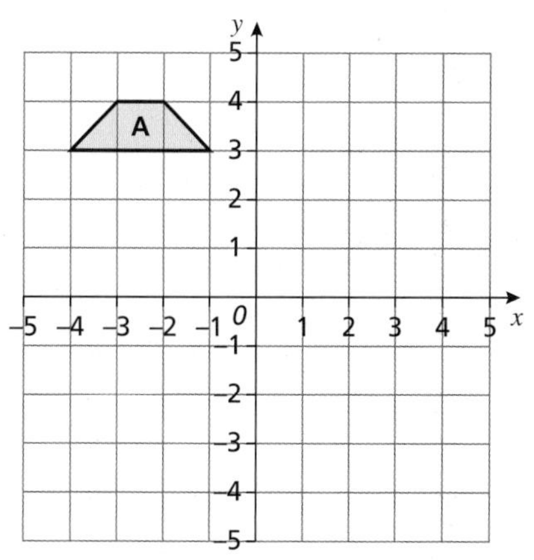

4. Junaid says that the translation from shape A to B is $\begin{pmatrix} 5 \\ 5 \end{pmatrix}$.

   Tiff says that the translation from shape A to B is $\begin{pmatrix} -5 \\ 5 \end{pmatrix}$.

   What mistakes have they made?

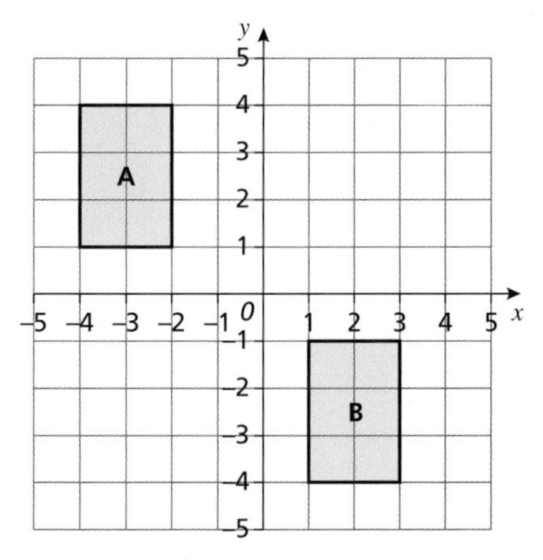

5. Which shapes are translations of each other?

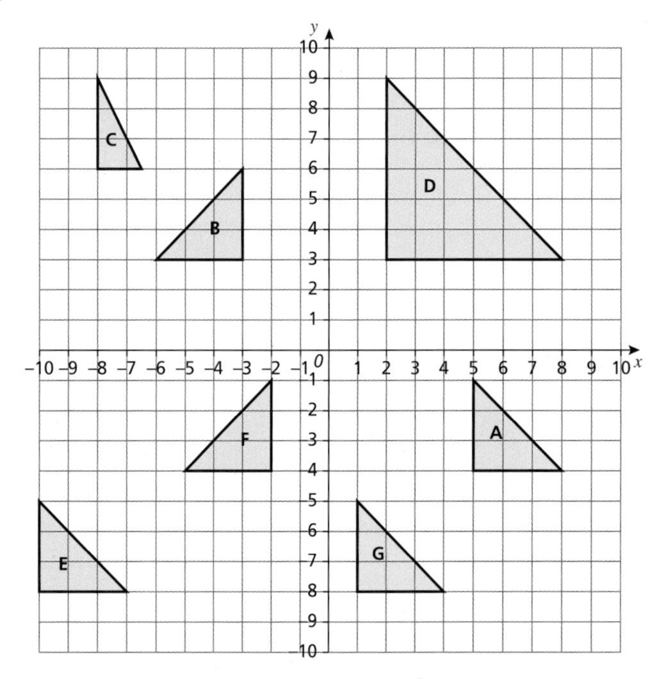

What do you think?

1. Is there more than one way to describe the transformation from shape A to shape B?

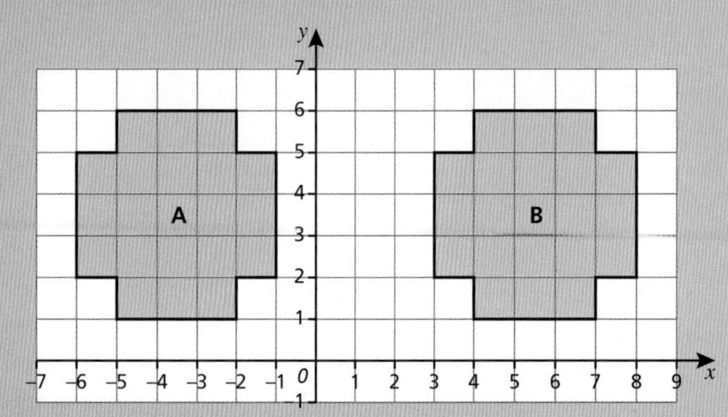

Consolidate – do you need more?

1 Copy the diagrams and translate each shape by the given vector.

a $\begin{pmatrix} 5 \\ -1 \end{pmatrix}$ **b** $\begin{pmatrix} -4 \\ 0 \end{pmatrix}$ **c** $\begin{pmatrix} -4 \\ -5 \end{pmatrix}$

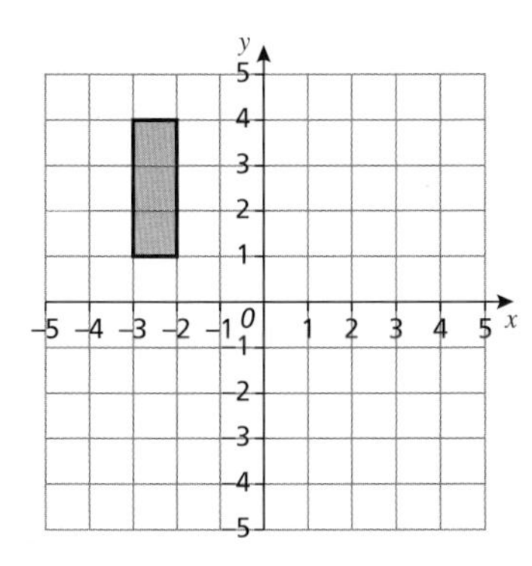

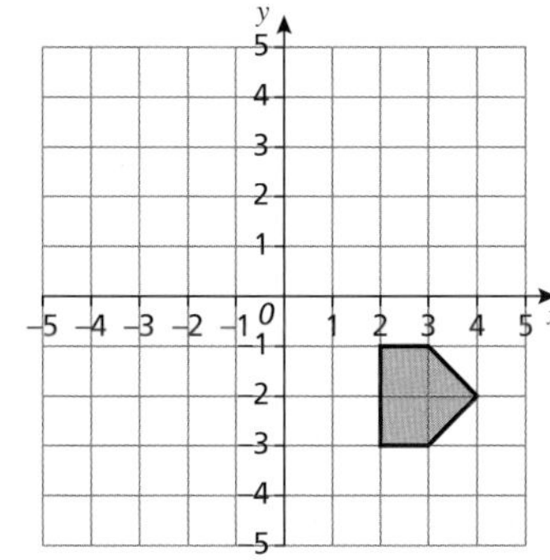

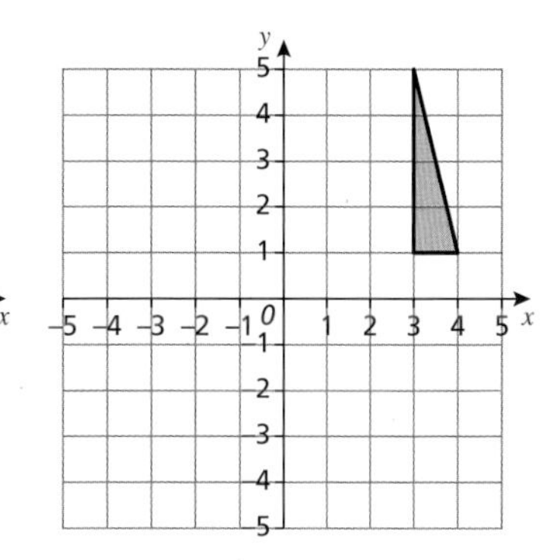

2 Describe fully the transformation from shape A to shape B in each diagram.

a

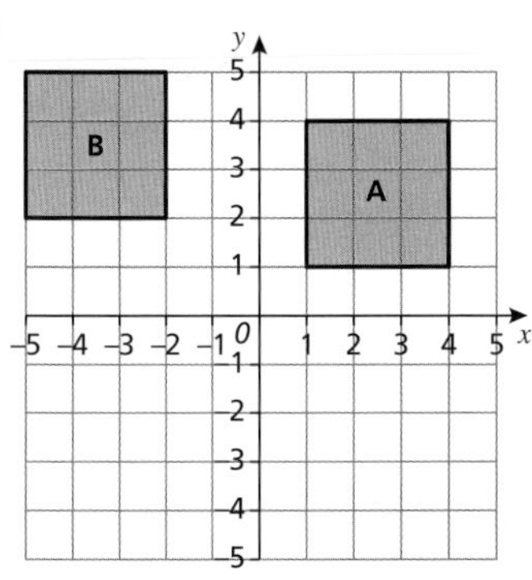

b

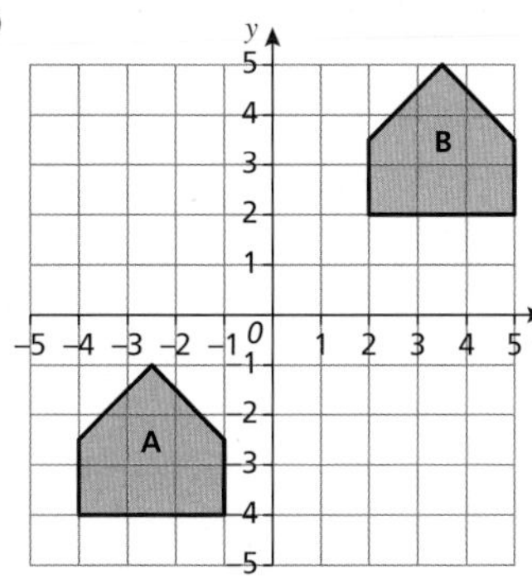

c

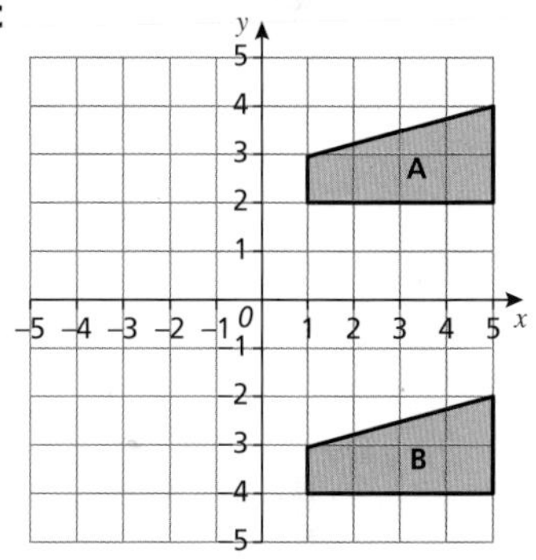

Stretch – can you deepen your learning?

1 The sketch shows a triangle that has been translated by the column vector $\begin{pmatrix} 7 \\ 10 \end{pmatrix}$.

What were the coordinates of each vertex before the translation?

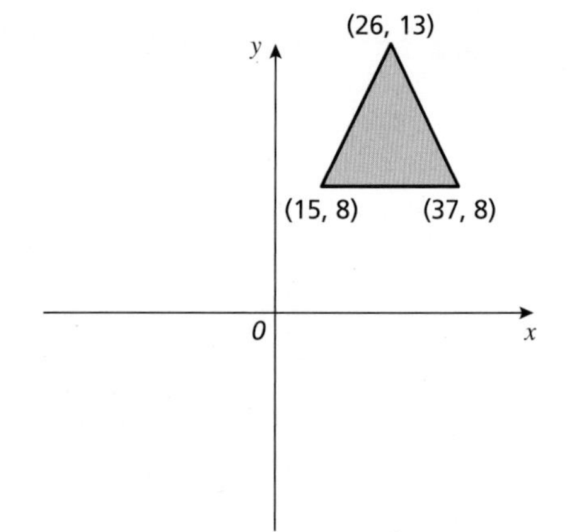

2 Shape A is translated by the vector $\begin{pmatrix} 3 \\ 7 \end{pmatrix}$, followed by $\begin{pmatrix} -6 \\ 12 \end{pmatrix}$, followed by $\begin{pmatrix} 2 \\ -19 \end{pmatrix}$, to give shape B.

Ed says, "Shape A has been translated by the vector $\begin{pmatrix} -1 \\ 0 \end{pmatrix}$."

Do you agree with Ed? Explain your answer.

3 Point A is translated by the vector $\begin{pmatrix} -4 \\ p \end{pmatrix}$ to give point B.

Point B has the coordinates (3, 5).

AB has a length of 12 units.

Work out the two possible values of p.

4 The point with coordinates (–5, 9) is translated by the vector $\begin{pmatrix} -8 \\ 4 \end{pmatrix}$ to the point $(3a + 8, 5 - 2b)$.

Work out the values of a and b.

5 Samira says, "Translations never result in some points being invariant."

Is Samira's statement **always, sometimes** or **never** true?

4.3 Enlargement

Are you ready?

1 Each pair of shapes is similar.

Work out the lengths of the lettered sides.

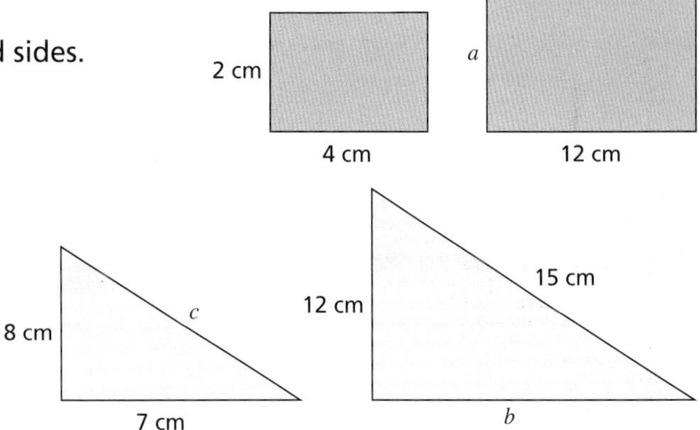

2 Enlarge this triangle by scale factor:

a 2

b $\frac{1}{2}$

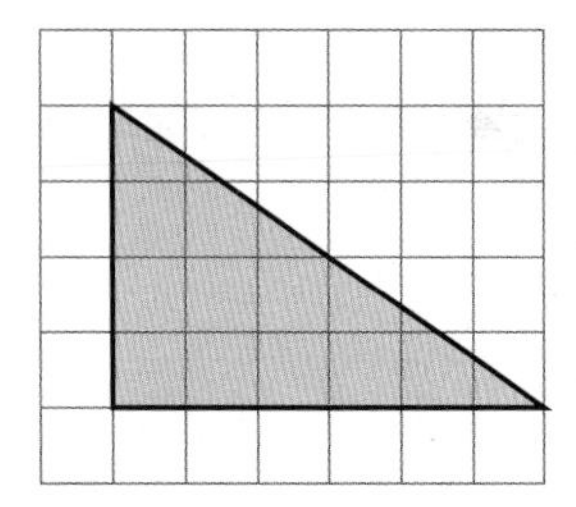

An **enlargement** is specified by its **scale factor** and its **centre of enlargement**.

- The scale factor of the enlargement can be found by comparing the lengths of the sides.
- The centre of enlargement can be found by joining corresponding vertices with line segments and seeing where these line segments meet.

Each side of rectangle Q is three times the length of the corresponding side in rectangle P. So the scale factor of enlargement is 3

Drawing ray lines through each vertex on the original shape and the corresponding vertex on the enlarged shape, and extending them so they meet at a point, finds the centre of enlargement at (–7, 1).

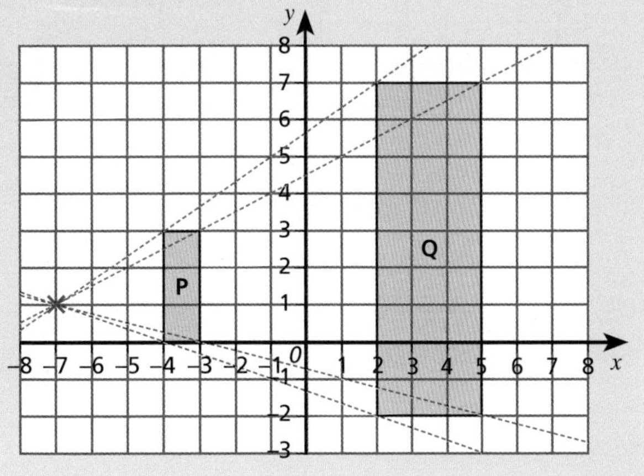

If you enlarge an object by:

- a scale factor greater than 1, the image will be larger than the object
- a scale factor less than 1, the image will be smaller than the object
- a negative scale factor, the image will be in a different orientation to the object.

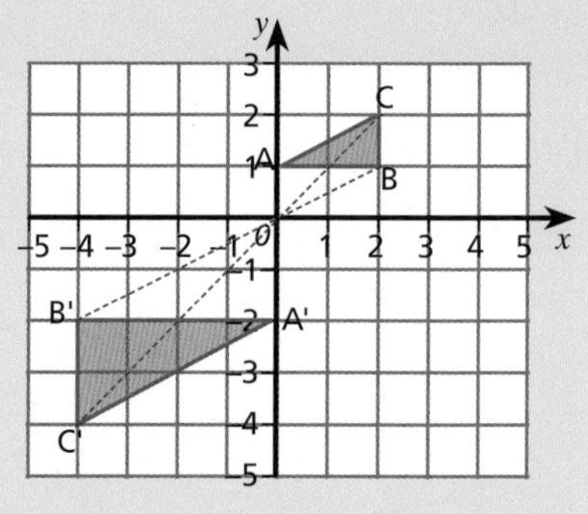

The triangle ABC is enlarged by scale factor –2 about the origin (0, 0). Drawing rays from each vertex of the triangle through the origin can help to identify the corresponding vertices of the image. The rays meet at the centre of enlargement. The image, triangle A'B'C', has side lengths twice as long as those of ABC. The distance from the centre of enlargement to a vertex on the image is twice the distance from the centre to the corresponding vertex on the object.

Example 1

Enlarge shape P by scale factor $\frac{1}{3}$, centre (–7, –4).

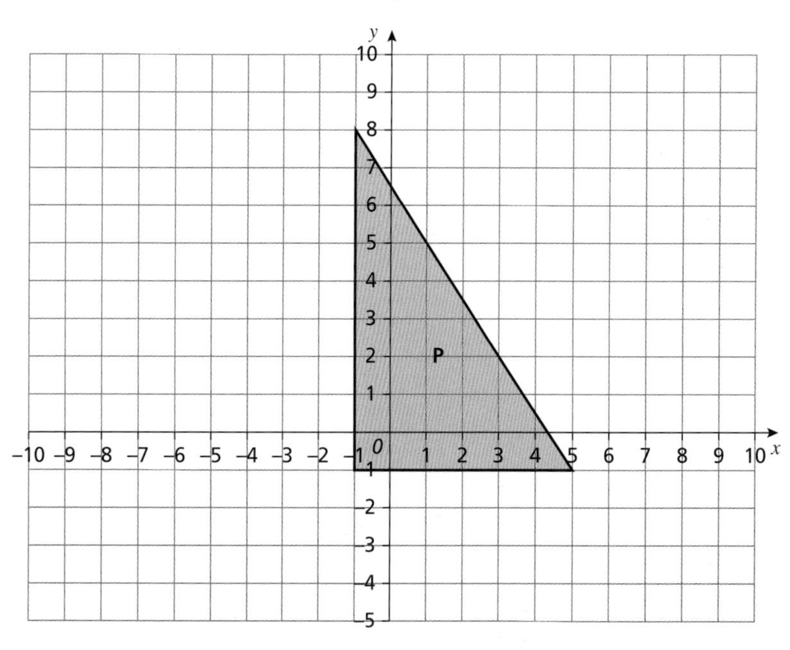

Method

Solution	Commentary
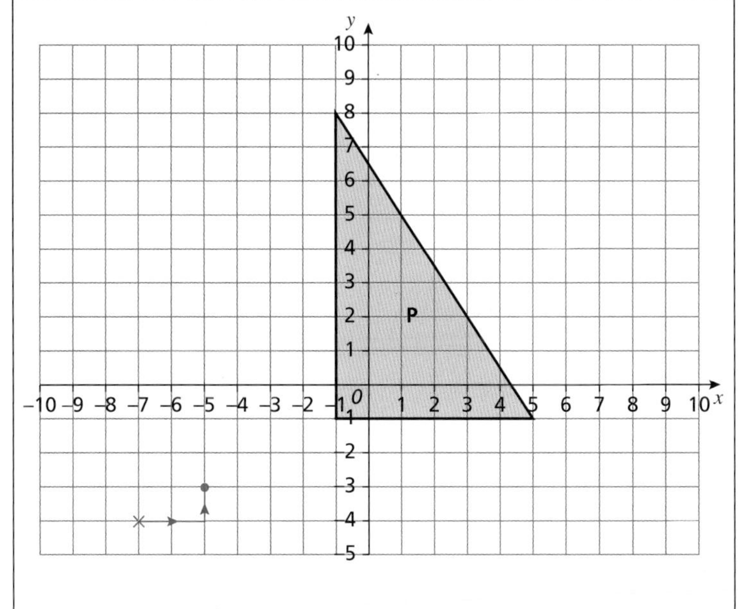	First plot the point (–7, –4) on the grid. This is the centre of enlargement. The bottom left vertex of P is 6 squares right and 3 squares up from the centre of enlargement. You can write this as the vector $\begin{pmatrix}6\\3\end{pmatrix}$. You need to enlarge the shape by scale factor $\frac{1}{3}$, so you can multiply this vector by $\frac{1}{3}$ $\begin{pmatrix}6\\3\end{pmatrix} \times \frac{1}{3} = \begin{pmatrix}2\\1\end{pmatrix}$ Use the vector $\begin{pmatrix}2\\1\end{pmatrix}$ from the centre of enlargement to draw the first vertex of the image.
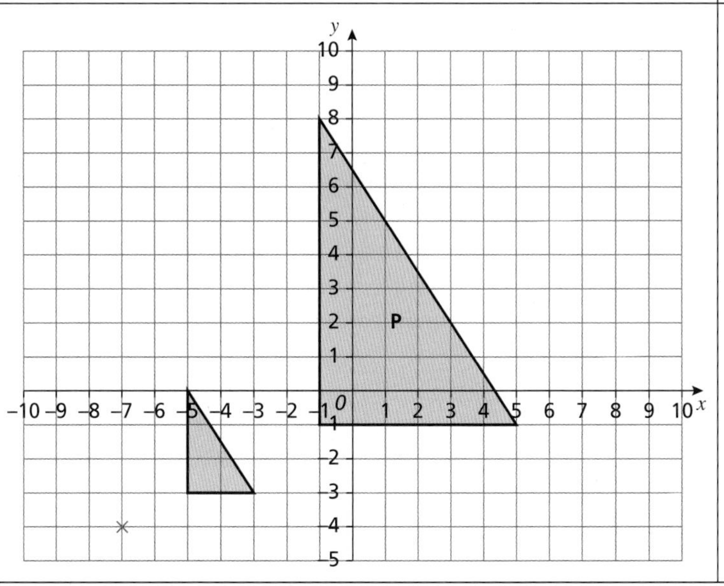	Repeat this for each vertex of the triangle and join the vertices to form the image of P.

Example 2

Enlarge the shape by scale factor –2, using the point marked with a cross as the centre of enlargement.

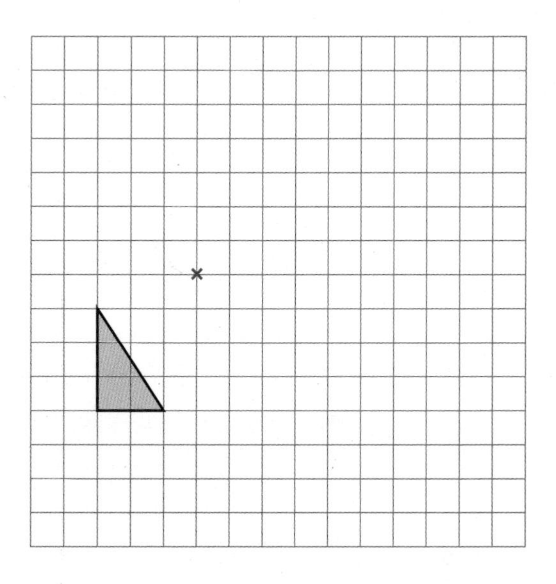

Method

Solution	Commentary
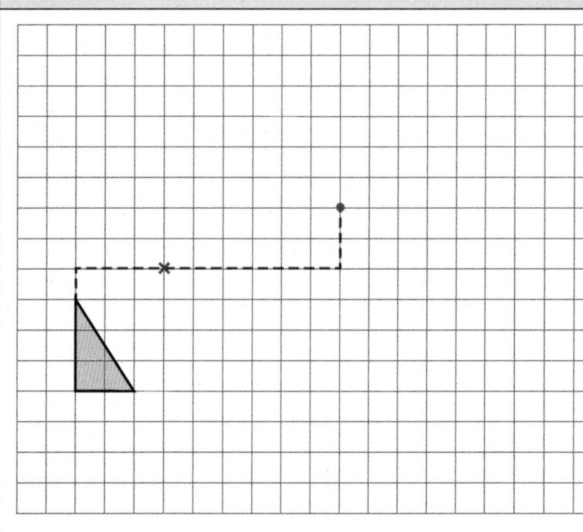	The top left vertex of the triangle is 3 squares left and 1 square down from the centre of enlargement. You can write this as the vector $\begin{pmatrix} -3 \\ -1 \end{pmatrix}$. You need to enlarge the shape by scale factor –2, so you can multiply this vector by –2: $\begin{pmatrix} -3 \\ -1 \end{pmatrix} \times -2 = \begin{pmatrix} 6 \\ 2 \end{pmatrix}$ Use the vector $\begin{pmatrix} 6 \\ 2 \end{pmatrix}$ from the centre of enlargement to draw the first vertex of the new shape.
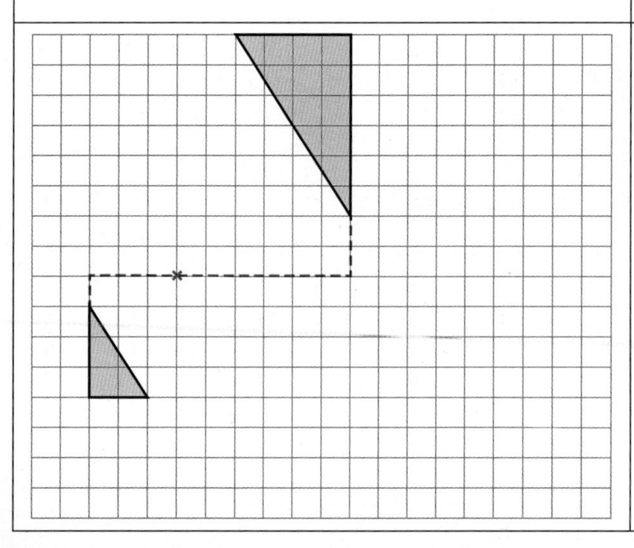	Repeat this for each vertex of the triangle and join the vertices to form the enlarged triangle. Note that the image has a different orientation to the object.

Practice

1. Copy each shape and enlarge it using the given scale factor and centre of enlargement (marked by the red cross).

   i Scale factor 2 **ii** Scale factor 3 **iii** Scale factor $\frac{1}{2}$

a

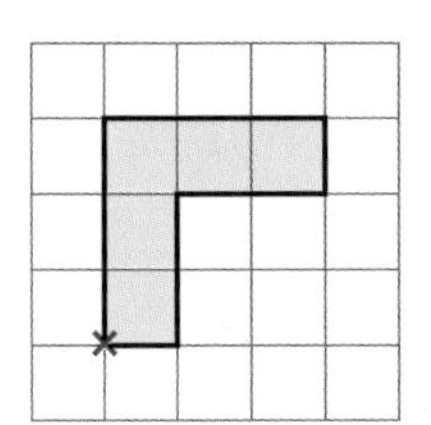

b

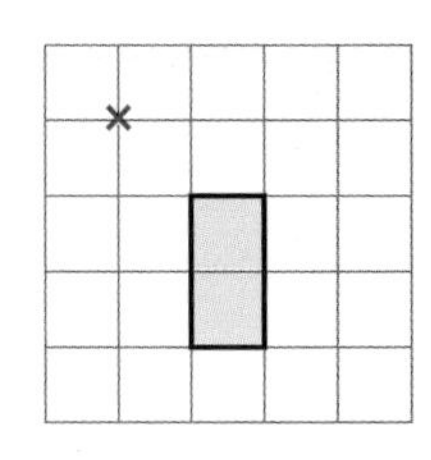

c

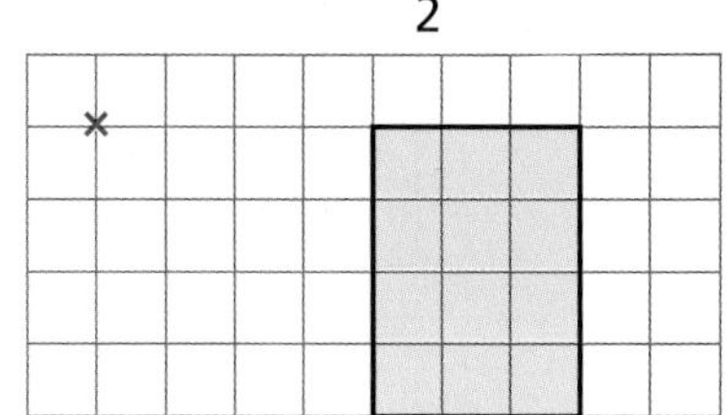

2. Copy and enlarge each shape using the given scale factor and centre of enlargement.

   a Scale factor 2 Centre of enlargement (0, 0)

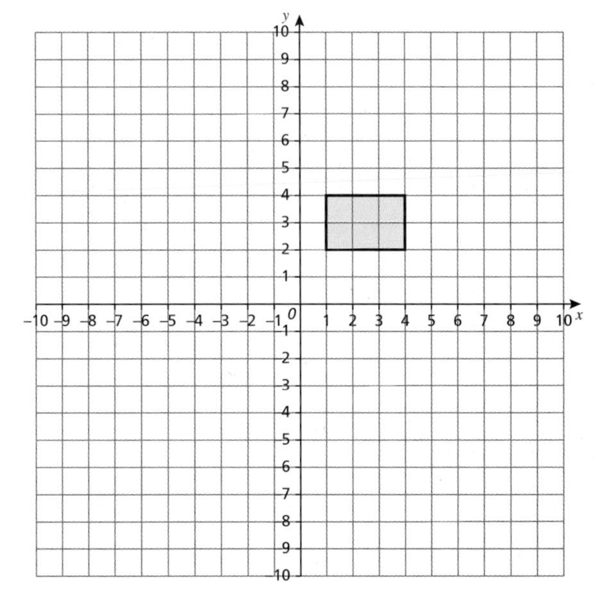

   b Scale factor 3 Centre of enlargement (–5, –4)

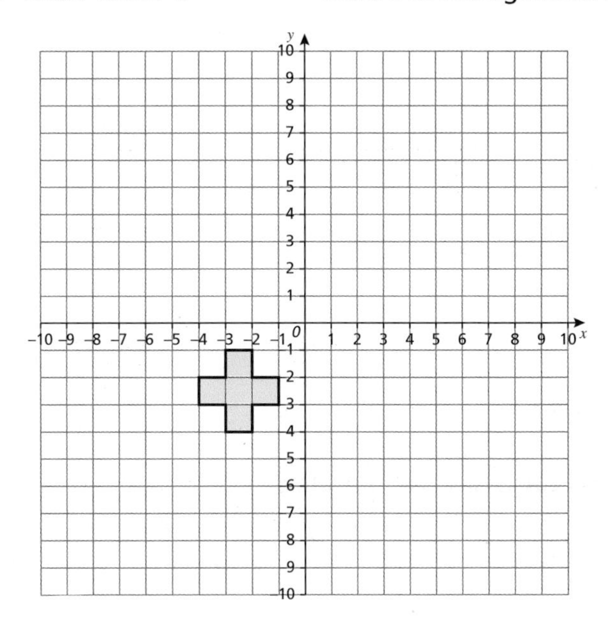

c Scale factor $\frac{1}{2}$ Centre of enlargement (0, 1)

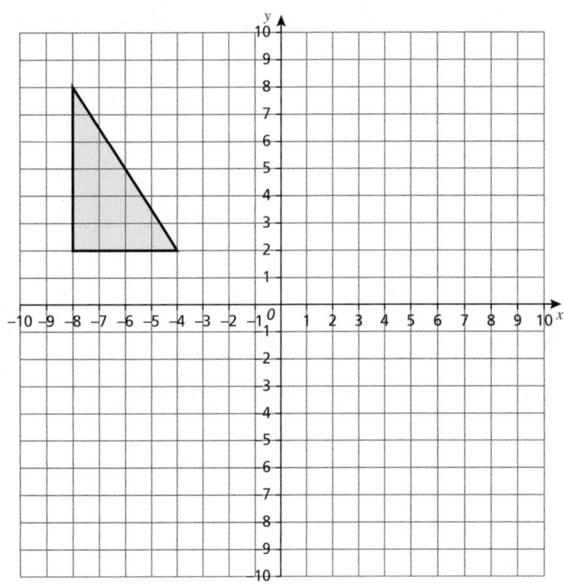

3 For each diagram, describe fully the transformation:

a A to B

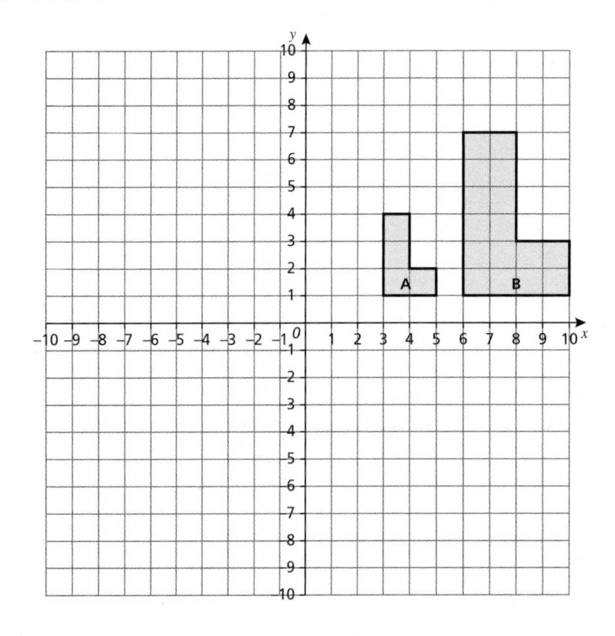

b B to A.

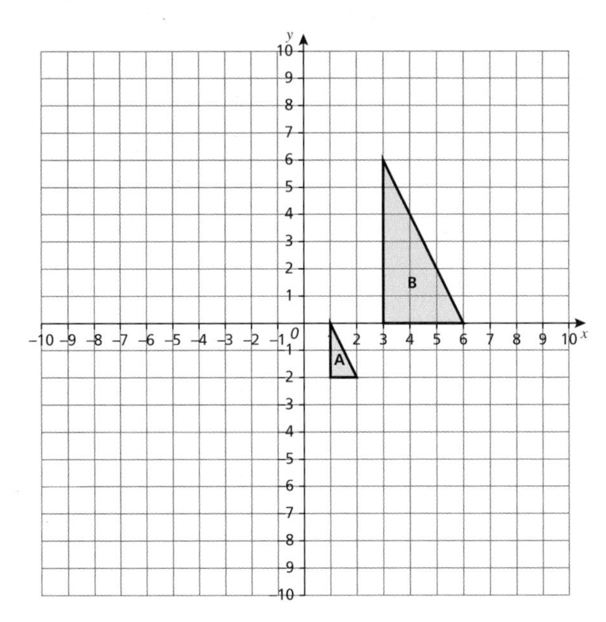

4 Copy and enlarge each shape by scale factor –2 using the given centre of enlargement (marked by the red cross).

a

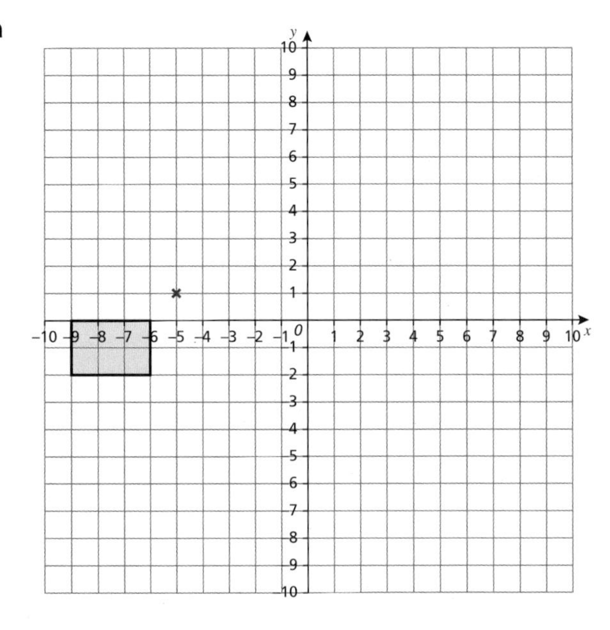

b

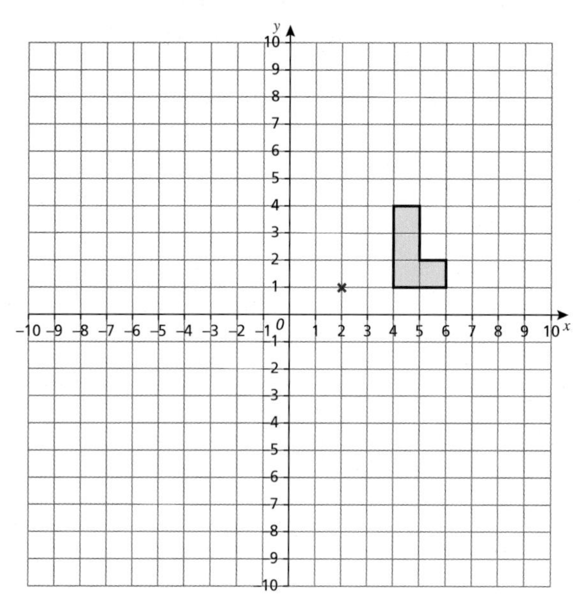
y
10
9
8
7
6
5
4
3
2
1
0
–1
–2
–3
–4
–5
–6
–7
–8
–9
–10
–10 –9 –8 –7 –6 –5 –4 –3 –2 –1
1 2 3 4 5 6 7 8 9 10
x

c

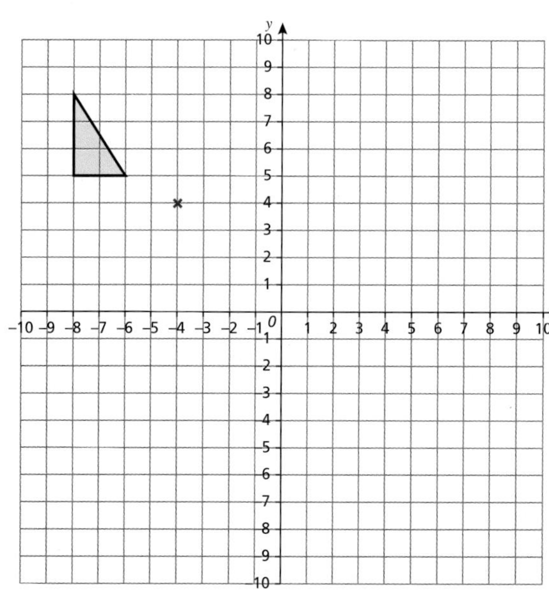
y
10
9
8
7
6
5
4
3
2
1
0
–1
–2
–3
–4
–5
–6
–7
–8
–9
–10
–10 –9 –8 –7 –6 –5 –4 –3 –2 –1
1 2 3 4 5 6 7 8 9 10
x

d

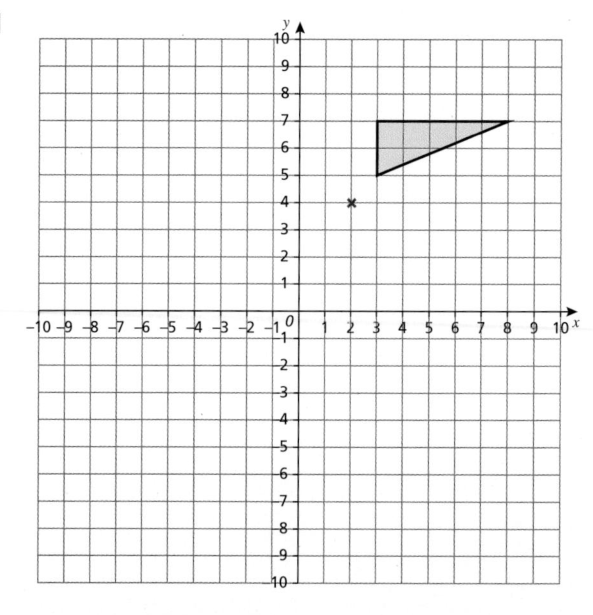
y
10
9
8
7
6
5
4
3
2
1
0
–1
–2
–3
–4
–5
–6
–7
–8
–9
–10
–10 –9 –8 –7 –6 –5 –4 –3 –2 –1
1 2 3 4 5 6 7 8 9 10
x

5 Copy each diagram and complete the enlargements.

a Scale factor −3, centre (6, 4)

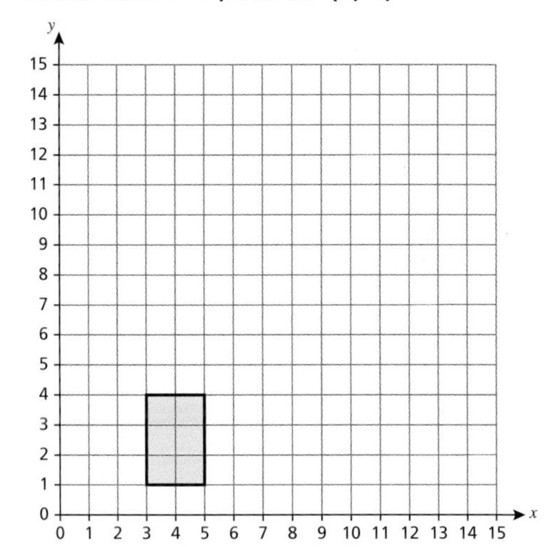

b Scale factor −2, centre (10, 3)

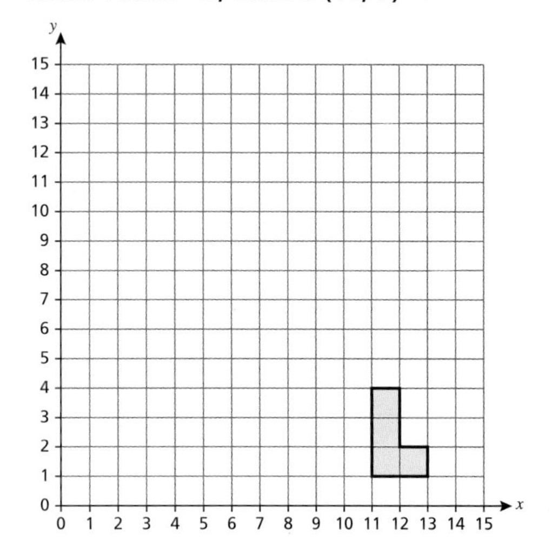

c Scale factor −2, centre (8, 8)

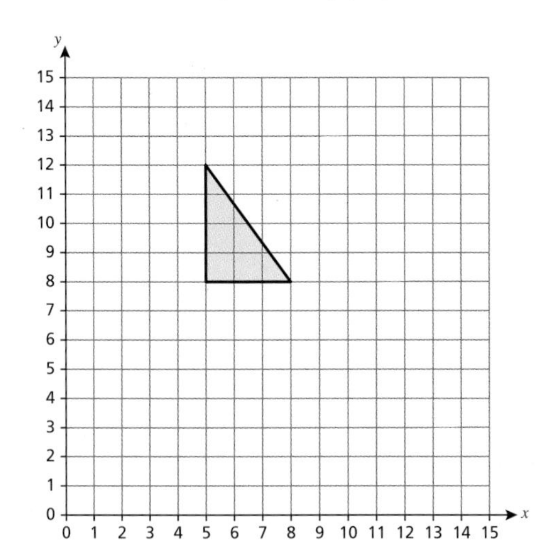

d Scale factor $-\frac{1}{2}$, centre (8, 5)

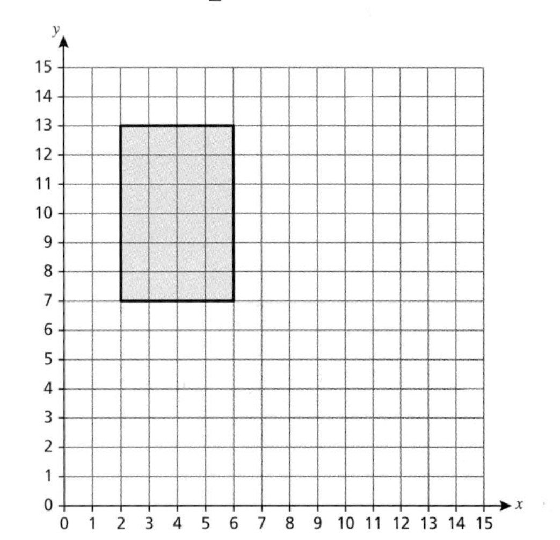

6 Copy each diagram and complete the enlargements.

a Scale factor –2, centre (0, 0)

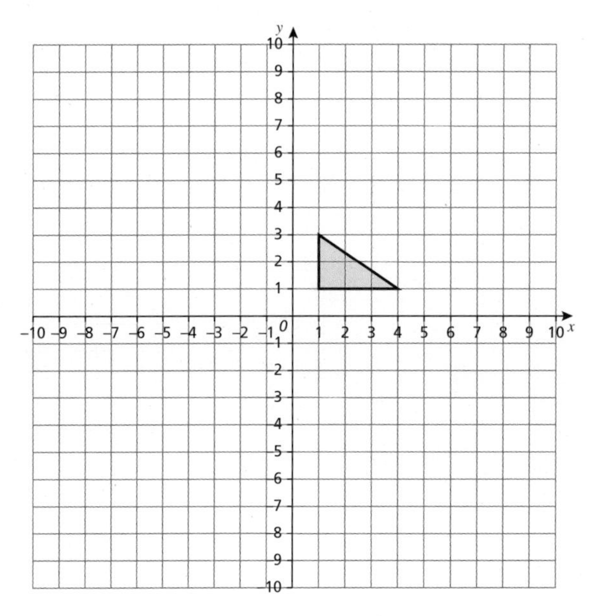

b Scale factor –5, centre (–5, –3)

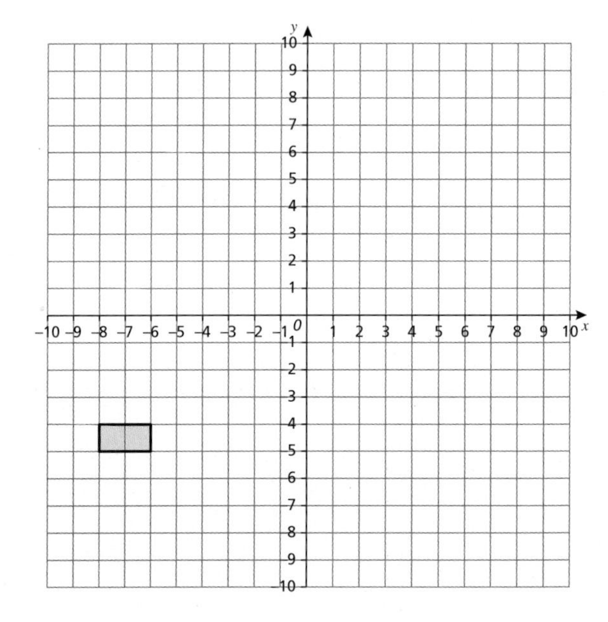

c Scale factor $-\frac{2}{3}$, centre (5, −1)

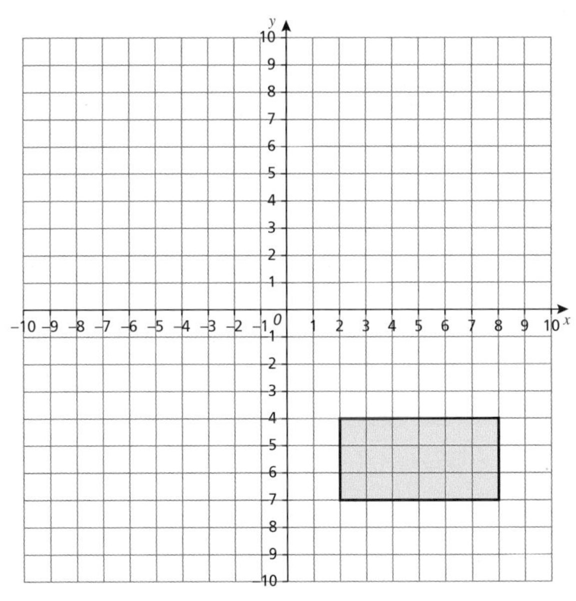

d Scale factor $-\frac{3}{2}$, centre (1, 4)

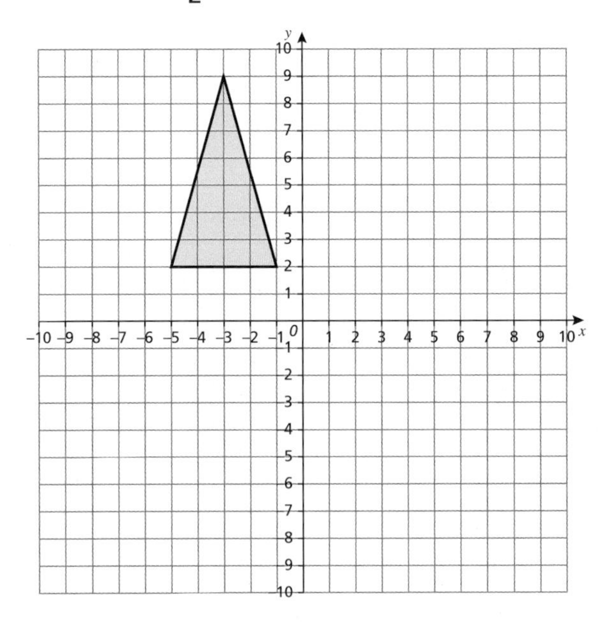

What do you think?

1 Ed enlarges a shape by a scale factor of $\frac{1}{3}$ and labels the shape P.

He then enlarges the original shape by a scale factor of $-\frac{1}{3}$ and labels the shape Q.

Ed says, "If I do this to any shape, then P and Q will always be congruent."

Do you agree with Ed? Explain your answer.

Consolidate – do you need more?

1 Copy each diagram and enlarge the shape by the given scale factor.

a Scale factor $\frac{1}{2}$

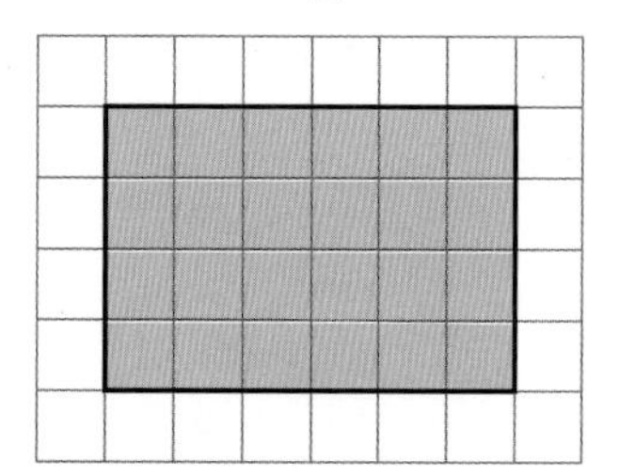

b Scale factor $\frac{1}{3}$

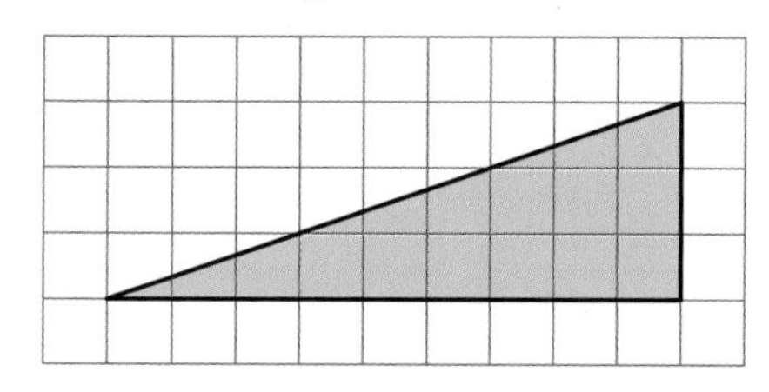

c Scale factor $\frac{1}{3}$

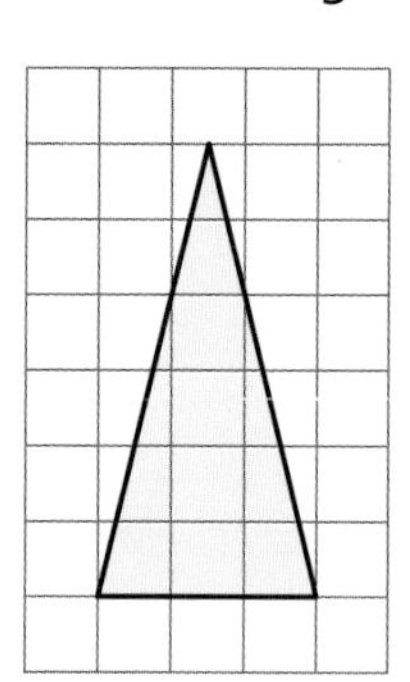

d Scale factor $\frac{1}{4}$

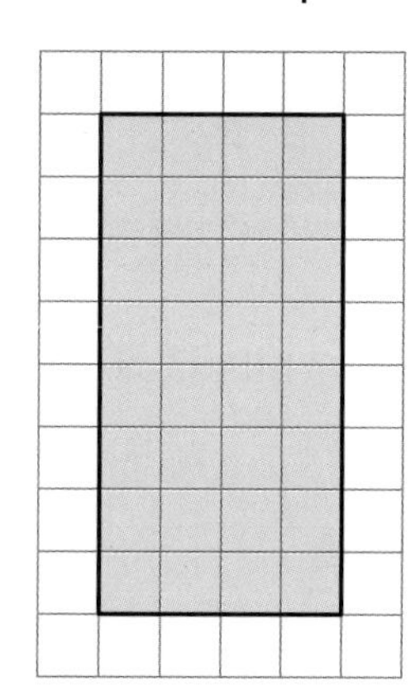

2 For each diagram, describe the enlargement from A to B.

a

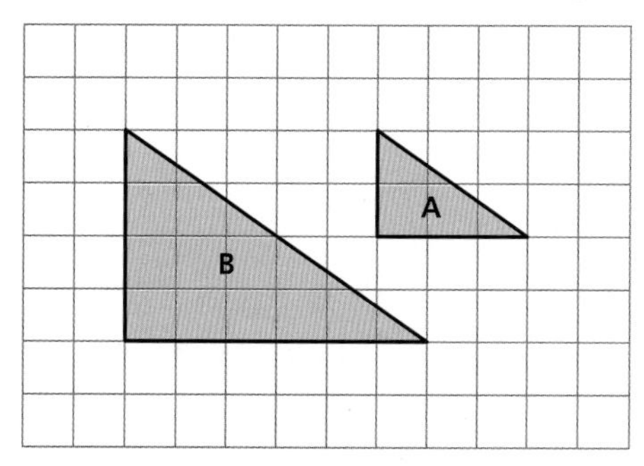

b

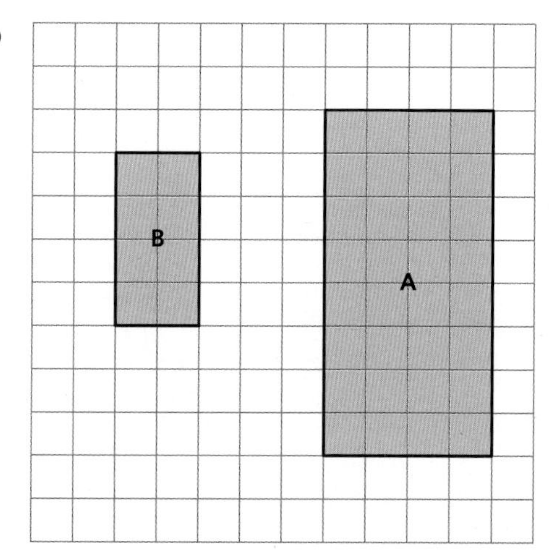

3 Copy each diagram and enlarge the shape by scale factor –2, centre O.

a

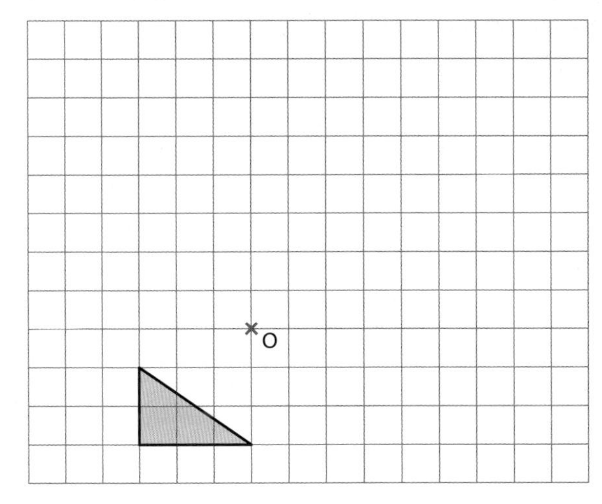

b

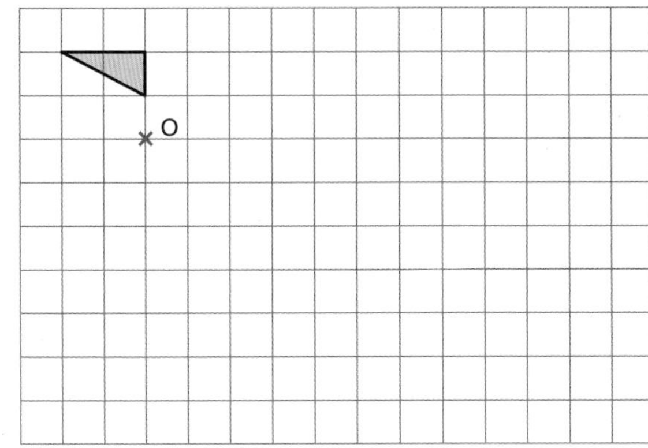

Stretch – can you deepen your learning?

1 Q is an enlargement of P by scale factor $\frac{1}{2}$, centre (0, 0).
R is an enlargement of P by scale factor –2, centre (0, 0).

Describe fully the single transformation that maps shape Q onto shape R.

2 A shape with an area of 20 cm^2 is enlarged by scale factor –3

Huda says, "The area of the new shape must be 60 cm^2. It can't be –60 cm^2 because area can't be measured with negative values."

Refer to Chapter 2.1 for help with this question.

Explain why the numerical part of Huda's statement is incorrect.

3 Shape P is enlarged by scale factor $-\frac{3}{2}$, centre (0, 0), to give shape Q.
Eva says, "If I enlarge shape Q by scale factor $\frac{2}{3}$, centre (0, 0), it will give shape P."

Do you agree with Eva? Explain your answer.

4 If I enlarge a shape by scale factor $\frac{a}{b}$ then the side lengths of the object and the image are in the ratio $b : a$.

Is this statement **true** or **false**?

4.4 Combinations of transformations

White Rose MATHS

Are you ready?

1 Copy the diagram four times and draw the result of each transformation of the triangle.

a Rotation 180° about the origin

b Reflection in the line $y = x$

c Translation by $\begin{pmatrix} 3 \\ -4 \end{pmatrix}$

d Enlargement by scale factor –2, centre (0, 0)

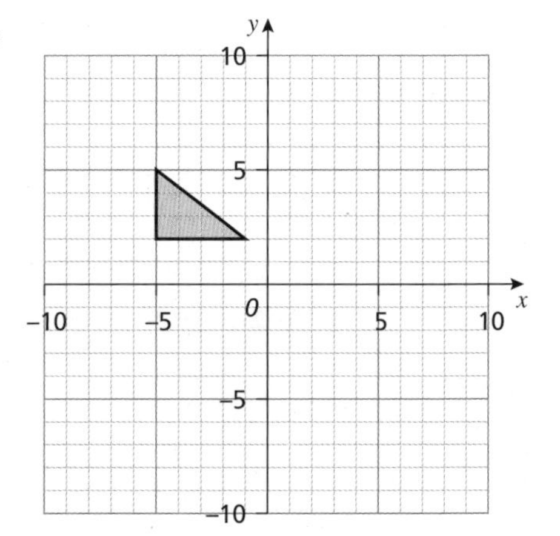

When describing transformations, you need to do so **fully**.

You should state the correct mathematical name of the transformation and:

- for a **reflection**, state the equation of the mirror line.

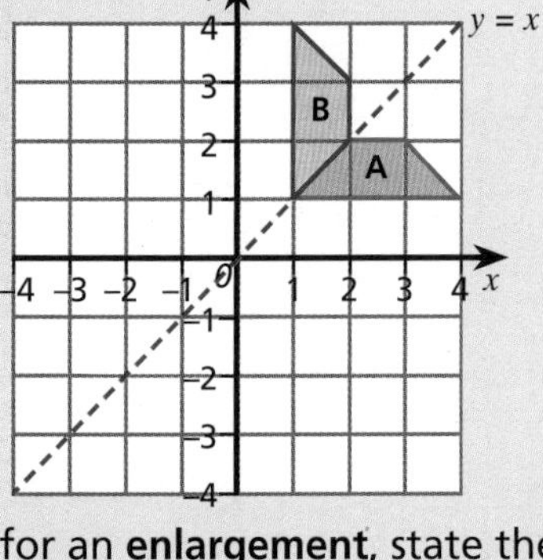
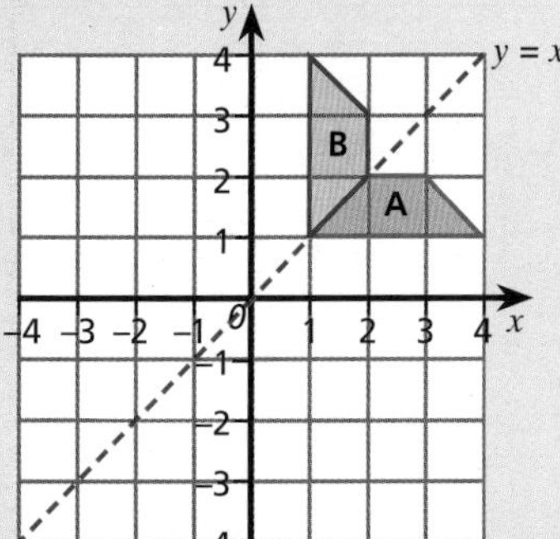

- for a **rotation**, state the angle of rotation, the direction and the centre of rotation.

90° clockwise about (0, 0)

- for an **enlargement**, state the scale factor and the centre of enlargement.

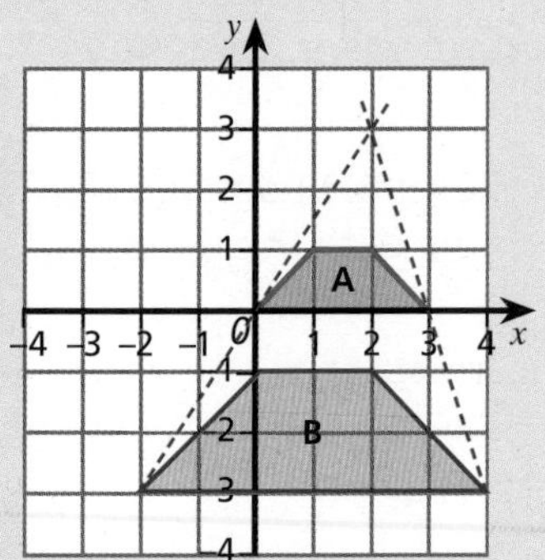

Scale factor 2, centre (2, 3)

- for a **translation**, state the vector the shape has been translated by.

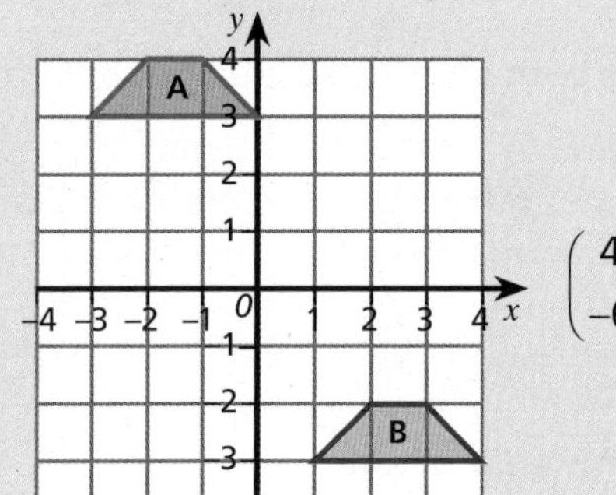

$\begin{pmatrix} 4 \\ -6 \end{pmatrix}$

You can perform a series of transformations to a shape; for example, applying a second transformation to the image of an initial transformation.

However, in an examination, you will often be required to describe a series of transformations as one single transformation.

Example

a Reflect the shape A in the y-axis.

Label the new shape B.

b Reflect the shape B in the line $y = 1$

Label the new shape C.

c There are two ways to describe fully the transformation which maps shape A to shape C as a single transformation.

State one of these ways.

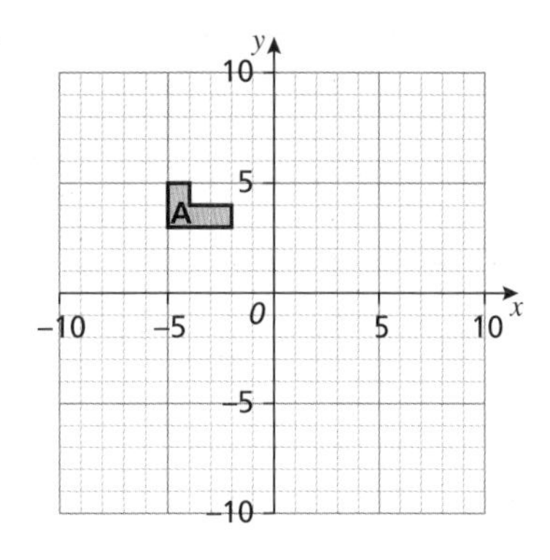

Method

Solution	Commentary
a	Reflect each vertex of the shape in the y-axis, ensuring that the corresponding vertices of the image are the same distance from the y-axis as those of the object. Label the new shape B.
b	Identify the line $y = 1$ and reflect shape B in the line. Label the new shape C.
c A rotation, 180° about point (0, 1)	The transformation from A to C cannot be a translation as the shapes have different orientations. You can also tell by eye that the transformation is not a reflection. Note that the transformation could also be described as an enlargement, scale factor −1, centre (0, 1).

Practice

1. Copy the shape A onto a grid.

   a Reflect shape A in the y-axis.

   Label the new shape B.

   b Translate shape B by $\begin{pmatrix} 2 \\ -3 \end{pmatrix}$.

   Label the new shape C.

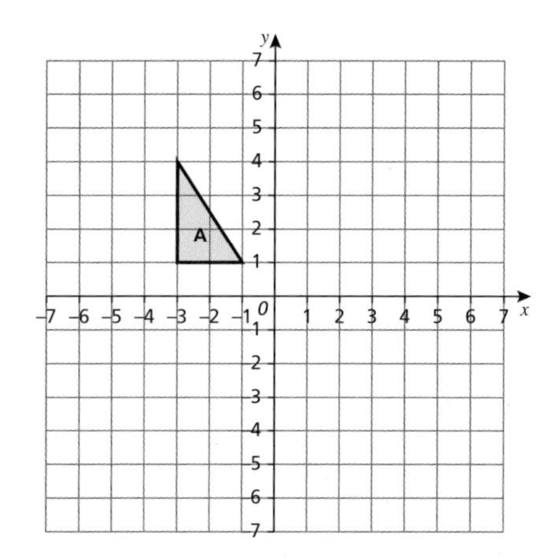

2. Copy the shape P onto a grid.

   a Translate shape P by $\begin{pmatrix} -4 \\ 0 \end{pmatrix}$.

   Label the new shape M.

   b Rotate shape M by 180° about the origin.

   Label the new shape N.

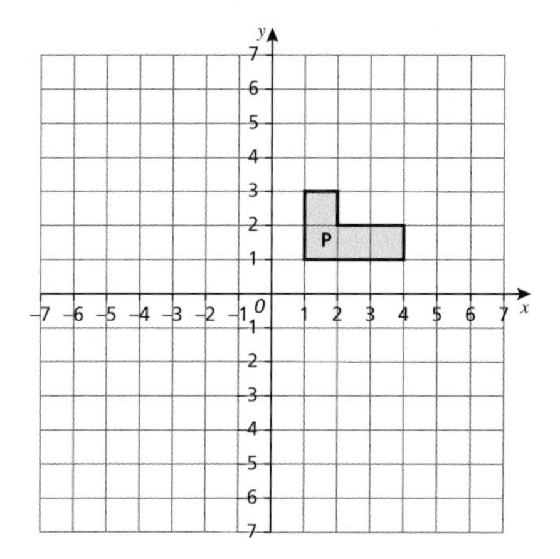

3. Copy the shape X onto a grid.

   a Enlarge shape X by scale factor 2, centre (6, 0).

   Label the new shape Y.

   b Reflect shape Y in the x-axis.

   Label the new shape Z.

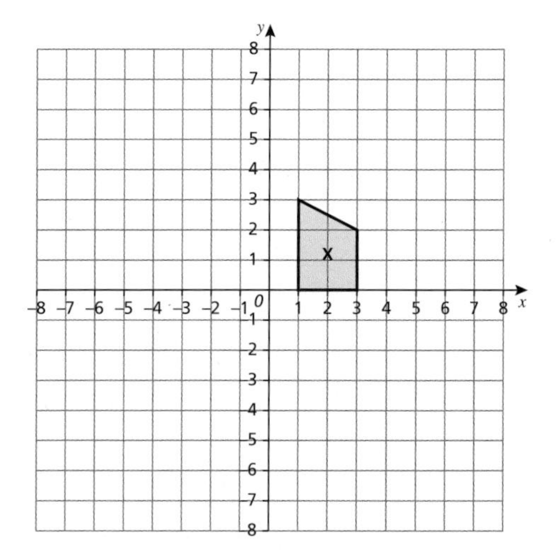

4 Copy the shape A onto a grid.

a Rotate shape A 90° clockwise about the origin.

Label the new shape B.

b Enlarge shape B by scale factor –2, centre (0, 0).

Label the new shape C.

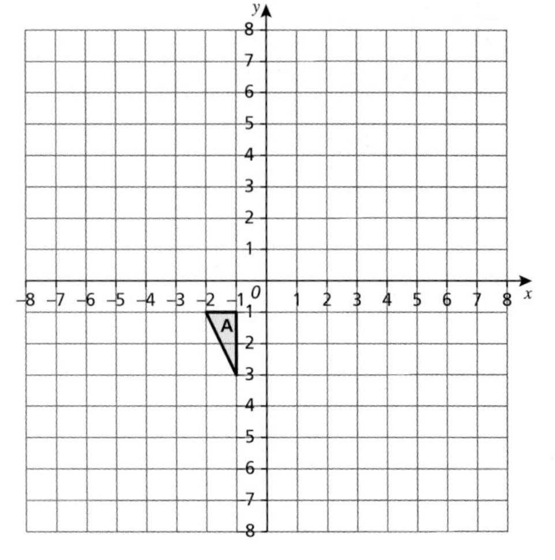

5 Shape K is reflected in the x-axis to give shape L.

Shape L is reflected in the y-axis to give shape M.

Describe **fully** the single transformation that maps shape K onto shape M.

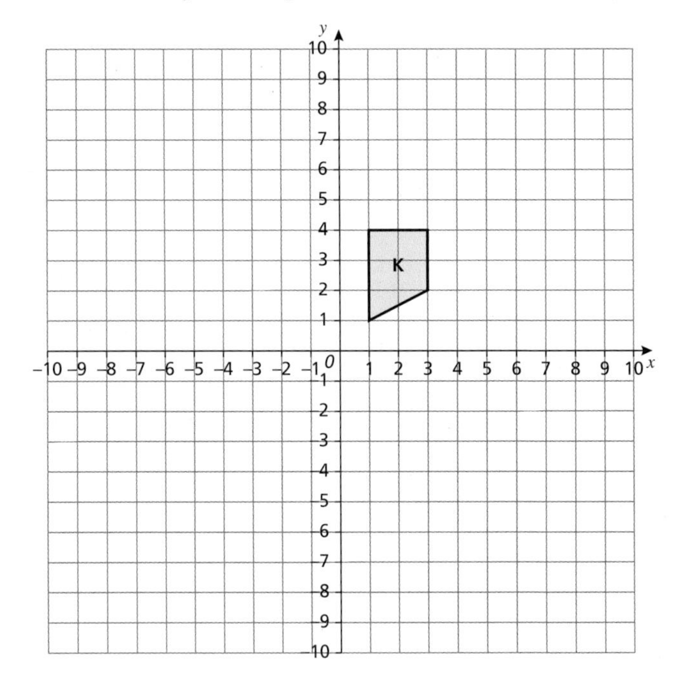

6 Triangle P is reflected in the y-axis to give triangle Q.

Triangle Q is then reflected in the line $y = -2$ to give triangle R.

Describe **fully** the single transformation that maps triangle P onto triangle R.

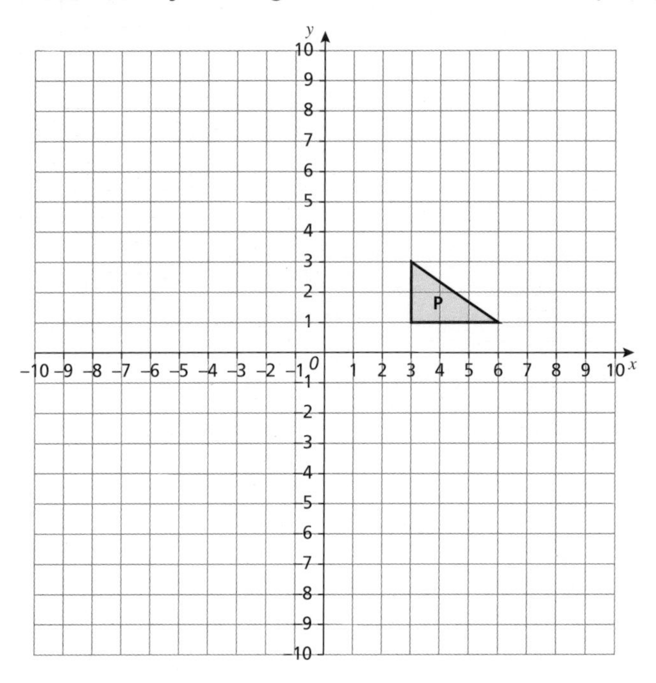

7 Triangle A is rotated 180° about (0, 0) to give triangle B.

Triangle B is translated by the vector $\begin{pmatrix} 6 \\ -2 \end{pmatrix}$ to give triangle C.

Describe **fully** the single transformation that maps triangle A onto triangle C.

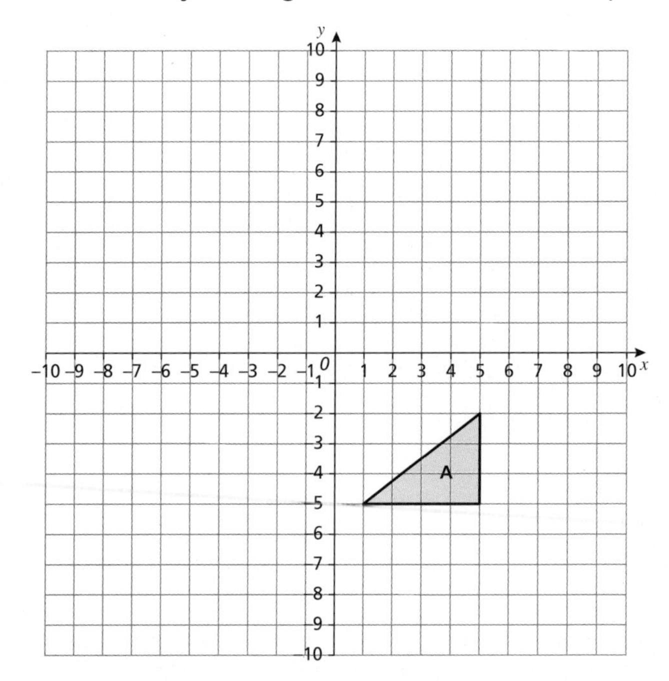

8 Shape A is rotated 90° clockwise about the point (0, –2) to give shape B.

Shape B is translated by the vector $\begin{pmatrix} -4 \\ 0 \end{pmatrix}$ to give shape C.

Describe **fully** the single transformation that maps shape A onto shape C.

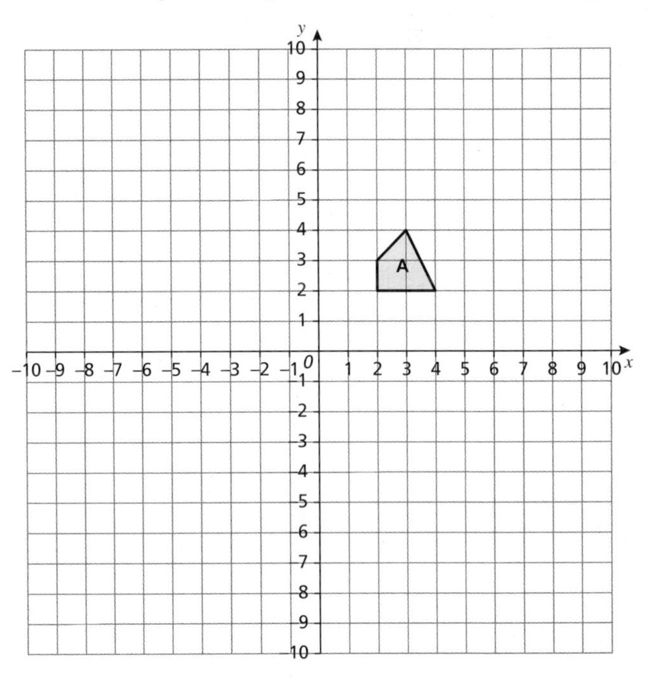

9 Shape A can be transformed to shape B by a reflection in the line $y = -1$, followed by a translation by the vector $\begin{pmatrix} p \\ q \end{pmatrix}$.

Work out the values of p and q.

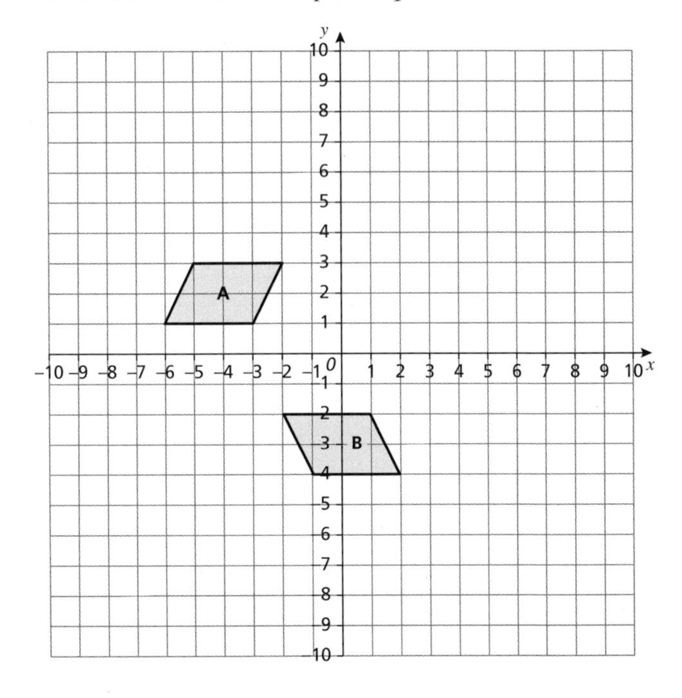

What do you think?

1. Draw a shape on a grid.

   a Rotate the shape 180° about a centre.

   b Enlarge the new shape, scale factor –1, using the centre of rotation from part **a** as the centre of enlargement.

   What do you notice?

2. Huda says, "When you carry out two transformations on a shape, it doesn't matter what order you do them in. The result will always be the same."

   Do you agree with Huda? Explain your answer.

3. Shape L is translated by the vector $\begin{pmatrix} 2 \\ -5 \end{pmatrix}$ to give shape M.

   Shape M is translated by the vector $\begin{pmatrix} -4 \\ 5 \end{pmatrix}$ to give shape N.

   Describe **fully** the single transformation that maps shape L onto shape N.

   What do you notice? Can you generalise your result?

Consolidate – do you need more?

1. **a** Rotate shape A 180° about the origin.

   Label the new shape B.

   b Reflect shape B in the x-axis.

   Label the new shape C.

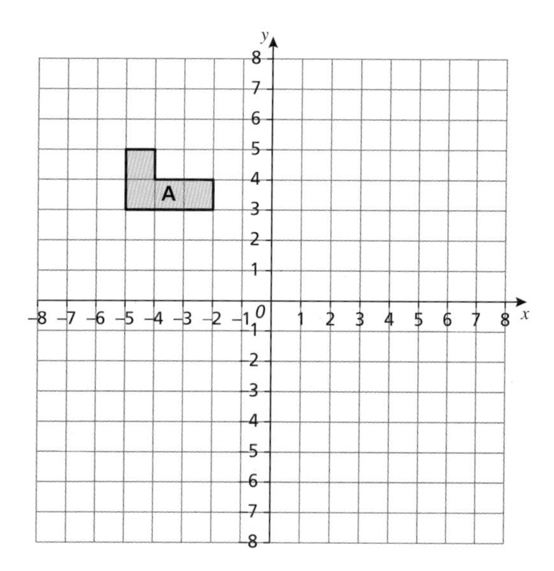

2 **a** Translate shape A by $\begin{pmatrix} 2 \\ -4 \end{pmatrix}$.

Label the new shape B.

b Rotate shape B 90° clockwise about the origin.

Label the new shape C.

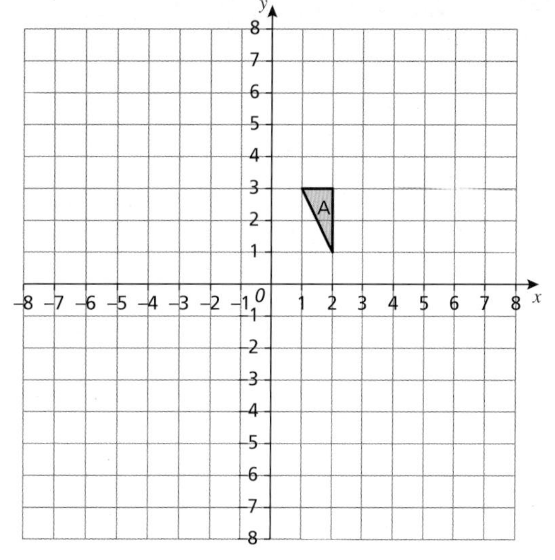

3 Describe **fully** the single transformation that maps shape A to shape B.

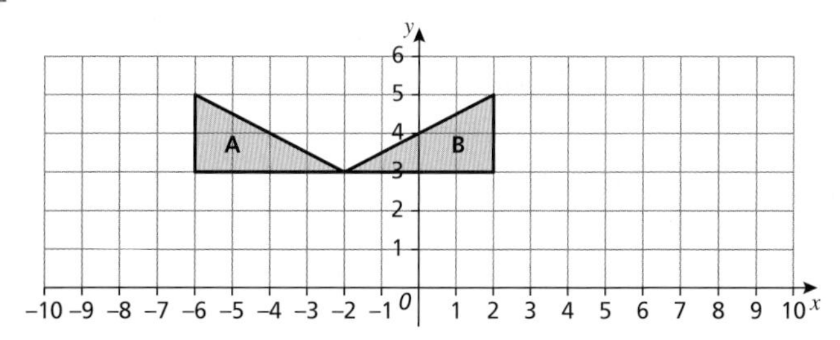

4 Describe **fully** the single transformation that maps shape X to shape Y.

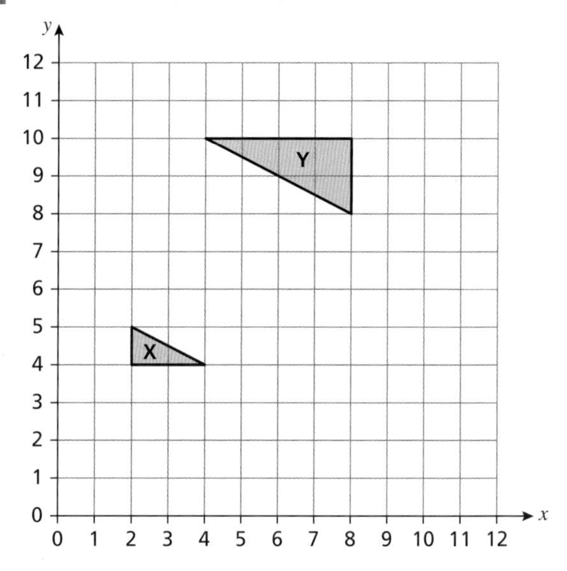

5 Describe **fully** the single transformation that maps shape P to shape Q.

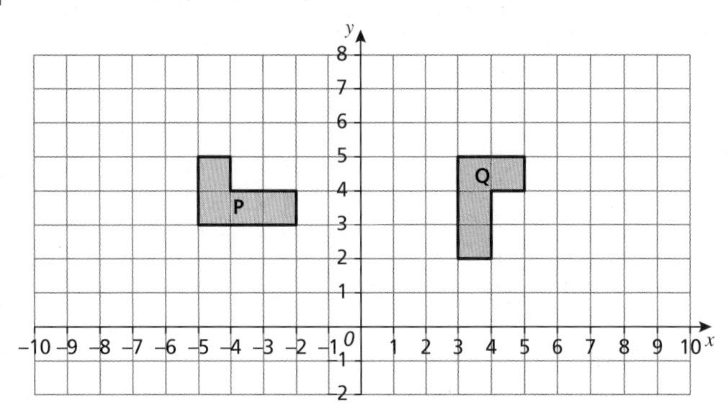

Stretch – can you deepen your learning?

1 Shape A is transformed to shape B by a reflection in the x-axis followed by another transformation.

a Describe **fully** the second transformation.

b Huda says, "Shape A could have been reflected in a vertical line, then rotated 180°."

Describe **fully** the two transformations that support Huda's statement.

Is there more than one solution?

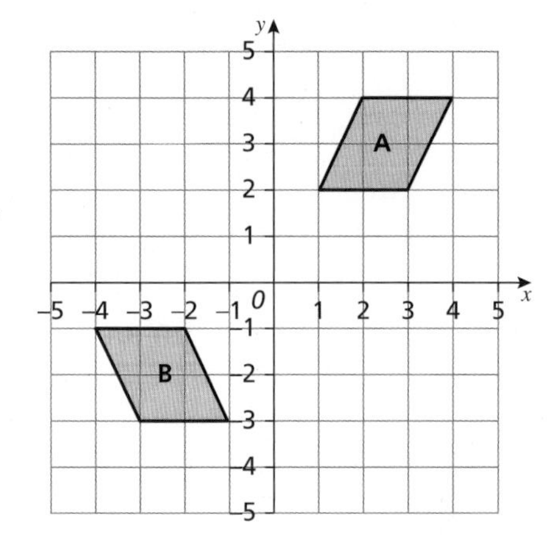

2 **a** Reflect the shape in the line $x = -1$

b Enlarge the new shape by scale factor $-\frac{1}{2}$, centre (0, 3).

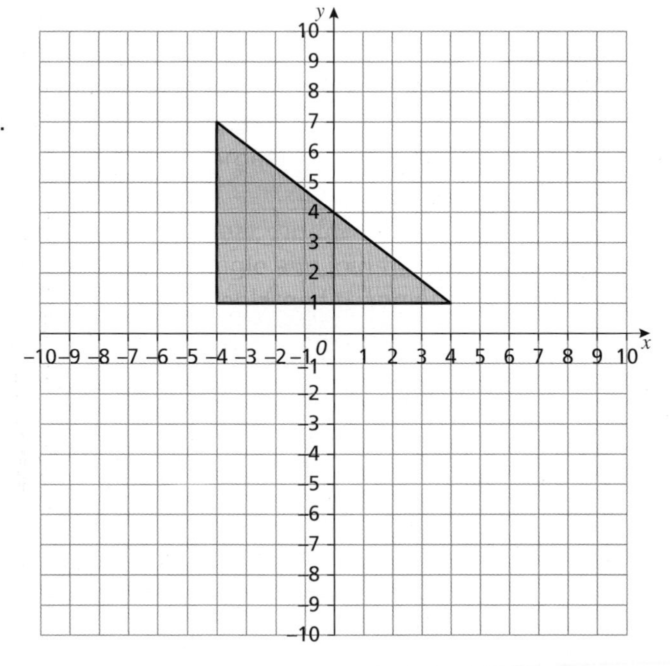

Transformations: exam practice

White Rose MATHS

4–6

1 Rotate the shape 90° anticlockwise about the point (2, 3) on a copy of the grid. **[2 marks]**

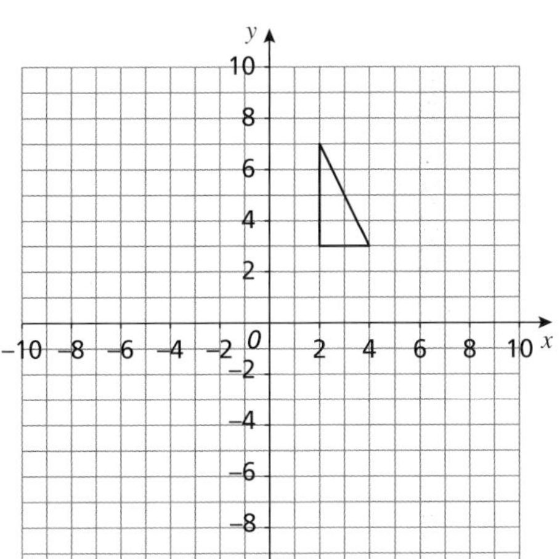

2 Reflect the shape in the line $y = x$ on a copy of the grid. **[3 marks]**

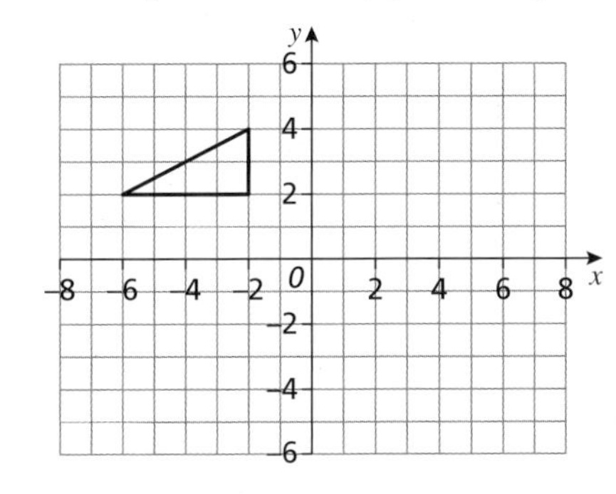

3 Describe **fully** the single transformation that maps triangle A to triangle B. **[3 marks]**

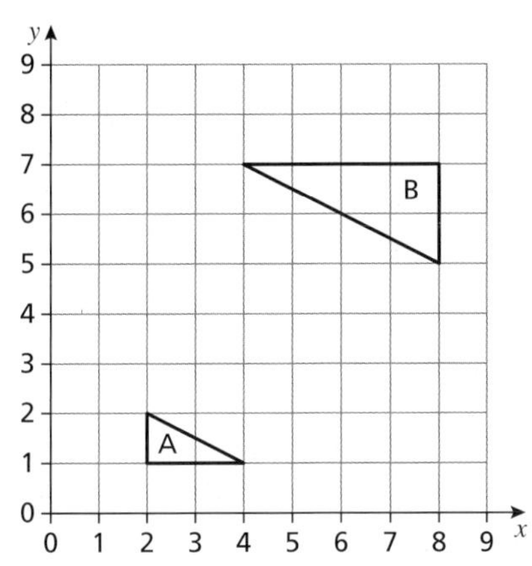

4–6

4 Part of a shape is drawn on the grid.

The dashed line is a line of symmetry of the shape.

Complete the shape on a copy of the grid and then rotate your completed shape through 270° anticlockwise about (0, 0). **[3 marks]**

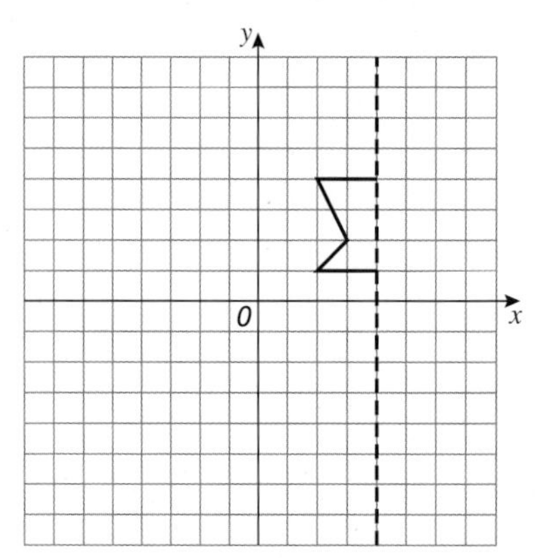

7–9

5 State a type of geometric transformation which produces an image that is **not** congruent to the original shape. **[1 mark]**

6 Write the coordinates of vertex B after the triangle is reflected in the line $x = -1$ and then translated by the vector $\begin{pmatrix} 2 \\ -3 \end{pmatrix}$. **[3 marks]**

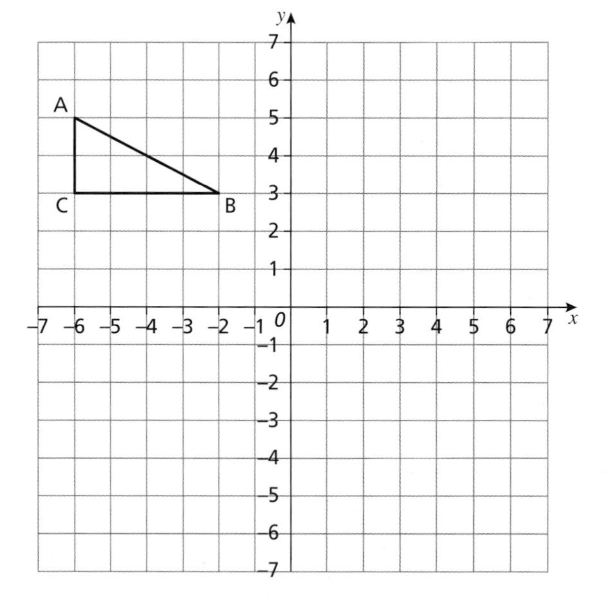

5 2D and 3D shapes

White Rose
MATHS

In this block, we will cover...

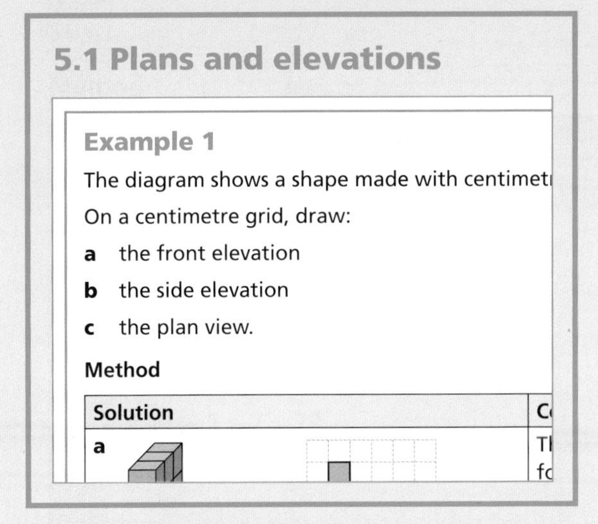
5.1 Plans and elevations
Example 1
The diagram shows a shape made with centimet
On a centimetre grid, draw:
a the front elevation
b the side elevation
c the plan view.
Method
Solution
a

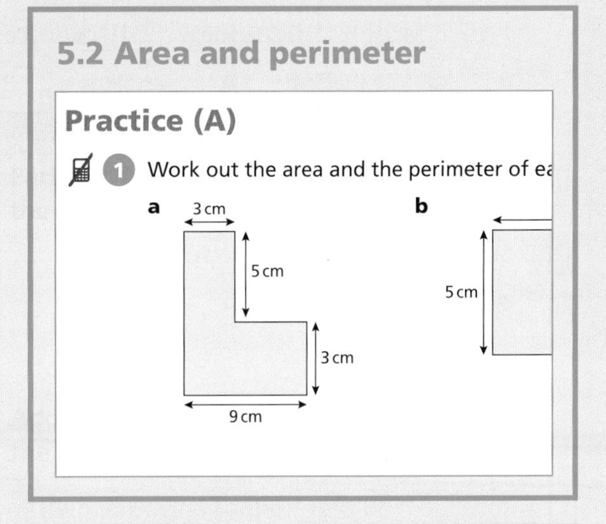
5.2 Area and perimeter
Practice (A)
1 Work out the area and the perimeter of ea
a
3 cm
5 cm
3 cm
9 cm
b
5 cm

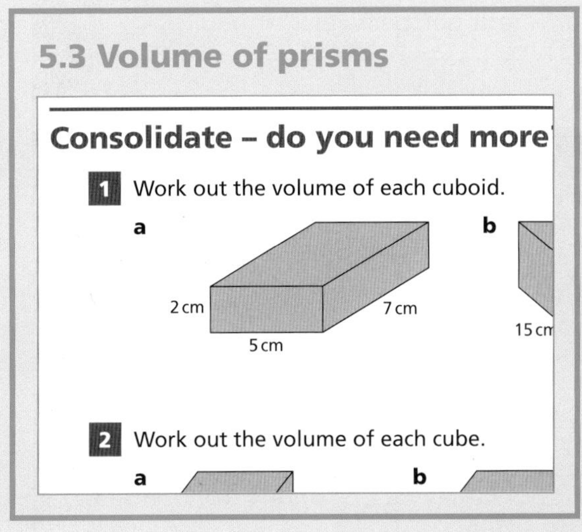
5.3 Volume of prisms
Consolidate – do you need more
1 Work out the volume of each cuboid.
a
2 cm
7 cm
5 cm
b
15 cm
2 Work out the volume of each cube.
a
b

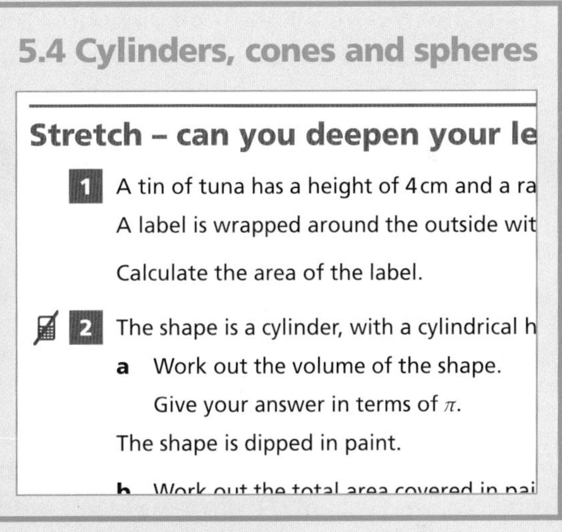
5.4 Cylinders, cones and spheres
Stretch – can you deepen your le
1 A tin of tuna has a height of 4 cm and a ra
A label is wrapped around the outside wit
Calculate the area of the label.
2 The shape is a cylinder, with a cylindrical h
a Work out the volume of the shape.
Give your answer in terms of π.
The shape is dipped in paint.

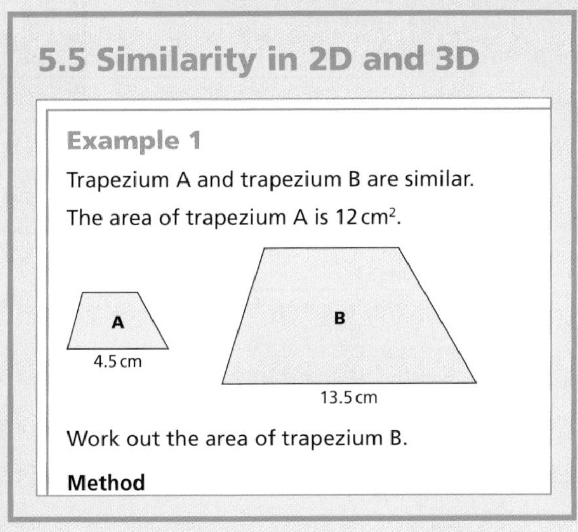
5.5 Similarity in 2D and 3D
Example 1
Trapezium A and trapezium B are similar.
The area of trapezium A is 12 cm².
A
4.5 cm
B
13.5 cm
Work out the area of trapezium B.
Method

5.1 Plans and elevations

Are you ready?

1 For each 3D shape, state the number of:

i faces **ii** edges **iii** vertices.

a

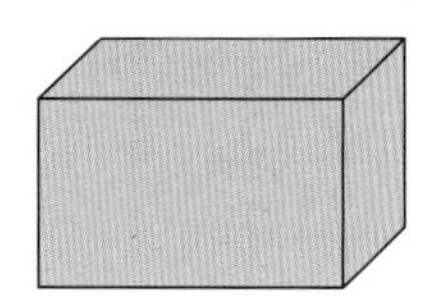

b

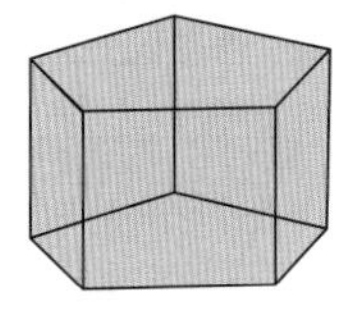

c

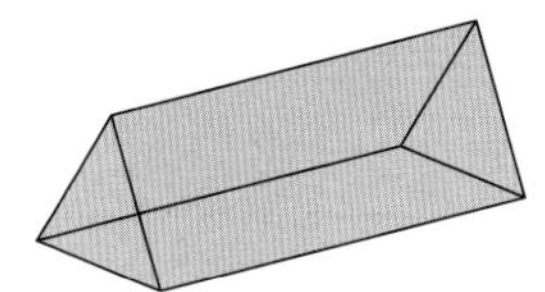

2 Sketch the net of each shape.

a

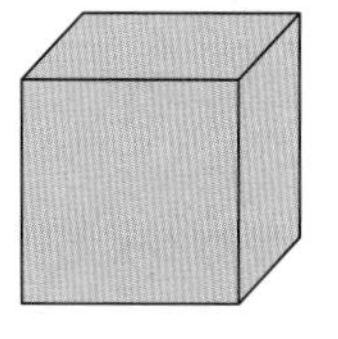

b

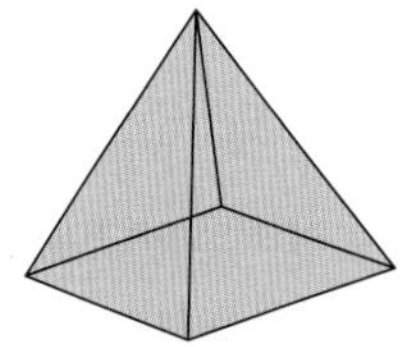

c

Plans and **elevations** show an object from three different perspectives.

Plan view

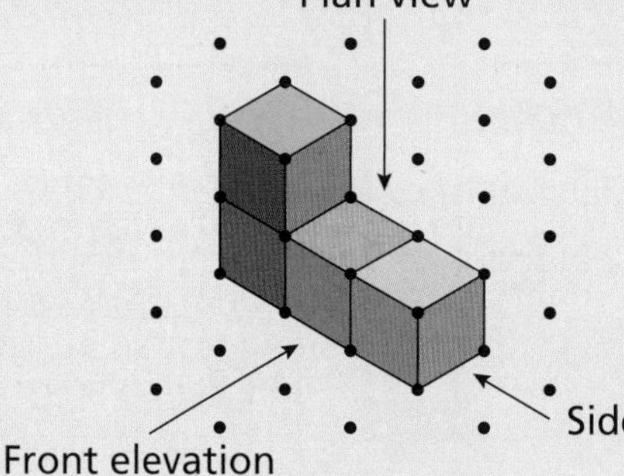

Side elevation

Front elevation

The **plan view** is the view when looking down on the object.

The **side elevation** is the view from one side.

The **front elevation** is the view from the front.

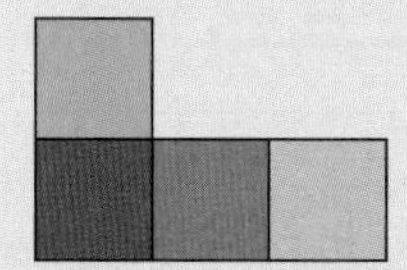

Example 1

The diagram shows a shape made with centimetre cubes.

On a centimetre grid, draw:

a the front elevation

b the side elevation

c the plan view.

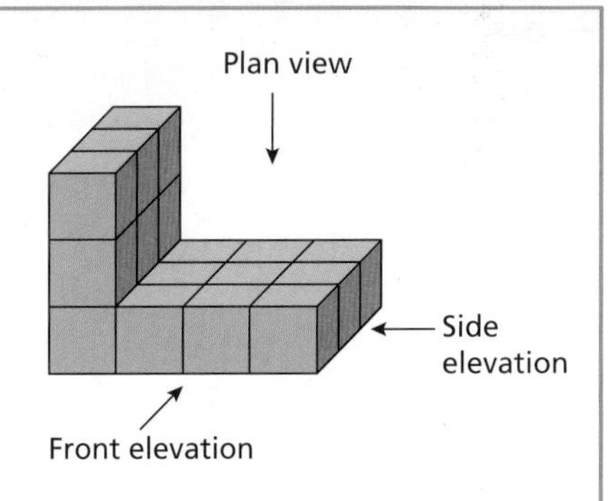

Method

Solution	Commentary
a	The front elevation is determined by the four cubes at the base and the cubes on the left-hand side.
b	For the side elevation, show the internal line that connects the two parts of the view.
c	Likewise for the plan view.

Example 2

The diagram shows the front elevation, the side elevation and a plan view of a 3D shape drawn on a centimetre grid.

Front Side Plan

Sketch the 3D shape. Give the dimensions of the shape on your sketch.

Method

Solution	Commentary
	From the top and front, you can see two identical rectangles. From the side, you can see a square. The shape to sketch will be a cuboid.
2 cm, 4 cm, 2 cm	From the grid, you can see that the length is 4 cm, the height is 2 cm and the depth is 2 cm. Show these dimensions on your sketch.

Practice

1. The diagrams show shapes made from 1 cm^3 cubes.

   For each shape, draw the plan view, front elevation and side elevation on a centimetre grid.

   a 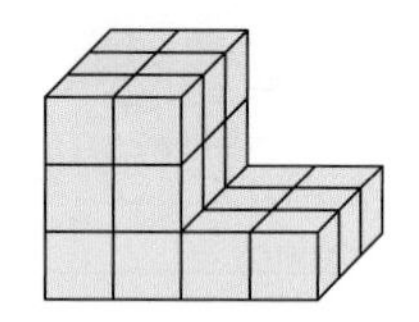

   b 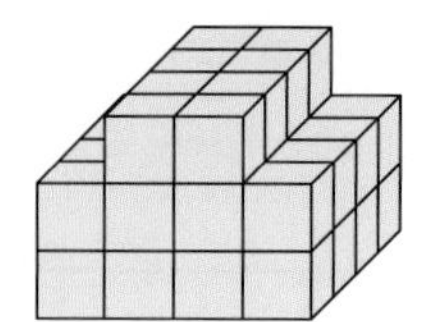

   c 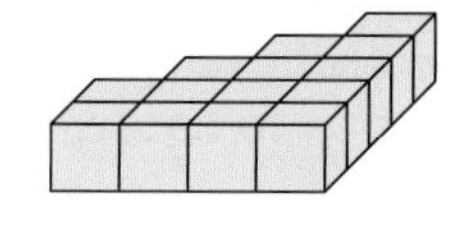

2. For each shape, draw the plan view, front elevation and side elevation on a centimetre grid.

   a 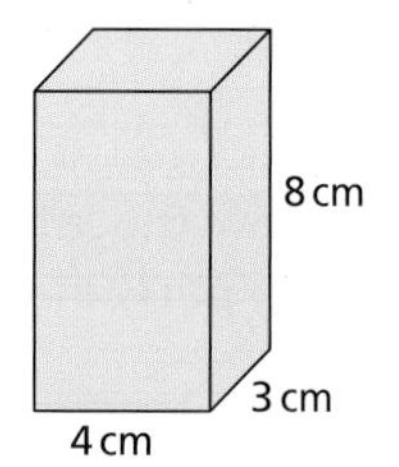

   b 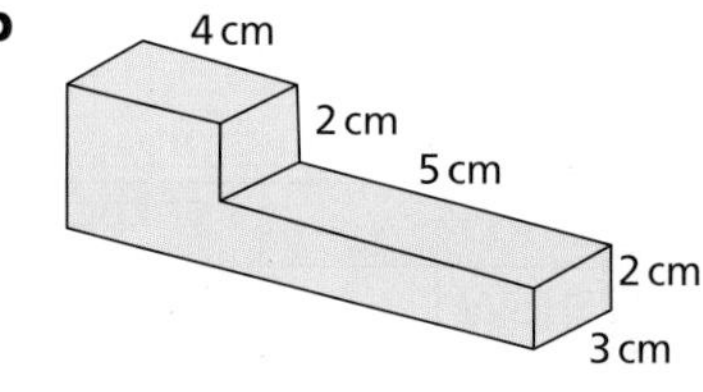

3. For each shape, draw the plan view, front elevation and side elevation on a centimetre grid.

   a 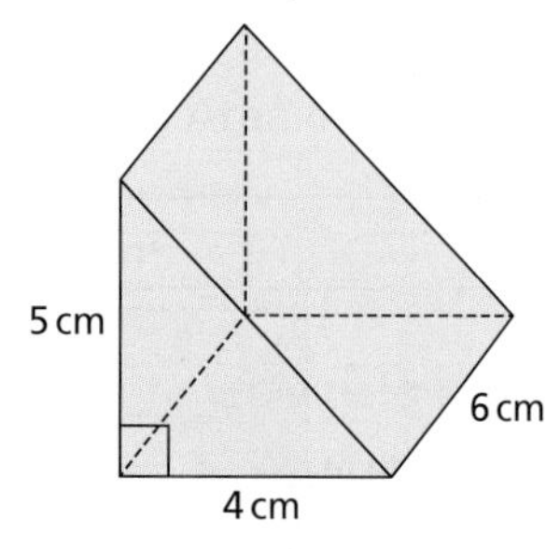

   b 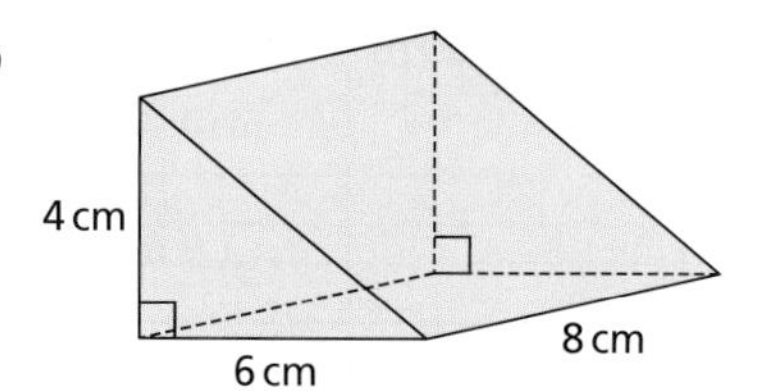

   c 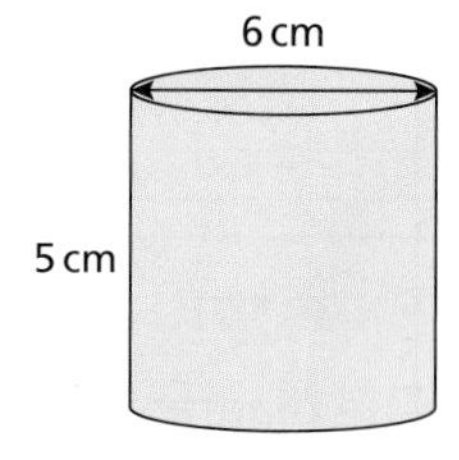

4. Name one possible 3D shape with the given elevations.

   a Plan view = rectangle   Front elevation = rectangle   Side elevation = rectangle

   b Plan view = circle   Front elevation = rectangle   Side elevation = rectangle

   c Plan view = square   Front elevation = triangle   Side elevation = triangle

   Compare your answers with a partner.

5 The diagrams show the plan view, front elevation and side elevation of 3D shapes on a centimetre grid.

Sketch each 3D shape, labelling its dimensions.

a

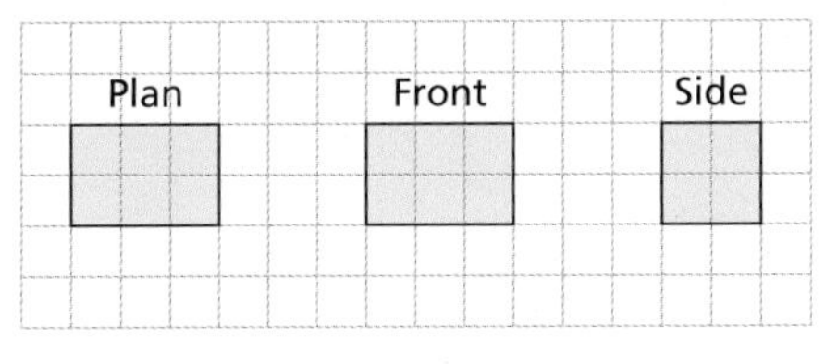

b

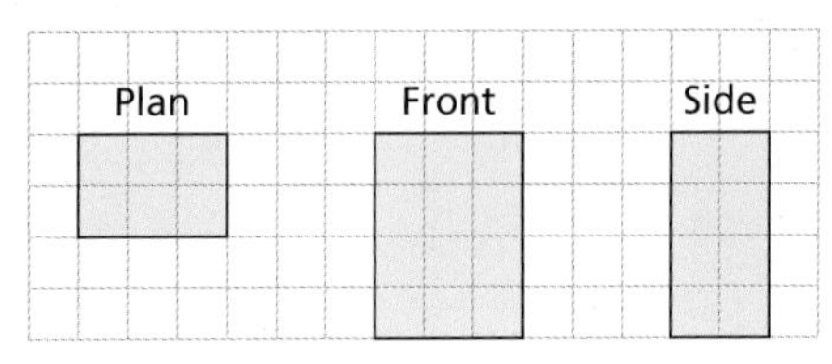

c

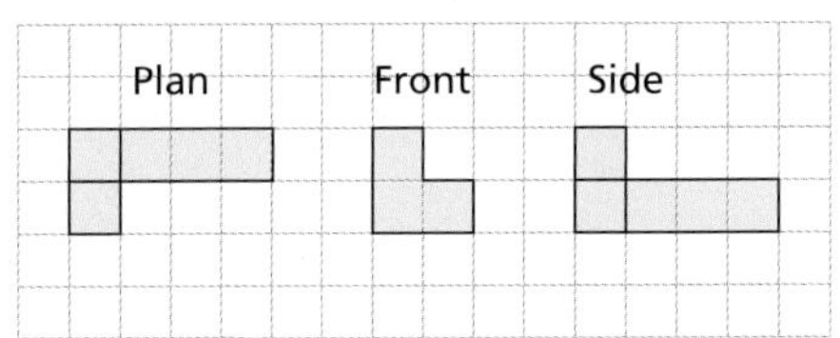

d

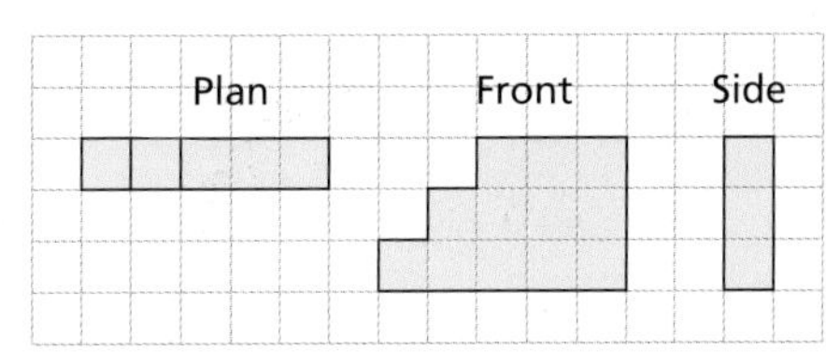

6 The diagrams show the plan view, front elevation and side elevation of 3D shapes.

Sketch each 3D shape.

a Plan Front Side

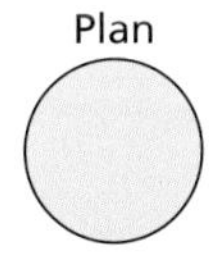

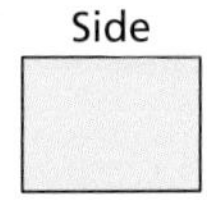

b Plan Front Side

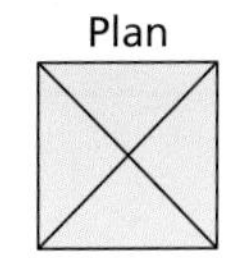

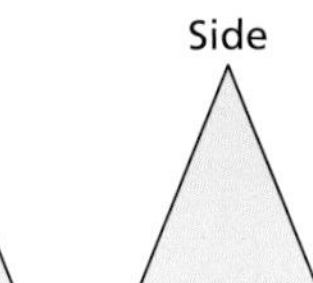

c Plan Front Side

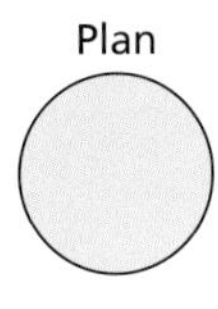

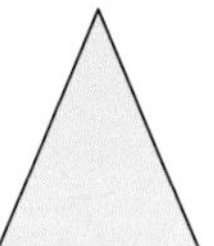

What do you think?

1. This shape is made from 12 cubes. Three cubes are not visible on the diagram.

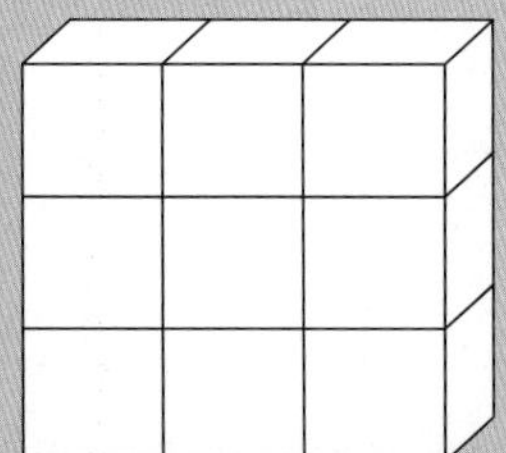

Draw a possible plan view, front elevation and side elevation of the shape.

Compare your answer with a partner's.

2. Here is the cuboid ABCDEFGH.

The shape is rotated 90° clockwise along the length EH.

On a centimetre grid, draw the plan view, front elevation and side elevation of the cuboid in its new orientation.

Investigate different plans and elevations formed by other rotations of the cuboid.

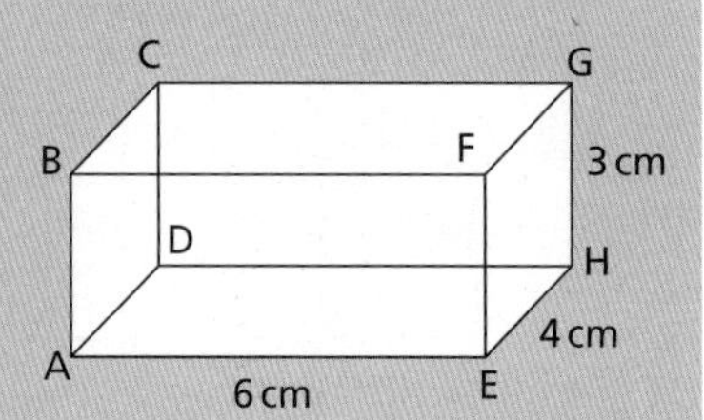

Consolidate – do you need more?

1. The plan view, front elevation and side elevation of a 3D shape are drawn on a centimetre grid.

Sketch the 3D shape, labelling its dimensions.

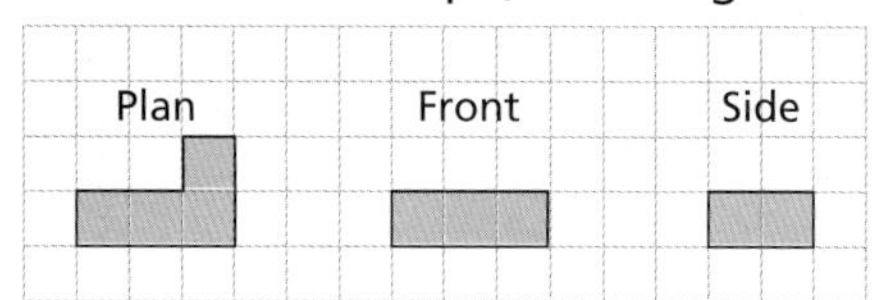

2. The diagrams show the plan view, front elevation and side elevation of a 3D shape.

Sketch the 3D shape.

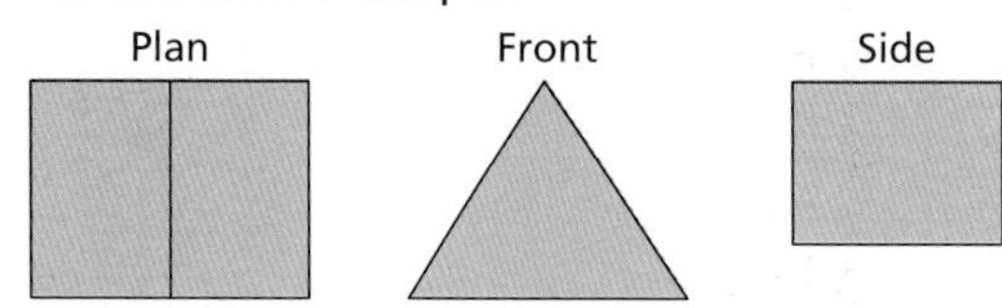

3. The plan view, front elevation and side elevation of a 3D shape are drawn on a centimetre grid.

Sketch the 3D shape, labelling its dimensions.

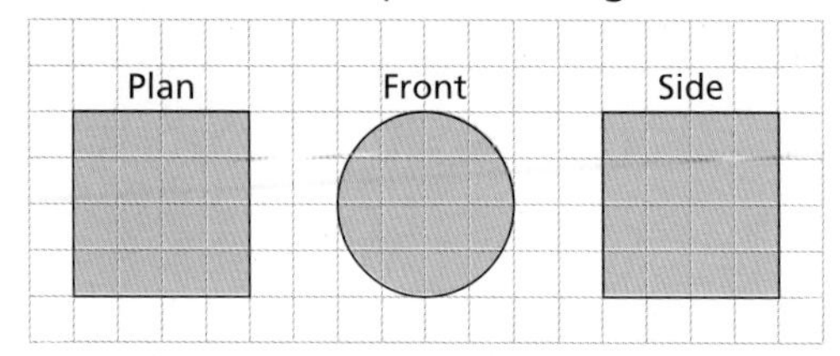

4 For each shape, draw the plan view, front elevation and side elevation on a centimetre grid.

a

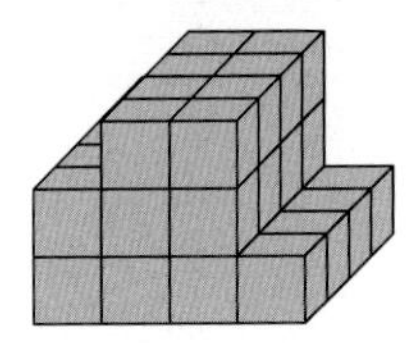

b

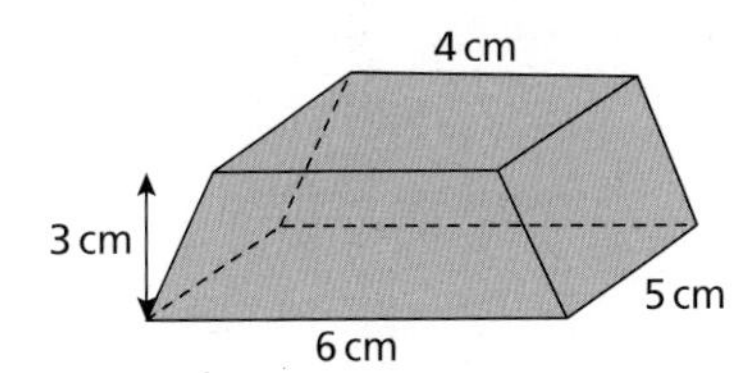

Stretch – can you deepen your learning?

1 The plan view, front elevation and side elevation of the replica of a monument are shown.

2 cm on the replica represents 1 m in real life.

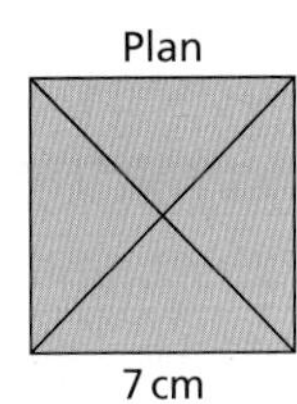

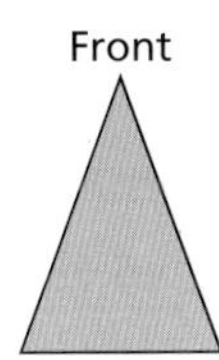

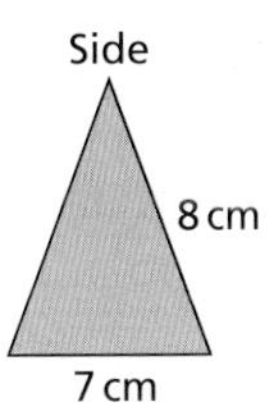

Work out the height of the actual monument.

2 The plan view, front elevation and side elevation of two 3D shapes are drawn on centimetre grids.

Work out the exact volume of each shape.

a

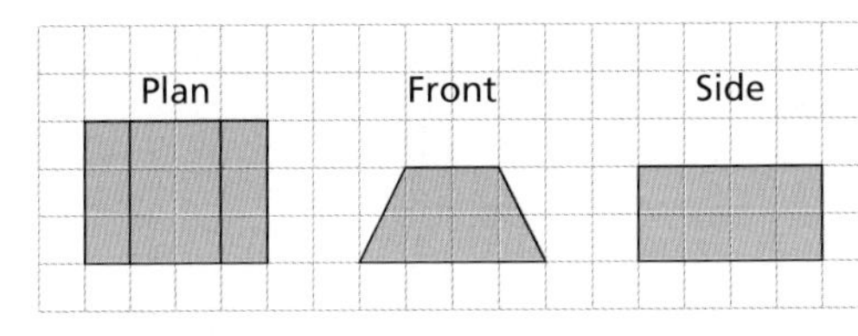

b

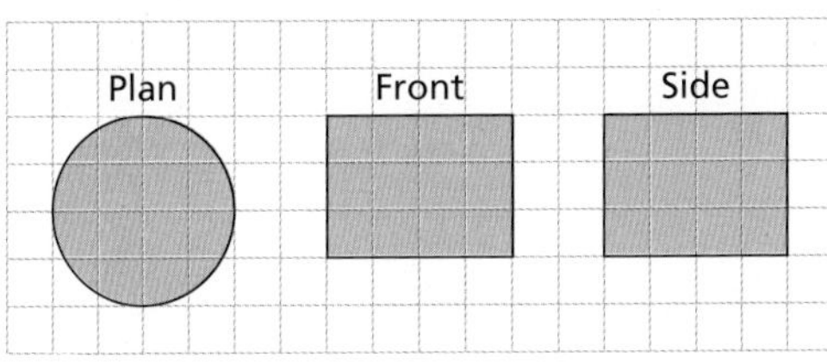

5.2 Area and perimeter

White Rose MATHS

Are you ready? (A)

1 Work out the area and perimeter of each shape.

a

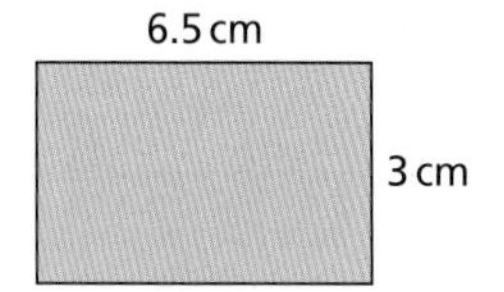

b

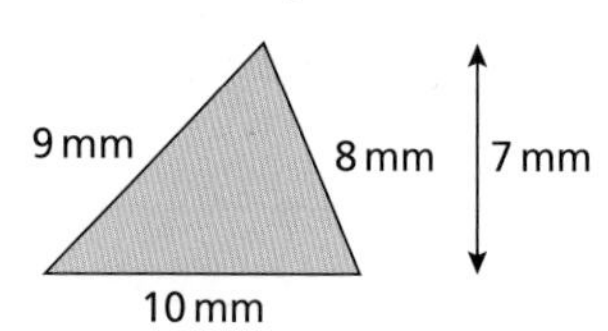

2 Work out the area and circumference of each circle.

a

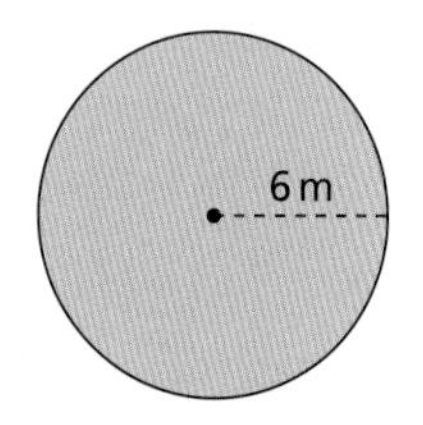

b

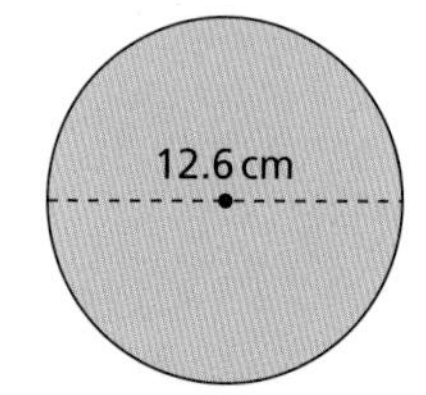

Area measures the space inside a 2D shape. Area is measured in squared units such as cm^2, mm^2 and m^2. You should know the formulae for working out A, the area of these shapes.

Rectangle

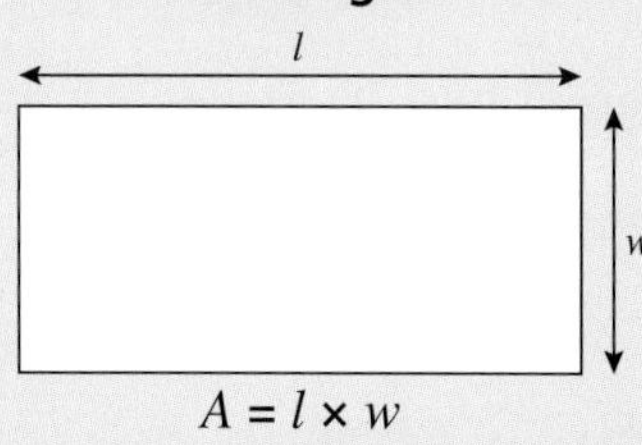

$A = l \times w$
$= lw$

Parallelogram

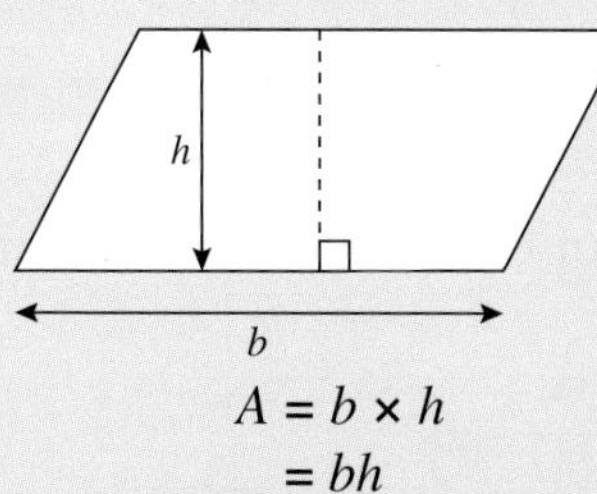

$A = b \times h$
$= bh$

Triangle

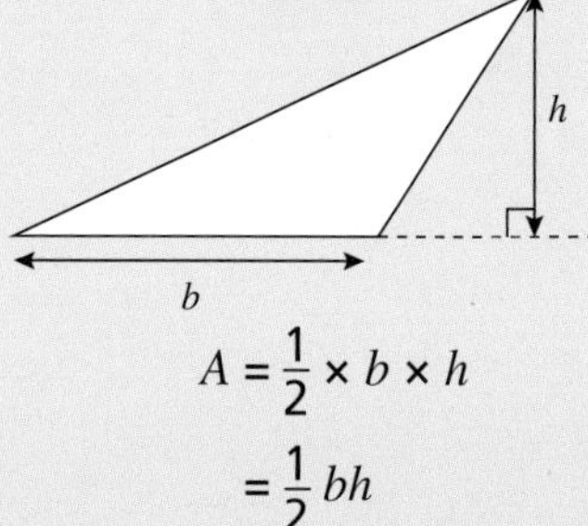

$A = \frac{1}{2} \times b \times h$
$= \frac{1}{2} bh$

Trapezium

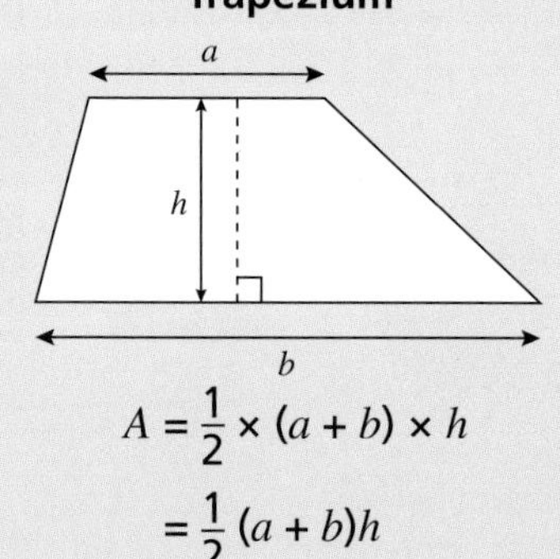

$A = \frac{1}{2} \times (a + b) \times h$
$= \frac{1}{2} (a + b)h$

Notice that h stands for the perpendicular height, which is at right angles to the base of the shape. The area, A, of a circle is given by the formula $A = \pi r^2$, where r is the radius of the circle.

To find the area of a **compound shape**, which is made by joining other shapes together, use addition or subtraction as appropriate.

Be careful not to confuse $A = \pi r^2$ with $C = \pi d$, which gives the circumference of a circle of diameter d. This can also be written as $C = 2\pi r$, giving the circumference in terms of a circle's radius, r.

Example

a Work out the area of this shape.

b Work out the perimeter.

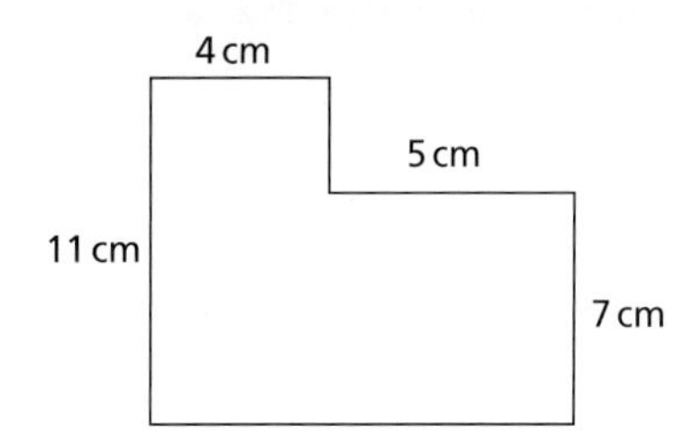

Method

Solution	Commentary
a	Partition the shape into two rectangles, then work out the area of each rectangle.
$4 \times 11 + 5 \times 7 = 79$ Area = $79\,\text{cm}^2$	Add the two areas to find the total area of the shape. The compound shape could be partitioned in a different way, giving the same answer. See if you can find this alternative method.
b	First work out the missing side lengths.
$4\,\text{cm} + 5\,\text{cm} = 9\,\text{cm}$ $11\,\text{cm} - 7\,\text{cm} = 4\,\text{cm}$ $11 + 4 + 4 + 5 + 7 + 9 = 40$ Perimeter = 40 cm	The total horizontal length of the shape must be the sum of 4 cm and 5 cm (9 cm). The total height is 11 cm. The unknown vertical length can be calculated by subtracting 7 cm from 11 cm. To work out the perimeter, calculate the sum of all six side lengths.

Practice (A)

1 Work out the area and the perimeter of each shape.

a

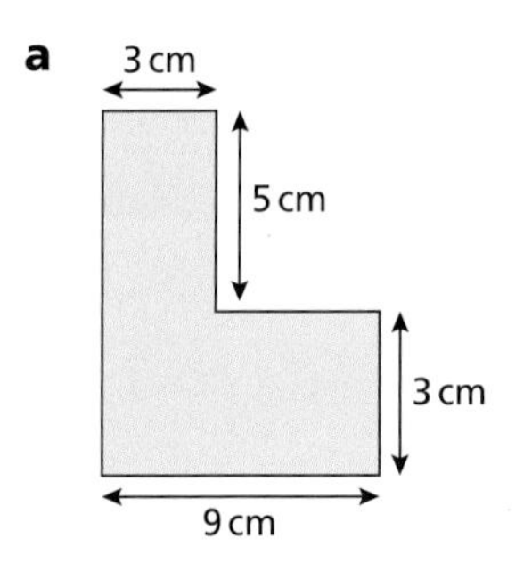

b

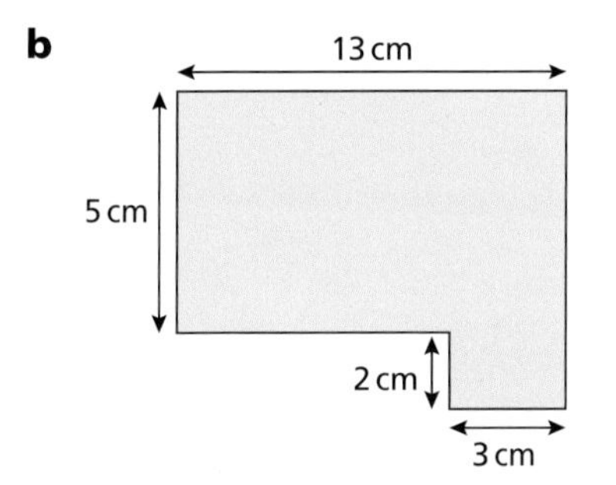

c

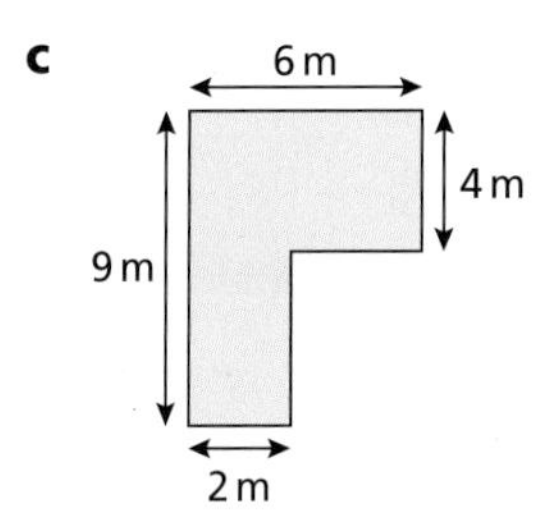

2 Work out the area of the shape.

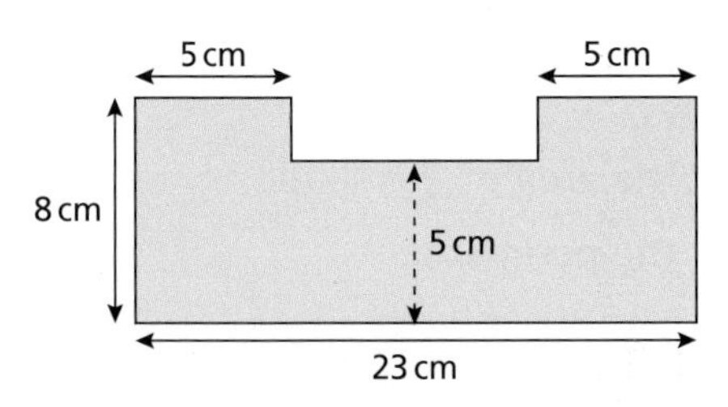

3 Work out the perimeter of the shape.

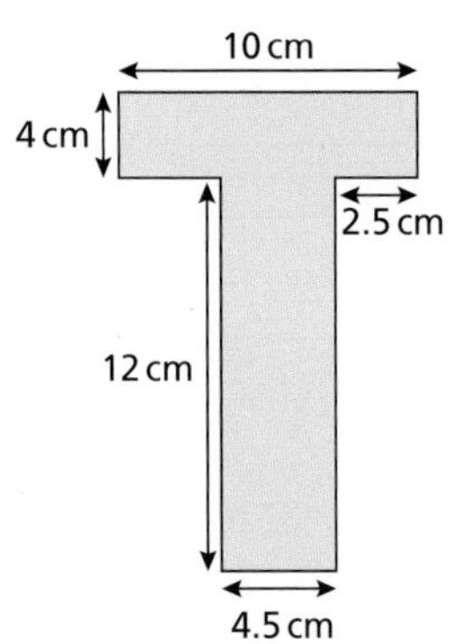

4 Work out the shaded areas.

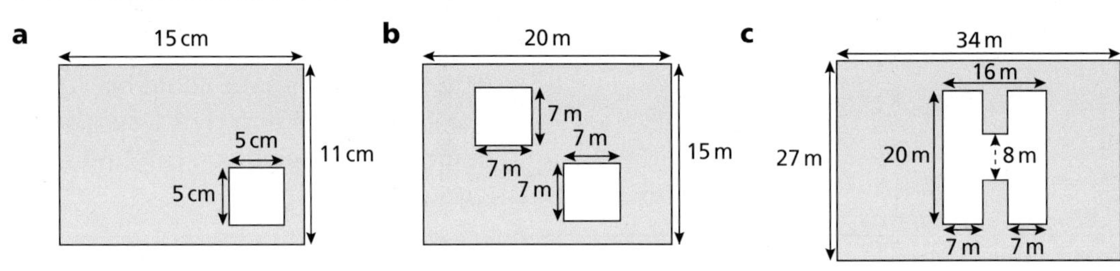

5 Work out the area of each shape.

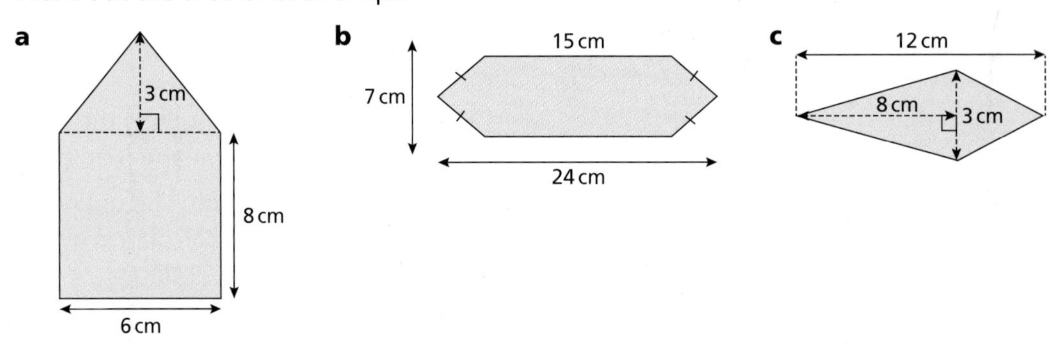

6 Work out the perimeter of the shape, giving your answer to 1 decimal place.

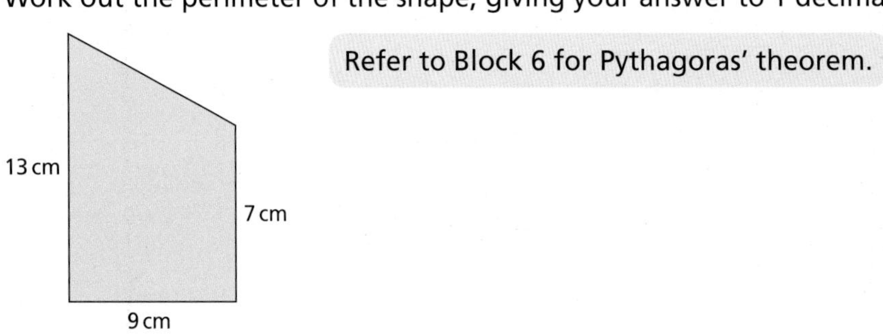

Refer to Block 6 for Pythagoras' theorem.

7 Work out the area of each shape.

a

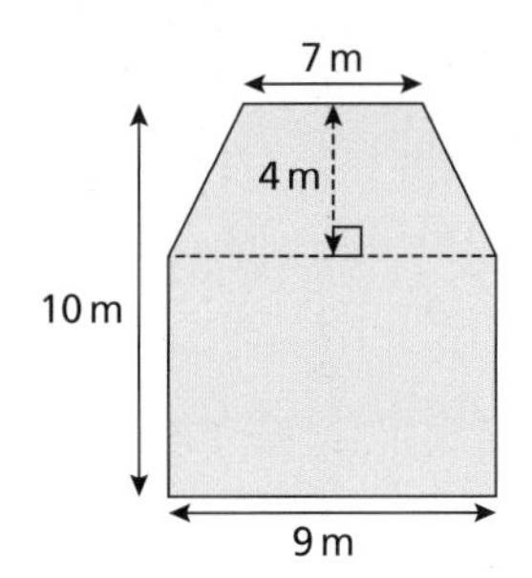

b

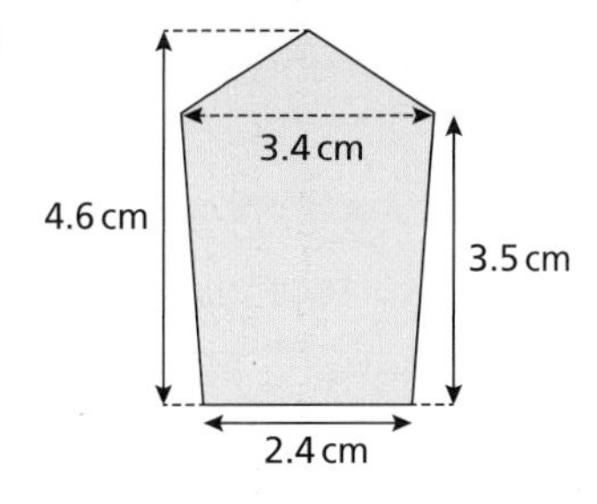

8 The shape is formed from a rectangle and a semicircle.

a Work out the area of the shape, giving your answer to 1 decimal place.

b Work out the perimeter of the shape, giving your answer to 1 decimal place.

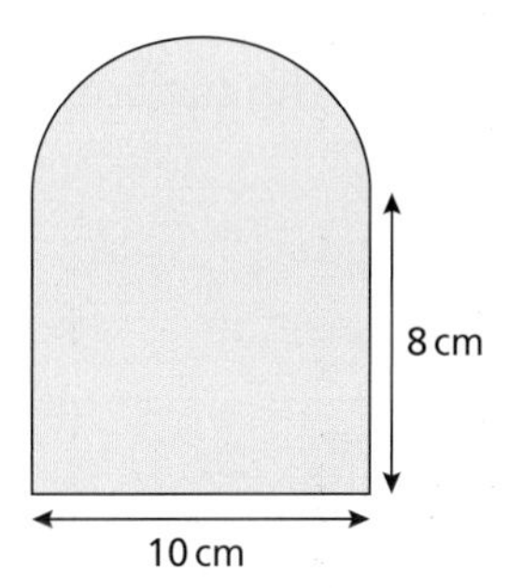

9 A company logo is formed from a triangle and a quarter of a circle, as shown.

a Work out the area of the shape, giving your answer in terms of π.

b Work out the perimeter of the shape, giving your answer in terms of π.

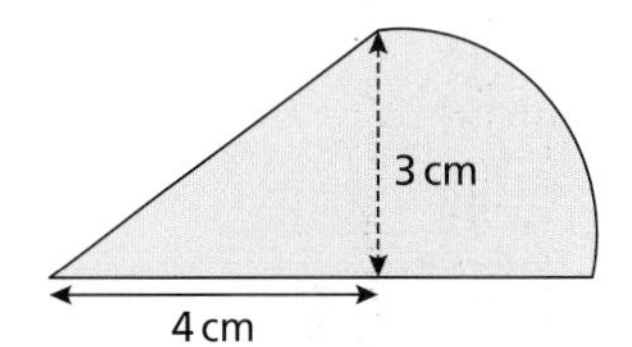

What do you think? (A)

1 A perfect shape is a shape that has a perimeter and area which are numerically equal.

A square with a side of 4 units is an example of a perfect shape as it has a perimeter of 16 units and an area of 16 units^2.

Can you find any other examples of perfect shapes?

2 Huda says, "It is impossible for a shape to have a perimeter numerically greater than its area."

Do you agree? Explain your answer.

3 A shape is formed by placing two overlapping, congruent rectangles.

Will the perimeter of the shape always be the same regardless of where the rectangles overlap?

Will the area of the shape always be the same regardless of where the rectangles overlap?

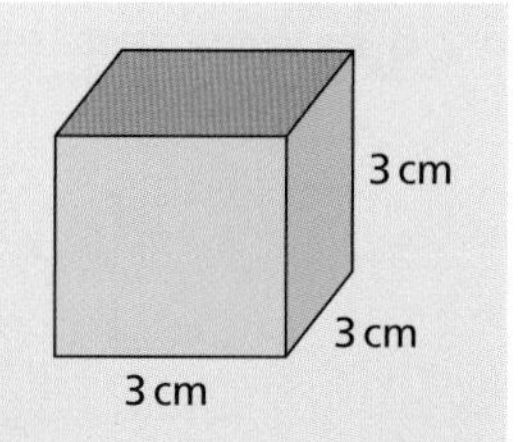

The **surface area** of a 3D shape is the sum of the areas of all the faces.

For example, each of the faces of the cube shown has a base and height of 3 cm.

Therefore each face has an area of $3 \times 3 = 9\,\text{cm}^2$

As the cube has six faces, the total surface area is $6 \times 9 = 54\,\text{cm}^2$

It is sometimes useful to draw the net of a 3D shape to work out the area of each face.

Example

Work out the surface area of the cuboid.

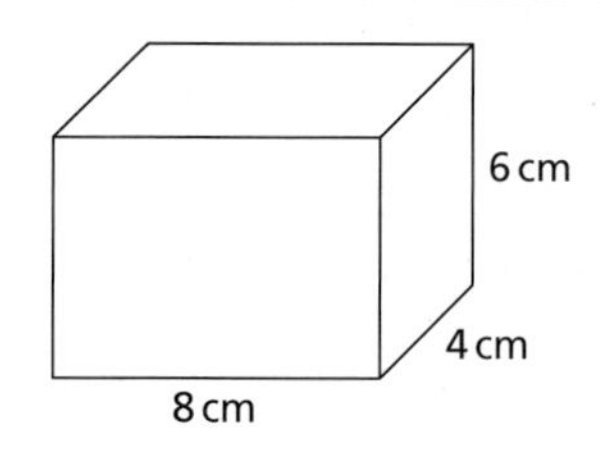

Method

Solution	Commentary
Area of front = Area of back $= 8 \times 6 = 48$ $48 \times 2 = 96\,\text{cm}^2$	Work out the area of the rectangular faces. A cuboid has three pairs of identical faces, and you can work out these areas in pairs.
Area of sides $= 4 \times 6 = 24$ $24 \times 2 = 48\,\text{cm}^2$	
Area of top = Area of bottom $= 8 \times 4 = 32\,\text{cm}^2$ $32 \times 2 = 64\,\text{cm}^2$	
Total surface area $= 96 + 48 + 64 = 208\,\text{cm}^2$	Find the sum of the areas of all six faces.

Practice (B)

1. Work out the surface area of the cube.

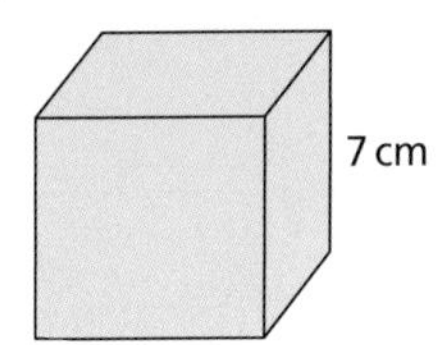

2. Work out the surface area of each cuboid.

a

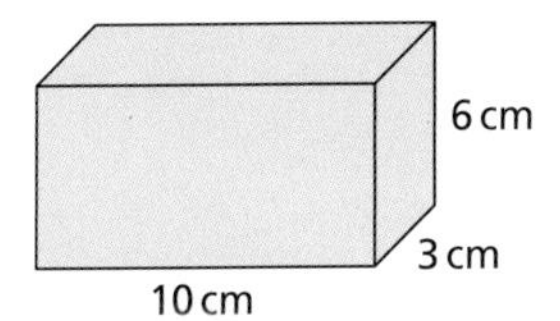

b

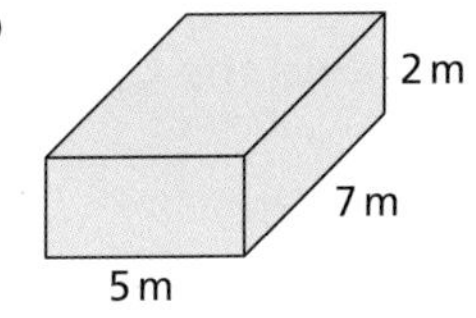

c

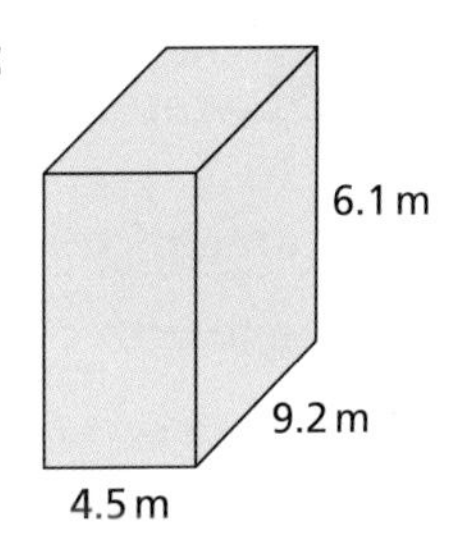

3. Work out the surface area of the triangular prism.

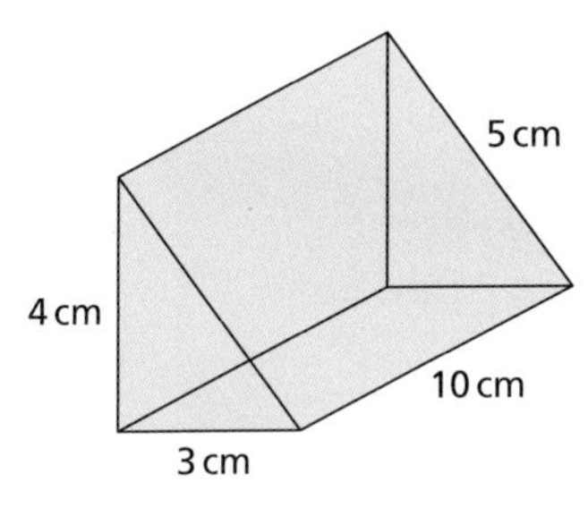

4. Work out the surface area of the shape.

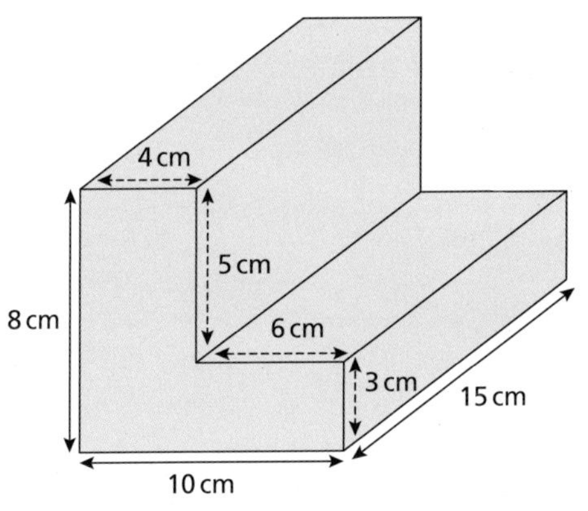

5. Work out the surface area of the prism.

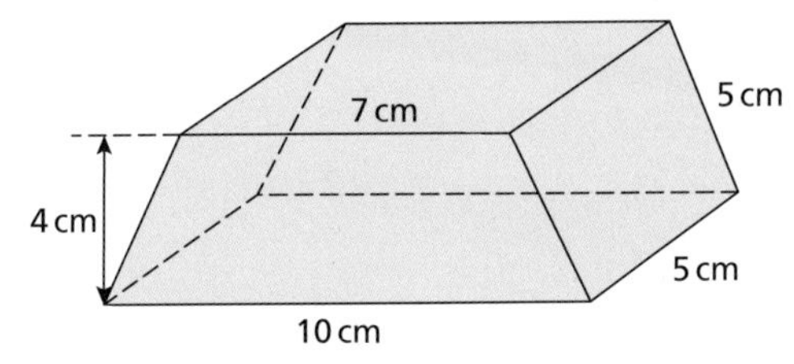

6. The cube has a surface area of $726\,m^2$.

   Work out the length of x.

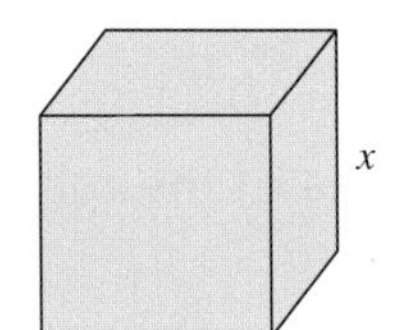

7. This cuboid is to be wrapped in paper.

   Wrapping paper costs £1.20 per sheet and can only be bought in whole sheets. Each sheet covers $1\,m^2$.

   How much will it cost to wrap the cuboid? Assume no overlap of wrapping paper around the cuboid.

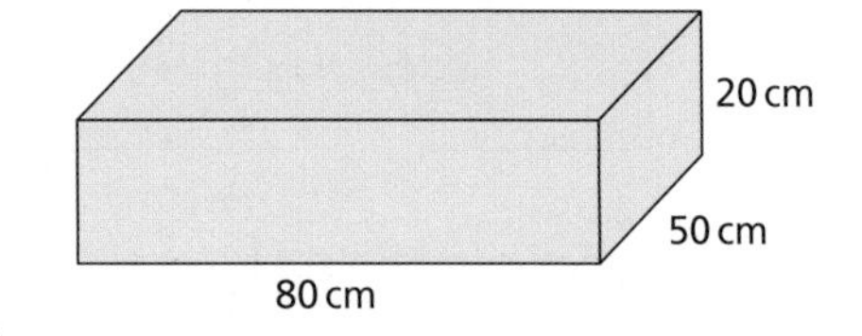

8. The surface area of this cuboid is $108\,m^2$.

   Work out the length of x.

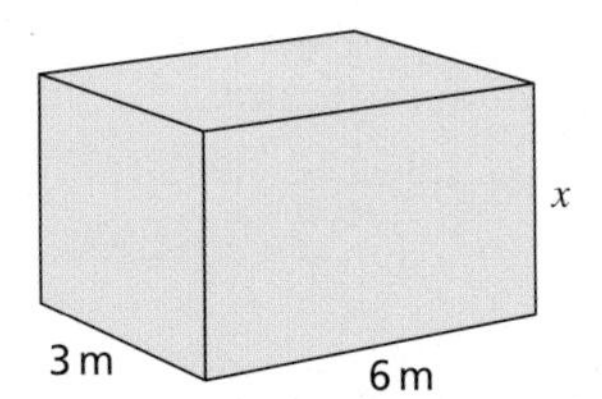

What do you think? (B)

1. Ed says, "If you place two identical cuboids on top of each other, the surface area will be twice the size of the surface area of one of the cuboids."

   Do you agree with Ed? Explain your answer.

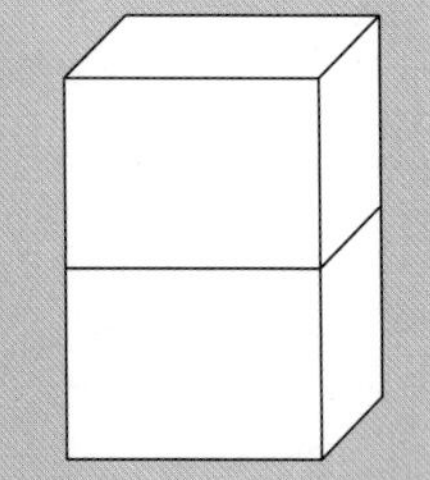

2. Mario has 10 identical cubes. He makes a shape using all of these cubes.

   What would a shape with the smallest possible surface area look like?

   What would a shape with the largest possible surface area look like?

Consolidate – do you need more?

1 Work out the area and perimeter of each shape.

a

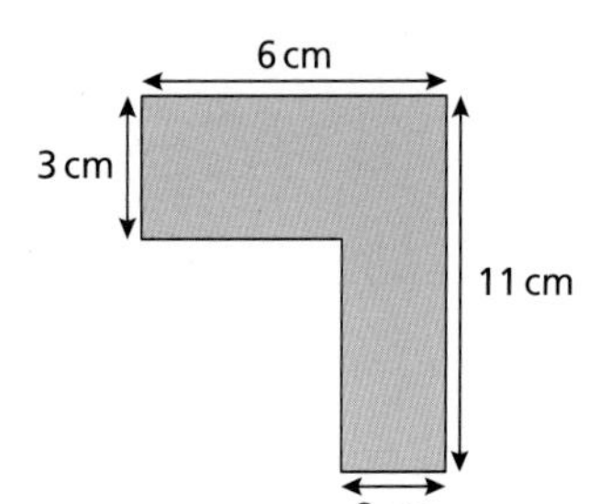

b

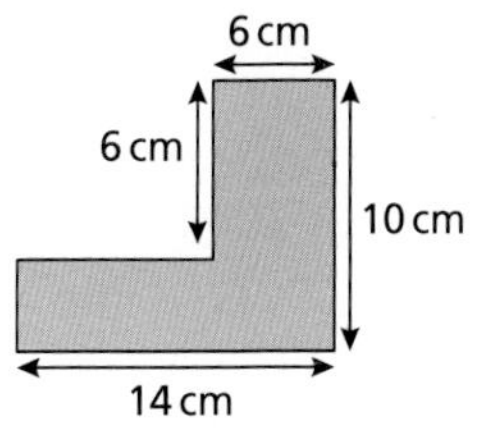

c

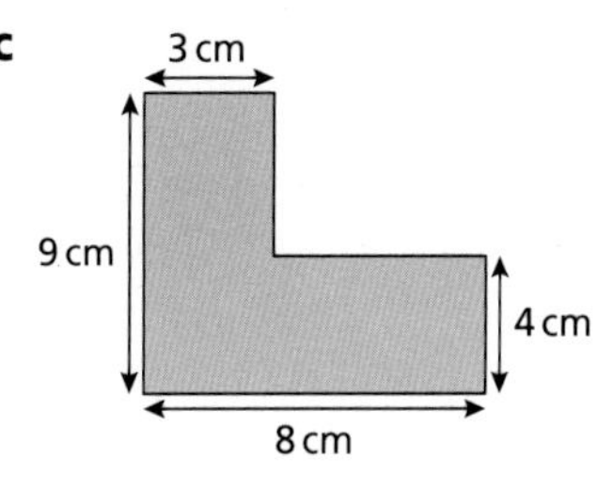

2 Work out the area of each shape.

a

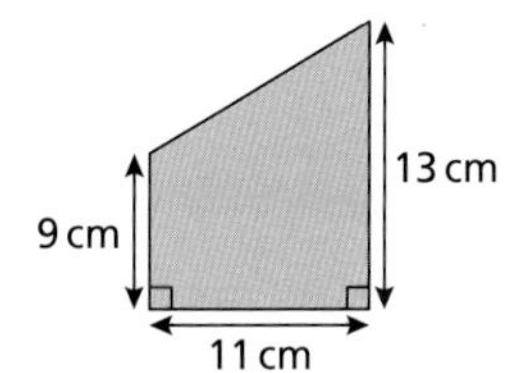

b

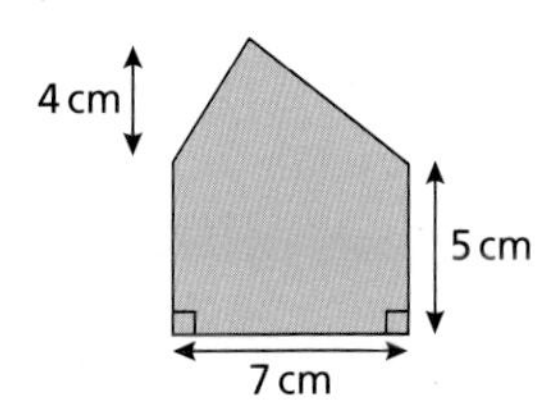

c

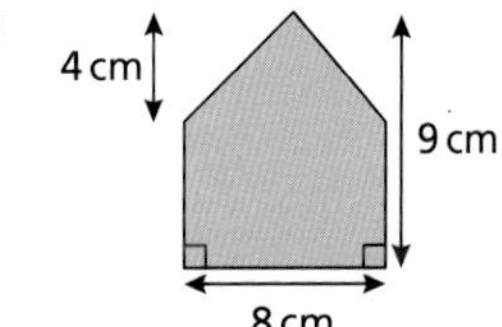

3 Work out the surface area of each cuboid.

a

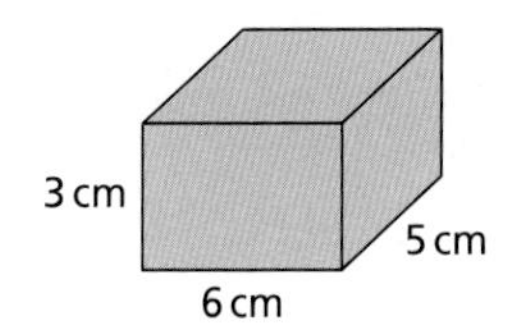

b

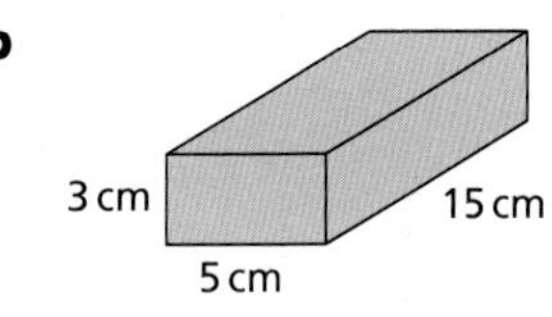

c

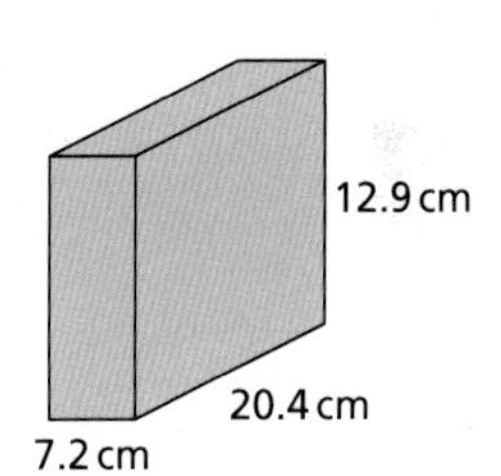

Stretch – can you deepen your learning?

1 The shape has a perimeter of 40 m.

Work out the area of the shape.

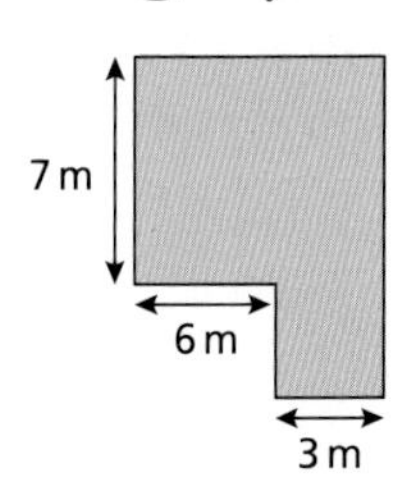

2 A regular hexagon has a side length of 3 cm.

Work out the area of the hexagon. Give your answer in the form $\frac{a\sqrt{b}}{c}$, where a, b and c are integers.

3 The diagram shows a rectangle.

Write and solve an equation to work out the value of x.

$(2x + 3)$ cm

Area: 20 cm^2

x cm

5.3 Volume of prisms

Are you ready?

1 Work out the area of each shape.

a

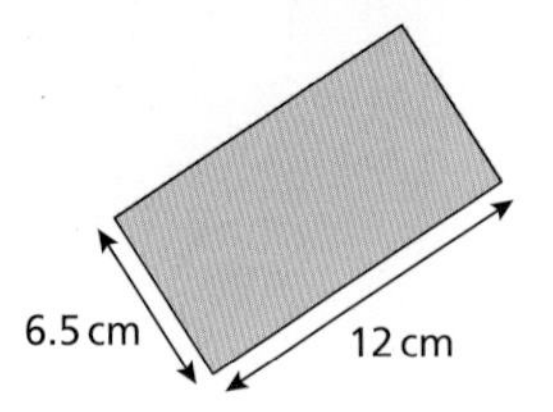

b

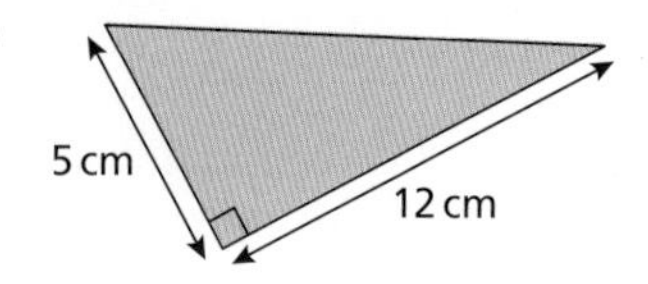

c

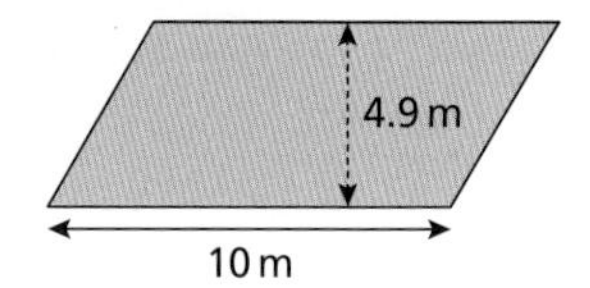

2 Work out the area of the shape.

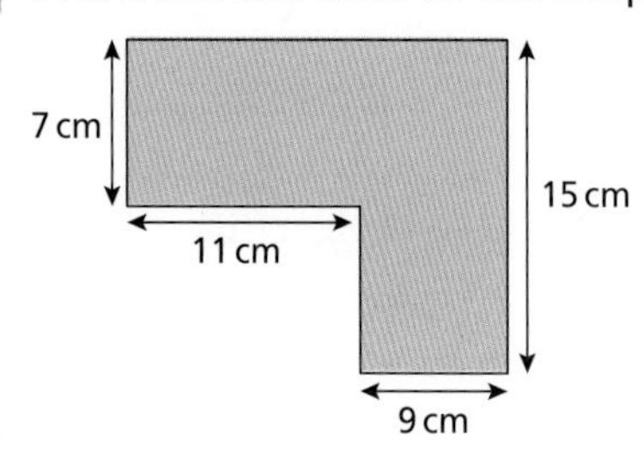

Volume is the amount of solid space taken up by a 3D shape. It is measured in cubic units such as cm^3 or m^3.

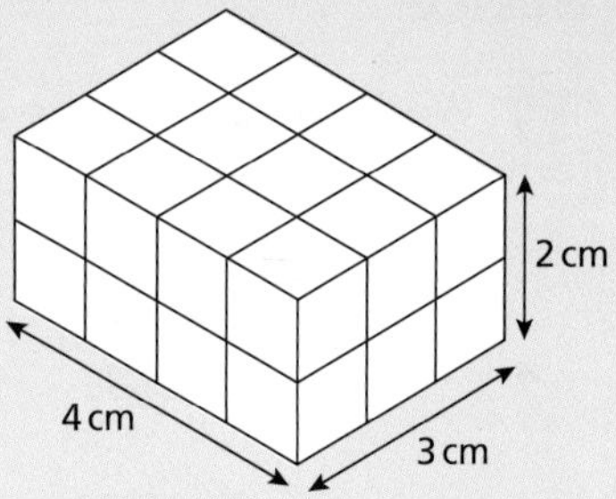

You can work out the volume of a simple shape by counting how many 1 cm cubes it is made up of.

To work out the volume of a cuboid, count the number of cubes in one layer, then multiply this by the number of layers.

This cuboid has four layers, each with six cubes, and therefore the volume is $4 \times 6 = 24\,cm^3$.

In general, the volume of a cuboid can be calculated by multiplying its length, width and height.

A **prism** is a 3D shape with a **uniform cross-section**. This means the same shape is made when the solid is cut through by a plane.

Below is a triangular prism.

It can be divided into layers containing centimetre cubes.

The number of cubes in one layer represents the cross-sectional area. There are 24 cubes in the first layer because $\frac{1}{2} \times 8 \times 6 = 24$ (area of a triangle).

There are five layers. So the volume is the total number of cubes $= 24 \times 5 = 120\,cm^3$.

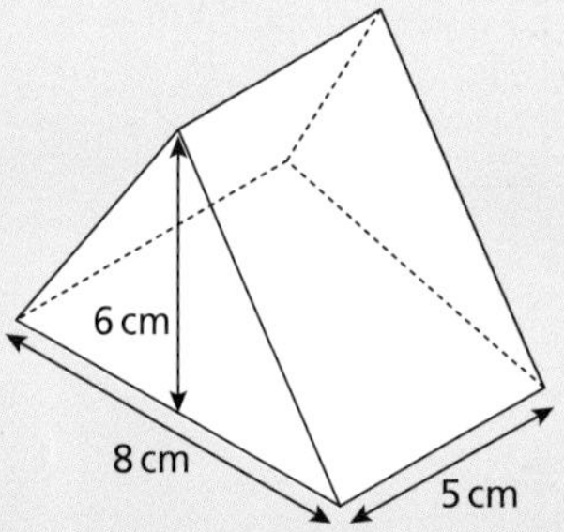

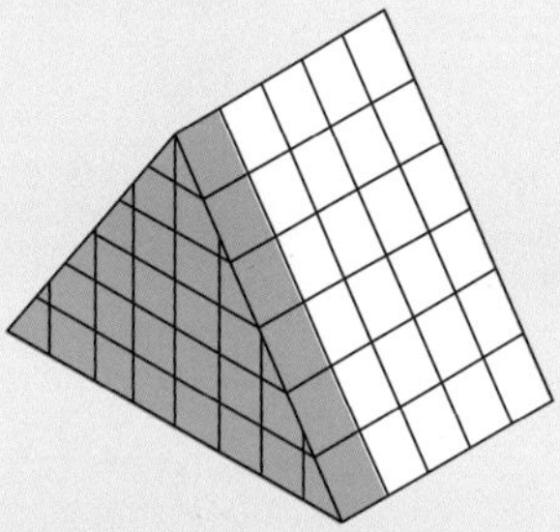

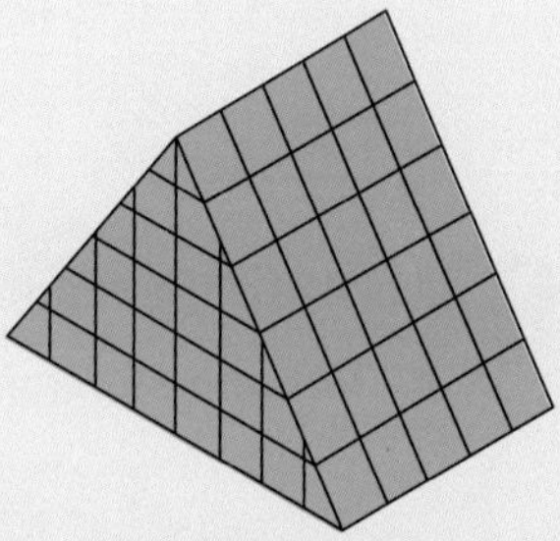

For any prism, **volume = area of cross-section × length**

This formula works whether or not unit cubes can be counted, as shown in the next example.

Example

Work out the volume of the prism.

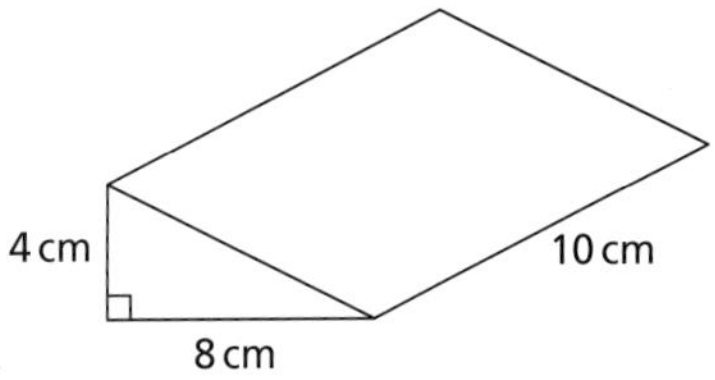

Method

Solution	Commentary
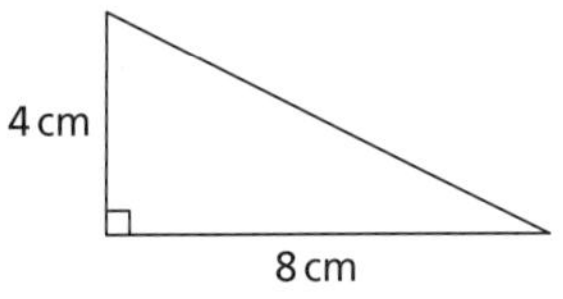	The cross-section is a right-angled triangle. Work out the area of the triangle, using the formula $\frac{1}{2}$ × base × perpendicular height.
Cross-sectional area $= \frac{1}{2} \times 4 \times 8 = 16$	
Volume $= 16 \times 10 = 160$ Volume $= 160\,cm^3$	Multiply the area of the cross-section by the length of the prism to calculate the volume.

Practice

1 Work out the volume of each cuboid.

a

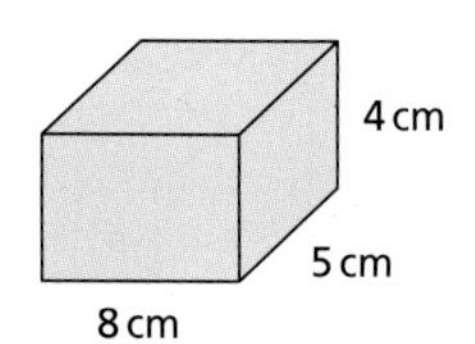

b

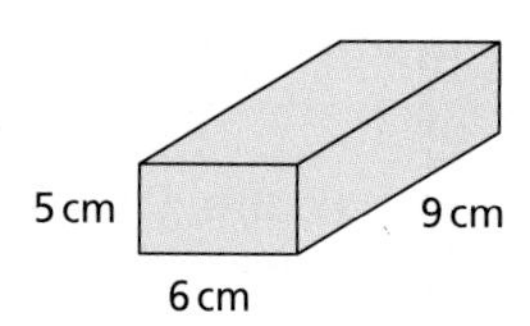

c

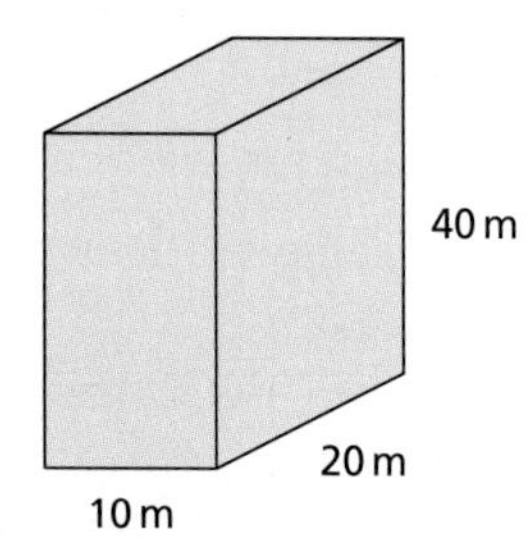

d

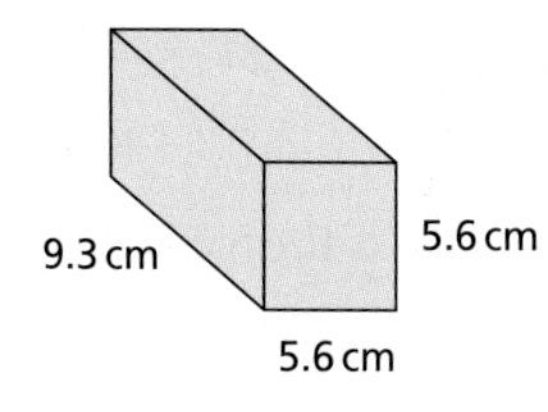

2 Work out the volume of each cube.

a

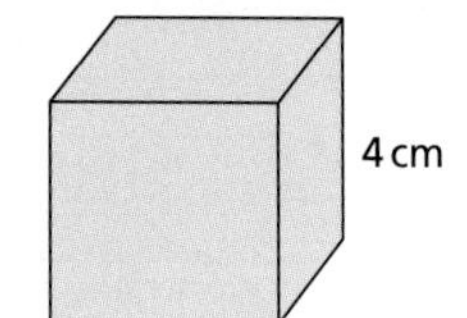

b

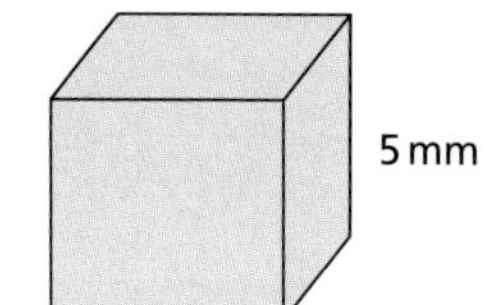

c

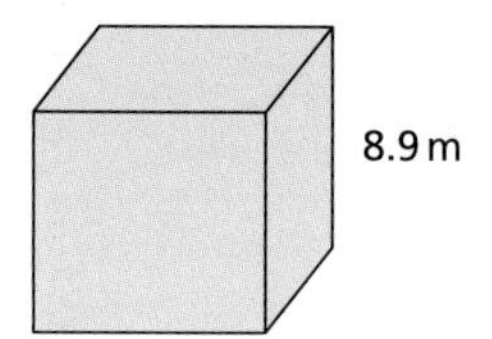

3 Work out the length of each lettered side.

a Volume 360 cm^3

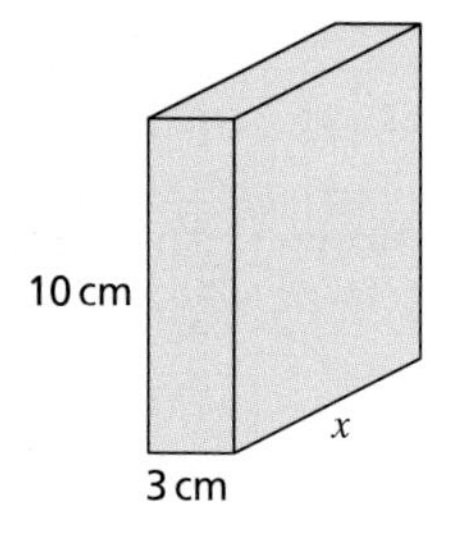

b Volume 280 m^3

5 m

y

7 m

4 Work out the volume of each triangular prism.

a

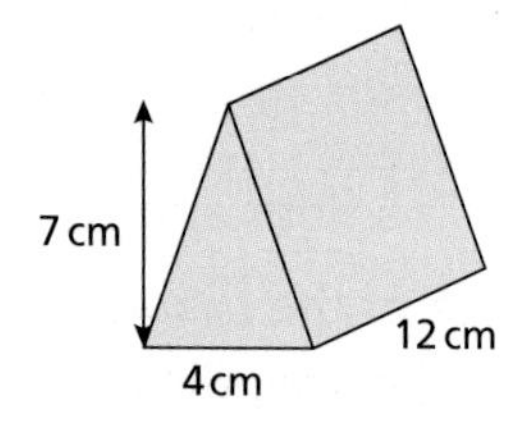

b

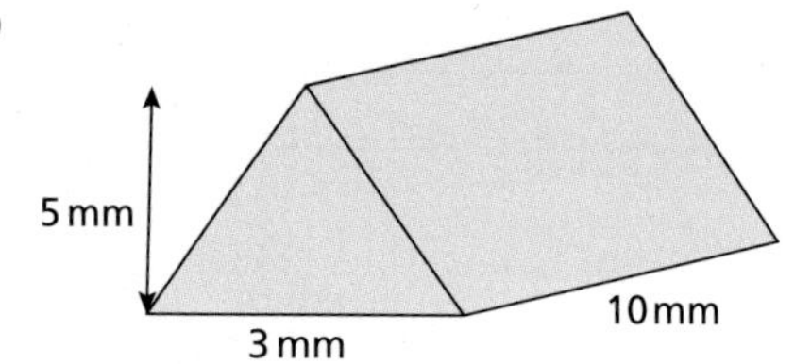

c

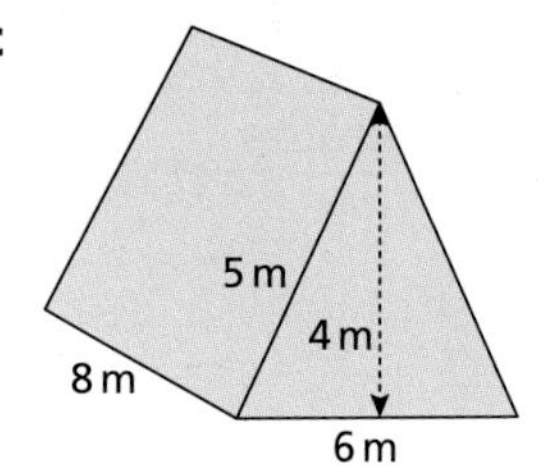

5 Work out the volume of the trapezoidal prism.

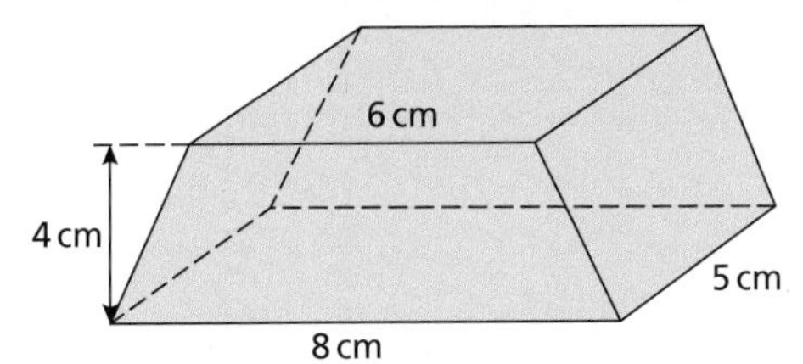

6 Work out the volume of each compound 3D shape.

a

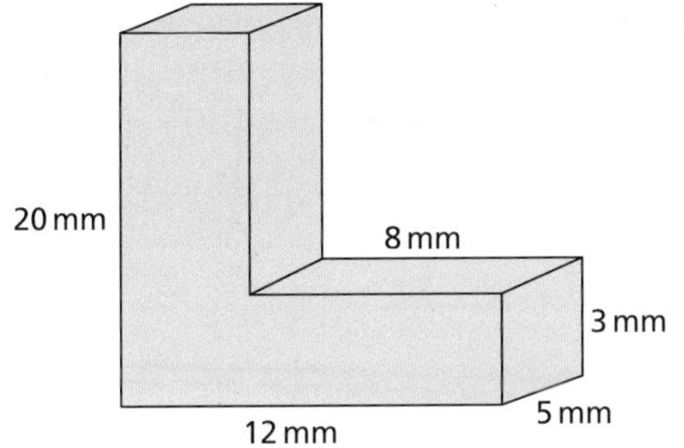

b

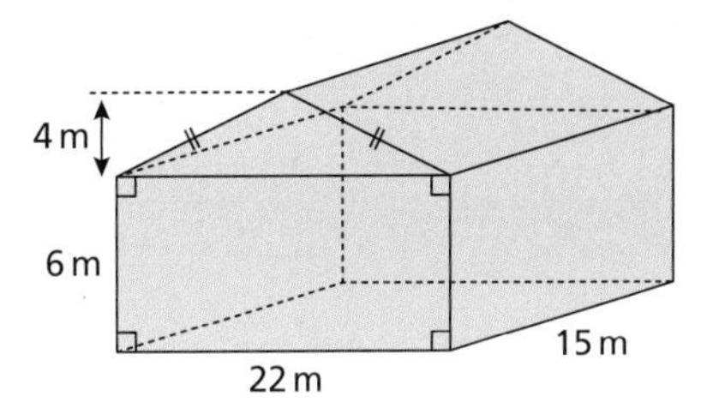

7 Work out the side length of the cube.

Volume = 729 cm^3

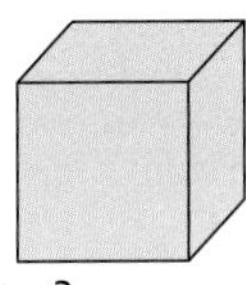

8 Work out the length of each lettered side in these triangular prisms.

a Volume = 37.5 m^3

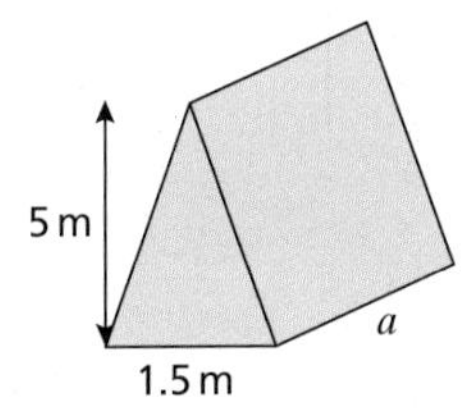

b Volume = 135 cm^3

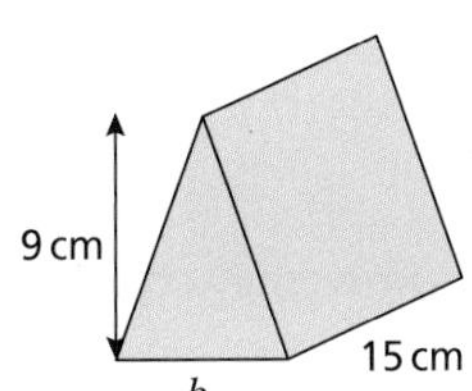

9 Rhys uses his van to deliver boxes. His van has the shape of a cuboid with length 3 m, width 2 m, and depth 5 m.

Rhys wants to fit some boxes in his van. Each box is a cuboid measuring 60 cm × 20 cm × 50 cm.

What is the greatest number of boxes Rhys can fit in his van?

10 A cuboidal container measures 20 cm × 65 cm × 1 m.

1 litre = 1000 cm^3

A tap can fill the container at a rate of 13 litres per minute.

How long will it take to fill the container?

What do you think?

1. Huda says, "I can draw a cuboid where the surface area is numerically equal to the volume."

   Sketch the cuboid.

   Is there more than one possible answer?

2. The volume of a cuboid is $200\,cm^3$.

   Two of the dimensions are equal.

   Sketch the cuboid.

   Is there more than one possible answer?

Consolidate – do you need more?

1 Work out the volume of each cuboid.

a

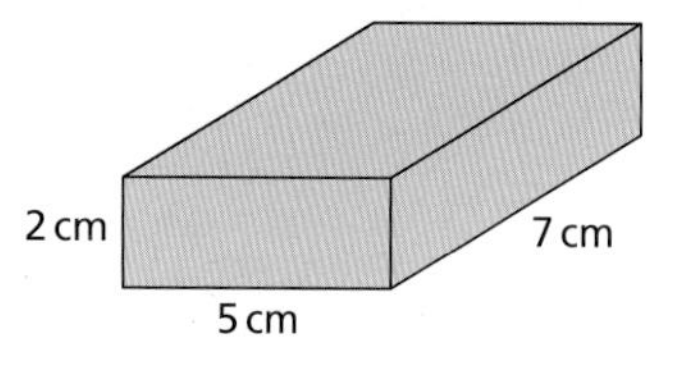

b

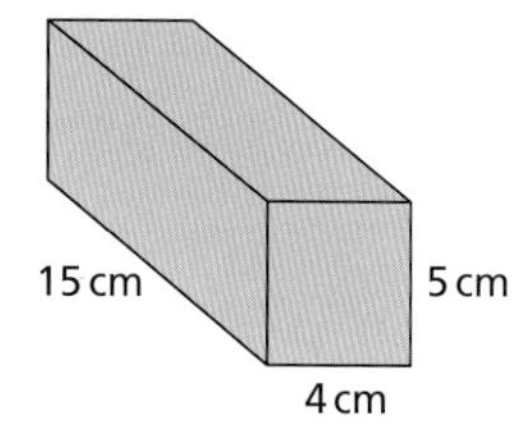

c

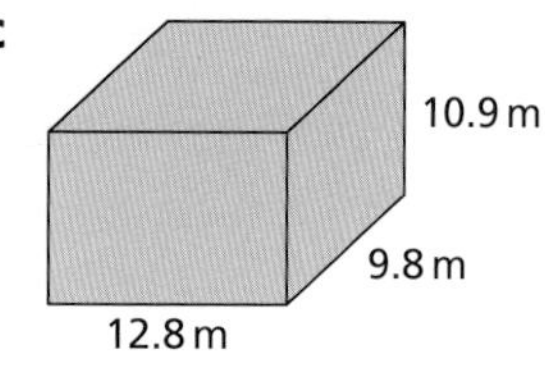

2 Work out the volume of each cube.

a

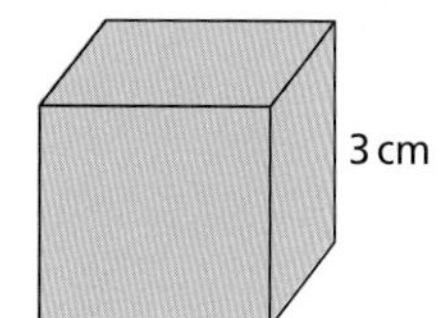

b

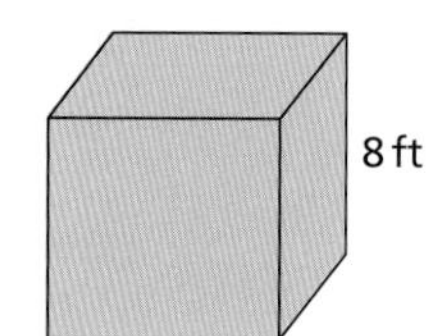

c

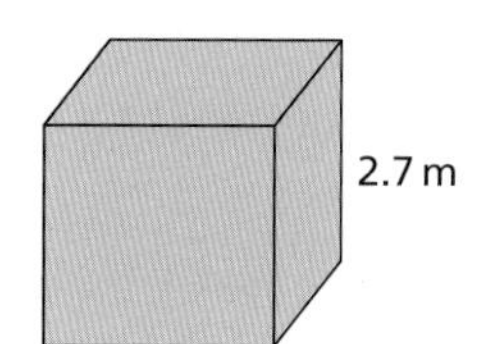

3 Work out the volume of each triangular prism.

a

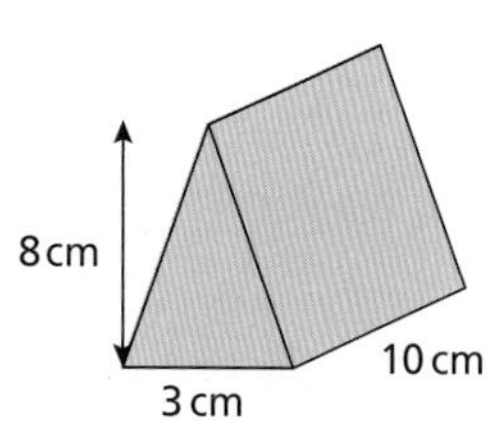

b

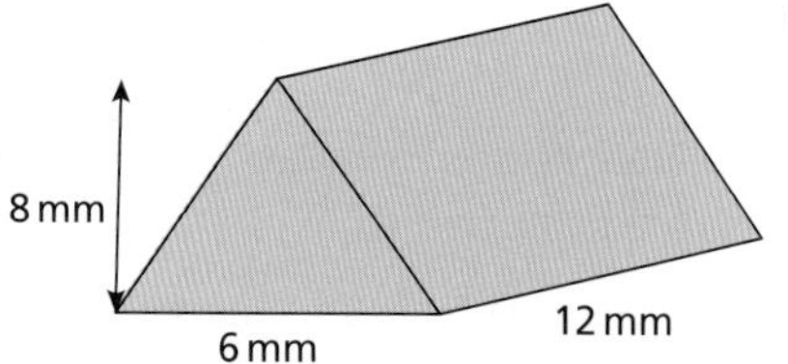

c

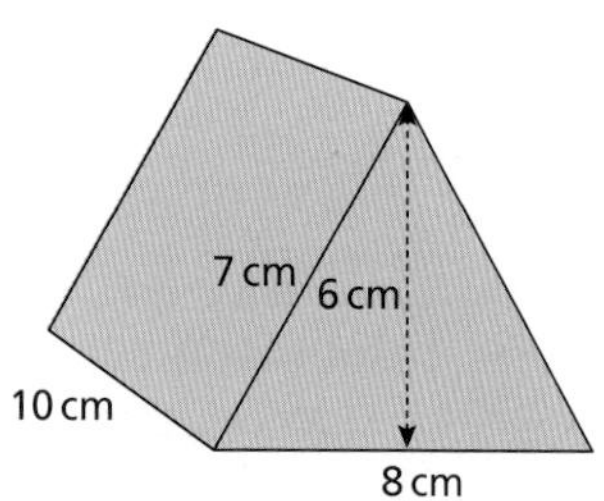

Stretch – can you deepen your learning?

1 A cube has a surface area of 726 cm^2.

Work out the volume of the cube.

2 A carton of juice measures 12 cm by 12 cm by 28 cm.

The depth of the juice in the carton is 10 cm.

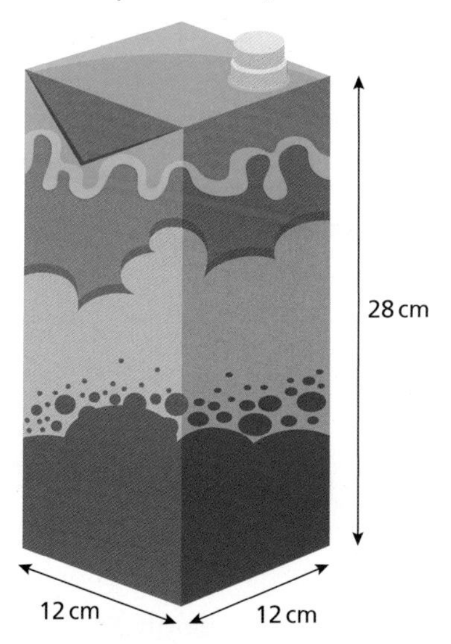

Mario flips the carton so it now sits like this:

Work out the depth of the juice in the carton now.

3 Here is a triangular prism.

Work out the volume of the prism.

Give your answer in the form $\frac{a\sqrt{3} + b}{c}$ where a, b and c are integers.

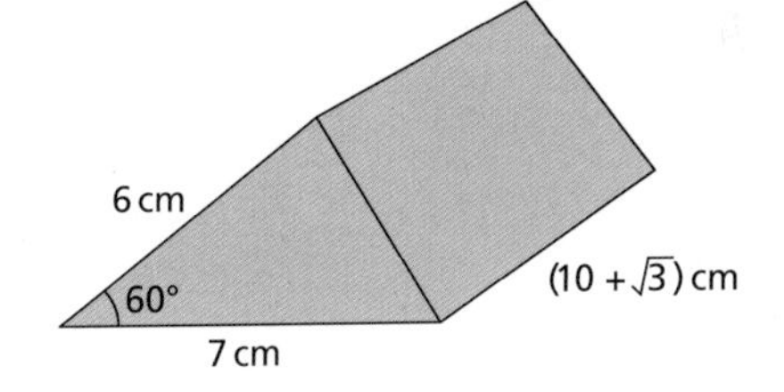

4 An electronic component has a square cross-section.

The component has a hole that runs the length of the component.

The cross-section of the hole is an equilateral triangle.

Work out the volume of the component. Give your answer in mm^3 to 3 significant figures.

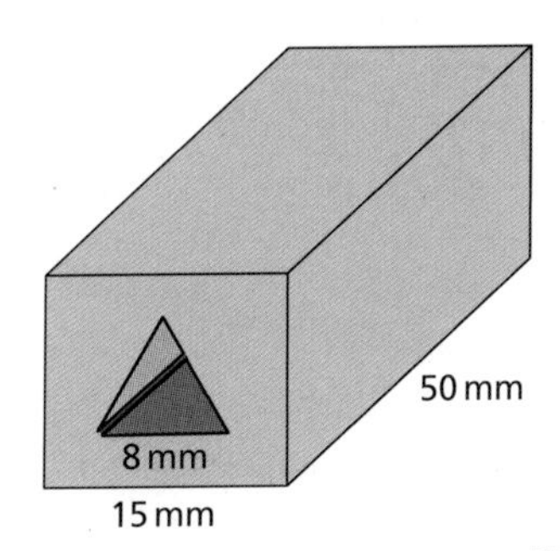

5.4 Cylinders, cones and spheres

Are you ready? (A)

1 For each circle, work out: **i** the area **ii** the circumference.

a

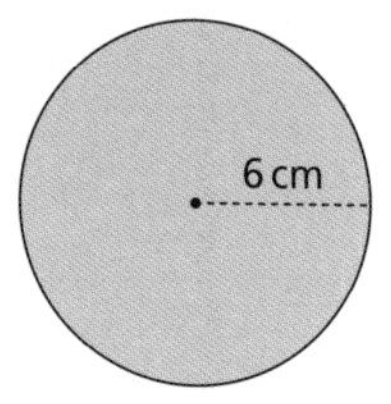

b

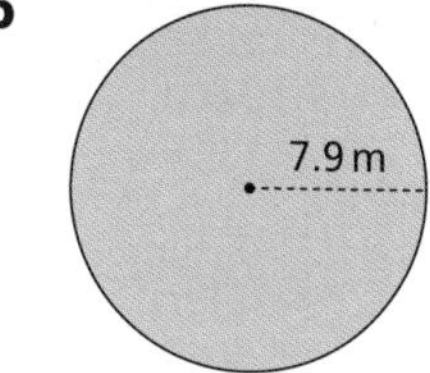

c

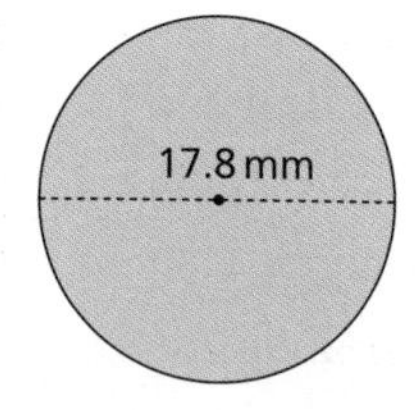

2 For each shape, work out: **i** the volume **ii** the surface area.

a

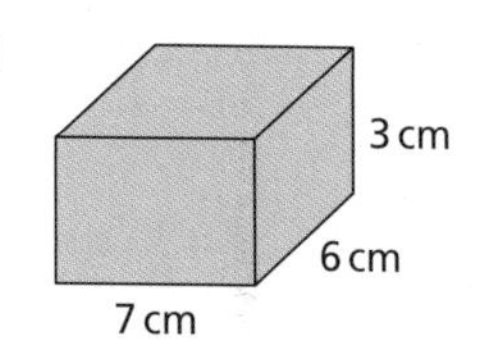

b

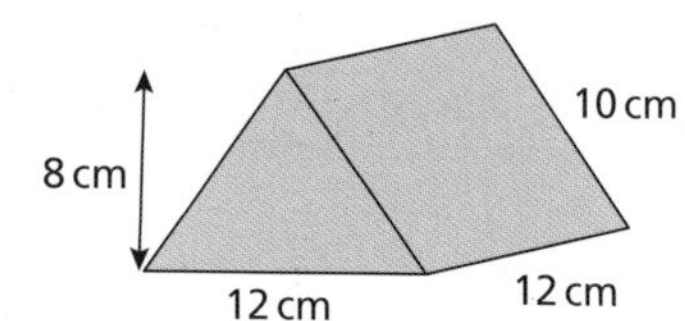

A **cylinder** is a 3D shape with two parallel circular bases, joined by a curved surface. As a cylinder has a curved surface, it is not a prism but the volume can be calculated in the same way:

Work out the area of the base and multiply the result by the length (or height) of the cylinder.

The volume, V, of a cylinder of radius r is therefore given by the formula $V = \pi r^2 h$

In this section, you will learn how to work out the surface area of a cylinder.

Here is the net of a cylinder. It is made up of two circles and a rectangle.

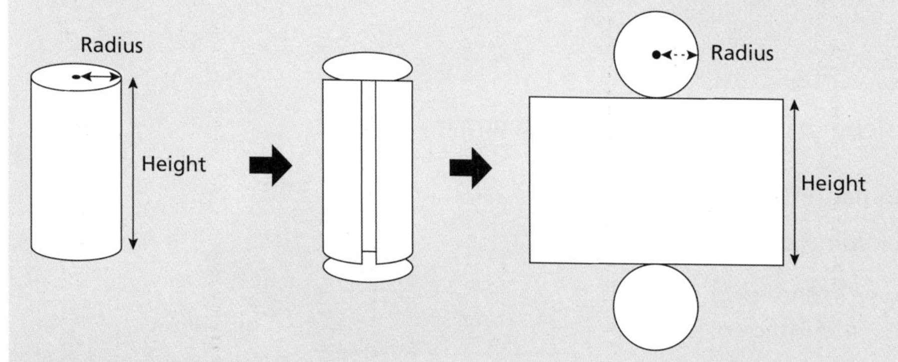

The area of each circle is given by $A = \pi \times r^2$

The area of a rectangle is length × width.

The length of the rectangle is the same as the circumference of the circle, which is given by $2\pi r$ or πd.

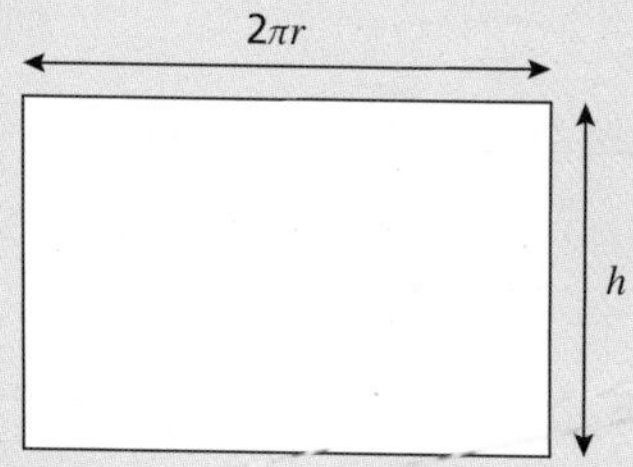

The width of the rectangle is the same as the height of the cylinder.

The area of the rectangle is equal to the curved surface area of the cylinder.

Total surface area = area of two circles + curved surface area

Total surface area = $2 \times \pi r^2 + 2\pi r \times h = 2\pi r^2 + 2\pi rh$

Example

A cylinder is shown.

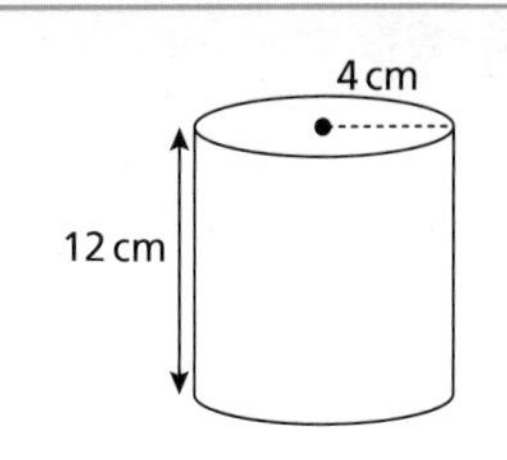

a Work out the volume of the cylinder, giving your answer to 1 decimal place.

b Work out the surface area of the cylinder, giving your answer in terms of π.

Method

Solution	Commentary
a $V = \pi r^2 h$ $= \pi \times 4^2 \times 12$ $= 603.1857894\ldots$ $= 603.2\text{ cm}^3$ (to 1 d.p.)	Round your answer to 1 decimal place and remember to give the units.
b Area of top = Area of bottom = πr^2 $= \pi \times 4^2$ $= 16\pi$ Curved surface area = $\pi d \times h$ $= \pi \times 8 \times 12$ $= 96\pi$	Work out the area of each part of the cylinder.
Total surface area = $16\pi \times 2 + 96\pi$ $= 128\pi\text{ cm}^2$	Add all three areas to find the total surface area.

Practice (A)

1 Work out the volume of each cylinder. Give your answers to 1 decimal place.

a

4 cm
10 cm

b

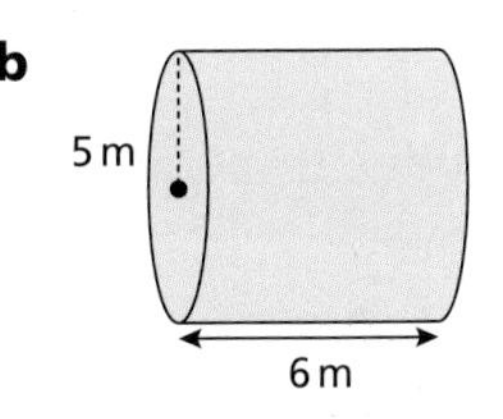

c

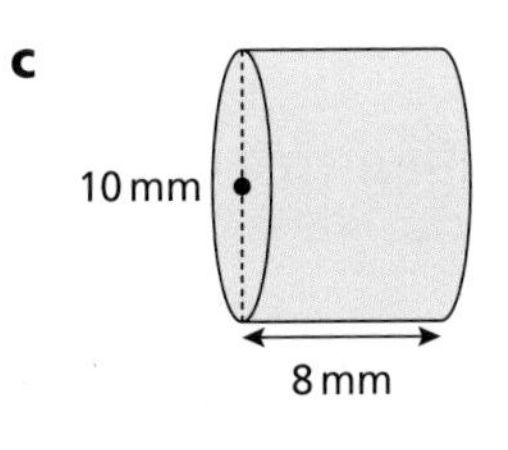

d

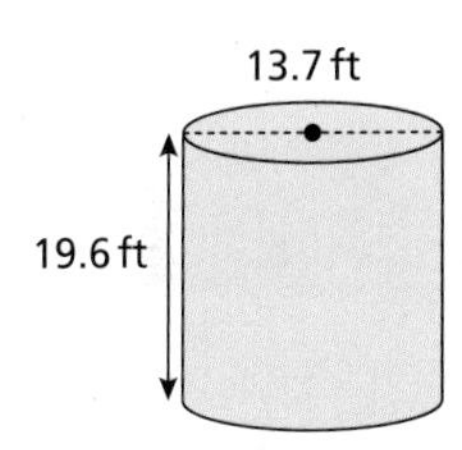

2 Work out the volume of each cylinder. Give your answers in terms of π.

a

2 cm
10 cm

b

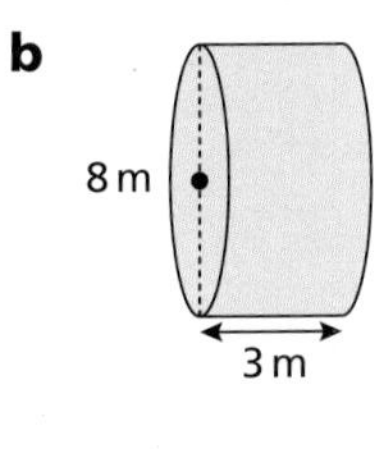

c

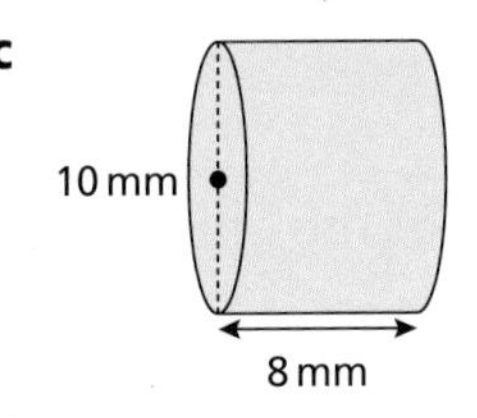

d

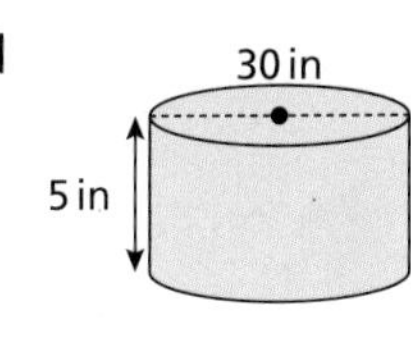

3 The volume of this cylinder is 200 cm³.

Work out the height of the cylinder.
Give your answer to 1 decimal place.

4 The volume of this cylinder is 450 m³.

Work out the radius (r) of the cylinder, giving your answer to 1 decimal place.

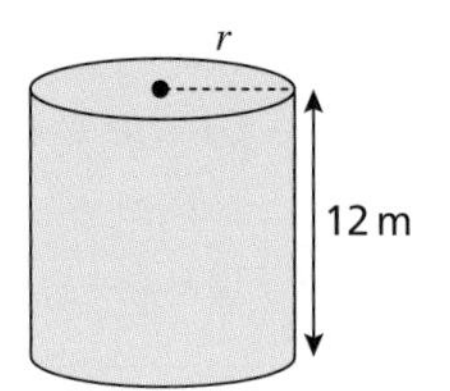

5 The volume of this cylinder is 1250 mm³.

Work out the diameter (d) of the cylinder, giving your answer to 1 decimal place.

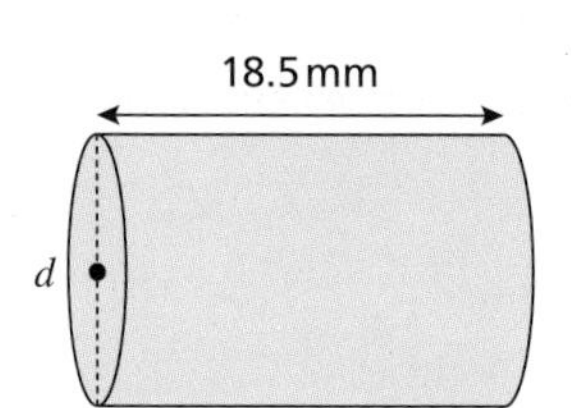

6 A cylinder and a cube are shown.

The cylinder and the cube have the same volume.

Work out the height (x) of the cylinder, giving your answer to 1 decimal place.

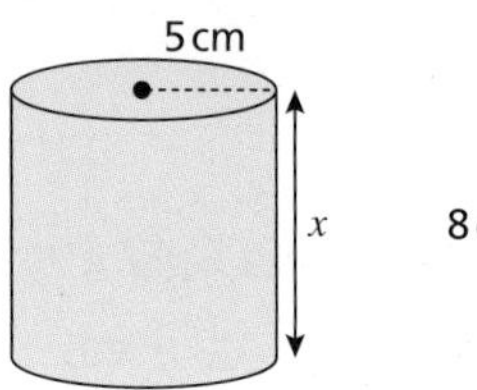

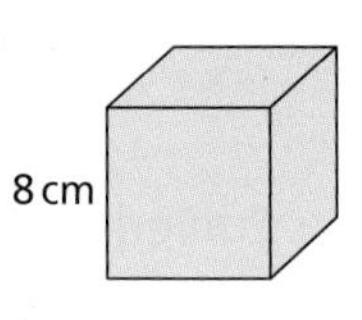

7 Work out the surface area of each cylinder. Give your answers to 1 decimal place.

a

b

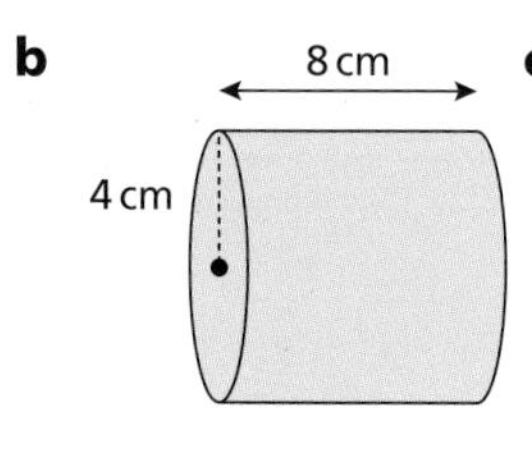

c

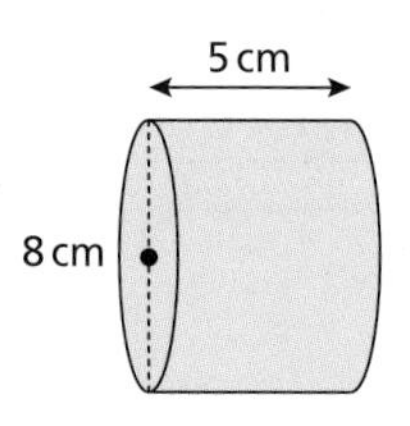

d

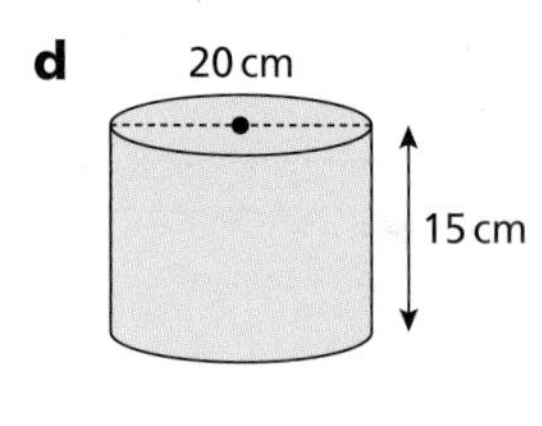

8 Work out the surface area of each cylinder. Give your answers in terms of π.

a

b

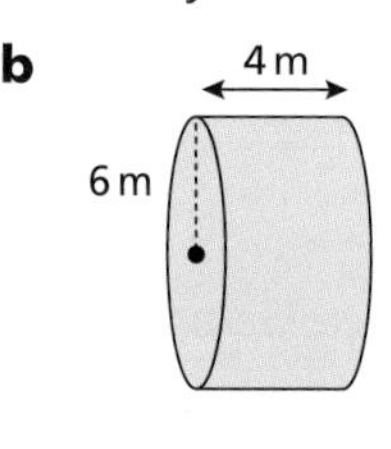

c

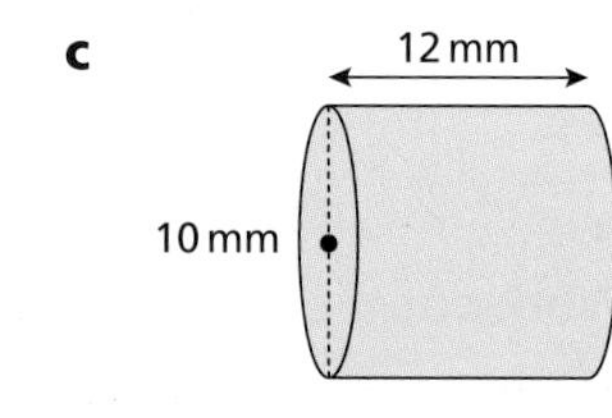

9 The surface area of this cylinder is 800 cm².

Work out the height (h) of the cylinder, giving your answer to 1 decimal place.

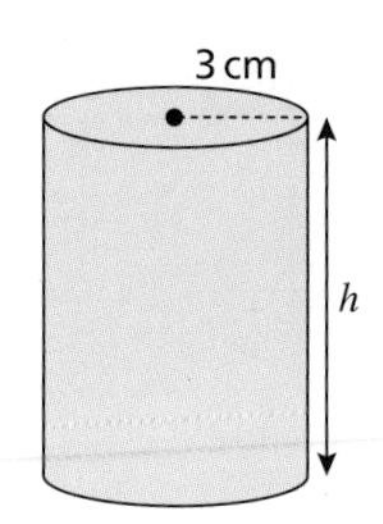

10 The surface area of this cylinder is 120π cm^2.

a Write an expression in terms of r for the surface area of the cylinder.

b Work out the length of the radius.

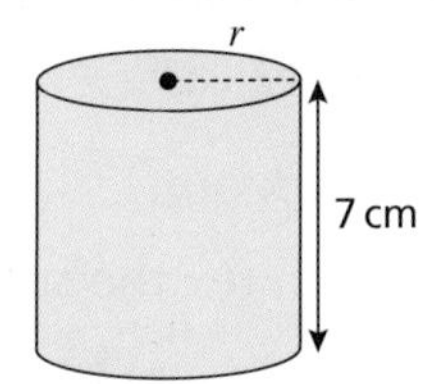

What do you think? (A)

1 Some cylindrical tubes are placed in a box.

The tubes come in two sizes, as shown below right.

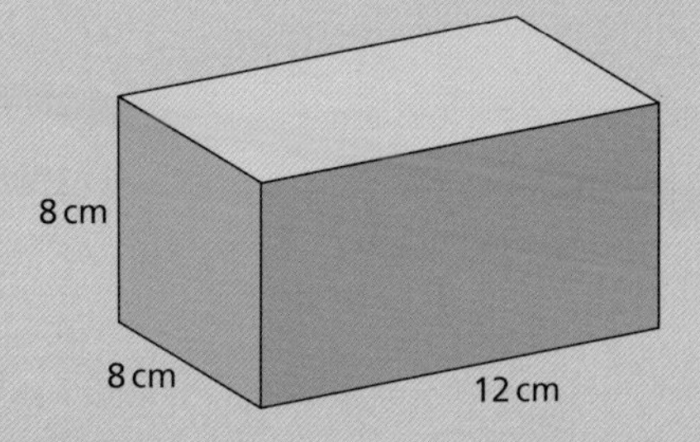

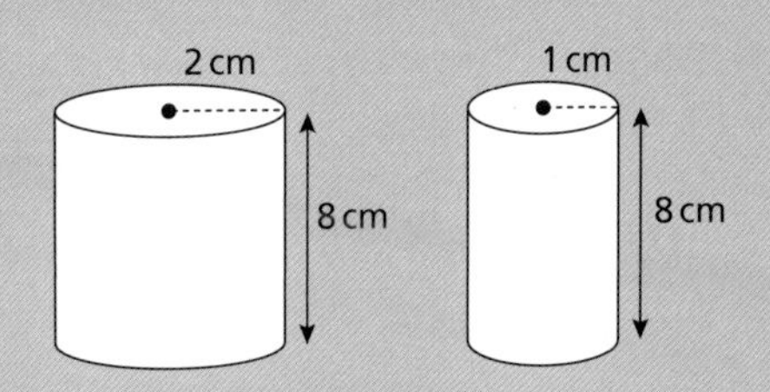

Huda says, "If I pack the box with the smaller tubes, they will take up more space than the larger tubes."

Is Huda correct? Explain your answer.

2 For each shape, work out: **i** the volume **ii** the surface area.

a

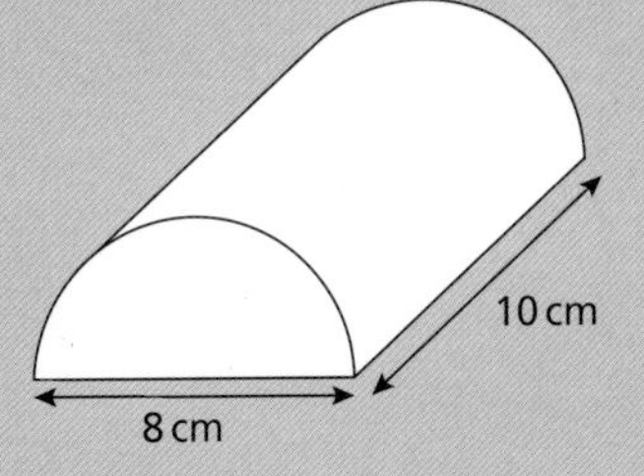

b

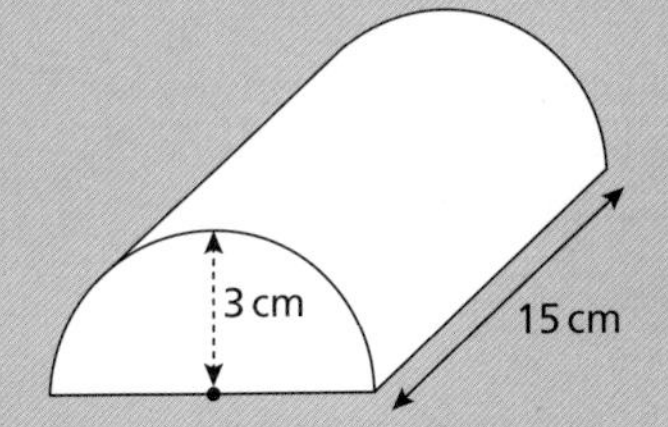

3 Work out the surface area of this shape.

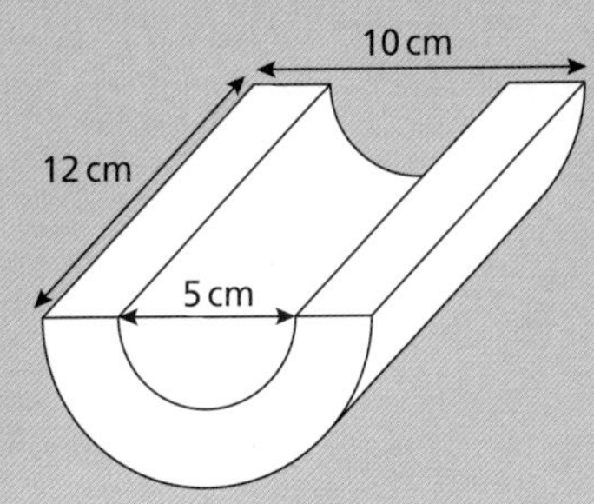

Are you ready? (B)

1 For each cylinder, work out:

i the volume **ii** the surface area.

a 3 cm

8 cm

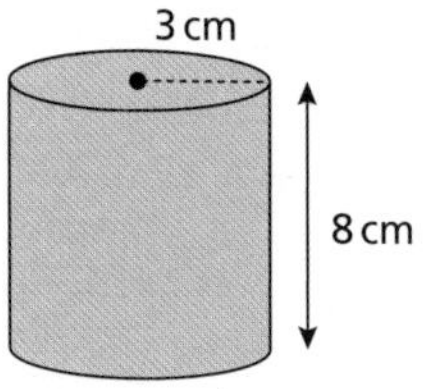

b

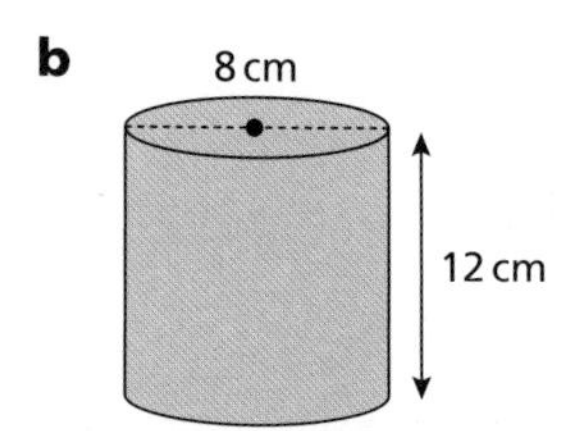

A **cone** is a 3D shape with one circular base, which narrows as it extends to a single point called its **apex**.

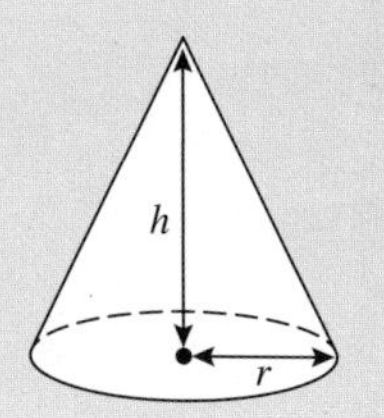

The volume of a cone is exactly $\frac{1}{3}$ of the volume of a cylinder with the same base and perpendicular height.

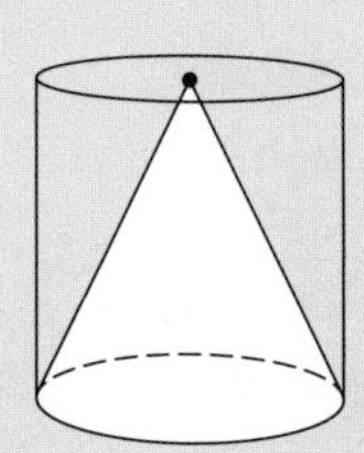

The volume of a cone is given by $V = \frac{1}{3}\pi r^2 h$, where r is the radius and h is the perpendicular height.

The **curved surface area** of a cone is given by the expression πrl, where r is the radius of the base and l is the **slant height**. This is the area of the curved surface.

The area of the base of the cone is given by πr^2, since it is a circle.

Therefore, the total surface area, A, of a cone can be found using the formula $A = \pi r^2 + \pi rl$

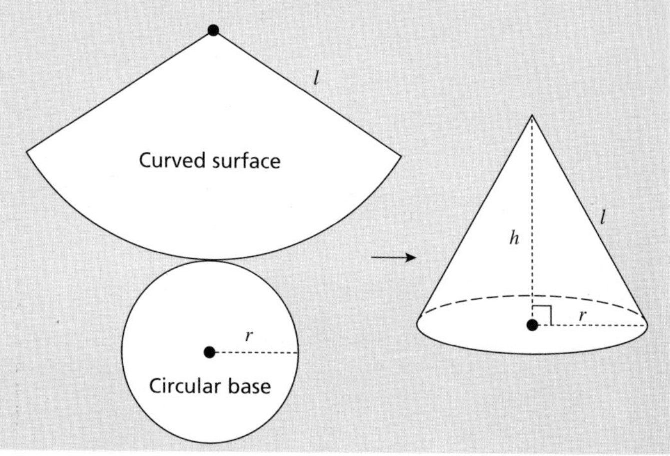

Example

A cone is shown.

a Work out the volume of the cone, giving your answer to 1 decimal place.

b Work out the surface area of the cone, giving your answer in terms of π.

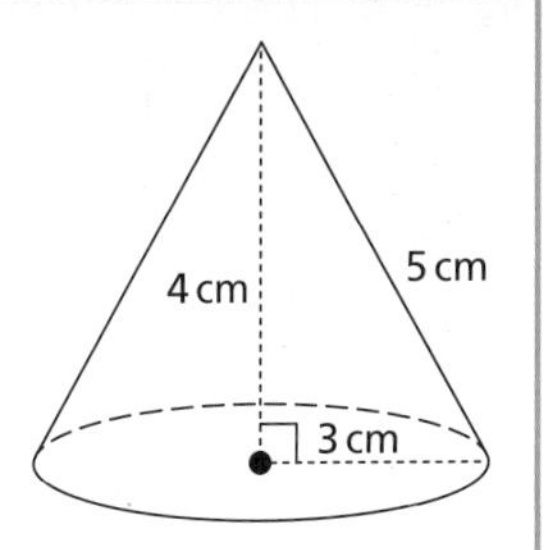

Method

Solution	Commentary
a $V = \frac{1}{3}\pi r^2 h = \frac{1}{3} \times \pi \times 3^2 \times 4$ $= 37.699\,111\,84\ldots$ $= 37.7\ \text{cm}^3$ (to 1 d.p.)	Use the formula for the volume of a cone and give the correct units with your answer.
b Area of base $= \pi r^2 = \pi \times 3^2 = 9\pi$ Curved surface area $= \pi r l = \pi \times 3 \times 5 = 15\pi$	Work out the area of the circular base and the curved surface.
Total surface area $= 9\pi + 15\pi = 24\pi\ \text{cm}^2$	Add the two areas to find the total surface area.

Practice (B)

1 Work out the volume of each cone. Give your answers to 1 decimal place.

a

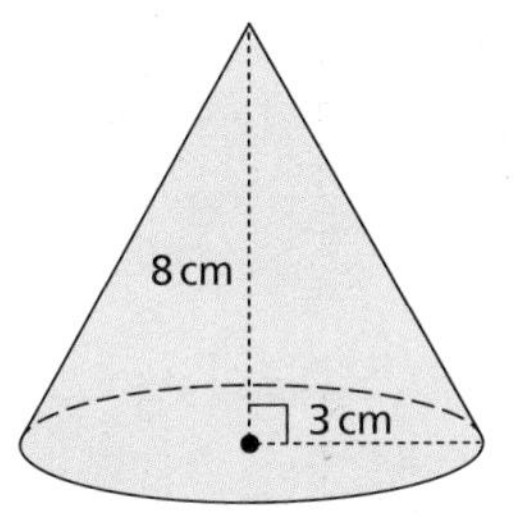

b

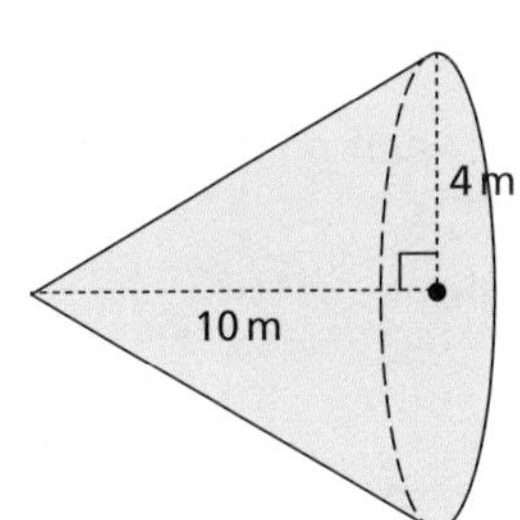

c

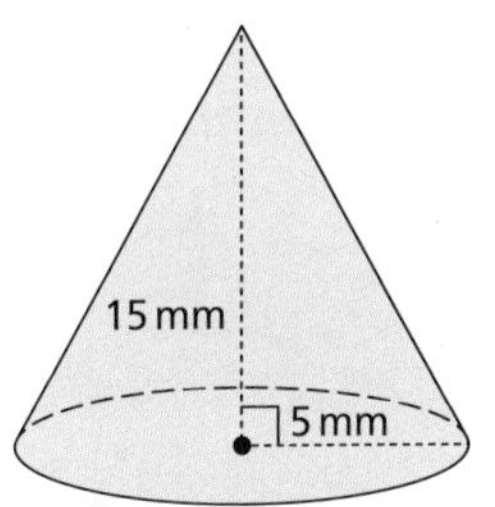

d

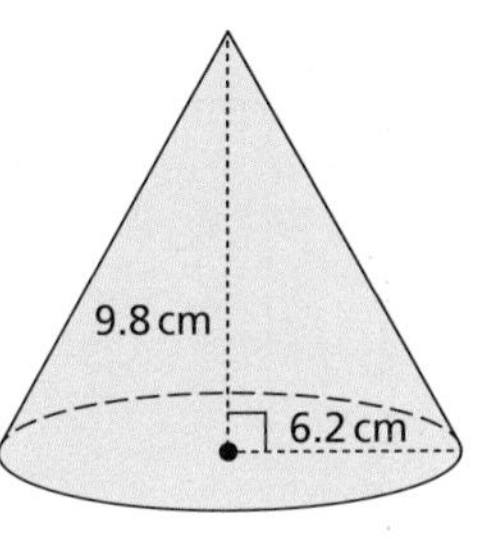

e

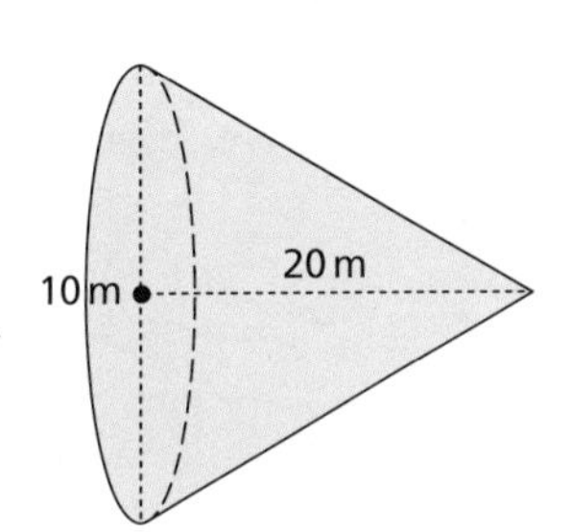

f

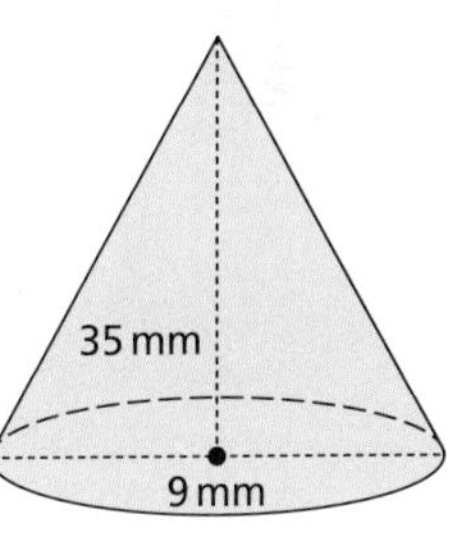

2 Work out the volume of each cone. Give your answers in terms of π.

a

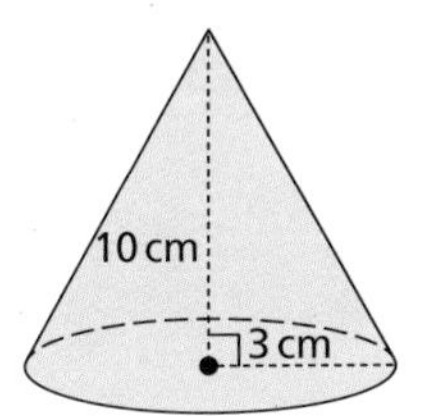

b

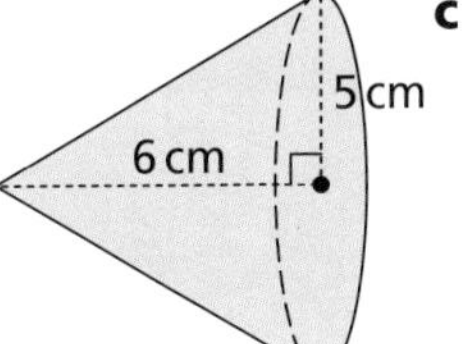

c

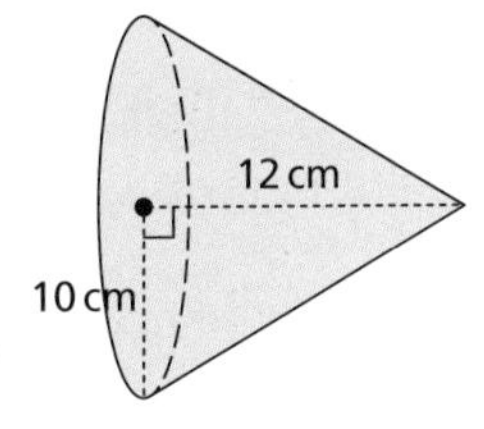

d

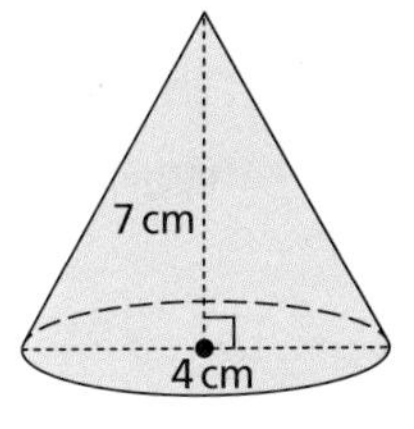

3 The volume of the cone is 500 cm³.

Work out the height, h, of the cone.
Give your answer to 1 decimal place.

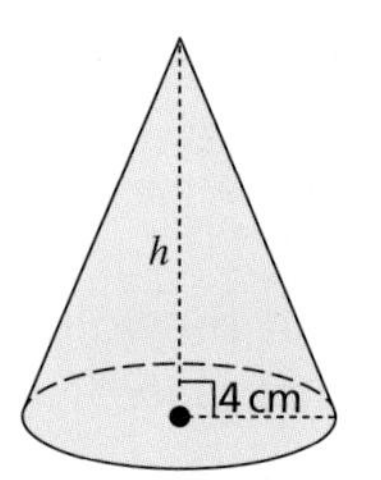

4 The volume of the cone is 2500 cm³.

Work out the radius, r, of the cone.
Give your answer to 1 decimal place.

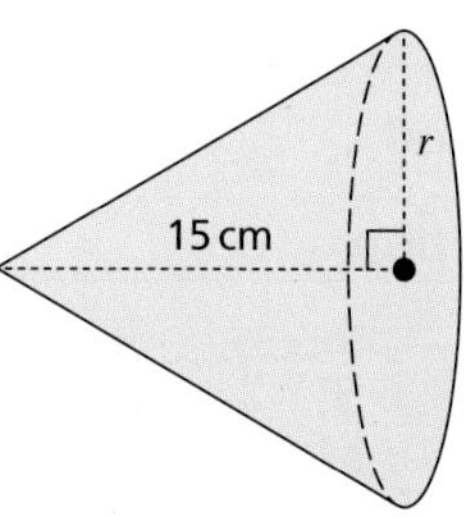

5 The shape is made from a cone and a cylinder.

Work out the volume of the shape.
Give your answer to 1 decimal place.

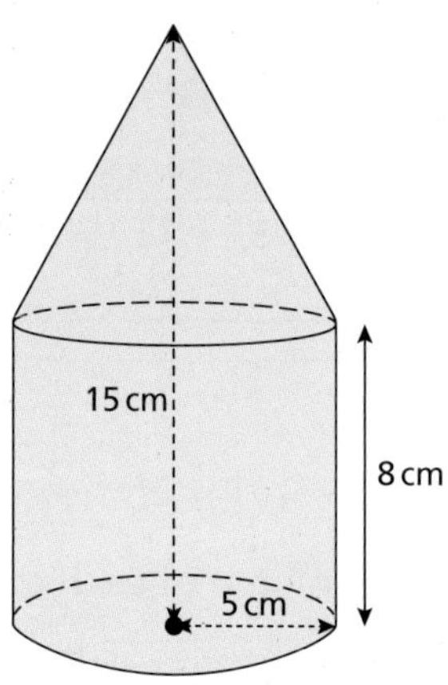

6 Work out the surface area of each cone. Give your answers to 1 decimal place.

a

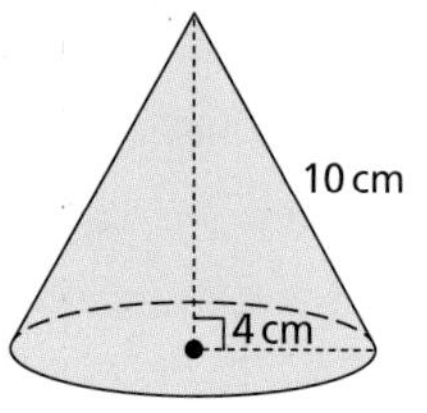

b

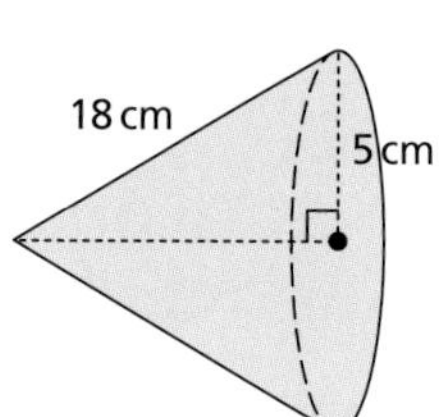

c

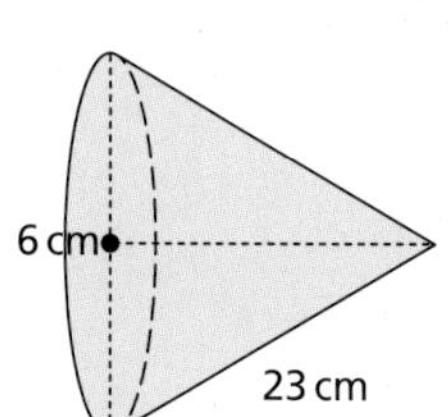

d

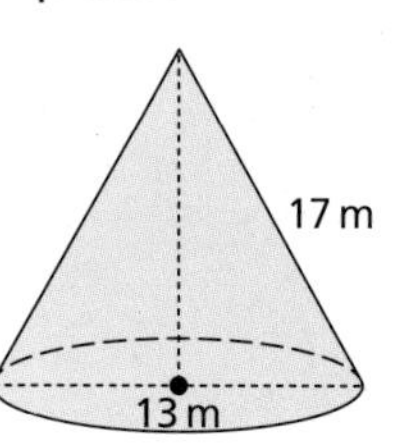

7 Work out the surface area of each cone. Give your answers in terms of π.

a

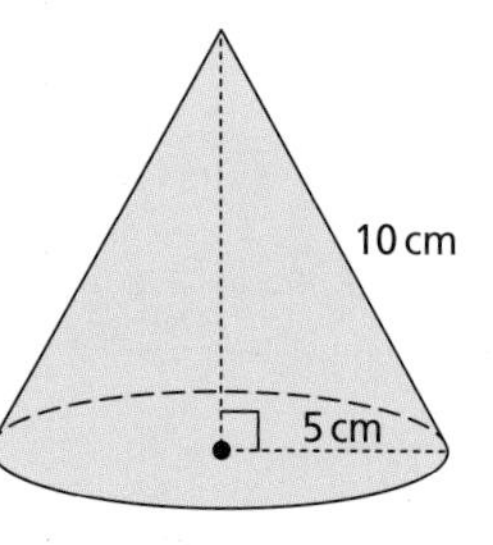

b

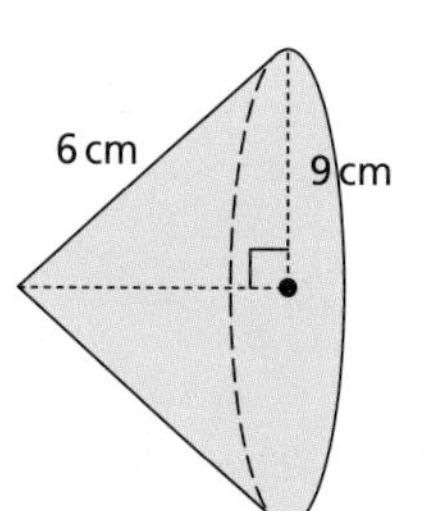

c

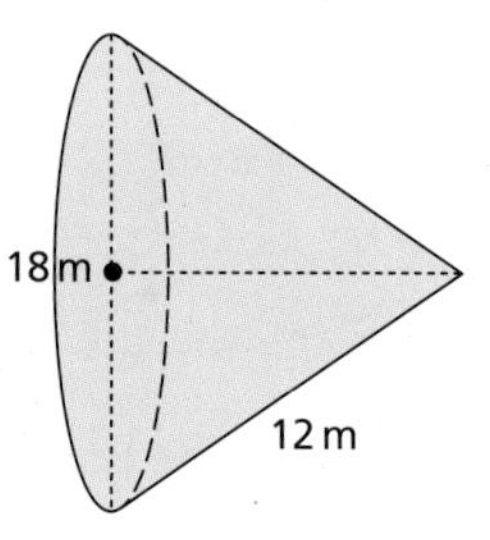

8 The cone has a surface area of 200 cm².

Work out the slant height, x, of the cone.

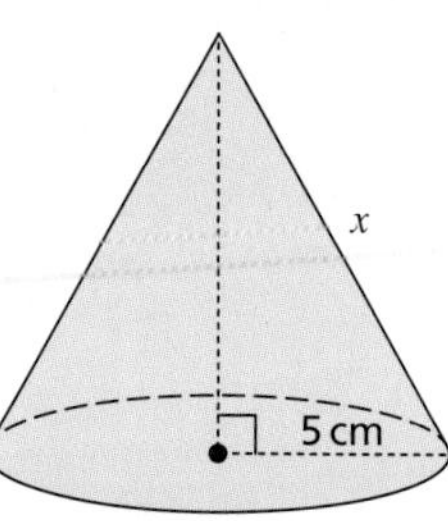

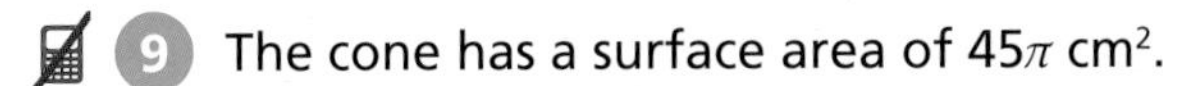

9 The cone has a surface area of 45π cm^2.

a Write an expression for the surface area of the cone in terms of r.

b Work out the radius of the cone.

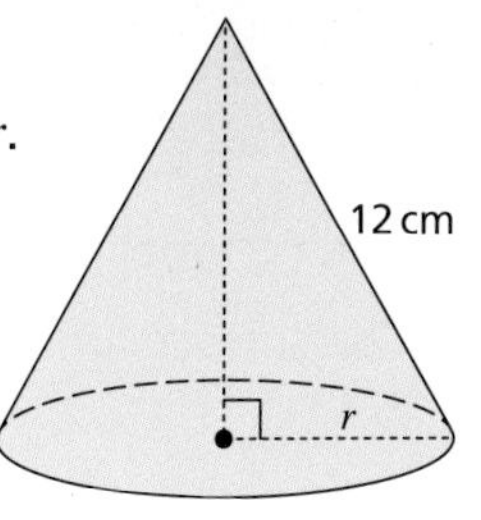

10 The cone and the cuboid have the same surface area.

Work out the slant height, x, of the cone.

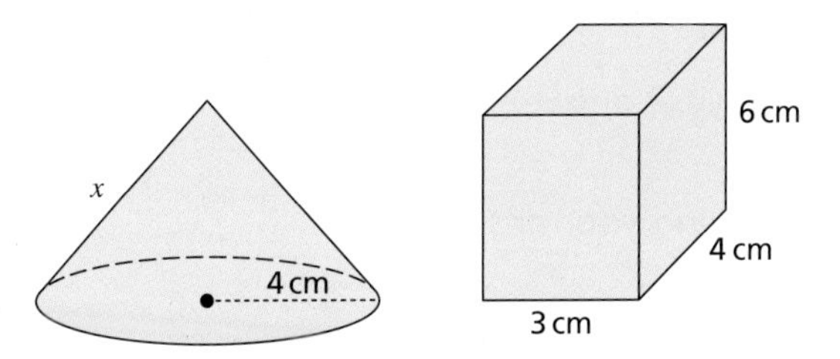

What do you think? (B)

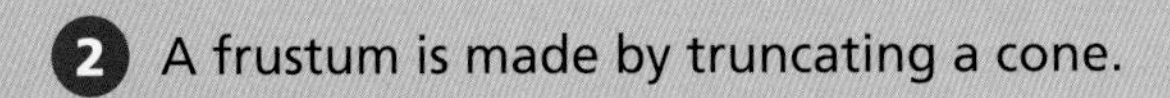

1 A cone is shown.

Huda says, "If the radius and the perpendicular height are doubled then the volume will double."

Show that Huda is incorrect.

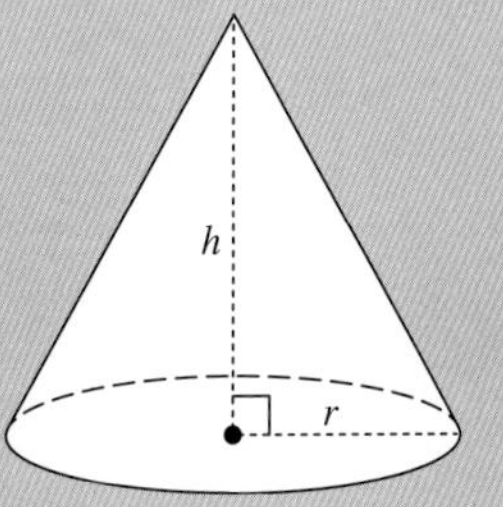

2 A frustum is made by truncating a cone.

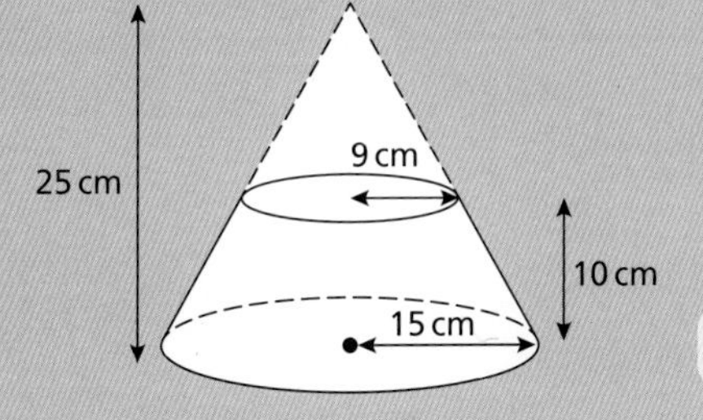

Truncating means removing a part of the shape.

Work out the volume of the frustum.

Are you ready? (C)

1 For each circle, work out the area:

i to 1 decimal place ii in terms of π.

a

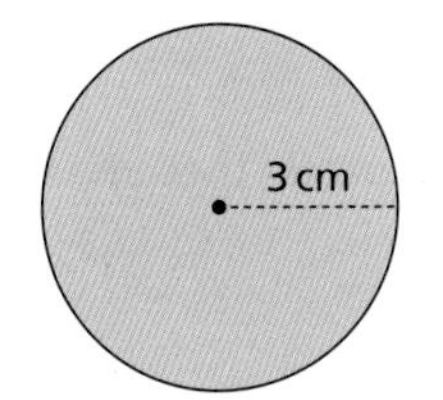

b

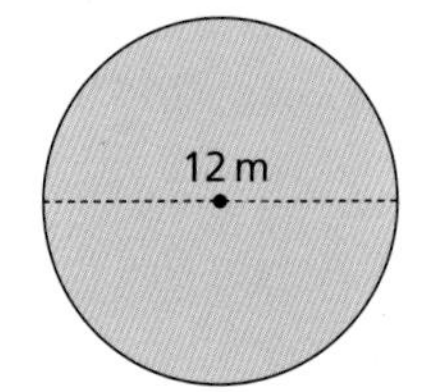

c

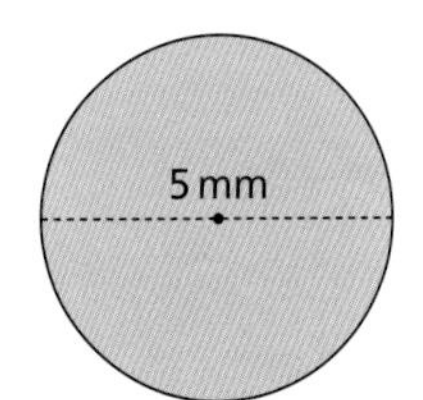

The radius of a **sphere** is the distance from the centre to any point on its surface.

The volume, V, of a sphere of radius r is given by the formula $V = \frac{4}{3}\pi r^3$

The surface area, A, of a sphere of radius r is given by the formula $A = 4\pi r^2$

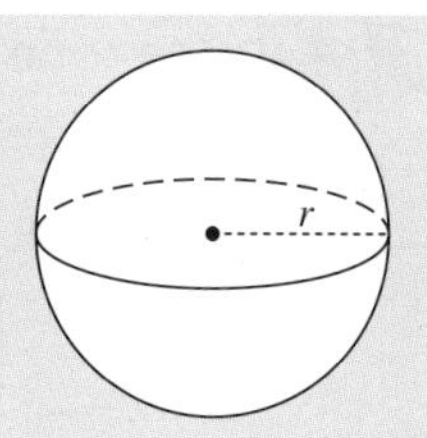

Example 1

This sphere has a diameter of 12 cm.

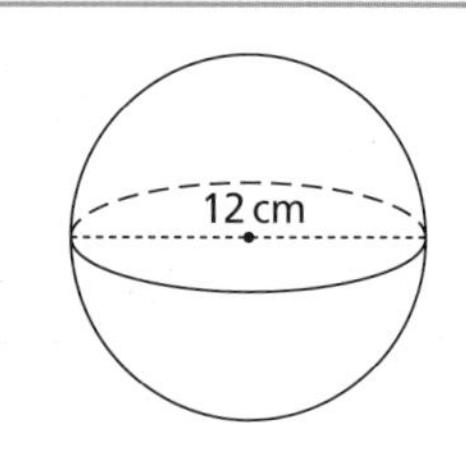

a Work out the volume of the sphere, giving your answer to 1 decimal place.

b Work out the surface area of the sphere, giving your answer in terms of π.

Method

Solution	Commentary
a Volume $= \frac{4}{3}\pi r^3 = \frac{4}{3} \times \pi \times 6^3$ $= 904.778\,6842\ldots$ $= 904.8\,\text{cm}^3$ (to 1 d.p.)	Remember that the radius is half the diameter. Use the formula for the volume of a sphere and give the correct units with your answer.
b $A = 4\pi r^2 = 4 \times \pi \times 6^2$ $= 144\pi\ \text{cm}^2$	Use the formula for the surface area of a sphere and give the correct units with your answer.

Example 2

The hemisphere has a radius of 5 cm.

a Work out the volume of the hemisphere.

b Work out the surface area of the hemisphere.

Give your answers in terms of π.

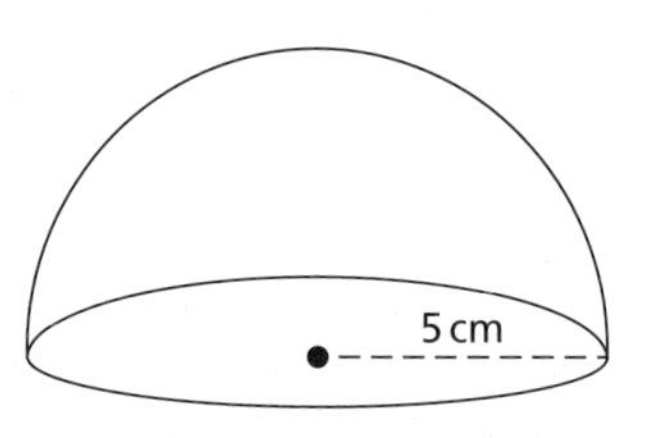

Method

Solution	Commentary
a Volume of a sphere $= \frac{4}{3} \times \pi \times 5^3 = \frac{500}{3}\pi$ Volume of the hemisphere $= \frac{500}{3}\pi \div 2 = \frac{250}{3}\pi\ \text{cm}^3$	Work out the volume of a sphere with the same radius using the formula $V = \frac{4}{3}\pi r^3$ and halve your answer.
b Curved surface area $= 4\pi r^2 \div 2 = 2 \times \pi \times 5^2 = 50\pi\ \text{cm}^2$	Find half the curved surface area of a full sphere.
Area of base $= \pi \times 5^2 = 25\pi\ \text{cm}^2$ Total surface area $= 50\pi + 25\pi = 75\pi\ \text{cm}^2$	Find the area of the circular base and add this to the curved surface area.

Practice (C)

1. Work out the volume of each sphere, giving your answers to 1 decimal place.

a 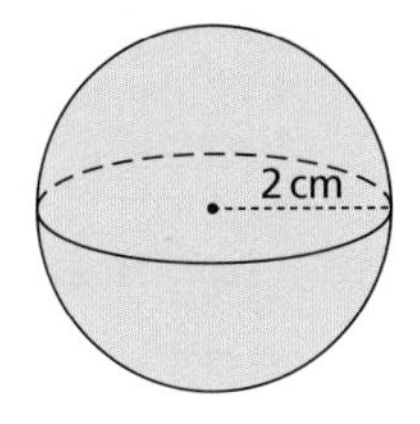

b 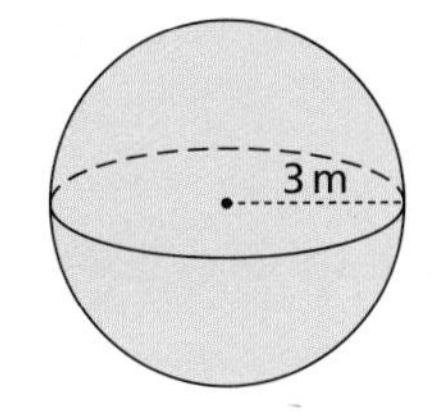

c 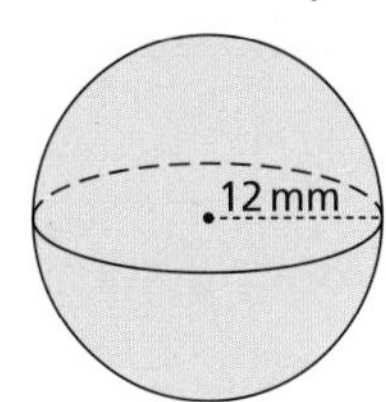

d 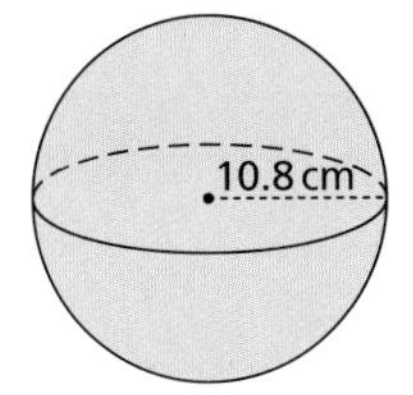

e 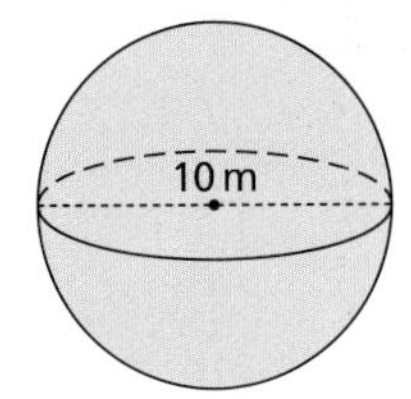

f 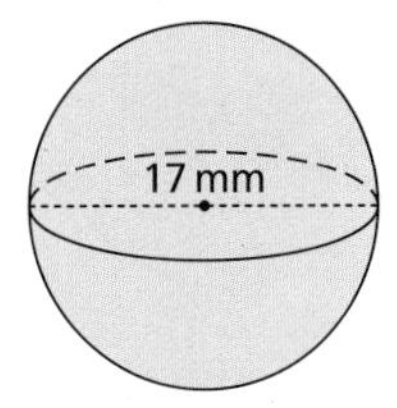

2. Work out the volume of each sphere. Give your answers in terms of π.

a 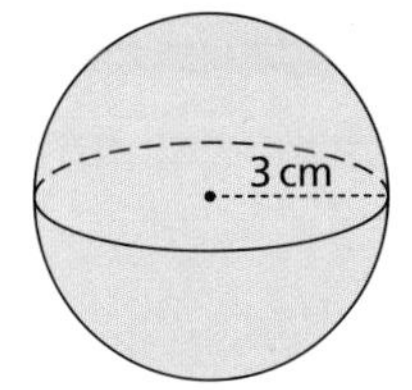

b 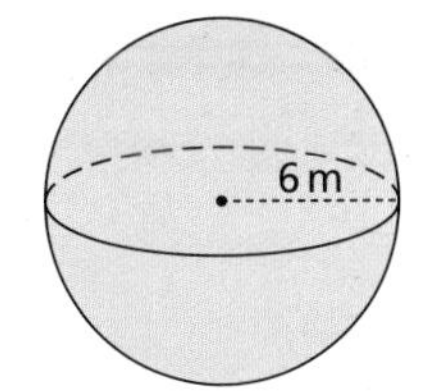

c 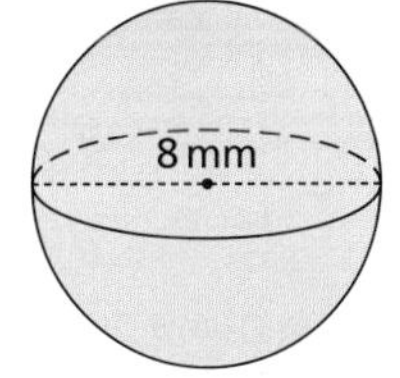

3. Work out the volume of this hemisphere, giving your answer to 1 decimal place.

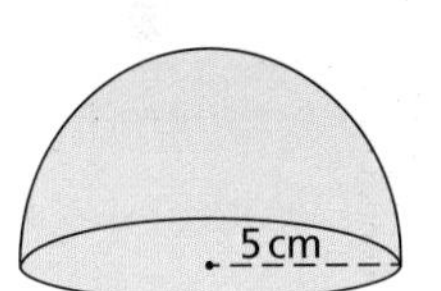

4. Work out the volume of the shape, giving your answer to 1 decimal place.

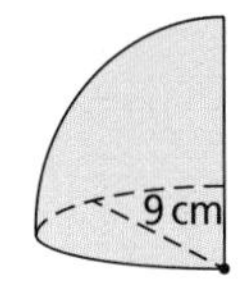

5. The volume of the sphere is 850 cm^3.

   Work out the radius of the sphere.
   Give your answer to 1 decimal place.

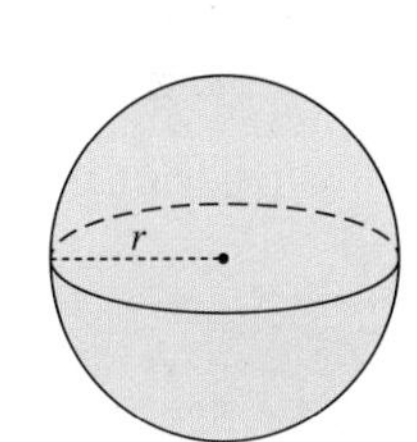

6. The volume of the sphere is 3000 cm^3.

   Work out the radius of the sphere.
   Give your answer in terms of π.

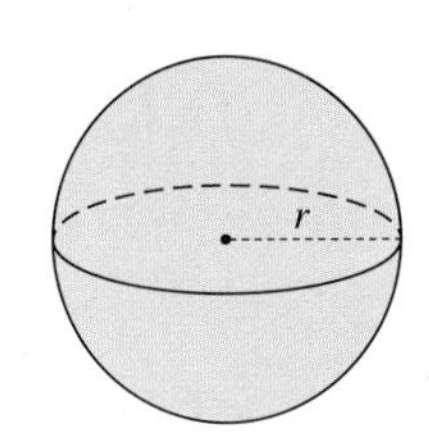

7 Here is a cuboid made from metal.

The cuboid is melted down and moulded to form a sphere.

Work out the radius of the sphere.

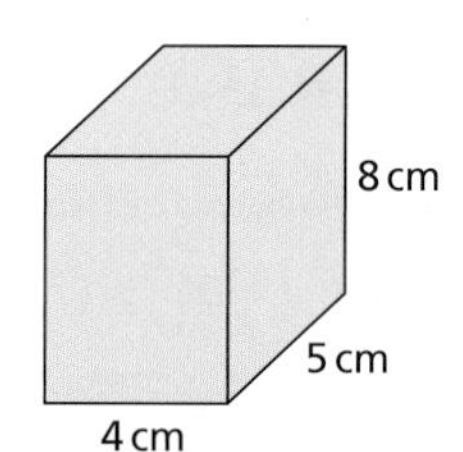

8 Work out the surface area of each sphere. Give your answers to 3 significant figures.

a 8 cm

b

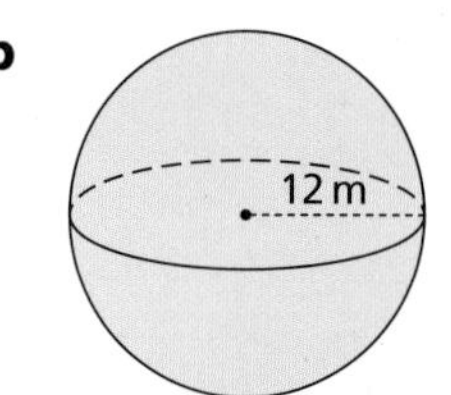

c

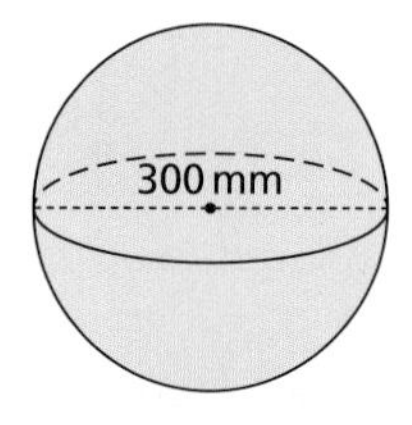

d

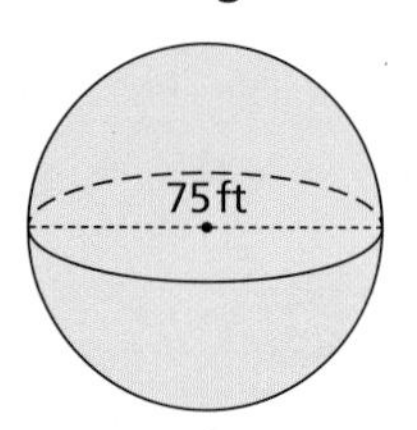

9 Work out the surface area of each sphere. Give your answers in terms of π.

a

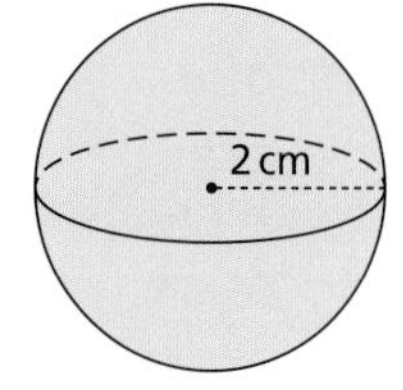

b

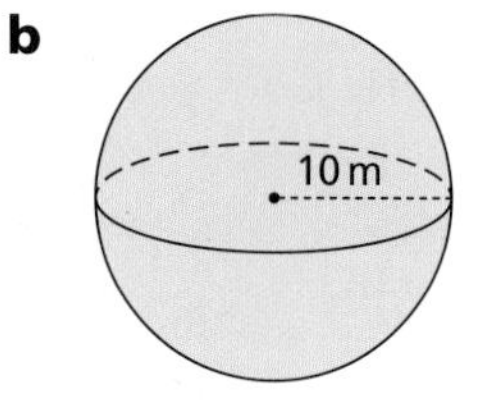

c

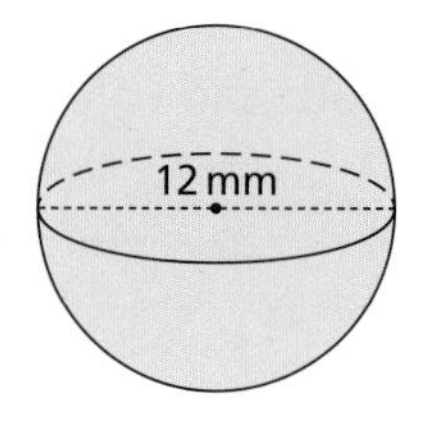

10 A sphere has a surface area of 325 cm^2.

Work out the radius of the sphere. Give your answer to 1 decimal place.

11 A sphere has a surface area of 900π cm^2.

Work out the radius of the sphere.

12 The sphere and the cube have the same surface area.

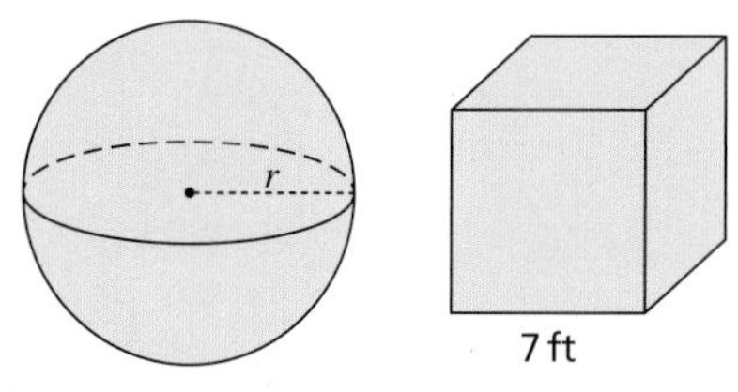

Work out the radius of the sphere.

What do you think? (C)

1 A sphere is placed tightly inside a cube so that it just touches each side.

Will a sphere always occupy the same proportion of space when placed inside a cube this way, regardless of how big the cube is? Explain how you know.

2 A hemisphere of radius 4 cm is shown.

Huda says, "I can calculate the surface area with the formula $A = 3\pi r^2$"

Is Huda correct?

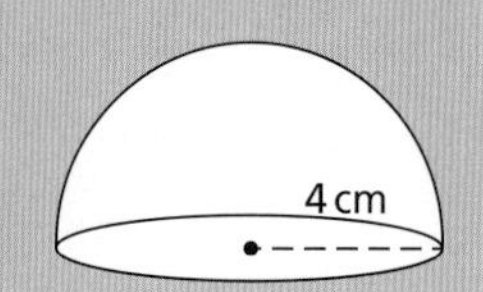

Consolidate – do you need more?

1 For each cylinder, work out: **i** the volume **ii** the surface area.

Give your answers to 3 significant figures.

a

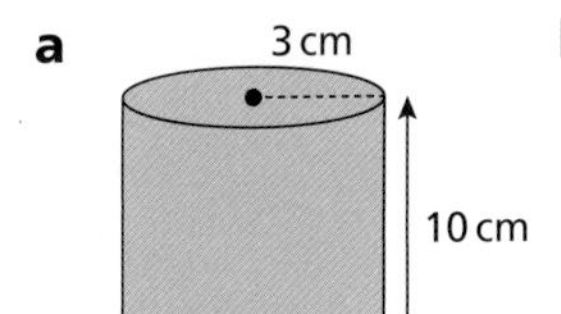

b

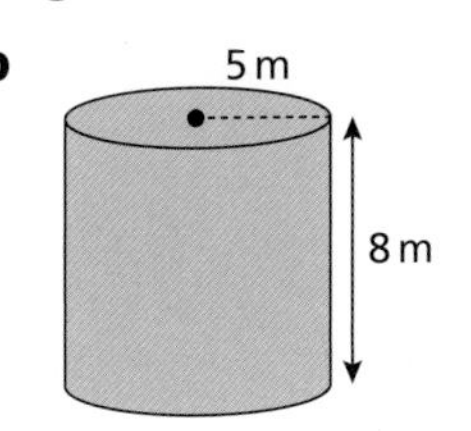

c

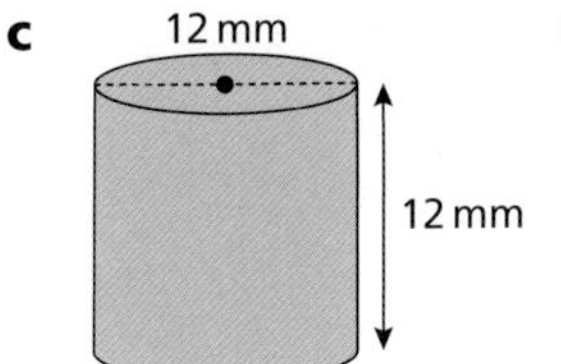

d

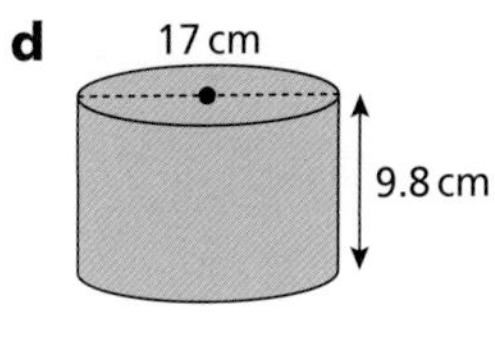

2 Work out the volume of each cone.

a Give your answer in terms of π.

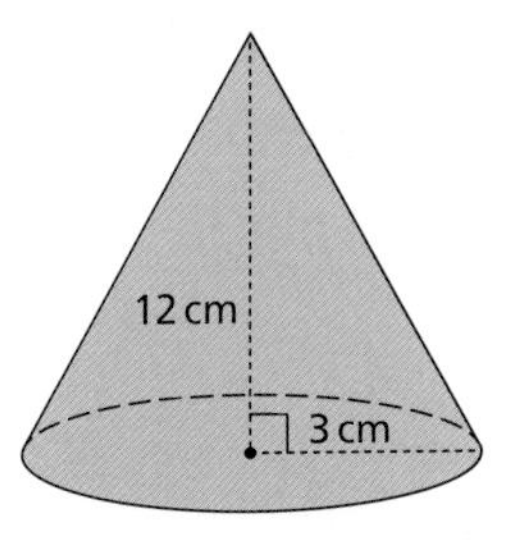

b Give your answer to 3 significant figures.

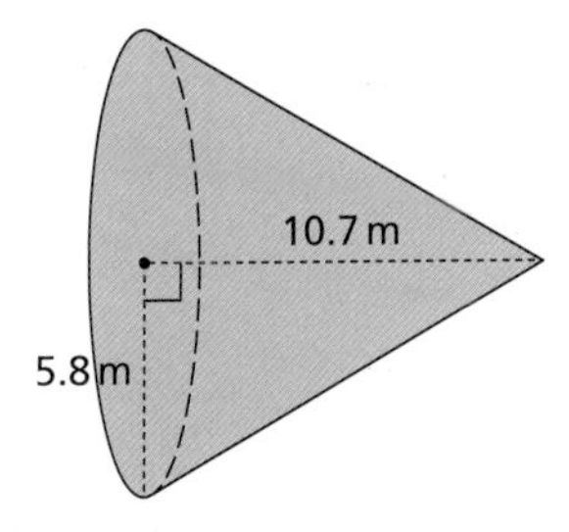

c Give your answer in terms of π.

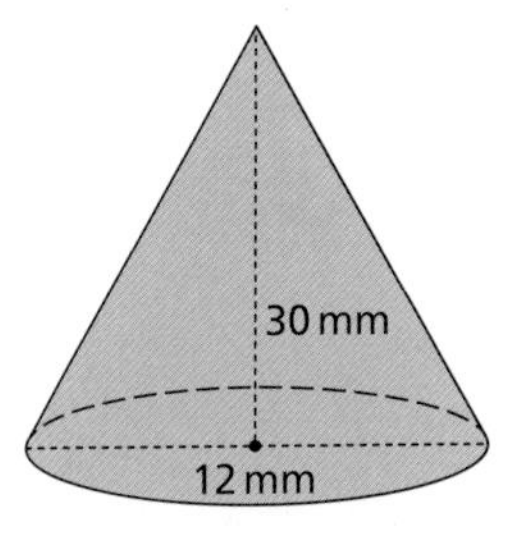

d Give your answer to 3 significant figures.

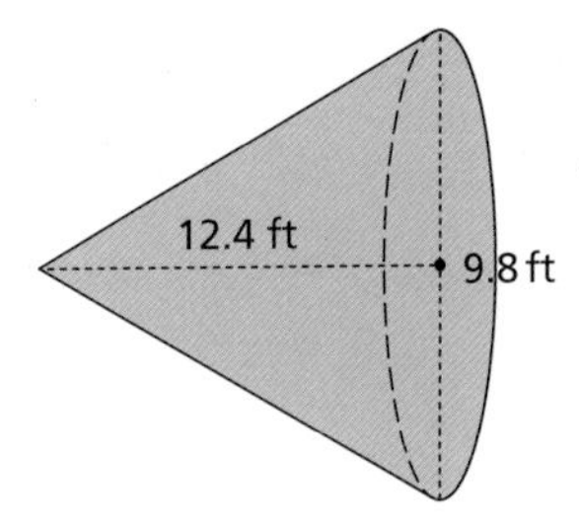

3 Work out the surface area of each cone. Give your answers to 1 decimal place.

a

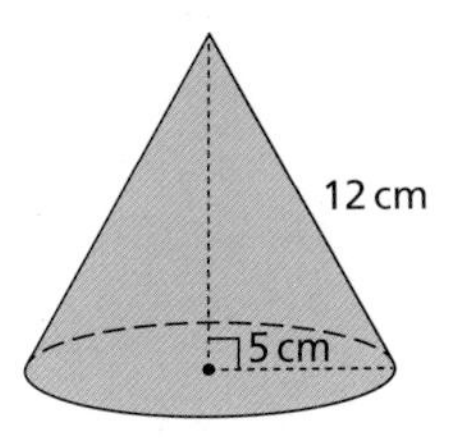

b

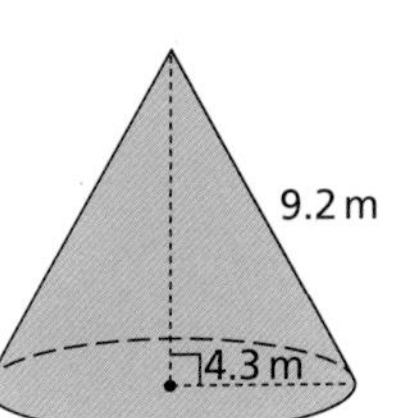

c

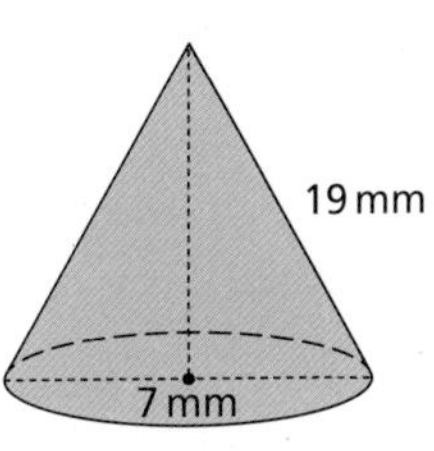

d

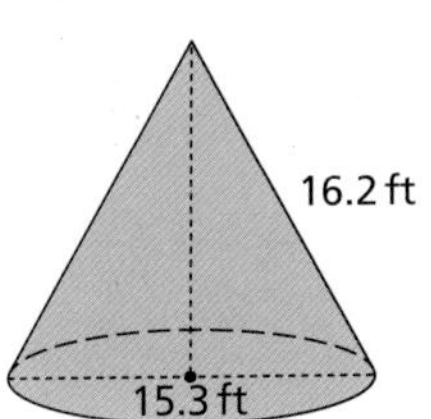

4 For each sphere, work out: **i** the volume **ii** the surface area.

Give your answers to 1 decimal place where appropriate.

a

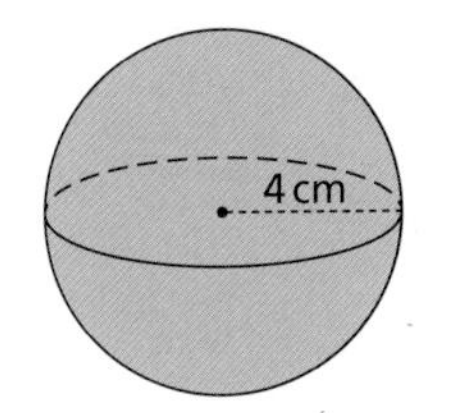

b

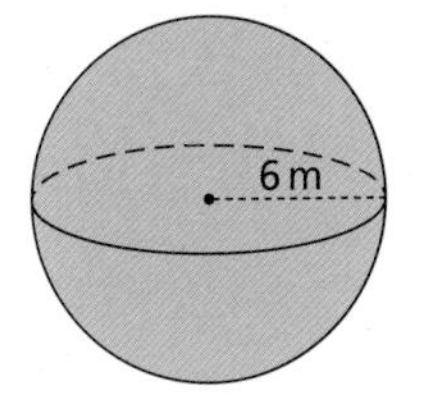

c

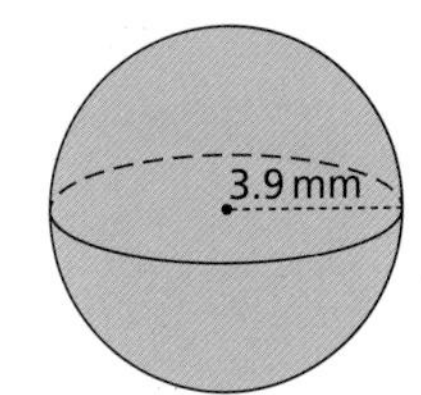

d

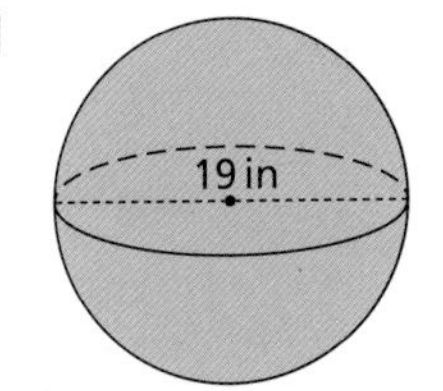

Stretch – can you deepen your learning?

1 A tin of tuna has a height of 4 cm and a radius of 3.75 cm.

A label is wrapped around the outside with a 1.5 cm overlap.

Calculate the area of the label.

2 The shape is a cylinder, with a cylindrical hole cut through its centre.

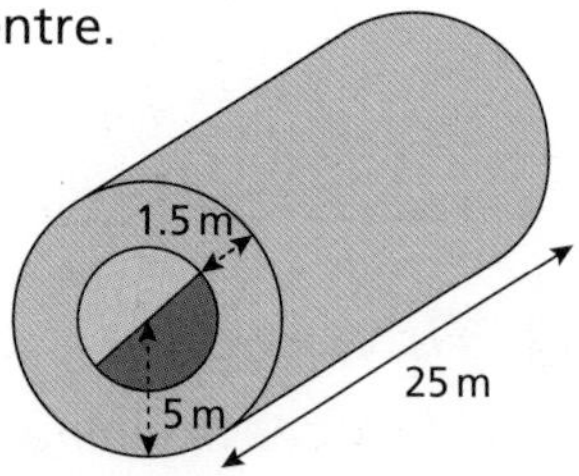

1.5 m
5 m
25 m

a Work out the volume of the shape.

Give your answer in terms of π.

The shape is dipped in paint.

b Work out the total area covered in paint.

Give your answer in terms of π.

3 The cylinder and the sphere have the same volume.

The radius of the cylinder is $\frac{1}{3}$ of the height.

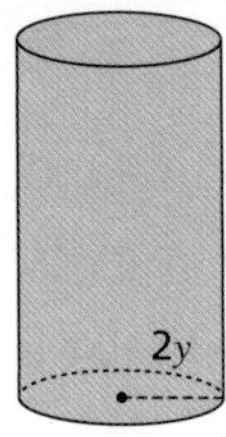

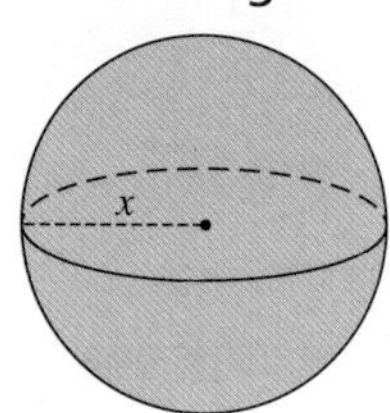

Express x in terms of y.

5.5 Similarity in 2D and 3D

Are you ready? (A)

1 Work out the area of each shape.

a

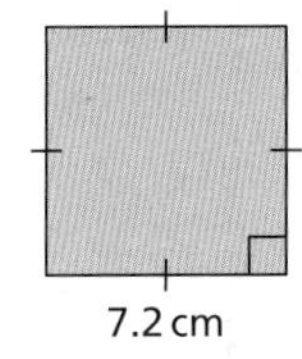

b

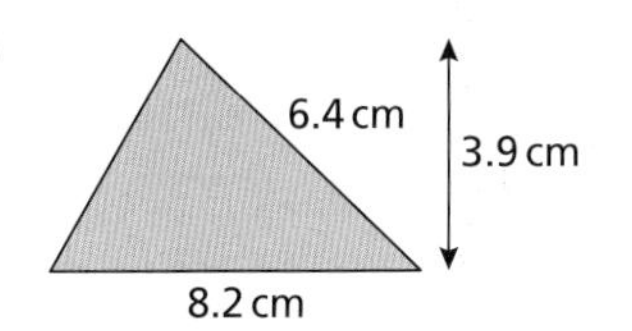

c

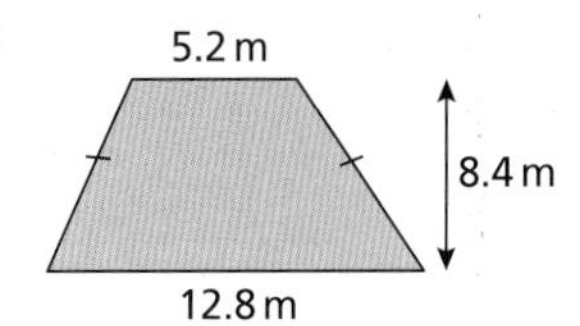

2 Each pair of shapes are similar.

Work out the length of each lettered side.

a

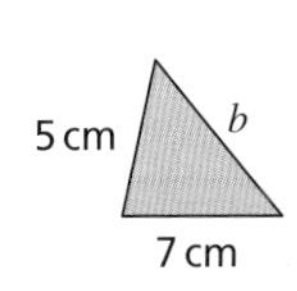

15 cm
21 cm
a

b

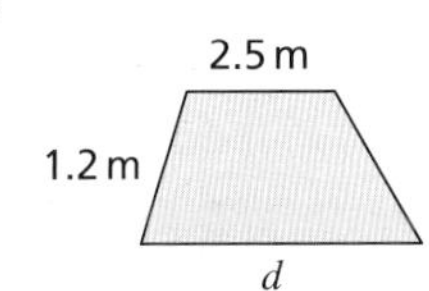

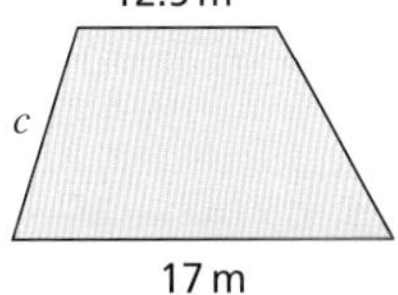

In **similar shapes,** the lengths of the corresponding sides are in the same ratio.

Here are two rectangles, with dimensions as shown and areas calculated using length × width.

3 cm
$A = 6\text{ cm}^2$
2 cm

6 cm
$A = 24\text{ cm}^2$
4 cm

You can work out the ratio of their dimensions.

Ratio of widths = 2 cm : 4 cm = 1 : 2

Ratio of lengths = 3 cm : 6 cm = 1 : 2

You can also work out the ratio of their areas.

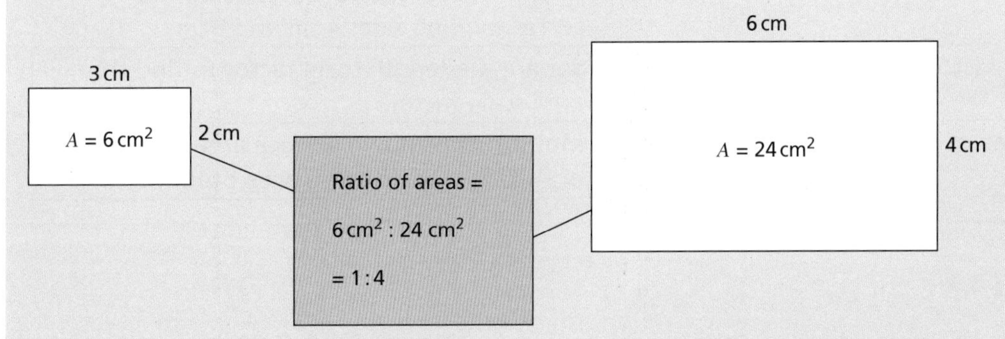

The area of the larger rectangle is $\frac{24}{6} = 4 = 2^2$ times the area of the smaller rectangle.

Generally:

If the lengths in a shape are multiplied by a scale factor of k, then the area will be multiplied by a scale factor of k^2.

Or, if the ratio of corresponding lengths is $a : b$, then the ratio of areas is $a^2 : b^2$.

Example 1

Trapezium A and trapezium B are similar.

The area of trapezium A is 12 cm².

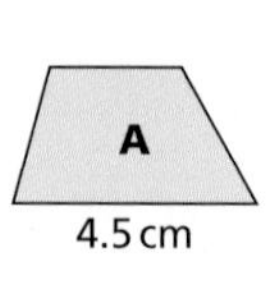

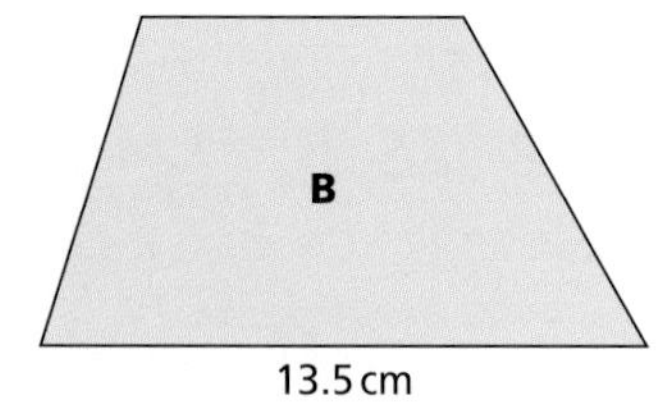

Work out the area of trapezium B.

Method

Solution	Commentary
Length scale factor = 13.5 ÷ 4.5 = 3	Find the length scale factor using the corresponding sides 4.5 cm and 13.5 cm.
Area scale factor = 3^2 = 9	Square the length scale factor to find the area scale factor.
Area of trapezium B = 12 × 9 = 108 cm²	Multiply the area of A by 9 to find the area of B.

Example 2

A 3D printer prints two similar cuboids.

The surface area of cuboid A is 76 cm².

Work out the surface area of cuboid B.

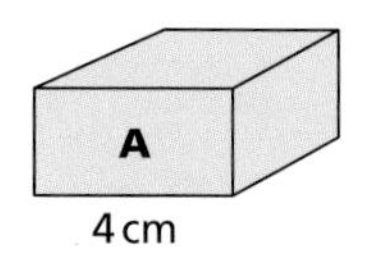

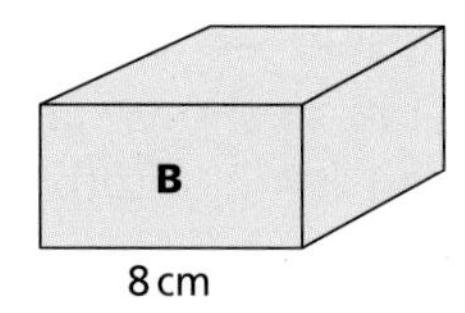

Method

Solution	Commentary
Length scale factor = 8 ÷ 4 = 2	Find the length scale factor using the corresponding sides 4 cm and 8 cm.
Area scale factor = 2^2 = 4	Square the length scale factor to find the area scale factor.
Surface area of cuboid B = 76 × 4 = 304 cm²	Multiply the known surface area of cuboid A by 4 to find the surface area of B.

Practice (A)

1. Rectangles A and B are similar.

   a Work out the scale factor of the lengths of rectangle A to B.

   b Work out the scale factor of the areas of rectangle A to B.

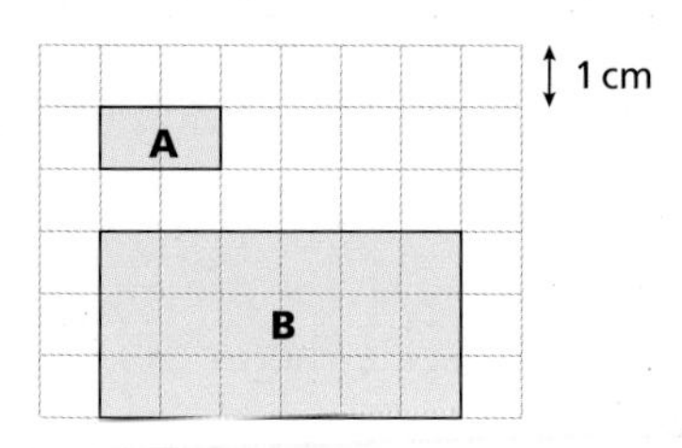

2 Triangles A and B are similar.

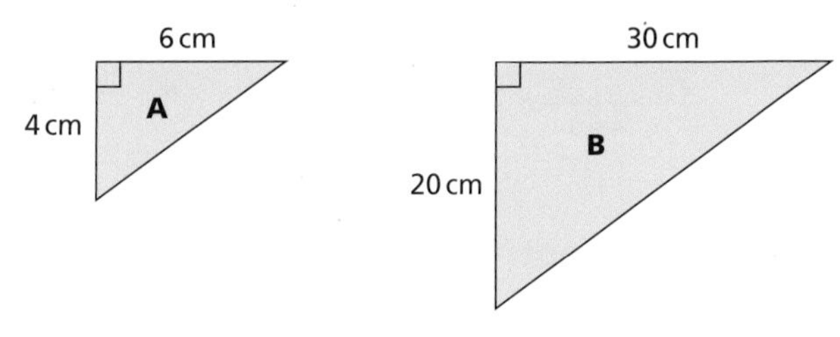

a Work out the length scale factor from A to B.

b Work out the area of triangle A.

c Work out the area of triangle B.

d Work out the area scale factor from A to B.

3 For each pair of similar shapes, work out:

i the scale factor of the lengths

ii the scale factor of the areas.

a

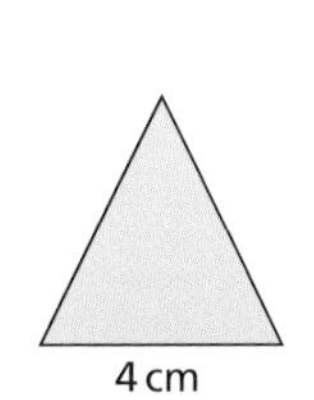

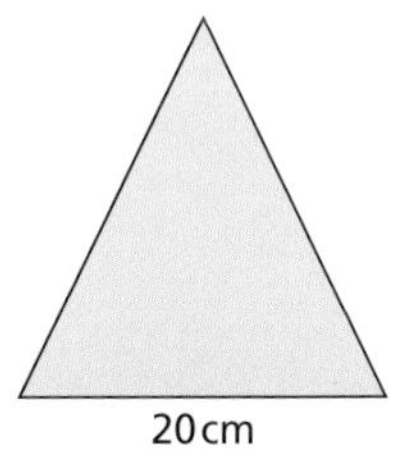

b

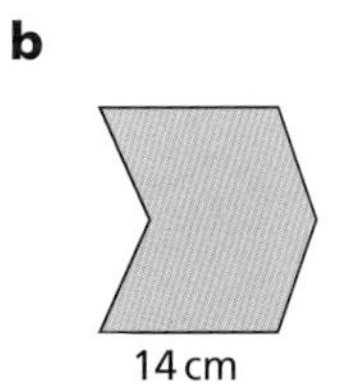

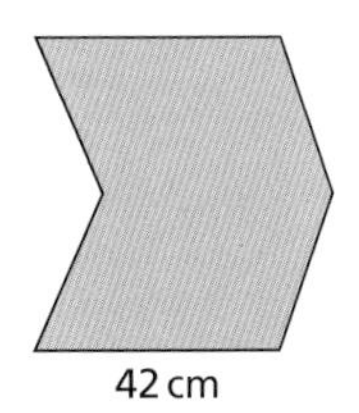

4 Triangles A and B are similar.

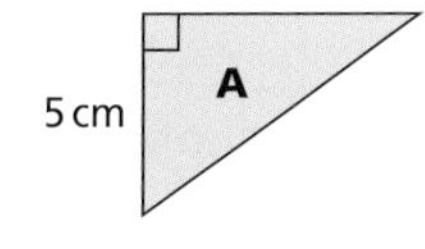

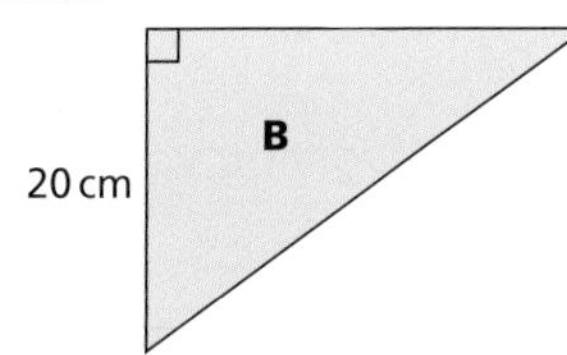

a Work out the length scale factor from A to B.

b Work out the area scale factor from A to B.

c Triangle A has an area of 15 cm^2.

Work out the area of triangle B.

5 Pentagons A and B are similar.

The area of pentagon A is 54 cm^2.

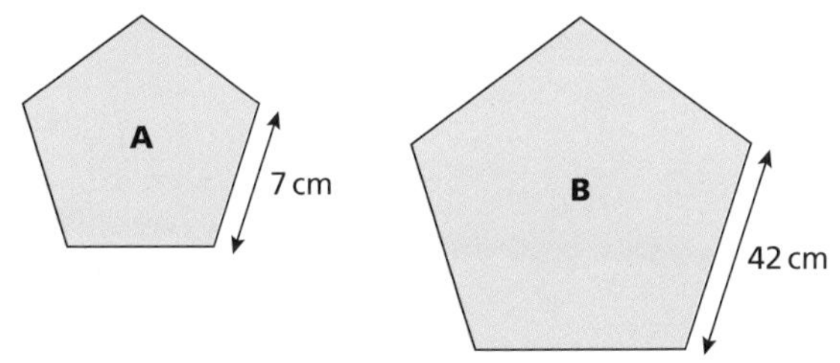

a Work out the area scale factor from A to B.

b Work out the area of B.

6 Triangles A and B are similar.

The area of triangle A is 15 cm^2.

Work out the area of triangle B.

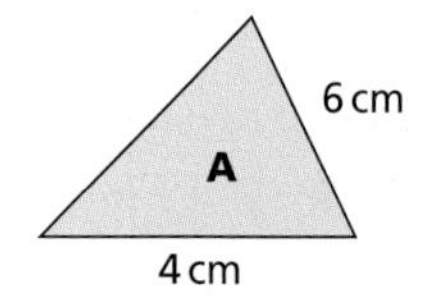

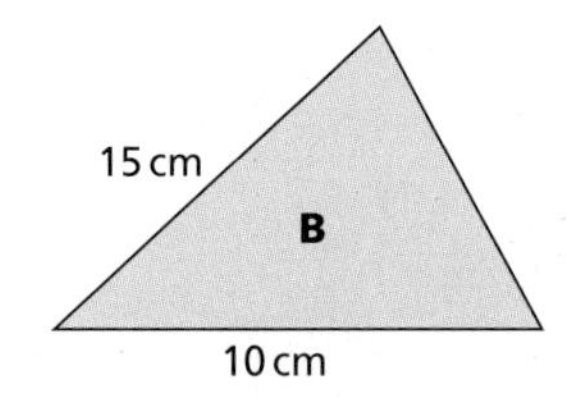

7 The pairs of shapes are similar. The area of shape A in each pair is given.

Work out the area of shape B in each pair.

a Area = 12 cm^2

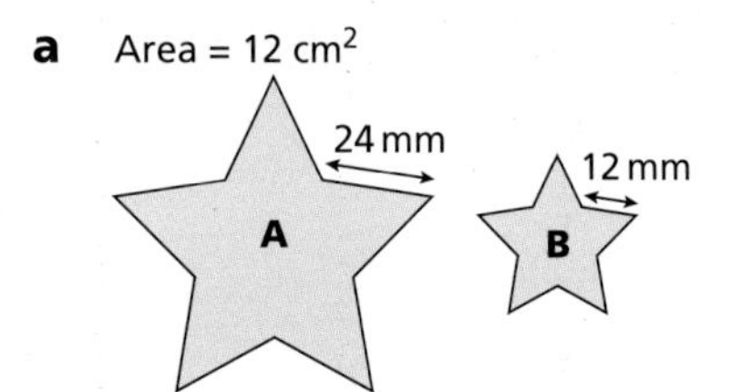

b Area = 200 cm^2

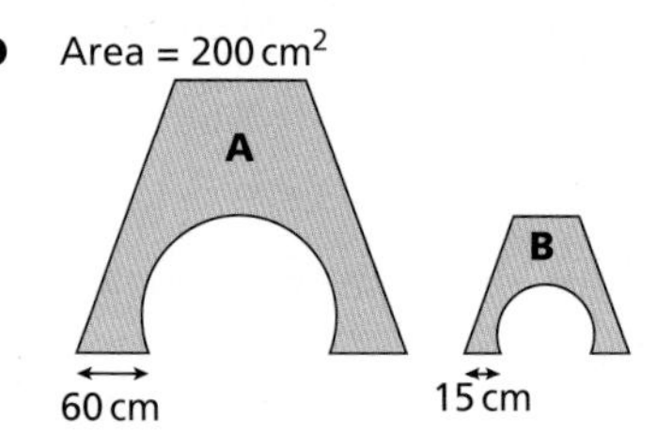

8 Pentagons A and B are similar.

The area of pentagon B is 144 m^2.

Work out the area of pentagon A.

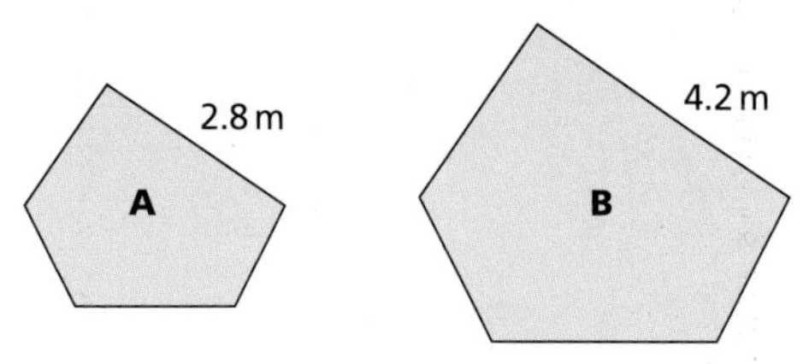

9 Cuboids A and B are similar.

Cuboid A has a surface area of 158 cm^2.

Work out the surface area of cuboid B, giving your answer to 2 decimal places.

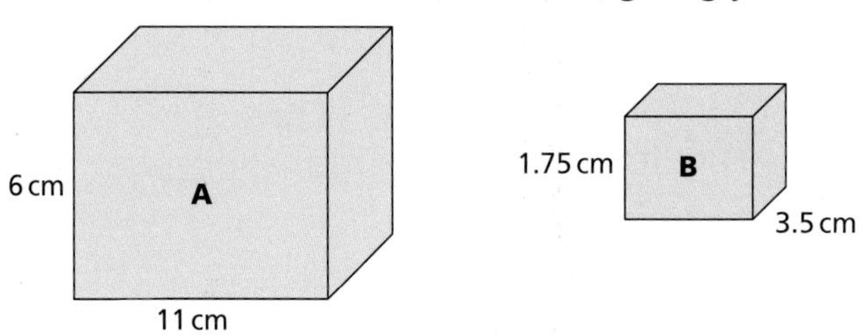

10 Cuboids A and B are similar.

The surface area of cuboid A is 60 cm^2 and the surface area of cuboid B is 2940 cm^2.

Work out the value of x.

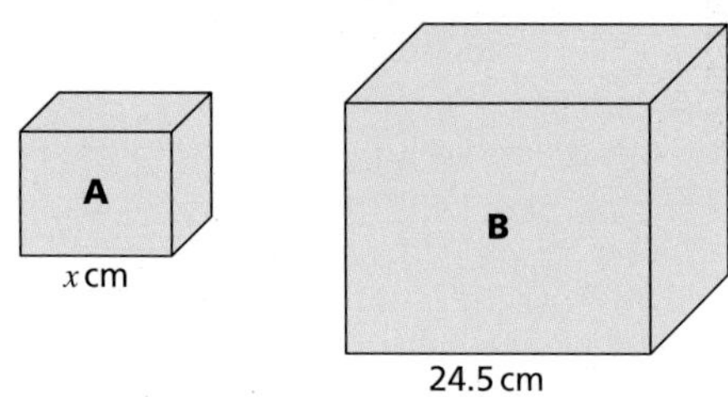

What do you think? (A)

1 Ed says, "The area of triangle BCD is $\frac{9}{25}$ of the area of triangle ABE."

Huda says, "The area of triangle BCD is $\frac{9}{16}$ of the area of triangle ABE."

Who do you agree with? Explain your answer.

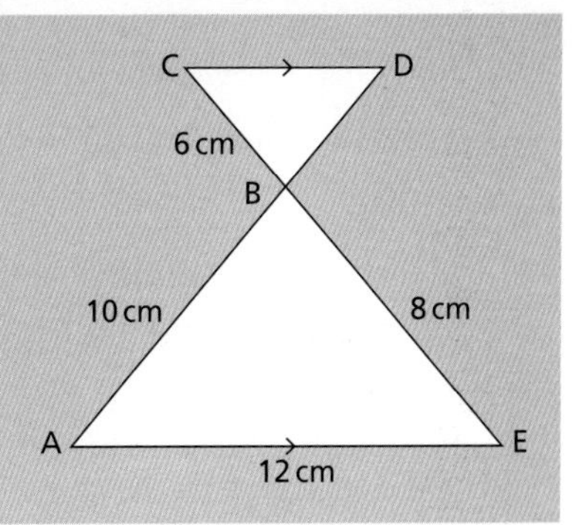

Are you ready? (B)

1 Work out the volume of each shape. Give your answers to 4 significant figures where appropriate.

a

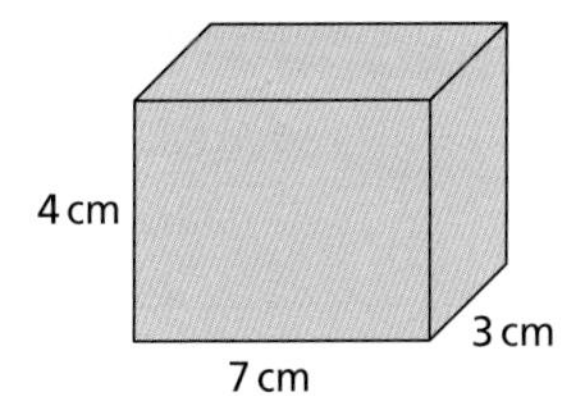

b

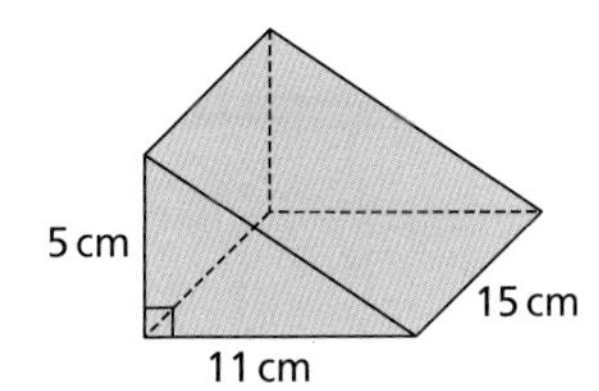

c

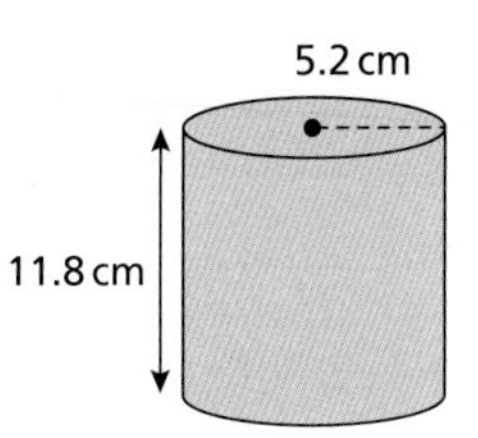

d

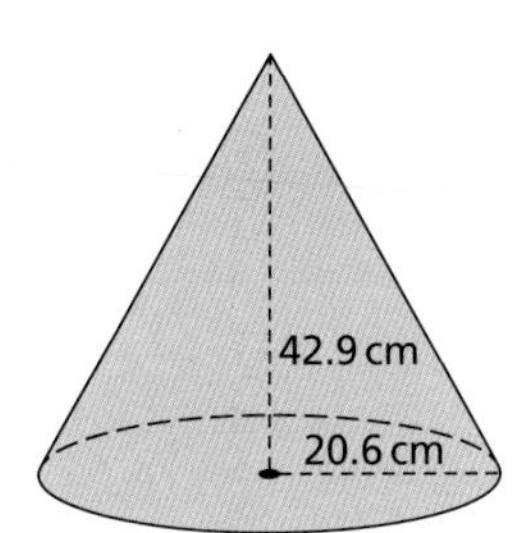

These two cubes are similar.

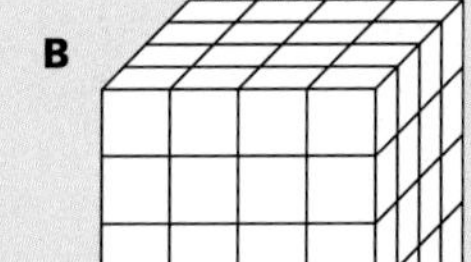

Dividing a length in B by its corresponding length in A gives a length scale factor of $\frac{4}{2} = 2$

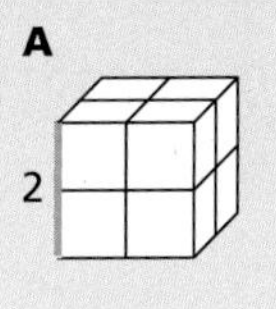

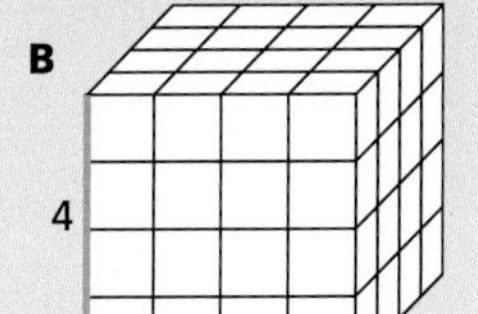

Dividing the area of a face on B by the area of a face on A gives an area scale factor of $\frac{16}{4} = 4$ or 2^2

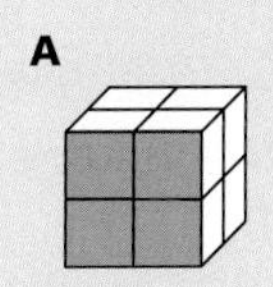

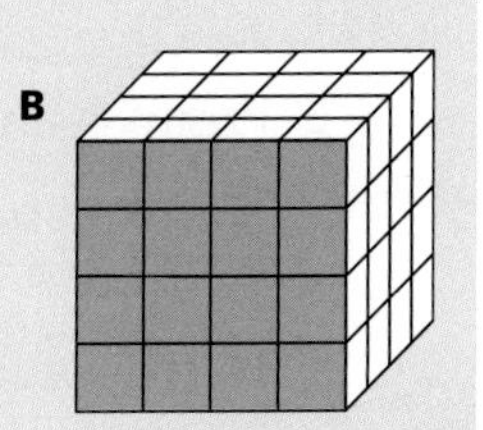

Dividing the volume of B by the volume of A gives a volume scale factor of $\frac{64}{8} = 8$ or 2^3

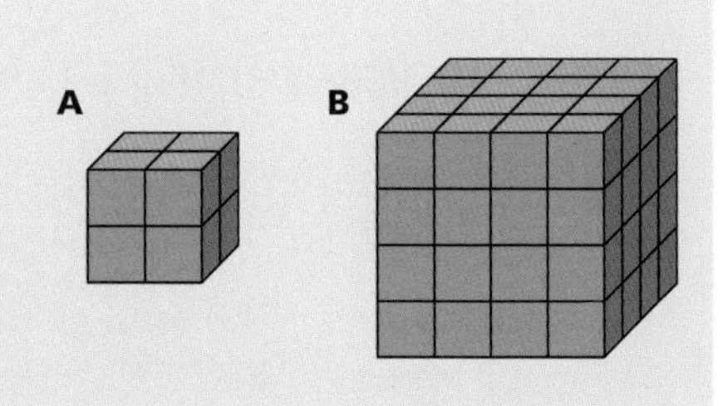

Generally,

If the lengths in a shape are multiplied by a scale factor of k, then the volume will be multiplied by a scale factor of k^3.

Or, if the ratio of corresponding lengths is $a : b$, then the ratio of the volumes is $a^3 : b^3$.

Example

Cuboids A and B are similar.

The volume of A is 72 cm³.

Work out the volume of B.

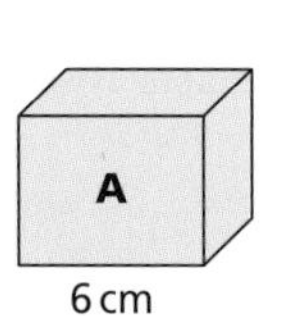

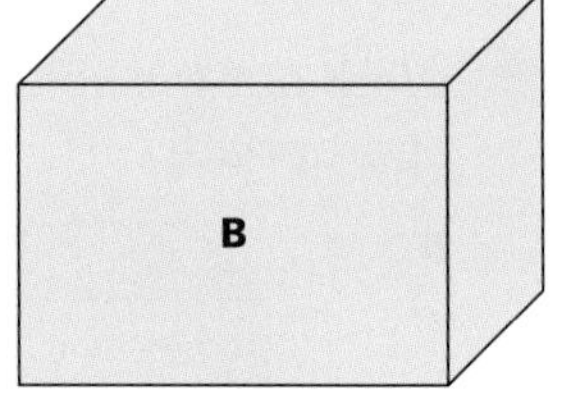

Method

Solution	Commentary
Length scale factor = 18 ÷ 6 = 3	Find the length scale factor using the corresponding sides 6 cm and 18 cm.
Volume scale factor = 3^3 = 27	Cube the length scale factor to find the volume scale factor.
Volume of B = 72 × 27 = 1944 cm³	Multiply the known volume of cuboid A by 27 to find the volume of cuboid B.

Practice (B)

1. Cubes A and B are shown.

   a Work out the length scale factor from A to B.

   b Work out the volume scale factor from A to B.

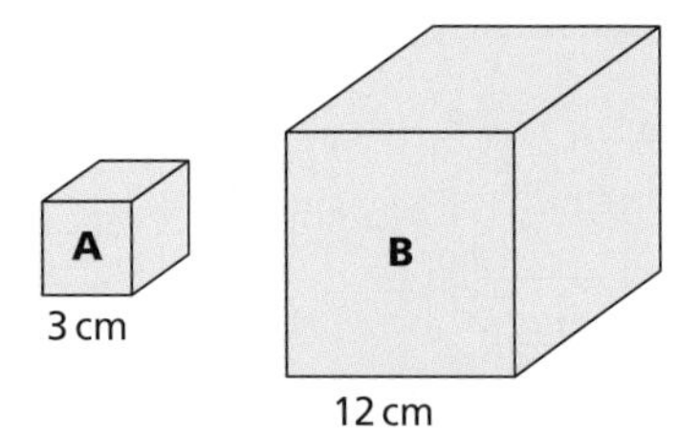

2. Pyramids A and B are similar.

   a Work out the length scale factor from A to B.

   b Work out the volume scale factor from A to B.

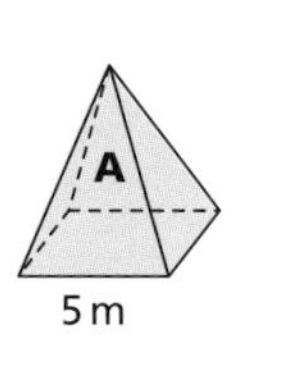

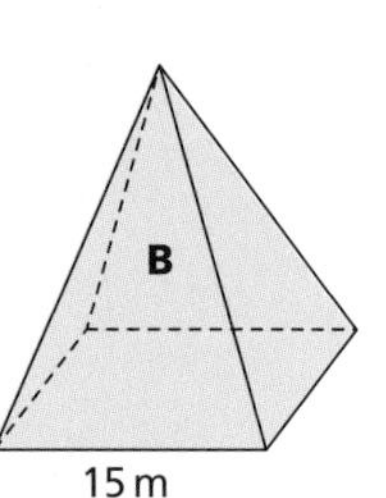

3. Prisms A and B are similar.

   a Work out the volume scale factor from A to B.

   b The volume of prism A is 35 cm³.

   Work out the volume of prism B.

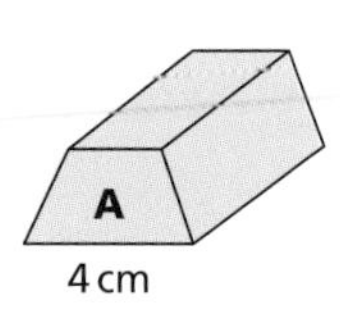

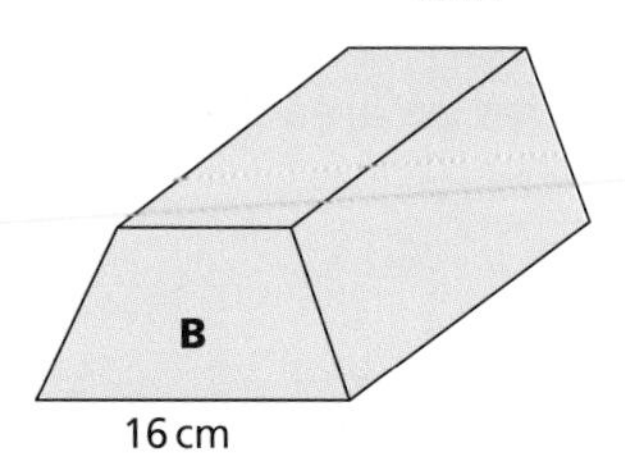

4 Shapes A and B are similar.

The volume of shape A is $35\,\text{cm}^3$.

Work out the volume of shape B.

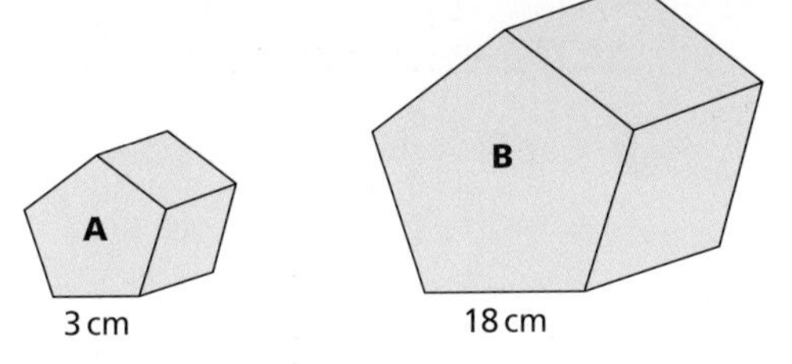

5 The pairs of shapes are similar. The volume of shape A in each pair is given.

Work out the volume of shape B in each pair. Give your answers to 4 significant figures.

a

Volume = $40\,\text{cm}^3$

A 7 cm

B 21 cm

b

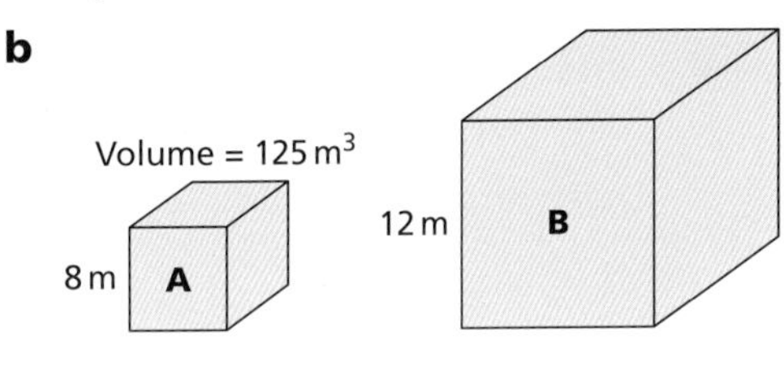

c

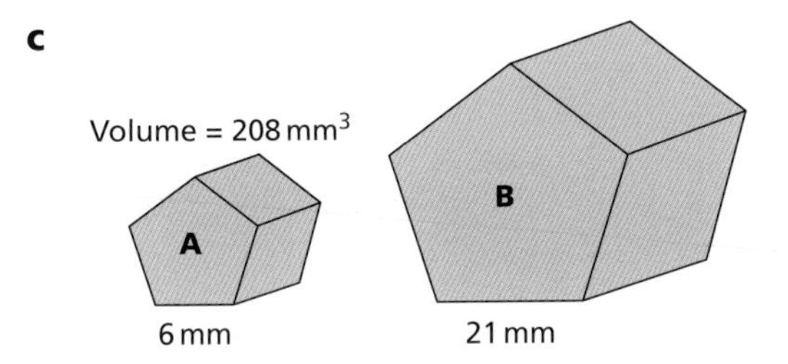

d

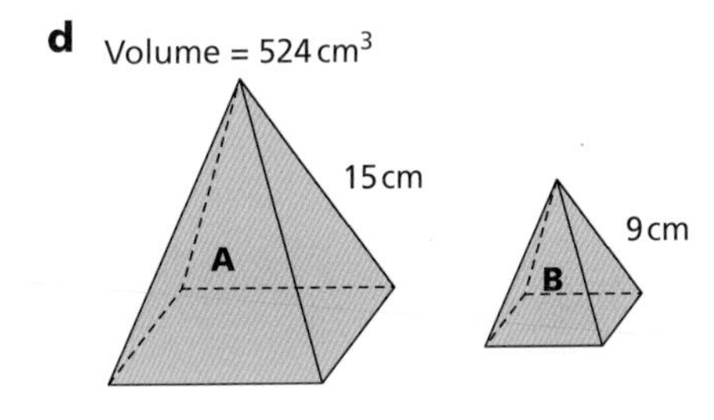

6 Cylinders A and B are similar.

The surface area of cylinder A is $30\,\text{m}^2$.

The surface area of cylinder B is $120\,\text{m}^2$.

a Work out the scale factor of the surface area from A to B.

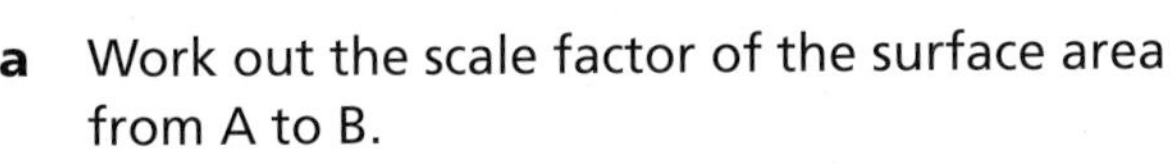

b Work out the scale factor of the length from A to B.

c The volume of cylinder A is $315\,\text{m}^3$.

Work out the volume of cylinder B.

A B

7 Prisms A and B are similar.

The surface area of prism A is $208\,\text{cm}^2$.

The surface area of prism B is $3328\,\text{cm}^2$.

The volume of prism A is $128\,\text{cm}^3$.

Work out the volume of prism B.

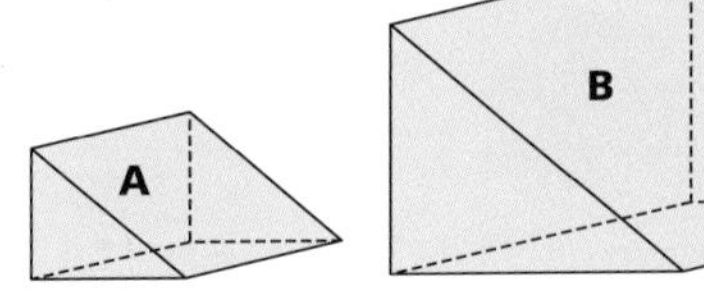

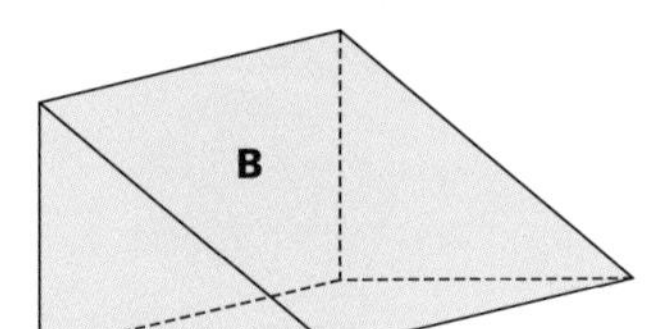

What do you think? (B)

1. Cuboids A and B are similar.

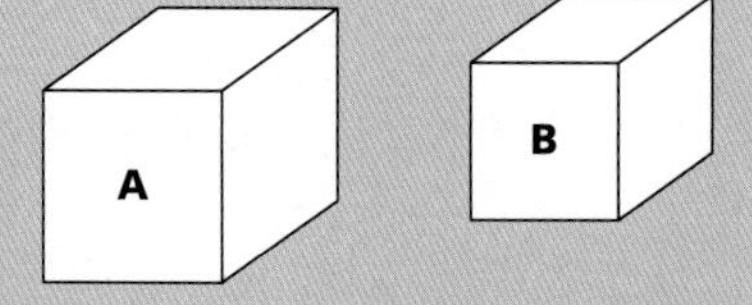

Cuboid A has a surface area of $514\,cm^2$.

Cuboid B has a surface area of $257\,cm^2$.

Huda says, "514 divided by 257 is 2, so the volume of cuboid A is 8 times larger than the volume of cuboid B."

Show that Huda is incorrect.

2. Investigate the sizes of different packages of the same item, for example cereal boxes.

Are the packages similar?

How does the ratio of prices compare to the ratios of lengths, areas and volumes?

Consolidate – do you need more?

1. The pairs of shapes are similar. The area of shape A in each pair is given.

Work out the area of shape B in each pair.

a Area = $32\,cm^2$ A 4 cm; B 12 cm

b Area = $20\,mm^2$ A 7 mm; B 28 mm

c Area = $756\,cm^2$ A 12.3 cm; B 4.1 cm

2. The pairs of shapes are similar. The volume of shape A in each pair is given.

Work out the volume of shape B in each pair. Give your answers to 3 significant figures.

a Volume = $25\,cm^3$

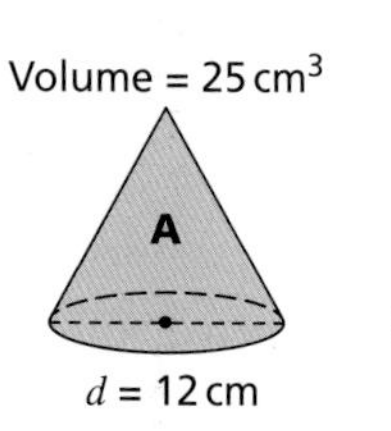

d = 12 cm

B d = 36 cm

b Volume = $155\,m^3$

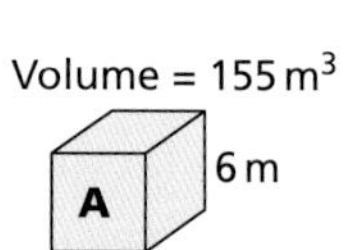

B 21 m

c Volume = $346\,mm^3$

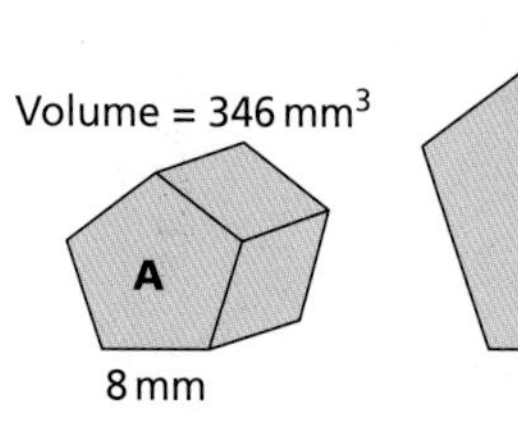

8 mm

B 12 mm

d Volume = $300\,cm^3$

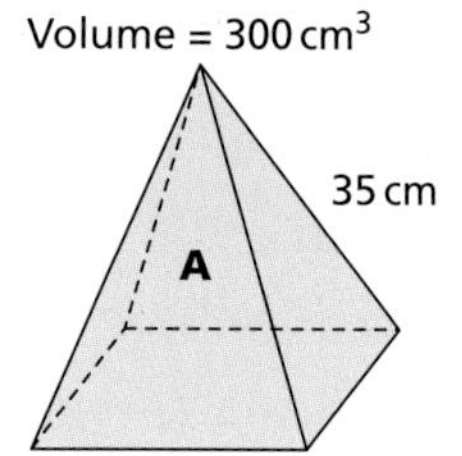

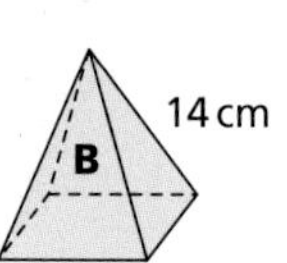

Stretch – can you deepen your learning?

1 The areas of two similar shapes are in the ratio 5 : 2

Show that the ratio of the volumes of the shapes is $a : b\sqrt{10}$, where a and b are integers.

2 Shapes A and B are similar.

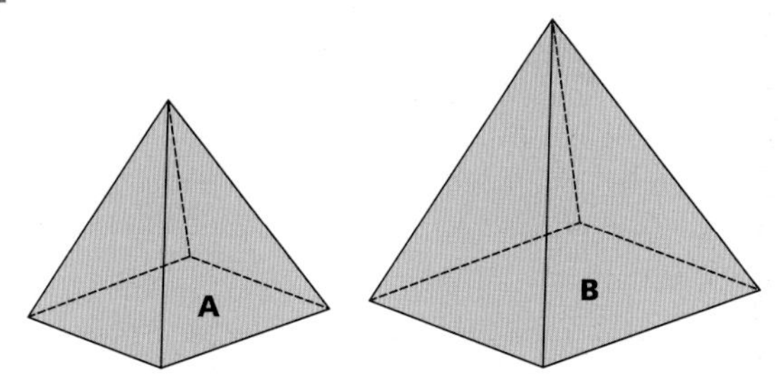

The surface area of shape A is 3 m².

The surface area of shape B is 7.5 m².

Work out the ratio of the volume of shape A to the volume of shape B.

Write your answer in the form 1 : n, giving n to 3 significant figures.

3 The larger cone is split into a smaller cone and a frustum.

The larger cone has a volume of 15 cm³.

Work out the volume of the frustum.

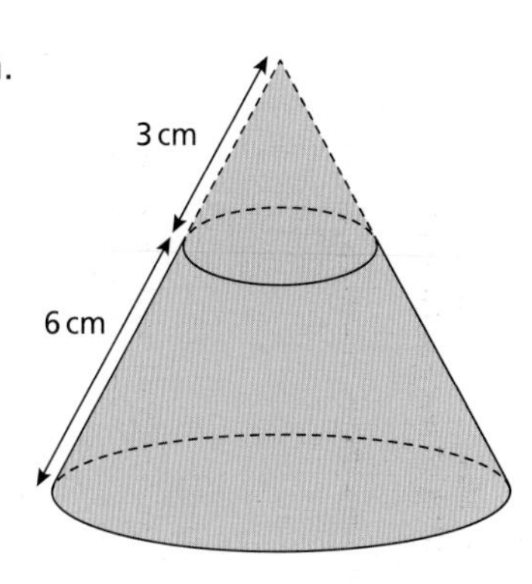

4 Two square-based pyramids, A and B, are shown.

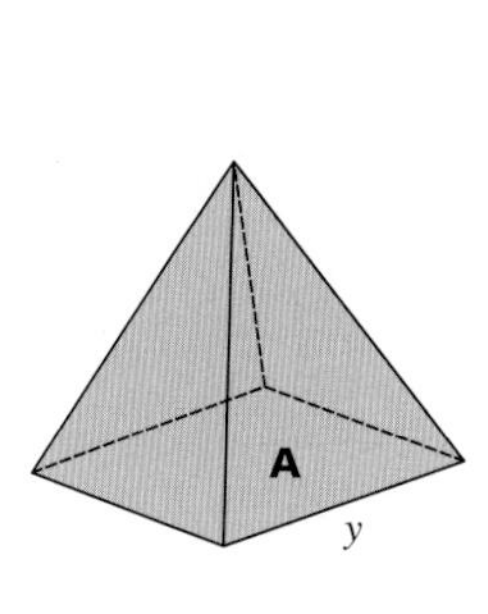

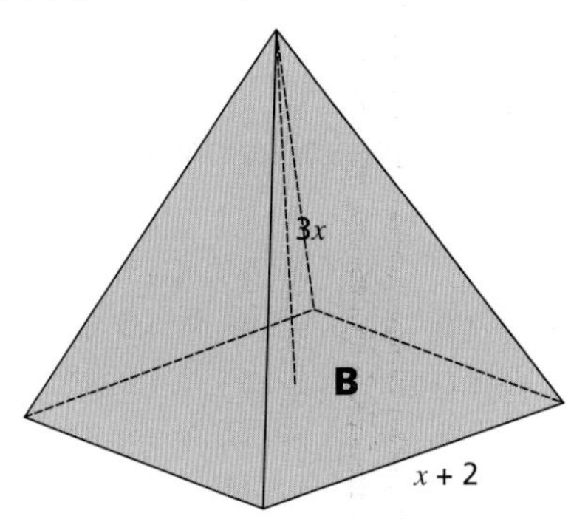

The surface area of pyramid A is 350 square units.

The surface area of pyramid B is 1400 square units. Its perpendicular height is $3x$ units.

a Write the length of y in terms of x.

b Write the volume of A in terms of x.

2D and 3D shapes: exam practice

* *Diagrams are not accurately drawn unless stated.*

4–6

1 The circumference of a circle is 12π cm.

Work out the area of the circle, giving your answer in terms of π. **[2 marks]**

2 The area of the rectangle is twice the area of the triangle.

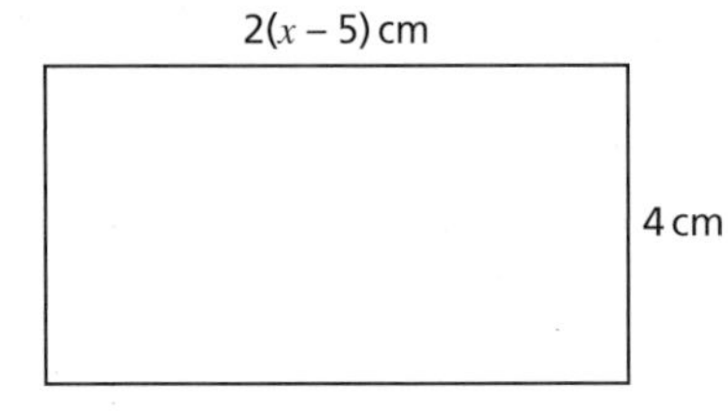

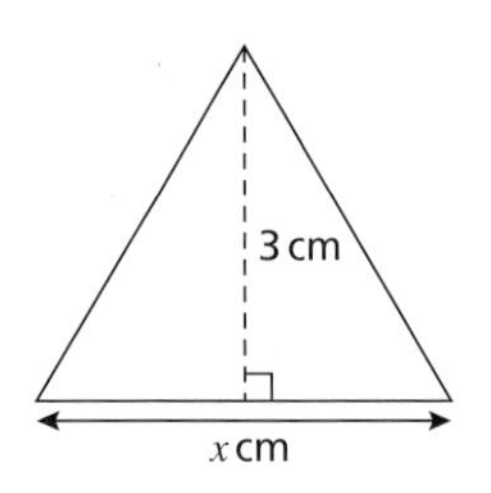

Work out the perimeter of the rectangle. **[4 marks]**

3 The volume, V, of a sphere of radius r is given by the formula $V = \frac{4}{3}\pi r^3$

Work out the volume of a hemisphere of diameter 10 cm. Give your answer correct to 3 significant figures. **[3 marks]**

4 This shape is made using 1 cm cubes.

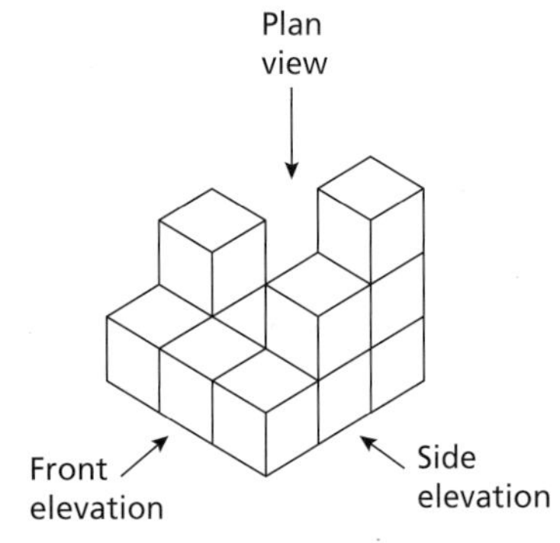

Diagram accurately drawn

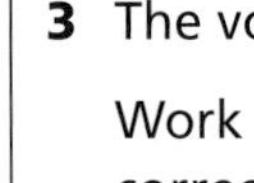

Draw the front and side elevations of the shape on centimetre squared paper. **[2 marks]**

5 The volume of the triangular prism is 120 cm^3.

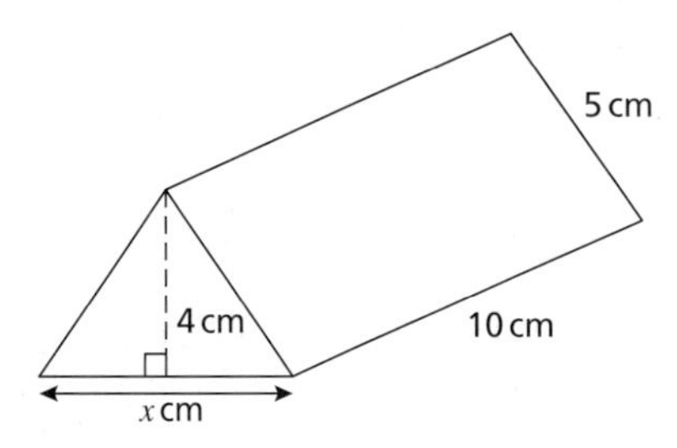

(a) Work out the value of x. **[3 marks]**

(b) Is the cross-section of the prism an isosceles triangle? Explain your answer. **[2 marks]**

7–9

6 A shop sells two mathematically similar posters.

The area of the larger poster is 44% greater than the area of the smaller poster.

The width of the larger poster is 42 cm.

Calculate the width of the smaller poster. **[3 marks]**

7 Two similar cones have volumes of $30\,cm^3$ and $3750\,cm^3$.

The radius of the base of the smaller cone is 2.6 cm.

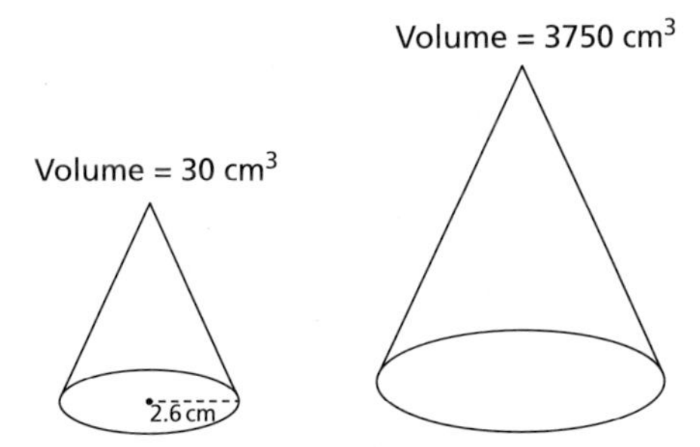

Calculate the radius of the base of the larger cone. **[3 marks]**

6 Right-angled triangles

White Rose
MATHS

In this block, we will cover...

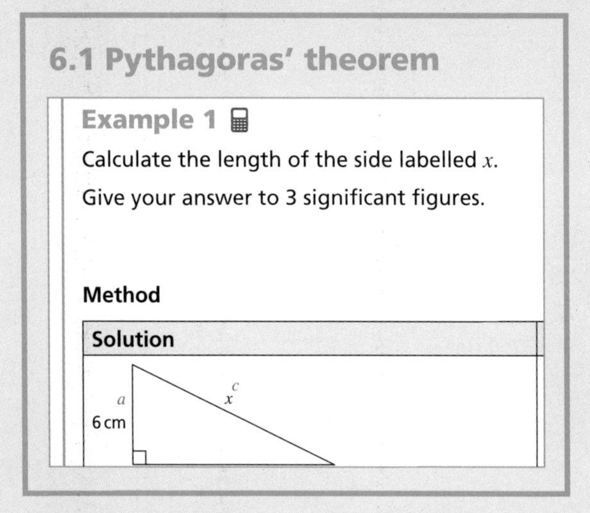
6.1 Pythagoras' theorem
Example 1
Calculate the length of the side labelled x.
Give your answer to 3 significant figures.
Method
Solution
a
6 cm
c
x

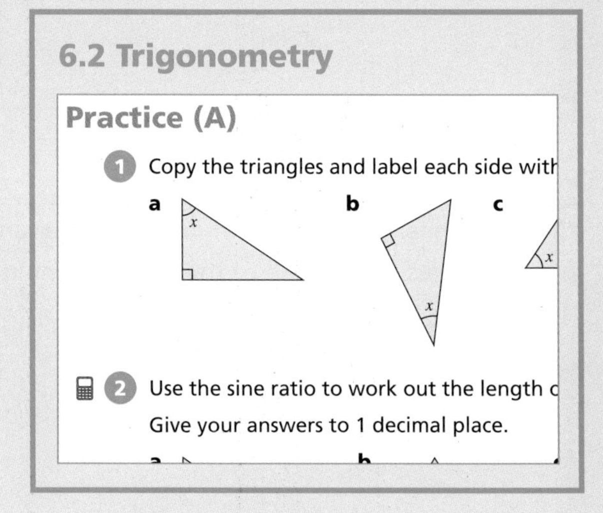
6.2 Trigonometry
Practice (A)
Copy the triangles and label each side with
a
b
c
Use the sine ratio to work out the length o
Give your answers to 1 decimal place.

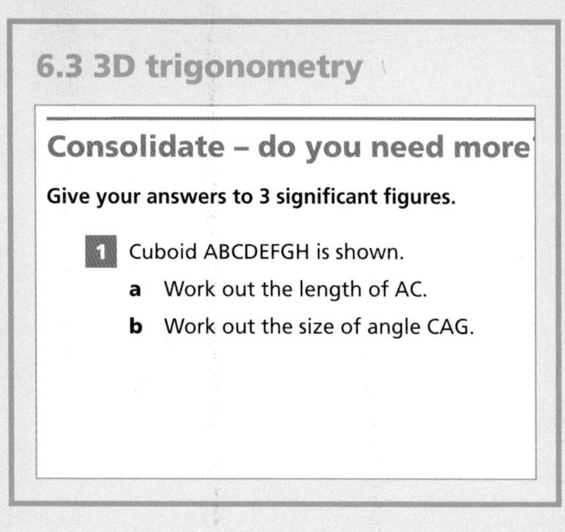
6.3 3D trigonometry
Consolidate – do you need more
Give your answers to 3 significant figures.
Cuboid ABCDEFGH is shown.
a Work out the length of AC.
b Work out the size of angle CAG.

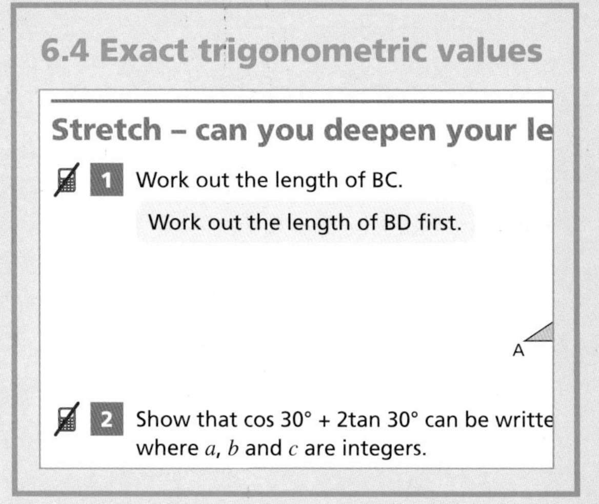
6.4 Exact trigonometric values
Stretch – can you deepen your le
Work out the length of BC.
Work out the length of BD first.
A
Show that cos 30° + 2tan 30° can be writte
where a, b and c are integers.

6.1 Pythagoras' theorem

Are you ready? (A)

1 Work out:

a 5^2 **b** 11^2 **c** 20^2 **d** 8.5^2

2 Work out:

a $\sqrt{36}$ **b** $\sqrt{37}$ **c** $\sqrt{47}$ **d** $\sqrt{4.7}$

3 Round each number to:

i 1 decimal place **ii** 2 decimal places **iii** 3 significant figures.

a 3.2456 **b** 12.6789 **c** 19.058 **d** 20.37298

Pythagoras' theorem states that 'in a right-angled triangle, the square on the hypotenuse is equal to the sum of the squares on the other two sides'.

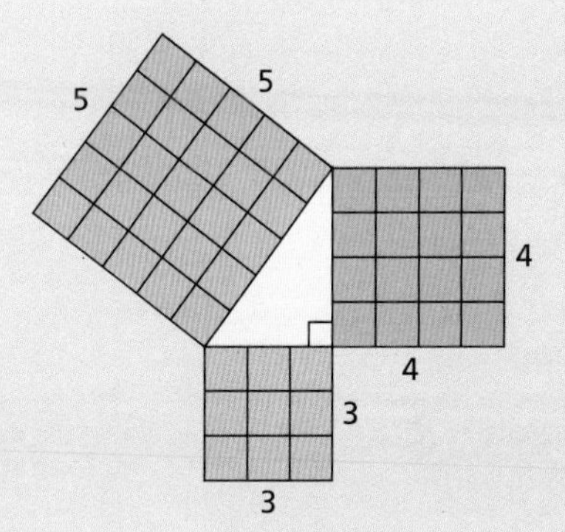

The **hypotenuse** in a right-angled triangle is the name given to the side opposite the right angle.

In this example, the sides of the triangle have lengths 3 units, 4 units and 5 units.

$3^2 + 4^2 = 5^2$

$9 + 16 = 25$

This can be generalised by labelling the sides of the triangle as a, b and c, where c is always the hypotenuse.

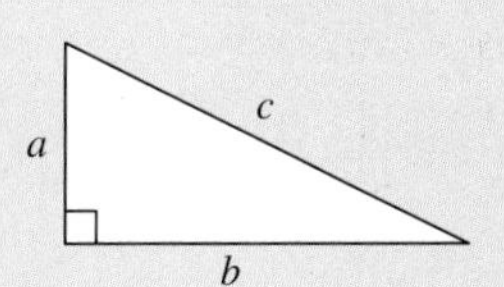

$a^2 + b^2 = c^2$

This can be written as $a^2 + b^2 = \text{hyp}^2$ if you prefer.

Example 1

Calculate the length of the side labelled x.

Give your answer to 3 significant figures.

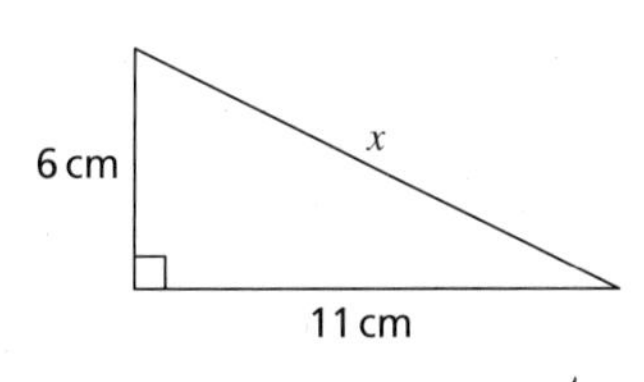

Method

Solution	Commentary
a 6 cm, c x, 11 cm b	Label the sides a, b and c. Make sure that c is the hypotenuse.

$a^2 + b^2 = c^2$	Or you can use $a^2 + b^2 = \text{hyp}^2$
$6^2 + 11^2 = x^2$ $36 + 121 = x^2$ $157 = x^2$ $x = \sqrt{157}$ $x = 12.5$ cm (to 3 s.f.)	Substitute the values you know into the formula for Pythagoras' theorem and solve for x.

Example 2

ABC is an isosceles triangle.

Work out its area. Give your answer to 3 significant figures.

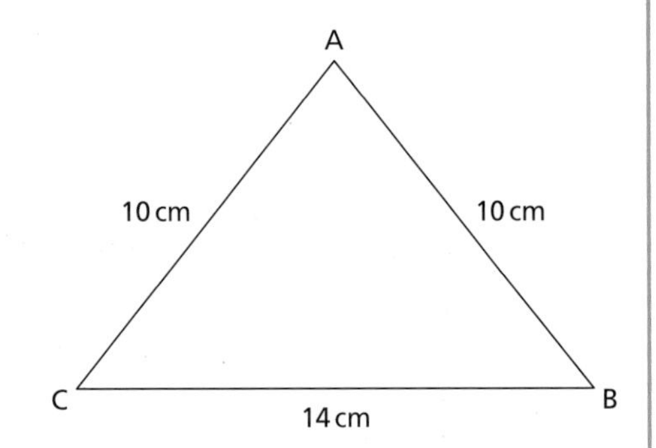

Method

Solution	**Commentary**
	Pythagoras' theorem only applies to right-angled triangles. Start by drawing a line to split the triangle into two right-angled triangles and label the sides of one of them.
$a^2 + b^2 = \text{hyp}^2$	You can use either triangle ABM or ACM as both are right-angled.
$x^2 + 7^2 = 10^2$ $x^2 + 49 = 100$ $x^2 = 100 - 49$ $x^2 = 51$ $x = 7.141\,42\ldots$ cm	Substitute the values you know into the formula for Pythagoras' theorem and solve for x.
Area $= \frac{1}{2} \times 14 \times 7.141\,42\ldots$ $= 50.0\ \text{cm}^2$ (to 3 s.f.)	Use the formula $\frac{1}{2} \times$ base $\times$ height to find the area of the isosceles triangle.

Practice (A)

1 Here is a right-angled triangle.

Work out the length of the hypotenuse.

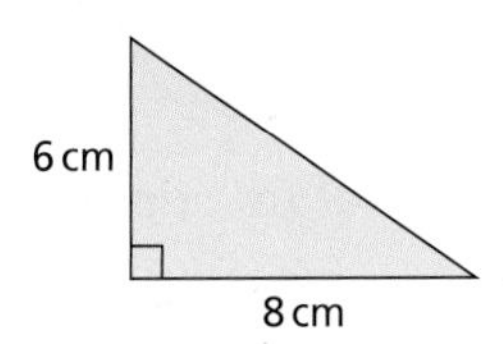

2 Calculate the length of the unknown side in each triangle.

Give your answers to 1 decimal place.

a

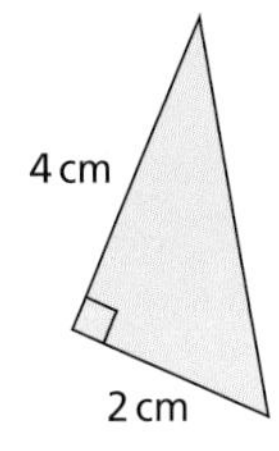

b

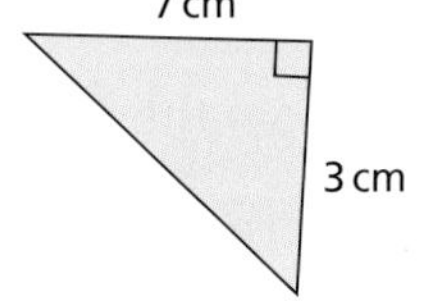

c

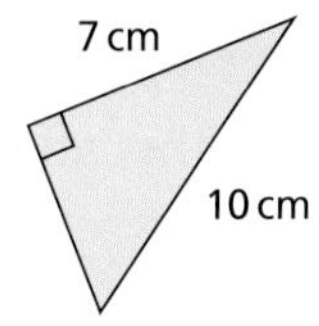

d

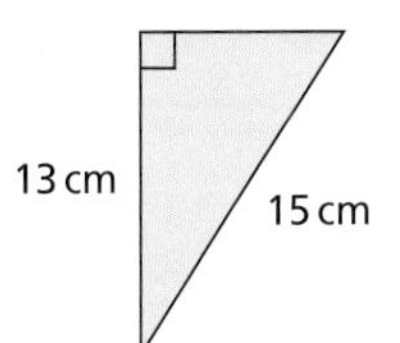

e

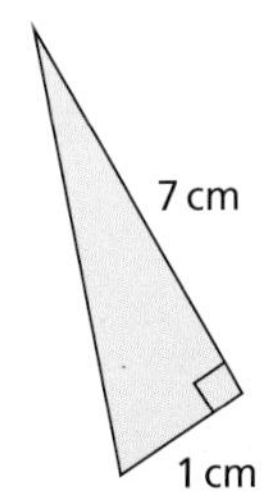

f

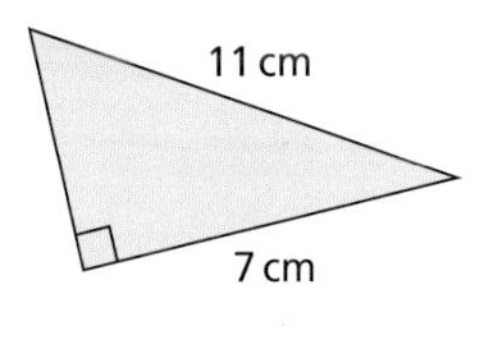

3 Calculate the length of the unknown side in each triangle.

Give your answers to 3 significant figures.

a

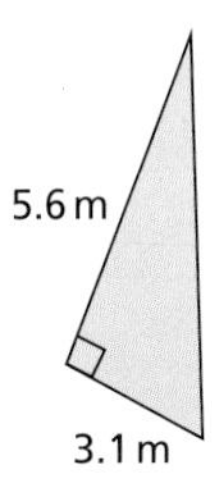

b

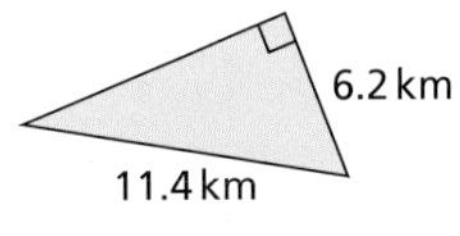

c

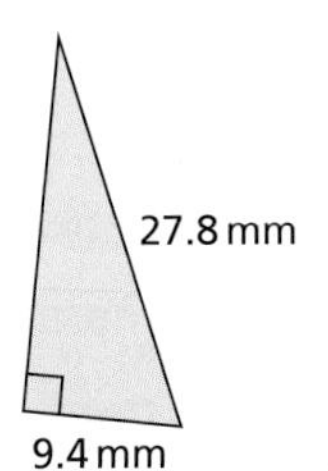

4 PQR is an isosceles triangle.

Work out the length of QR.

Give your answer to 1 decimal place.

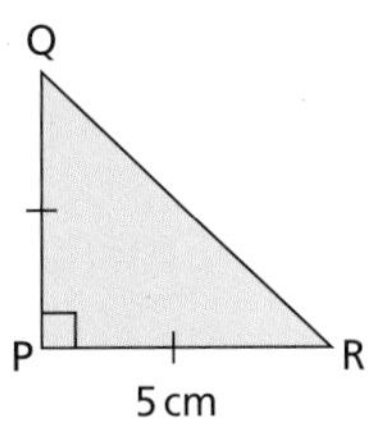

5 ABCD is a rectangle.

a Calculate the length of AB.

b Work out the area of the rectangle.

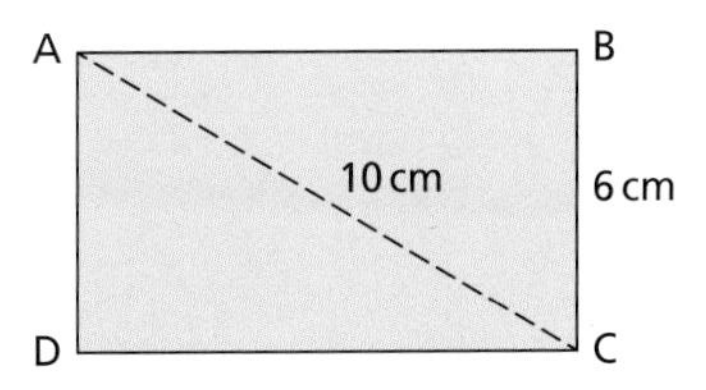

6 ABC is an isosceles triangle.

a Calculate the perpendicular height, h, of the triangle.

Give your answer to 1 decimal place.

b Calculate the area of the triangle.

Give your answer to 1 decimal place.

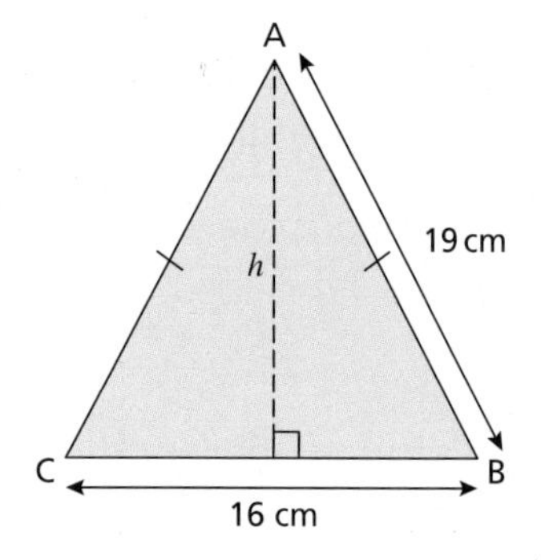

7 A 6-metre long ladder is placed against a wall.

The foot of the ladder is 2 metres from the wall.

How far up the wall does the ladder reach?

8 Calculate the length of BD.

Give your answer to 2 decimal places.

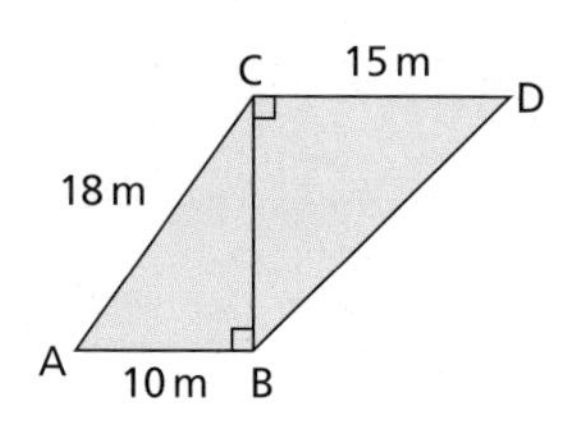

9 ABC is a right-angled triangle.

Work out the length of BC. Give your answer in the form $\sqrt{a}$, where a is an integer.

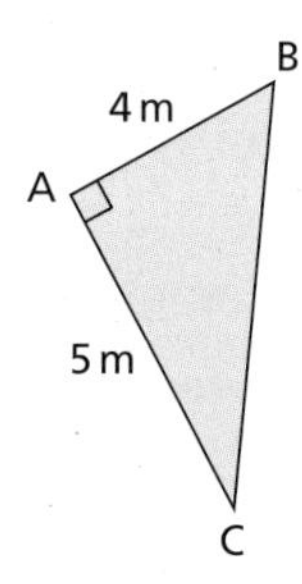

10 Two points are shown on the coordinate grid.

a Work out the distance between the two points.

b Work out the distance between the points (–3, 5) and (6, –1). Give your answer to 1 decimal place.

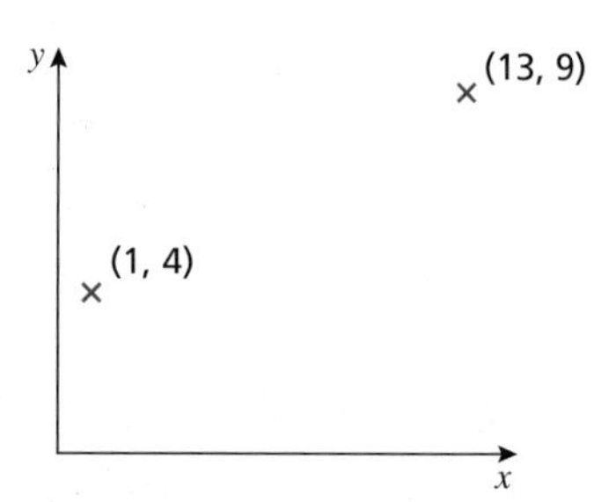

11 Here is a right-angled isosceles triangle.

Explain why the hypotenuse must be less than 10 cm.

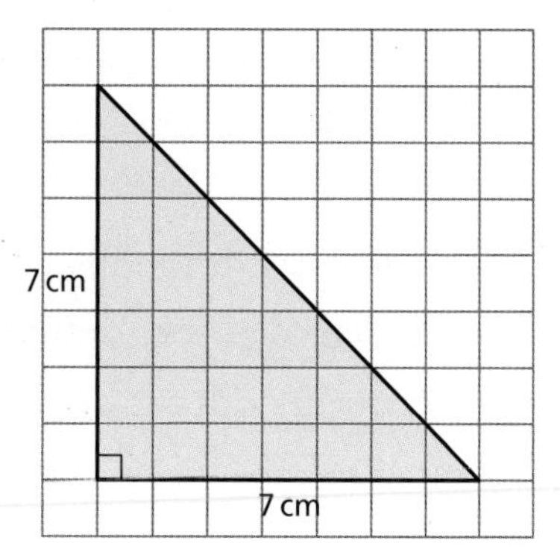

What do you think? (A)

1 Here are three triangles, not drawn accurately.

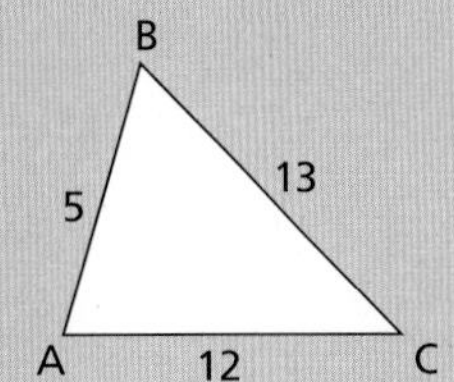

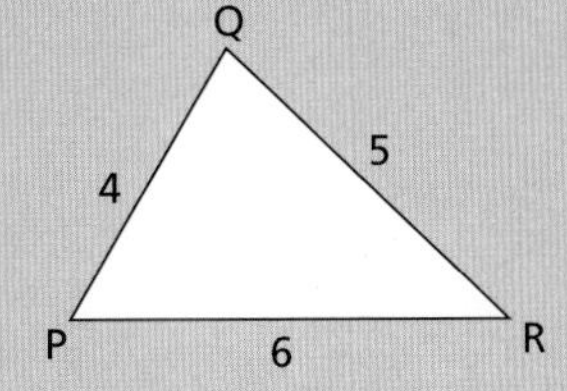

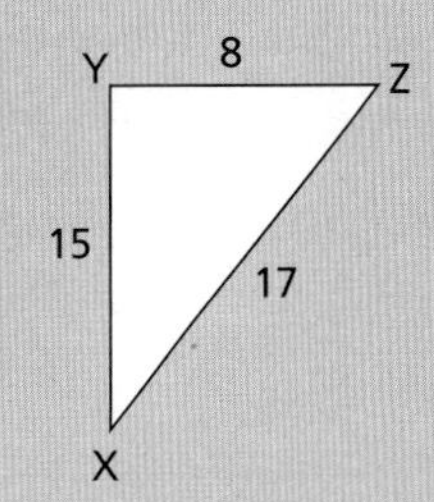

a Which of the triangles are right-angled? Justify your answers.

b Do any of the non-right-angled triangles contain an obtuse angle? Explain how you know.

2 ABD and BCD are right-angled triangles.

Work out the length of CD.

Investigate the results with different lengths of AB, AD and BC.

Can you generalise your findings?

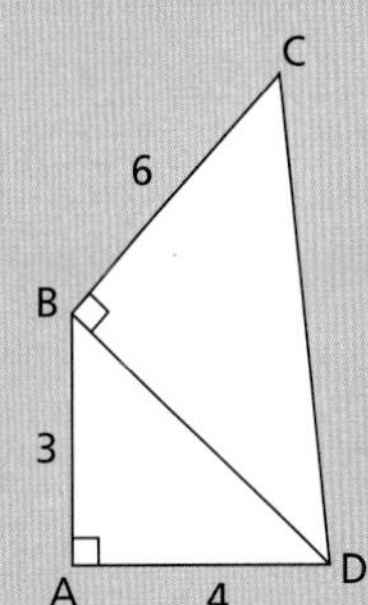

3 A Pythagorean triple is a set of three positive integers a, b and c that are related by the formula $a^2 + b^2 = c^2$.

For example, 3, 4, 5 is a Pythagorean triple, as well as 8, 15, 17

Investigate other Pythagorean triples.

Are you ready? (B)

1 Here is the cuboid ABCDEFGH.

a Name the three sides with lengths that are equal in length to AD.

b Name the three sides with lengths that are equal in length to EF.

c Name a line segment with the same length as CH.

d Name a line segment with the same length as BG.

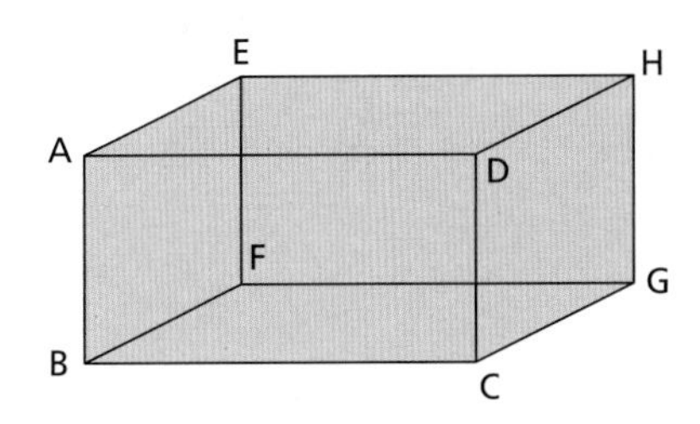

2 Here is the cube ABCDEFGH.

ABC is a right-angled triangle.

Name two more right-angled triangles that can be formed using vertices of the cuboid.

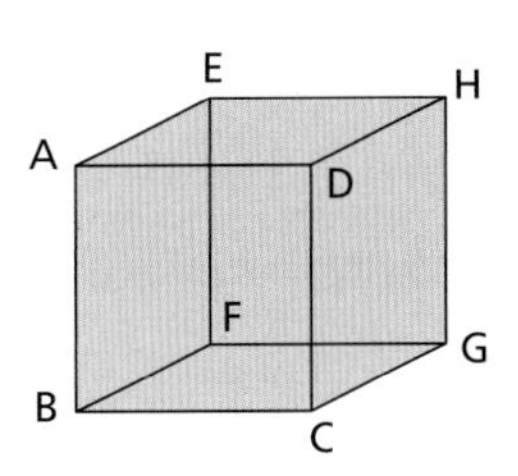

Example

ABCDEFGH is a cuboid.

a Work out the length of BG.

b Work out the length of BH.

Give your answer to 3 significant figures.

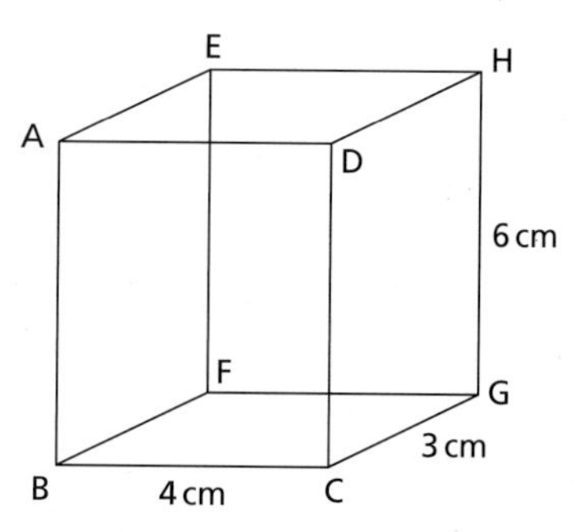

Method

Solution	Commentary
a $a^2 + b^2 = \text{hyp}^2$ $4^2 + 3^2 = BG^2$ $BG^2 = 25$ $BG = 5$ $BG = 5\text{ cm}$	Consider the right-angled triangle BCG. The two shorter sides are 4 cm and 3 cm. Use Pythagoras' theorem to find the length of the hypotenuse.
b $a^2 + b^2 = \text{hyp}^2$ $5^2 + 6^2 = BH^2$ $BH^2 = 61$ $BH = 7.810\,24\ldots$ $BH = 7.81\text{ cm}$ (to 3 s.f.)	Now consider the right-angled triangle BGH. The two shorter sides are 5 cm and 6 cm. Use Pythagoras' theorem to find the length of the hypotenuse and round to 3 significant figures.

Practice (B)

1 ABCDEFGH is a cuboid.

a Calculate the length FC.

Give your answer to 2 decimal places.

b Calculate the length EC.

Give your answer to 2 decimal places.

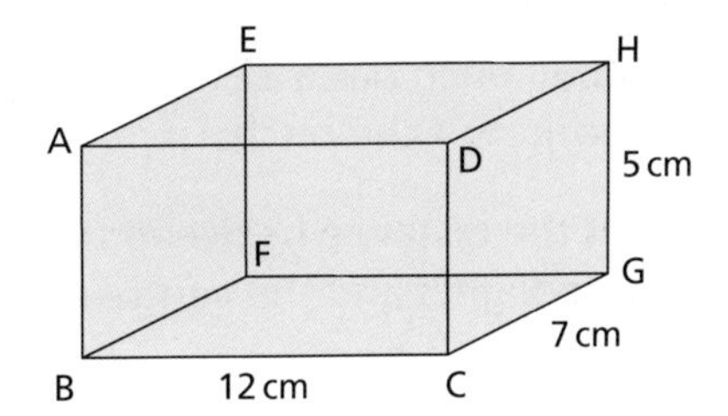

2 ABCDEFGH is a cube.

a Work out the length of BG.

Give your answer to 3 significant figures.

b Work out the length of AG.

Give your answer to 3 significant figures.

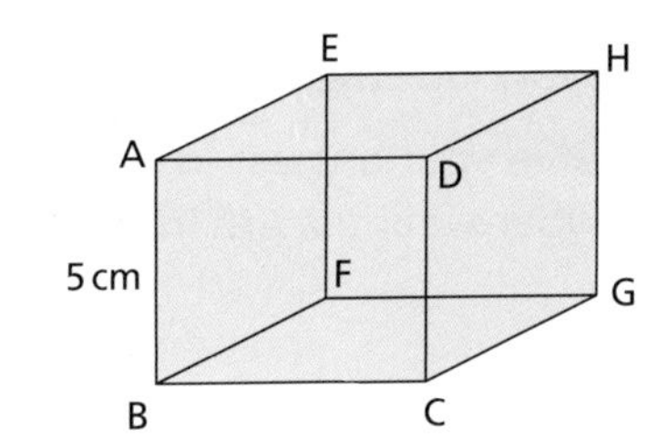

3 Here is a cone.

Work out the slant height of the cone.

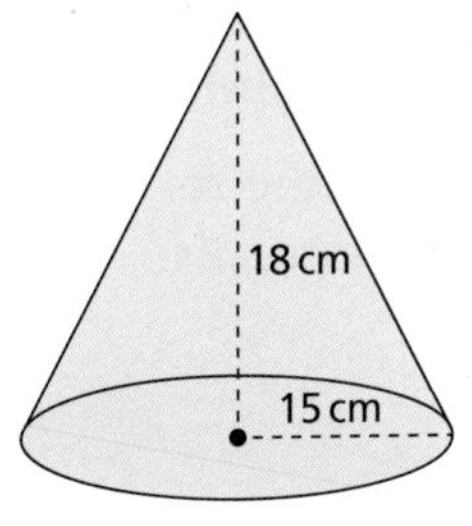

4 ABCDE is a square-based pyramid.

a Work out the length of AC.

Give your answer to 3 significant figures.

b Work out the perpendicular height of the pyramid.

Give your answer to 3 significant figures.

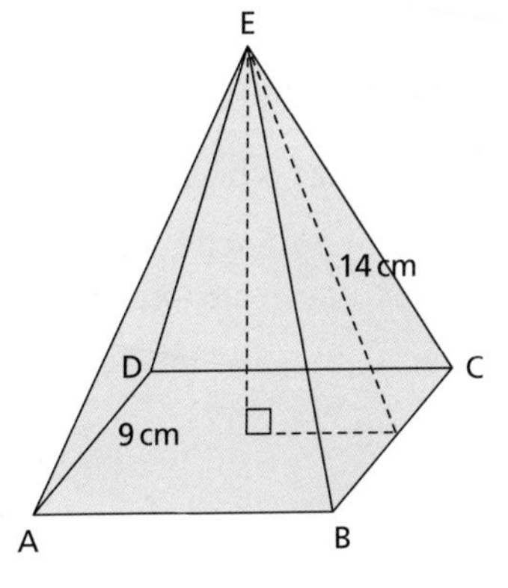

5 ABCDEF is a triangular prism.

Work out the length of BF.

Give your answer to 3 significant figures.

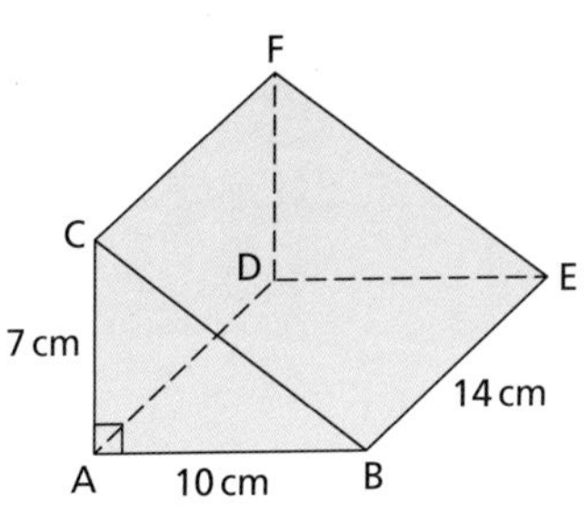

6 A rectangular football pitch measures 105 m by 65 m.

The football pitch has a 12 m floodlight standing in each corner of the pitch.

Work out the distance between the top of one of the floodlights and the centre of the pitch.

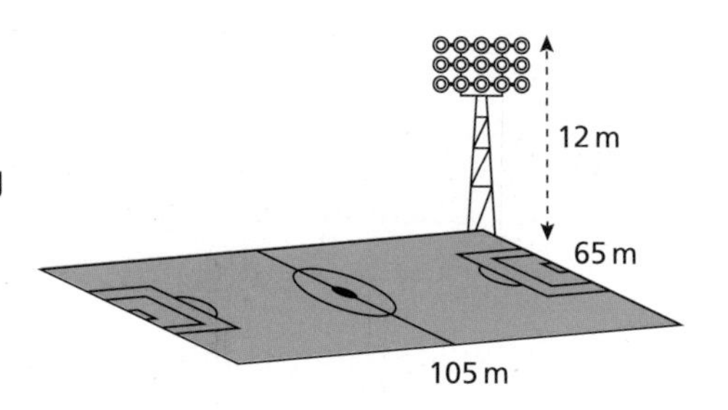

7 ABCDEFGH is a cube.

Work out the length of the longest diagonal.

Give your answer in the form $a\sqrt{b}$, where a and b are integers.

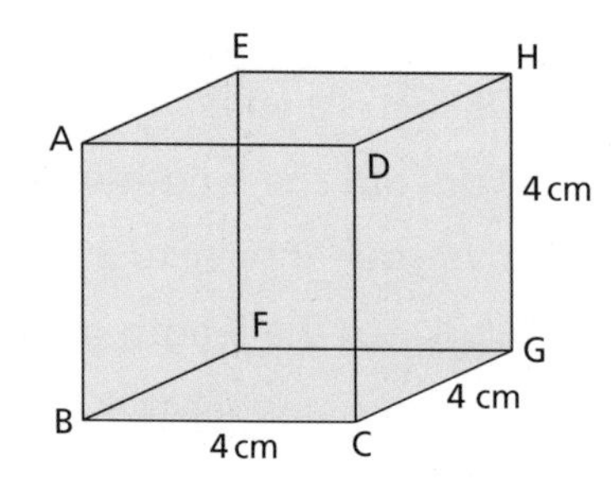

8 ABCDEFGH is a cuboid.

Work out the length of the longest diagonal.

Give your answer in the form $a\sqrt{b}$, where a and b are integers.

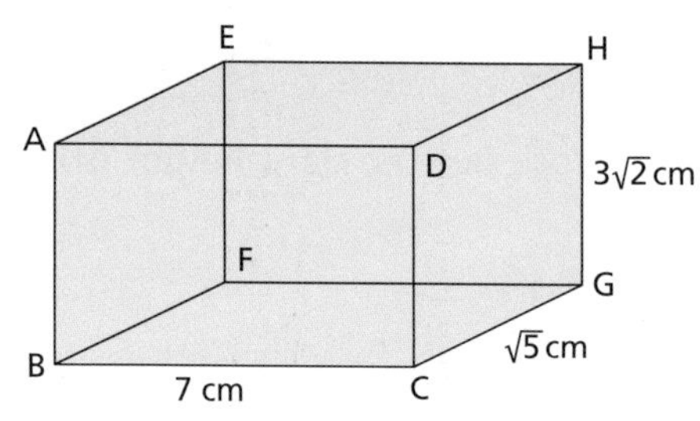

What do you think? (B)

1 Flo has a box to keep pencils in.

Will a pencil of length 12 cm fit inside the box? Explain your answer.

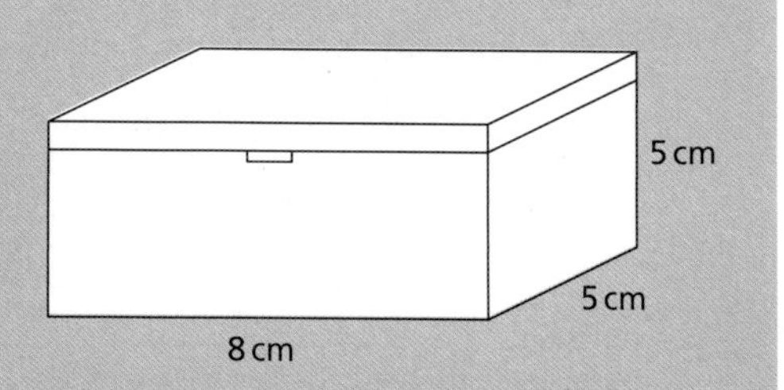

2 A cuboid is shown.

a Show that the longest diagonal of the cuboid is $\sqrt{a^2 + b^2 + c^2}$

b Show that the longest diagonal in a cube with side length a cm is $\sqrt{3a^2}$ cm.

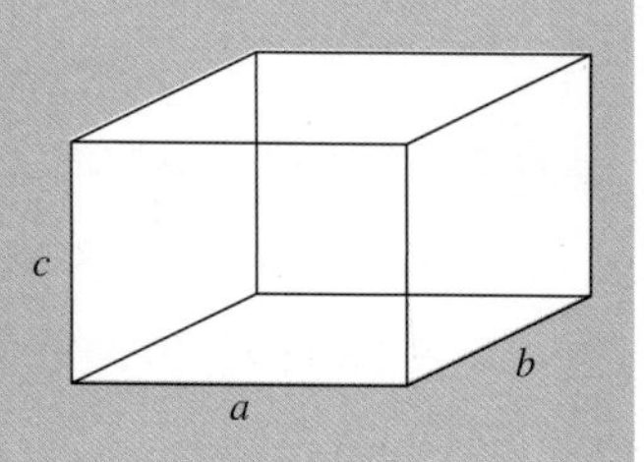

Consolidate – do you need more?

1 Work out the length of the hypotenuse in each triangle.

Give your answers to 1 decimal place.

a

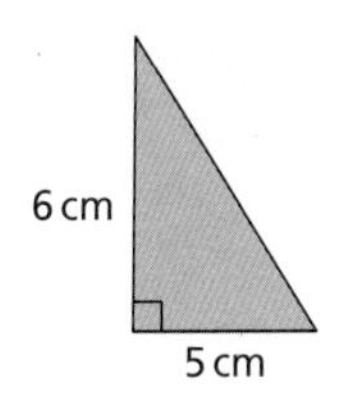

b

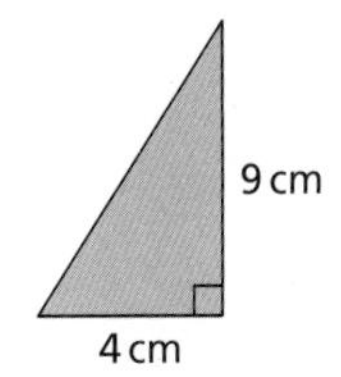

c

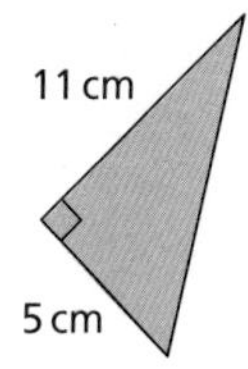

d

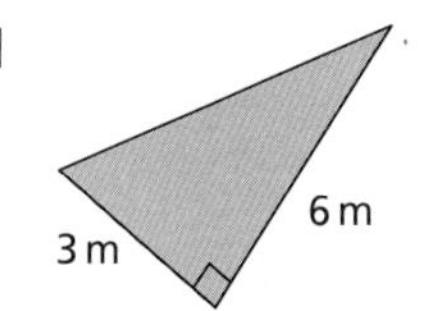

e

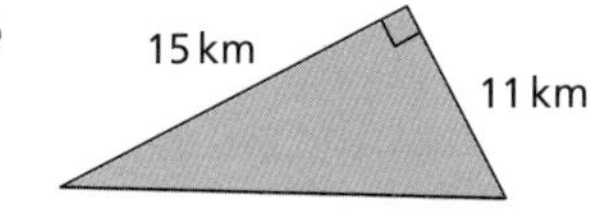

f

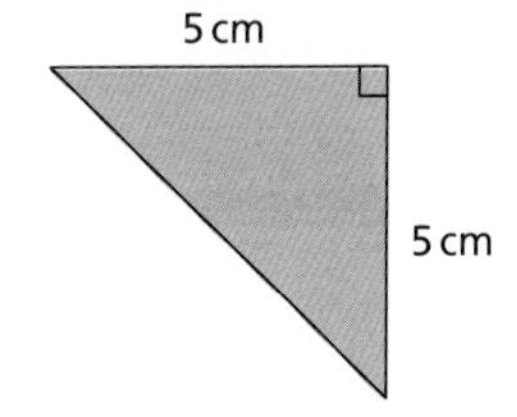

2 Work out the length of the unknown side in each triangle.

Give your answers to 1 decimal place.

a

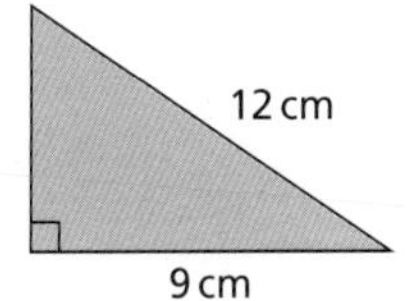

b

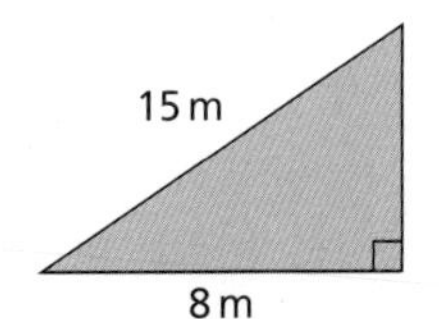

c

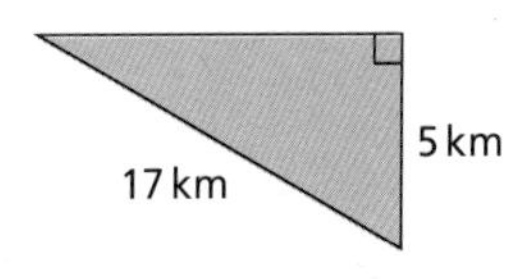

d

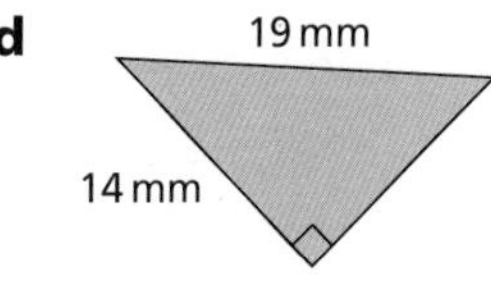

e

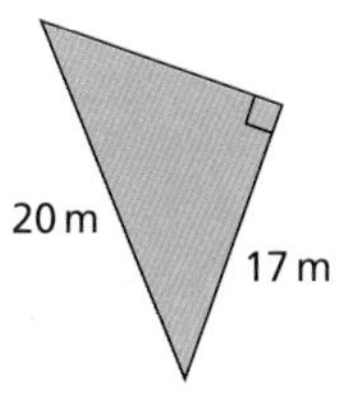

f

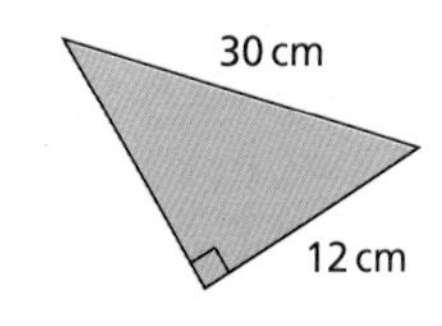

3 A bird flies due East 9 km from a nest.

It then flies due South 11 km.

Calculate the distance between the bird and the nest.

4 Two points are marked on the grid.

Work out the distance between the two points.

Give your answer to 1 decimal place.

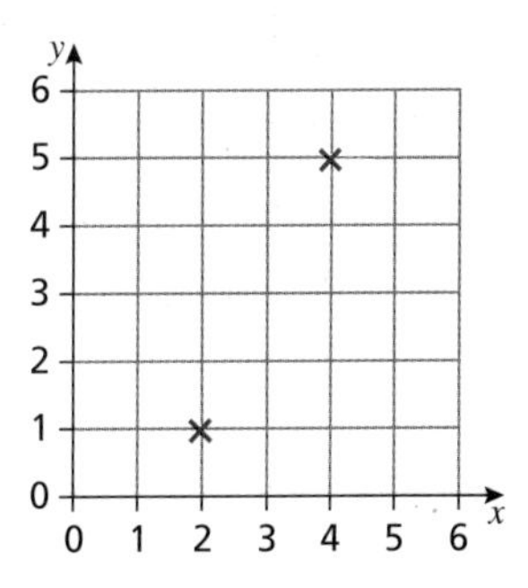

Stretch – can you deepen your learning?

1 The length AB can be written in the form $a\sqrt{b}$, where a and b are integers.

Work out the values of a and b.

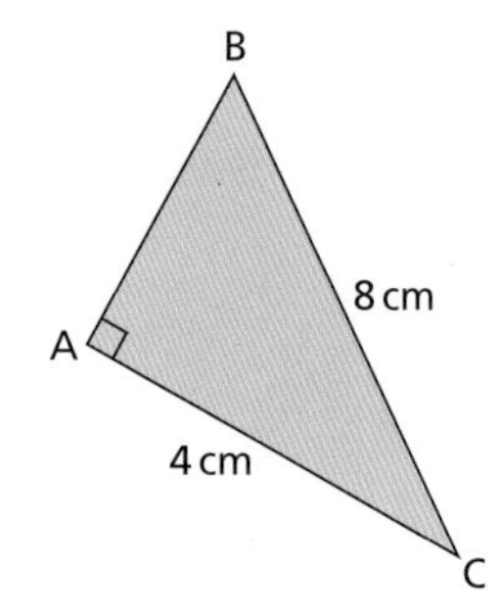

Simplifying surds is covered in *Collins White Rose Maths AQA GCSE 9–1 Higher Student Book 1*, Chapter 3.1

2 Triangle PQR is shown.

The lengths of PQ and QR have both been rounded to 3 significant figures.

Work out the error interval for the length of PR.

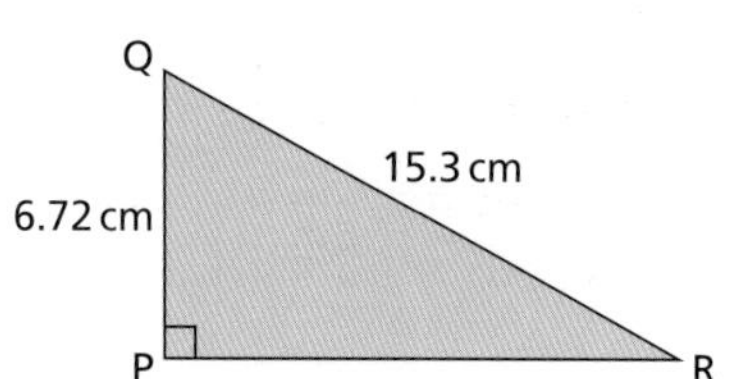

If you need help with rounding and error intervals, see *Collins White Rose Maths AQA GCSE 9–1 Higher Student Book 1*, Block 5.

3 Show that the sum of the areas of the two smaller semicircles is equal to the area of the larger semicircle.

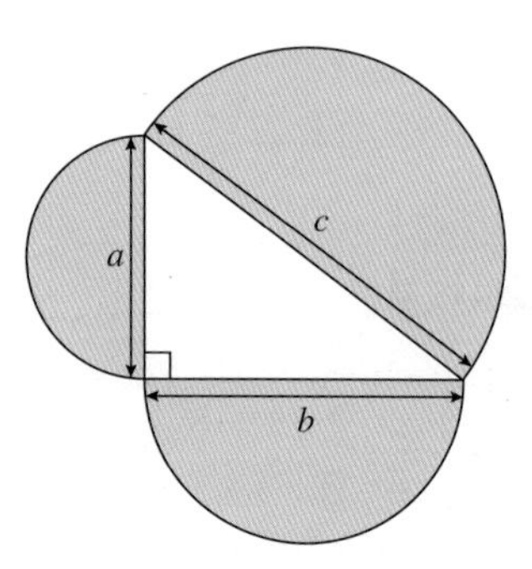

4 The points M and N are exactly 13 units apart.

Point M is (–2, 5) and point N is (3, k).

Work out the two possible values of k.

5 Work out the volume of the cone.

Give your answer to 3 significant figures.

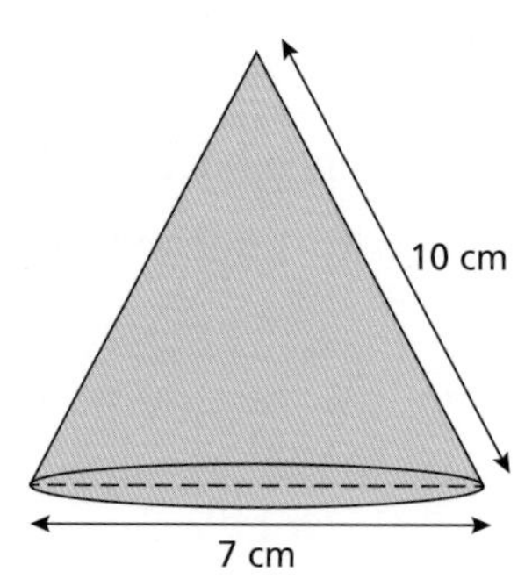

6 Write an expression for the distance between the points with coordinates (a, b) and (c, d).

6.2 Trigonometry

White Rose MATHS

Are you ready? (A)

1 Solve:

a $\frac{m}{2} = 7$ **b** $7 = \frac{t}{2}$ **c** $7 = \frac{2}{y}$

2 Solve:

a $\frac{p}{2.5} = 8$ **b** $6.5 = \frac{k}{3.5}$ **c** $11.2 = \frac{6.4}{r}$

3 Round each number below to:

i 1 decimal place **ii** 2 decimal places **iii** 3 significant figures.

a 6.7367 **b** 12.252 94 **c** 10.0736 **d** 200.972

4 Copy each triangle and label the hypotenuse.

a

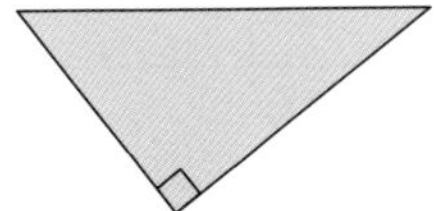

b

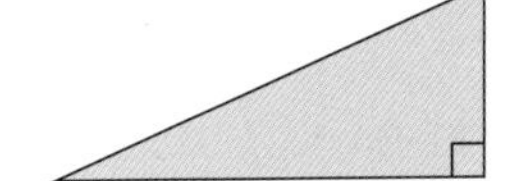

c

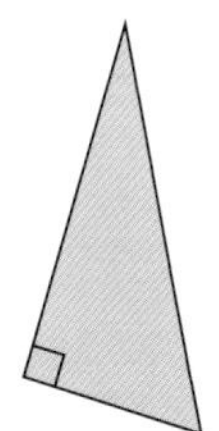

Using your calculator

In this chapter, you will learn about trigonometric ratios, which are very useful in solving problems in geometry. The values of these ratios are stored on your scientific calculator, and can be accessed by using the keys 'sin', 'cos' and 'tan'.

As you work through the chapter, make sure you know how to use your calculator to calculate with these ratios.

You can label the sides of a right-angled triangle as follows:

- The **hypotenuse** is the longest side and is opposite the right angle.
- The **opposite** is the side directly opposite an angle of interest, denoted x in this diagram.
- The **adjacent** is the side 'touching' the angle x that is not the hypotenuse.

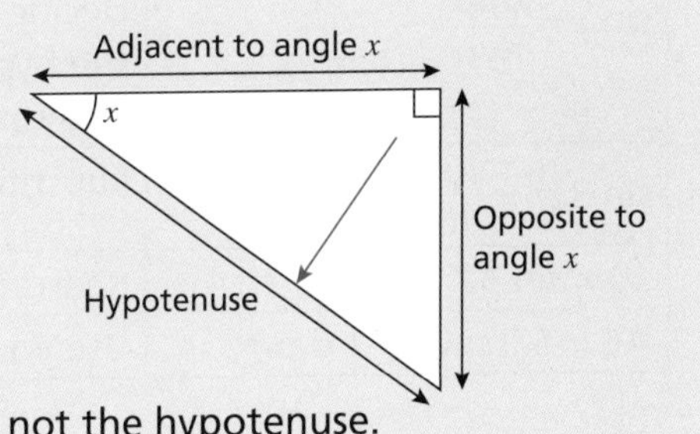

6.2 Trigonometry

For any given angle x, the ratios of the lengths of the sides are constant. It doesn't matter how big or small the triangle is.

These two triangles are similar. The ratios of the lengths of their sides will be the same because they have the same angles.

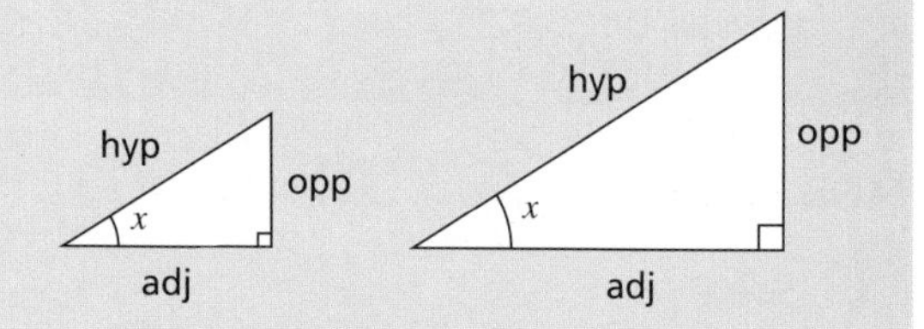

Hypotenuse, opposite and adjacent can be abbreviated to hyp, opp and adj respectively.

These trigonometric ratios are called the **sine, cosine** and **tangent** ratios and are defined as follows:

$$\sin x = \frac{\text{opposite}}{\text{hypotenuse}} \qquad \cos x = \frac{\text{adjacent}}{\text{hypotenuse}} \qquad \tan x = \frac{\text{opposite}}{\text{adjacent}}$$

These values are stored in your calculator. You can use these ratios to work out missing sides and angles in right-angled triangles.

Note that sine is shortened to 'sin', cosine is shortened to 'cos' and tangent is shortened to 'tan'.

Example 1

Work out the length of side AC.

Give your answer to 1 decimal place.

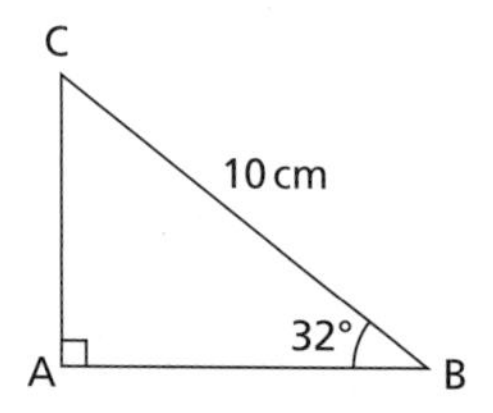

Method

Solution	Commentary
C, opp, 10 cm hyp, 32°, A, adj, B	First, label the sides of the triangle as opposite (opp), adjacent (adj) and hypotenuse (hyp), using the given angle of 32° to decide which is which.
$\sin x = \frac{\text{opp}}{\text{hyp}}$	Decide which ratio to use and write it down. Here you know the hypotenuse and want to find the opposite, so use the sine ratio.
$\sin 32° = \frac{\text{AC}}{10}$	Substitute in the information you know.
$10 \times \sin 32° = \text{AC}$	Rearrange the equation to find AC.
AC = 5.3 cm (to 1 d.p.)	Use your calculator to find the answer.

Make sure your calculator is set up to work in degrees.

Example 2

Work out the length of the side PR.

Give your answer to 1 decimal place.

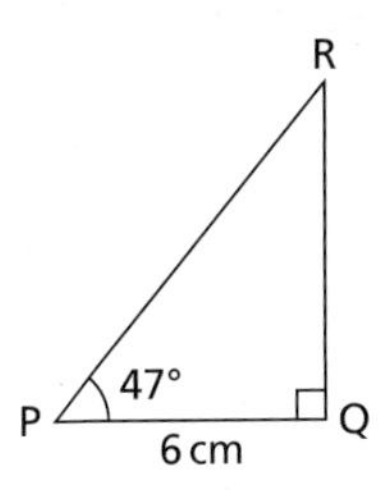

Method

Solution	Commentary
R, hyp, opp, P, 47°, Q, 6 cm, adj	First, label the sides of the triangle as opposite (opp), adjacent (adj) and hypotenuse (hyp), using the given angle of 47° to decide which is which.
$\cos x = \frac{\text{adj}}{\text{hyp}}$	Decide which ratio to use and write it down. Here you know the adjacent and want to find the hypotenuse, so use the cosine ratio.
$\cos 47° = \frac{6}{\text{PR}}$	Substitute in the information you know.
$\frac{6}{\cos 47°} = \text{PR}$	Rearrange the equation to find PR.
PR = 8.8 cm (to 1 d.p)	Use your calculator to find the answer and round as required.

Practice (A)

1. Copy the triangles and label each side with hypotenuse, opposite and adjacent.

a

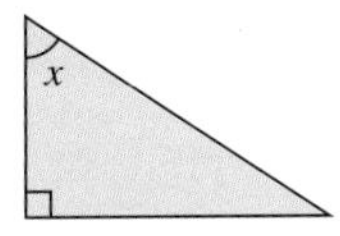

b

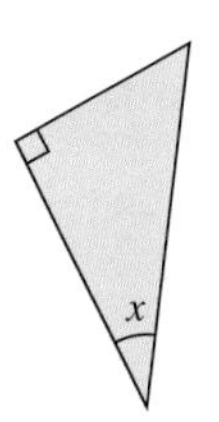

c

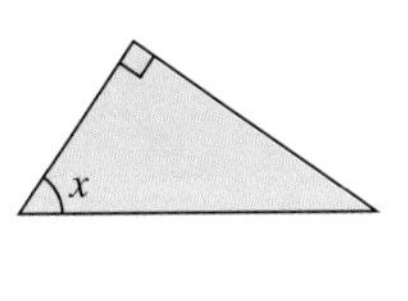

d

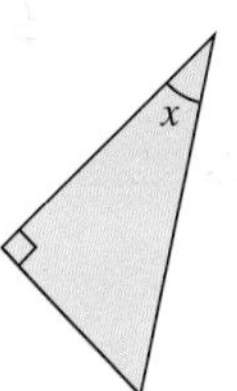

2. Use the sine ratio to work out the length of each lettered side.

Give your answers to 1 decimal place.

a

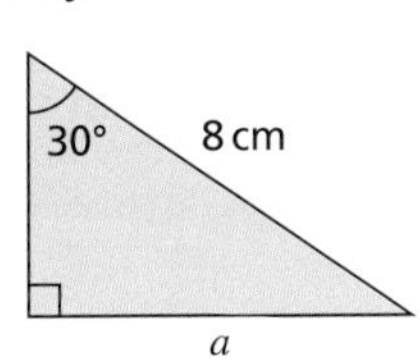

b

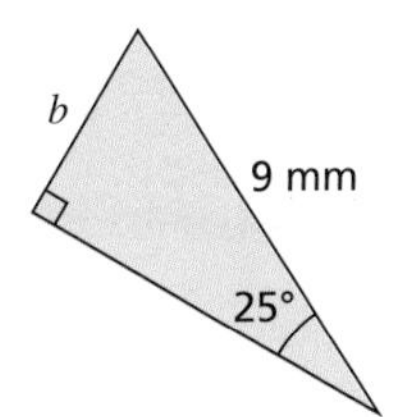

c

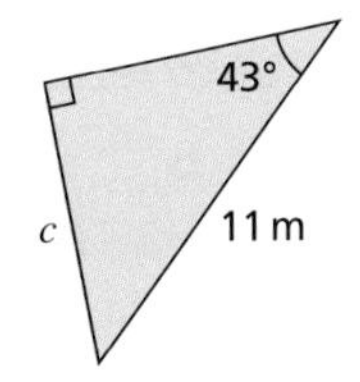

d

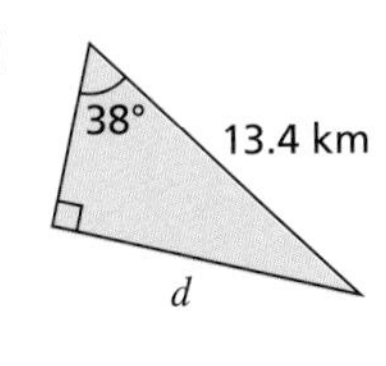

3 Use the cosine ratio to work out the length of each lettered side.

Give your answers to 1 decimal place.

a

30°
6 cm
a

b

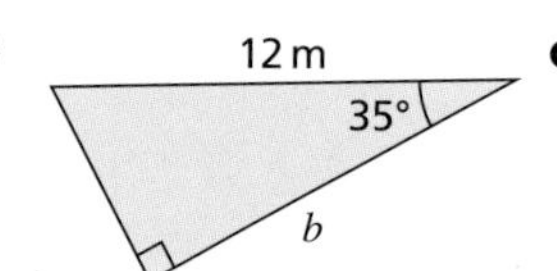

c

51°
10.8 cm
c

d

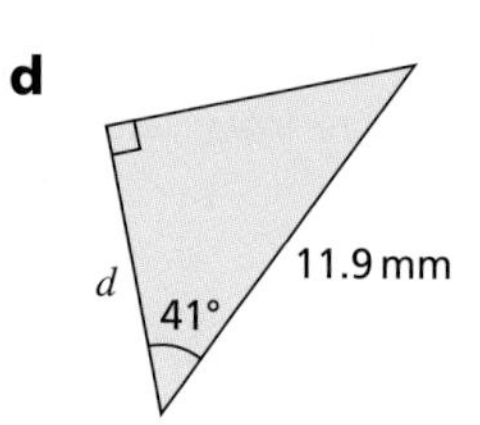

4 Use the tangent ratio to work out the length of each lettered side.

Give your answers to 1 decimal place.

a

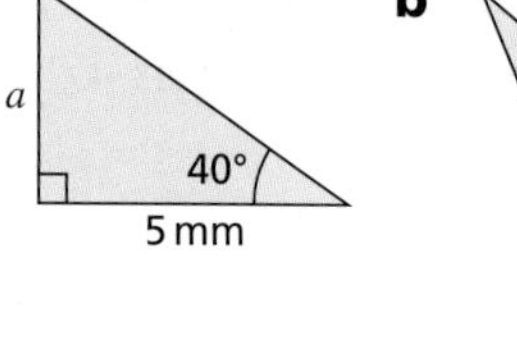

b

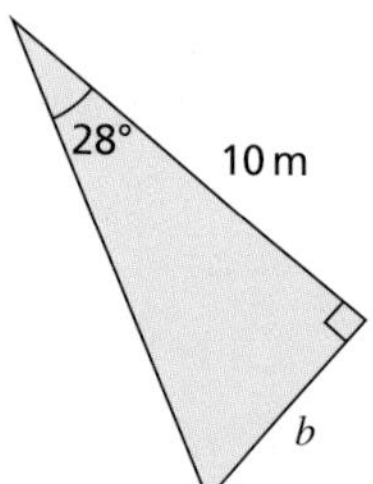

c

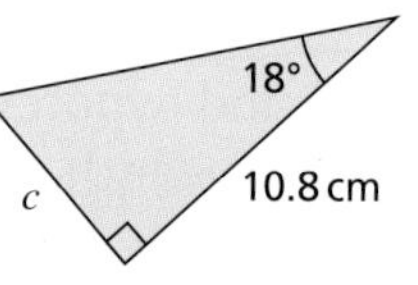

d

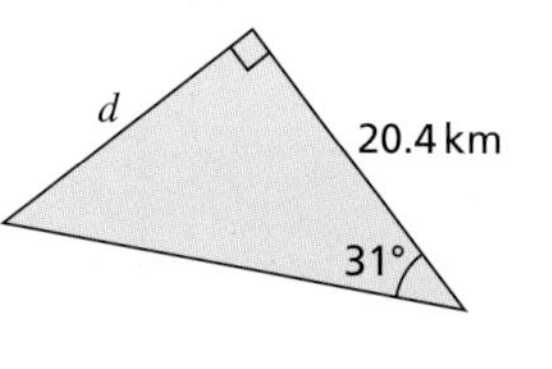

5 For each triangle, state whether to use the sine (sin), cosine (cos) or tangent (tan) ratio to work out the length of the lettered side.

a

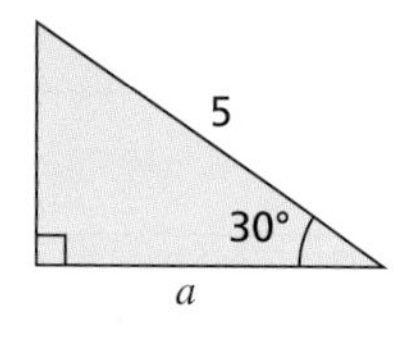

b

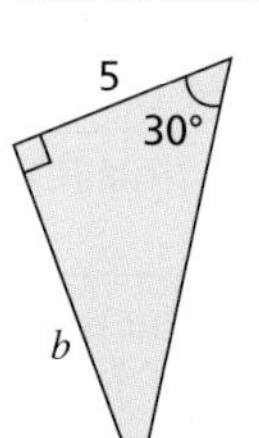

c

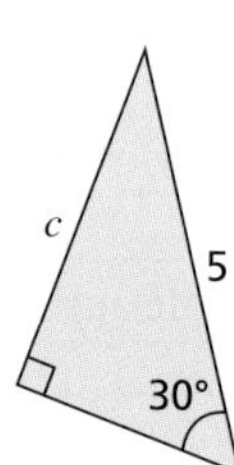

d

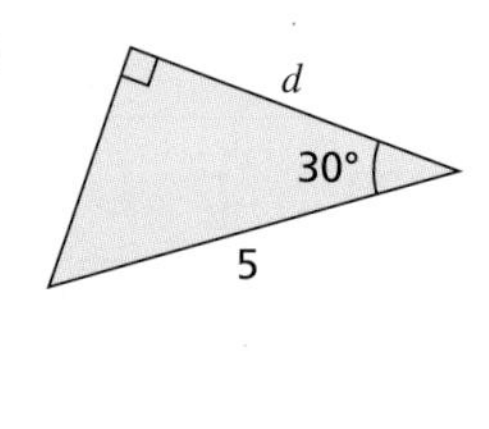

6 Work out the length of each lettered side. Give your answers to 1 decimal place.

a

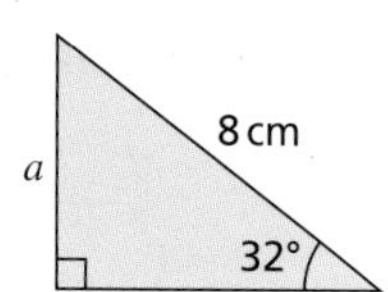

b

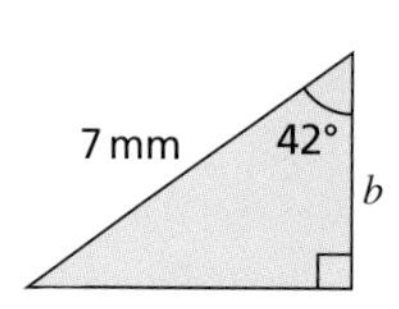

c

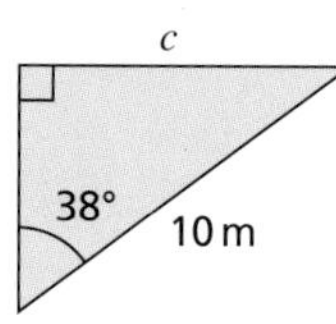

d

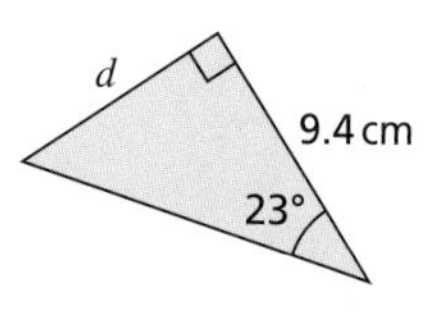

e

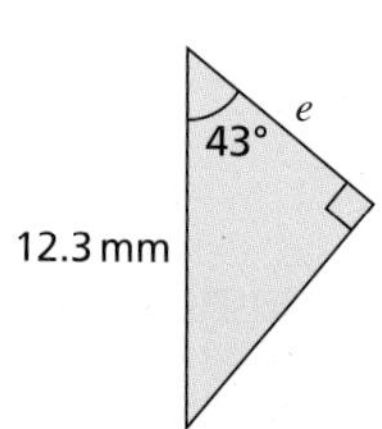

f

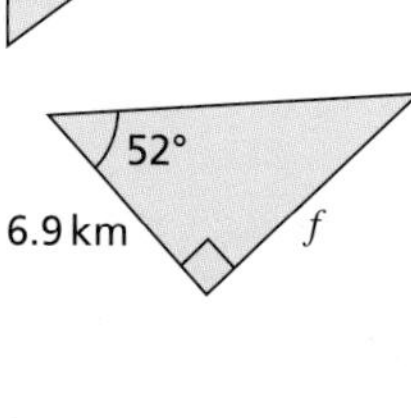

7 Work out the length of each lettered side. Give your answers to 1 decimal place.

a

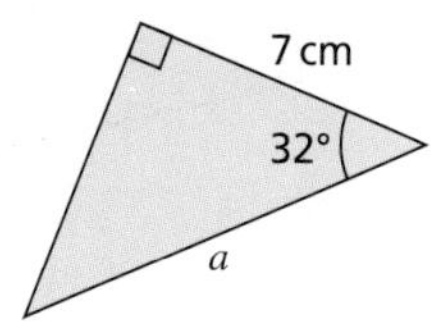

b

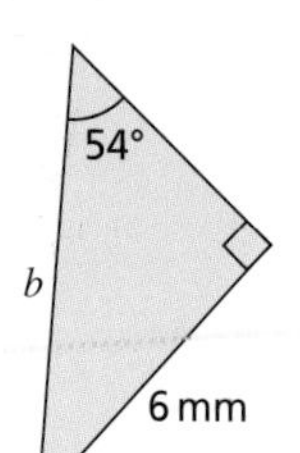

c

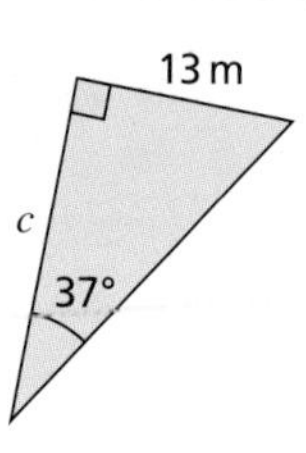

8 Marta is working out the length of PQ.

Her working out is shown.

$$\tan 28° = \frac{11}{PQ}$$

$$PQ = \frac{11}{\tan 28°}$$

$$PQ = 20.687\,99\ldots \text{cm}$$

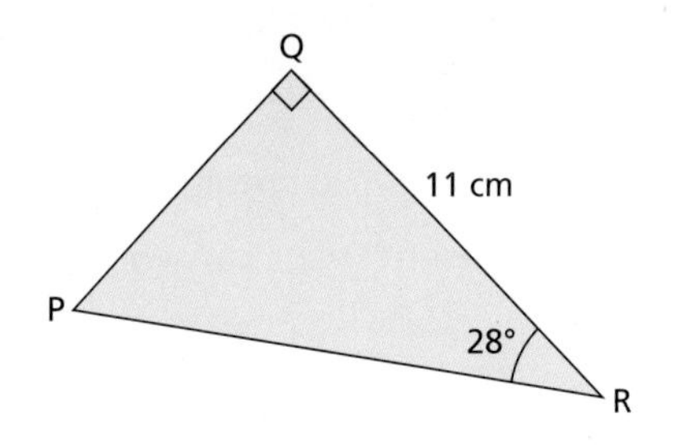

a What mistake has Marta made?

b Work out the length of PQ. Give your answer to 1 decimal place.

9 Work out the length of each lettered side. Give your answers to 2 decimal places.

a

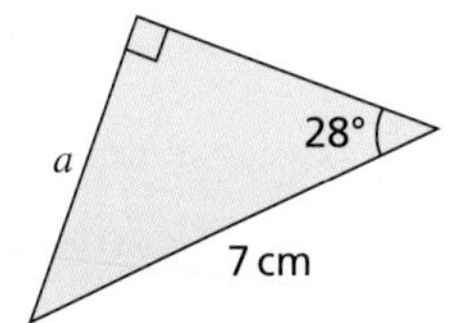

b

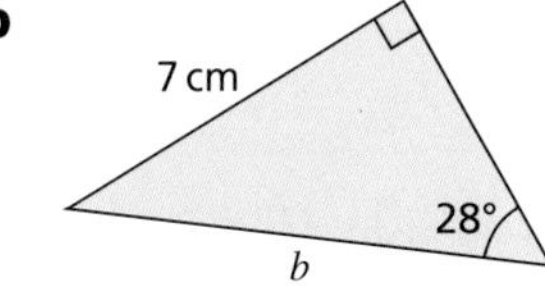

c

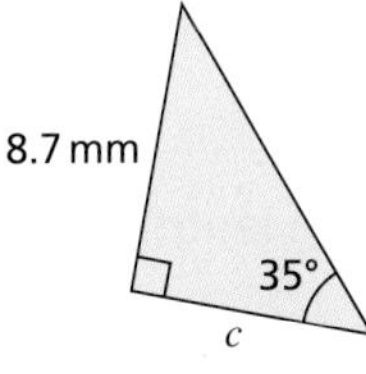

d

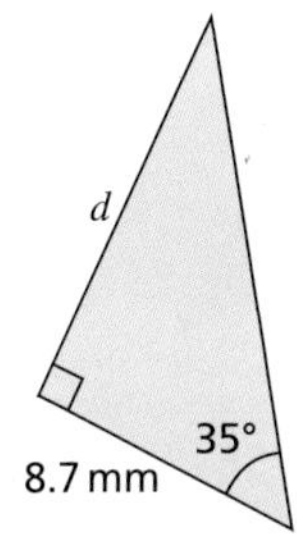

e

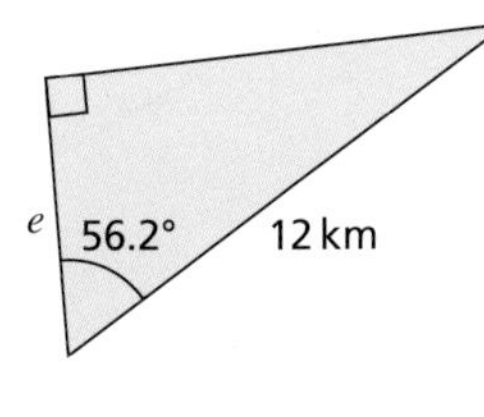

f

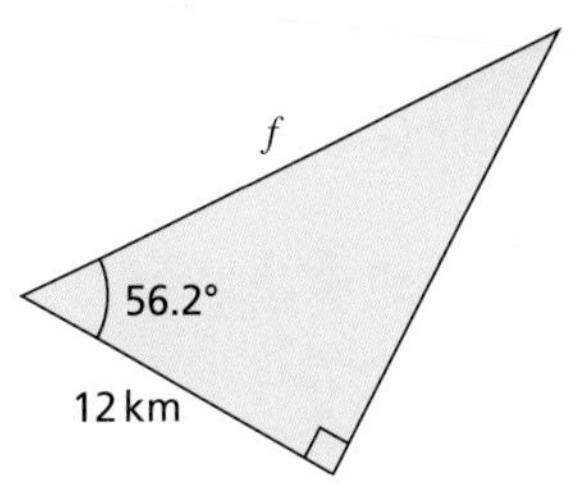

10 A 5 m ladder rests against a wall.

The angle between the ground and the foot of the ladder is 53°.

Work out the distance between the foot of the ladder and the wall.

Give your answer to 3 significant figures.

11 Work out the area of the triangle.

Give your answer to 3 significant figures.

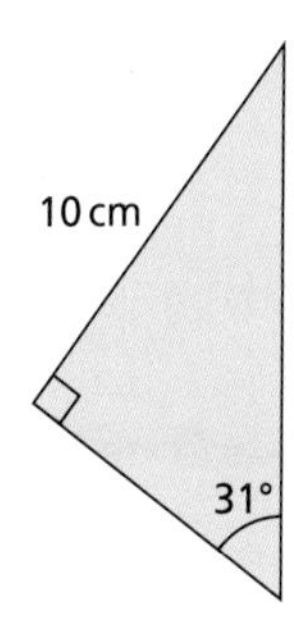

What do you think? (A)

1. Here is a right-angled triangle.

   How many different ways can you work out the length of x?

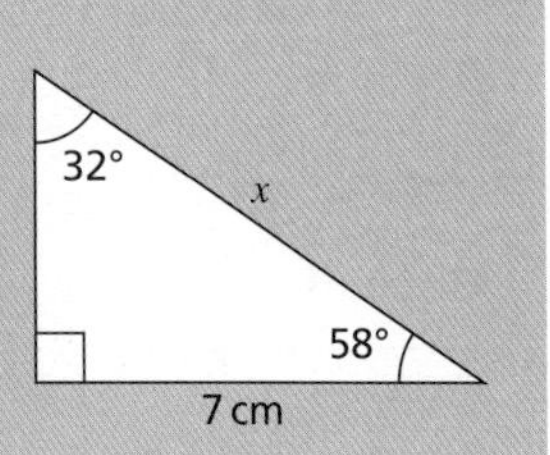

2. Samira says, "sin 30° is equal to cos 60°."

   Work out other values of x and y such that $\sin x = \cos y$

   Explain why this is true.

Are you ready? (B)

1 Copy and complete the formulae using the words hypotenuse, adjacent and opposite.

$\sin x =$ ☐ $\cos x =$ ☐ $\tan x =$ ☐

2 Work out the length of each lettered side. Give your answers to 3 significant figures.

a

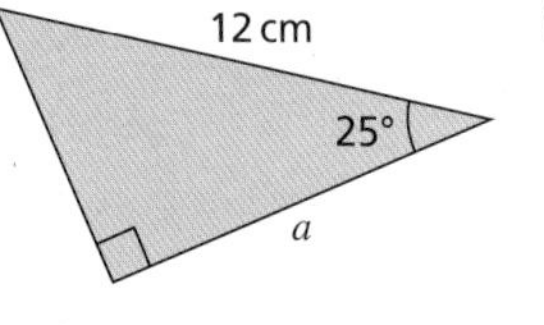

b

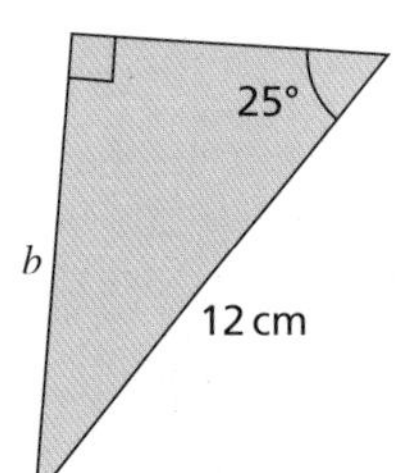

c

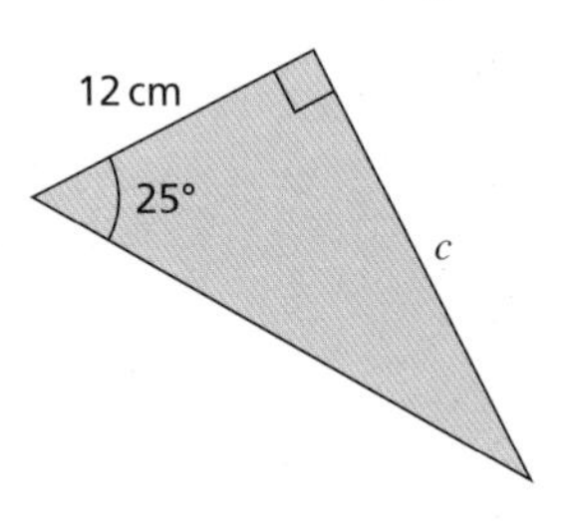

d

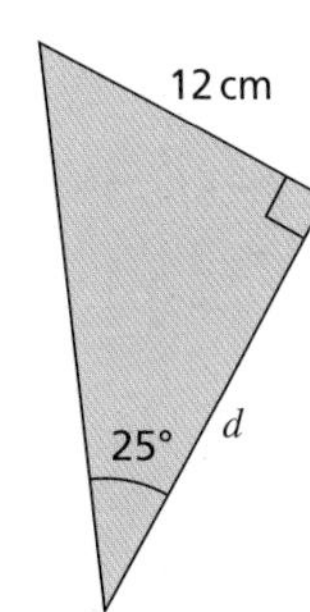

Using your calculator

You can also use your calculator to find unknown angles, using the inverse trigonometric functions. These are usually situated above the sin, cos and tan keys and can be accessed using the 'shift', 'inverse' or '2nd function' key. Again, as you work through the next section of this chapter, ensure that you are familiar with the workings of your own calculator.

$\sin^{-1}$ **sin** $\cos^{-1}$ **cos** $\tan^{-1}$ **tan**

As seen earlier, for any given angle, the ratios of the lengths of the sides in right-angled triangles are constant.

When you know the ratio and need to work out the angle, you can use the **inverse** of the ratio.

For example, if $\sin 30° = 0.5$, then $\sin^{-1}(0.5) = 30°$

Example 1

Work out the size of angle x.

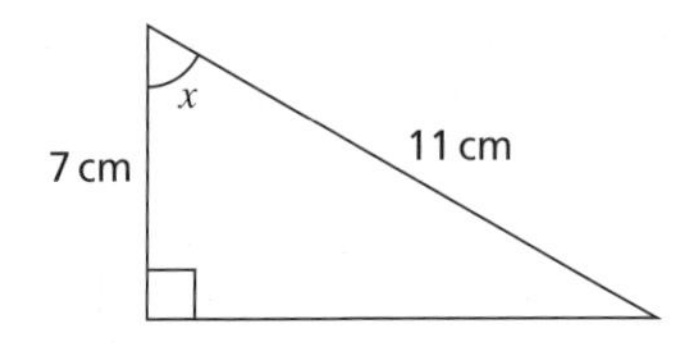

Method

Solution	Commentary
adj 7 cm, 11 cm hyp, opp, x	First, label the triangle using angle x to determine the labels.
$\cos x = \frac{\text{adj}}{\text{hyp}}$	Use the labelling to decide which ratio to use. Here you know the adjacent and hypotenuse, so use the cos ratio.
$\cos x = \frac{7}{11}$	Substitute in the information you know.
$x = \cos^{-1}\left(\frac{7}{11}\right)$ $x = 50.5°$	If you know a ratio, use the inverse function on your calculator to find the angle. Here the inverse of cosine is denoted as $\cos^{-1}$.

Example 2

A ladder of length 5 m rests against a wall.

The base of the ladder is 1.4 m from the wall, as shown in the diagram.

Calculate the size of the angle the ladder makes with the ground.

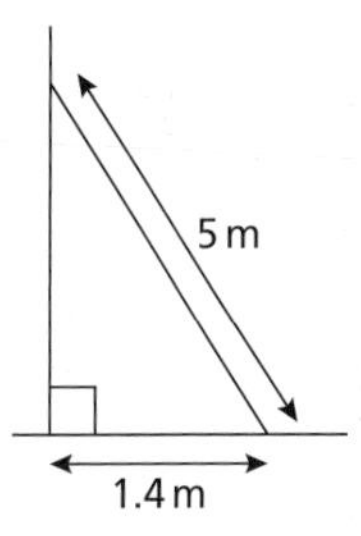

Method

Solution	Commentary
opp, hyp 5 m, x, 1.4 m adj	First, identify the angle you have been asked to find and label the sides of the triangle relative to that angle.
$\cos x = \frac{\text{adj}}{\text{hyp}}$	Decide which ratio you need to use and write it out. Here, you know the adjacent and hypotenuse so use the cosine ratio.
$\cos x = \frac{1.4}{5}$	Substitute in the information you know.
$x = \cos^{-1}\left(\frac{1.4}{5}\right)$ $x = 73.739\ 795\ 2917\ldots$ $x = 74°$ (to the nearest degree)	Use the inverse function on your calculator to find x.

Practice (B)

1 Work out the value of x. Give your answers to 1 decimal place.

a $\sin x = 0.2$ **b** $\cos x = 0.35$ **c** $\tan x = \frac{5}{8}$

d $\cos x = \frac{3}{10}$ **e** $\sin x = \frac{12.5}{15}$ **f** $\tan x = \frac{11.2}{5.7}$

2 Use the sine ratio to work out the size of each lettered angle.

a

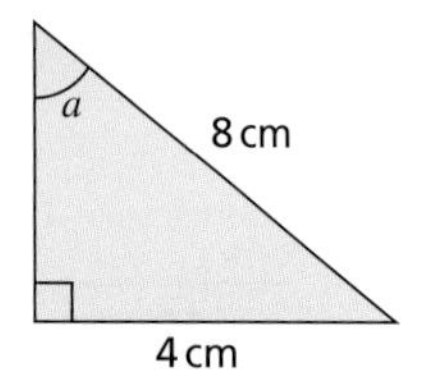

b

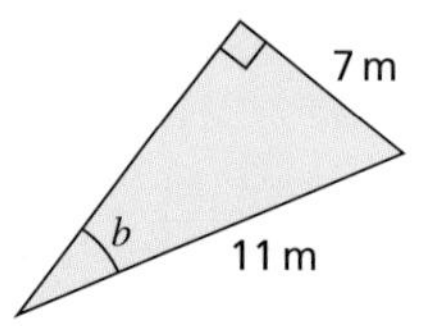

c

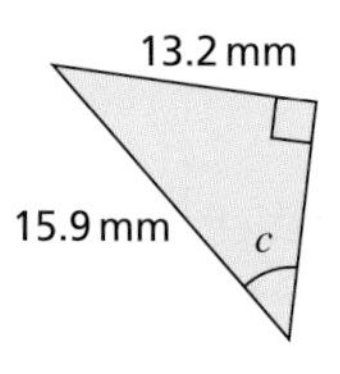

d

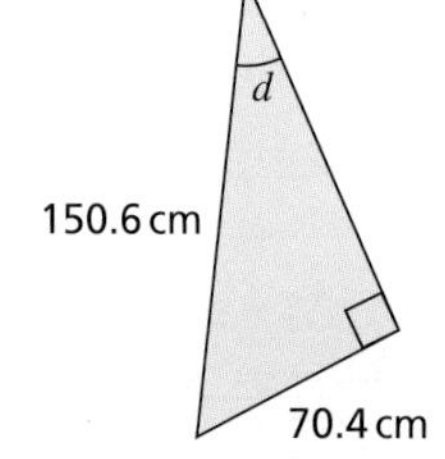

3 Use the cosine ratio to work out the size of each lettered angle.

a

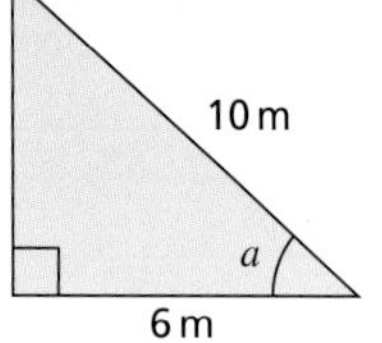

b

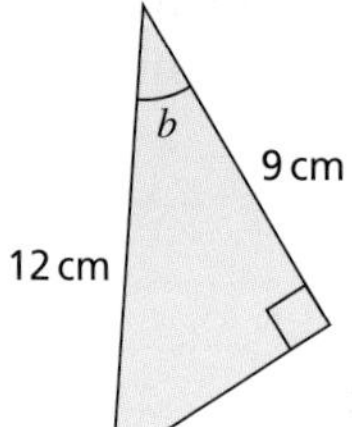

c

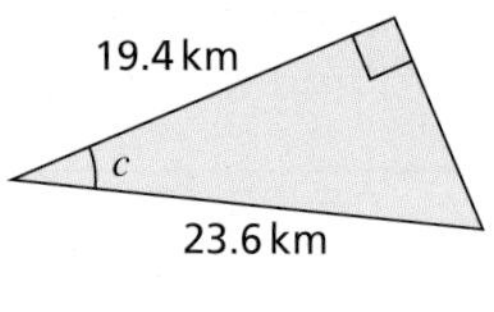

d

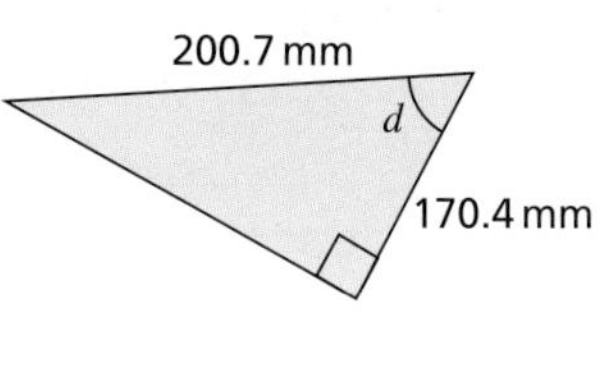

4 Use the tangent ratio to work out the size of each lettered angle.

a

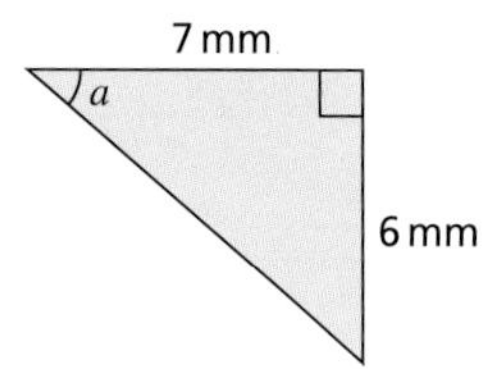

b

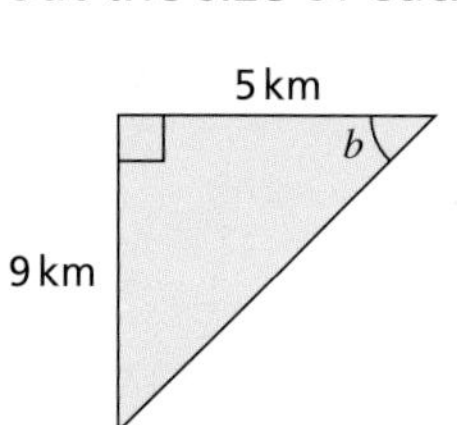

c

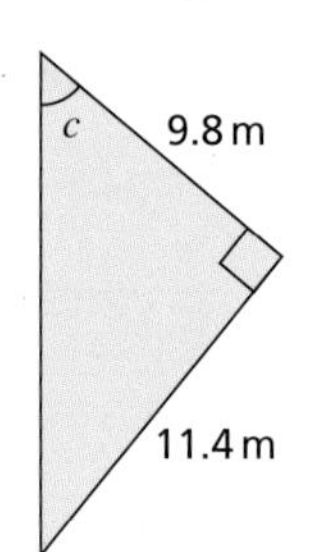

d

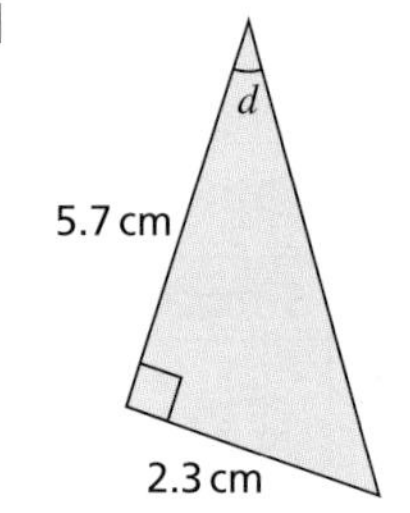

5 Calculate the size of the angle θ in each triangle. Give your answers to 1 decimal place.

a

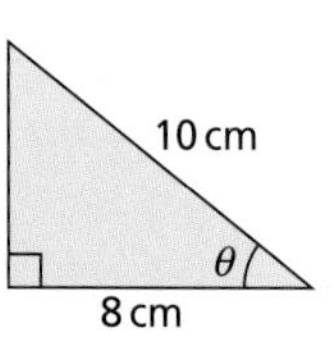

b

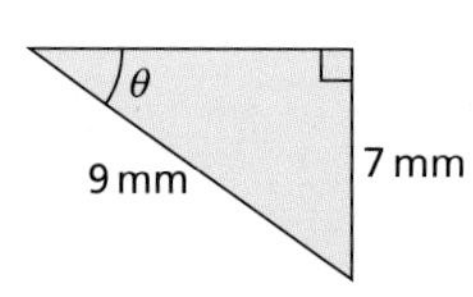

c

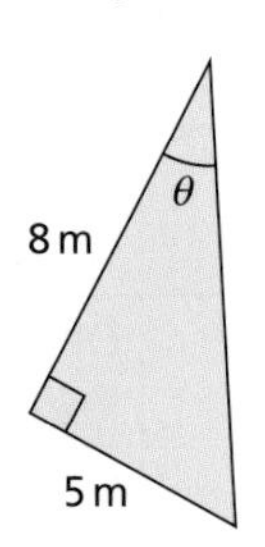

d

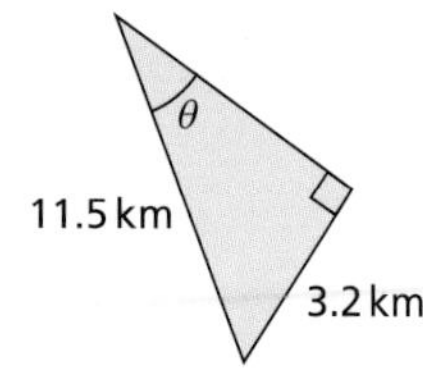

e

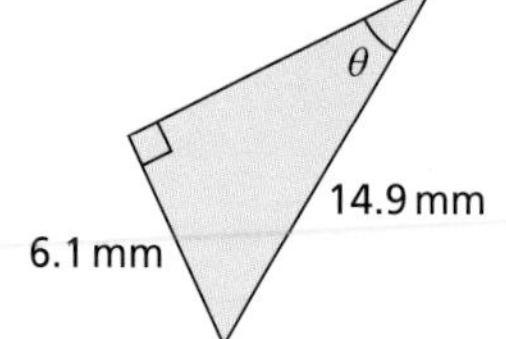

f

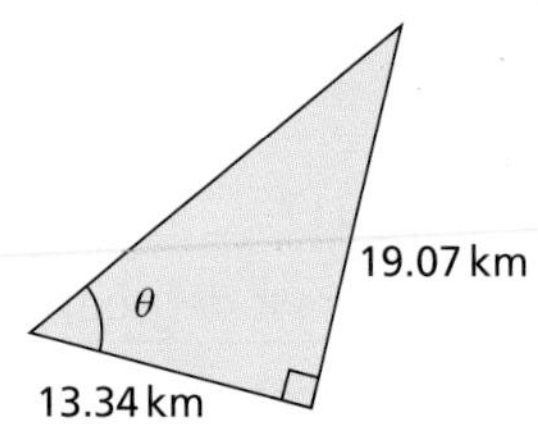

6 Work out the size of angle ACB.

Give your answer to 3 significant figures.

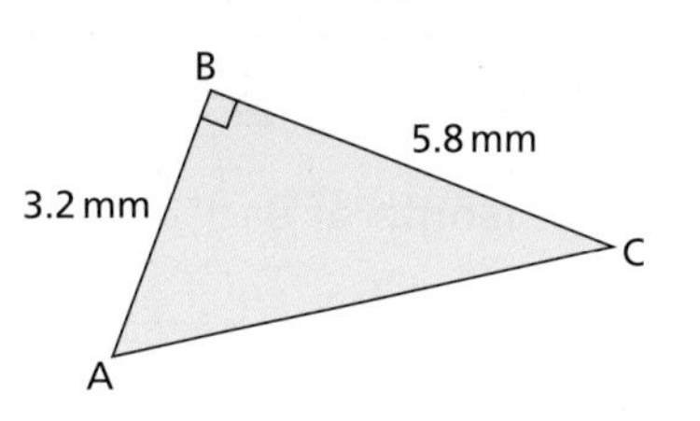

7 Here is the rectangle ABCD.

Work out the size of angle ACD.

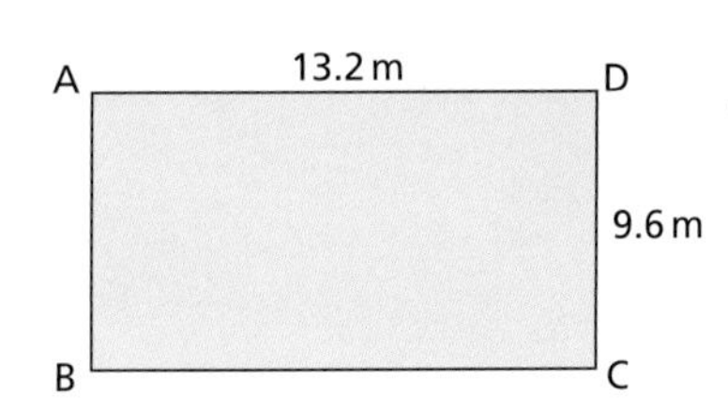

What do you think? (B)

1 A triangle is shown.

Both measurements have been rounded to the nearest integer.

What is the largest possible value for x?

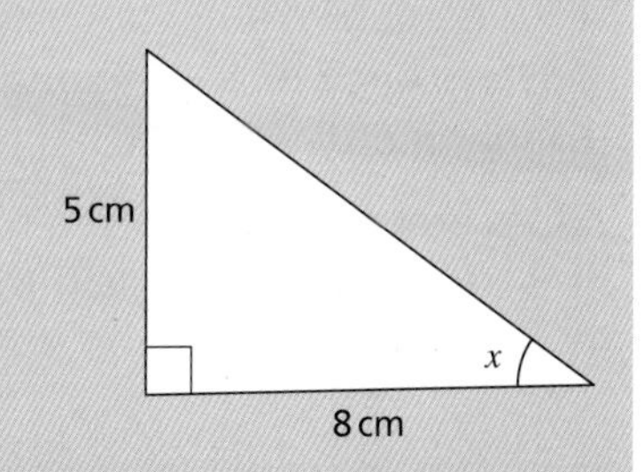

2 Triangle ABC is shown.

Sven says, "If you double the length of AB and the length of AC, the angle ACB will also double."

Explain why Sven is incorrect.

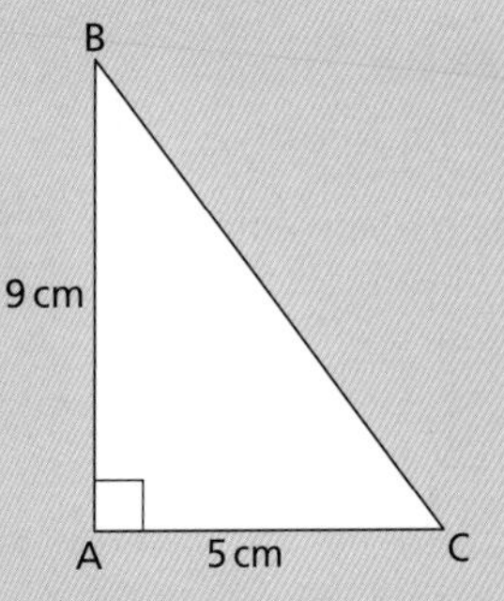

3 An isosceles triangle has a base of 7 cm and a height of 10 cm.

Ed says, "The two equal angles are 55° to the nearest degree because $\tan^{-1}\left(\frac{10}{7}\right) = 55$ degrees."

What mistake has Ed made?

Work out the size of the two equal angles.

Are you ready? (C)

1 Work out the length of each lettered side.

a

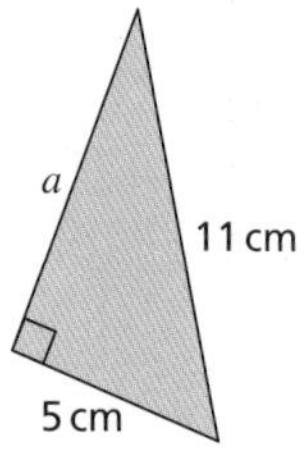

b

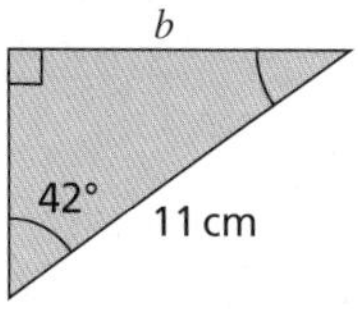

c

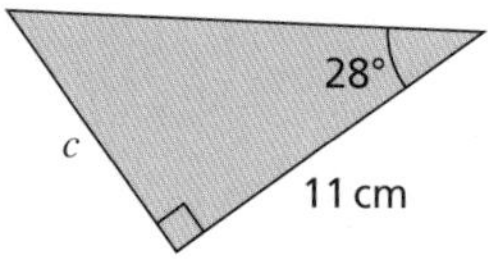

d

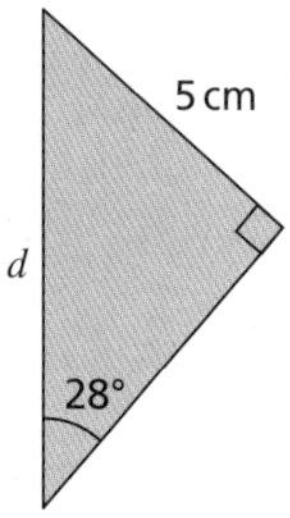

Pythagoras' theorem can be used to work out a missing length in a right-angled triangle, if the other two sides are known.

Trigonometric ratios can be used to work out a missing length when an angle and another side are known. Missing angles in a right-angled triangle can be calculated using trigonometry when two of the sides are known.

Multi-step problems often require the use of both Pythagoras' theorem and trigonometry. It is often useful to sketch the known information in order to apply a correct strategy.

Example

a Work out the length of BD.

b Calculate the size of angle CBD.

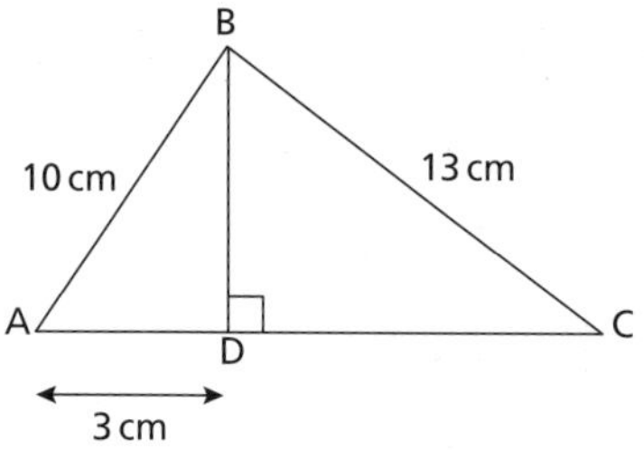

Method

Solution	Commentary
a 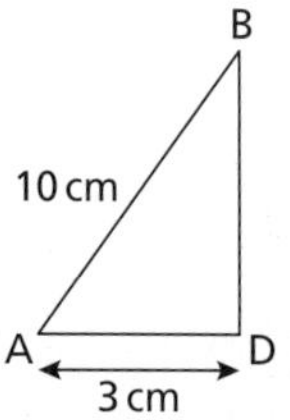	Consider triangle ABD first. The triangle is right-angled and you know two of the three sides.
$a^2 + b^2 = \text{hyp}^2$ $10^2 - 3^2 = 91$ $\sqrt{91} = 9.5393...$ BD = 9.54 cm (to 2 d.p.)	You can use Pythagoras' theorem to work out the length of BD.

b

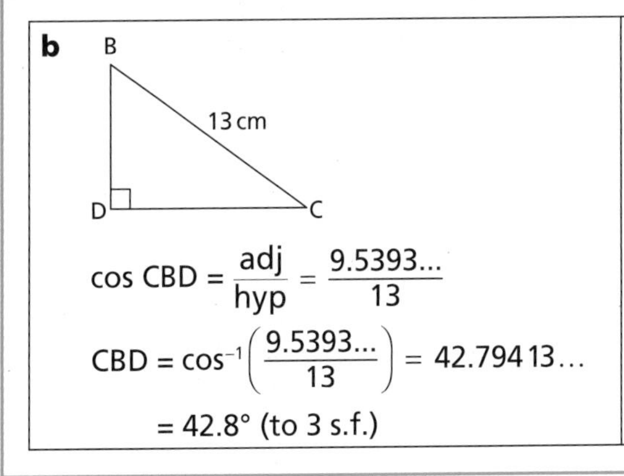

$\cos \text{CBD} = \frac{\text{adj}}{\text{hyp}} = \frac{9.5393...}{13}$

$\text{CBD} = \cos^{-1}\left(\frac{9.5393...}{13}\right) = 42.79413...$

$= 42.8°$ (to 3 s.f.)

Now consider the triangle BDC, where you know the length of the side adjacent to angle CBD (9.5393... cm) and the hypotenuse (13 cm).

You can use the cosine ratio to calculate the size of angle CBD.

As you are calculating the size of an angle, use the inverse cosine function.

Practice (C)

1 Here are two triangles.

a Work out the length of BD.

Give your answer to 2 decimal places.

b Work out the length of CD.

Give your answer to 2 decimal places.

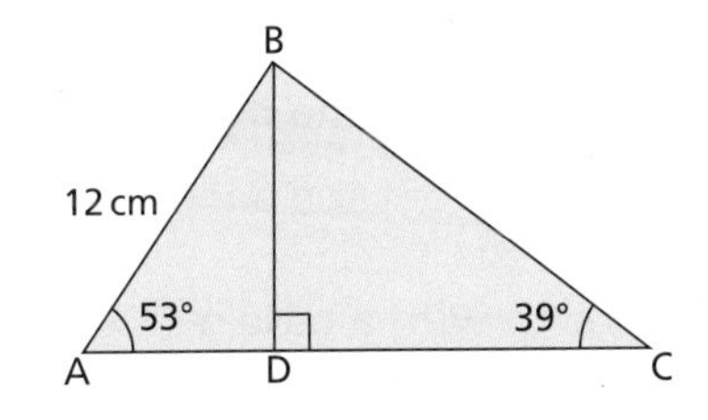

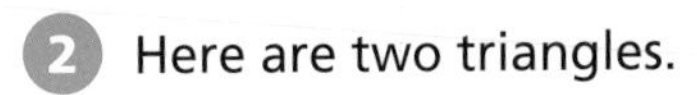

2 Here are two triangles.

a Work out the length of QS.

Give your answer to 3 significant figures.

b Work out the size of angle QPS.

Give your answer to 3 significant figures.

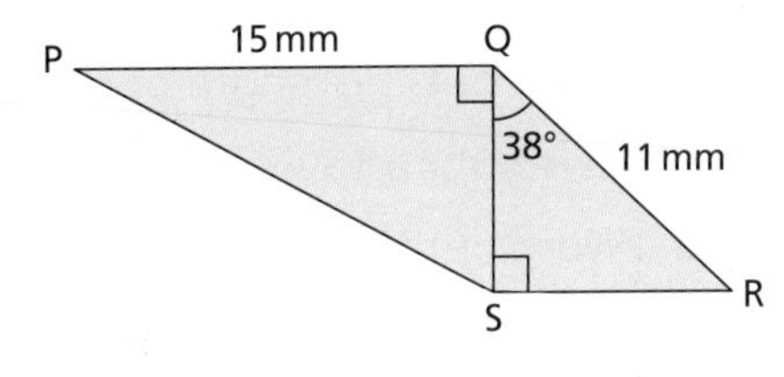

3 Here are two triangles.

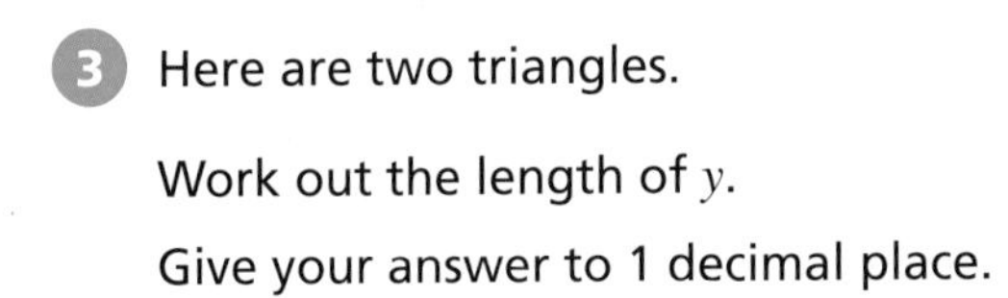

Work out the length of y.

Give your answer to 1 decimal place.

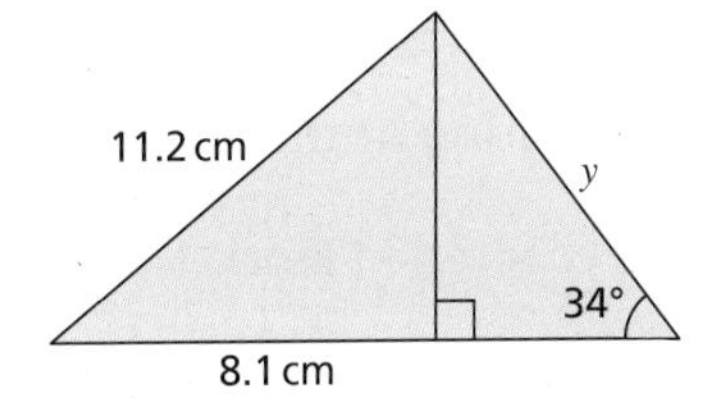

4 Here are two right-angled triangles.

Work out the size of angle ADC.

Give your answer to 3 significant figures.

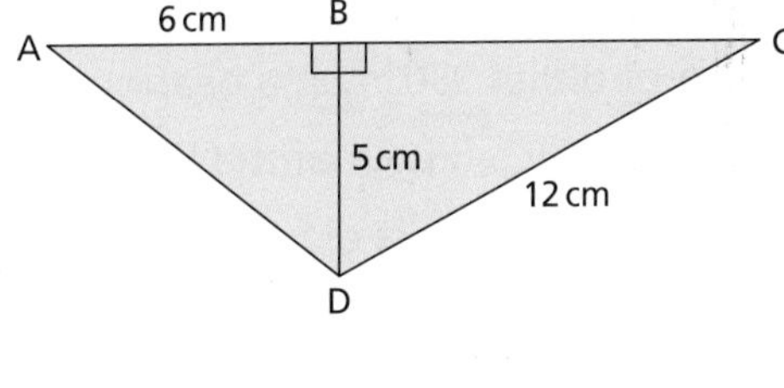

5 The rectangle has an area of $60\,\text{cm}^2$.

Work out the size of angle ABD.

Give your answer to 1 decimal place.

6 The triangle has an area of 25 cm².

Work out the size of angle x.

Give your answer to 3 significant figures.

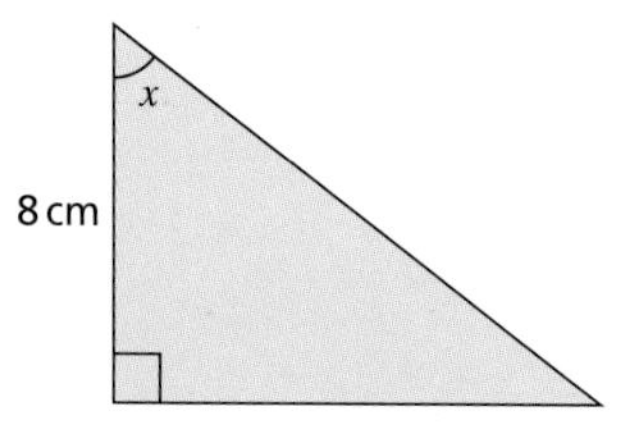

7 Work out the size of angle x.

Give your answer to 1 decimal place.

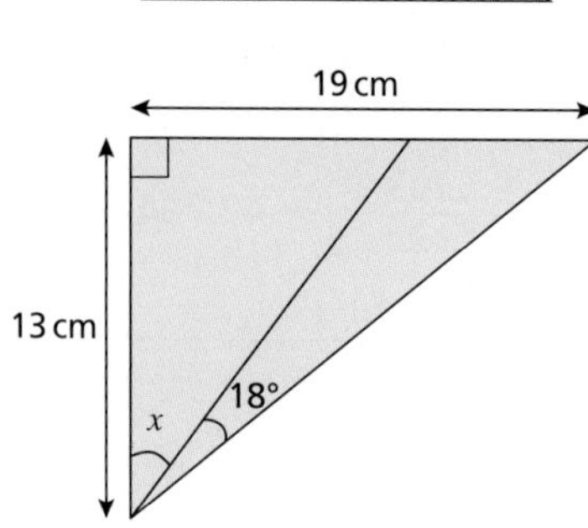

8 A 6 m ladder rests 5 m up a vertical wall.

a Work out the distance from the foot of the ladder to the wall.

b Work out the angle between the ladder and the wall.

9 A ramp with a base of 130 cm and a height of 65 cm lies on a horizontal surface.

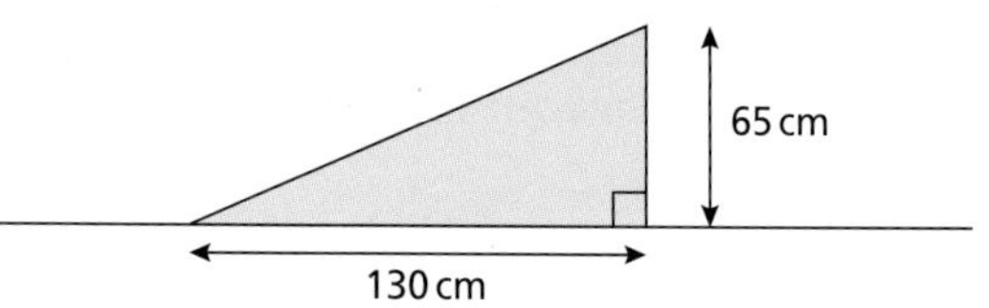

a Work out the length of the ramp.

Give your answer to 3 significant figures.

b Work out the angle between the ramp and the horizontal surface.

Give your answer to 3 significant figures.

10 A boat leaves a port and travels 8 km due East, then 5 km due South to a lighthouse.

Work out the bearing of the lighthouse from the port.

Bearings are covered in Block 1 (Constructions).

11 A bird flies 350 m on a bearing of 154° from a nest to a cliff top.

a Draw a sketch of the bird's journey.

b How far East did the bird travel? Give your answer to 3 significant figures.

c How far South did the bird travel? Give your answer to 3 significant figures.

Consolidate – do you need more?

1 Work out the length of each lettered side. Give your answers to 1 decimal place.

a

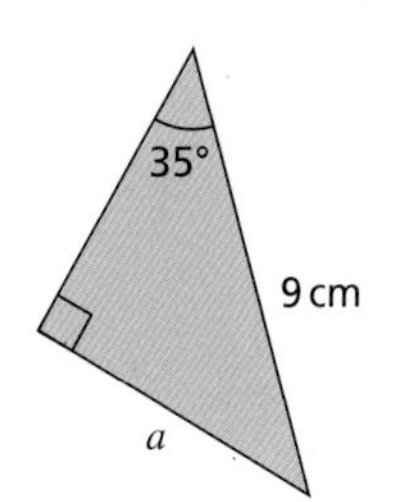

b

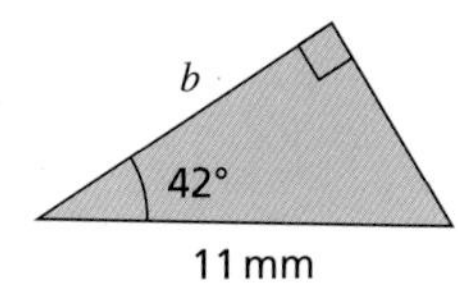

c

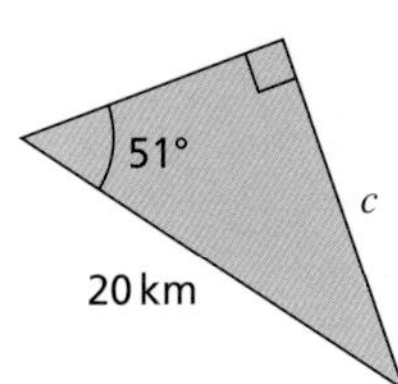

d

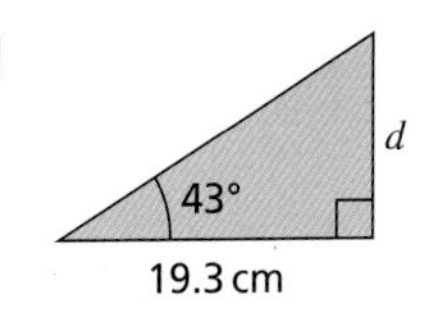

e

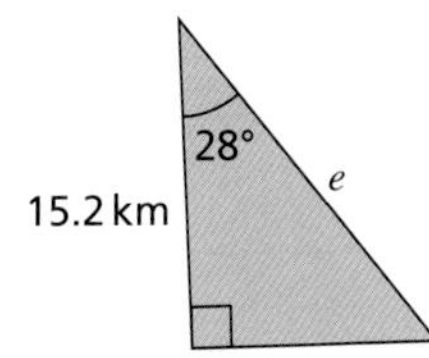

f

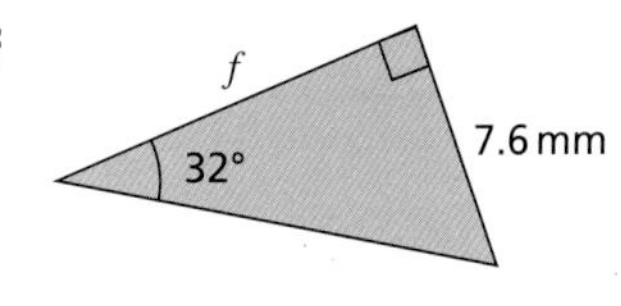

2 Work out the size of each lettered angle. Give your answers to 1 decimal place.

a

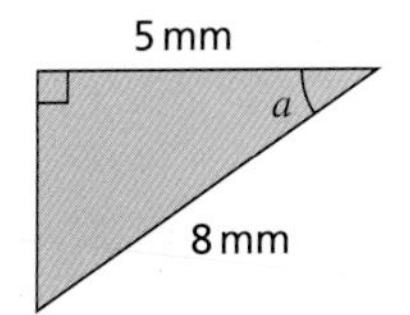

b

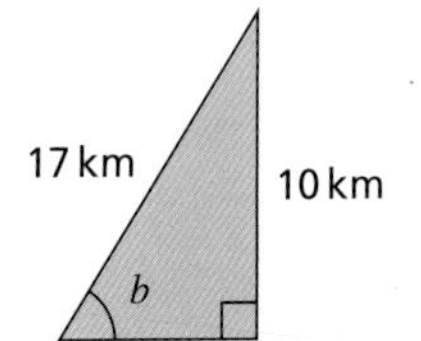

c

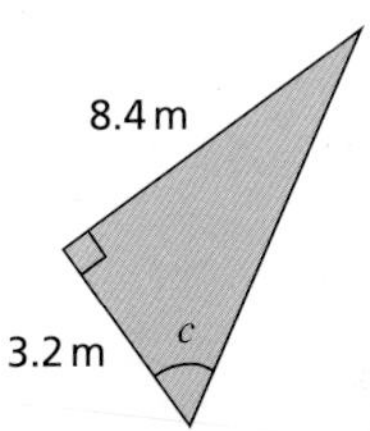

d

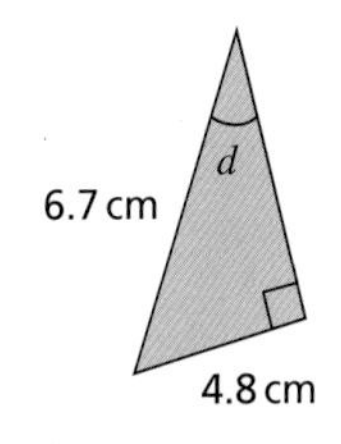

e

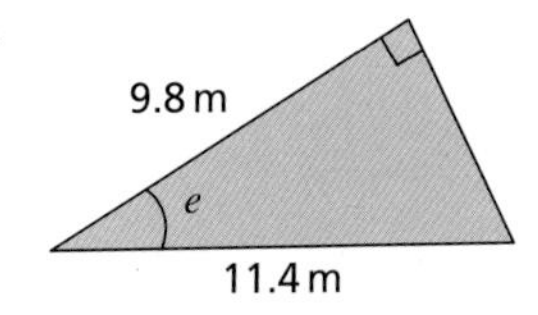

f

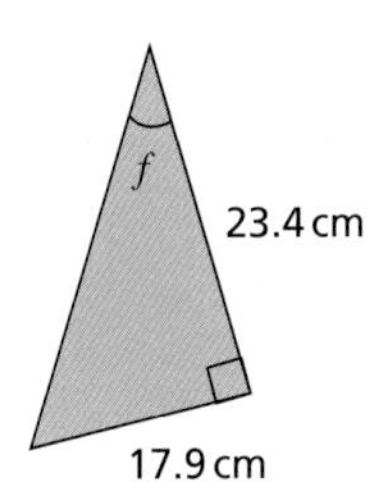

3 Work out angle DEF.

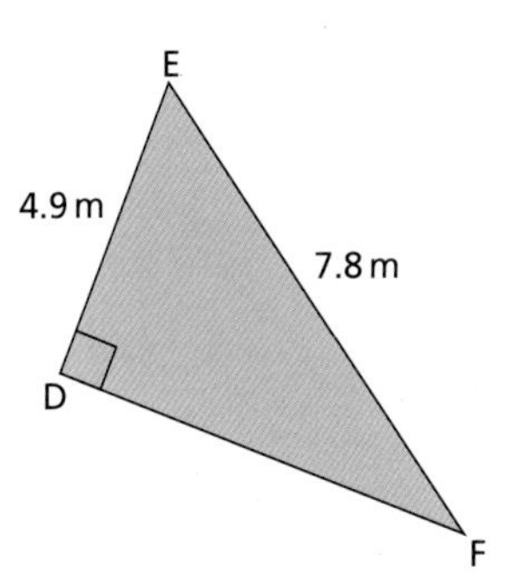

4 Work out the length of PR.

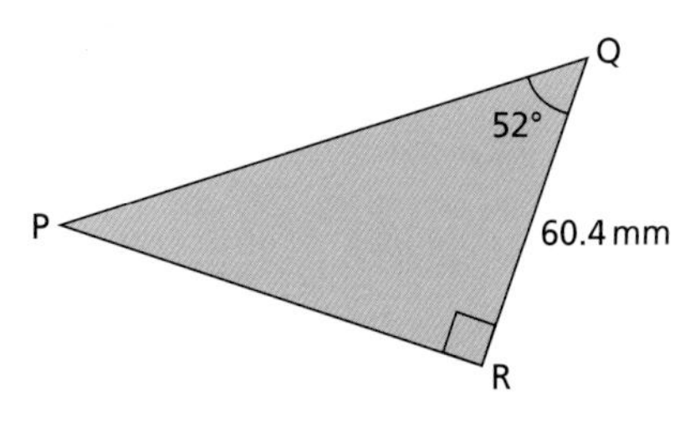

5 Here are two triangles.

a Work out the length of MN.

Give your answer to 2 decimal places.

b Work out the length of KL.

Give your answer to 2 decimal places.

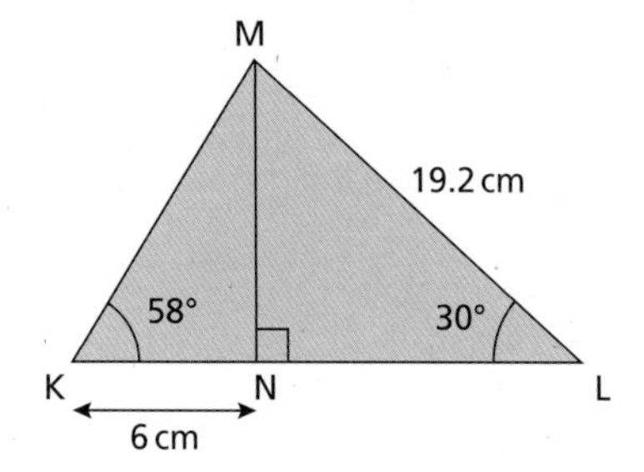

Stretch – can you deepen your learning?

1 ABCD is a rhombus with sides of length 8 cm. Angle ABC is 65°.

Calculate the lengths of the diagonals of rhombus ABCD.

2 If you divide the sine of an angle by the cosine of the same angle, the result is the tangent of that angle.

Is the statement **true** or **false**? Explain your answer.

3 $\sin^2 x$ means the square of $\sin x$. For example, $\sin 30° = 0.5$ and $\sin^2 30° = 0.25$

Investigate values of $\sin^2 x + \cos^2 x$ for different values of x.

What do you notice? Explain your answer.

4 **a** Given that $x^2 + 5x + 6 \equiv (x + 3)(x + 2)$, factorise $\sin^2 x + 5\sin x + 6$

b Factorise:

i $\sin^2 x + 8\sin x + 12$ **ii** $\cos^2 x + 2\cos x - 15$ **iii** $2\tan^2 x + 9\tan x + 10$

5 Sketch an equilateral triangle with a side length of 2 units.

How can you use the triangle to work out the value of sin 30°?

Are there any other trigonometric values you can work out?

You will explore this idea in Chapter 6.4

6.3 3D trigonometry

White Rose MATHS

Are you ready?

1 Work out the length of each lettered side. Give your answers to 1 decimal place.

a

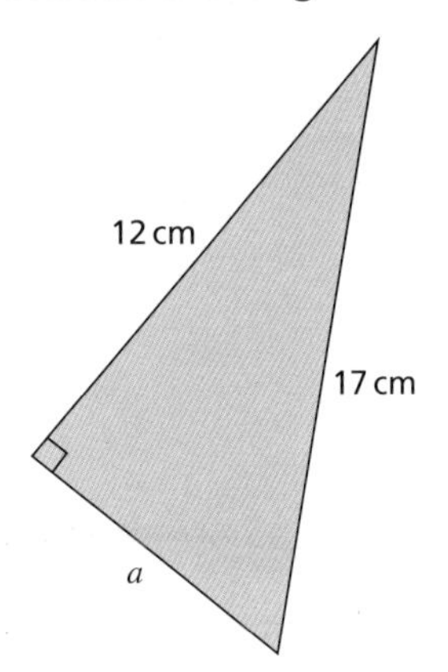

b

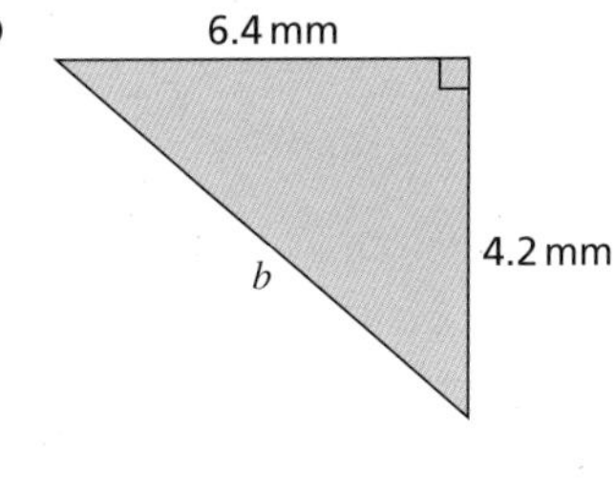

c

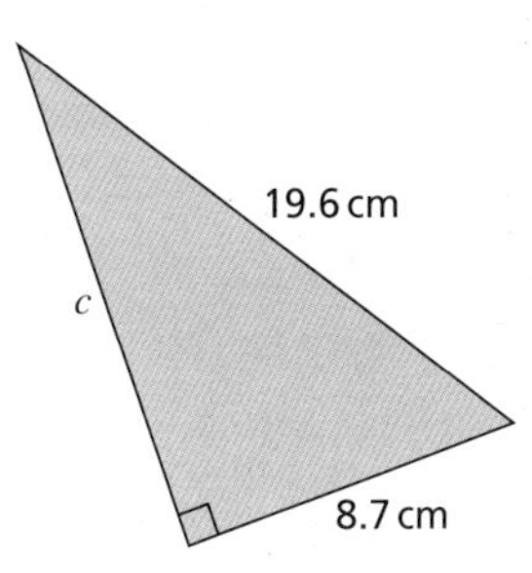

2 Work out the length of each lettered side. Give your answers to 1 decimal place.

a

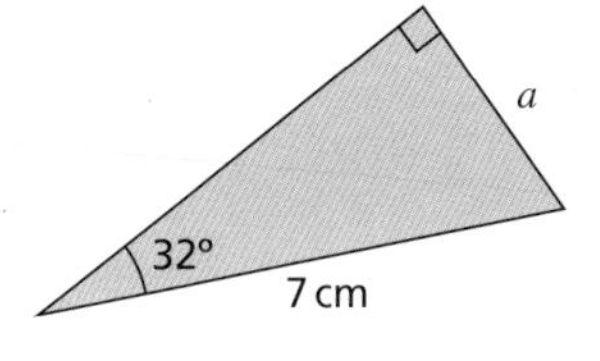

b

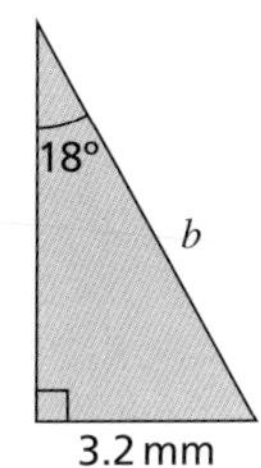

c

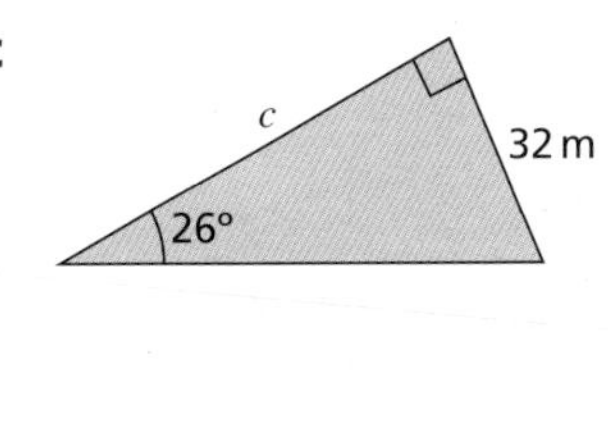

3 Work out the size of each lettered angle. Give your answers to 1 decimal place.

a

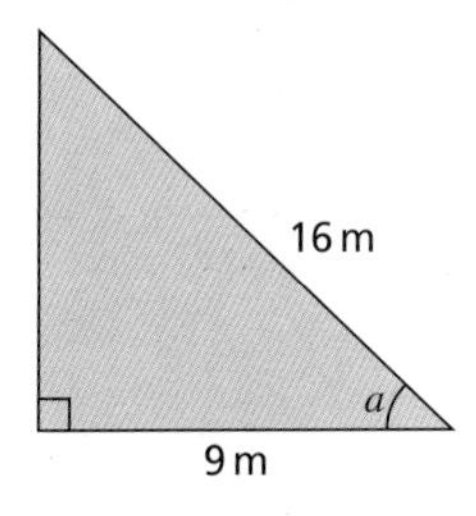

b

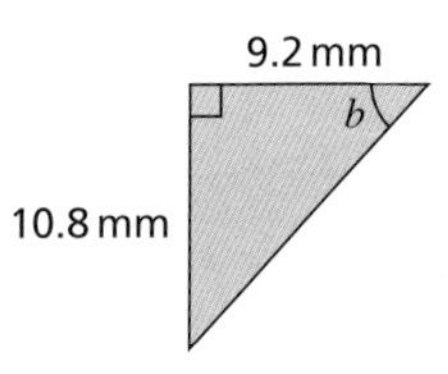

c

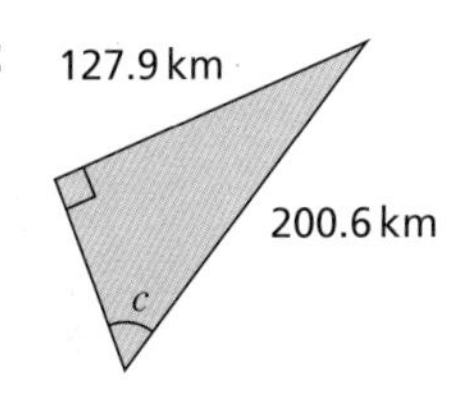

The rules and theorems that are used to solve problems in 2D shapes can also be used for right-angled triangles in 3D shapes.

Pythagoras' theorem: $a^2 + b^2 = \text{hypotenuse}^2$

Trigonometry: $\sin\theta = \frac{\text{opp}}{\text{hyp}}$ $\cos\theta = \frac{\text{adj}}{\text{hyp}}$ $\tan\theta = \frac{\text{opp}}{\text{adj}}$

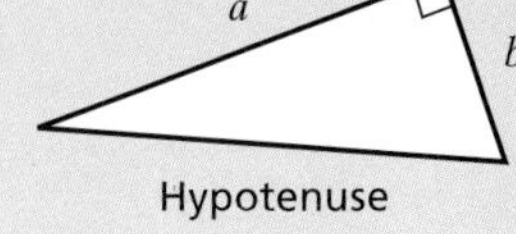

You can identify right-angled triangles in 3D shapes.

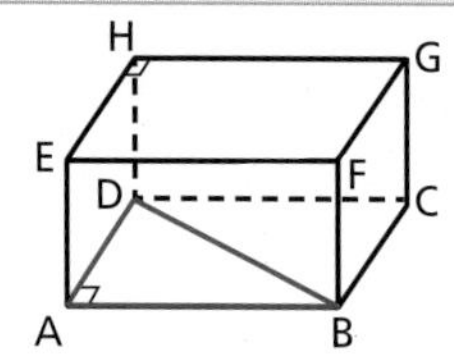

Triangles ADB, BDC, EFH and GHF are congruent right-angled triangles.

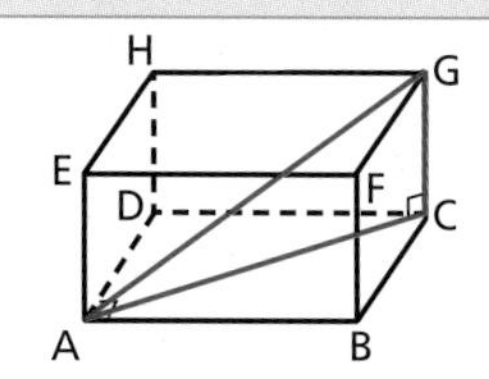

Triangles CAG, CAE, BDF and HFD are congruent right-angled triangles.

When working with 3D figures, a flat surface is sometimes referred to as a **plane**.

In the cube ABCDEFGH, EFGH is a square.

EFGH is an example of a plane.

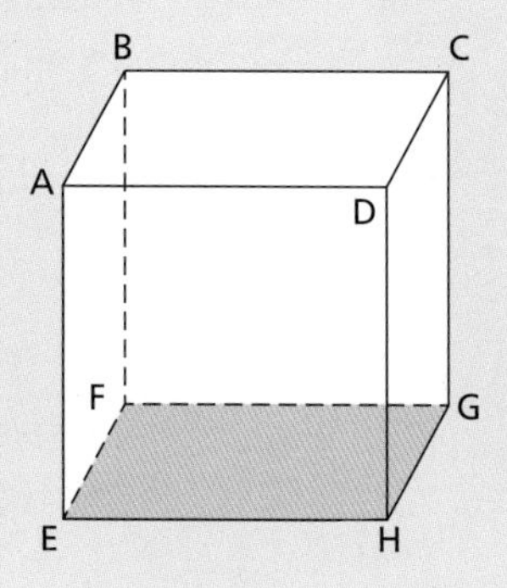

Example

ABCDEFGH is a cuboid.

a Calculate length AG, giving your answer to 3 significant figures.

b Calculate angle CAG to 1 decimal place.

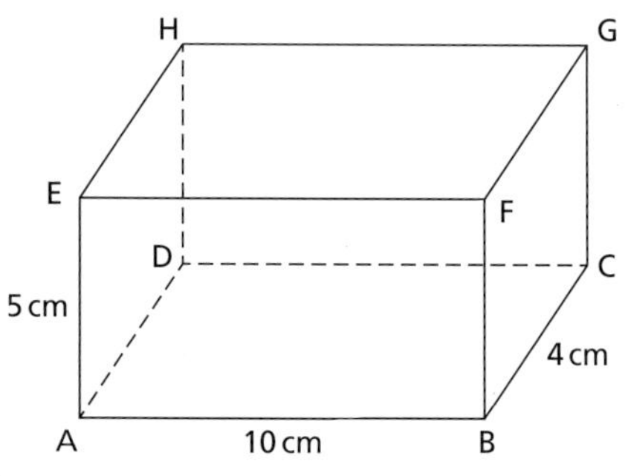

Method

Solution	Commentary
a $AC = \sqrt{10^2 + 4^2}$ $= \sqrt{116}$ cm	Work out the diagonal AC using Pythagoras' theorem.
$AG^2 = AC^2 + CG^2$ $= 116 + 25$ $= 141$ $AG = \sqrt{141} = 11.9$ cm (to 3 s.f.)	Using the theorem again in triangle ACG, work out length AG using length AC; you know $AC^2 = 116$ cm.

b G, opp 5 cm, A, C, $\sqrt{116}$ cm adj	From the diagram of the cuboid, you can see the right-angled triangle ACG. You may find it useful to draw this separately. Label the sides so you know which trigonometric ratio to use.
$\tan\theta = \frac{\text{opposite}}{\text{adjacent}}$	As you know the opposite side and the adjacent side, use the tangent ratio to work out angle CAG.
$\angle CAG = \tan^{-1}\left(\frac{5}{\sqrt{116}}\right)$ $= 24.9°$ (to 1 d.p.)	You know that GC = 5 cm and AC = $\sqrt{116}$ cm from part **a**. It is important to use the exact value, not the rounded value.
	As you also know the length of AG from part **a**, you could use any of the three trigonometric ratios and get the same answer.

Practice

1 Cuboid ABCDEFGH is shown.

The diagonal AC is 8 cm and the diagonal AG is 10 cm.

Work out the size of the angle between AC and AG.

Give your answer to 1 decimal place.

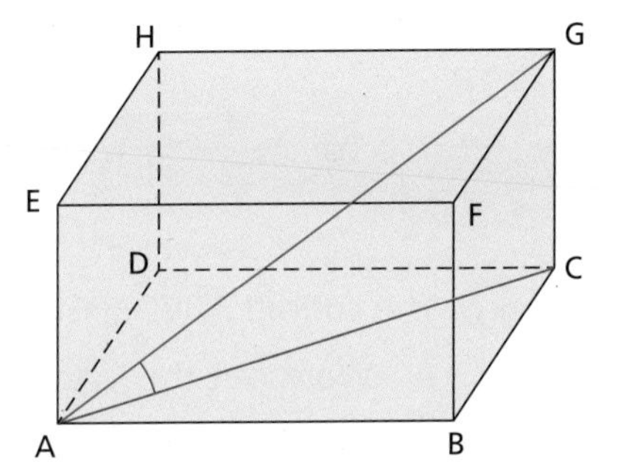

2 Here is a cuboid.

a Work out the length of AC.

Give your answer to 3 significant figures.

b Work out the size of angle CAG.

Give your answer to 3 significant figures.

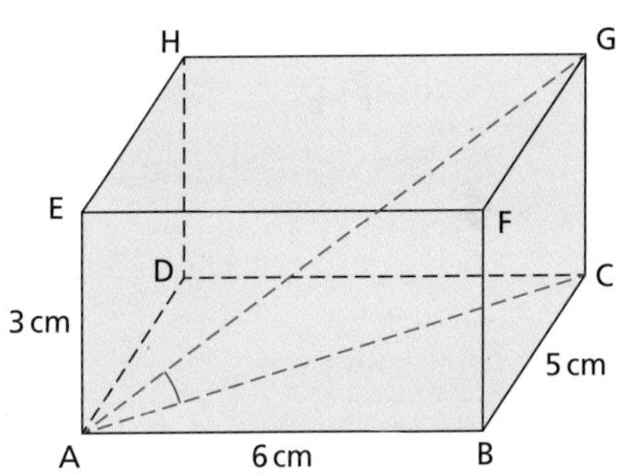

3 ABCDEFGH is a cuboid.

a Work out the length of AC.

Give your answer to 3 significant figures.

b Work out the size of angle CAG.

Give your answer to 3 significant figures.

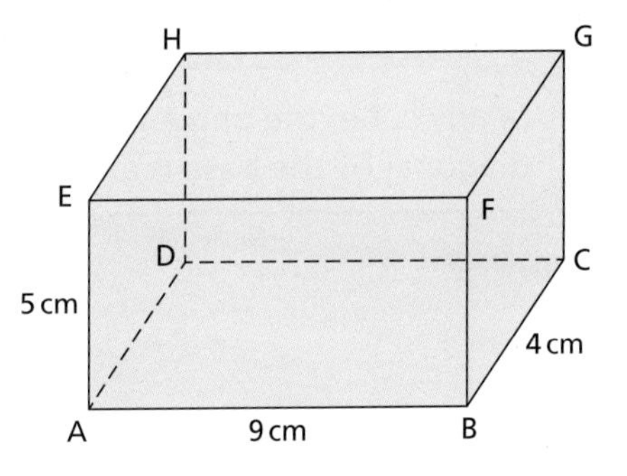

4 ABCDEFGH is a cube.

a Work out the length of BD.

Give your answer to 3 significant figures.

b Work out the size of angle BHD.

Give your answer to 3 significant figures.

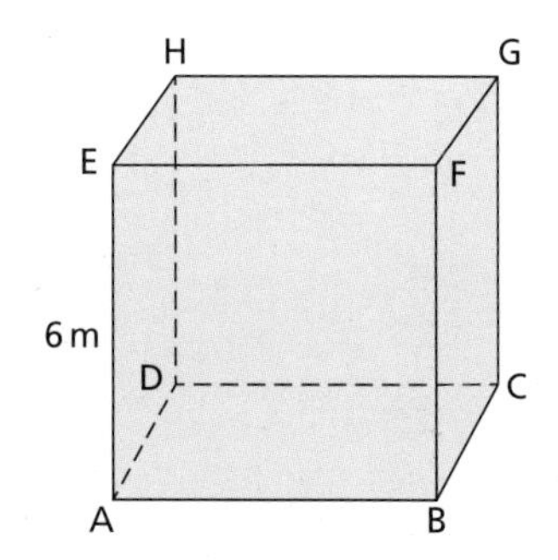

5 The cuboid ABCDEFGH is shown.

AB = 7 cm, BC = 9 cm and BF = 5 cm.

Work out the size of the angle between the plane ABCD and BH.

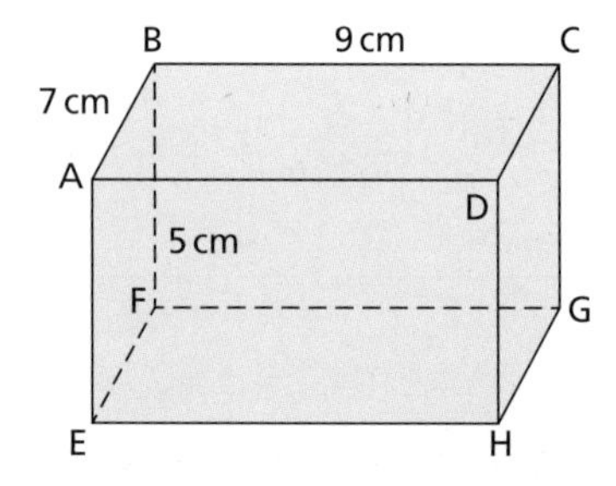

6 A triangular prism is shown.

Work out the size of the angle between the plane ABCD and the line AF.

Give your answer to 3 significant figures.

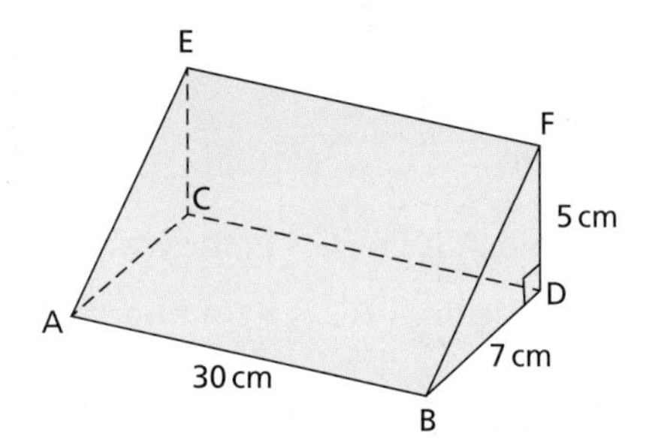

7 The cuboid ABCDEFGH has a volume of 350 cm^3.

a Work out the length of x.

b Work out the length of AG.

c Work out the angle between AG and the plane ABCD.

Sketch right-angled triangle ACG to help you.

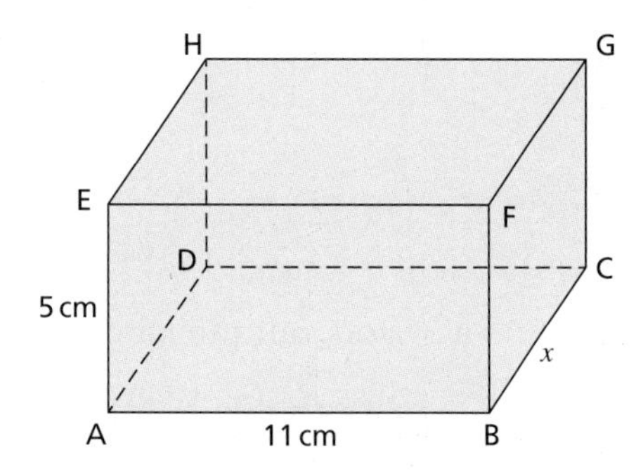

What do you think?

1 Is the statement **true** or **false**?

In any cube, the angle between the longest diagonal and the diagonal of the base is always the same.

Justify your answer.

Consolidate – do you need more?

Give your answers to 3 significant figures.

1 Cuboid ABCDEFGH is shown.

a Work out the length of AC.

b Work out the size of angle CAG.

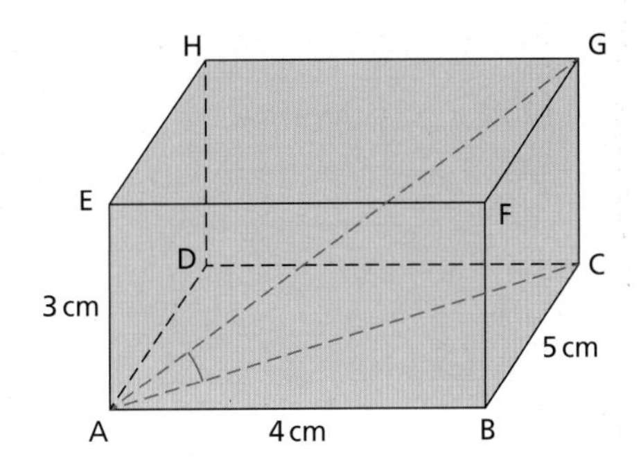

2 Cuboid ABCDEFGH is shown.

a Work out the length of AC.

b Work out the size of angle CAG.

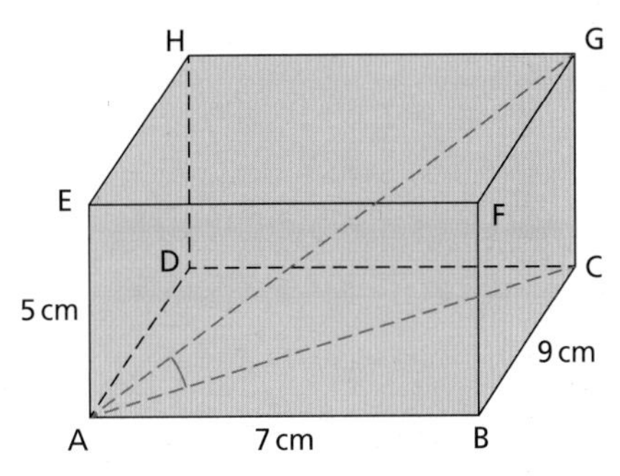

3 Cuboid ABCDEFGH is shown.

a Work out the length of AC.

b Work out the size of angle CAG.

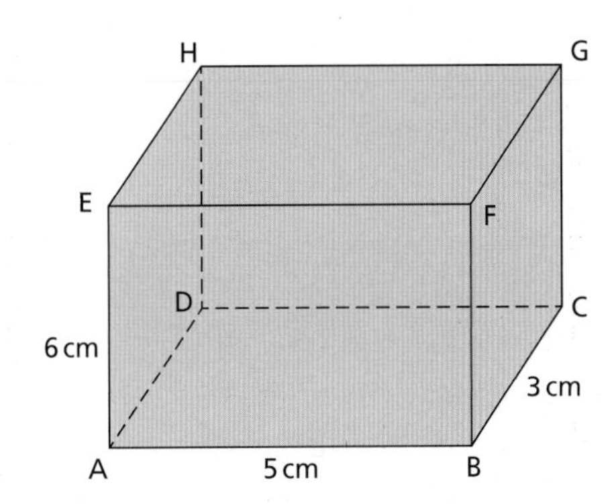

4 Cube ABCDEFGH is shown.

a Work out the length of BD.

b Work out the size of angle BHD.

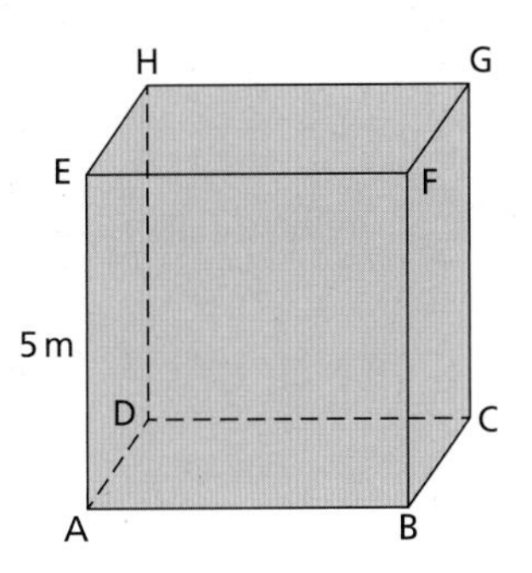

Stretch – can you deepen your learning?

Give your answers to 3 significant figures.

1 A square-based pyramid is shown.

> Volume of a pyramid
> $= \frac{1}{3} \times$ area of the base $\times$ perpendicular height

Work out the volume of the pyramid.

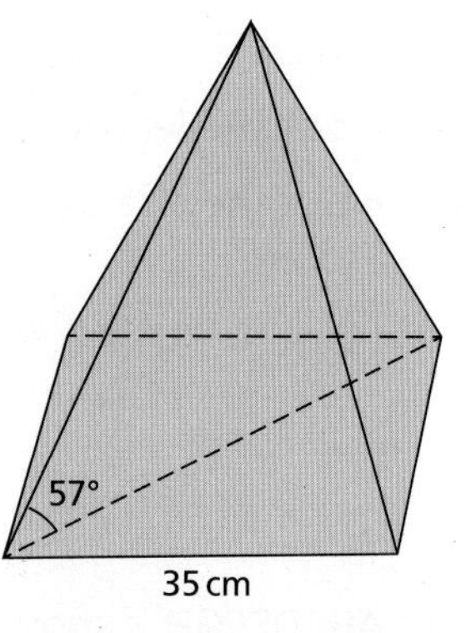

2 The diagram shows two cubes.

The cubes have a total volume of 169 cm^3.

Work out the length of AB.

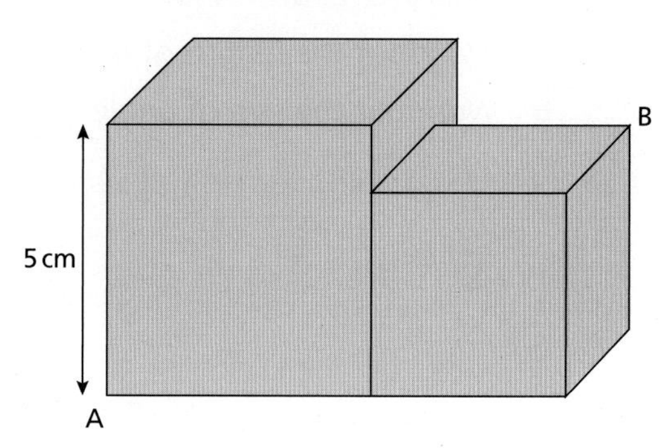

3 The diagram shows the plan of a house.

EFGH is the floor of the loft.

The apex point I lies vertically above the point M, which is the centre of the square base ABCD.

Work out the angle between the floor of the loft and the edge of the roof, HI.

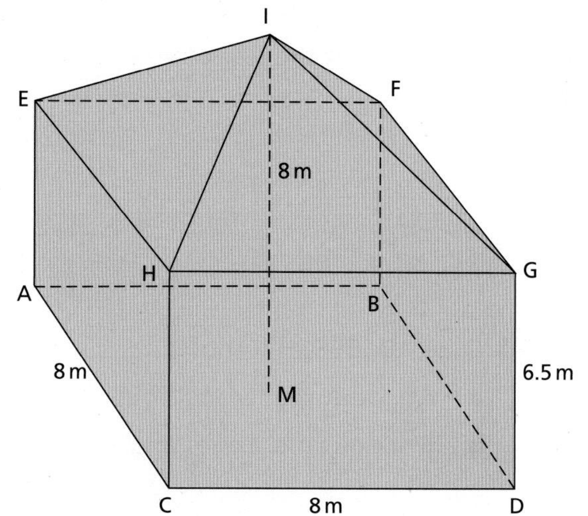

6.4 Exact trigonometric values

White Rose MATHS

Are you ready?

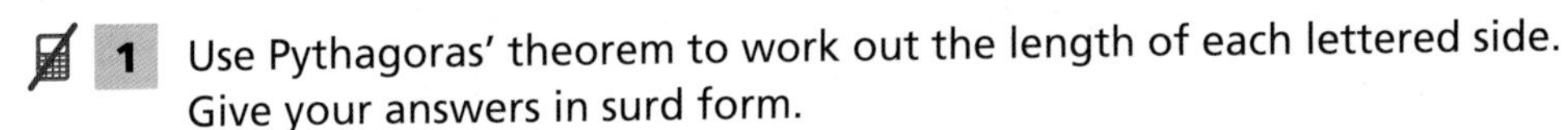

1 Use Pythagoras' theorem to work out the length of each lettered side.
Give your answers in surd form.

a

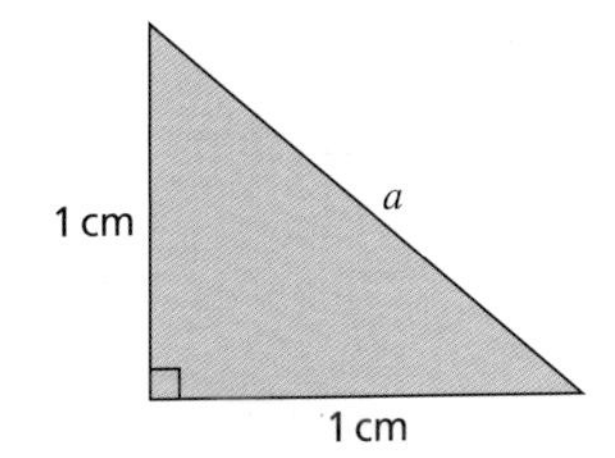

b

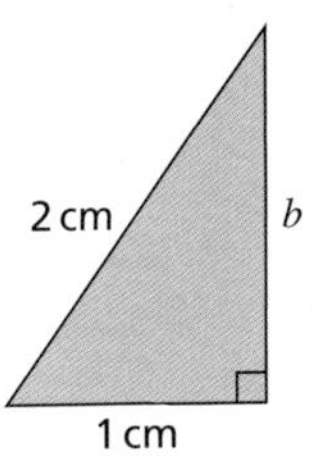

2 Use trigonometry to work out the length of each lettered side.

a

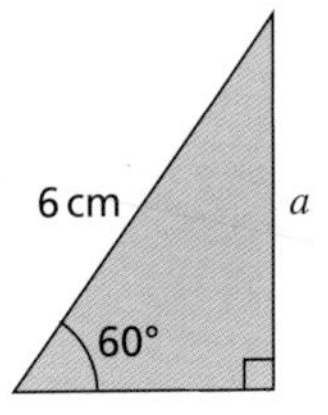

b

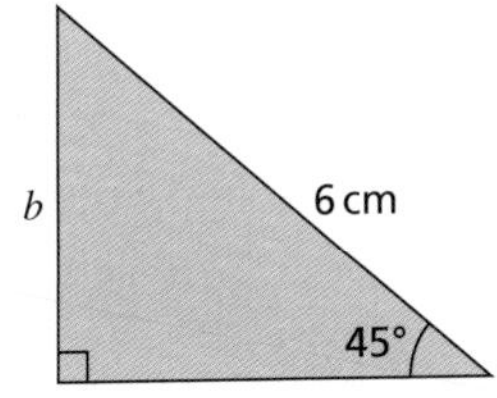

3 Simplify:

a $\frac{2\sqrt{3}}{2}$ **b** $\frac{8\sqrt{3}}{2}$ **c** $\frac{10\sqrt{2}}{2}$

4 Solve:

a $\frac{a}{5} = 10$ **b** $\frac{1}{2} = \frac{b}{3}$ **c** $\frac{\sqrt{3}}{2} = \frac{c}{5}$ **d** $\frac{\sqrt{2}}{2} = \frac{10}{d}$

The trigonometric ratios for some angles can be worked out using special triangles.

In a right-angled isosceles triangle with two sides of 1 unit, you can use Pythagoras' theorem to calculate the length of AC to be $\sqrt{2}$ units.

Then you can label the sides of the triangle relative to one of the 45° angles.

Use the definitions of sine, cosine and tangent ratios to find:

$\sin 45° = \frac{\text{opp}}{\text{hyp}} = \frac{1}{\sqrt{2}}$ $\cos 45° = \frac{\text{adj}}{\text{hyp}} = \frac{1}{\sqrt{2}}$ $\tan 45° = \frac{\text{opp}}{\text{adj}} = \frac{1}{1} = 1$

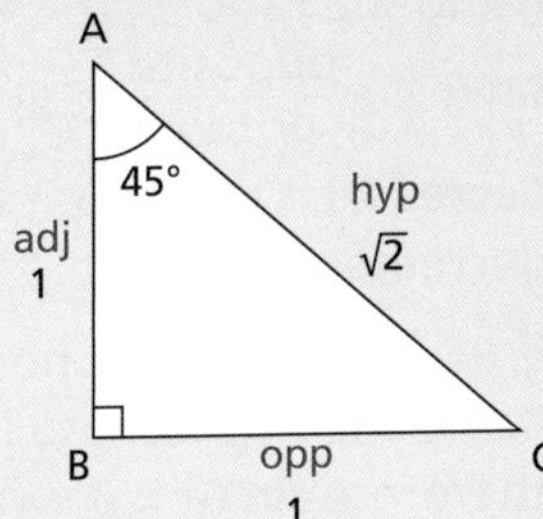

Another special triangle is a right-angled triangle formed with half of an equilateral triangle with a side length of 2 units.

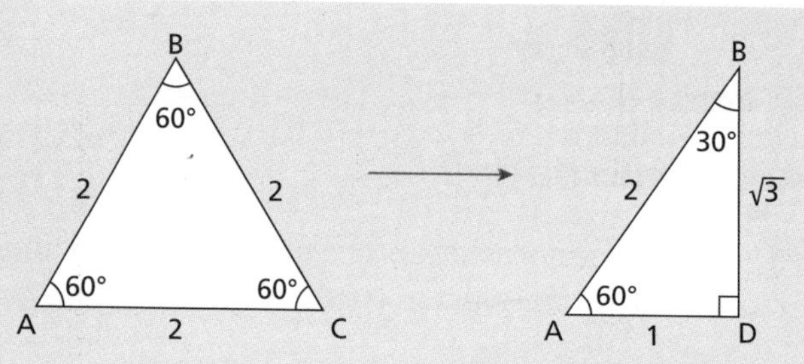

Bisecting the angle ABC will form two right-angled triangles with a base of 1 and a hypotenuse of 2. You can calculate the length of BD to be $\sqrt{3}$ units using Pythagoras' theorem.

You can use the definitions of sine, cosine and tangent ratios to find:

$\sin 30° = \frac{1}{2}$ $\quad \cos 30° = \frac{\sqrt{3}}{2}$ $\quad \tan 30° = \frac{1}{\sqrt{3}} = \frac{\sqrt{3}}{3}$

$\sin 60° = \frac{\sqrt{3}}{2}$ $\quad \cos 60° = \frac{1}{2}$ $\quad \tan 60° = \frac{\sqrt{3}}{1} = \sqrt{3}$

A circle of radius 1, called a **unit circle,** can also be used to determine exact values.

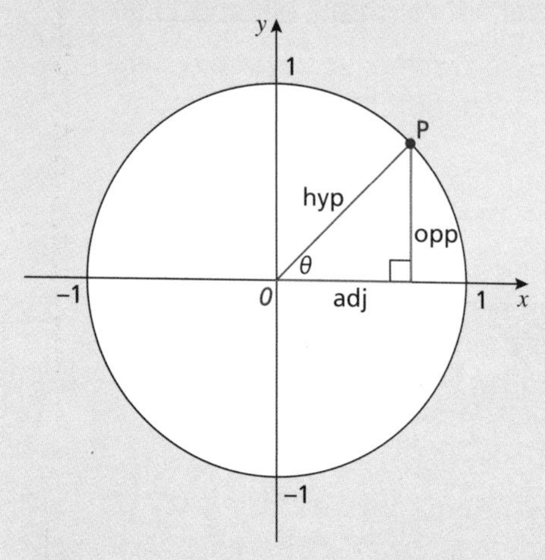

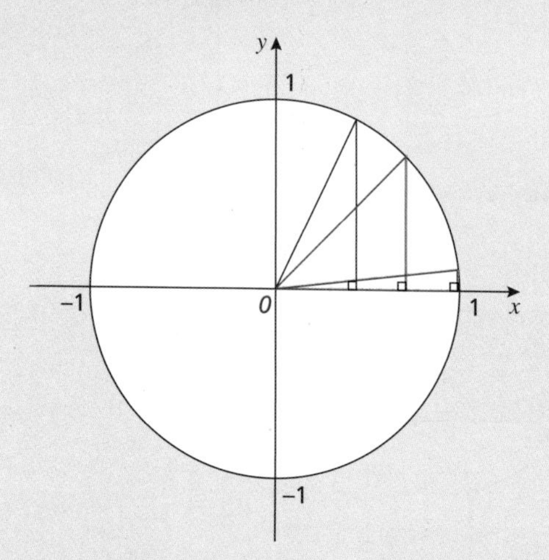

Here is a unit circle. It traces the position of a point P relative to the origin, *0*. A right-angled triangle can be drawn, with angle θ.

As P moves around the circle, the hypotenuse always has the value 1, but the trigonometric ratios change. The following shows how sine, cosine and tangent change from 0° to 90°.

Sine	**Cosine**	**Tangent**
Using the ratio $\sin \theta = \frac{\text{opposite}}{\text{hypotenuse}}$, as θ decreases, the opposite decreases. If $\theta = 0°$ then the opposite is 0 and the hypotenuse is 1, therefore $\sin 0° = 0$	Using the ratio $\cos \theta = \frac{\text{adjacent}}{\text{hypotenuse}}$, as θ decreases, the adjacent increases. If $\theta = 0°$ then the adjacent side and the hypotenuse are both 1, therefore $\cos 0° = 1$	Using the ratio $\tan \theta = \frac{\text{opposite}}{\text{adjacent}}$, as θ decreases, the opposite decreases and the adjacent increases. If $\theta = 0°$ then the opposite is 0 and the adjacent is 1, therefore $\tan 0° = 0$

Using a similar approach when $\theta = 90°$, you can determine $\sin 90° = 1$, and $\cos 90° = 0$

When $\theta = 90°$, the opposite side is 1 and the adjacent side is 0. As it is not possible to divide a value by 0, we say that tan 90° is undefined.

The graphs of $y = \sin x$, $y = \cos x$ and $y = \tan x$ can also be used to determine these values.

For more on this, see *Collins White Rose Maths AQA GCSE 9–1 Higher Student Book 1* (Chapter 15.1, Trigonometric graphs).

Here are the values you need to know:

	0°	30°	45°	60°	90°
$\sin x$	0	$\frac{1}{2}$	$\frac{1}{\sqrt{2}} = \frac{\sqrt{2}}{2}$	$\frac{\sqrt{3}}{2}$	1
$\cos x$	1	$\frac{\sqrt{3}}{2}$	$\frac{1}{\sqrt{2}} = \frac{\sqrt{2}}{2}$	$\frac{1}{2}$	0
$\tan x$	0	$\frac{1}{\sqrt{3}} = \frac{\sqrt{3}}{3}$	1	$\sqrt{3}$	undefined

You need to know these values to be able to solve problems without a calculator.

Example

Calculate the exact length of AB.

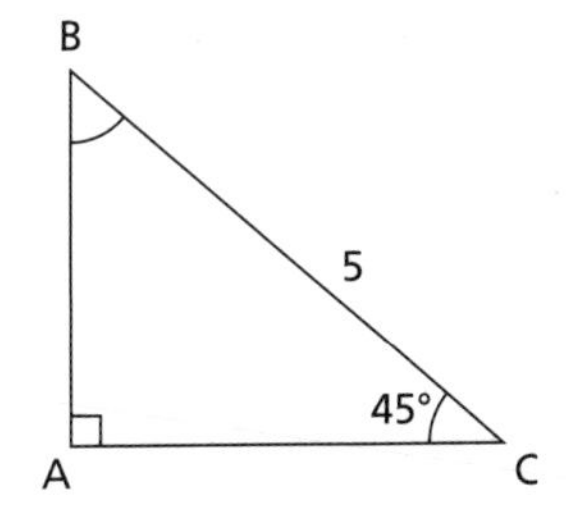

Method

Solution	**Commentary**
B opp hyp 5 45° A adj C	First, label the triangle relative to the given 45° angle.
$\sin x = \frac{\text{opp}}{\text{hyp}}$	Then decide which ratio you need to use and write it down. Here, you know the hypotenuse and you want to find the opposite, so you need the sine ratio.
$\sin 45° = \frac{\text{AB}}{5}$	Substitute in the information you know.
$\frac{1}{\sqrt{2}} = \frac{\text{AB}}{5}$	Since you know $\sin 45° = \frac{1}{\sqrt{2}}$, substitute this into the equation.
$\frac{1}{\sqrt{2}} \times 5 = \text{AB}$	Rearrange to solve for AB.
$\text{AB} = \frac{5}{\sqrt{2}}$	This could also be written as $\frac{5\sqrt{2}}{2}$ If the question asks for an **exact** answer, you will lose marks if you only give a rounded answer.

Practice

1. Use the triangle to work out the value of:

   a cos 30° **b** sin 30° **c** tan 30°

   d cos 60° **e** sin 60° **f** tan 60°

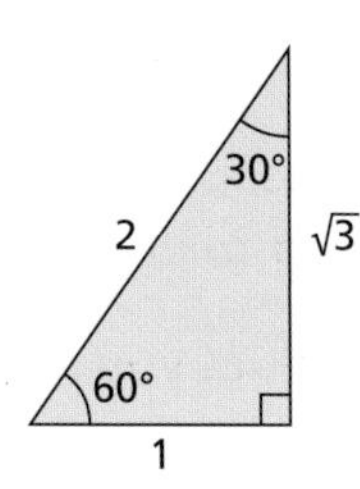

2. Work out the value of cos 60° + sin 30°

3. Use the triangle to work out the value of:

   a sin 45°

   b cos 45°

   c tan 45°

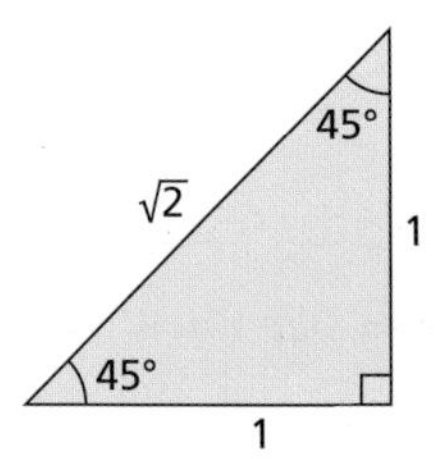

4. Write the value of:

   a sin 0° **b** sin 90° **c** cos 0° **d** cos 90° **e** tan 0° **f** tan 90°

5. Work out:

   a 2sin 30° **b** 2cos 30° **c** 2tan 30°

   d 3tan 30° **e** 8tan 45° **f** 8sin 45°

   g 10sin 60° **h** 20cos 60° **i** 9cos 90°

6. Work out:

   a $\sqrt{2}$sin 30° **b** $\sqrt{3}$tan 60° **c** $2\sqrt{3}$cos 30°

7. Solve:

   a $\sin 30° = \frac{p}{10}$ **b** $\cos 60° = \frac{q}{6}$ **c** $\tan 60° = \frac{3}{r}$

8. Work out the length of each lettered side.

   a

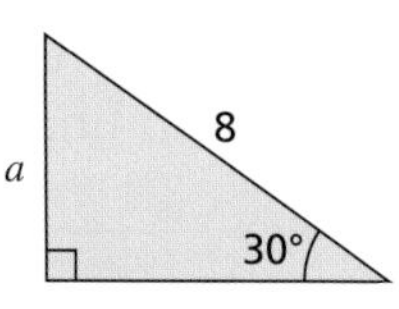

   b

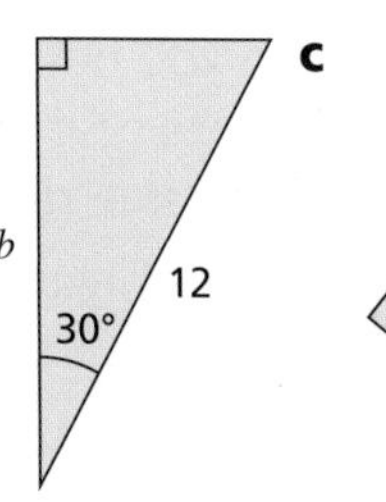

   c 12, 30°, c

   d

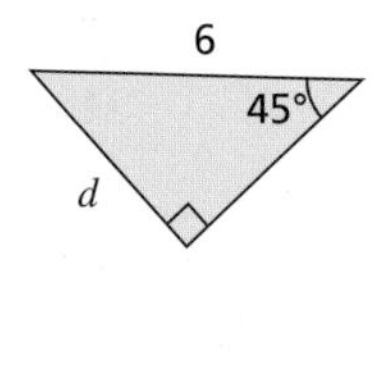

   e

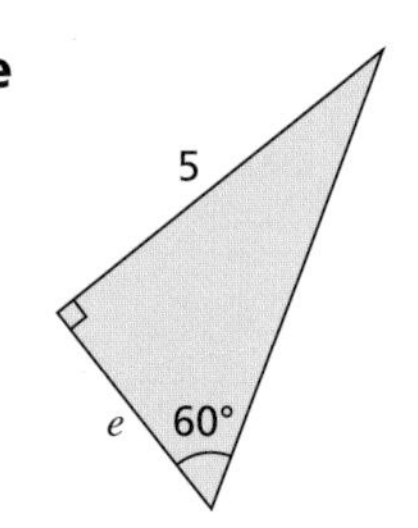

9. Work out the area of the triangle.

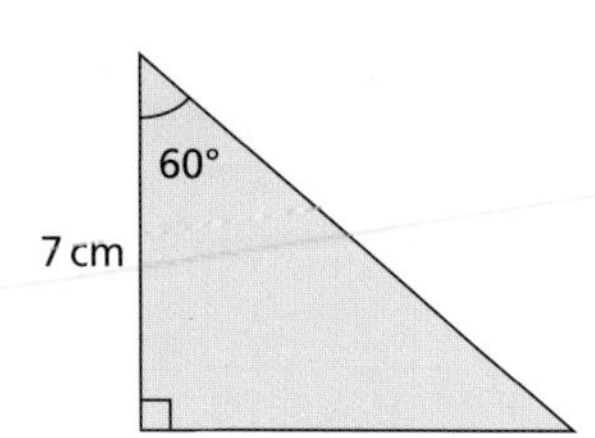

10 Work out the size of each lettered angle.

a

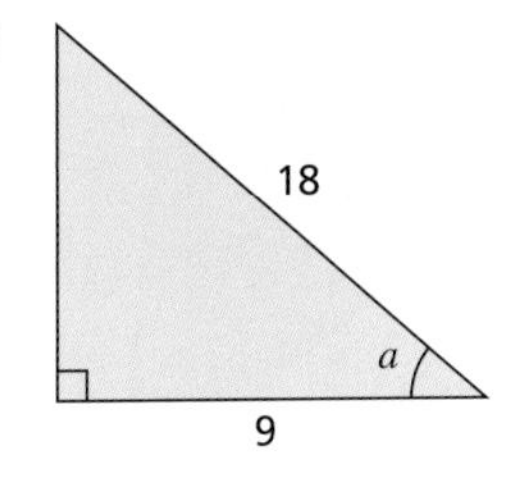

b

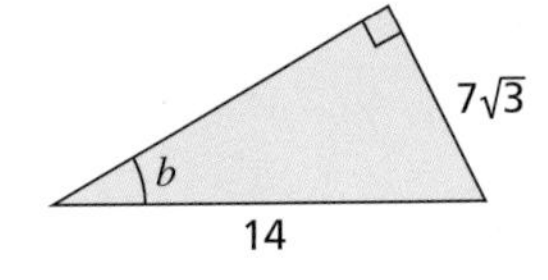

c

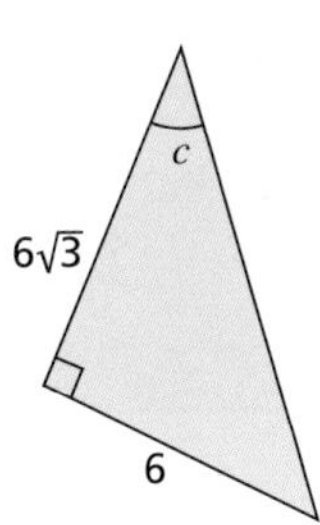

11 Work out:

sin 30° + 2cos 60° + tan 45°

12 ABCDEFGH is a cuboid.

∠GBH = 30°

GH = 4 cm

Work out the length of BG.

Give your answer in the form $a\sqrt{b}$, where a and b are integers.

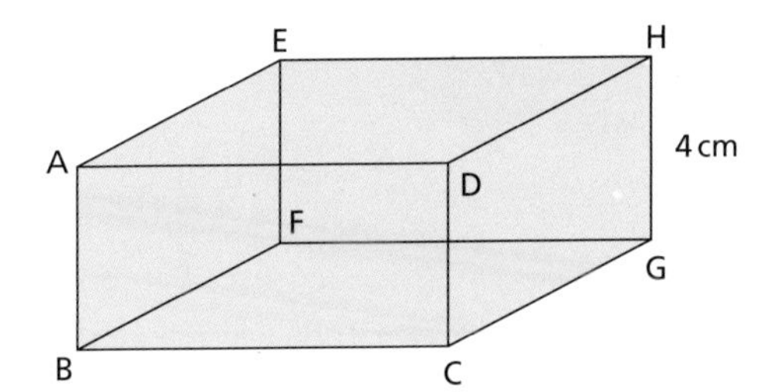

What do you think?

1 Is the statement **true** or **false**? Explain how you know.

sin 30° + cos 30° = cos 60° + sin 60°

2 Write the values in ascending order.

sin 30° | cos 30° | tan 60° | cos 45°

Consolidate – do you need more?

1 Write down the value of:

a sin 30° **b** sin 60° **c** sin 45° **d** cos 30° **e** cos 60°

f cos 45° **g** tan 30° **h** tan 60° **i** tan 45°

2 Work out the length of each lettered side.

a

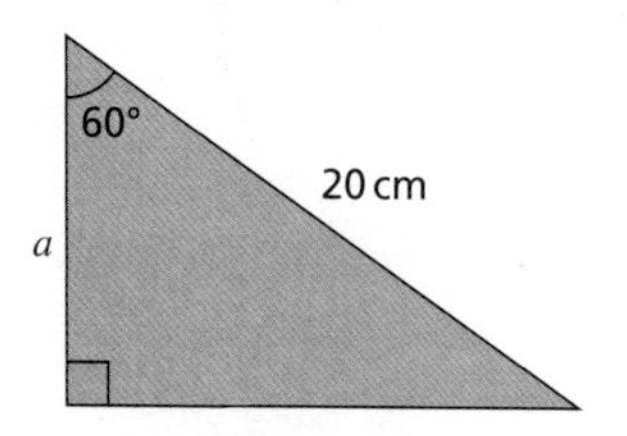

b

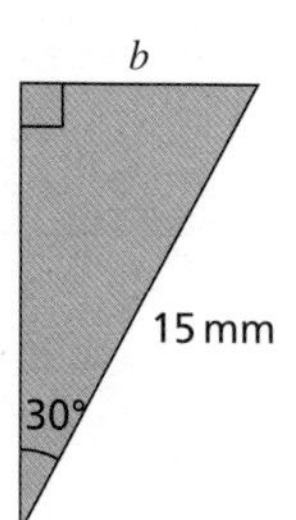

c

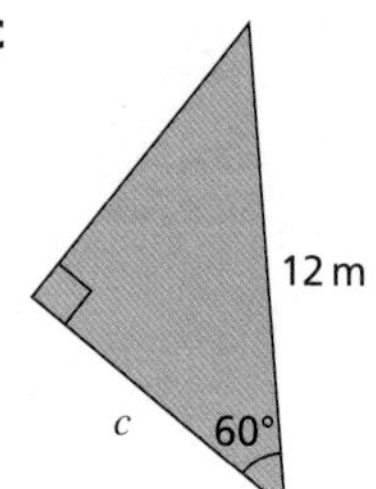

d

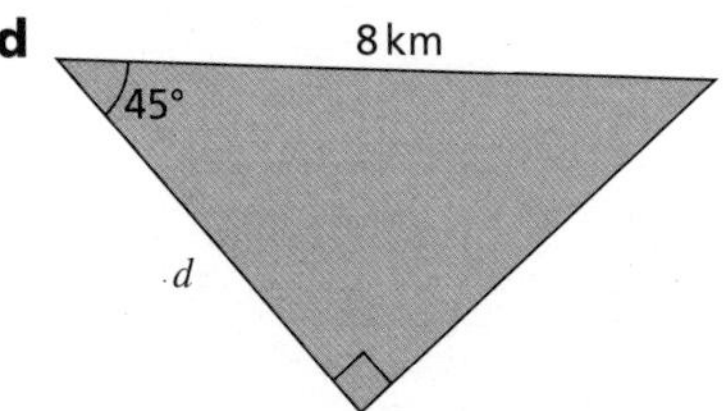

e

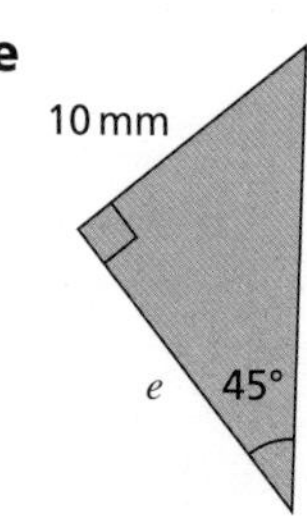

Stretch – can you deepen your learning?

1 Work out the length of BC.

Hint: Work out the length of BD first.

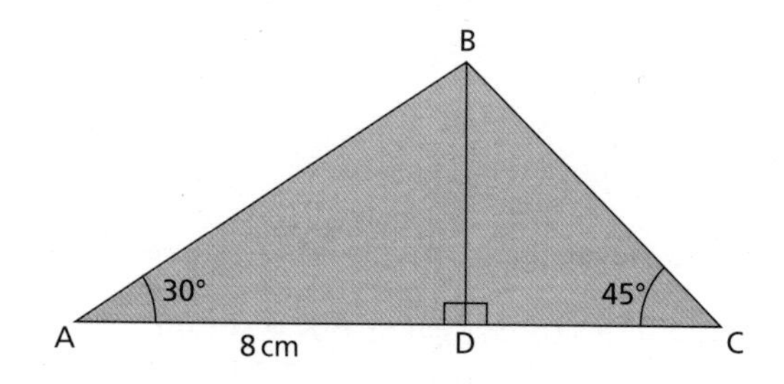

2 Show that $\cos 30° + 2\tan 30°$ can be written in the form $\frac{a\sqrt{b}}{c}$ where a, b and c are integers.

3 Show that $\frac{\sin 60°}{\tan 30°}$ can be written in the form $\frac{a}{b}$ where a and b are integers.

4 Work out the size of angle x.

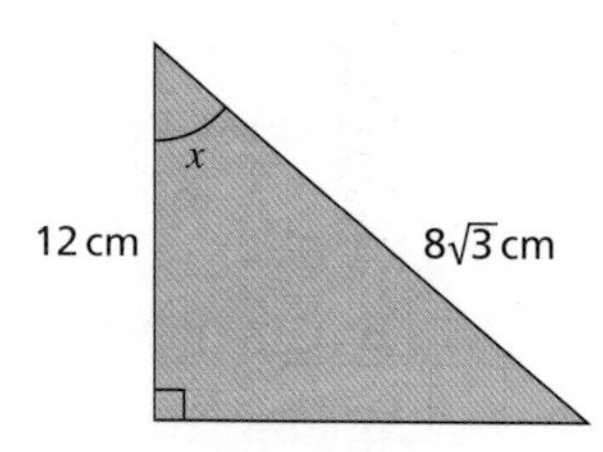

5 Show that the ratio $\cos 30° : \tan 60° : \sin 45°$ can be written in the form $\sqrt{a} : b\sqrt{c} : d$, where a, b, c and d are integers.

6 y and x are connected by the formula $y = a\sin x + b$.

When $x = 30°$, $y = 3.5$

When $x = 0°$, $y = 2$

Work out the values of a and b.

Right-angled triangles: exam practice

White Rose MATHS

* *Diagrams are not accurately drawn.*

4–6

1 Calculate the size of angle m. **[3 marks]**

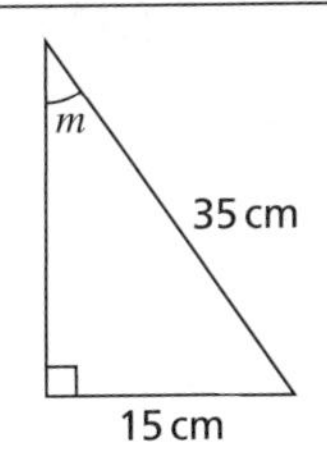

2 The sides of a triangle have lengths of 36 cm, 77 cm and 85 cm.

Is the triangle right-angled? Show working to justify your answer. **[3 marks]**

3 Calculate the length of the side labelled x. **[3 marks]**

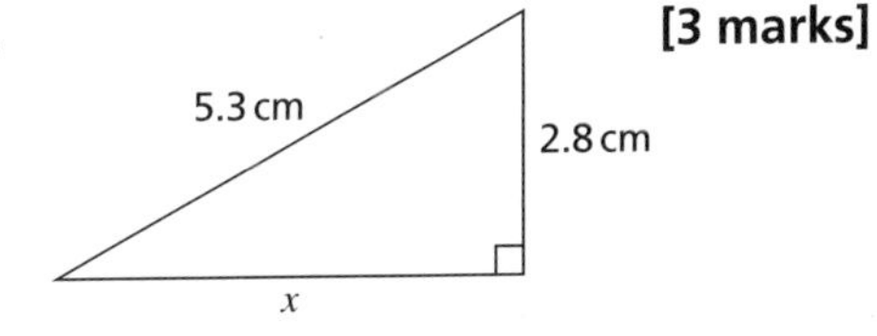

4 The perimeter of a square is 60 cm.

Calculate the length of its diagonal. **[3 marks]**

5 The area of the triangle is 60 cm^2.

Calculate the perimeter of the triangle. **[4 marks]**

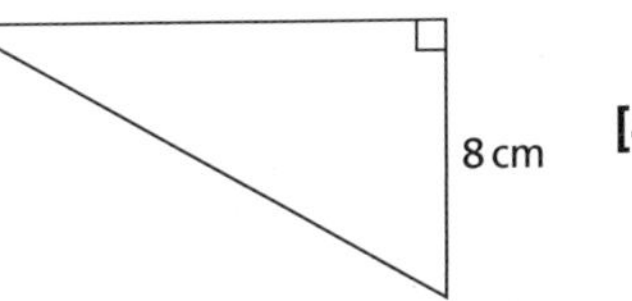

6 Write down the values of:

(a) cos 0° **[1 mark]**

(b) sin 30° **[1 mark]**

7–9

7 Here is a cuboid with HG = 4 cm, EH = 6 cm and AE = 5 cm.

Calculate the angle AG makes with the base of the cuboid. **[3 marks]**

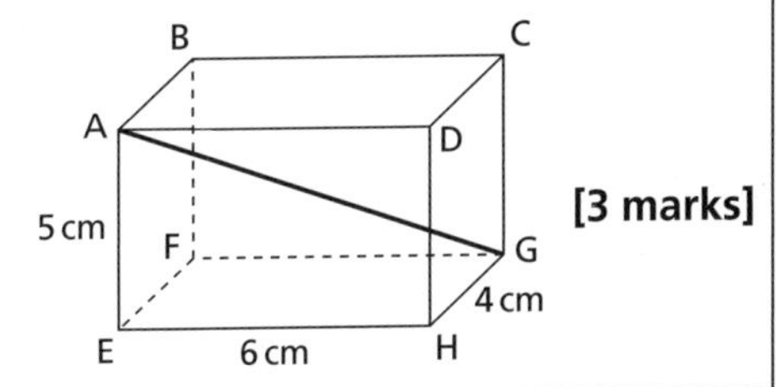

8 Here is a sketch of a square-based pyramid.

Work out the length of the sloping edge of the pyramid.

Give your answer in the form $a\sqrt{b}$, where a and b are integers. **[4 marks]**

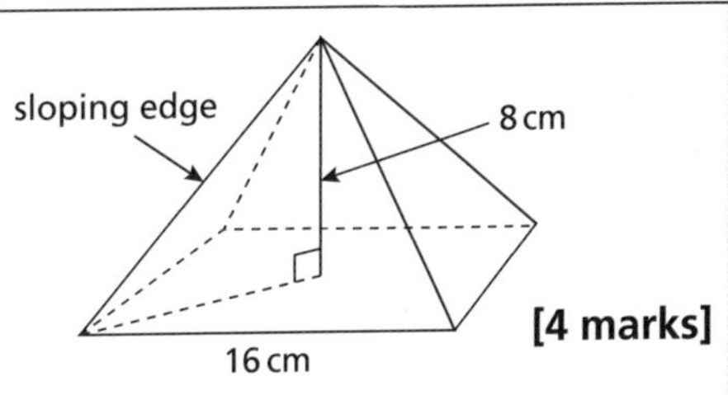

7 Non-right-angled triangles

White Rose MATHS

In this block, we will cover...

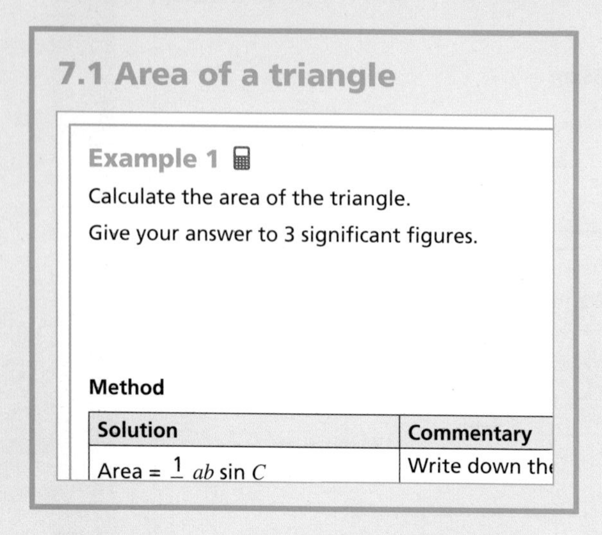
7.1 Area of a triangle
Example 1
Calculate the area of the triangle.
Give your answer to 3 significant figures.
Method
Solution
Commentary
Area = 1 ab sin C
Write down th

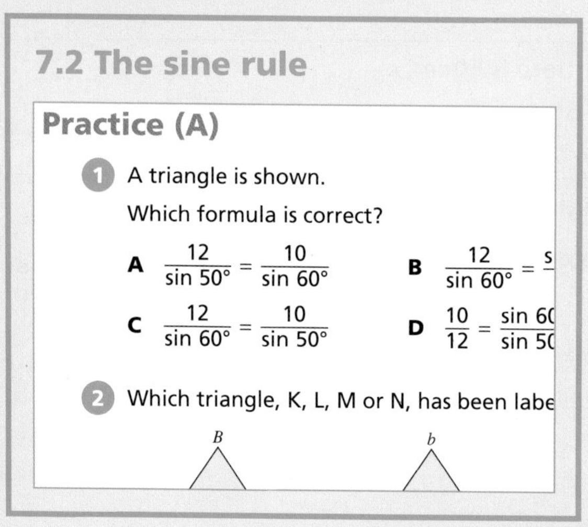
7.2 The sine rule
Practice (A)
A triangle is shown.
Which formula is correct?
Which triangle, K, L, M or N, has been labe

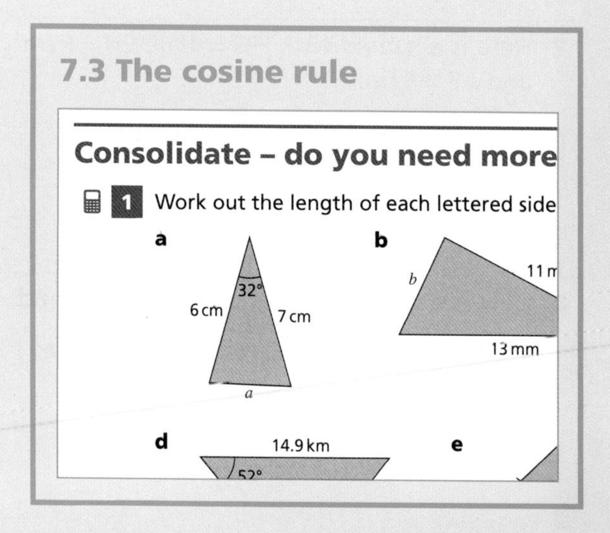
7.3 The cosine rule
Consolidate – do you need more
Work out the length of each lettered side
6 cm
7 cm
32°
13 mm
14.9 km

7.1 Area of a triangle

White Rose MATHS

Are you ready? 🖩

1 Work out the value of each expression when $a = 4$, $b = 5$ and $\theta = 35°$

a ab **b** $\frac{1}{2}ab$ **c** $\sin\theta$ **d** $\frac{1}{2}ab\sin\theta$

2 Solve the equations, giving your solutions to 2 decimal places.

a $6a\sin 35° = 70$ **b** $32\sin\theta = 20$

Trigonometry can be used to solve problems with non-right-angled triangles.

As shown on the triangle, the general convention is to label each angle as A, B and C and each opposing side as a, b and c respectively.

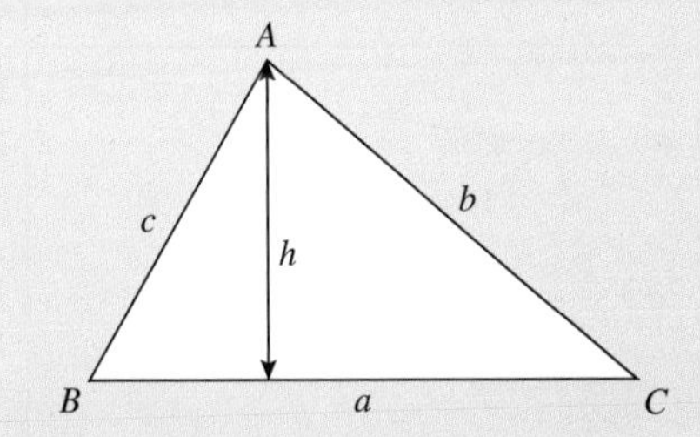

$$\sin C = \frac{h}{b}, \text{ and therefore } h = b\sin C$$

The area of any triangle can be found by calculating $\frac{1}{2}$ × base × height.

Substituting, the area of a triangle can be found using the formula:

$$\text{Area} = \frac{1}{2}ab\sin C$$

The area could also be found using:

$$\text{Area} = \frac{1}{2}bc\sin A \text{ or Area} = \frac{1}{2}ac\sin B$$

The formula can be used to find the area of any triangle if two sides and the angle between them (the **included angle**) are known.

You can also rearrange the formula to find side lengths or angles if you are given the area.

Example 1

Calculate the area of the triangle.

Give your answer to 3 significant figures.

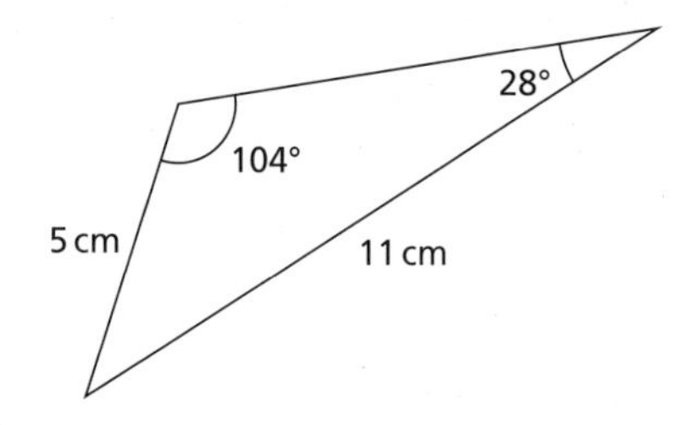

Method

Solution	Commentary
Area = $\frac{1}{2}ab \sin C$	Write down the formula.
28° 104° b 5 cm 11 cm a C	Label the triangle with a, b and C. Remember C needs to be the included angle, so the angle between 5 cm and 11 cm. It doesn't matter which side is a and which is b.
180° – 104° – 28° = 48°	The included angle is not given so you need to work it out using the sum of the angles in a triangle.
Area = $\frac{1}{2} \times 11 \times 5 \times \sin 48°$ = 20.4 cm^2 (to 3 s.f.)	Substitute the values into the formula. Give the answer to 3 significant figures.

Example 2

The area of the triangle is 2.7 m^2.

Calculate the size of angle θ.

Give your answer to 1 decimal place.

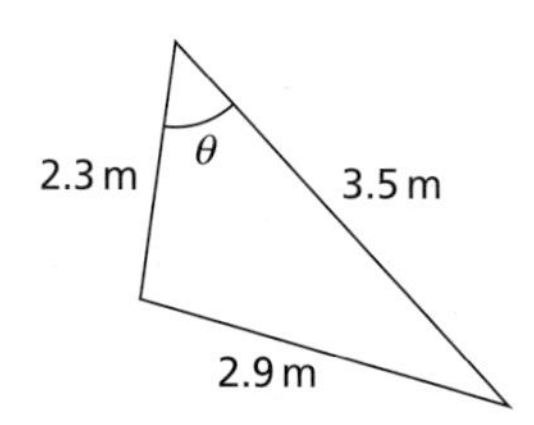

Method

Solution	Commentary
Area = $\frac{1}{2}ab \sin C$	Start by writing the formula and substituting in what you know.
$2.7 = \frac{1}{2} \times 2.3 \times 3.5 \sin \theta$	θ is between the sides of 2.3 m and 3.5 m, so use those for your lengths of a and b.
$2.7 = 4.025 \sin \theta$	Simplify the equation.
$\frac{2.7}{4.025} = \sin \theta$	Rearrange.
$\theta = \sin^{-1}\left(\frac{2.7}{4.025}\right)$ $\theta = 42.1°$ (to 1 d.p.)	Use the inverse function to find the value of θ.

Practice

1. Calculate the area of each triangle. Give your answers to 1 decimal place.

a

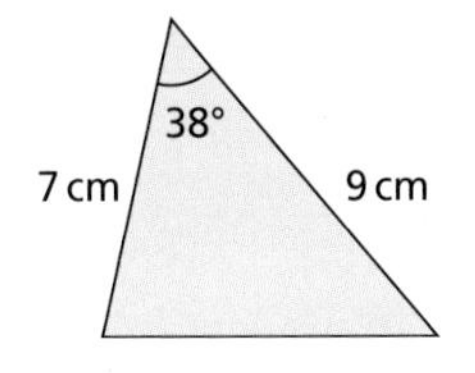

b

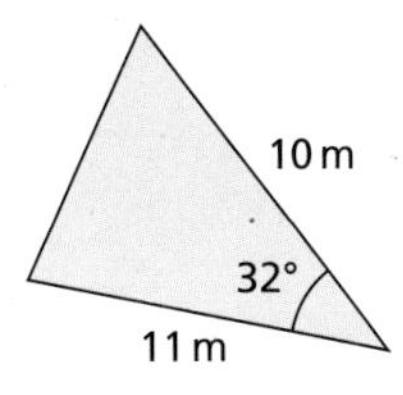

c

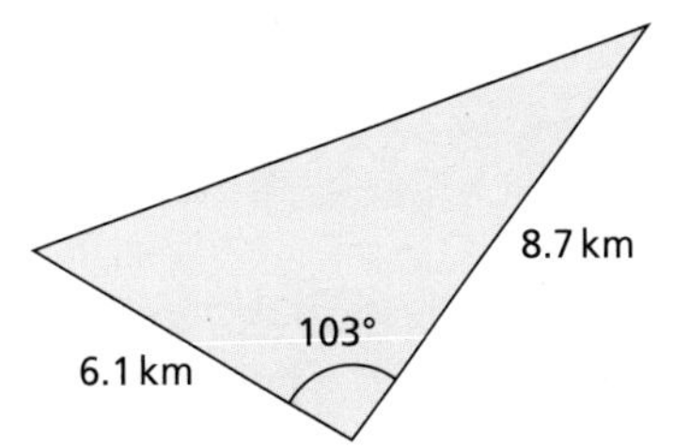

2. Calculate the area of triangle ABC.

Give your answer to 3 significant figures.

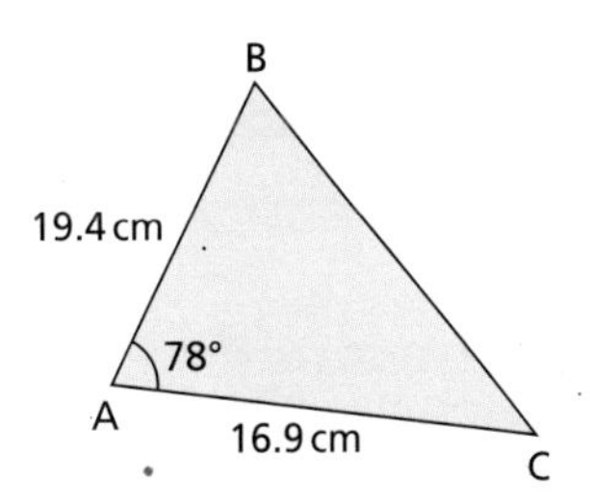

3. Calculate the area of each triangle. Give your answers to 1 decimal place.

a

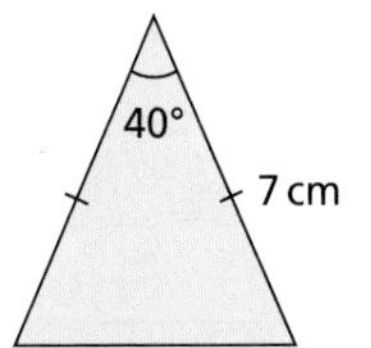

b

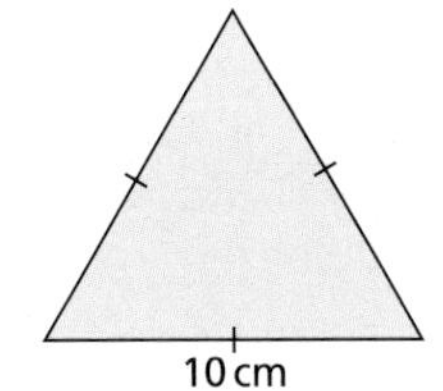

4. Calculate the area of triangle PQR.

Give your answer to 3 significant figures.

Hint: Work out the size of angle PRQ first.

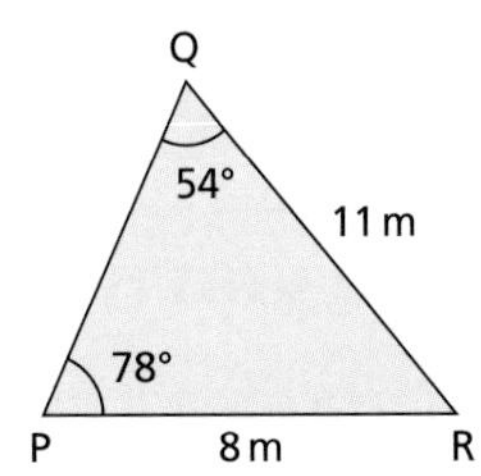

5. Work out the length of each lettered side. Give your answers to 1 decimal place.

a Area = 80 cm^2

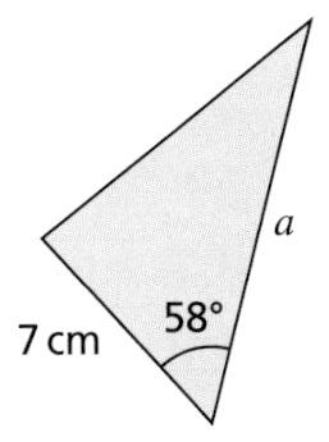

b Area = 10 cm^2

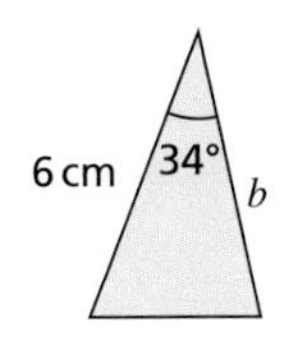

c Area = 58.5 cm^2

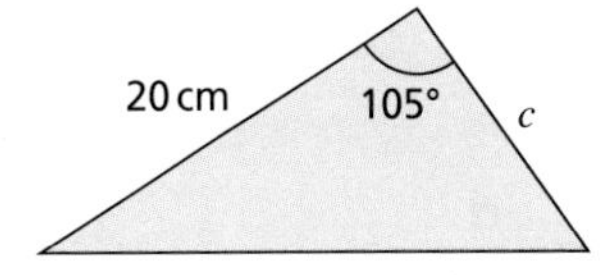

6. The triangle has an area of 60 cm^2.

Work out the value of x.

Give your answer to 3 significant figures.

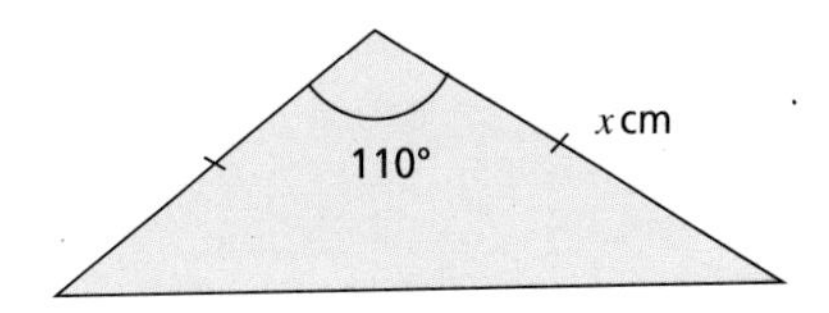

7 The triangle has an area of $100\,\text{cm}^2$.

Calculate the acute angle θ.

Give your answer to 3 significant figures.

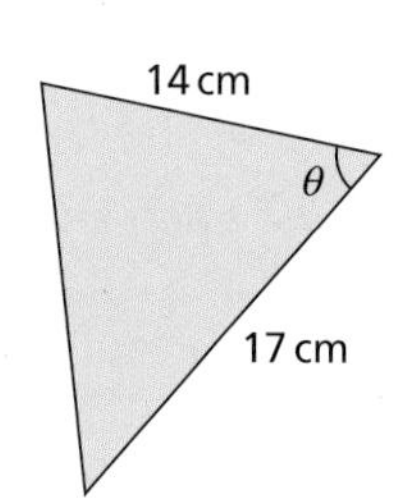

What do you think?

1 Lida and Benji are working out the area of the triangle shown.

Here is their working out:

Explain why both methods work.

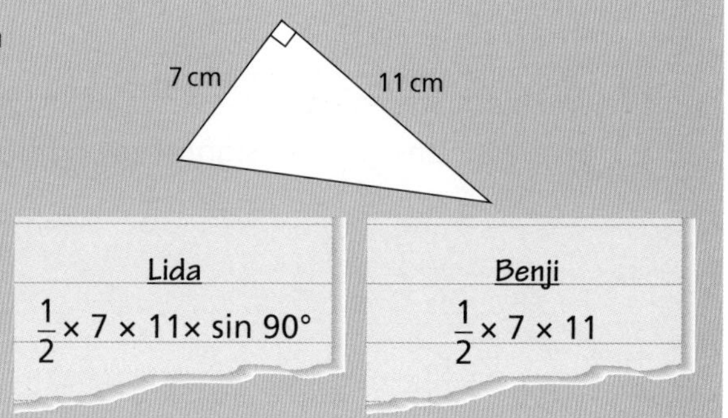

2 Triangle PQR has an area of $30\,\text{cm}^2$.

Huda says, "There are two possible values for θ."

Draw two different constructions of PQR, marking the size of the angle PQR clearly.

Are there always two possible angles if the area and two sides forming an angle are known?

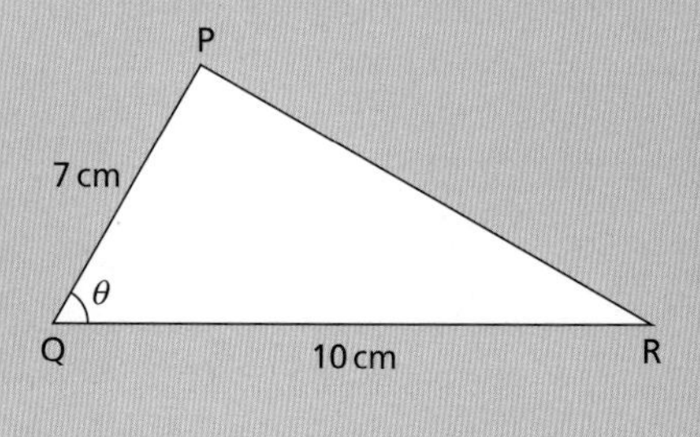

Consolidate – do you need more?

1 Work out the area of each triangle. Give your answers to 1 decimal place.

a

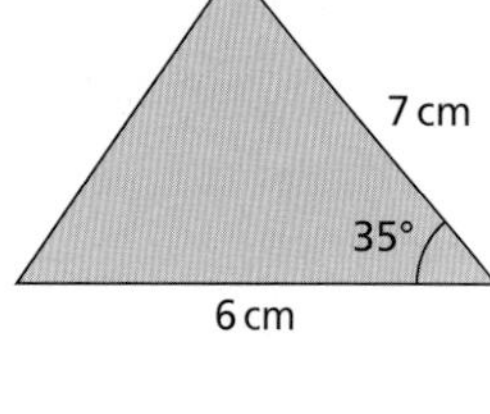

b

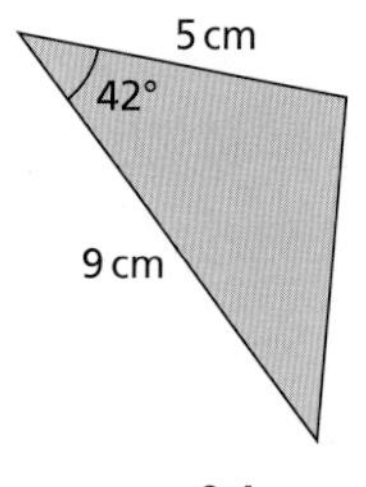

c

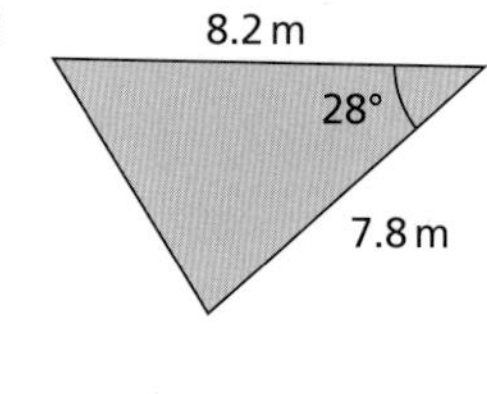

d

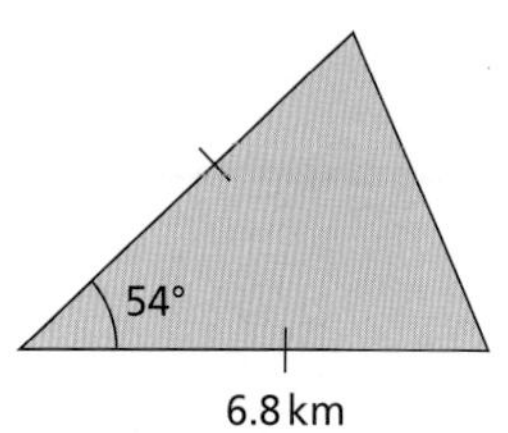

e

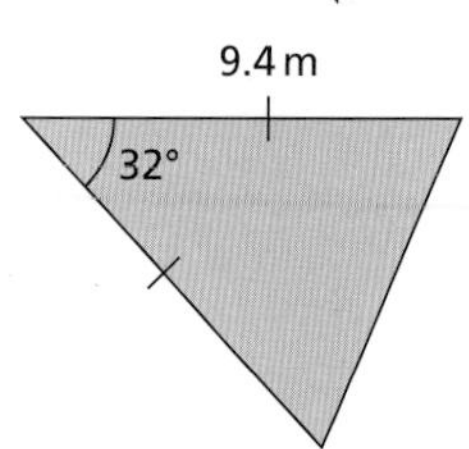

f

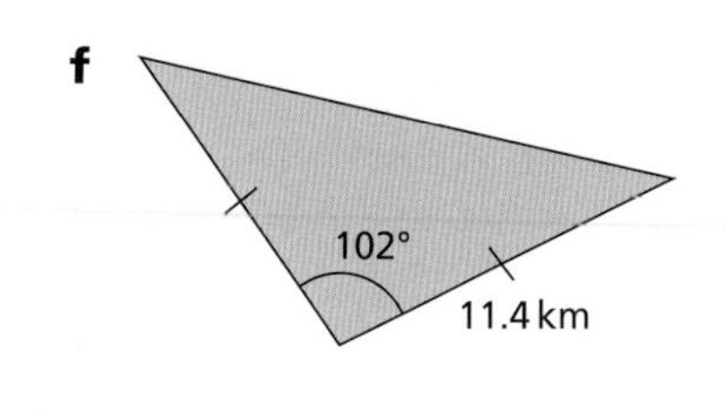

2 Calculate the size of each lettered angle. Give your answers to 1 decimal place.

a Area = 30 cm²

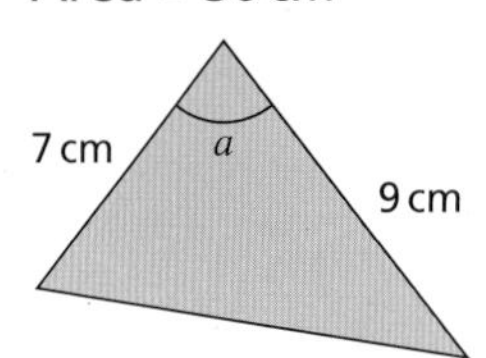

b Area = 25 cm²

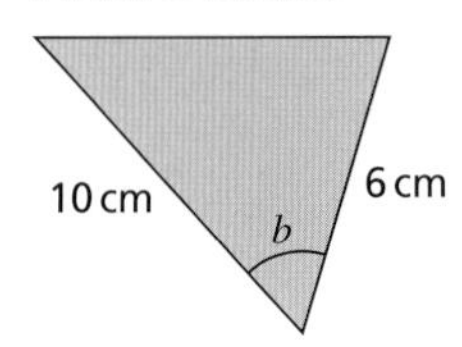

c Area = 32.5 cm²

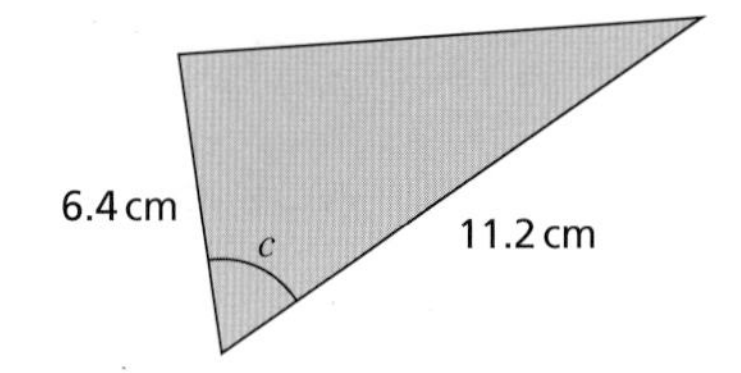

Stretch – can you deepen your learning?

1 Work out the exact area of each triangle.

a

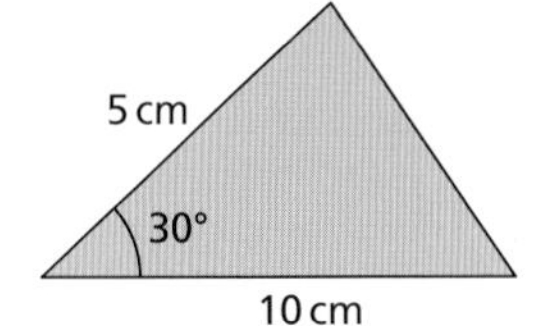

b

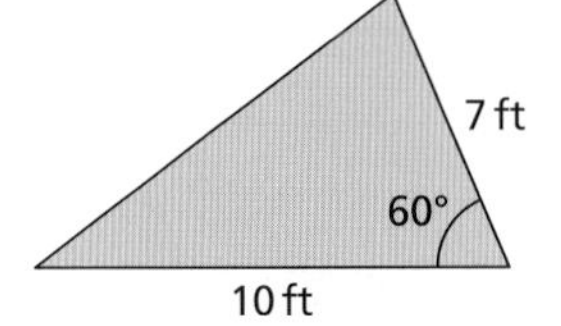

c

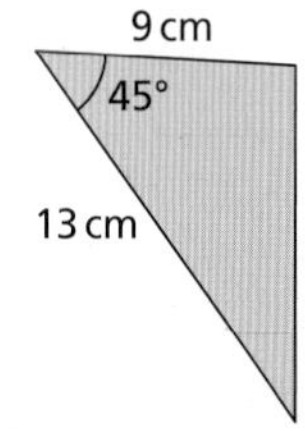

2 The diagram shows sector POQ of a circle with centre O.

PQ is a chord.

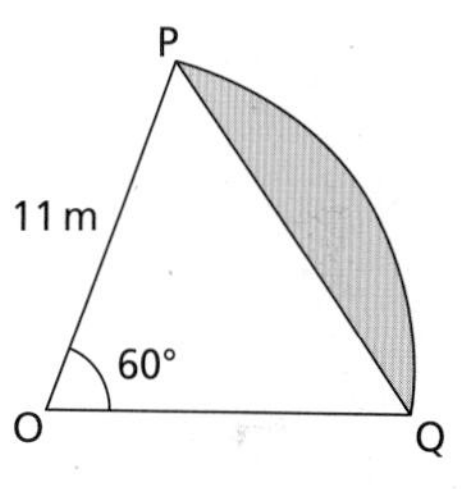

a Work out the area of the shaded segment.

b Ed says, "sin 60° = sin 120°, so if the angle was 120°, the area of the sector would be the same."

Show that Ed is wrong.

Parts of circles are covered in Chapter 3.4

3 Show that the area of the triangle can be written in the form $\frac{\sqrt{a}(x^2 + bx)}{c}$ where a, b and c are integers.

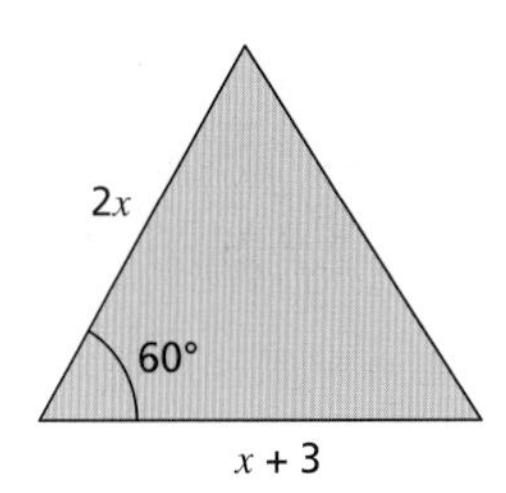

7.2 The sine rule

White Rose MATHS

Are you ready? (A)

1 Solve the equations, giving your solutions to 1 decimal place.

a $\frac{a}{3} = \frac{2}{7}$ **b** $\frac{3}{\sin 40°} = \frac{b}{6}$ **c** $\frac{4}{\sin 35°} = \frac{c}{\sin 45°}$ **d** $\frac{7.5}{\sin 42°} = \frac{d}{\sin 53°}$

A triangle is shown, with a perpendicular height of h.

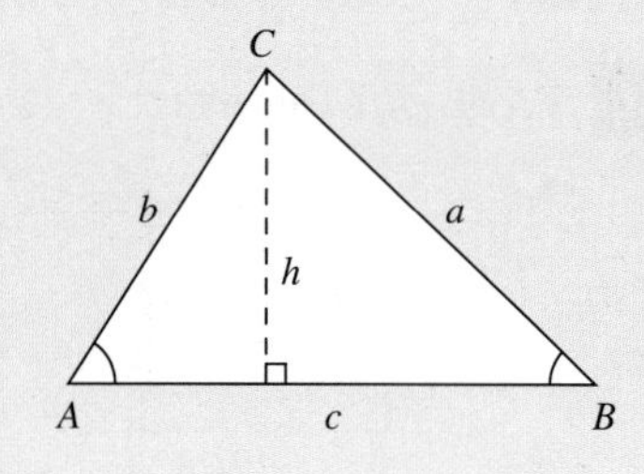

It has been split into two right-angled triangles.

As $\sin B = \frac{h}{a}$, then $h = a \sin B$

And as $\sin A = \frac{h}{b}$, then $h = b \sin A$.

It follows that $a \sin B = b \sin A$ and therefore $\frac{a}{\sin A} = \frac{b}{\sin B}$

Drawing another perpendicular height through A gives $c \sin B = b \sin C$

And therefore $\frac{b}{\sin B} = \frac{c}{\sin C}$

This gives the sine rule:

$$\frac{a}{\sin A} = \frac{b}{\sin B} = \frac{c}{\sin C}$$

The sine rule helps you to find angles and sides in triangles that are not right-angled.

Consider the triangle labelled as shown.

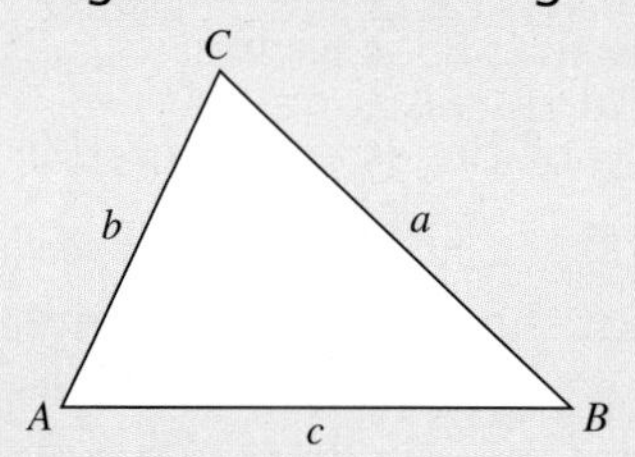

The sine rule states:

$$\frac{a}{\sin A} = \frac{b}{\sin B} = \frac{c}{\sin C} \quad \text{or} \quad \frac{\sin A}{a} = \frac{\sin B}{b} = \frac{\sin C}{c}$$

To use the sine rule, you need an angle and the side opposite it, and one other angle or side.

Example

Calculate the length of YZ.

Give your answer to 3 significant figures.

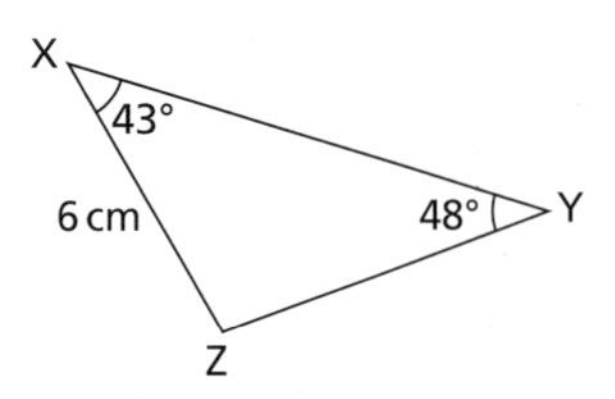

Method

Solution	Commentary
$\frac{a}{\sin A} = \frac{b}{\sin B}$	Since you have two angles and a length opposite, use the sine rule to find the missing length YZ.
$\frac{\text{YZ}}{\sin 43°} = \frac{6}{\sin 48°}$	Substitute the values into the formula. You only need to use two of the three fractions from the formula.
$\text{YZ} = \frac{6}{\sin 48°} \times \sin 43° = 5.50631588...$	Rearrange the formula.
YZ = 5.51 cm (to 3 s.f.)	Round your answer to 3 significant figures.

Practice (A)

1. A triangle is shown.

   Which formula is correct?

   A $\frac{12}{\sin 50°} = \frac{10}{\sin 60°}$ **B** $\frac{12}{\sin 60°} = \frac{\sin 50°}{10}$

   C $\frac{12}{\sin 60°} = \frac{10}{\sin 50°}$ **D** $\frac{10}{12} = \frac{\sin 60°}{\sin 50°}$

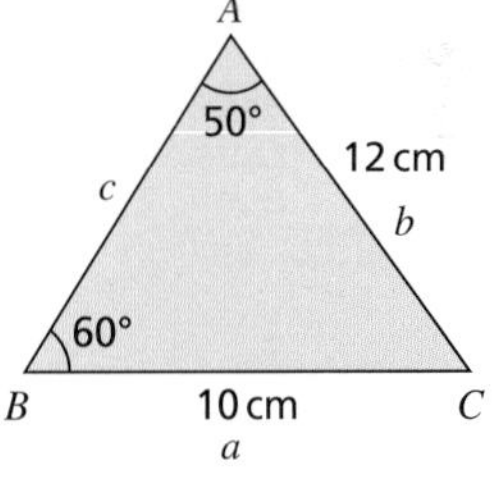

2. Which triangle, K, L, M or N, has been labelled correctly to use the sine rule?

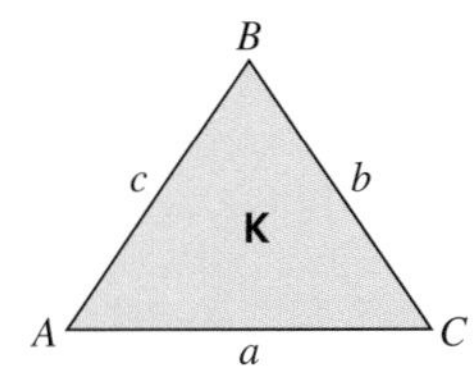

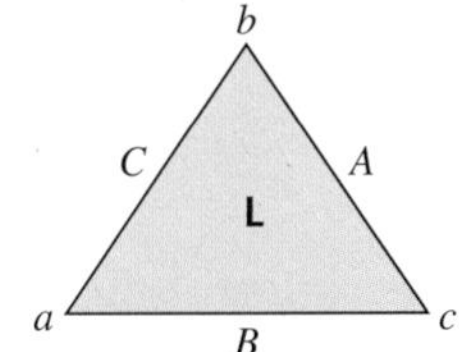

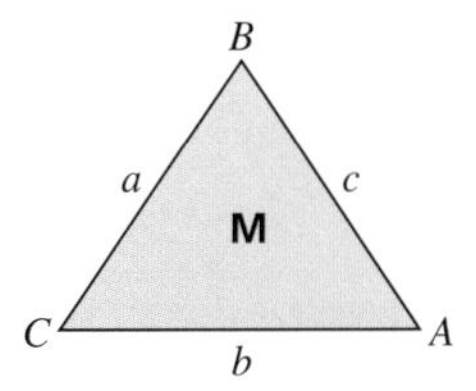

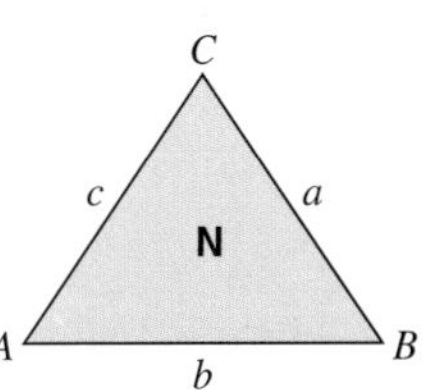

3. Copy the triangle.

   a Label the angles with A, B and C and the corresponding sides with a, b and c to use the sine rule.

   b Work out the length of x, giving your answer to 3 significant figures.

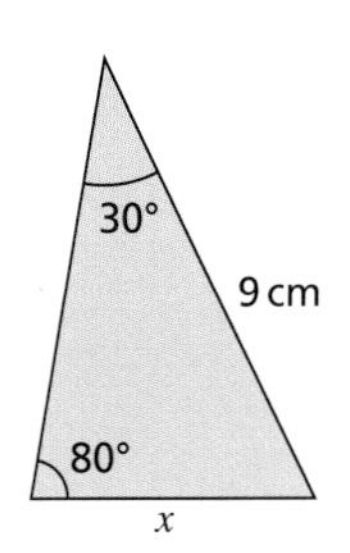

4 Work out the length of each lettered side. Give your answers to 1 decimal place.

a

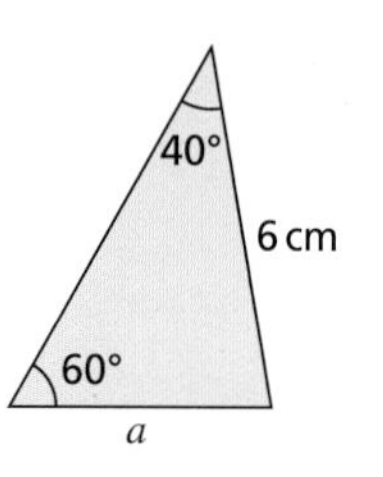

b

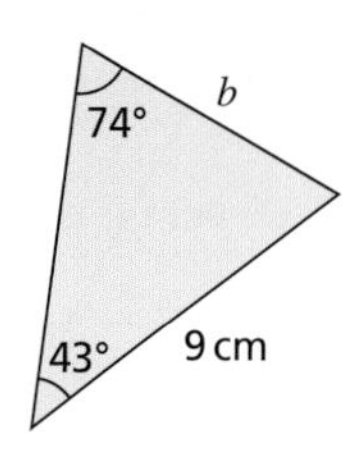

c

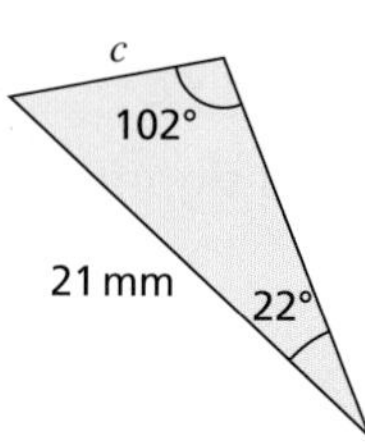

d

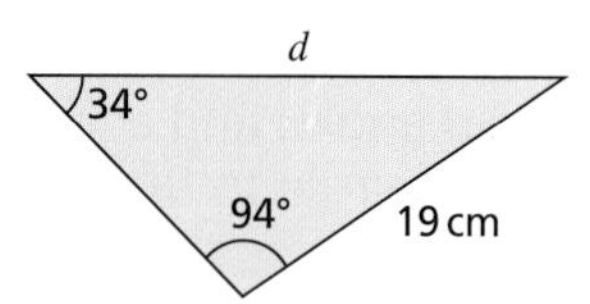

e

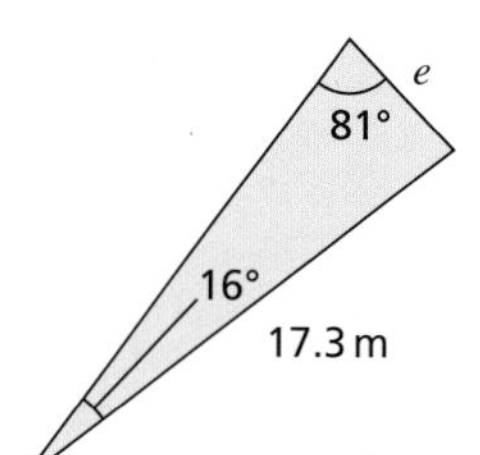

f

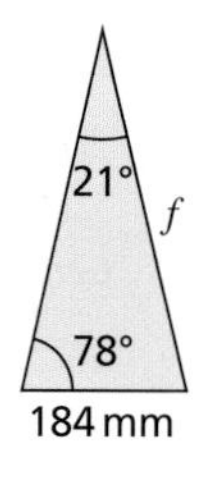

5 Work out the length of PQ.

Give your answer to 1 decimal place.

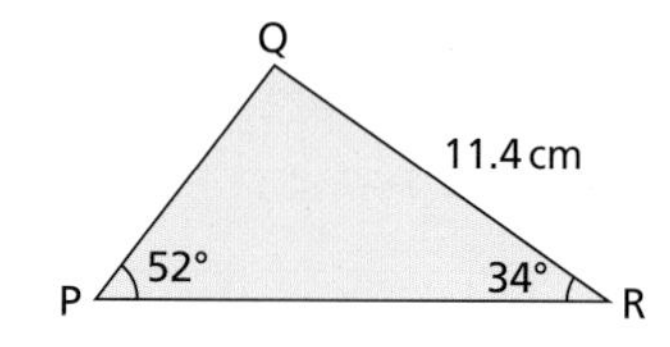

6 Work out the length of XY.

Give your answer to 2 decimal places.

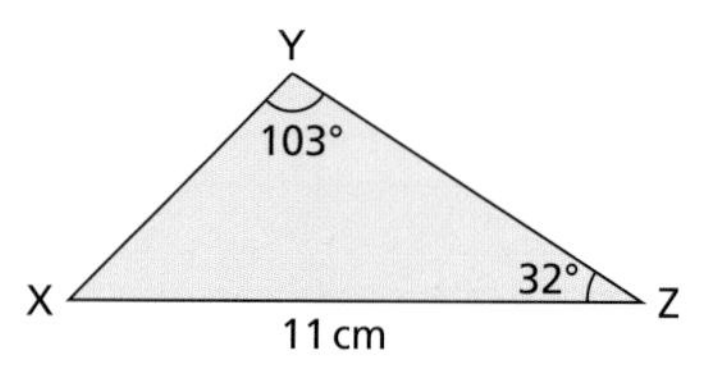

7 Work out the length of MN.

Give your answer to 3 significant figures.

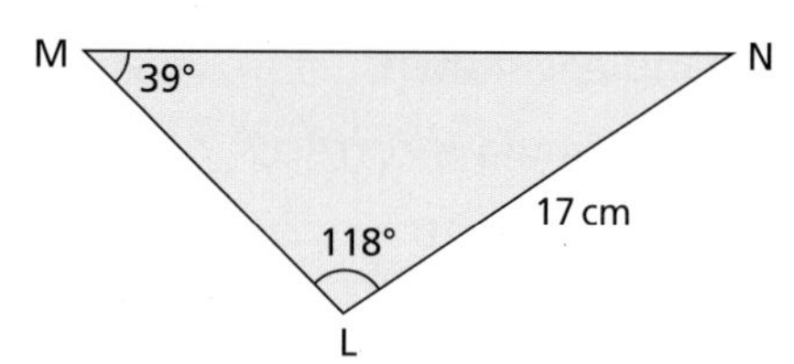

8 Work out the length of each lettered side. Give your answers to 3 significant figures.

a

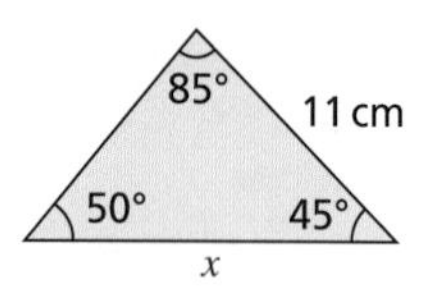

b

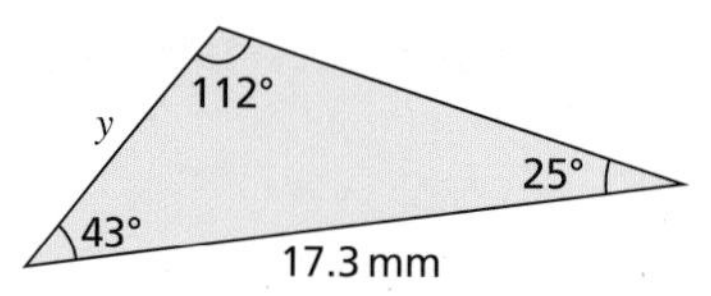

9 Here is the triangle LMN.

a Work out the missing angle.

b Work out the length of MN.

Give your answer to 2 decimal places.

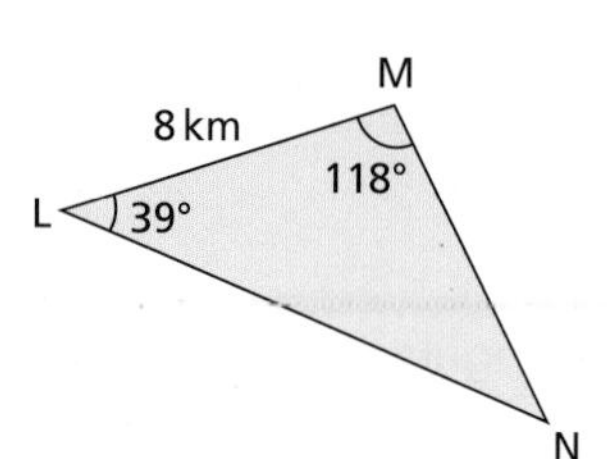

10 Work out the length of each lettered side.

Give your answers to 1 decimal place.

a

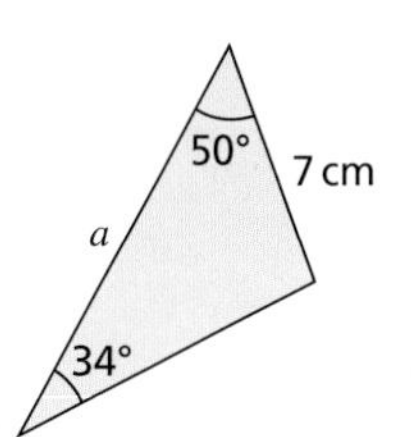

b

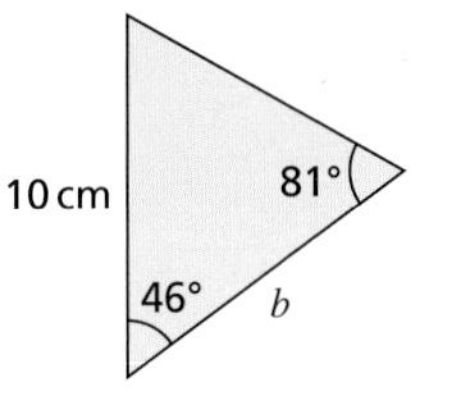

c

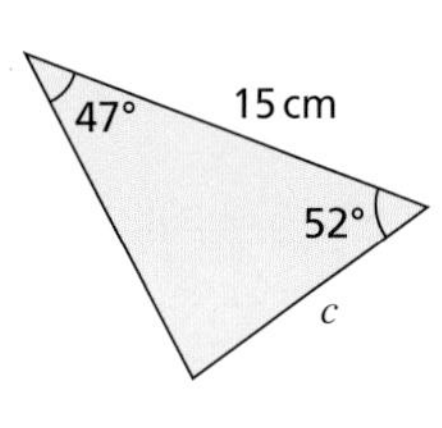

11 Calculate the perimeter of the triangle.

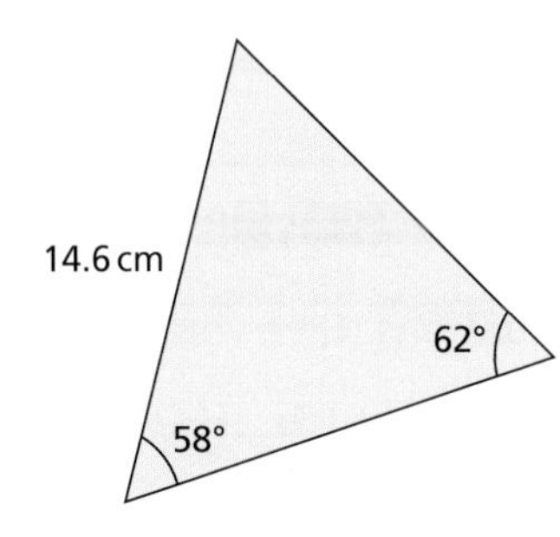

12 DEF is an isosceles triangle.

Work out the length of DF.

Give your answer to 1 decimal place.

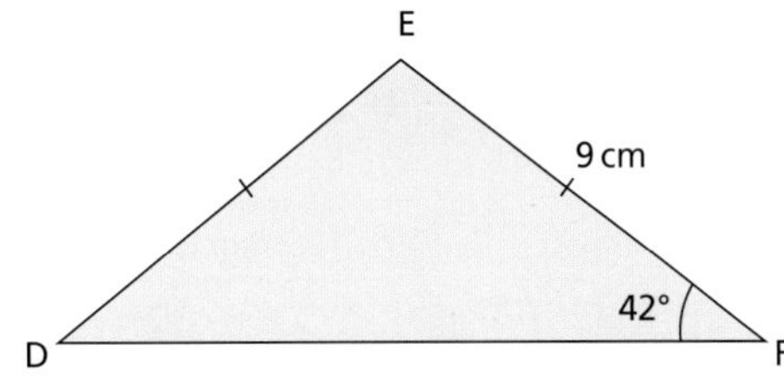

13 Work out the length of QR.

Give your answer to 3 significant figures.

Hint: Work out the length of QS first.

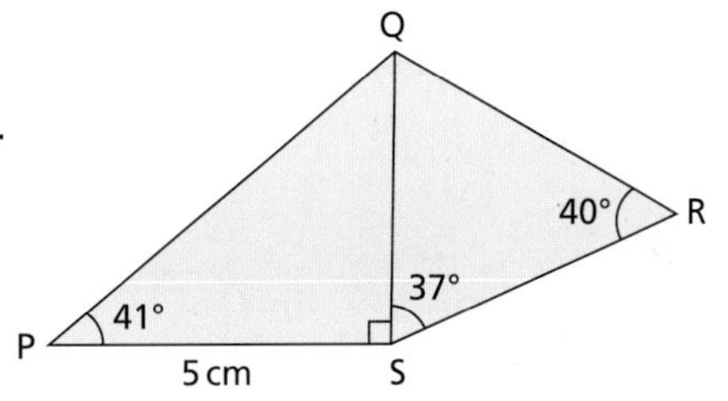

14 In triangle PQR, QR = 11 cm, angle PQR = 113° and angle QPR = 28°

Work out the length of PR.

What do you think? (A)

1 Jakub and Chloe are trying to work out the value of x in this diagram.

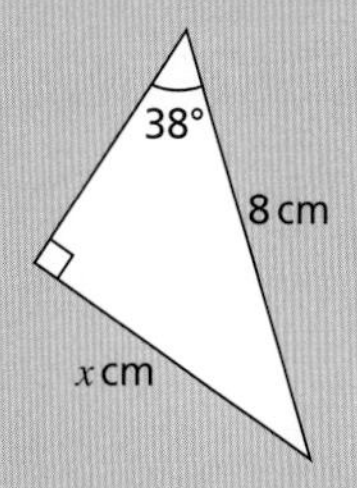

<u>Jakub</u>

$$\frac{8}{\sin 90°} = \frac{x}{\sin 38°}$$

<u>Chloe</u>

$$\sin 38° = \frac{x}{8}$$

Whose working out is correct? Choose option A, B or C.

A Jakub **B** Chloe **C** Both

Explain your answer.

2 Draw a circle with a diameter of 10 cm.

Mark three points, A, B and C, on the circumference.

Join the points to make a triangle and label the sides a, b and c.

Work out $\frac{a}{\sin A}, \frac{b}{\sin B}$ and $\frac{c}{\sin C}$

What do you notice?

Investigate with other circles and triangles.

Are you ready? (B)

1 Solve the equations. Give your solutions to 1 decimal place.

a $\sin x = 0.6$ **b** $\sin x = \frac{4}{5}$ **c** $\sin x = \frac{4.6}{8.9}$

2 Work out the size of each lettered angle. Give your answers to 1 decimal place.

a

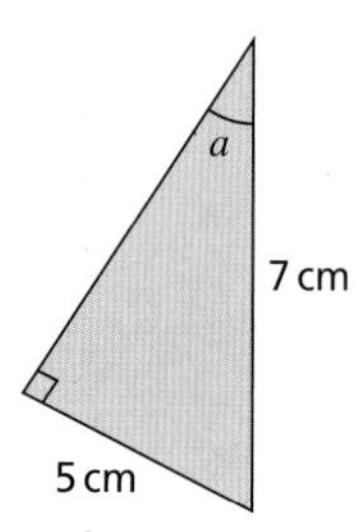

b

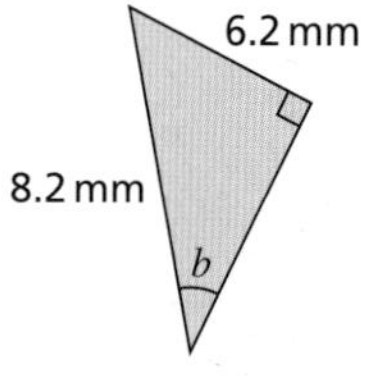

The sine rule $\frac{a}{\sin A} = \frac{b}{\sin B} = \frac{c}{\sin C}$ can be rearranged to $\frac{\sin A}{a} = \frac{\sin B}{b} = \frac{\sin C}{c}$

This rearranged form of the sine rule is useful when trying to work out angles.

To do this, you need to know the side length opposite the unknown angle, and another corresponding side length and angle.

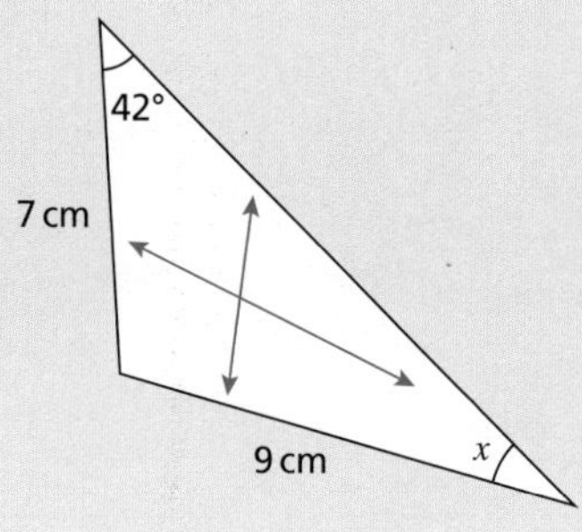

Ambiguous case

In some cases, there is more than one solution to problems with unknown angles using the sine rule.

Consider the triangle with sides of length 9 cm and 8 cm, and a non-included angle of 52°.

The triangle could look like this:

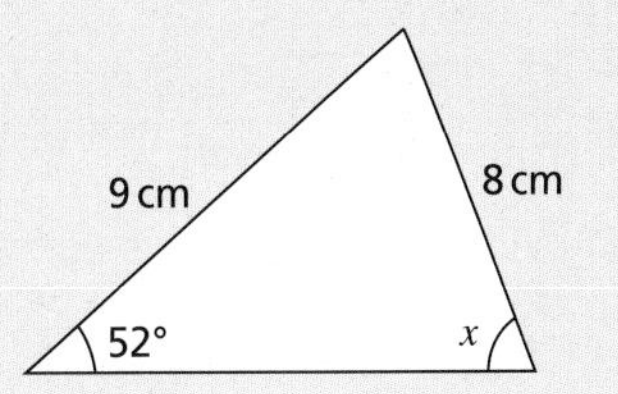

You could then calculate with the sine rule as normal.

$$\frac{\sin x}{9} = \frac{\sin 52°}{8}$$

$$\sin x = \frac{\sin 52°}{8} \times 9$$

$x = 62.4°$ (to 1 d.p.)

So the acute angle x is 62.4°

However, it is possible to draw a different triangle with the same given measurements, by moving the 8 cm side as shown.

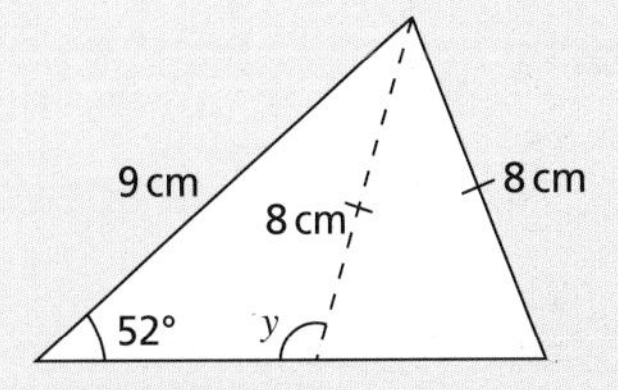

Since you know the acute angle x is 62.4°, and base angles in an isosceles triangle are equal, then $y = 180° - 62.4° = 117.6°$

This means that in this case the unknown angle in this triangle could be 62.4° or 117.6°

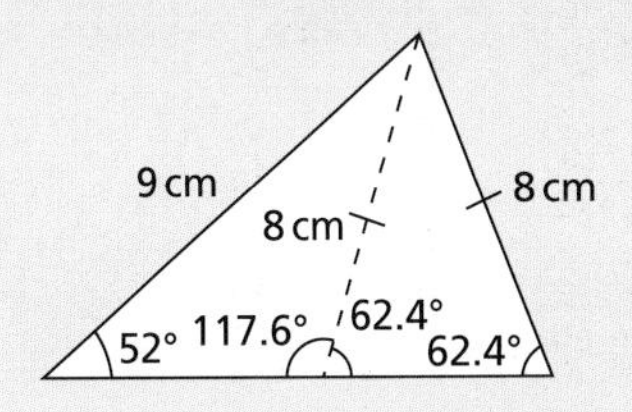

You may be told that the unknown angle is acute or obtuse.

Example

Work out the acute angle y.

Give your answer to the nearest integer.

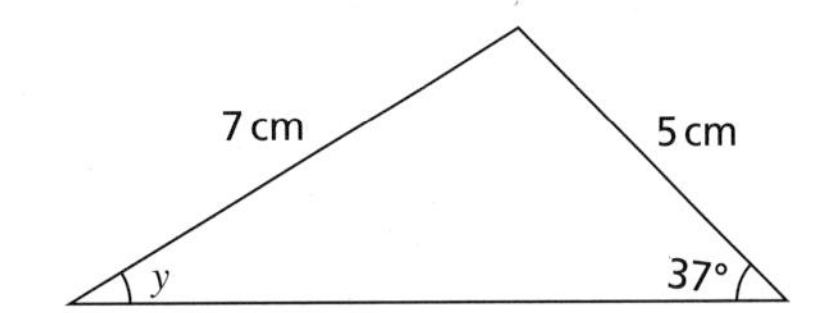

Method

Solution	Commentary
$\frac{\sin A}{a} = \frac{\sin B}{b}$	Since you are finding an angle, you can use the formula written in this format. Label the triangle and substitute the values into the formula.
$\frac{\sin y}{5} = \frac{\sin 37°}{7}$ $\sin y = \frac{\sin 37°}{7} \times 5$ $\sin y = 0.429\,867\,87\ldots$ $\sin^{-1}(0.429\,867\ldots) = 25.459\,1753\ldots°$	Rearrange the equation and use the inverse sine function to find the value of y.
$y = 25°$	Give the answer to the nearest integer.

Practice (B)

1 Solve the equations.

a $\frac{\sin 35°}{7} = \frac{\sin x}{5}$ **b** $\frac{\sin x}{11} = \frac{\sin 54°}{16}$ **c** $\frac{\sin x}{13.1} = \frac{\sin 81°}{19.2}$

2 A triangle is shown.

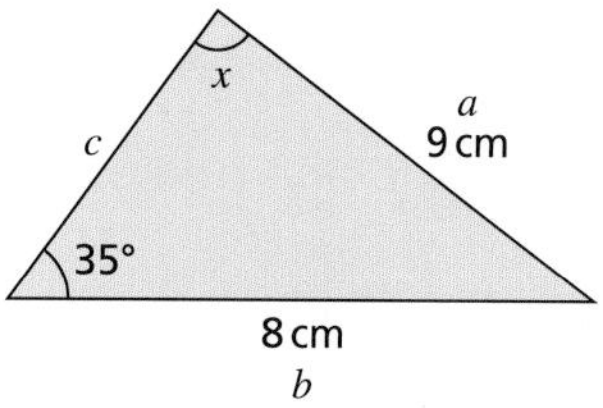

a Which equation is correct?

A $\frac{\sin 35°}{8} = \frac{\sin x}{9}$

B $\frac{\sin x}{8} = \frac{\sin 35°}{9}$

C $\frac{\sin x}{9} = \frac{\sin 35°}{8}$

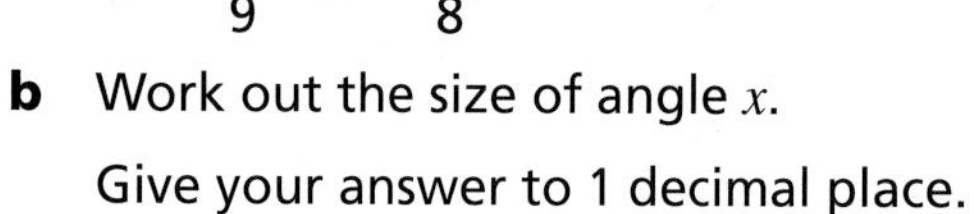

b Work out the size of angle x.

Give your answer to 1 decimal place.

3 Work out the size of each lettered angle. Give your answers to 1 decimal place.

a

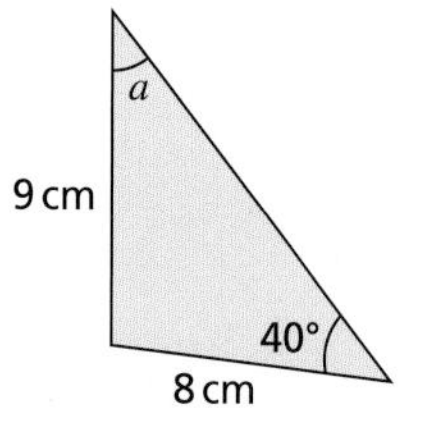

b

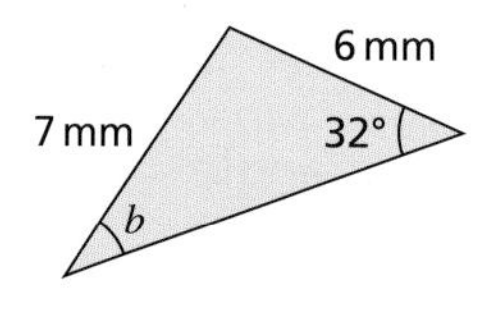

c

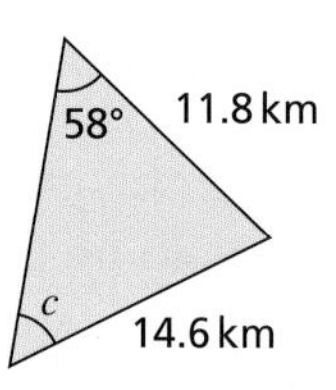

d

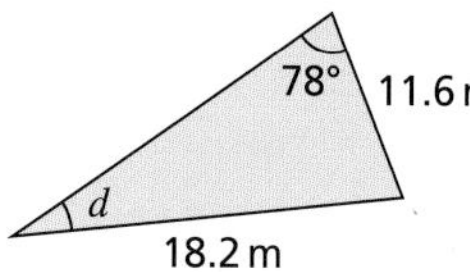

e

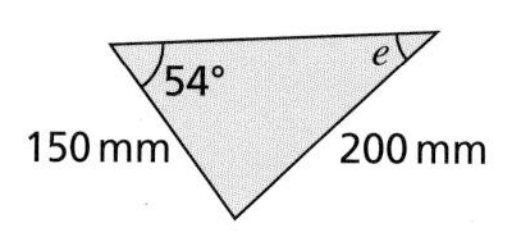

f

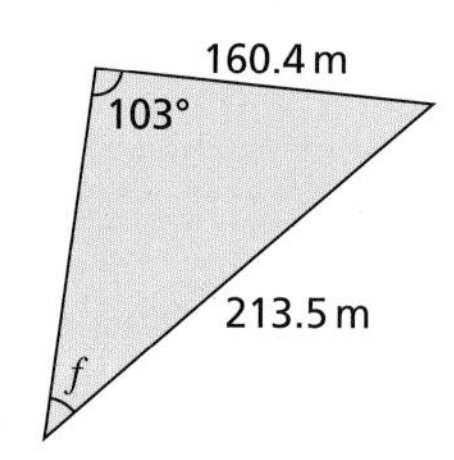

4 Triangle DEF is shown.

Work out the size of angle EFD.

Give your answer to 1 decimal place.

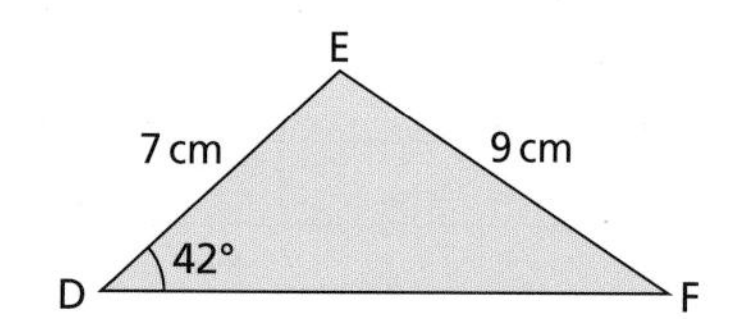

5 Triangle ABC is shown.

Work out the size of angle BCA.

Give your answer to 3 significant figures.

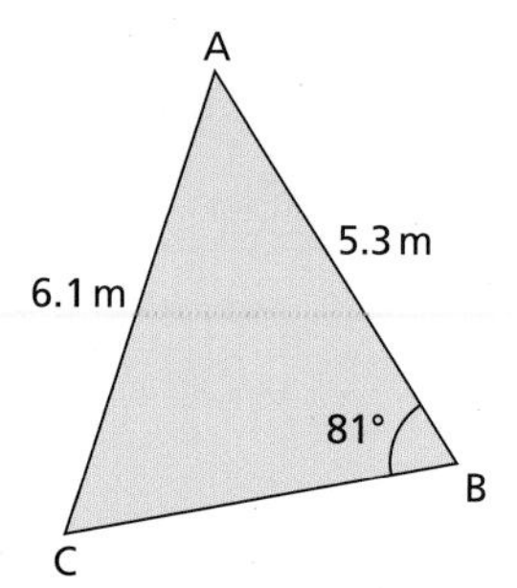

6 On each diagram, PQR is a straight line.

Work out the size of each lettered angle. Give your answers to 1 decimal place.

a

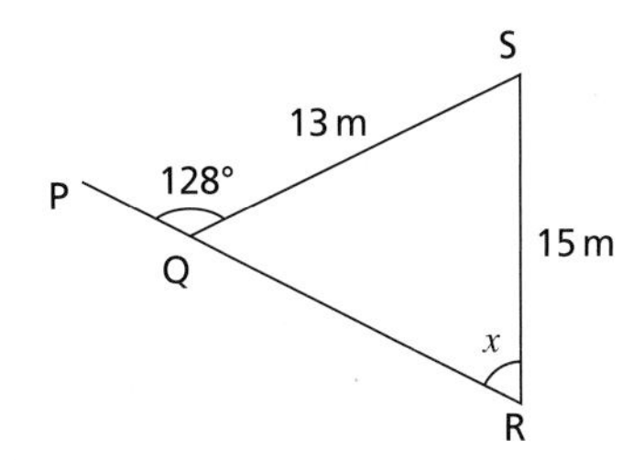

b

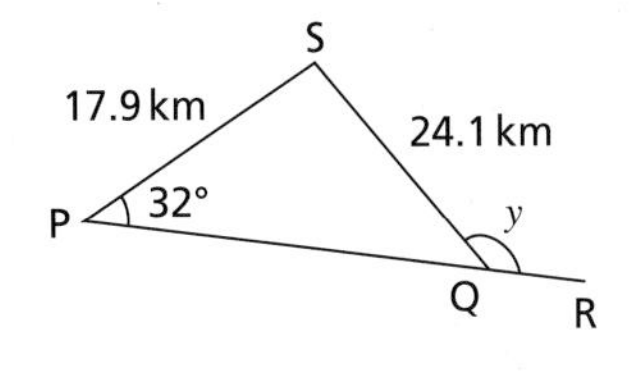

7 ABCD is a parallelogram.

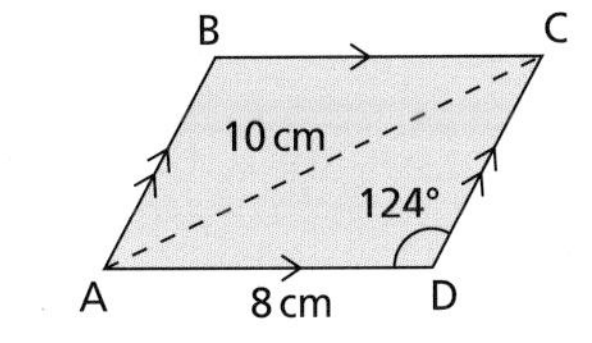

a Calculate the size of angle ACD.

Give your answer to 1 decimal place.

b Calculate the size of angle BCA.

Give your answer to 1 decimal place.

8 Work out the size of angle QPS.

Give your answer to 2 decimal places.

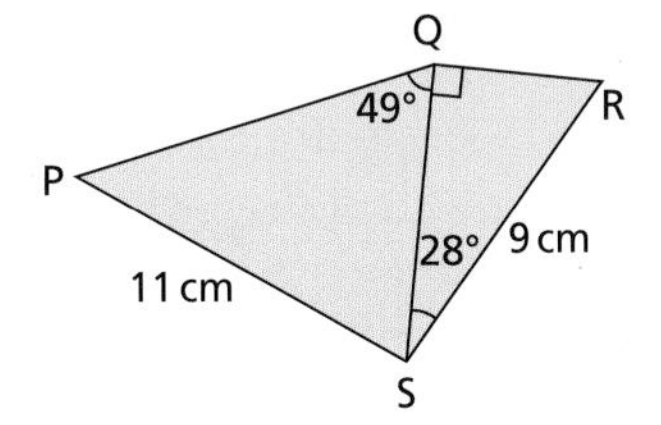

9 The triangle has **not** been drawn accurately.

There are two possible values for θ.

Work out the value of:

a the acute angle, θ

b the obtuse angle, θ

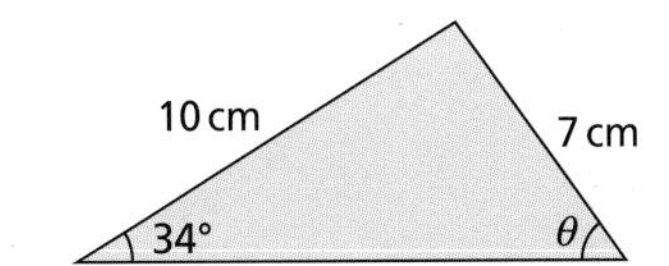

10 The triangle has **not** been drawn accurately.

Work out the two possible values for θ.

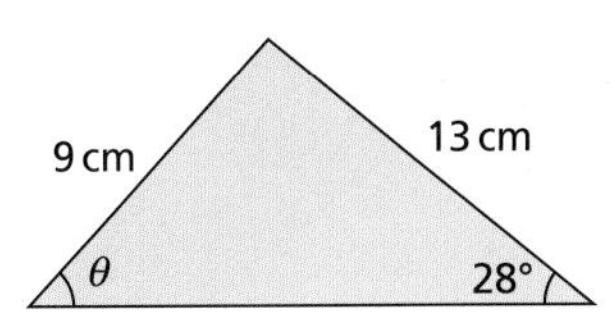

What do you think? (B)

1 Ed and Mo are discussing this triangle.

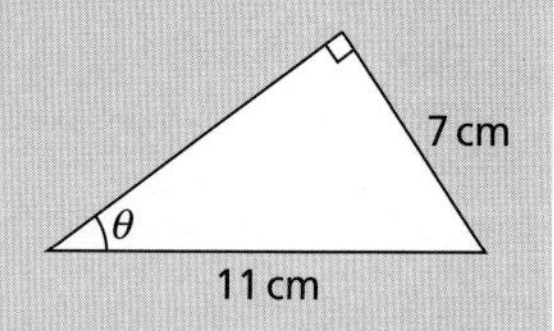

Ed says, "You have to use the sine rule to find the value of θ"

Mo says, "You don't have to use the sine rule."

Who do you agree with? Explain your answer.

2 B is due North of A.

C is 90 m from B on a bearing of 140°.

A is 60 m from C and lies north of C.

Copy the diagram, show the information and calculate the bearing of C from A.

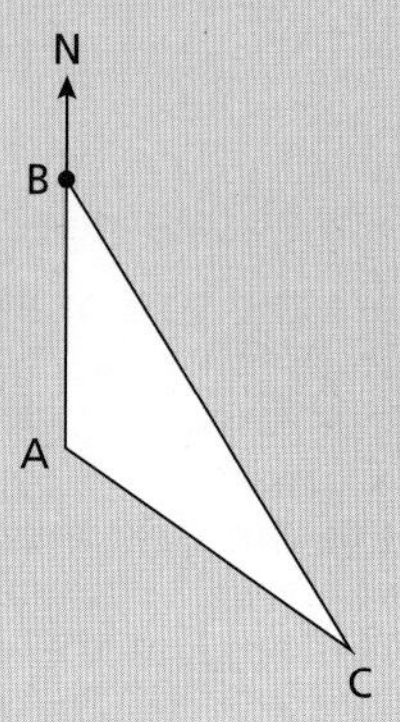

3 In the triangle ABC, AB = 10 cm, BC = 9 cm and $\angle BAC = 55°$

Abdullah says, "There are two possible triangles that I could draw."

a Draw two different triangles to show that Abdullah is correct.

b Work out all of the unknown angles in each of your triangles.

4 Decide if the triangles have **one**, **two** or **no** possible values for θ.

Explain how you know.

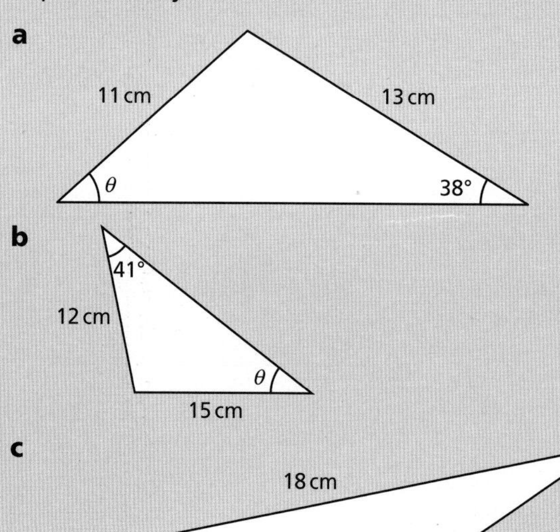

Consolidate – do you need more?

1 Work out the length of each lettered side. Give your answers to 1 decimal place.

a

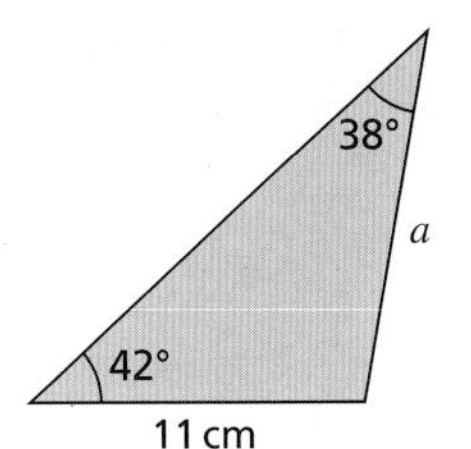

b

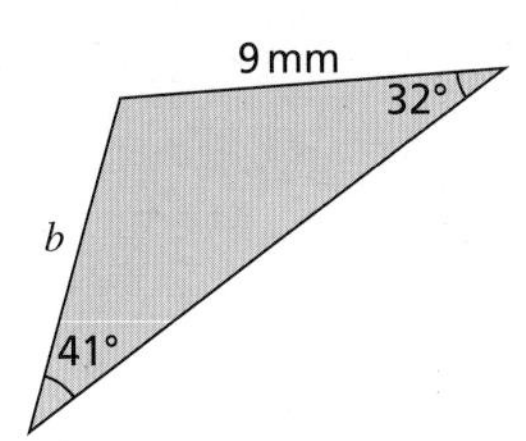

c

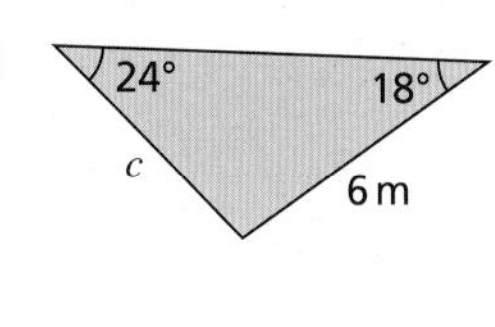

d

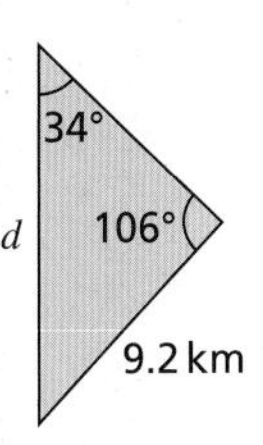

2 Work out the size of each lettered angle. Give your answers to 1 decimal place.

a

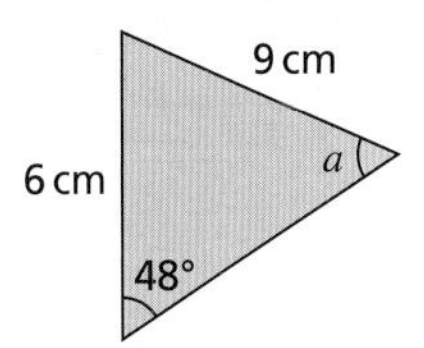

b

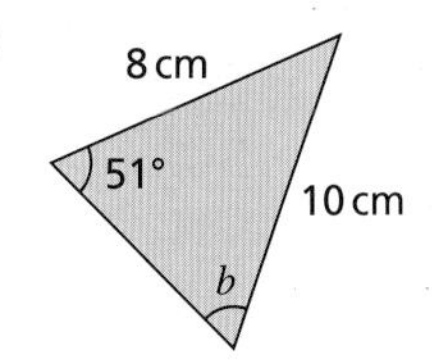

c

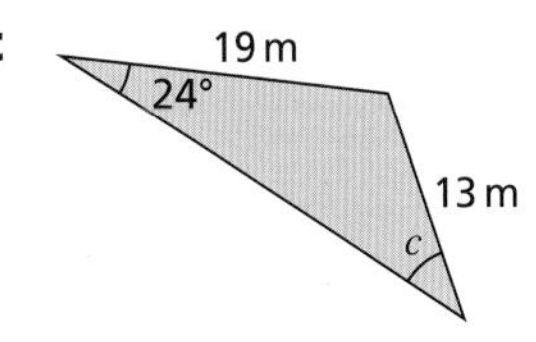

d

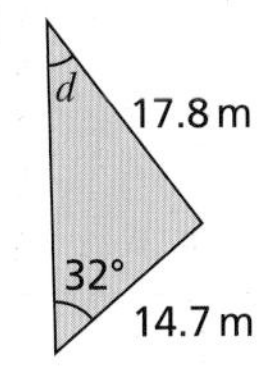

Stretch – can you deepen your learning?

1 Work out the length of the side labelled a.

Give your answer to 3 significant figures.

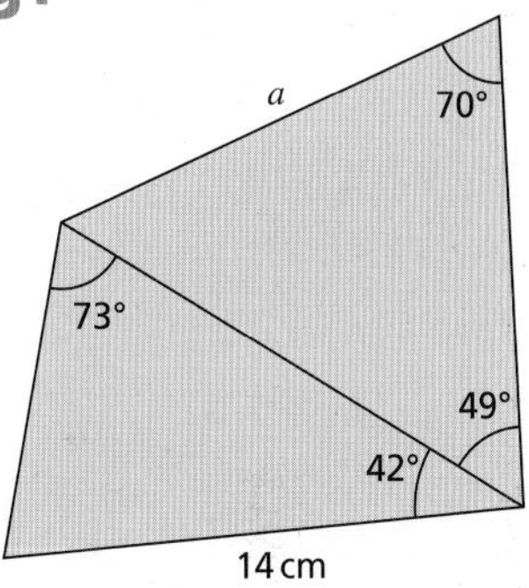

2 Write the value of y in terms of x.

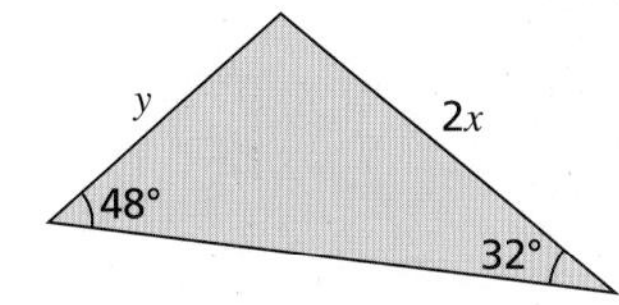

3 Write the value of x. Give your answer in the form $a\sqrt{b}$, where a and b are integers.

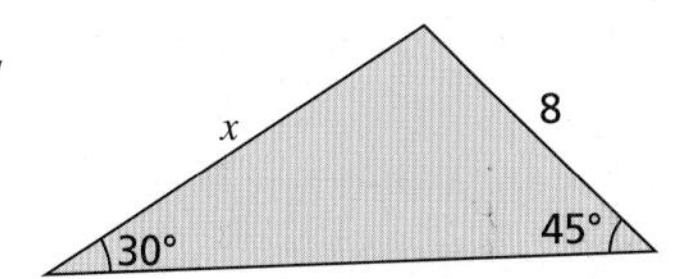

4 The triangle PQR is shown. All values have been rounded to the nearest integer.

a Work out the upper bound for the size of angle PRQ. Give your answer to 5 decimal places.

b Work out the lower bound for the size of angle PRQ. Give your answer to 5 decimal places.

P, Q, R; 19 cm; 38°; 16 cm; x

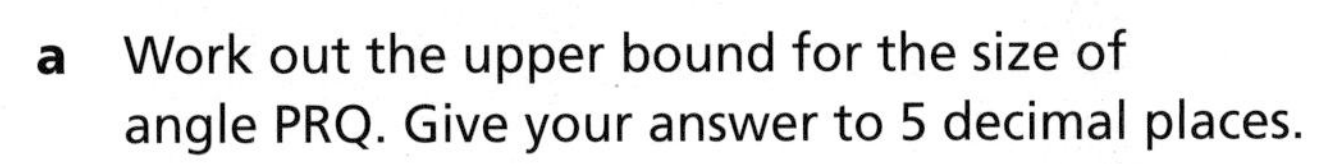
Bounds are covered in *Collins White Rose Maths AQA GCSE 9–1 Higher Student Book 1*, Chapter 5.2

7.3 The cosine rule

White Rose MATHS

Are you ready? (A)

1 Calculate the value of each expression, when $b = 5$, $c = 3$ and $A = 55°$

Give your solutions to 1 decimal place where appropriate.

a b^2 **b** c^2 **c** $2bc$ **d** $2bc\cos A$ **e** $b^2 + c^2 - 2bc\cos A$

2 Copy the triangles and label the sides of each triangle with a, b and c.

a

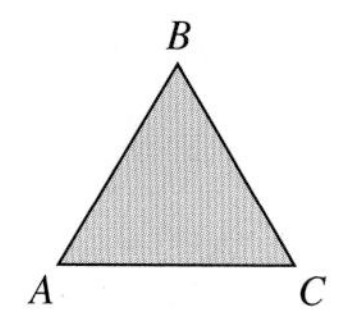

b

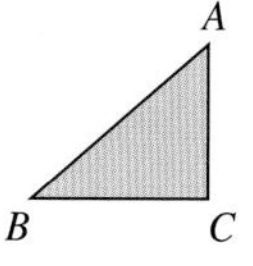

c

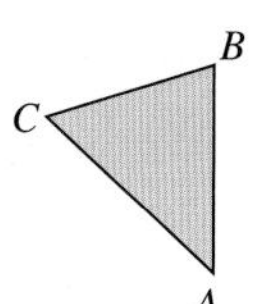

See Chapter 7.2 (The sine rule).

The **cosine rule** is another rule that helps you find angles and sides in triangles that are not right-angled.

Consider the triangle shown.

Using Pythagoras' theorem in the triangle ACD,

$b^2 = h^2 + x^2$ and therefore $h^2 = b^2 - x^2$

From triangle BCD, $a^2 = h^2 + (c - x)^2$

and therefore $h^2 = a^2 - (c - x)^2$

Equating the expressions:

$b^2 - x^2 = a^2 - (c - x)^2$

Which gives $a^2 = b^2 + c^2 - 2cx$

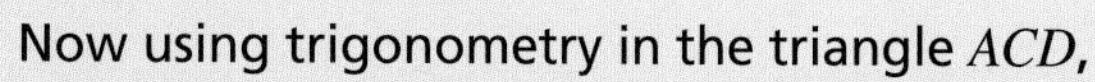

Now using trigonometry in the triangle ACD,

$\cos A = \frac{x}{b}$ and therefore $b\cos A = x$

Substituting $b\cos A$ for x in the formula $a^2 = b^2 + c^2 - 2cx$

gives $a^2 = b^2 + c^2 - 2bc\cos A$

The cosine rule states: $a^2 = b^2 + c^2 - 2bc\cos A$

A is the **included angle**, which is the angle between the two given sides.

To use the cosine rule, you need all three sides and one angle, or two sides and the angle between them.

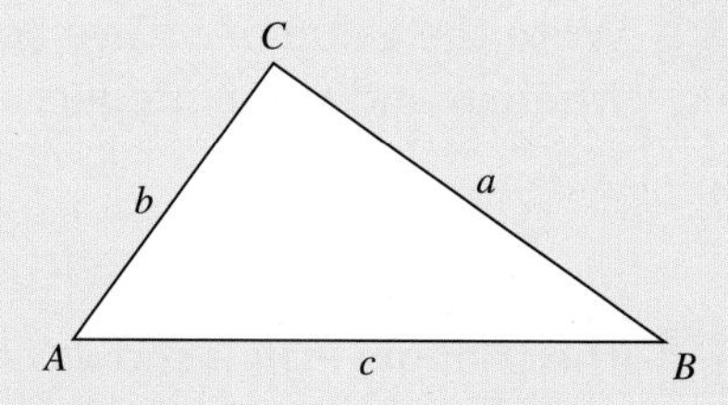

Remember that to use the sine rule, you need an angle, the side opposite it and one other angle or side. You will often need to decide which rule to use when solving problems involving triangles that are not right-angled.

Example

Work out the length of DY.

Give your answer to 1 decimal place.

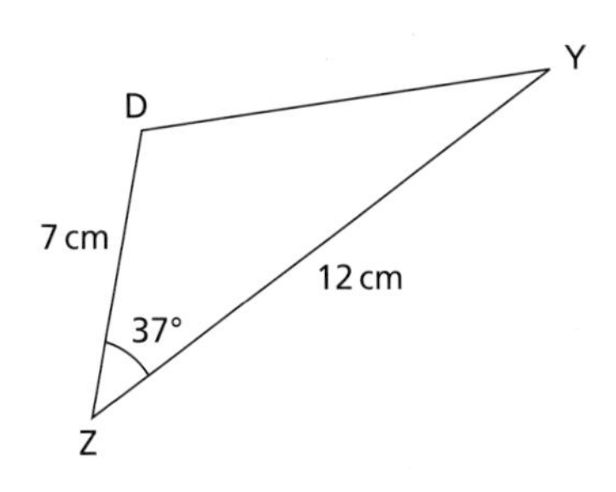

Method

Solution	Commentary
$a^2 = b^2 + c^2 - 2bc\cos A$	Since you have two lengths and the included angle, you can use the cosine rule to find the required length.
B Y, a, C D, b 7 cm, c 12 cm, 37°, Z A	Label the triangle and substitute the values into the formula.
$a^2 = 7^2 + 12^2 - 2 \times 7 \times 12\cos 37°$ $a^2 = 58.829234...$ $a = \sqrt{58.829234...}$ DY = 7.7 cm (to 1 d.p.)	This gives the value of a^2, so you need to find the square root.

Practice (A)

1 A triangle is shown.

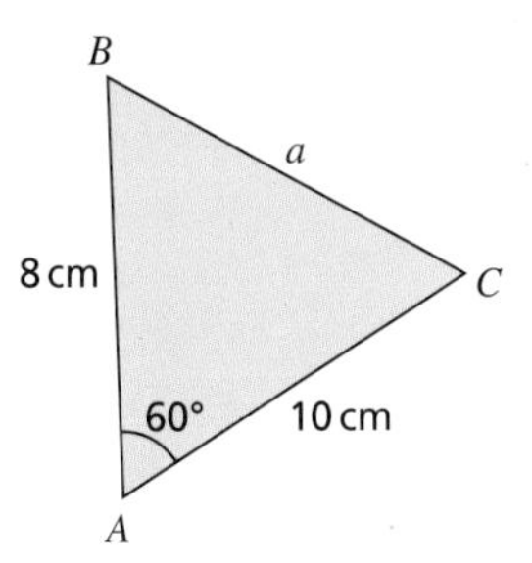

a Copy the triangle and label the corresponding sides b and c.

b Which equation is correct?

A $a^2 = 8^2 + 10^2 + 2 \times 8 \times 10\cos 60°$

B $a^2 = 8 + 10 + 2 \times 8 \times 10\cos 60°$

C $a^2 = 8^2 + 10^2 - 2 \times 8 \times 10\cos 60°$

c Work out the length of side a.

Give your answer to 1 decimal place.

2 Work out the length of each lettered side. Give your answers to 1 decimal place.

a

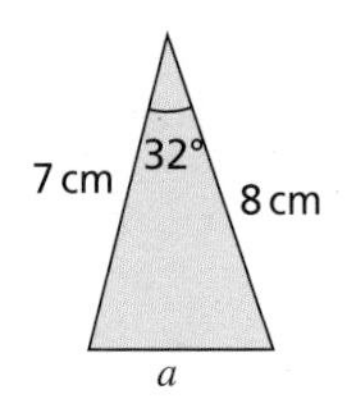

b

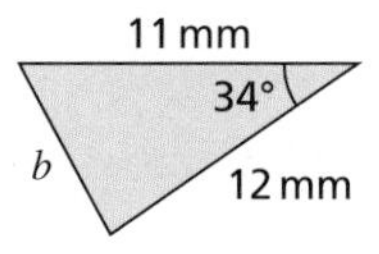

c

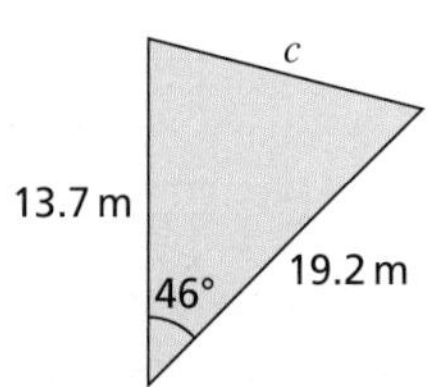

d

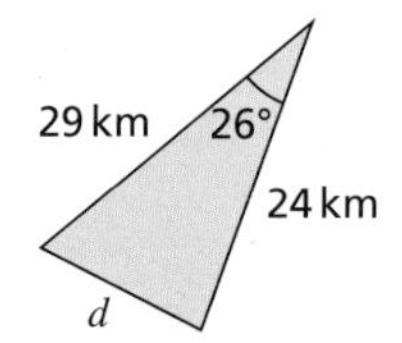

3 Calculate the length of QR.

Give your answer to 1 decimal place.

Q
13.2 m
37°
P
11.4 m
R

4 Calculate the value of x.

Give your answer to 2 decimal places.

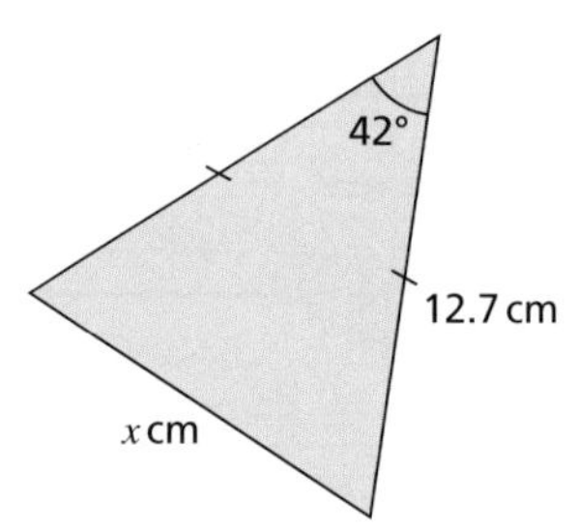

5 Calculate the length of RS.

Give your answer to 1 decimal place.

Hint: Work out the length of QS first.

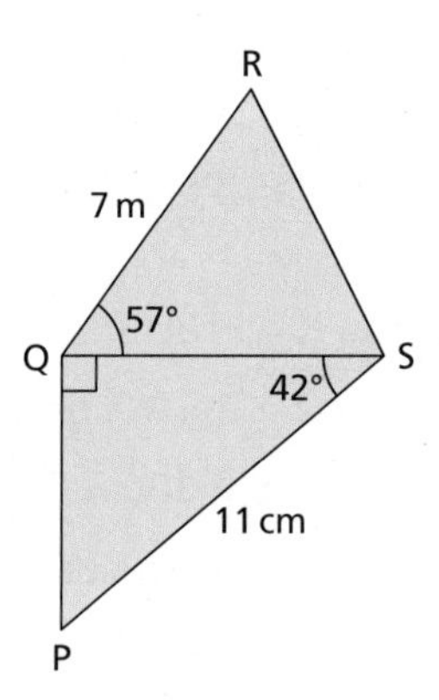

6 Work out the perimeter of ABC.

Give your answer to 3 significant figures.

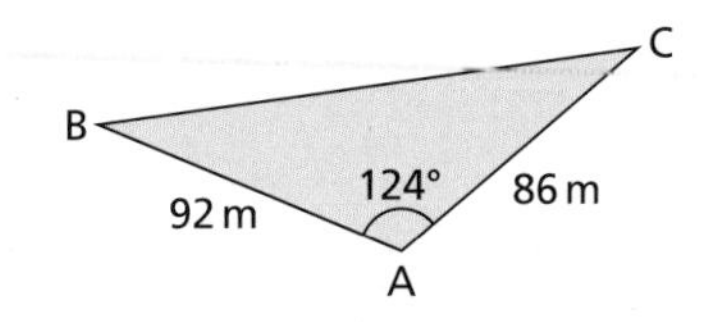

7 Work out the value of x. Give your answer in the form $a\sqrt{b}$, where a and b are integers.

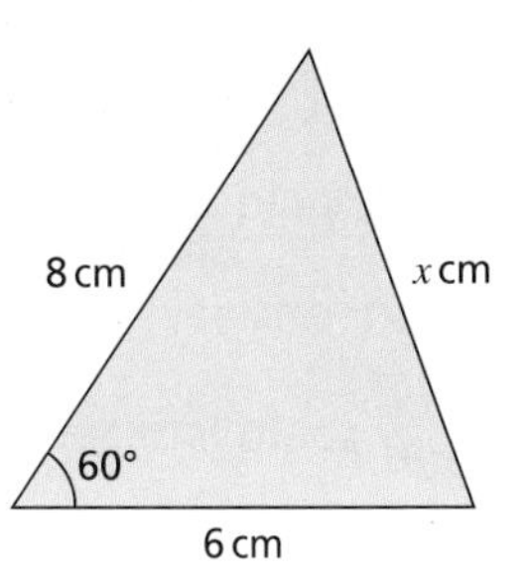

8 Work out the value of y. Give your answer in the form $a\sqrt{b}$, where a and b are integers.

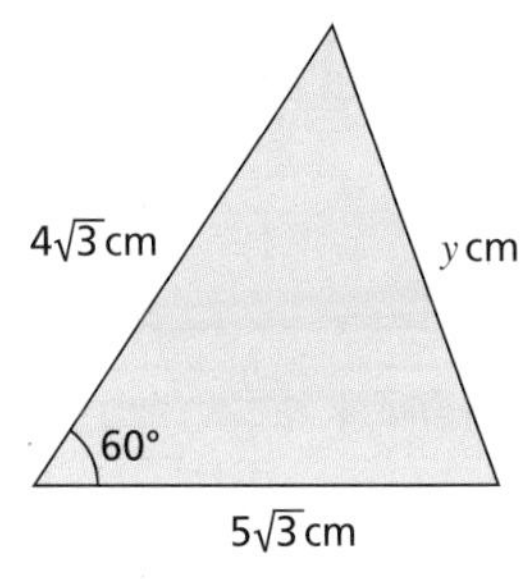

9 PQR is a triangle.

PQ = 9 cm, QR = 11 cm and ∠PQR = 42°

Work out the length of PR. Give your answer to 3 significant figures.

10 A boat sails 12 km North from a harbour.

The boat then travels on a bearing of 055° for another 9 km.

a Sketch the journey of the boat.

b Calculate the distance between the harbour and the boat.

Give your answer to 3 significant figures.

What do you think? (A)

1 What is the connection between the cosine rule and Pythagoras' theorem?

Are you ready? (B)

1 Solve each equation, giving your solutions to 1 decimal place where appropriate.

a $\cos x = \frac{7}{10}$ **b** $\cos y = \frac{13.5}{15.6}$ **c** $2\cos x = 0.6$ **d** $\cos x + 0.7 = 1.2$

Example

Calculate the size of angle TRS.

Give your answer to 1 decimal place.

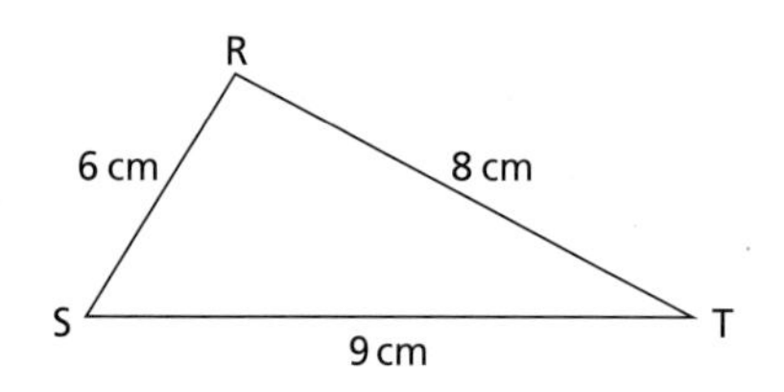

Method A

Solution	Commentary
R A ? 6 cm c b 8 cm S T 9 cm a	Since you have three lengths, use the cosine rule to find the required angle. Label the triangle and substitute the values that you know into the formula.
$a^2 = b^2 + c^2 - 2bc\cos A$ $9^2 = 8^2 + 6^2 - 2 \times 8 \times 6 \times \cos A$ $81 = 100 - 96\cos A$ $+ 96\cos A$ $+ 96\cos A$ $96\cos A + 81 = 100$ $- 81$ $- 81$ $96\cos A = 19$ $\div 96$ $\div 96$ $\cos A = \frac{19}{96}$ $\cos^{-1}\left(\frac{19}{96}\right) = 78.58484...$ Angle TRS = 78.6° (to 1 d.p.)	Rearrange the equation and use the inverse cosine function to find the size of the angle. Depending on your calculator, you may get a fraction like this or a decimal equivalent 0.197916666...

Method B

Solution	Commentary
$a^2 = b^2 + c^2 - 2bc\cos A$ $+ 2bc\cos A$ $+ 2bc\cos A$ $2bc\cos A + a^2 = b^2 + c^2$ $- a^2$ $- a^2$ $2bc\cos A = b^2 + c^2 - a^2$ $\div 2bc$ $\div 2bc$ $\cos A = \frac{b^2 + c^2 - a^2}{2bc}$	Rearrange the cosine rule formula to make $\cos A$ the subject.
$\cos A = \frac{8^2 + 6^2 - 9^2}{2 \times 8 \times 6} = \frac{19}{96}$ $\cos^{-1}\left(\frac{19}{96}\right) = 78.58484....$ Angle TRS = 78.6° (to 1 d.p.)	Substitute the values for a, b and c into the formula, and use the inverse cosine function to find the size of the angle.

Practice (B)

1 Here is a triangle.

a Copy the triangle and label the angles to use the cosine rule.

b Which equation is correct?

A $9^2 = 7^2 + 11^2 - 2 \times 7 \times 11\cos x$

B $8^2 = 9^2 + 7^2 - 2 \times 9 \times 7\cos x$

C $7^2 = 9^2 + 11^2 - 2 \times 9 \times 11\cos x$

c Work out the size of angle x. Give your answer to 1 decimal place.

2 Work out the size of each lettered angle, accurate to 1 decimal place.

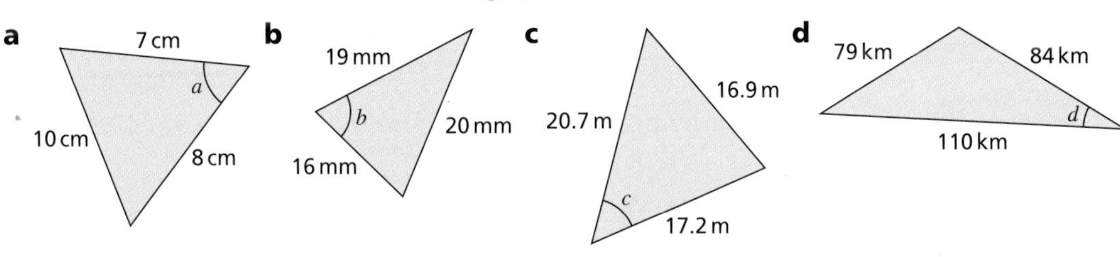

3 Work out the size of angle XYZ.

Give your answer to 1 decimal place.

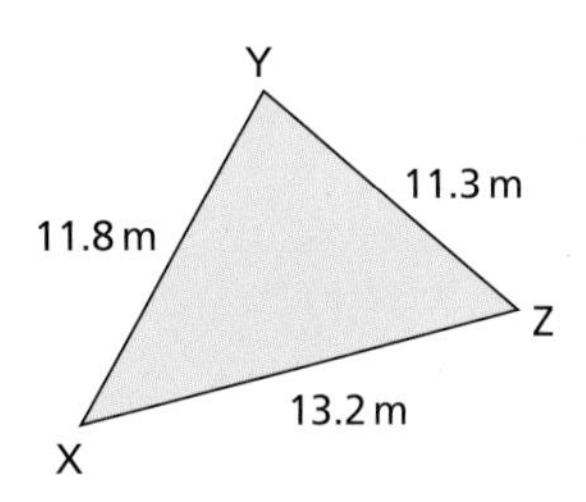

4 Work out the size of angle x.

Give your answer to 3 significant figures.

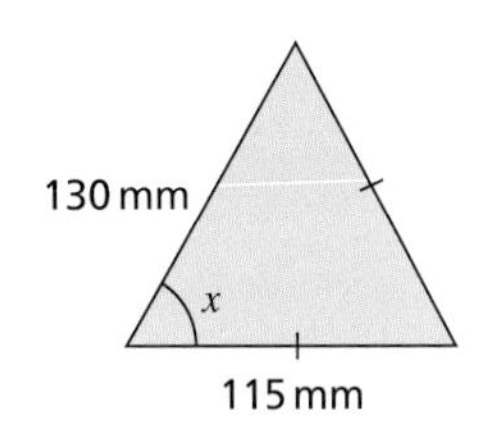

5 Work out the size of angle TUV.

Give your answer to 3 significant figures.

Hint: Work out the length of TV first.

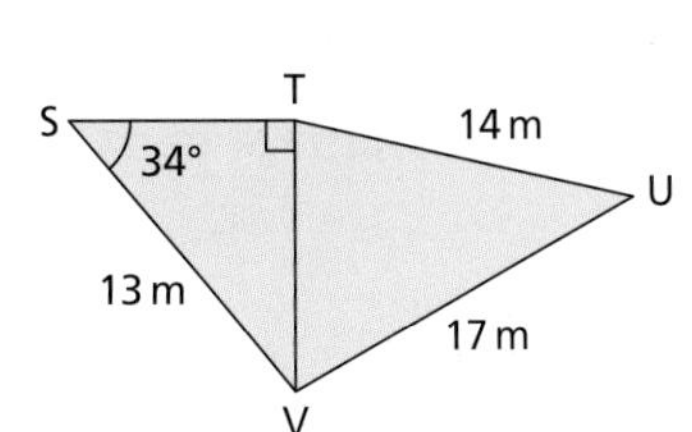

6 Work out the size of angle ACD.

Give your answer to 3 significant figures.

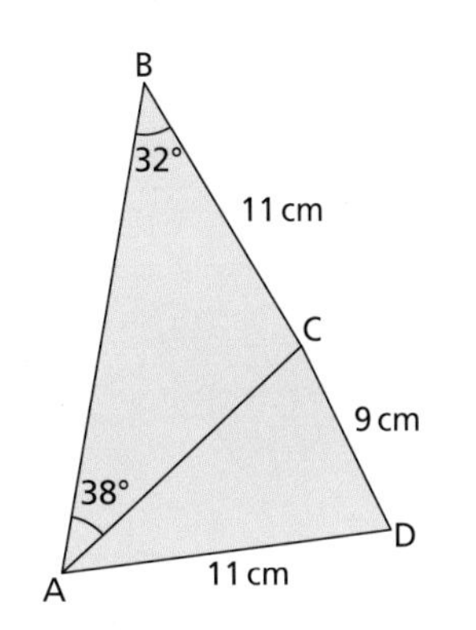

7 A triangle is shown.
Show that $\cos A = \frac{13}{14}$

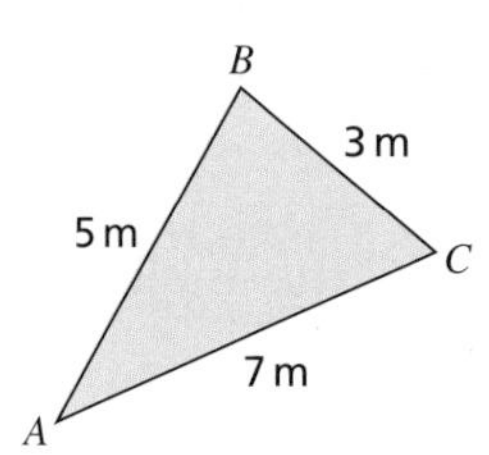

What do you think? (B)

1 Work out the size of the largest angle in the triangle.

Hint: The largest angle of a triangle is opposite the longest side.

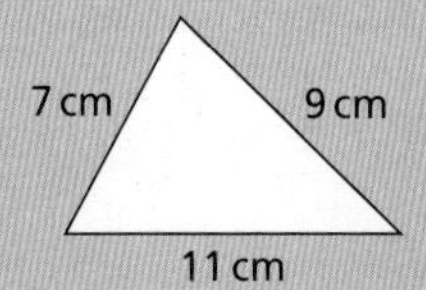

2 Use the cosine rule to show that this triangle is right-angled.

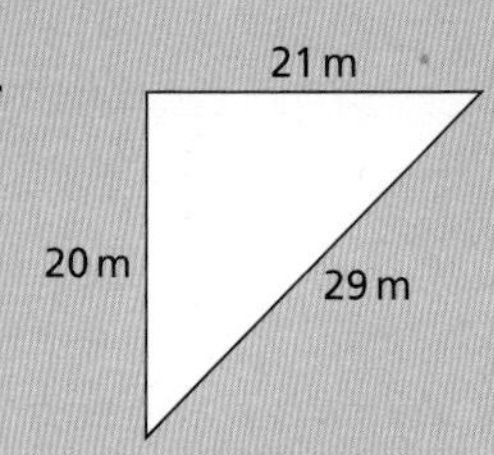

3 For each triangle, state whether you would use the sine rule or the cosine rule to find the angle marked θ. You do not need to work out the value of θ.

a

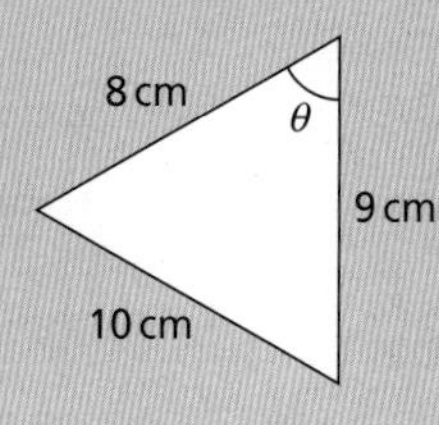

b

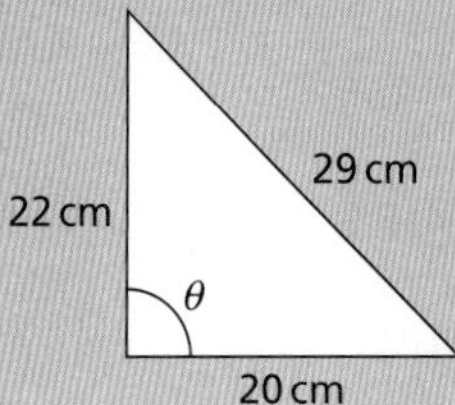

c

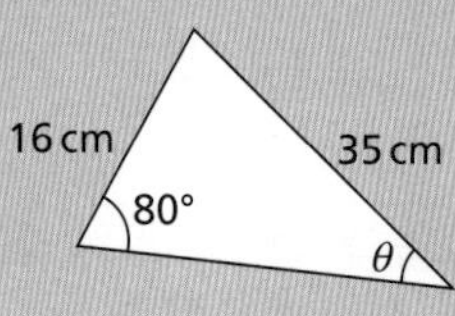

d

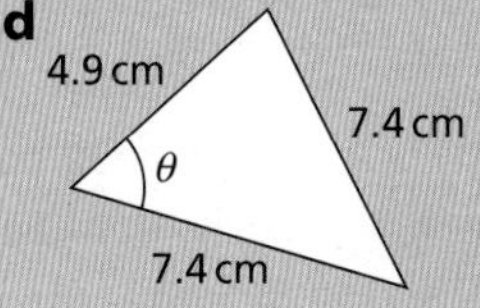

Consolidate – do you need more?

1 Work out the length of each lettered side. Give your answers to 1 decimal place.

a

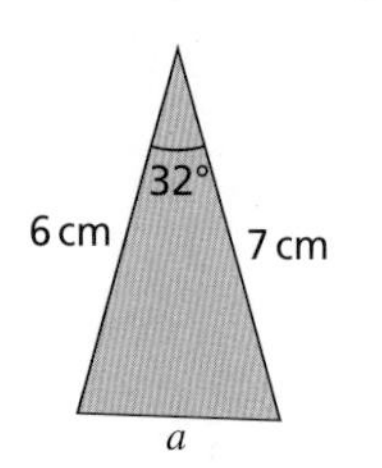

b

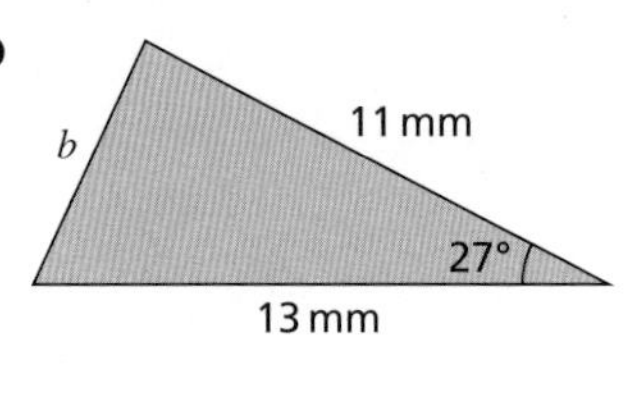

c

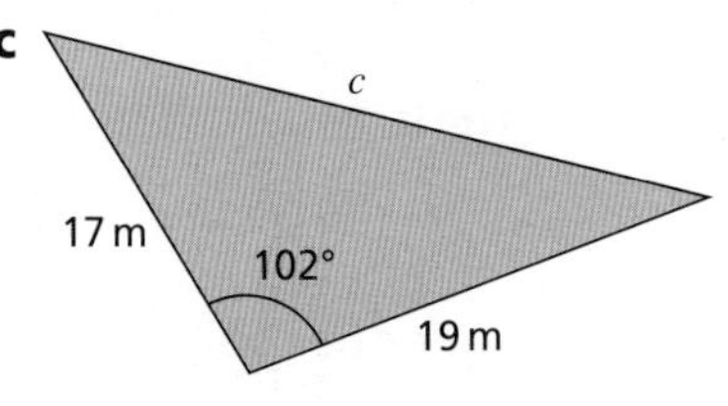

d

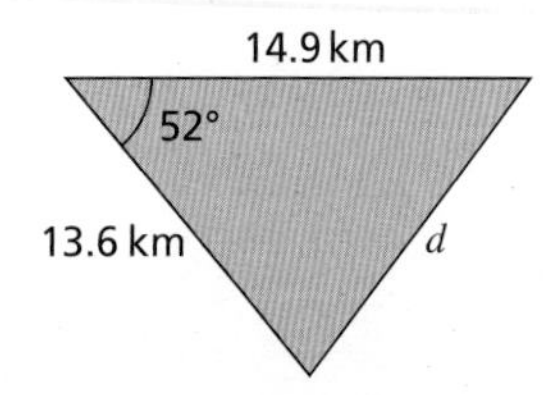

e

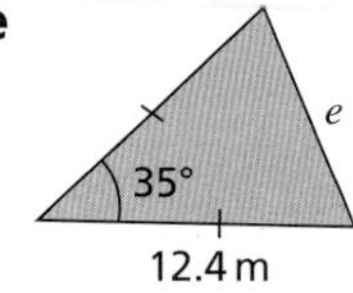

f

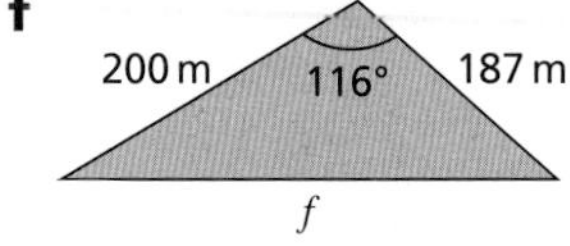

2 Work out the size of each lettered angle, accurate to 1 decimal place.

a 7 cm, 9 cm, 11 cm, angle a

b 11 m, 14 m, 16 m, angle b

c 10.4 km, 12.4 km, 9.8 km, angle c

d 20 m, 19 m, 24 m, angle d

e 380 mm, 400 mm, 415 mm, angle e

f 16.2 cm, 21.8 cm, angle f

3 Work out the value of x. Give your answer in the form $\sqrt{a}$, where a is an integer.

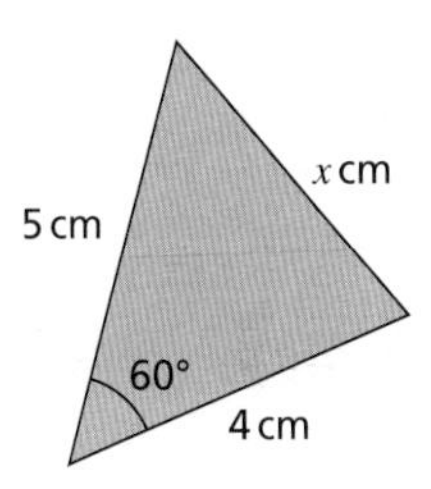

4 A triangle is shown.

Show that $\cos A = \dfrac{41}{55}$

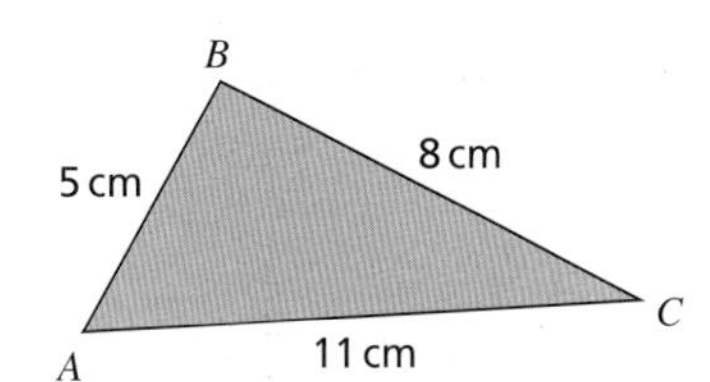

Stretch – can you deepen your learning?

1 Triangle EFG is shown.

Work out the length FG in centimetres. Give your answer in the form $a\sqrt{b}$, where a and b are integers.

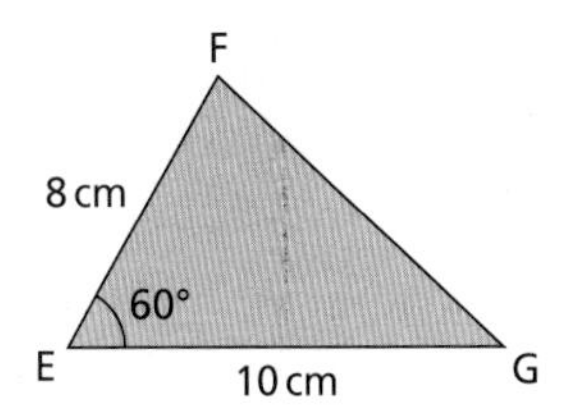

2 Here is a triangle.

Work out an expression for y^2 in terms of x. Give your answer in the form $ax^2 + bx + c$, where a, b and c are integers.

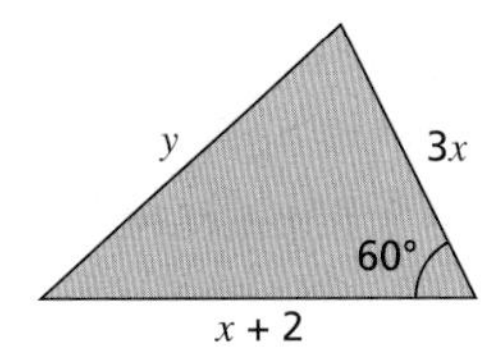

Non-right-angled triangles: exam practice

White Rose MATHS

7–9

* *Diagrams are not accurately drawn.*

1 Calculate the area of the triangle.
Give your answer to 3 significant figures. **[2 marks]**

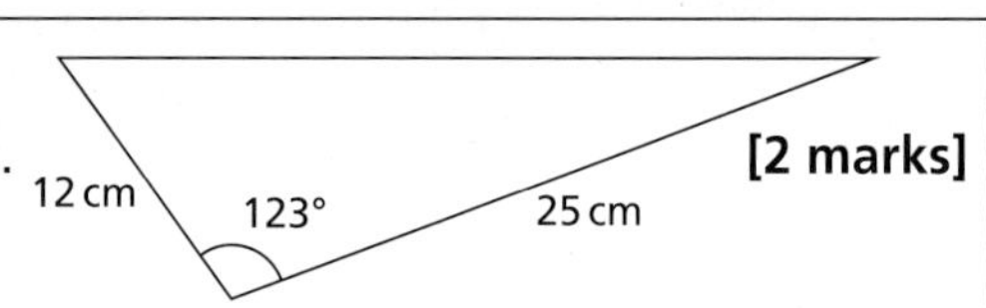

2 Calculate the size of ∠BAC.
Give your answer to 3 significant figures. **[4 marks]**

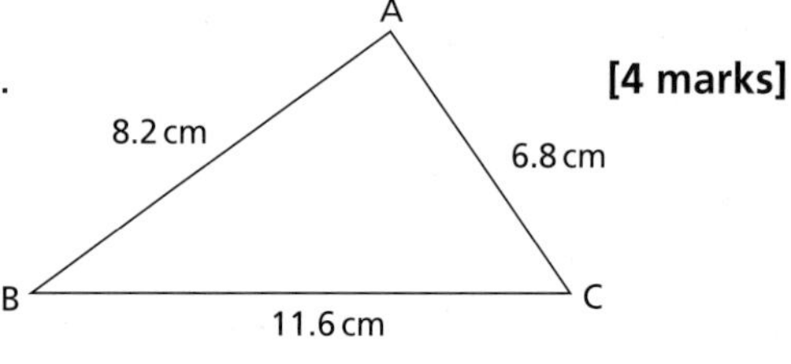

3 The area of the triangle is 42 cm².
Work out the length of BC. **[3 marks]**

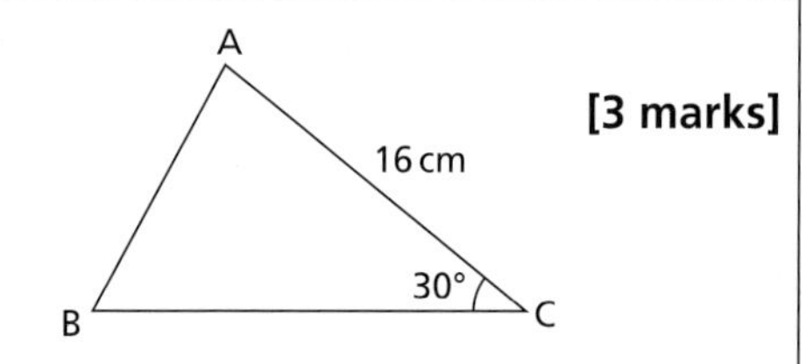

4 A kite has sides of length 5 cm and 12 cm.
Two of the angles in the kite are 140°.
Work out the area of the kite. Give your answer to 3 significant figures. **[3 marks]**

5 Calculate the size of ∠XYZ.
Give your answer to 3 significant figures. **[3 marks]**

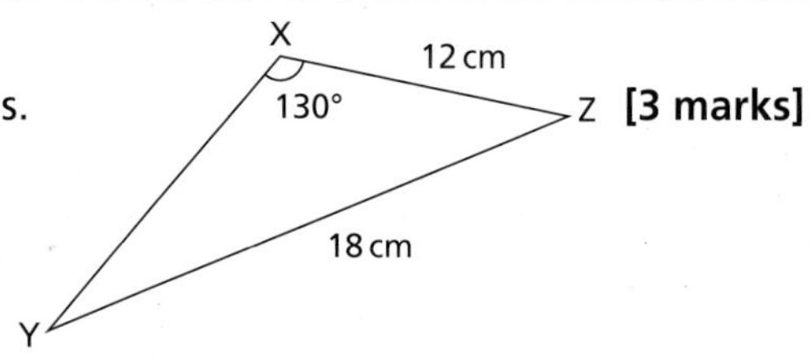

6 Here is quadrilateral ABCD.
Calculate the area of triangle BCD.
Give your answer to 3 significant figures. **[5 marks]**

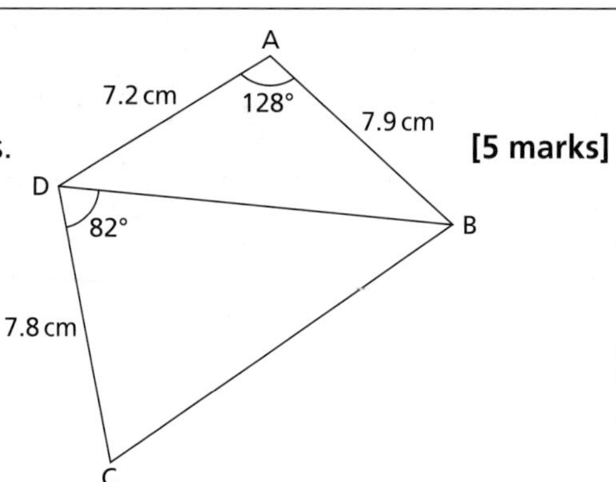

8 Vectors

White Rose
MATHS

In this block, we will cover...

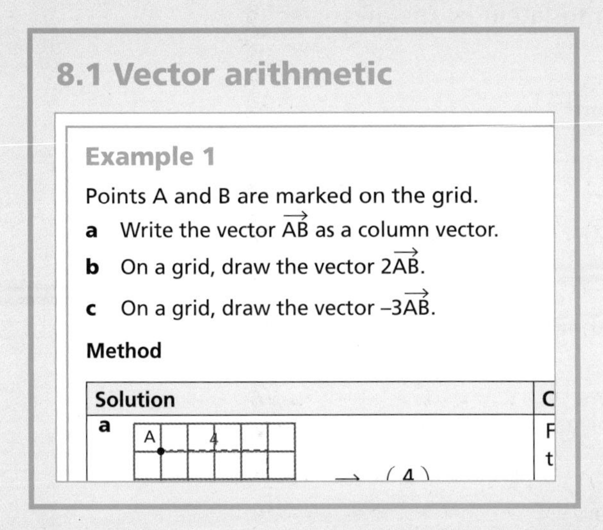
8.1 Vector arithmetic
Example 1
Points A and B are marked on the grid.
a Write the vector $\overrightarrow{AB}$ as a column vector.
b On a grid, draw the vector $2\overrightarrow{AB}$.
c On a grid, draw the vector $-3\overrightarrow{AB}$.
Method
Solution
a
A
4

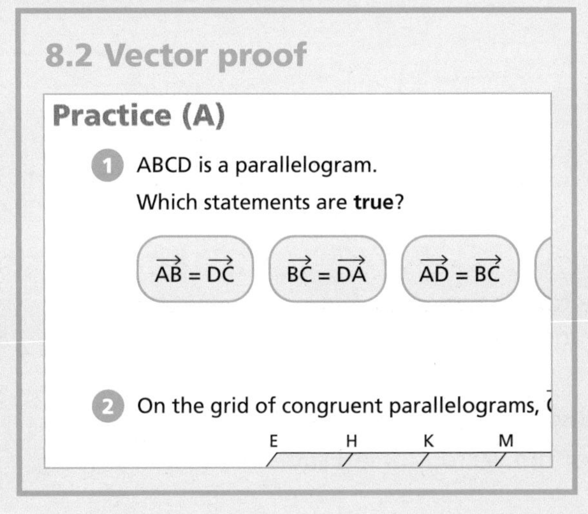
8.2 Vector proof
Practice (A)
1 ABCD is a parallelogram.
Which statements are true?
$\overrightarrow{AB} = \overrightarrow{DC}$
$\overrightarrow{BC} = \overrightarrow{DA}$
$\overrightarrow{AD} = \overrightarrow{BC}$
2 On the grid of congruent parallelograms,
E
H
K
M

8.1 Vector arithmetic

White Rose MATHS

Are you ready?

1 Copy each shape onto squared paper and translate it by the given vector.

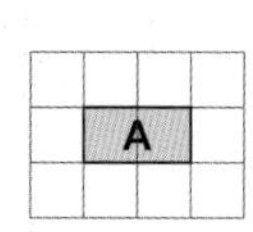

$\begin{pmatrix} 3 \\ -2 \end{pmatrix}$

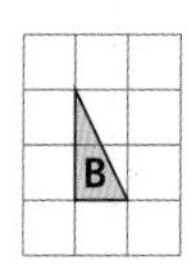

$\begin{pmatrix} -1 \\ 3 \end{pmatrix}$

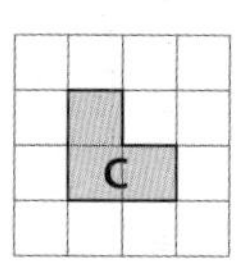

$\begin{pmatrix} 0 \\ 2 \end{pmatrix}$

2 Describe each translation from shape A to shape B.

a

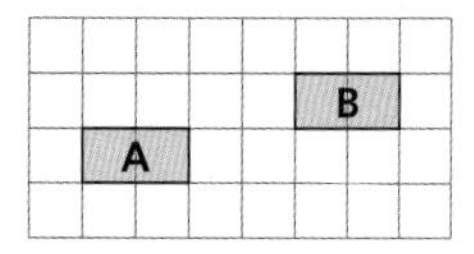

b

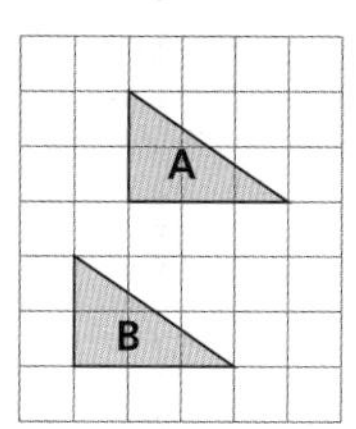

3 Work out:

a $3 - 7$ **b** $-3 - 7$ **c** $3 - -7$ **d** $-3 - -7$

e 3×-7 **f** -3×7 **g** -3×-7 **h** -7×-3

The diagram shows the translation of a square by the **vector** $\begin{pmatrix} 2 \\ -1 \end{pmatrix}$. The square has been translated two units to the right and one unit down.

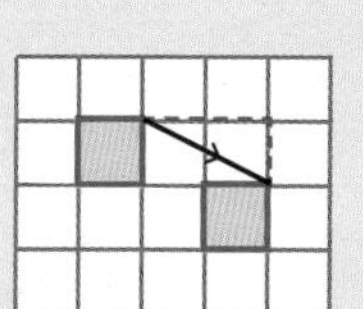

Vectors are usually represented using a line segment with an arrow showing the direction of the vector. $\overrightarrow{AB}$ indicates the vector starting at A and ending at B.

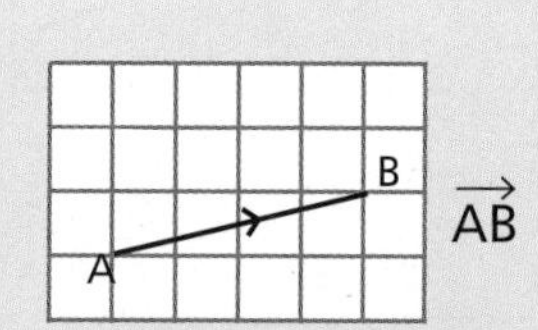

Alternatively, a vector can be denoted using a bold lowercase letter such as **a**.

Two vectors are **equal** if they have the same length and are in the same direction.

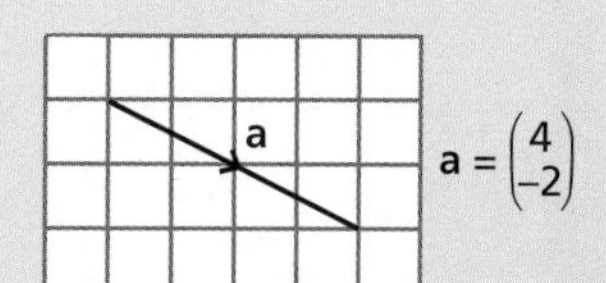

Two vectors are parallel if one vector is a multiple of the other.

For example, the vector 3**a** is three times as long as the vector **a** in the same direction.

You can work out the components of the vector 3**a** by multiplying the components of vector **a** by 3

$3\mathbf{a} = 3 \times \begin{pmatrix} 4 \\ -2 \end{pmatrix} = \begin{pmatrix} 12 \\ -6 \end{pmatrix}$

Notice that the result is parallel to **a**.

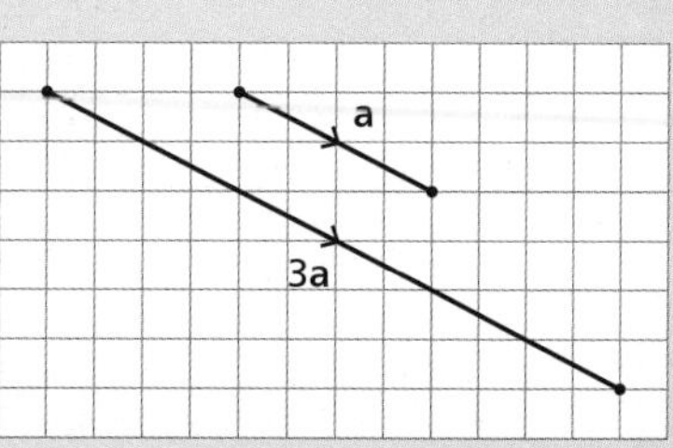

The diagram shows the vectors $\overrightarrow{AB} = \begin{pmatrix} 4 \\ 2 \end{pmatrix}$ and $\overrightarrow{BC} = \begin{pmatrix} 5 \\ -3 \end{pmatrix}$.

The vector $\overrightarrow{AB}$ followed by $\overrightarrow{BC}$ is equivalent to the vector $\overrightarrow{AC}$.

$\overrightarrow{AC}$ can be calculated by adding the horizontal and vertical components of $\overrightarrow{AB}$ and $\overrightarrow{BC}$.

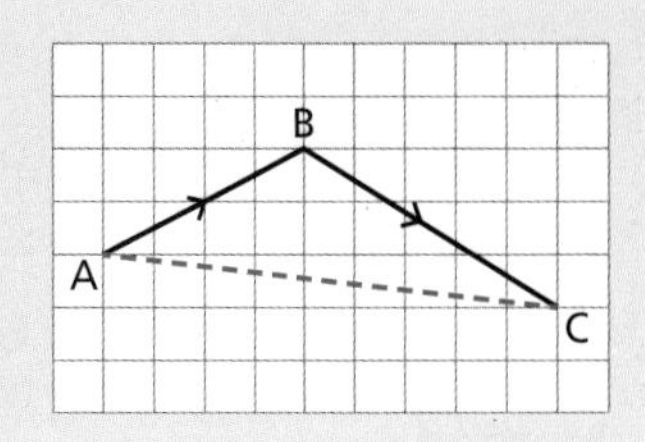

Example 1

Points A and B are marked on the grid.

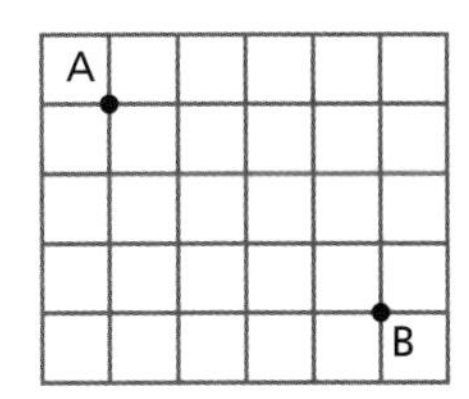

a Write the vector $\overrightarrow{AB}$ as a column vector.

b On a grid, draw the vector $2\overrightarrow{AB}$.

c On a grid, draw the vector $-3\overrightarrow{AB}$.

Method

Solution	Commentary
a $\overrightarrow{AB} = \begin{pmatrix} 4 \\ -3 \end{pmatrix}$	From A to B is four units to the right and three units down. Now write this using a column vector.
b $2 \times \overrightarrow{AB} = \begin{pmatrix} 8 \\ -6 \end{pmatrix}$	$2\overrightarrow{AB}$ would be $2 \times \begin{pmatrix} 4 \\ -3 \end{pmatrix}$ Now draw this on squared paper. Remember to use the arrow to give the direction of the vector.
c $-3 \times \overrightarrow{AB} = \begin{pmatrix} -12 \\ 9 \end{pmatrix}$	$-3\overrightarrow{AB}$ would be $-3 \times \begin{pmatrix} 4 \\ -3 \end{pmatrix}$ Now draw this on squared paper. Notice that the vector has now changed to the opposite direction.

Example 2

You are given these vectors: $\mathbf{a} = \begin{pmatrix} 1 \\ -3 \end{pmatrix}$ $\mathbf{b} = \begin{pmatrix} 4 \\ 2 \end{pmatrix}$ $\mathbf{c} = \begin{pmatrix} 2 \\ 5 \end{pmatrix}$

Work out:

a vector **a** + **b** **b** vector 2**a** – **c** **c** vector **b** – **c**

Method

Solution	Commentary
a $\begin{pmatrix} 1 \\ -3 \end{pmatrix} + \begin{pmatrix} 4 \\ 2 \end{pmatrix} = \begin{pmatrix} 5 \\ -1 \end{pmatrix}$	You can add vectors by adding each of the components separately: 1 + 4 = 5 and –3 + 2 = –1
b $2\mathbf{a} - \mathbf{c} = 2 \times \begin{pmatrix} 1 \\ -3 \end{pmatrix} - \begin{pmatrix} 2 \\ 5 \end{pmatrix}$ $= \begin{pmatrix} 2 \\ -6 \end{pmatrix} - \begin{pmatrix} 2 \\ 5 \end{pmatrix} = \begin{pmatrix} 0 \\ -11 \end{pmatrix}$	First multiply **a** by 2, by multiplying both components by 2 Then subtract the components of **c**.
c $\begin{pmatrix} 4 \\ 2 \end{pmatrix} - \begin{pmatrix} 2 \\ 5 \end{pmatrix} = \begin{pmatrix} 2 \\ -3 \end{pmatrix}$	On the top row, 4 – 2 = 2 On the bottom row, 2 – 5 = –3

Practice

1 Write down the column vectors **a** to **h**.

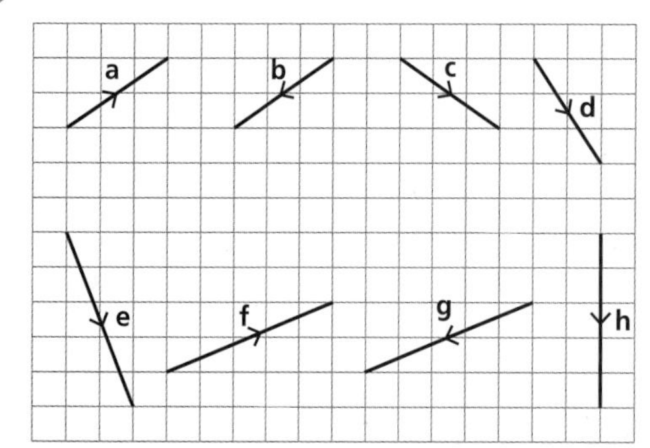

2 Draw each vector on squared paper.

a $\begin{pmatrix} 2 \\ 4 \end{pmatrix}$ **b** $\begin{pmatrix} 3 \\ 4 \end{pmatrix}$ **c** $\begin{pmatrix} 4 \\ 2 \end{pmatrix}$ **d** $\begin{pmatrix} 4 \\ -2 \end{pmatrix}$

e $\begin{pmatrix} -4 \\ -2 \end{pmatrix}$ **f** $\begin{pmatrix} -4 \\ 0 \end{pmatrix}$ **g** $\begin{pmatrix} 0 \\ -4 \end{pmatrix}$ **h** $\begin{pmatrix} 0 \\ 4 \end{pmatrix}$

3 Vector $\mathbf{a} = \begin{pmatrix} 5 \\ 2 \end{pmatrix}$

Draw each vector on a grid.

a a **b** 2a **c** 3a **d** –3a

4 The vector **b** is $\begin{pmatrix} 4 \\ -3 \end{pmatrix}$.

Write as a column vector:

a 2b **b** 5b **c** –4b **d** $\frac{1}{2}$b

5 The vectors **a**, **b** and **a** + **b** are shown on the grid.

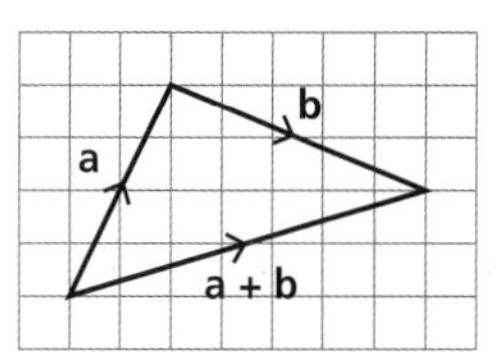

a Write **a** as a column vector.

b Write **b** as a column vector.

c Write **a** + **b** as a column vector.

6 a = $\begin{pmatrix} 1 \\ 3 \end{pmatrix}$ b = $\begin{pmatrix} 2 \\ -4 \end{pmatrix}$ c = $\begin{pmatrix} -1 \\ 4 \end{pmatrix}$

Work out the vector calculations.

a a + b **b** a – b **c** a – c **d** a + c

e b + c **f** b – c **g** c – a **h** c – b

7 a = $\begin{pmatrix} 2 \\ 4 \end{pmatrix}$ b = $\begin{pmatrix} -3 \\ 5 \end{pmatrix}$

Write as column vectors:

a a + b **b** 2a + b **c** a + 2b **d** 3a + 2b

e –3a + 2b **f** 2b – 3a **g** 5b – 3a **h** –5b – 3a

8 The vectors **a** and **b** are shown on the grid.

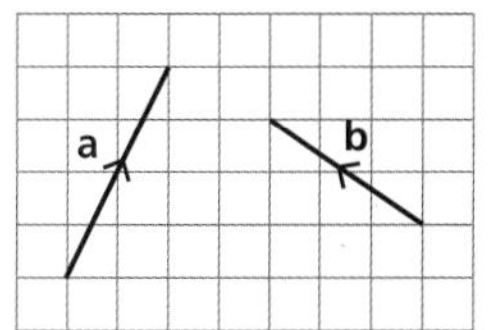

a Draw the vector **a** + **b** on a grid.

b Work out 2**a** – 3**b** as a column vector.

9 The vectors **a** and **b** are defined as follows:

a = $\begin{pmatrix} 2 \\ 4 \end{pmatrix}$ **b** = $\begin{pmatrix} k \\ 5 \end{pmatrix}$

Given that 3**a** + **b** = $\begin{pmatrix} 10 \\ 17 \end{pmatrix}$, work out the value of k.

10 A is the point (2, –1) and B is the point (5, 4).

Write each of these as a column vector.

a $\overrightarrow{AB}$ **b** $\overrightarrow{BA}$ **c** $3\overrightarrow{AB}$

11 P is the point (1, 5), Q is the point (3, –4) and R is the point (–2, 1).

Write each of these as a column vector.

a $\overrightarrow{PQ}$ **b** $\overrightarrow{QP}$ **c** $\overrightarrow{QR}$ **d** $2\overrightarrow{QR}$

e $\overrightarrow{QR} + \overrightarrow{PQ}$ **f** $\overrightarrow{QR} - \overrightarrow{PQ}$ **g** $2\overrightarrow{QR} - \overrightarrow{PQ}$

12 Given that $\begin{pmatrix} 2 \\ q \end{pmatrix} + 3\begin{pmatrix} p \\ -10 \end{pmatrix} = \begin{pmatrix} 2 \\ -15 \end{pmatrix}$, work out the values of p and q.

What do you think?

1 The vector **a** is shown.

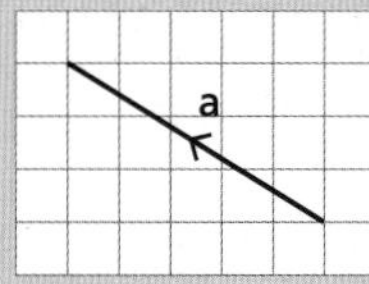

Which of these column vectors are parallel to **a**?

$\begin{pmatrix} 5 \\ -3 \end{pmatrix}$ $\begin{pmatrix} -5 \\ -3 \end{pmatrix}$ $\begin{pmatrix} -10 \\ 6 \end{pmatrix}$ $\begin{pmatrix} 20 \\ -12 \end{pmatrix}$

Write three more vectors that are parallel to **a**.

2 The points A, B and C are plotted on a grid.

$\overrightarrow{AB} = \begin{pmatrix} 0 \\ -5 \end{pmatrix}$ and $\overrightarrow{BC} = \begin{pmatrix} 4 \\ 2 \end{pmatrix}$

What type of triangle is created when you join the points A, B and C?

3 Ed has drawn the vector $\begin{pmatrix} 5 \\ -3 \end{pmatrix}$ in the way shown.

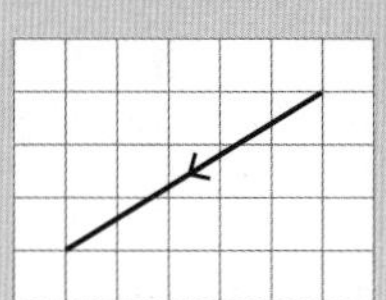

Ed's diagram is incorrect.

What mistake has he made?

Correctly draw the vector $\begin{pmatrix} 5 \\ -3 \end{pmatrix}$.

4 The diagram shows the vectors **a**, **b**, **c** and **d**.

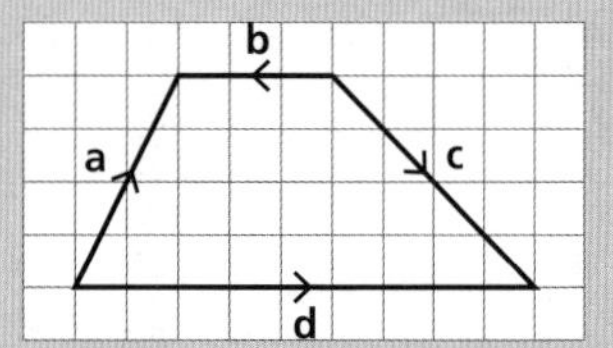

Decide if the statements below are **true** or **false**.

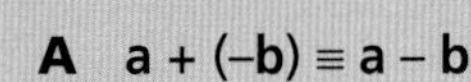

A $\mathbf{a} + (-\mathbf{b}) \equiv \mathbf{a} - \mathbf{b}$

B $\mathbf{a} - \mathbf{b} \equiv \mathbf{d} - \mathbf{c}$

C $\mathbf{a} + \mathbf{d} \equiv \mathbf{b} + \mathbf{c}$

D $\mathbf{a} + \mathbf{c} \equiv \mathbf{b} + \mathbf{d}$

E $\mathbf{b} - \mathbf{c} \equiv \mathbf{d} - \mathbf{a}$

Consolidate – do you need more?

1 Write each of these as a column vector.

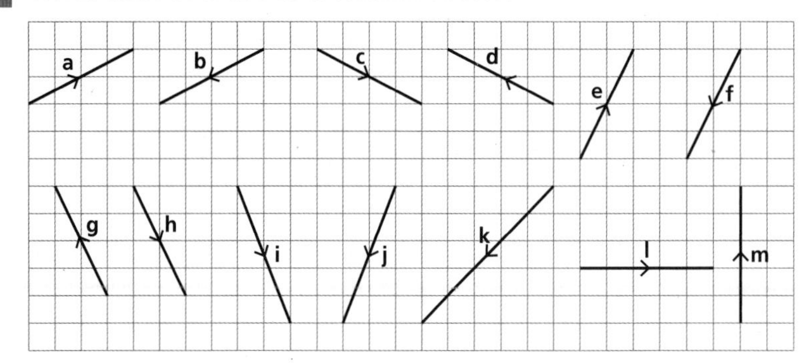

2 Draw each vector on a grid.

a $\begin{pmatrix} 3 \\ 1 \end{pmatrix}$ **b** $\begin{pmatrix} 3 \\ 2 \end{pmatrix}$ **c** $\begin{pmatrix} 5 \\ 2 \end{pmatrix}$ **d** $\begin{pmatrix} 5 \\ -2 \end{pmatrix}$

e $\begin{pmatrix} 5 \\ -3 \end{pmatrix}$ **f** $\begin{pmatrix} -5 \\ -3 \end{pmatrix}$ **g** $\begin{pmatrix} -5 \\ 3 \end{pmatrix}$ **h** $\begin{pmatrix} -5 \\ 0 \end{pmatrix}$

3 The vector **a** is $\begin{pmatrix} 2 \\ 3 \end{pmatrix}$.

Draw these vectors on a grid.

a a **b** 2a **c** 3a **d** 5a

4 The vector **b** is $\begin{pmatrix} -2 \\ 3 \end{pmatrix}$.

Write each of these as a column vector.

a 2b **b** 3b **c** 4b **d** 5b

5 a = $\begin{pmatrix} 2 \\ 4 \end{pmatrix}$ b = $\begin{pmatrix} -3 \\ 5 \end{pmatrix}$

Write each of these as a column vector.

a 2a **b** 2a + b **c** 2a + 2b **d** 2a + 3b

e 2a – 3b **f** 3b – 2a **g** 3b – a **h** –3b + a

Stretch – can you deepen your learning?

1 Given that $m\begin{pmatrix} 3 \\ 2 \end{pmatrix} + n\begin{pmatrix} 2 \\ -4 \end{pmatrix} = \begin{pmatrix} 18 \\ -20 \end{pmatrix}$, work out the values of m and n.

2 On a grid, $\overrightarrow{AB} = \begin{pmatrix} -2 \\ 8 \end{pmatrix}$.

Work out the distance between the points A and B. Give your answer in the form $a\sqrt{b}$, where a and b are integers.

3 On a grid, $\overrightarrow{PQ} = \begin{pmatrix} m \\ -3 \end{pmatrix}$. The distance between P and Q is $3\sqrt{5}$

Work out all the possible values for m.

8.2 Vector proof

White Rose MATHS

Are you ready? (A)

1 Simplify:

a $a + 3a - b$ **b** $-a + 3a - b$ **c** $-a - 3a - b$ **d** $-b - 3a - b$

2 Multiply out:

a $-2(a + b)$ **b** $-2(3a + b)$ **c** $-2(3a - b)$ **d** $-2(3a - 5b)$

3 Expand:

a $-(a + b)$ **b** $-(a - b)$ **c** $-(2a - b)$ **d** $-(-2a - b)$

4 Point X lies on the line segment AB so that AX : XB = 4 : 3

What fraction of AB is AX?

A **vector** has both magnitude (size) and direction.

Vectors can be written using the notation $\overrightarrow{XY}$ to show the start and end points, in this case 'the vector from X to Y'.

Vectors can also be expressed using single letters, such as **a**, and this is useful for comparing them.

For example, 3**a** is the vector in the same direction as **a** but three times as long.

–**a** is the same length as **a** but in the opposite direction.

Example 1

ABC is a triangle.

$\overrightarrow{AB} = \mathbf{p}$ and $\overrightarrow{BC} = 2\mathbf{p} - \mathbf{q}$

a Write the vector $\overrightarrow{AC}$ in terms of **p** and **q**.

b Write the vector $\overrightarrow{CA}$ in terms of **p** and **q**.

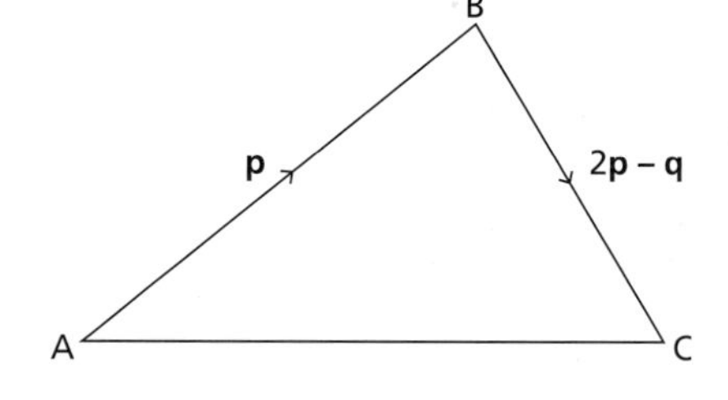

Method

Solution	Commentary
a $\overrightarrow{AC} = \overrightarrow{AB} + \overrightarrow{BC}$ $= \mathbf{p} + 2\mathbf{p} - \mathbf{q}$ $= 3\mathbf{p} - \mathbf{q}$	Write the journey from A to C in terms of vectors you know. Write the vectors $\overrightarrow{AB}$ and $\overrightarrow{BC}$ in terms of **p** and **q**. Add the two expressions and simplify in the same way as collecting like terms in algebraic expressions.
b $\overrightarrow{CA} = \overrightarrow{CB} + \overrightarrow{BA}$ $= -(2\mathbf{p} - \mathbf{q}) + -(\mathbf{p})$ $= -2\mathbf{p} + \mathbf{q} - \mathbf{p}$ $= -3\mathbf{p} + \mathbf{q}$	Write the vectors $\overrightarrow{CB}$ and $\overrightarrow{BA}$ in terms of **p** and **q**. $\overrightarrow{CB} = -\overrightarrow{BC}$ and $\overrightarrow{BA} = -\overrightarrow{AB}$. Add the two expressions and simplify. Note also that $\overrightarrow{CA} = -\overrightarrow{AC}$, so you could just use your answer to part **a** as $-(3\mathbf{p} - \mathbf{q}) = -3\mathbf{p} + \mathbf{q}$

Example 2

PQR is a triangle.

M is the midpoint of QR.

a Write $\overrightarrow{QR}$ in terms of **a** and **b**.

b Write $\overrightarrow{QM}$ in terms of **a** and **b**.

c Write $\overrightarrow{PM}$ in terms of **a** and **b**.

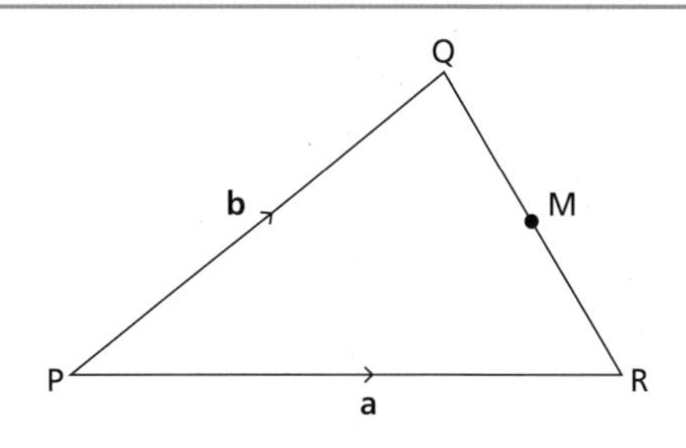

Method

Solution	Commentary
a $\overrightarrow{QR} = \overrightarrow{QP} + \overrightarrow{PR}$ $= -\mathbf{b} + \mathbf{a}$ $= \mathbf{a} - \mathbf{b}$	Add the vectors $\overrightarrow{QP}$ and $\overrightarrow{PR}$ to work out $\overrightarrow{QR}$. Note that $-\mathbf{b} + \mathbf{a}$ is the same as $\mathbf{a} - \mathbf{b}$, so you could give the answer in either form.
b $\overrightarrow{QM} = \frac{1}{2}\overrightarrow{QR}$ $= \frac{1}{2}(\mathbf{a} - \mathbf{b})$	M is the midpoint of QR, so $\overrightarrow{QM}$ is half of $\overrightarrow{QR}$. Multiply the expression from part **a** by $\frac{1}{2}$ You could also write the answer as $\frac{1}{2}\mathbf{a} - \frac{1}{2}\mathbf{b}$.
c $\overrightarrow{PM} = \overrightarrow{PQ} + \overrightarrow{QM}$ $= \mathbf{b} + \frac{1}{2}\mathbf{a} - \frac{1}{2}\mathbf{b}$ $= \frac{1}{2}\mathbf{a} + \frac{1}{2}\mathbf{b}$	Add the vectors $\overrightarrow{PQ}$ (= **b**) and $\overrightarrow{QM}$ $\left(\frac{1}{2}\mathbf{a} - \frac{1}{2}\mathbf{b}\right)$. Simplify the expression.

Example 3

ABC is a triangle.

The point P lies on the line BC such that BP : PC = 2 : 1

a Write $\overrightarrow{BP}$ in terms of **a** and **b**.

b Write $\overrightarrow{AP}$ in terms of **a** and **b**.

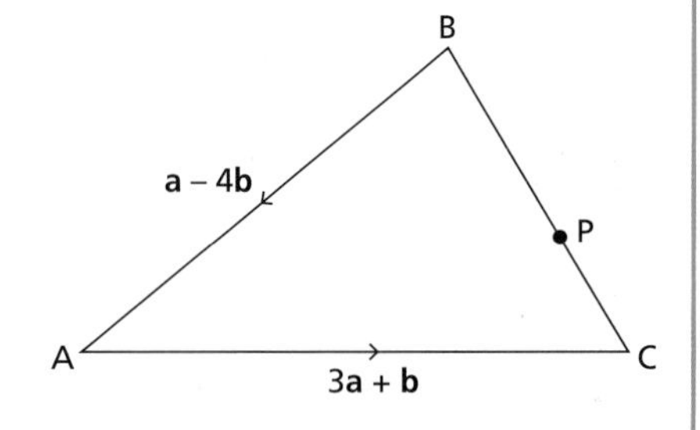

Method

Solution	Commentary
a $\overrightarrow{BP} = \frac{2}{3}\overrightarrow{BC}$ $\overrightarrow{BC} = \overrightarrow{BA} + \overrightarrow{AC}$ $= \mathbf{a} - 4\mathbf{b} + 3\mathbf{a} + \mathbf{b}$ $= 4\mathbf{a} - 3\mathbf{b}$ $\overrightarrow{BP} = \frac{2}{3}(4\mathbf{a} - 3\mathbf{b}) = \frac{8}{3}\mathbf{a} - 2\mathbf{b}$	Using the ratio 2 : 1, $\overrightarrow{BP}$ is $\frac{2}{3}\overrightarrow{BC}$. Write $\overrightarrow{BC}$ in terms of **a** and **b**. Multiply the expression for $\overrightarrow{BC}$ by $\frac{2}{3}$ to work out $\overrightarrow{BP}$.
b $\overrightarrow{AP} = \overrightarrow{AB} + \overrightarrow{BP}$ $\overrightarrow{AB} = -(\mathbf{a} - 4\mathbf{b}) = -\mathbf{a} + 4\mathbf{b}$ $\overrightarrow{AP} = -\mathbf{a} + 4\mathbf{b} + \frac{2}{3}(4\mathbf{a} - 3\mathbf{b})$ $= \frac{5}{3}\mathbf{a} + 2\mathbf{b}$	Add the vectors $\overrightarrow{AB}$ and $\overrightarrow{BP}$. Use the fact that $\overrightarrow{AB} = -\overrightarrow{BA}$. Simplify the expression.

Practice (A)

1 ABCD is a parallelogram.

Which statements are **true**?

$\overrightarrow{AB} = \overrightarrow{DC}$ $\overrightarrow{BC} = \overrightarrow{DA}$ $\overrightarrow{AD} = \overrightarrow{BC}$ $\overrightarrow{AC} = \overrightarrow{DB}$

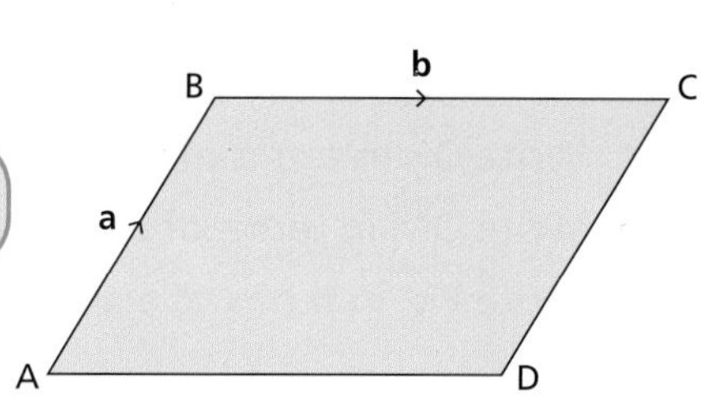

2 On the grid of congruent parallelograms, $\overrightarrow{OA} = \mathbf{a}$ and $\overrightarrow{OB} = \mathbf{b}$.

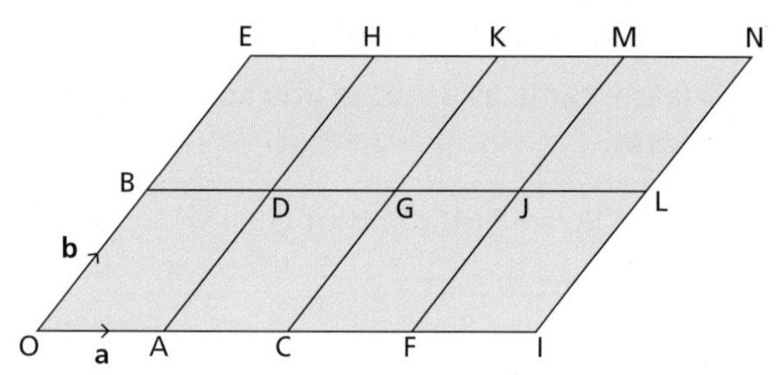

Write the vectors in terms of **a** and **b**.

a $\overrightarrow{OA}$ **b** $\overrightarrow{OB}$ **c** $\overrightarrow{OE}$ **d** $\overrightarrow{OF}$

e $\overrightarrow{OD}$ **f** $\overrightarrow{OH}$ **g** $\overrightarrow{HJ}$ **h** $\overrightarrow{JA}$

3 Here is the quadrilateral ABCD.

a Write the vector $\overrightarrow{AC}$ in terms of **n** and **o**.

b Write the vector $\overrightarrow{AC}$ in terms of **m** and **p**.

c Copy and complete the vector journeys for $\overrightarrow{DB}$.

$\overrightarrow{DB} = \overrightarrow{DA} + \overrightarrow{AB} =$ ______

$\overrightarrow{DB} = \overrightarrow{DC} + \overrightarrow{CB} =$ ______

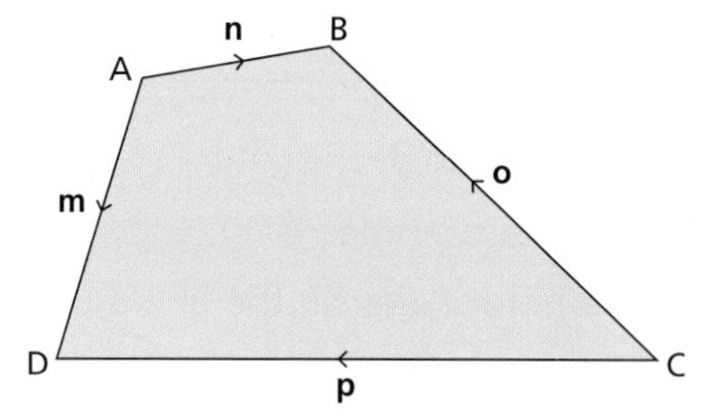

4 PQR is a triangle.

$\overrightarrow{PR} = 2\mathbf{a}$ and $\overrightarrow{RQ} = \mathbf{b}$

Write in terms of **a** and **b**:

a $\overrightarrow{PQ}$ **b** $\overrightarrow{QP}$

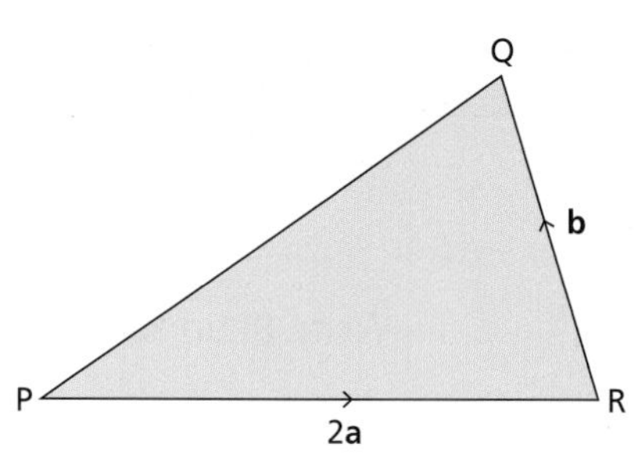

5 ABC is a triangle.

$\overrightarrow{AB} = 2\mathbf{a} + \mathbf{b}$ and $\overrightarrow{BC} = \mathbf{a} - 3\mathbf{b}$

Write in terms of **a** and **b**:

a $\overrightarrow{AC}$ **b** $\overrightarrow{CA}$

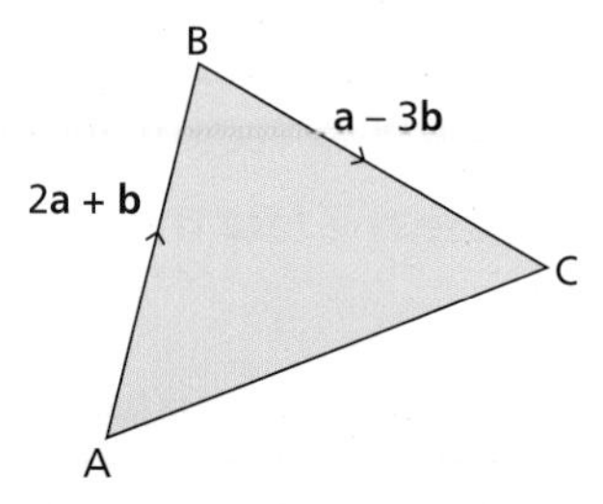

6 Write the vectors $\overrightarrow{AC}$ in terms of $\mathbf{a}$ and $\mathbf{b}$.

a

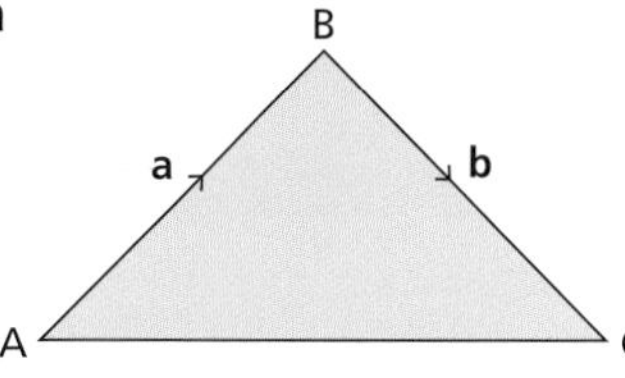

b

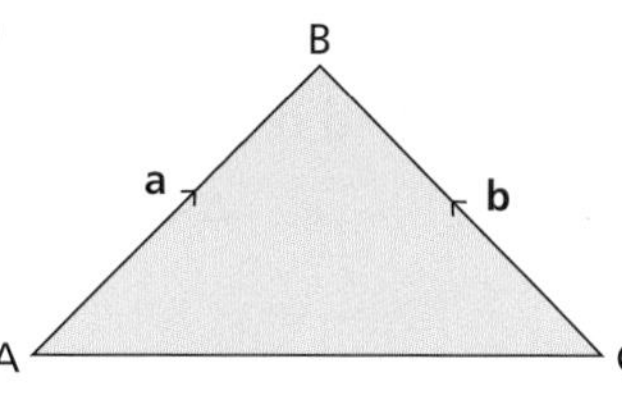

c

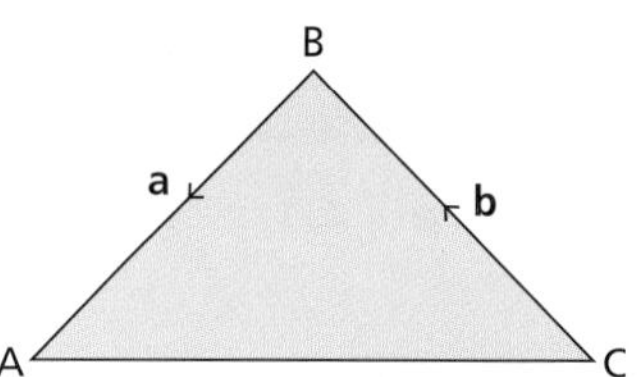

7 LMN is a triangle.

$\overrightarrow{ML} = 3\mathbf{a}$ and $\overrightarrow{NL} = \mathbf{a} + \mathbf{b}$

Write in terms of $\mathbf{a}$ and $\mathbf{b}$:

a $\overrightarrow{LM}$ **b** $\overrightarrow{NM}$

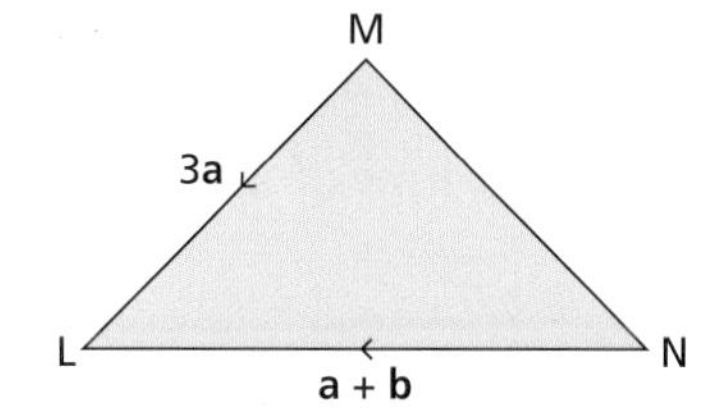

8 WXYZ is a quadrilateral.

Write the vector $\overrightarrow{WZ}$ in terms of $\mathbf{a}$ and $\mathbf{b}$.

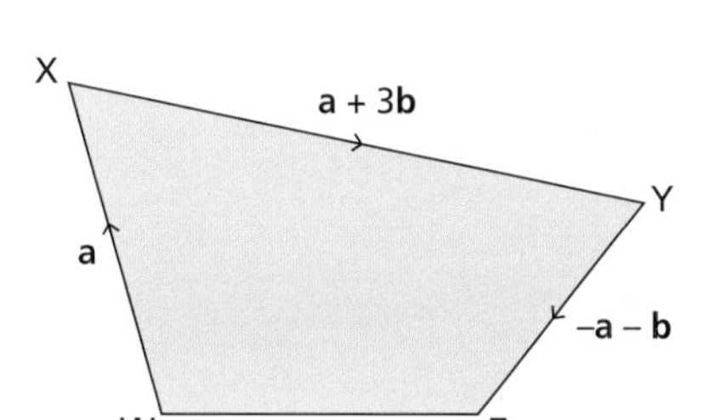

9 ABC is a triangle.

The point M is the midpoint of BC.

Write the vectors in terms of $\mathbf{a}$ and $\mathbf{b}$.

a $\overrightarrow{BC}$ **b** $\overrightarrow{BM}$ **c** $\overrightarrow{AM}$

$\overrightarrow{BM}$ is $\frac{1}{2}$ of $\overrightarrow{BC}$.

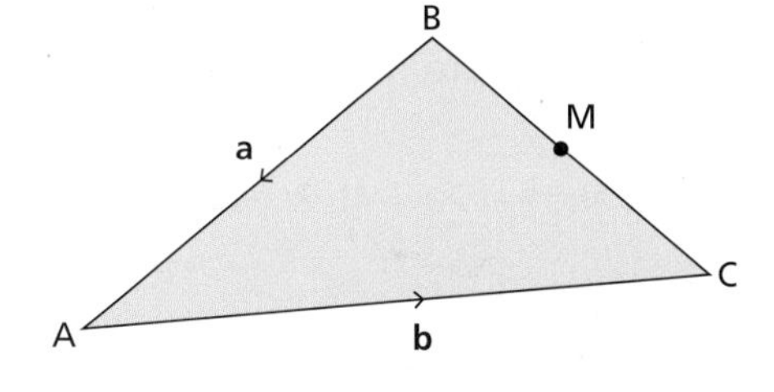

10 PQR is a triangle.

The point M is the midpoint of QR.

a Write $\overrightarrow{QM}$ in terms of $\mathbf{a}$ and $\mathbf{b}$.

b Write $\overrightarrow{PM}$ in terms of $\mathbf{a}$ and $\mathbf{b}$.

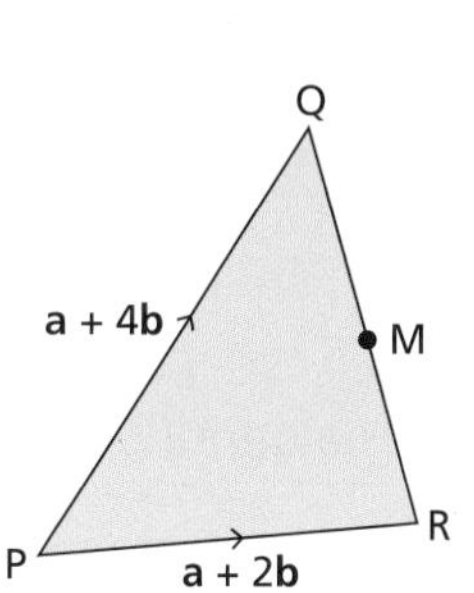

11 ABC is a triangle.

The point M is the midpoint of BA.

a Write $\overrightarrow{BM}$ in terms of $\mathbf{a}$ and $\mathbf{b}$.

b Write $\overrightarrow{CM}$ in terms of $\mathbf{a}$ and $\mathbf{b}$.

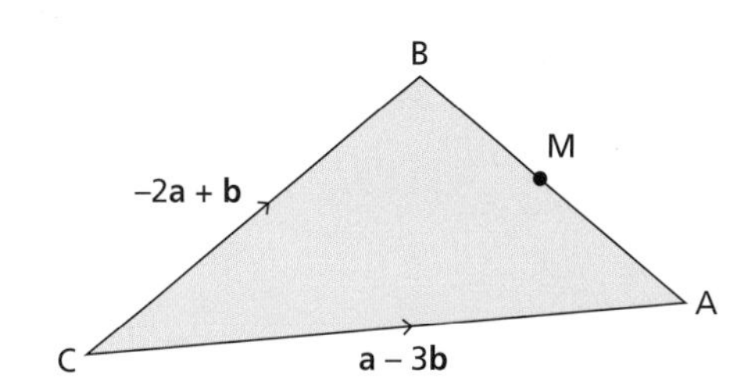

12 ABCD is a trapezium.

$\overrightarrow{AD} = 3\overrightarrow{BC}$

Write the vectors in terms of **a** and **b**.

a $\overrightarrow{AD}$ **b** $\overrightarrow{AC}$ **c** $\overrightarrow{AB}$

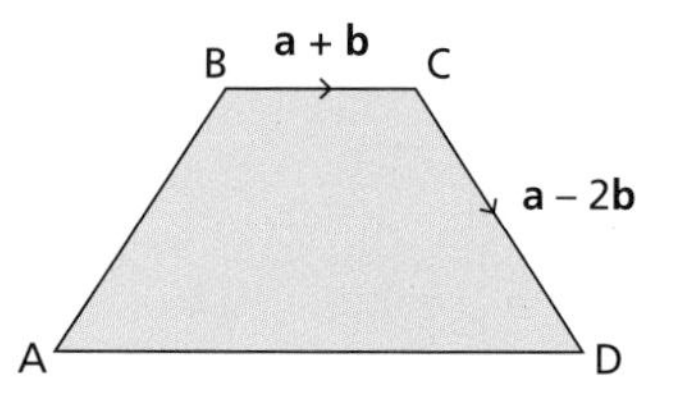

13 $\overrightarrow{OA} = \mathbf{x}$

$\overrightarrow{OB} = \mathbf{y}$

N is the point on AB such that AN : NB = 1 : 3

Write $\overrightarrow{ON}$ in terms of **x** and **y**.

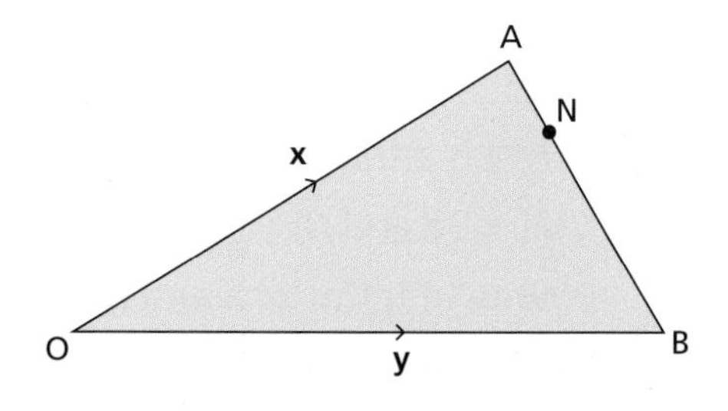

Are you ready? (B)

1 Factorise:

a $3a + 3b$ **b** $10a - 16b$ **c** $3a - 27b$ **d** $-20a + 12b$

2 Expand:

a $2(a + b)$ **b** $3(2a - b)$ **c** $-2(2a + 4b)$ **d** $-3(12a - 15b)$

Vectors that are multiples of each other are parallel, even if they involve more than one term.

For example, 2**a** + 4**b** is parallel to **a** + 2**b** as 2**a** + 4**b** = 2(**a** + 2**b**)

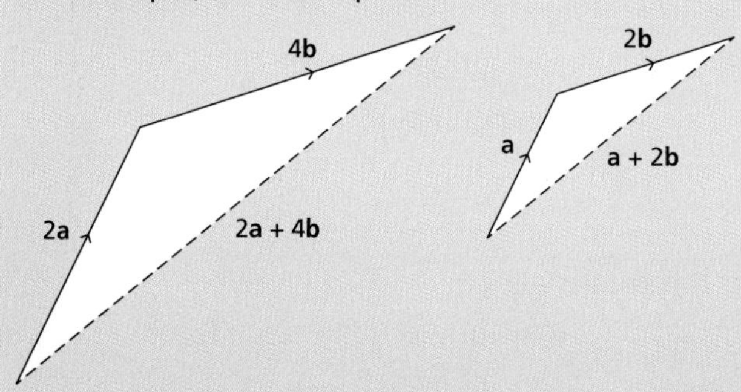

Factorising vectors expressed as algebraic expressions is useful when working out if two (or more) vectors are parallel.

For example, 4**a** + 6**b** = 2(2**a** + 3**b**)

Therefore, the vector 4**a** + 6**b** is parallel to the vector 2**a** + 3**b**.

Example

$\overrightarrow{AB} = \mathbf{a} + 2\mathbf{b}$ and $\overrightarrow{BC} = \mathbf{a} - 4\mathbf{b}$

Show that $\overrightarrow{AC}$ is parallel to $\mathbf{a} - \mathbf{b}$.

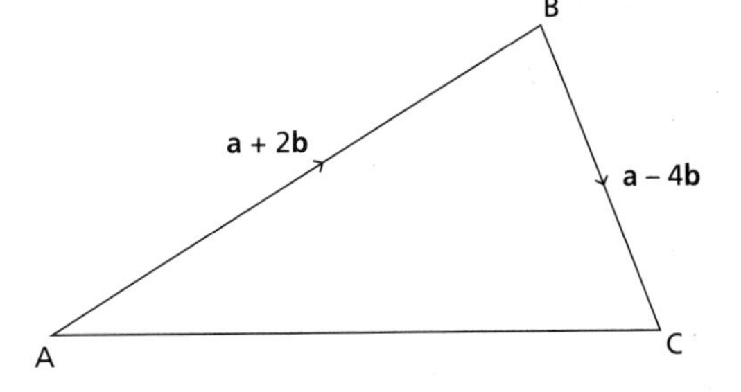

Method

Solution	Commentary
$\overrightarrow{AB} + \overrightarrow{BC} = \overrightarrow{AC}$ $\mathbf{a} + 2\mathbf{b} + \mathbf{a} - 4\mathbf{b} = 2\mathbf{a} - 2\mathbf{b}$	Work out $\overrightarrow{AC}$, by adding $\overrightarrow{AB}$ and $\overrightarrow{BC}$ in terms of **a** and **b**.
$2\mathbf{a} - 2\mathbf{b} \equiv 2(\mathbf{a} - \mathbf{b})$	Factorise $2\mathbf{a} - 2\mathbf{b}$ with a common factor of 2 Writing as $2(\mathbf{a} - \mathbf{b})$ proves that $\overrightarrow{AC}$ is parallel to $\mathbf{a} - \mathbf{b}$, since $2(\mathbf{a} - \mathbf{b})$ is a multiple of $\mathbf{a} - \mathbf{b}$.

Practice (B)

1. $\overrightarrow{AB} = 4\mathbf{a} + 6\mathbf{b}$ and $\overrightarrow{CD} = 10\mathbf{a} + 15\mathbf{b}$

   Show that $\overrightarrow{AB}$ and $\overrightarrow{CD}$ are parallel.

2. $\overrightarrow{PQ} = \mathbf{a} + 3\mathbf{b}$ and $\overrightarrow{RS} = 4\mathbf{a} + 15\mathbf{b}$

   Show that $\overrightarrow{PQ}$ and $\overrightarrow{RS}$ are **not** parallel.

3. ABC is a triangle.

   a Write $\overrightarrow{AC}$ in terms of **a** and **b**.

   b Show that $\overrightarrow{AC}$ is parallel to $2\mathbf{a} - \mathbf{b}$.

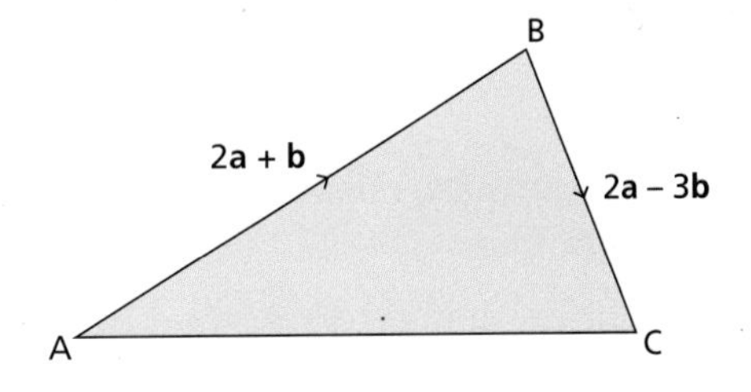

4. PQR is a triangle.

   Show that $\overrightarrow{PR}$ is parallel to $\mathbf{a} - \mathbf{b}$.

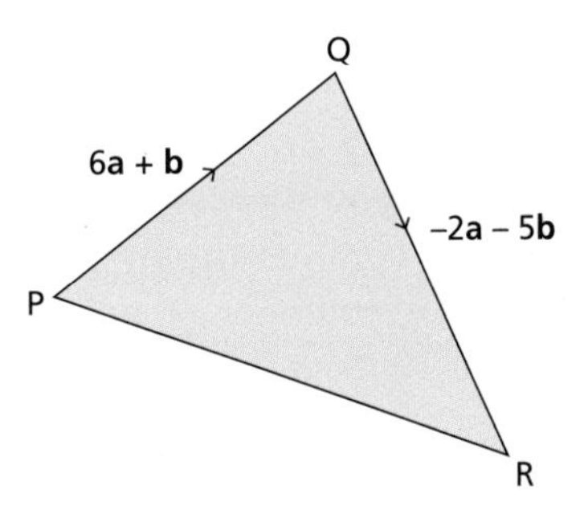

5 OAB is a triangle.

M is the midpoint of OA.

N is the midpoint of OB.

$\overrightarrow{OM} = \mathbf{a}$

$\overrightarrow{ON} = \mathbf{b}$

Show that AB is parallel to MN.

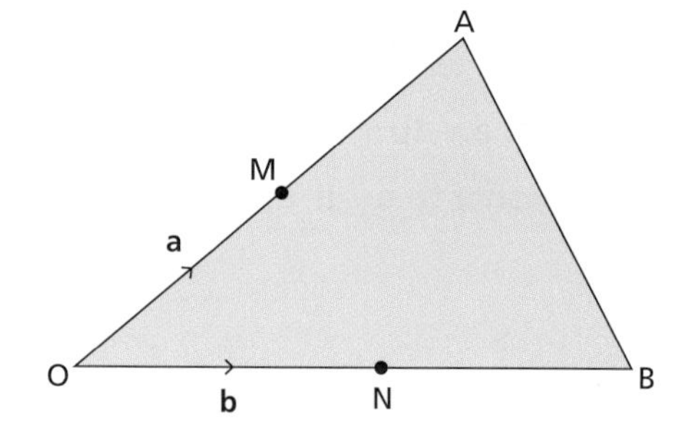

What do you think?

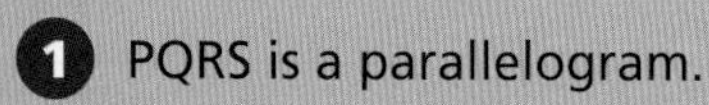

1 PQRS is a parallelogram.

M is the midpoint of PS and N is the midpoint of RS.

Is PR parallel to MN? Use vectors to prove your answer.

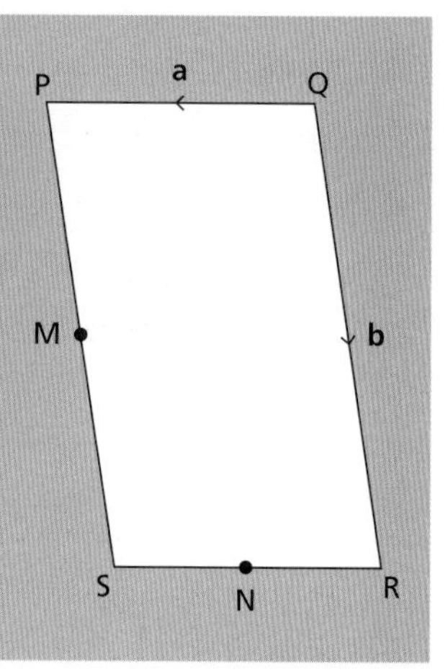

Are you ready? (C)

1 $\overrightarrow{AB} = 2\mathbf{a} - 3\mathbf{b}$

Write the vectors in terms of **a** and **b**.

a $2\overrightarrow{AB}$ **b** $5\overrightarrow{AB}$ **c** $-3\overrightarrow{AB}$ **d** $\frac{1}{2}\overrightarrow{AB}$

Three or more points are **collinear** if they all lie on the same straight line.

The diagram shows the vectors $\overrightarrow{AB}$ and $\overrightarrow{BC}$.

$\overrightarrow{AB} = \begin{pmatrix} 3 \\ 2 \end{pmatrix}$ and $\overrightarrow{BC} = \begin{pmatrix} 6 \\ 4 \end{pmatrix}$

The vectors are parallel as $2 \times \begin{pmatrix} 3 \\ 2 \end{pmatrix} = \begin{pmatrix} 6 \\ 4 \end{pmatrix}$

The vectors also share a common point B, therefore the points A, B and C are collinear.

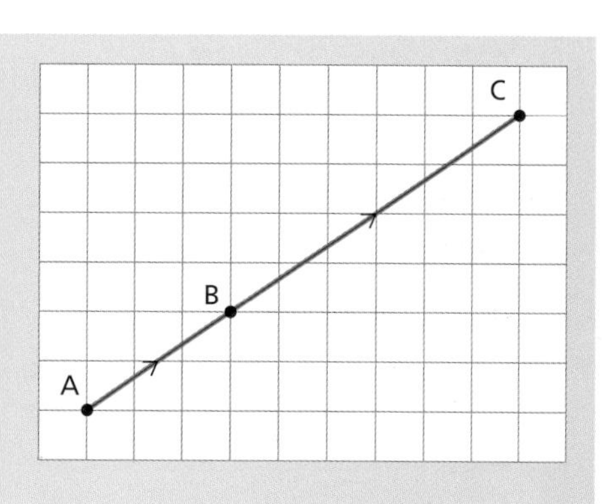

Example 1

$\overrightarrow{AC} = 4\mathbf{a} - 8\mathbf{b}$ and $\overrightarrow{BC} = 6\mathbf{a} - 12\mathbf{b}$

Prove that the points A, B and C lie on a straight line.

Method

Solution	Commentary
$\overrightarrow{AC} = 4\mathbf{a} - 8\mathbf{b} \equiv 4(\mathbf{a} - 2\mathbf{b})$ $\overrightarrow{BC} = 6\mathbf{a} - 12\mathbf{b} \equiv 6(\mathbf{a} - 2\mathbf{b})$	Factorise $\overrightarrow{AC}$ and $\overrightarrow{BC}$.
$\overrightarrow{AC}$ and $\overrightarrow{BC}$ are both multiples of $\mathbf{a} - 2\mathbf{b}$ and both pass through the point C.	Doing this will prove that the vectors are parallel as they are both multiples of $\mathbf{a} - 2\mathbf{b}$. As both vectors pass through the point C, this proves that they are both on the same straight line. Therefore, the points A, B and C are collinear.

Example 2

In this diagram of congruent parallelograms, $\overrightarrow{BE} = \mathbf{a}$ and $\overrightarrow{BC} = \mathbf{b}$

Prove that A, E and I are collinear.

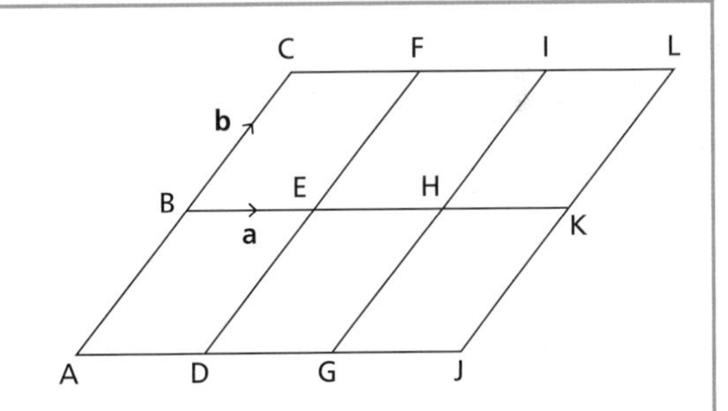

Method

Solution	Commentary
$\overrightarrow{AE} = \overrightarrow{AB} + \overrightarrow{BE}$ $= \mathbf{b} + \mathbf{a}$ $\overrightarrow{AI} = \overrightarrow{AC} + \overrightarrow{CI}$ $= 2\mathbf{b} + 2\mathbf{a}$	Write $\overrightarrow{AE}$ and $\overrightarrow{AI}$ in terms of **a** and **b**.
$2\mathbf{b} + 2\mathbf{a} = 2(\mathbf{b} + \mathbf{a})$ $\overrightarrow{AE}$ and $\overrightarrow{AI}$ are parallel and both pass through the point A.	Factorise the expression for $\overrightarrow{AI}$ to show that $\overrightarrow{AE}$ and $\overrightarrow{AI}$ are parallel.
Therefore, A, E and I are collinear.	The points A, E and I are collinear as both vectors are parallel, and both vectors pass through the point A.

Practice (C)

1. $\overrightarrow{AC} = 2\mathbf{a} - 3\mathbf{b}$ and $\overrightarrow{BC} = 8\mathbf{a} - 12\mathbf{b}$

   Prove that the points A, B and C lie on a straight line.

2 The vectors **x** and **y** are shown.

$\overrightarrow{AB} = \mathbf{x} + 2\mathbf{y}$

a Copy the grid and draw the vector $\overrightarrow{AB}$.

b Write $\overrightarrow{AB}$ as a column vector.

$\overrightarrow{AC} = 2\mathbf{x} + 4\mathbf{y}$

c Copy the grid and draw the vector $\overrightarrow{AC}$.

d Write $\overrightarrow{AC}$ as a column vector.

e Prove that the points A, B and C lie on a straight line.

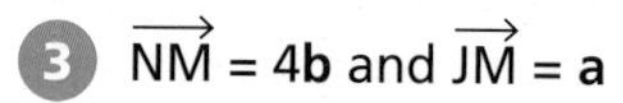

3 $\overrightarrow{NM} = 4\mathbf{b}$ and $\overrightarrow{JM} = \mathbf{a}$

a Write $\overrightarrow{NJ}$ in terms of **a** and **b**.

N, J and K are collinear.

$\overrightarrow{NK} = 3 \times \overrightarrow{NJ}$

b Write $\overrightarrow{NK}$ in terms of **a** and **b**.

c Write $\overrightarrow{MK}$ in terms of **a** and **b**.

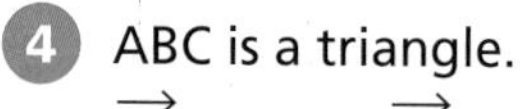

4 ABC is a triangle.

$\overrightarrow{AB} = \mathbf{a}$ and $\overrightarrow{AC} = 2\mathbf{b}$

M is the midpoint of AB.

N is the point on BC such that BN : NC = 3 : 1

D is the point such that $\overrightarrow{CD} = \frac{1}{2}\overrightarrow{AC}$

a Write the vector $\overrightarrow{MN}$ in terms of **a** and **b**.

b Use vectors to prove that M, N and D lie on the same straight line.

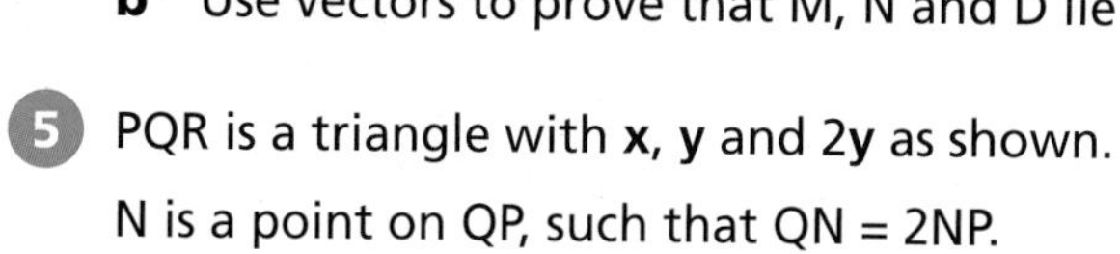

5 PQR is a triangle with **x**, **y** and 2**y** as shown.

N is a point on QP, such that QN = 2NP.

a Write $\overrightarrow{PR}$ in terms of **x** and **y**.

R is the midpoint of QS.

M is the midpoint of PR.

b Write $\overrightarrow{NM}$ in terms of **x** and **y**.

c Show that NMS is a straight line.

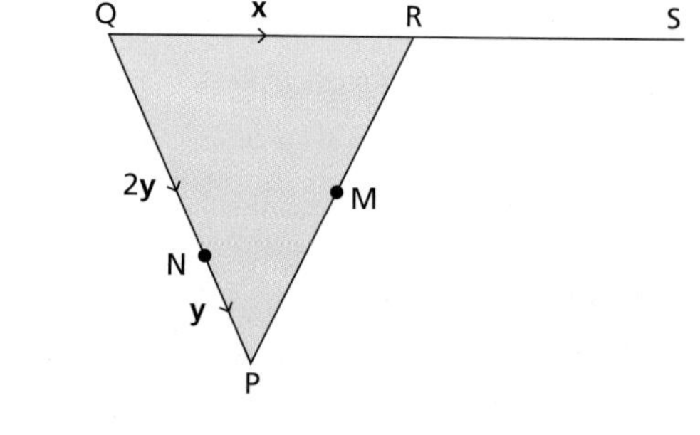

6 OAB is a triangle.

A is the midpoint of OC.

N is the midpoint of AB.

M is on OB such that OM : MB = 2 : 1

$\overrightarrow{NB} = \mathbf{x}$

$\overrightarrow{OA} = \mathbf{y}$

Prove that MNC is a straight line.

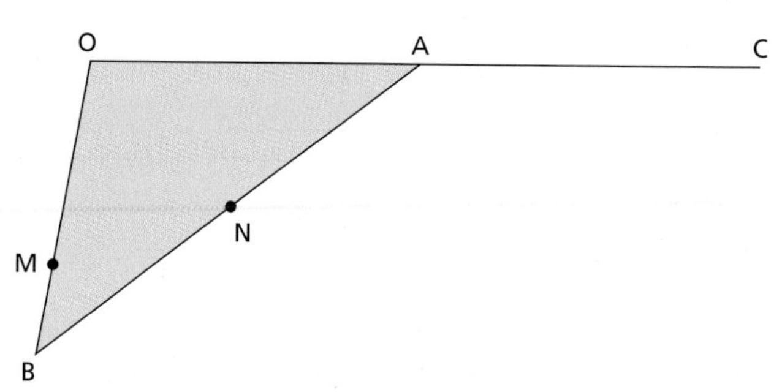

Consolidate – do you need more?

1 On the grid of congruent equilateral triangles, $\overrightarrow{AB} = \mathbf{x}$ and $\overrightarrow{AD} = \mathbf{y}$.

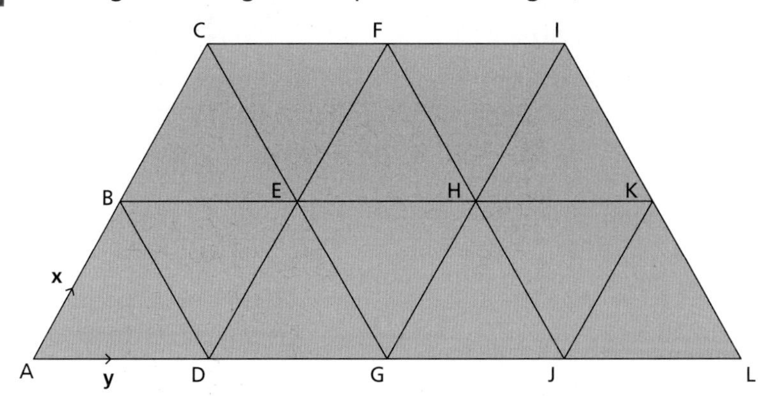

Write the vector journeys in terms of **x** and **y**.

a $\overrightarrow{AB}$ **b** $\overrightarrow{AD}$ **c** $\overrightarrow{AG}$ **d** $\overrightarrow{AC}$

e $\overrightarrow{BE}$ **f** $\overrightarrow{BH}$ **g** $\overrightarrow{BF}$ **h** $\overrightarrow{BI}$

2 For each triangle, write the vector $\overrightarrow{AC}$ in terms of **x** and **y**.

a

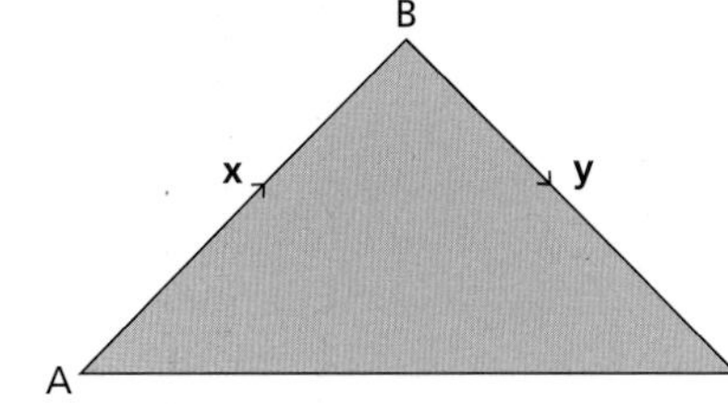

b

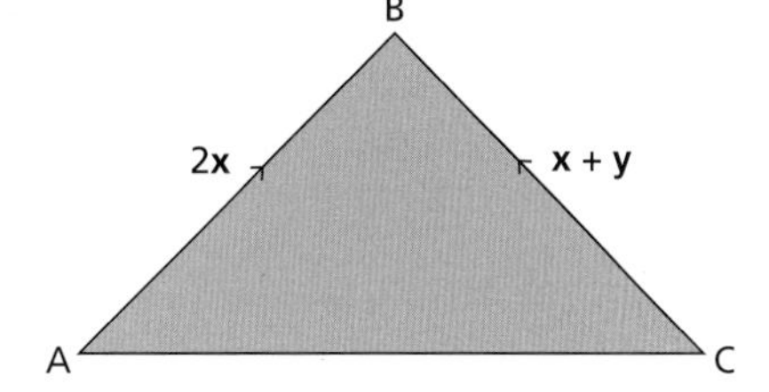

c

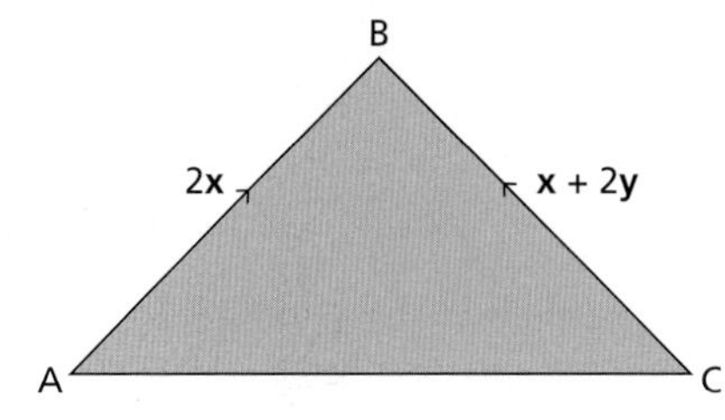

3 $\overrightarrow{AB} = \mathbf{a} + 3\mathbf{b}$ and $\overrightarrow{CD} = 4\mathbf{a} + 12\mathbf{b}$

Show that $\overrightarrow{AB}$ and $\overrightarrow{CD}$ are parallel.

4 $\overrightarrow{AC} = \begin{pmatrix} 1 \\ 2 \end{pmatrix}$ and $\overrightarrow{BC} = \begin{pmatrix} 3 \\ 6 \end{pmatrix}$

Prove that the points A, B and C lie on a straight line.

5 $\overrightarrow{PQ} = \begin{pmatrix} -3 \\ 4 \end{pmatrix}$ and $\overrightarrow{PR} = \begin{pmatrix} -6 \\ 8 \end{pmatrix}$

Prove that the points P, Q and R lie on a straight line.

6 $\overrightarrow{AC} = \mathbf{a} - 2\mathbf{b}$ and $\overrightarrow{BC} = 3\mathbf{a} - 6\mathbf{b}$

Prove that the points A, B and C lie on a straight line.

Stretch – can you deepen your learning?

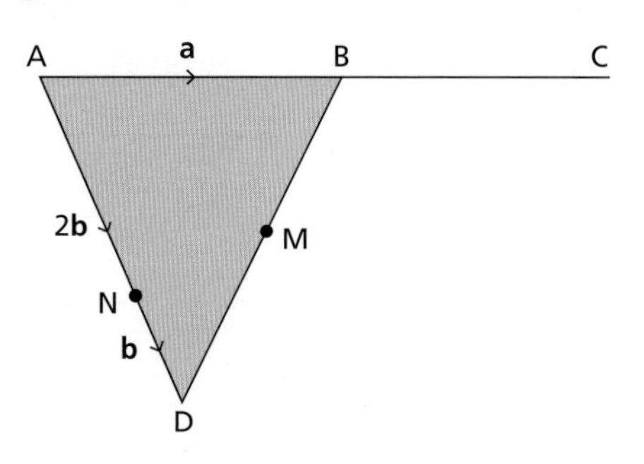

1 ABD is a triangle.

N is a point on AD such that AN : ND = 2 : 1

a Write $\overrightarrow{DB}$ in terms of **a** and **b**.

B is the midpoint of AC.

M is the midpoint of BD.

b Show that NMC is a straight line.

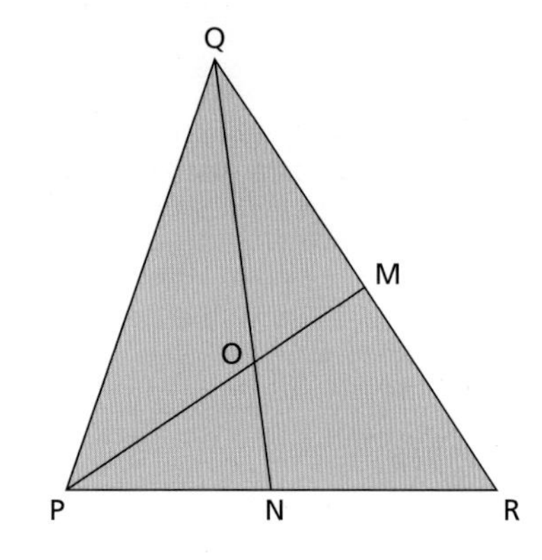

2 The triangle PQR is shown.

M is the midpoint of QR.

$\overrightarrow{PR} = \mathbf{x}$

$\overrightarrow{PQ} = \mathbf{y}$

$\overrightarrow{PO} : \overrightarrow{OM} = 3 : 2$

$\overrightarrow{PN}$ can be written in the form of $k\mathbf{x}$.

Work out the value of k.

Vectors: exam practice

4–6

1 $\mathbf{a} = \begin{pmatrix} 3 \\ 2 \end{pmatrix}$ and $\mathbf{b} = \begin{pmatrix} 5 \\ -1 \end{pmatrix}$

Work out $\mathbf{a} - \mathbf{b}$. **[2 marks]**

2 $\mathbf{p} = \begin{pmatrix} -1 \\ 2 \end{pmatrix}$ $\mathbf{q} = \begin{pmatrix} 2 \\ -3 \end{pmatrix}$ $\mathbf{r} = \begin{pmatrix} -4 \\ 12 \end{pmatrix}$

Show that $\mathbf{r}$ is parallel to $3\mathbf{p} + \mathbf{q}$. **[3 marks]**

3 In a quadrilateral PQRS, $\overrightarrow{PQ} = 2\overrightarrow{RS}$.

What type of quadrilateral must PQRS be? **[1 mark]**

4 A is the point with coordinates (4, –2) and B is the point with coordinates (7, –4).

Work out the vector $3\overrightarrow{AB}$. **[2 marks]**

7–9

5 Given that $\overrightarrow{OP} = 2\mathbf{a} + 3\mathbf{b}$, $\overrightarrow{OQ} = 5\mathbf{a} + 2\mathbf{b}$ and $\overrightarrow{XY} = 9\mathbf{a} - 3\mathbf{b}$, explain fully the geometrical relationships between $\overrightarrow{PQ}$ and $\overrightarrow{XY}$. **[3 marks]**

6 OABC is a parallelogram.

$\overrightarrow{OA} = \mathbf{a}$ and $\overrightarrow{OC} = \mathbf{c}$.

X is the midpoint of AB, and Y is the point that divides CB such that CY : YB = 2 : 1

Show that XY is **not** parallel to AC. **[4 marks]**

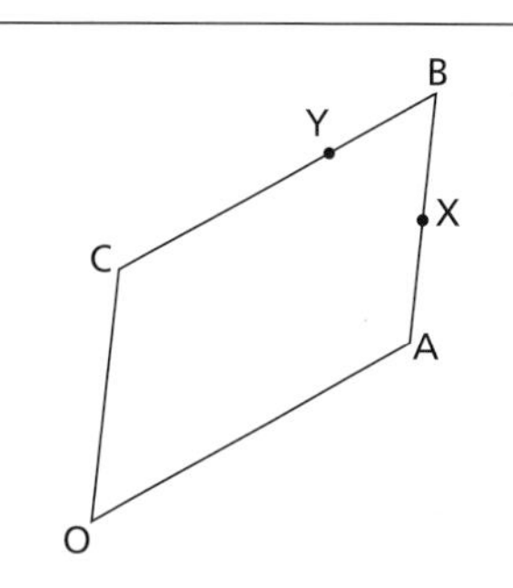

7 $\overrightarrow{OA} = 2\mathbf{a}$ and $\overrightarrow{OB} = 3\mathbf{b}$

X is the point on AB such that AX : XB = 3 : 1

Show that $\overrightarrow{OX} = k(2\mathbf{a} + 9\mathbf{b})$ and work out the value of k. **[4 marks]**

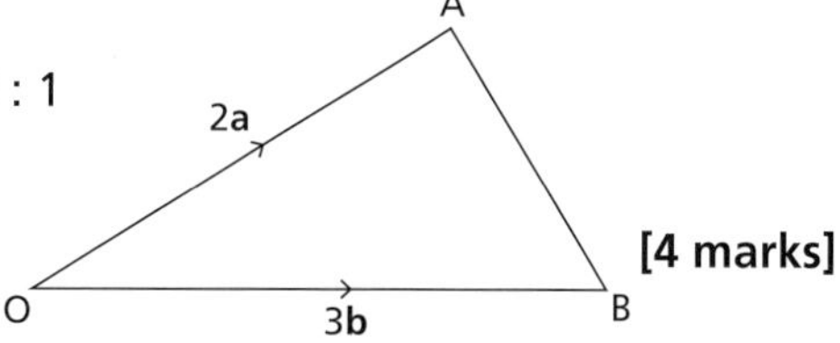

8 $\overrightarrow{OB} = \mathbf{b}$, and $\overrightarrow{OA} = 3\mathbf{a}$

OB : BC = 2 : 1, and OA : AX = 3 : 1

Y is the midpoint of BX.

Show that AYC is a straight line. **[3 marks]**

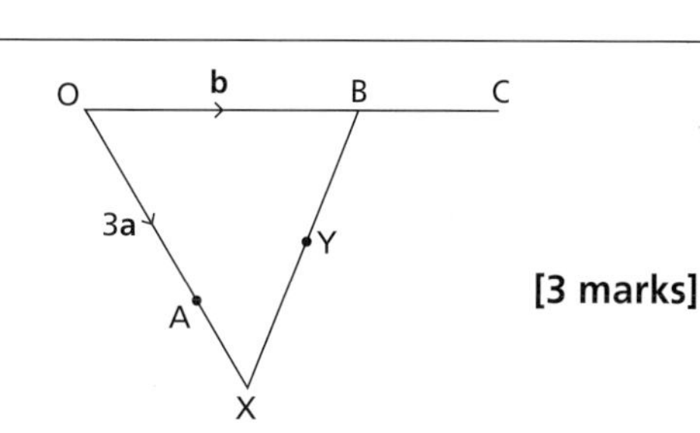

Geometry and measures: exam practice

** Diagrams are not accurately drawn unless stated.*

1 Shade the region that satisfies both conditions on a copy of the diagram.

- The points that are less than 5 cm from B.
- The points that are nearer to A than B.

A

Diagram accurately drawn

B

[3 marks]

2 Two similar triangles are shown.

Calculate the lengths of the sides labelled x and y.

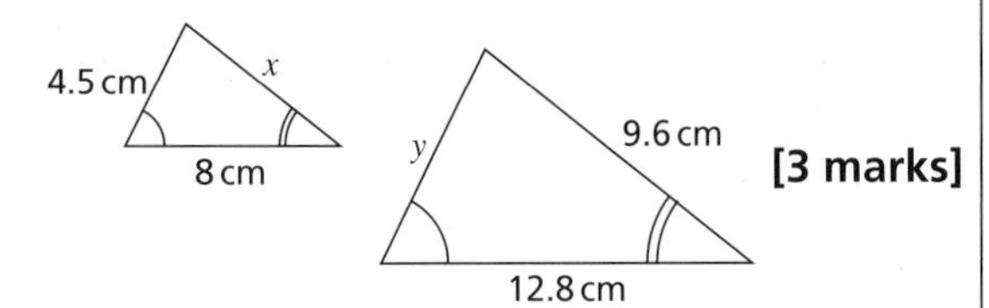

[3 marks]

3 $\mathbf{p} = \begin{pmatrix} 3 \\ -2 \end{pmatrix}$ and $\mathbf{q} = \begin{pmatrix} 5 \\ -1 \end{pmatrix}$

Work out $\mathbf{q} - 2\mathbf{p}$. **[2 marks]**

4 Draw an enlargement of the shape by scale factor 3 on squared paper.

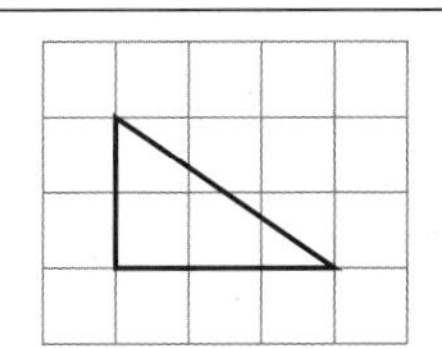

[2 marks]

5 Work out the size of angle EFG.

Give reasons at each stage of your working.

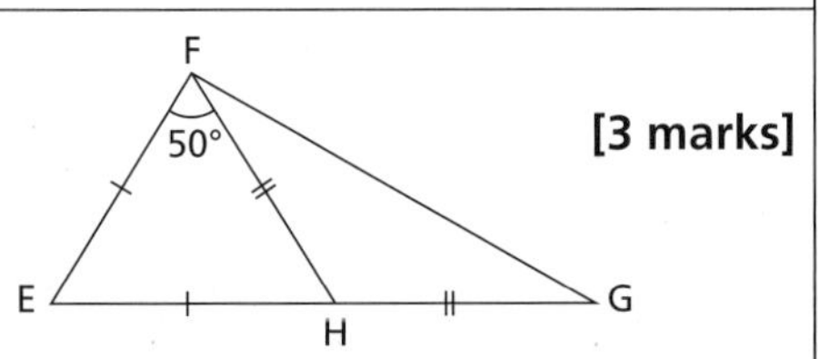

[3 marks]

6 A square and a circle are shown.

The diagonal of the square is equal to the diameter of the circle.

Calculate the area of the circle, giving your answer to 3 significant figures.

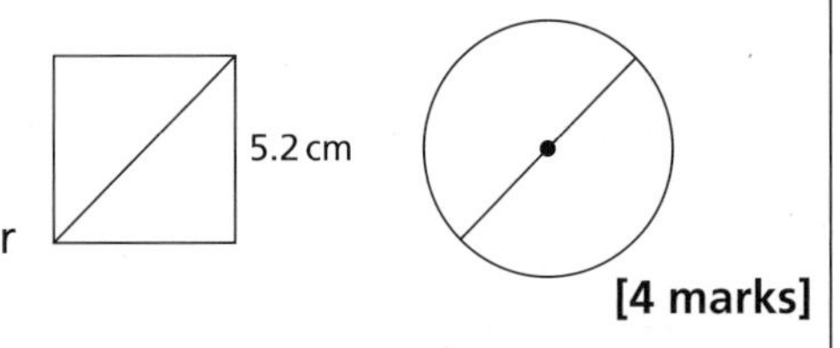

[4 marks]

7 A small boat is 5.3 km from a lighthouse.

The bearing of the small boat from the lighthouse is 090°.

The small boat is 7 km North of a ship.

Calculate the bearing of the ship from the lighthouse, giving your answer to the nearest degree. **[3 marks]**

8 The diagram shows triangles ABC and CDE.

BC = CE

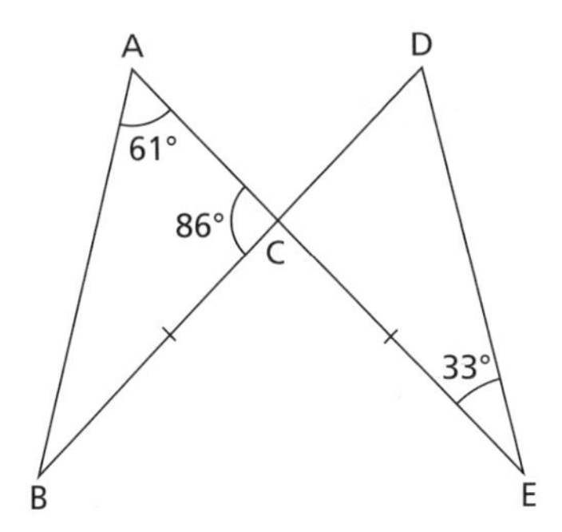

Prove that triangles ABC and CDE are congruent. **[3 marks]**

9 The diagram shows a circle with centre O.

Points A and B lie on the circumference of the circle.

$\angle BOC = 67°$, AC = 19 cm and the area of triangle ABC is 35 cm^2.

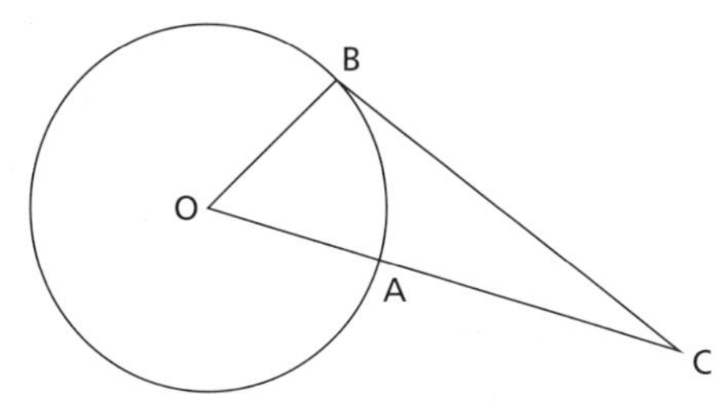

Show that the area of the circle is approximately 16π cm^2. **[4 marks]**

10 The diagram shows a square-based pyramid.

AB = 6 cm and AP = BP = CP = DP = 8 cm

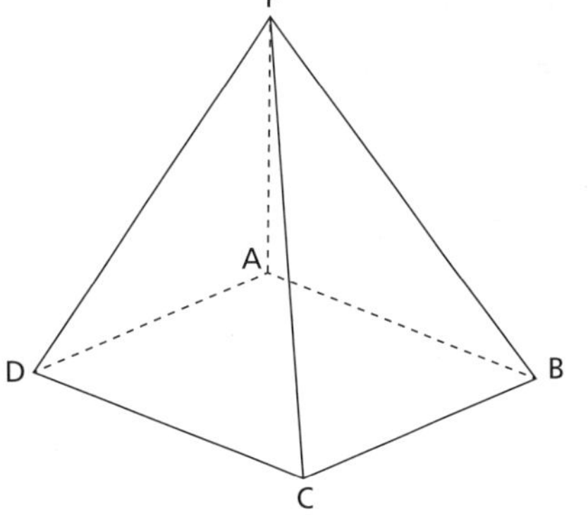

Work out the angle between CP and AC. **[5 marks]**

9 Basic probability

In this block, we will cover...

9.1 Equally likely outcomes

Example 1

A bag contains yellow counters, white counters, p

A counter is taken from the bag at random.

The table shows the probability of selecting each c

Colour	Yellow	White	Purple
Probability	15%	$\frac{1}{5}$	35%

Calculate the probability that the counter is:

a yellow or white **b** purple or orange.

9.1 Equally likely outcomes

Are you ready?

1 Write 0.35 as a fraction.

2 There are 15 teaspoons, 17 tablespoons and 9 dessert spoons in a kitchen drawer.

What fraction of the spoons are:

a teaspoons

b teaspoons or dessert spoons

c teaspoons, tablespoons, or dessert spoons?

3 Which of the following values could represent probabilities?

50 1 −0.8 $\frac{3}{7}$ 55% $\frac{3}{2}$ 0.35

Probability measures how likely an **event** is to occur. When all possible **outcomes** are **equally likely** to occur (such as when spinning a spinner with equally sized regions, or rolling a fair dice), then:

$$\text{Probability of event} = \frac{\text{Number of favourable outcomes}}{\text{Total number of possible outcomes}}$$

For example, the probability of rolling a '3' on a fair, six-sided dice is $\frac{1}{6}$

The probability of rolling 'not a 3' is $\frac{5}{6}$ as 1, 2, 4, 5 and 6 are all favourable outcomes.

P(A) is the notation used to denote 'the probability that event A happens'.

Notice that:

P(event happens) + P(event doesn't happen) = 1

The general result is 'The probabilities of all possible events sum to 1'

Probabilities can be expressed as fractions, decimals or percentages.

Example 1

A bag contains yellow counters, white counters, purple counters, and orange counters.

A counter is taken from the bag at random.

The table shows the probability of selecting each colour of counter.

Colour	Yellow	White	Purple	Orange
Probability	15%	$\frac{1}{5}$	35%	0.3

Calculate the probability that the counter is:

a yellow or white **b** purple or orange.

Method

Solution	Commentary
a $15\% + \frac{1}{5}$ $= 15\% + 20\%$ $= 35\%$	To add the probabilities, you need to make them both fractions, or both decimals, or both percentages. $\frac{1}{5} = 20\%$ so you can rewrite $\frac{1}{5}$ as 20%.
b $35\% + 0.3$ $= 0.35 + 0.3$ $= 0.65$	To add the probabilities, put them into the same form. 35% = 0.35

In this example, you can find the probabilities by adding the probabilities of the individual events because the events are **mutually exclusive**. This means they cannot occur at the same time.

Example 2

A spinner has red sections, green sections and blue sections. It is spun and the colour it lands on is noted.

P(blue) = $\frac{7}{10}$ and P(red) 0.15

Show that P(green) = P(red)

Method

Solution	Commentary
P(blue or red) = $\frac{7}{10} + 0.15$ $= 0.7 + 0.15$ $= 0.85$	Start by adding together the probabilities you are given.
So P(green) = 1 – 0.85 = 0.15 = P(red) Therefore P(green) = P(red)	The sum of all the probabilities is 1 so to work out P(green), subtract 0.85 from 1

Example 3

A crate of fruit contains apples, kiwis, mangoes and peaches. A fruit is selected at random.

The table shows the probability of selecting each type of fruit.

Fruit	Apple	Kiwi	Mango	Peach
Probability	$\frac{1}{4}$			

The chance of choosing a kiwi is 50% more likely than choosing an apple.

The probability of choosing a mango is twice the probability of choosing a peach.

Complete the table.

Method

Solution	**Commentary**
$\frac{1}{4} = 0.25$ P(kiwi) $= 0.25 \times 1.5 = 0.375$	Find P(kiwi) by increasing P(apple) by 50%. Remember, to increase by 50% multiply by 1.5
$1 - (0.25 + 0.375) = 0.375$	Add P(apple) and P(kiwi) and subtract from 1 to find the sum of P(mango) and P(peach).
P(peach) $= x$ $2x + x = 0.375$ P(mango) $= 2x$ $3x = 0.375$ $x = 0.125$	Let P(peach) $= x$ so P(mango) $= 2x$ Form and solve an equation to find the value of x.
(see table below)	Complete the table.

Fruit	Apple	Kiwi	Mango	Peach
Probability	$\frac{1}{4}$	$\frac{3}{8}$	$\frac{1}{4}$	$\frac{1}{8}$

Practice

1 A spinner has 13 equal parts as shown.

The spinner is spun once.

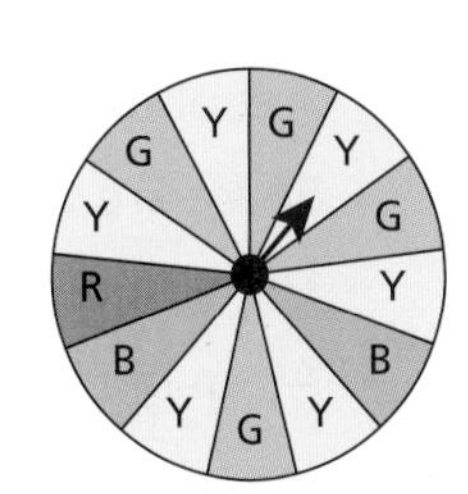

Work out the probability that the spinner lands on:

a red **b** blue **c** purple

d blue or green **e** not yellow.

2 Here is a list of numbers:

15 8 21 14 7 24 36

A number is selected at random.

Calculate the probability that it is:

a a multiple of 7 **b** prime **c** a multiple of 2 and 3 **d** a multiple of 2 or 4

3 There are 30 marbles in a bag. All the marbles are black, blue or red.

P(blue) = 0.2 P(red) = $\frac{1}{3}$

a Work out how many of each colour marble are in the bag.

A marble is chosen at random.

b Write down the probability of choosing a black marble.

c Write down the probability of choosing a blue marble or a red marble.

4 In a DIY shop, the probability of choosing a new lightbulb with a fault is 0.03

Work out the probability of choosing a new light bulb without a fault.

5 The probability that a darts player misses the bullseye is $\frac{11}{15}$

Write down the probability that the player hits the bullseye.

6 Amina and Benji are playing a game.

P(Amina wins) = P(Benji wins)

P(a tie or no result) = 0.2

Work out the probability that Amina does **not** lose.

7 Filipo is watching a football match.

He says that the probability of winning a football match is $\frac{1}{3}$ because the only possible outcomes of the result are win, draw or lose.

Explain why Filipo may be incorrect.

8 A drawer contains spotty, stripey, and plain pairs of socks.

The table shows the probability of choosing each pair of socks.

Copy and complete the table.

Pair of socks	Spotty	Stripey	Plain
Probability	25%		$\frac{1}{8}$

9 A cupboard has yellow cups, purple cups and orange cups.

There are twice as many purple cups as yellow cups.

The number of orange cups is half the number of yellow cups.

A cup is chosen at random.

Work out the probability of randomly choosing each colour of cup from the cupboard.

10 The table shows the number of Year 10 students at a school who have chosen to study German, French or Spanish. Each student studies exactly one of the languages.

Language	German	French	Spanish
Number of students	61	114	

The probability of randomly selecting a student who studies Spanish is $\frac{3}{10}$

Work out the total number of students in Year 10 who study Spanish.

What do you think?

1 There are only blue counters, green counters or yellow counters in a bag.

P(blue counter) = 15% P(green counter) = 0.225 P(yellow counter) = $\frac{5}{8}$

Abdullah says, "The number of counters in the bag must be a multiple of 8"

a Do you agree with Abdullah? Explain your reasoning.

b Work out the smallest possible number of counters in the bag.

2 The spinner shown has red, green and blue sections and is numbered 1, 2, 3 and 4

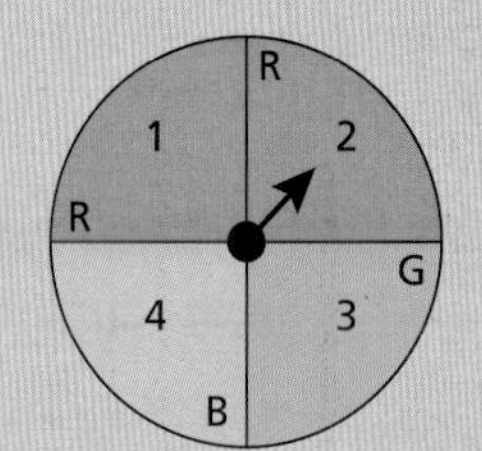

Seb says, "P(1 or red) = $\frac{1}{4} + \frac{1}{2} = \frac{3}{4}$"

Explain why Seb is wrong.

3 A fair, six-sided dice numbered 1 to 6 is rolled twice.

Lida says, "The probability of rolling 4 twice is $\frac{2}{12}$"

Rob says, "The probability of rolling 4 twice is $\frac{1}{36}$"

Who do you agree with? Give a reason for your answer.

Consolidate – do you need more?

1 Here is a list of numbers:

24 4 29 42 5 30 49

A number is selected at random.

Calculate the probability that it is:

a a prime number **b** a multiple of 6 **c** a square number

d a multiple of 6 or a square number **e** a multiple of 6 and a square number.

2 There are 20 straws in a box. All the straws are pink, yellow or green.

P(green) = 0.15 P(yellow) = $\frac{1}{4}$

a Work out how many of each colour of straw are in the box.

A straw is selected at random.

b Write down the probability of choosing a pink straw.

c Write down the probability of choosing a pink or yellow straw.

3 The probability that Junaid passes his music exam is 0.71

Work out the probability that Junaid does **not** pass his music exam.

4 Students A, B and C are running for head of student council.

The table shows the number of votes for each student.

Student	A	B	C
Number of votes	21	43	16

A voter is selected at random. Work out the probability that they voted for Student C.

Stretch – can you deepen your learning?

1 Amina makes up this probability question.

Rhys and Mario both sit the same test.
The probability of Rhys passing the test is twice the probability of Mario passing the test.
The probability of Mario not passing the test is 50% greater than the probability of him passing the test.
Work out the probability of Rhys not passing the test.

Discuss why Amina's question could be ambiguous.

How could she phrase the question better?

2 A spinner can land on either white, black, or red.

The table shows the probability of the spinner landing on white or black.

Colour	White	Black	Red
Probability	$\frac{1}{a}$	$\frac{1}{b}$	

Write an expression for P(red) as a single fraction.

3 A pack contains only yellow cards, red cards and green cards.

There are x green cards in the pack.

There are twice as many red cards as green cards.

There are two more yellow cards than green cards.

Write an expression for the probability of randomly selecting a yellow card from the pack.

Basic probability: exam practice

4–6

1 A fair spinner has six sections.

Each section must have one number from 1 to 4 written on it so that:

- the probability of spinning a 1 is equal to the probability of spinning a 3
- the probability of spinning a 4 is greater than the probability of spinning a 1
- the probability of spinning a 2 is less than the probability of spinning a 4

Write the numbers on a copy of the spinner. **[3 marks]**

2 The pie chart shows information about the languages studied by Year 11 students at a school. Each student studies exactly one language.

One student is selected at random.

What is the probability they study French? **[2 marks]**

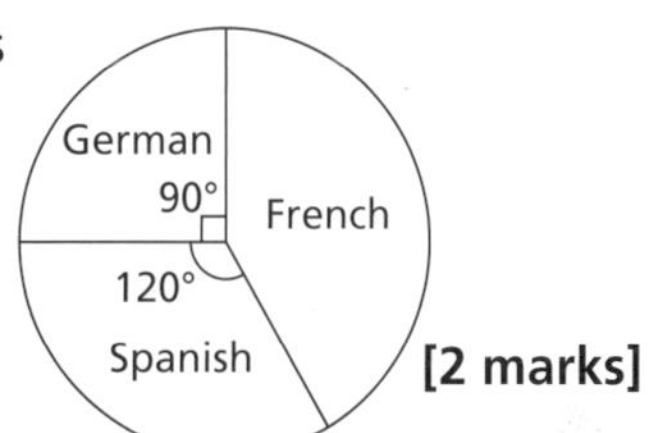

3 A bag contains only blue counters, green counters, yellow counters and red counters.

A counter is selected from the bag at random.

The probability of picking a blue counter is $\frac{2}{5}$

The probability of picking a green counter is 15%

The probability of picking a yellow counter is twice the probability of picking a red counter.

What is the probability of picking a yellow counter? Give your answer as a decimal. **[3 marks]**

4 The numbers 1 to 20 are written on cards. One card is selected at random.

What is the probability that the card shows a prime number? **[2 marks]**

5 In a bag there are only red counters, blue counters and green counters.

The ratio

number of red counters : number of blue counters : number of green counters = 5 : 8 : 7

A counter is taken at random from the bag.

Complete a copy of the table to show the probabilities, written as decimals, of the counter being red, blue or green.

Colour	Red	Blue	Green
Probability			

[3 marks]

6 There are n coins in a bag. Five of them are 10p coins.

A coin is taken at random from the bag.

Write an expression for the probability that the coin is **not** a 10p coin. **[2 marks]**

10 Probability diagrams

In this block, we will cover...

10.1 Lists and counting

Example 1

There are 30 students in Class A and 27 students

Two students are randomly selected to win a pri

a How many different pairs of students could b

i from different classes **ii** both from

b Huda is in Class A and Zach is in Class B.

Work out the probability of Huda and Zach b

Method

Solution	Commentary

10.2 Sample space diagrams

Practice

1 Two spinners A and B are spun and the res

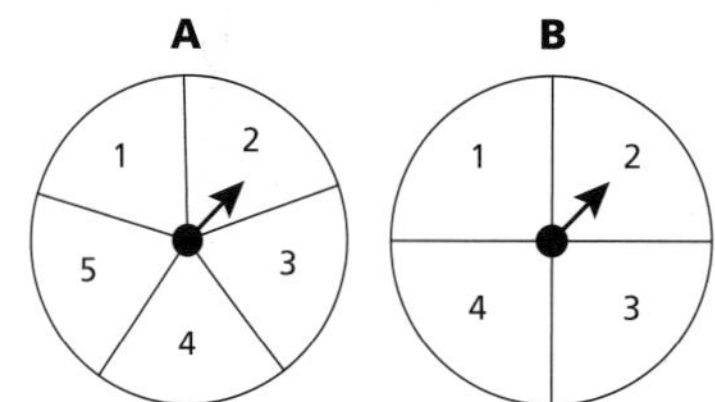

a Copy and complete the sample space d

10.3 Venn diagrams

Consolidate – do you need more

1 $\xi = \{1, 2, 3, 4, 5, 6, 7, 8, 9, 10\}$

$X = \{2, 4, 5, 6, 8, 9\}$

$Y = \{1, 5, 10\}$

a Copy and complete the Venn diagram to represent the information.

A number from the universal set ξ is select at random.

b Work out $P(X \cap Y)$. **c** Work out

10.1 Lists and counting

Are you ready?

1 A restaurant offers choices for both starters and mains.

A customer decides to order one starter and one main course.

List all the possible outcomes for the customer's order.

2 A fair coin is flipped twice. The coin can land on heads or tails.

a Copy and complete the table to show all the possible outcomes.

		2nd flip	
		H	T
1st flip	H		
	T		

b Work out the probability that the coin lands on:

i tails both times **ii** tails at least once.

3 The letters of the word 'DIVISION' are each written on a card and put into a bag.

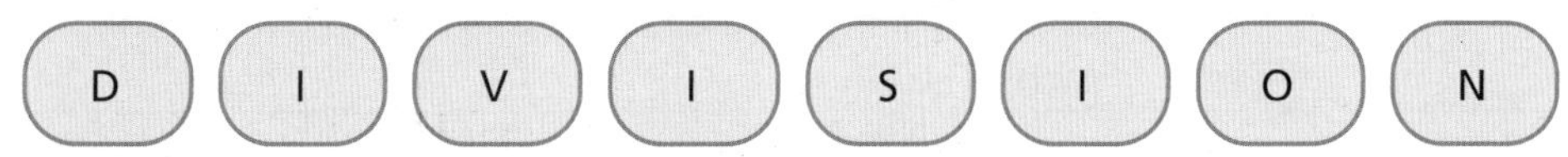

A card is selected at random from the bag.

Write down the probability of choosing a card with 'I' on it.

Consider a wardrobe that has 3 pairs of different coloured jeans and 4 different coloured shirts.

To work out how many possible unique outfit combinations there are, one way is to systematically match a pair of jeans with a shirt. Here you can see that the number of possible outfits is 12

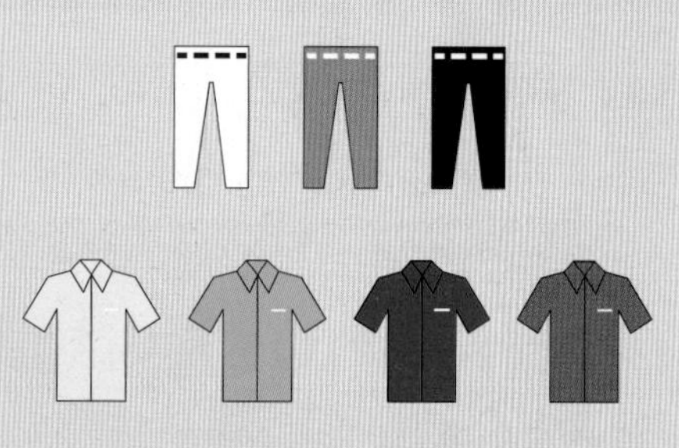

(White, Yellow), (White, Green), (White, Red), (White, Purple), (Blue, Yellow), (Blue, Green), (Blue, Red), (Blue, Purple), (Black, Yellow), (Black, Green), (Black, Red), (Black, Purple)

Another way to do this is to multiply the number of jeans options by the number of shirt options.

Multiplying 3 by 4 gives 12 unique outfit combinations. This is an example of the **product rule for counting**.

In general, to calculate the total number of possible outcomes of two or more **independent events** occurring, multiply together the total number of possible outcomes for each event.

If there are m ways of one event occurring and n ways of a different event occurring, then the total number of ways of both events occurring is $m \times n$.

Example 1

There are 30 students in Class A and 27 students in Class B.

Two students are randomly selected to win a prize.

a How many different pairs of students could be selected if the two students are:

i from different classes **ii** both from Class B?

b Huda is in Class A and Zach is in Class B.

Work out the probability of Huda and Zach both being selected.

Method

Solution	Commentary
a **i** $30 \times 27 = 810$ There are 810 different pairs of students.	Multiply together the number of options for each class.
ii $27 \times 26 = 702$ There are 702 different pairs of students.	If both students are from Class B then there are 27 ways to choose the first student. After the first student has been chosen, there are only 26 ways to choose the second student.
b $\frac{1}{810}$	Huda and Zach being selected is a unique combination of choosing one student from Class A and one student from Class B.

Example 2

Four friends, Ali, Jakub, Bev and Chloe, sit next to each other at a concert.

a How many different ways could the four friends be arranged to sit next to each other at the concert?

500 people attend the concert. Two people from the audience are going to be selected to win a prize.

b How many different pairs of people could be selected?

Method

Solution	Commentary
a $4 \times 3 \times 2 \times 1 = 24$ There are 24 possible arrangements.	There are four seats available to sit in. Any of the four friends can sit in the first seat. After one friend has sat in the first seat, any of the remaining three friends can sit in the second seat. After the first two seats have been occupied, either of the two remaining friends can sit in the third seat. This leaves only one friend to sit in the fourth seat. You can now use the product rule for counting to find the total number of possible arrangements.
b $\frac{500 \times 499}{2} = 124\,750$	There are 500 people to choose from for the first prize winner. Therefore, there are 499 people to choose from for the second prize winner. Use the product rule for counting to work out how many possible pairs. However, the order in which you select the pairs doesn't matter (choosing person A and then person B is the same as choosing person B and then person A). So to avoid 'double counting', divide your answer by 2

Practice

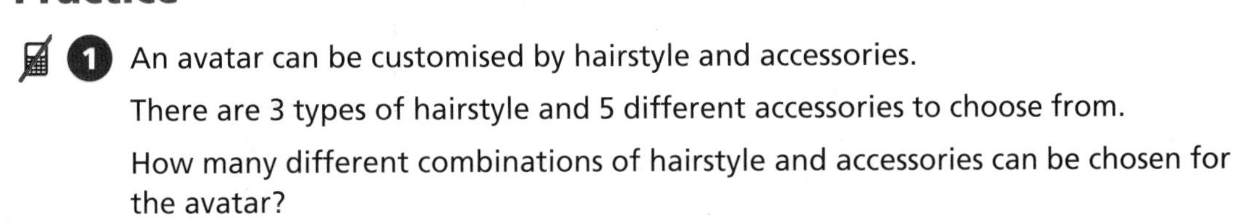

1 An avatar can be customised by hairstyle and accessories.

There are 3 types of hairstyle and 5 different accessories to choose from.

How many different combinations of hairstyle and accessories can be chosen for the avatar?

2 There are 8 fiction books and 3 non-fiction books on a bookshelf.

Jackson will choose a fiction book and a non-fiction book from the bookshelf.

How many different pairs of books could Jackson choose?

3 A sewing box has equal numbers of blue, black and white buttons.

There is also an equal amount of red, green, orange and brown thread.

A button and a thread colour are chosen at random.

Work out the probability of choosing a black button and orange thread.

4 A supermarket meal deal consists of buying a sandwich, a snack and a drink.

There 9 sandwich options, 6 snack options and 7 drink options to choose from.

Work out the total number of different meal deal combinations.

5 Here are three number cards:

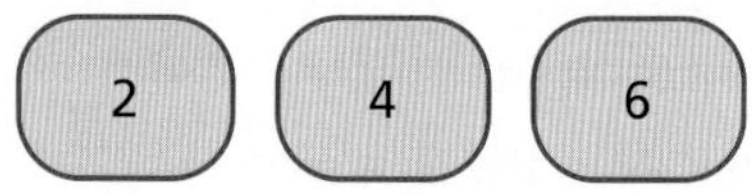

Kath is making three-digit numbers using each card once.

She says, "There are three different ways to choose each digit so the total number of different three-digit numbers is 3 × 3 × 3 = 27"

Explain why Kath is wrong.

6 An ice cream parlour has 38 different flavours.

Sven orders one scoop each of three different flavours.

Show that there are over 50 000 different possible flavour combinations that Sven could order.

7 Amina has 9 different pairs of socks and x different pairs of shoes.

The probability that she wears a unique combination of socks and shoes is $\frac{1}{36}$

Work out how many pairs of shoes Amina has.

8 Three fair, six-sided dice numbered 1 to 6 are rolled.

a How many possible outcomes are there altogether?

b Work out the probability of all three dice showing the same number.

9 Rhys, Benji, and Flo stand next to each other for a group photo.

a How many different ways can they be arranged in a line for the photo?

Beca and Rob join in the photo.

b How many different ways can the five students be arranged in a line for the photo?

10 Five friends sit around a circular table.

How many different ways can they be arranged around the table?

11 Kath is planning a holiday.

She decides she must visit **at least** two out of these three cities:

- New York San Francisco Chicago

If she chooses two cities, she must book tickets for two out of the three available shows:

- Broadway Musical Comedy Club Jazz Concert

If she decides to visit three cities, she will only have time for one show.

After choosing the cities and shows, Kath plans two activities from this list:

- Hiking Museum Tour Shopping Bike Riding

Finally, Kath wants to try one cuisine from each of these two lists:

- Italian Japanese Mexican
- Indian French Thai Greek

How many different combinations of cities, shows, activities, and cuisines could Kath plan for her holiday?

What do you think?

1 A social media platform allows friends to make 'connections' with each other.

Work out how many connections there would be if:

a 20 friends connect with each other

b 200 friends connect with each other

c n friends connect with each other.

Consolidate – do you need more?

1 In a supermarket, there are 7 varieties of pasta and 5 different sauces.

How many different possible pasta and sauce combinations are there altogether?

2 A restaurant offers the following menu options.

Starters	Main meals	Desserts
Lentil soup	Chickpea curry	Fruit sorbet
Caprese salad	Margherita pizza	Classic cheesecake
Chicken satay skewers	Grilled salmon fillet	Chocolate mousse
Roasted vegetable bruschetta	Beef lasagna	
	Vegetable stir-fry	

A customer randomly selects a starter, a main meal and a dessert.

Work out the probability that the customer orders this combination:

Lentil soup Beef lasagna Chocolate mousse

3 This fair spinner is spun four times.

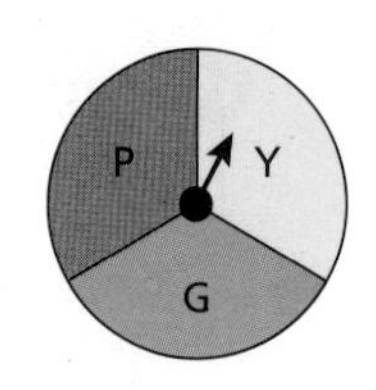

a How many possible colour combinations are there altogether?

b Work out the probability of all four spins landing on the same colour.

4 Chloe and Lida are on the school football team.

There are 15 people altogether on the school team.

A captain and a vice-captain are to be selected from the team.

The names of the team members are put into a hat and two are selected at random.

a How many different possible captain and vice-captain combinations are there?

b Work out the probability of Chloe being selected as captain and Lida as vice-captain.

c Work out the probability of Lida being selected as captain.

One of the other team members asks for their name to be removed from the hat.

d Work out the probability of Chloe being selected as captain and Lida as vice-captain.

Stretch – can you deepen your learning?

1 In a science class, there are 300 different ways to choose a pair of students.

How many students are in the class?

2 Marta, Abdullah, Ed, Tiff and Samira volunteer to help at a school fair.

Two of them are chosen at random by their teacher to help on the raffle.

Work out the probability of Tiff and Samira being selected.

3 Class X has one more student than Class Y.

Class Z has one fewer student than Class Y.

One student from Class X and one student from Class Z are selected.

Show that the total number of ways of selecting these students is 1 less than a square number.

4 A four-digit PIN code can be made using the digits 0 to 9

a Without listing, show that there are 10 000 different possible PIN codes.

Work out how many possible PIN codes can be made:

b with no repeated digits

c using just odd digits

d using just square digits

e where the sum of the digits is equal to 4

5 **a** How many different three-digit prime numbers can you make using only the digits 3, 4, 5, 6, 7 once?

b How many different three-digit prime numbers can you make from the digits 3, 4, 5, 6, 7 if you can reuse the digits?

10.2 Sample space diagrams

Are you ready?

1 The table shows information about a group of karate students.

	Left-hand dominant	Right-hand dominant	Total
Left-foot dominant	18		50
Right-foot dominant		46	
Total	22		

a Copy and complete the table.

b One student is selected at random. What is the probability that this student is left-hand and left-foot dominant?

2 The two-way table shows how some students from Year 9, Year 10 and Year 11 travel to school.

A student is selected at random.

	Bus	Bicycle	Walk	Total
Year 9	40	25	15	80
Year 10	35	30	20	85
Year 11	20	15	25	60
Total	95	70	60	225

a What is the probability of choosing a Year 9 student who uses a bicycle?

b What is the probability of choosing a student who walks to school, and is from Year 10 or Year 11?

3 Kath rolls two six-sided dice numbered 1 to 6 and adds the results together to get a score.

List all the possible scores.

A sample space diagram shows all the possible **outcomes** or results of an **experiment**. It is often in the form of a **table** or an **array** rather than just a long list.

For example, a spinner is spun and a six-sided dice is rolled. Here are the possible outcomes:

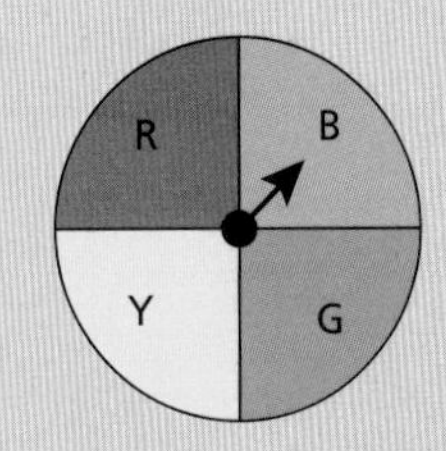

		Dice					
		1	**2**	**3**	**4**	**5**	**6**
Spinner	**Blue (B)**	(B, 1)	(B, 2)	(B, 3)	(B, 4)	(B, 5)	(B, 6)
	Green (G)	(G, 1)	(G, 2)	(G, 3)	(G, 4)	(G, 5)	(G, 6)
	Yellow (Y)	(Y, 1)	(Y, 2)	(Y, 3)	(Y, 4)	(Y, 5)	(Y, 6)
	Red (R)	(R, 1)	(R, 2)	(R, 3)	(R, 4)	(R, 5)	(R, 6)

Example 1

There are three cards inside Box X and four cards inside Box Y. The cards each have a number on them.

A card is taken at random from each box and the difference between the numbers on the cards is found.

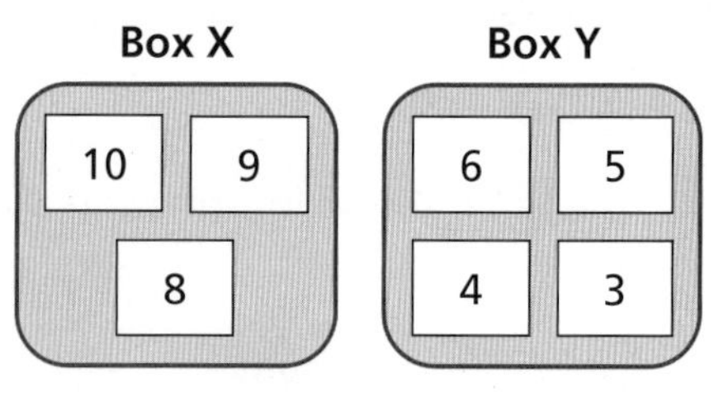

a Complete the sample space diagram to show all the possible outcomes.

		Box Y			
		6	**5**	**4**	**3**
Box X	**10**	4	5	6	7
	9	3			
	8				

b Work out the probability that the difference between the two cards is 3

c Work out the probability that the difference between the two cards is greater than 3

Method

<table>
<tr><th>Solution</th><th>Commentary</th></tr>
<tr><td>a
<table>
<tr><td></td><td></td><td colspan="4">Box Y</td></tr>
<tr><td></td><td></td><td>6</td><td>5</td><td>4</td><td>3</td></tr>
<tr><td>Box X</td><td>10</td><td>4</td><td>5</td><td>6</td><td>7</td></tr>
<tr><td></td><td>9</td><td>3</td><td>4</td><td>5</td><td>6</td></tr>
<tr><td></td><td>8</td><td>2</td><td>3</td><td>4</td><td>5</td></tr>
</table></td><td>Remember that you will not have negative answers when ‘finding the difference’.</td></tr>
<tr><td>b P(difference = 3) = $\frac{2}{12}$</td><td>Two of the outcomes give a difference of 3 and there are 12 outcomes in total, so the probability is $\frac{2}{12}$</td></tr>
<tr><td>c P(difference > 3) = $\frac{9}{12}$</td><td>There are 9 pairs of cards where the difference is greater than 3

You don’t have to simplify your answers unless you are asked to.</td></tr>
</table>

Example 2

Two fair, six-sided dice numbered 1 to 6 are rolled and the numbers obtained are multiplied together.

Work out the probability that the result is:

a a multiple of 5

b a square number

c neither a multiple of 5 nor a square number.

Method

Solution	Commentary

a

×	1	2	3	4	5	6
1	1	2	3	4	(5)	6
2	2	4	6	8	(10)	12
3	3	6	9	12	(15)	18
4	4	8	12	16	(20)	24
5	(5)	(10)	(15)	(20)	(25)	(30)
6	6	12	18	24	(30)	36

P(a multiple of 5) = $\frac{11}{36}$

Commentary: Draw a sample space diagram to show all the possible outcomes.

Of the 36 outcomes, 11 of them are multiples of 5

b

×	1	2	3	4	5	6
1	(1)	2	3	4	5	6
2	2	(4)	6	8	10	12
3	3	6	(9)	12	15	18
4	4	8	12	(16)	20	24
5	5	10	15	20	(25)	30
6	6	12	18	24	30	(36)

P(a square number) = $\frac{6}{36} = \frac{1}{6}$

Commentary: Of the 36 outcomes, 6 of them are square numbers.

You do not need to simplify your answer unless the question asks you to, but you can do so if you wish.

c

×	1	2	3	4	5	6
1	(1)	2	3	4	(5)	6
2	2	(4)	6	8	(10)	12
3	3	6	(9)	12	(15)	18
4	4	8	12	(16)	(20)	24
5	(5)	(10)	(15)	(20)	(25)	(30)
6	6	12	18	24	(30)	(36)

P(multiple of 5 or a square number) = $\frac{16}{36}$

P(neither a multiple of 5 nor a square number)

= 1 – P(multiple of 5 or a square number)

$= 1 - \frac{16}{36} = \frac{20}{36}$

Commentary: Of the 36 outcomes, 16 of them are multiples of 5 or a square number.

Subtract the probability from 1

Practice

1 Two spinners A and B are spun and the results are added together.

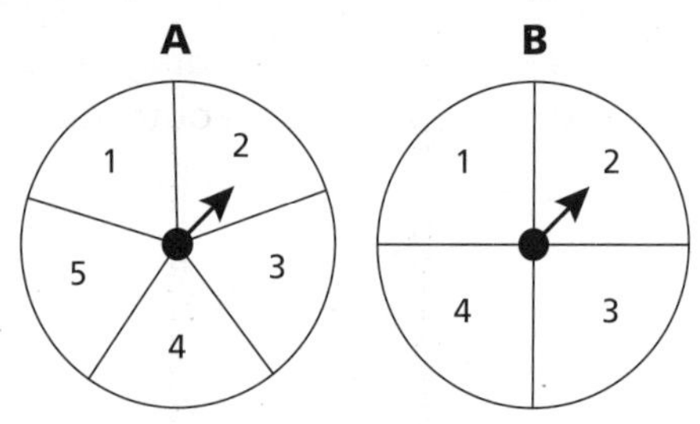

a Copy and complete the sample space diagram.

		Spinner A				
		1	2	3	4	5
Spinner B	1					
	2					
	3					
	4					

b Work out the probability that the sum of the results is an even number.

c Work out the probability that the sum of the results is a square number.

2 A fair coin is flipped and a fair, six-sided dice numbered 1 to 6 is rolled.

a Draw a sample space diagram to show all the possible outcomes.

b Work out the probability that the outcome includes heads or a factor of 6

3 Flo and Ed each have a bag of fruit.

Flo has a banana, an orange, a pear and a grapefruit.

Ed has an orange, an apple and a mango.

They each select one piece of fruit at random. Assume each fruit is equally likely to be chosen.

a Draw a sample space diagram to show all the possible outcomes.

b Work out the probability that:

i both Flo and Ed select an orange **ii** exactly one of them chooses an orange.

4 Samira rolls two fair, six-sided dice numbered 1 to 6 and the sum of their scores is noted. One dice is coloured blue and the other is coloured red.

a Draw a sample space diagram to show all the possible outcomes.

b Work out the probability of:

i getting a sum of 8 when rolling the two dice

ii getting a number on the red dice that is greater than the number on the blue dice

iii getting a prime number on the red dice and an even number on the blue dice

iv the sum of the numbers being less than 5 or the number on the red dice being odd.

Consolidate – do you need more?

1 Mario and Jackson each have a spinner.

		Mario's spinner				
		1	3	5	7	9
Jackson's spinner	2	(2, 1)				
	4					
	6					
	8					

a Copy and complete the sample space diagram.

b Work out the probability that Mario and Jackson both roll a multiple of 3

c Work out the probability that either Mario or Jackson roll a prime number.

2 Filipo and Chloe each have a box of chocolates containing milk, white and dark chocolates in equal number. They each randomly select a chocolate from their box.

a Draw a sample space diagram to show all the possible outcomes.

b Calculate:

i the probability that both select a dark chocolate

ii the probability that exactly one of them selects a milk chocolate

iii the probability that neither of them selects a white chocolate.

3 A spinner with four equal sections labelled A, B, C and D is spun, and a fair, six-sided dice numbered 1 to 6 is rolled.

a Draw a sample space diagram to show all the possible outcomes.

b Work out the probability of spinning 'A' and rolling a factor of 20

c Work out the probability of spinning 'not A' or rolling a factor of 24

Stretch – can you deepen your learning?

1 An arcade game dispenses tokens with values of 1, 2, 3, or 4

Each token is equally likely to be dispensed.

Lida plays the game three times, receiving one token each time, and she adds the numbers shown on her tokens.

a Draw a sample space diagram showing all possible outcomes.

b Given that the first two tokens were 3 and 4, work out the probability that the sum of Lida's three tokens is less than 10

c Consider how you could use a tree diagram to work out part **b**.

2 Marta rolls two fair, six-sided dice numbered 1 to 6 and adds up the numbers obtained.

She then rolls both dice again but this time multiplies the scores obtained.

Work out the probability of:

a both scores being equal

b both scores being square

c the sum of the numbers obtained on the first roll being greater than the product of the numbers obtained on the second roll

d the sum of the numbers obtained on the first roll being a factor of the product of the numbers obtained on the second roll.

3 Five fair coins are flipped.

a What is the probability of obtaining exactly three heads and two tails? Explain how you approached the question.

b Ed says, "If you flip a coin n times, where n is a very large integer, the probability of the coin landing on heads every single time is 0"

Show that Ed is wrong by writing an expression for P(n heads from n flips).

10.3 Venn diagrams

White Rose MATHS

Are you ready? (A)

1 Write down the first three prime numbers greater than 40

2 ξ = {1, 2, 3, 4, 5, 6, 7, 8, 9, 10}

A = {even numbers}   B = {multiples of 4}

a Write down the members of both A and B.

b Write down the members of A′.

3 The Venn diagram shows information about the extra-curricular club choices of some students.

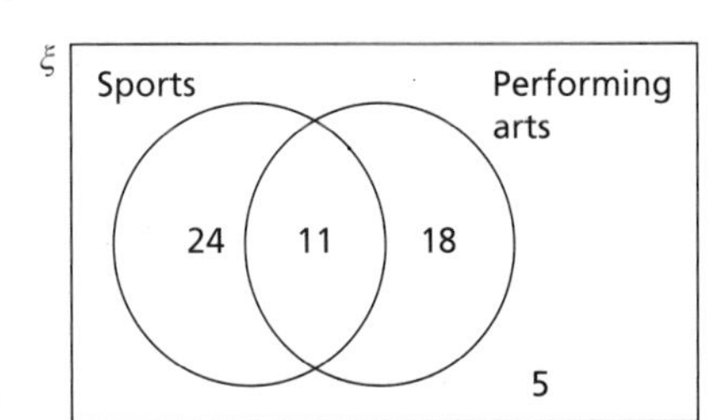

a How many students are there altogether?

b How many students chose a sports club and a performing arts club?

c How many students chose a performing arts club?

d How many students did not choose a sports club?

A **Venn diagram** is a visual way of representing the elements of the **universal set**, which is the set containing all objects or elements, often denoted by ξ. The Venn diagram also shows the relationships between two or more **sets**.

Each circle in a Venn diagram is a set. The elements that belong to that set are written within the circle.

The overlapping regions show the elements that belong to more than one set. These are also known as **intersections**.

Elements of the universal set, ξ, that are not a member of any of the sets are written outside of the circles.

This diagram highlights the elements of A.

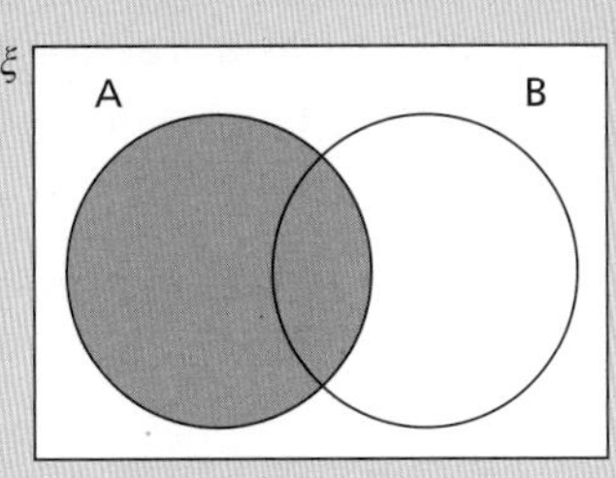

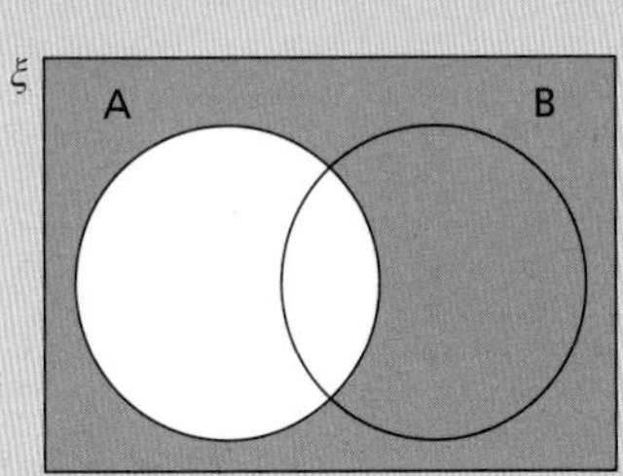

A′ is 'A **complement**' – this means all the elements that are not in A.

$A \cup B$ is 'A **union** B' – this means elements that are in A or B.

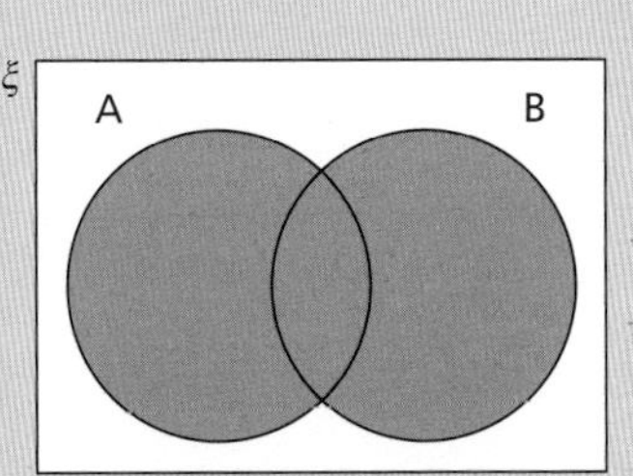

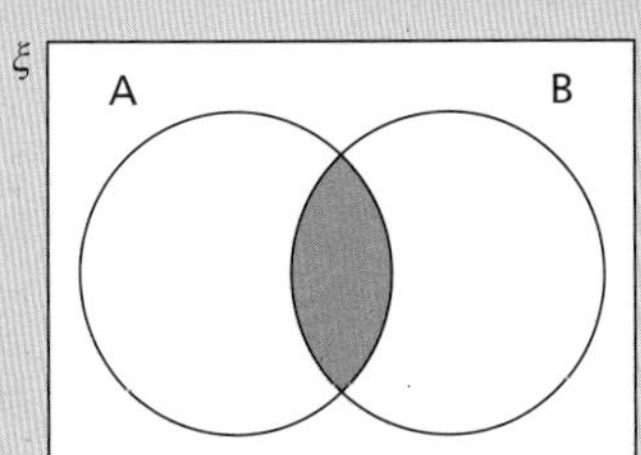

$A \cap B$ is 'A **intersection** B' – this means the elements that are in both A and B.

Example 1

In a survey of 120 students: 80 students study Computing, 60 students study Textiles, and 15 students study neither.

a Draw a Venn diagram to represent this information.

b A student is selected at random. Work out the probability that the student studies Computing but not Textiles. Give your answer as a fraction in its simplest form.

Method

Solution	Commentary
a $120 - 15 = 105$ so 105 students study Computing or Textiles or both	Start with the section outside the sets, as you are told that 15 students study neither Computing or Textiles. Subtracting this from the total means that there are 105 students who chose to study Computing or Textiles.
$80 + 60 - 105 = 35$ 35 students study both Computing and Textiles	80 students study Computing and 60 study Textiles. Some of the students will study both as the sum of these is 140, which is 35 more than 105
$80 - 35 = 45$ students study only Computing	Subtract 35 from 80 to find the number of students who study only Computing.
ξ Computing: 45; Computing and Textiles: 35; Textiles: 25; outside: 15	Show the information in a Venn diagram.
b P(Computing only) $= \frac{45}{120} = \frac{3}{8}$	Write your answer as a fraction: $\frac{\text{Number of students studying only computing}}{\text{Total number of students}}$ Simplify your answer as instructed.

Example 2

The Venn diagram shows the number of people who chose to have ice cream (I) and custard (C) with their pudding.

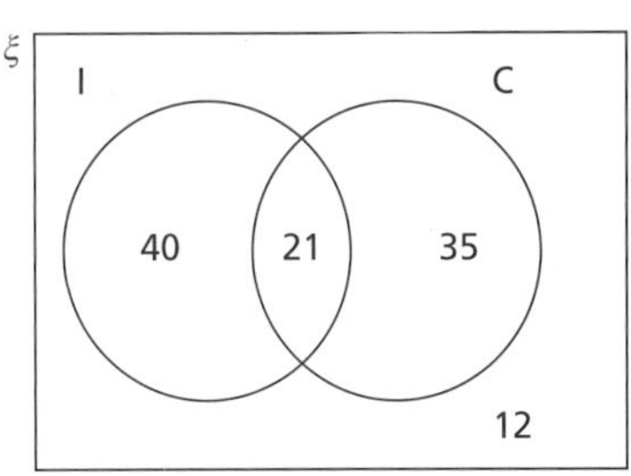

A person is picked at random.

Work out the probabilities:

a $P(I \cap C)$ **b** $P(C')$ **c** $P(I' \cap C')$

Method

Solution	Commentary
a $I \cap C = 21$	$I \cap C$ represents the intersection of the Venn diagram, i.e. the people who chose ice cream and custard.
$40 + 21 + 35 + 12 = 108$	Find the total number of people who had a pudding.
$P(I \cap C) = \frac{21}{108}$	21 out of 108 people had ice cream and custard.
b $C' = 40 + 12 = 52$	C′ means the complement of C, which represents the people who did not have custard.
	Find the number of people who did not have custard.
$P(C') = \frac{52}{108}$	Write down the probability.
c $P(I' \cap C') = \frac{12}{108}$	$I' \cap C'$ means the intersection of the complement of I and the complement of C. This represents the people who did not have ice cream and did not have custard.

Example 3

In a bag of 100 counters, 60 are red (R) and 40 have spots (S).
25 counters are red and have spots.

Work out: **a** $P(R \cup S)$ **b** $P(R' \cap S)$ **c** $P(R \cup S')$

Method

Solution	Commentary
a ξ: R only 35, R ∩ S 25, S only 15, outside 25 (Venn diagram with circles R and S)	Draw a Venn diagram to represent the information given. Start by putting 25 in the intersection, and work out the remaining values by subtraction.
$35 + 25 + 15 = 75$ $P(R \cup S) = \frac{75}{100}$	Find the number of counters that are red or have spots.
b $P(R' \cap S) = \frac{15}{100}$	Find the number of counters that are not red and have spots.
c $35 + 25 = 60$ $P(R \cup S') = \frac{60}{100}$	Find the number of counters that are red or have no spots. There are 35 counters that are red with no spots and 25 counters that are not red and have no spots.

Practice (A)

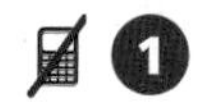

1 There are 240 students in Year 11:

- 75 study History and Geography
- 40 study only History
- 105 study Geography.

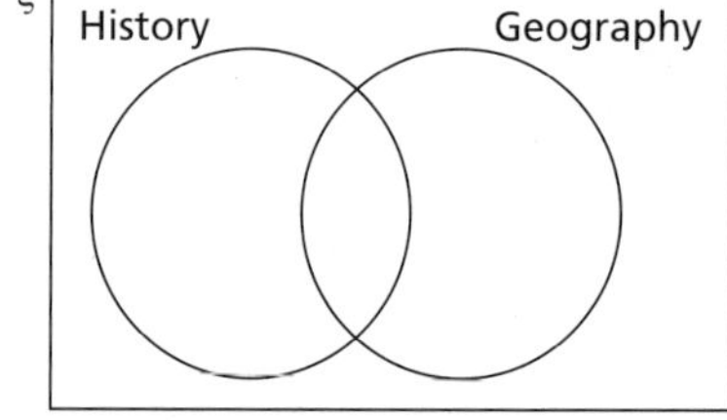

a Copy and complete the Venn diagram.

b A student is selected at random.
Work out the probability that the student studies History or Geography.

2 80 people in a library were asked whether they enjoy mystery (M) or science fiction (S) books:

- 15 people enjoy both
- 35 people enjoy mystery
- 40 enjoy science fiction.

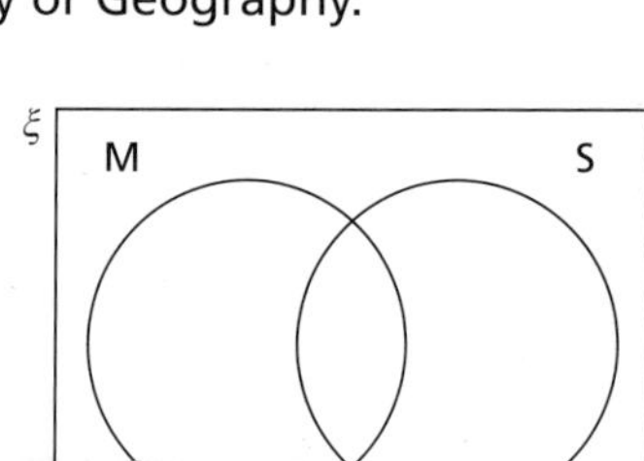

a Use the Venn diagram to represent this information.

b How many people do not enjoy either mystery or science fiction books?

c A person is selected at random. Work out the probability that the person enjoys mystery or science fiction, or both.

3 In a market research study with 180 participants, 100 participants use Product X and 70 participants use Product Y. If a participant is chosen at random, the probability that they use Product X and Product Y is $\frac{1}{3}$

a Draw a Venn diagram to represent this information.

b Work out how many participants use neither Product X nor Product Y.

c Work out the probability that a participant chosen at random uses only Product Y.

4 In a box of 200 buttons, 100 buttons have two holes, and 80 buttons are green.

The probability that a button has both two holes and is green is $\frac{1}{4}$

Calculate the probability that a randomly selected button has two holes or is green.

5 A tennis club has 60 members. 40 members play right-handed, 30 play left-handed, and 20 play either-handed.

What is the probability that a randomly chosen member plays right- or left-handed?

6 Match the set notation to its correct description, and write a suitable description for the blank card.

$X \cap Y$	The union of X and the complement of Y
Y'	The union of X and Y
$X \cup Y$	The intersection of X and Y
$X \cup Y'$	The complement of Y
$X' \cap Y'$	...

7 $\xi = \{1, 2, 3, 4, 5, 6, 7, 8, 9, 10, 11, 12\}$

The Venn diagram shows two sets A and B where:

A = {factors of 12} and

B = {square numbers}

Cards with the numbers 1 to 12 are placed in a bag.

A card is selected at random.

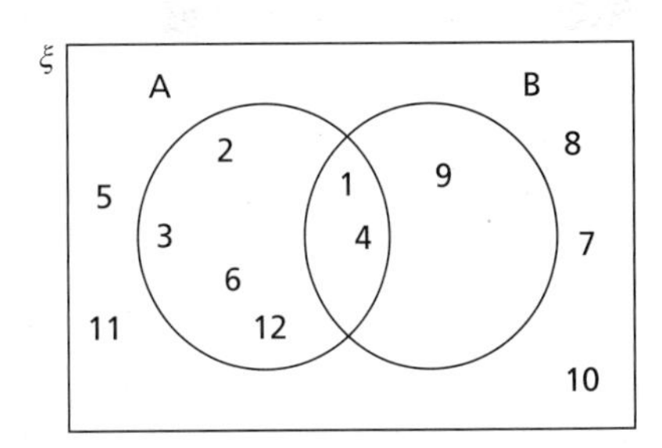

Work out:

a the probability of choosing a factor of 12

b P(B)

c $P(A \cap B)$

d the probability of not choosing a factor of 12

e P(B′)

f the probability of choosing a factor of 12 or a square number

g $P(A' \cup B')$

h the probability of choosing a factor of 12 that is not a square number

i $P(A' \cap B')$

What do you think? (A)

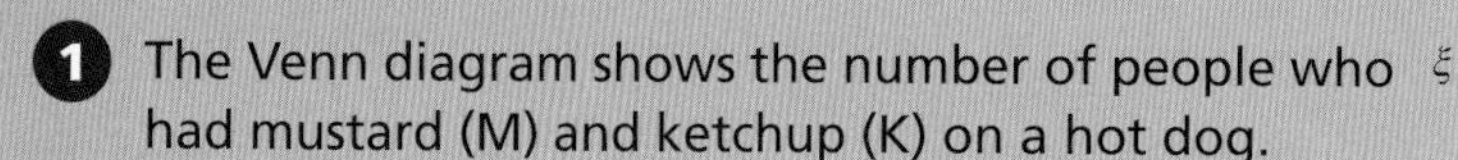

1 The Venn diagram shows the number of people who had mustard (M) and ketchup (K) on a hot dog.

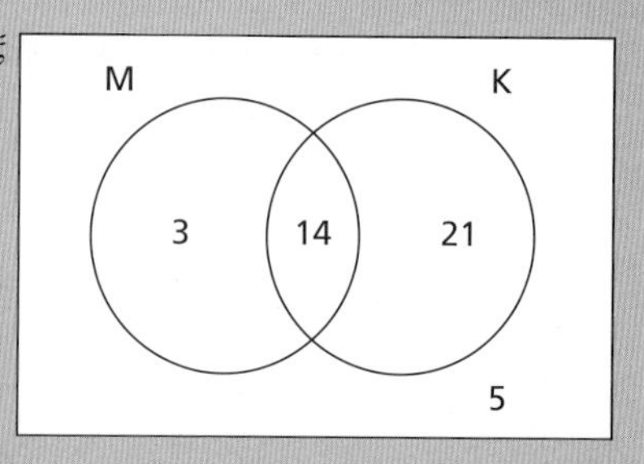

a Work out the probability that a randomly selected person had ketchup on their hot dog.

b Work out the probability that a randomly selected person had ketchup on their hot dog, given that it is known that they had mustard on their hot dog.

2 In the universal set, ξ, there are three elements that do not belong in either set A or B.

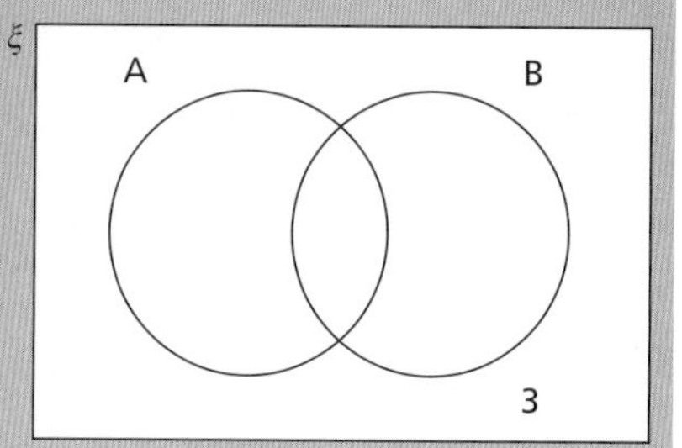

An element is chosen at random from ξ.

$P(A) = \frac{8}{15}$ $\quad P(A \cup B) = \frac{4}{5}$ $\quad P(B) = \frac{3}{4} P(A)$

Copy and complete the Venn diagram representing ξ using the given probabilities.

3 Describe each shaded region shown using set notation.

a

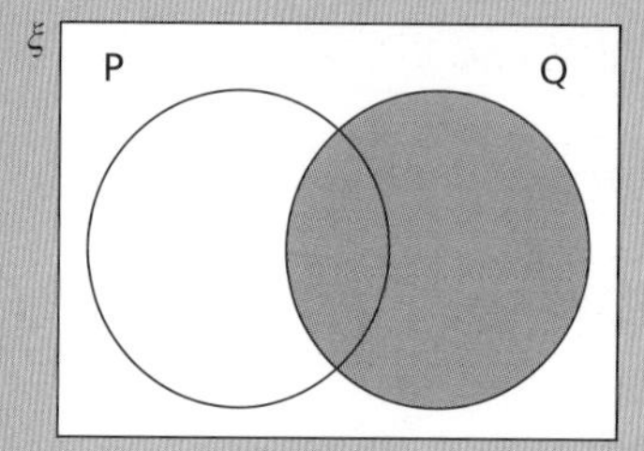

b

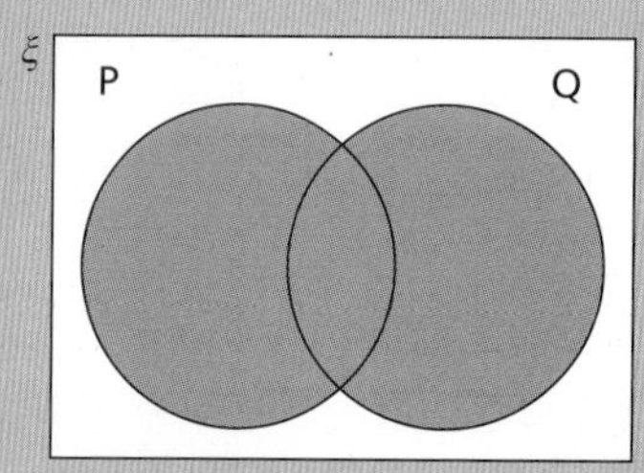

c

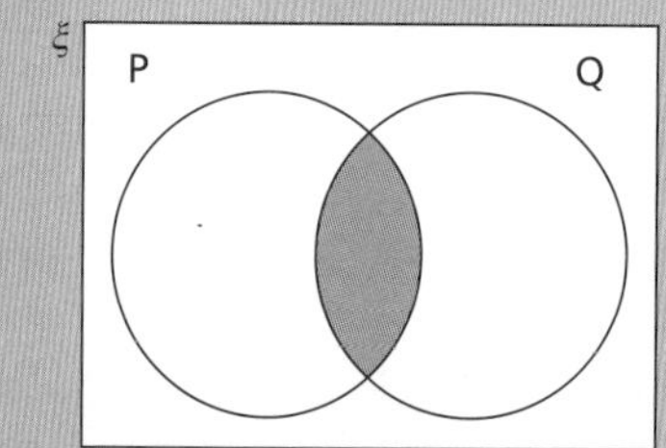

d

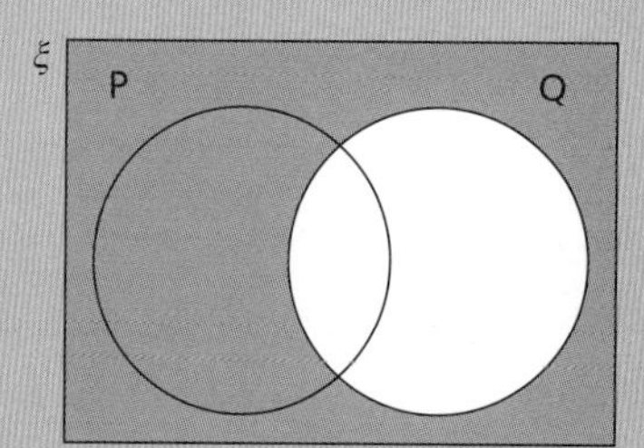

4 The Venn diagram represents the universal set, ξ.

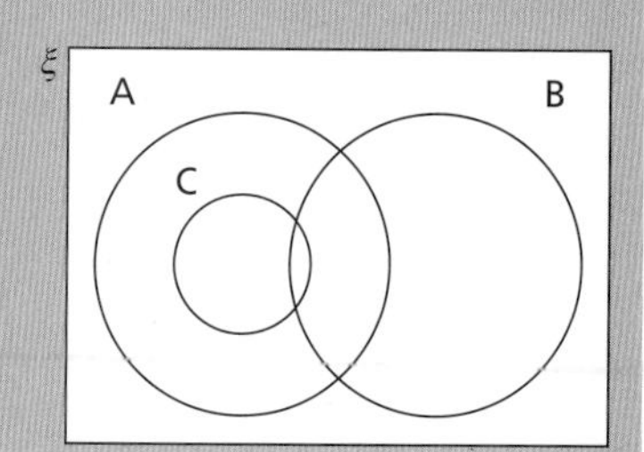

ξ = {positive integers}

A = {multiples of 2}

B = {multiples of 3}

C = {multiples of 4}

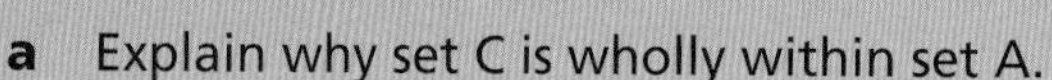

a Explain why set C is wholly within set A.

b Describe the elements of $A \cap B$.

c Describe the elements of $B \cap C$.

Are you ready? (B)

1 Some people were asked if they like coffee (C), tea (T) and hot chocolate (H).

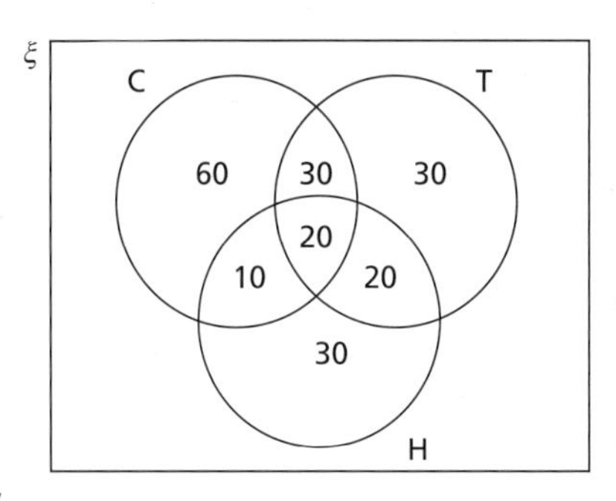

The Venn diagram represents the information from the responses.

a How many people like all three drinks?

b How many people were asked in total?

c How many people like only coffee?

d How many more people like 'tea and hot chocolate' than those that like 'coffee and hot chocolate'?

e How many people like exactly two drinks (not all three)?

$A \cap B \cap C$ means the **intersection** of sets A, B and C.

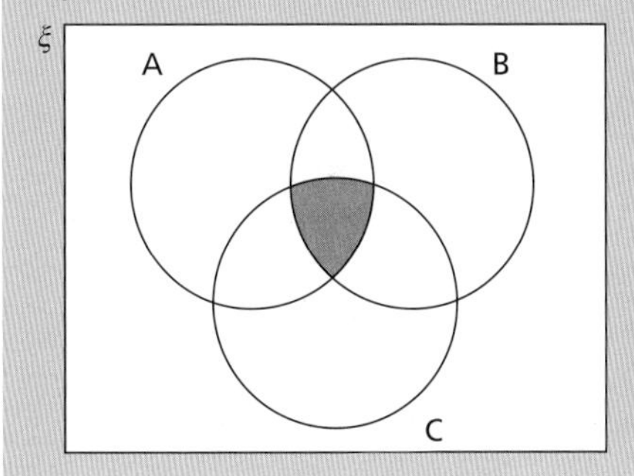

$A \cup B \cup C$ means the **union** of sets A, B and C.

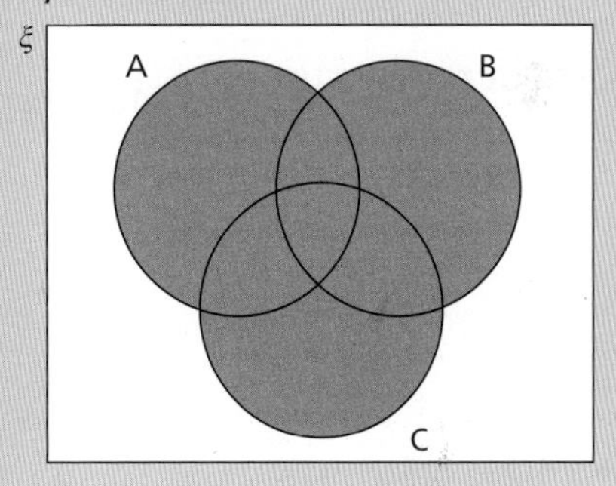

Example 1

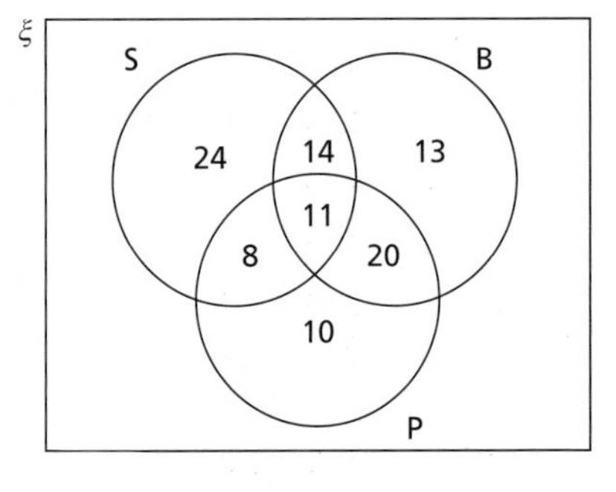

The Venn diagram shows the instruments played by 100 musicians in an orchestra.

S = musicians who play string instruments

B = musicians who play brass instruments

P = musicians who play percussion instruments

One of the musicians is selected at random.

a Work out the probability that the musician plays a brass instrument but not a percussion instrument.

b It is given that the randomly chosen musician plays a string instrument. Work out the probability that the musician plays a percussion instrument but not a brass instrument.

Method

Solution	Commentary
a $14 + 13 = 27$ P(brass but not percussion) $= \frac{27}{100}$	Work out the number of musicians who play a brass instrument but not a percussion instrument.
b 57 musicians play a string instrument. $\frac{8}{57}$	You only need to look at musicians who play a string instrument. Then within that set, find the number of musicians who play a percussion instrument but not a brass instrument.

Example 2

The universal set, ξ, consists of the integers 1 to 20 inclusive.

X = {factors of 36} Y = {triangular numbers} Z = {two-digit multiples of 3}

An element from ξ is selected at random.

Work out: **a** $P(X \cap Y)$ **b** $P(Y' \cap Z')$.

Method

Solution	Commentary
a The elements in $X \cap Y$ are {1, 3, 6} So $P(X \cap Y) = \frac{3}{20}$	Draw a Venn diagram to show the information about the sets. List the elements of each set and write them in their correct place in the Venn diagram.
b 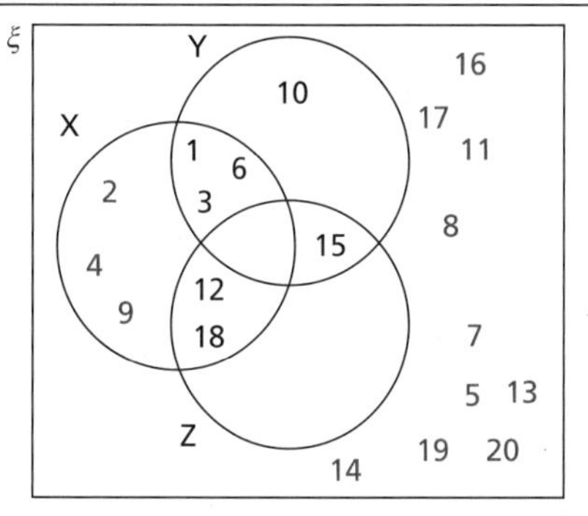 $Y' \cap Z'$ = {2, 4, 5, 7, 8, 9, 11, 13, 14, 16, 17, 19, 20} $P(Y' \cap Z') = \frac{13}{20}$	Use your Venn diagram to help identify the elements of $Y' \cap Z'$. $P(Y' \cap Z')$ $= \frac{\text{number of elements } Y' \cap Z'}{\text{total number of elements in } \xi}$

Practice (B)

1 Ed conducted a survey with students from his year group who own a smartphone (S), tablet (T) or a laptop (L). The Venn diagram shows his results.

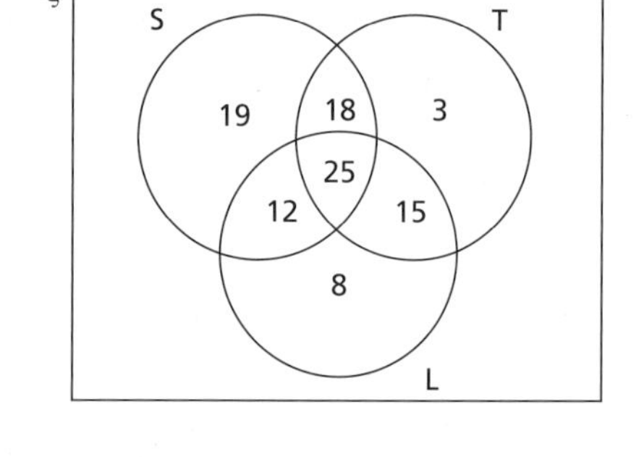

a How many people did Ed ask?

One of the students is picked at random.

b Work out the probability of choosing a student who owns:

i a laptop

ii a smartphone, tablet and laptop

iii a smartphone or laptop.

c Describe the set $S' \cap T$.

2 Visitors to a conference have a choice of three workshops to attend: A, B and C.

The Venn diagram shows information about the workshops attended by 1500 people at the conference.

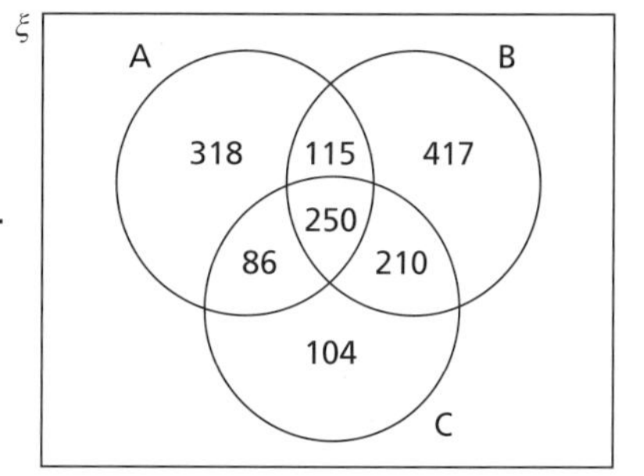

a How many people attended workshop A?

b What percentage of people at the conference attended workshop B? Give your answer to 1 decimal place.

c A person from the conference is chosen at random. Work out the probability that they attended workshop A but not workshop B.

d A person who attended workshop C is chosen at random. Work out the probability that this person only attended workshop C.

3 In a bookshop, 140 people were asked which type of books they read.

90 people read fiction (F), 70 people read non-fiction (N) and 60 people read poetry (P).

Of these, 40 people read both fiction and non-fiction, 30 people read both non-fiction and poetry, 20 people read both fiction and poetry, and 5 people read all three genres.

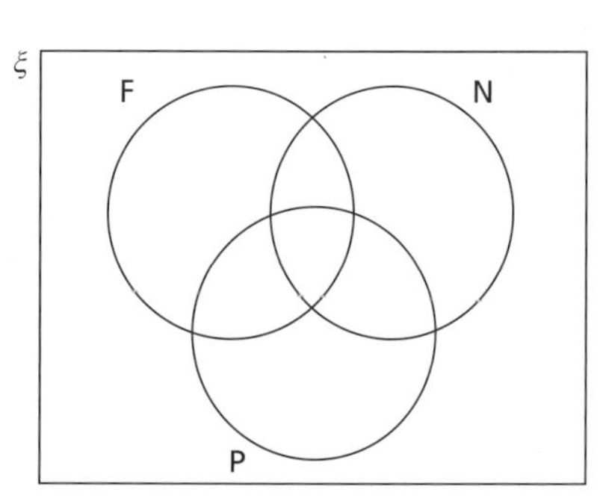

a Use the information to copy and complete the Venn diagram.

b Work out how many people do not read any of the genres.

c Work out the probability of choosing a person who reads non-fiction or poetry.

4 The Venn diagram shows the universal set ξ of positive integers from 1 to 20 inclusive.

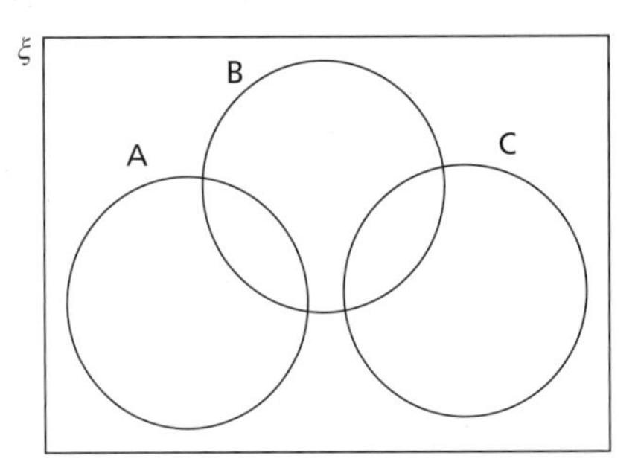

A = {prime numbers}

B = {integers between 10 and 20 inclusive}

C = {square numbers}

a Copy and complete the Venn diagram.

A number is chosen at random.

b Work out: **i** $P(B \cap C)$ **ii** $P(A \cup B \cup C)$.

5 The universal set ξ contains integers from 1 to 20 inclusive.

X = {factors of 12} Y = {odd numbers} Z = {multiples of 3}

a Represent this information in a Venn diagram.

An element from ξ is randomly selected.

b Work out: **i** $P(Y \cap Z)$ **ii** $P(X \cap Y \cap Z)$ **iii** $P(X \cup Y \cup Z)$ **iv** $P(X \cup Z')$.

6 In a survey of 120 students:

70 students play basketball (B),

60 students play football (F),

and 50 students play volleyball (V).

30 students play both basketball and football,

25 students play both football and volleyball,

20 students play both basketball and volleyball,

and 10 students play all three sports.

a Draw a Venn diagram to show this information.

One of the students is picked at random.

b Work out the probability that the student plays only basketball.

c Work out the probability that the student plays exactly two of the sports.

d Given that the student does **not** play football, work out probability that they play both basketball and volleyball.

7 160 people have brown hair, brown eyes or wear glasses.

40% of the group wear glasses.

20% of the group have brown hair and brown eyes.

15% of the group have brown hair and wear glasses.

10% of the group have brown eyes and wear glasses.

5% of the group have brown hair, brown eyes and wear glasses.

A person is picked at random from the group.

The probability of selecting a person with brown eyes is the same as the probability of selecting a person with brown hair.

Work out the number of people who have brown eyes but do not have brown hair nor wear glasses.

8 260 people at an activity centre are asked if they enjoy hiking, swimming and cycling.

The ratio of people who enjoy hiking to those who enjoy swimming is 1 : 3

The ratio of people who enjoy swimming to those who enjoy cycling is 2 : 1

30 people enjoy swimming and cycling.

Half of all the people who enjoy hiking also enjoy swimming.

The number of people who only enjoy hiking is the same as the number of people who enjoy hiking and cycling.

10 people enjoy all three activities.

Work out the probability that a person picked at random enjoys cycling but **not** swimming.

What do you think? (B)

1 Describe each shaded region shown using set notation.

a ξ F G H

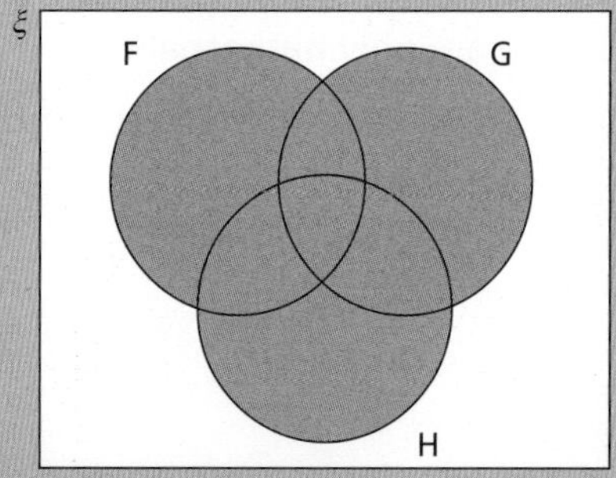

b ξ F G H

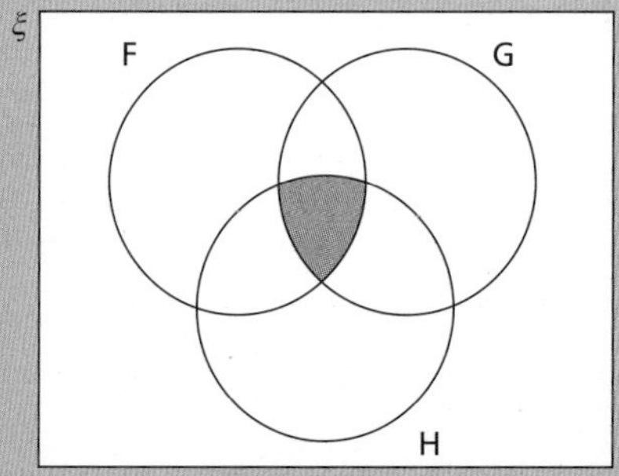

c ξ F G H

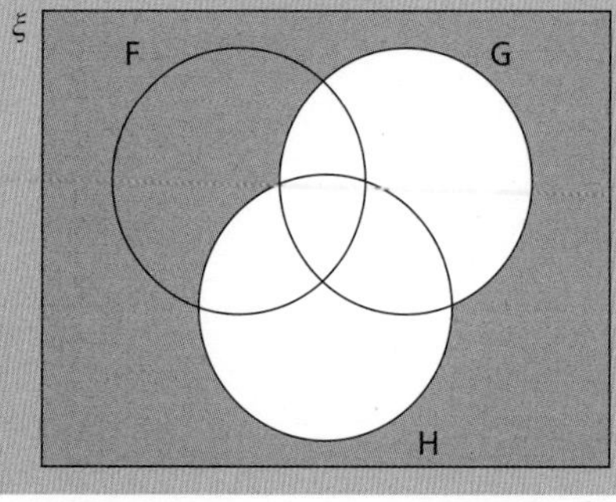

Consolidate – do you need more?

1 $\xi = \{1, 2, 3, 4, 5, 6, 7, 8, 9, 10\}$

$X = \{2, 4, 5, 6, 8, 9\}$

$Y = \{1, 5, 10\}$

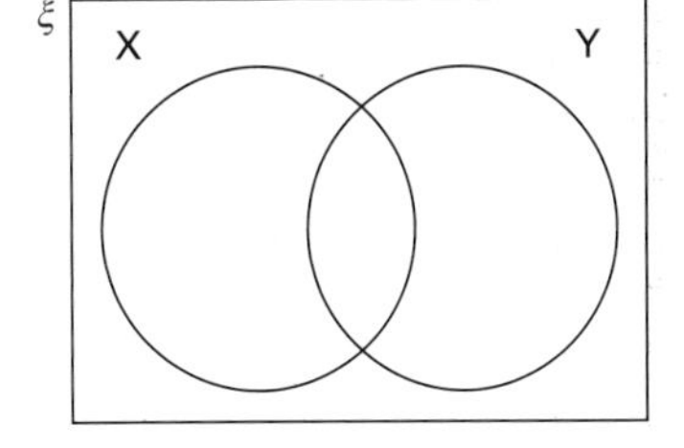

a Copy and complete the Venn diagram to represent the information.

A number from the universal set ξ is selected at random.

b Work out $P(X \cap Y)$. **c** Work out $P(Y')$.

2 50 people are surveyed to see if they watch the news on Channel A or Channel B.

30 people watch Channel A.

18 people watch both channels and 12 people watch neither.

a Draw a Venn diagram to represent this information.

b Work out the probability that a person chosen at random watches only Channel B.

c Work out the probability that a person chosen at random watches only one of the channels.

3 A school has 180 students in Year 11.

53 students study Food Technology, 48 students study Textiles, and 67 students study Electronics.

13 students study both Food Technology and Textiles.

12 students study both Food Technology and Electronics.

11 students study both Textiles and Electronics.

4 students study all three subjects.

a Draw a Venn diagram to represent this information.

A Year 11 student is picked at random.

b Work out the probability that the student studies Textiles but **not** Food Technology.

c Work out the probability that the student studies only Electronics.

It is known that the randomly picked student studies Food Technology.

d Work out the probability that the student studies all three subjects.

4 ξ = {odd numbers between 1 and 21 inclusive}

C = {factors of 60}

D = {multiples of 3}

E = {prime numbers}

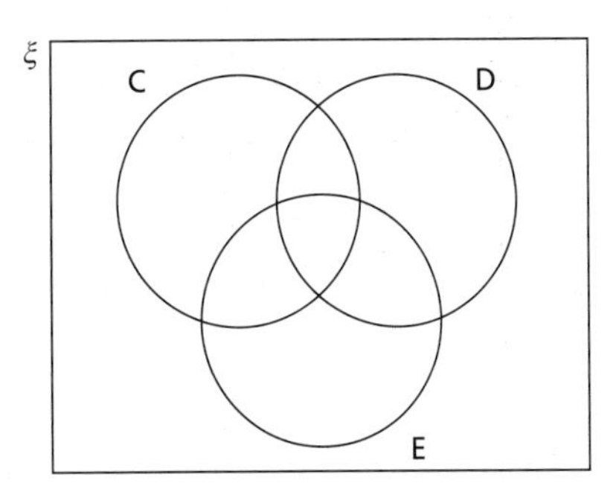

a Copy and complete the Venn diagram.

A number is selected at random.

b Work out:

i $P(C)$ **ii** $P(C \cap D \cap E)$ **iii** $P(D \cup E)$.

Stretch – can you deepen your learning?

1 A survey is conducted with people who drink tea. They are asked if they drink tea with milk or sugar.

9 people have both milk and sugar.

34% of people do not have milk or sugar.

17 people do not have milk or sugar.

a Work out how many people were in the survey.

b The probability of randomly selecting a person who has milk only is 44%.

Work out the probability of choosing someone who has sugar.

2 In a survey, 92 people like music (M), 105 people like novels (N) and x people like both music and novels.

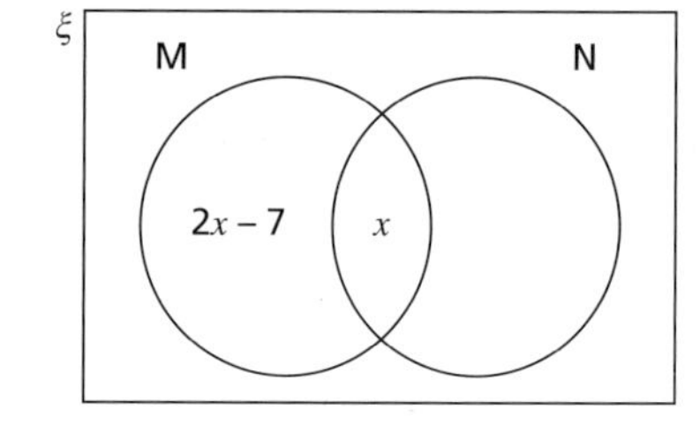

a Work out the value of x in the Venn diagram.

The number of people who do not like music or novels is $\frac{3}{11}x$.

b Work out the total number of people who took part in the survey.

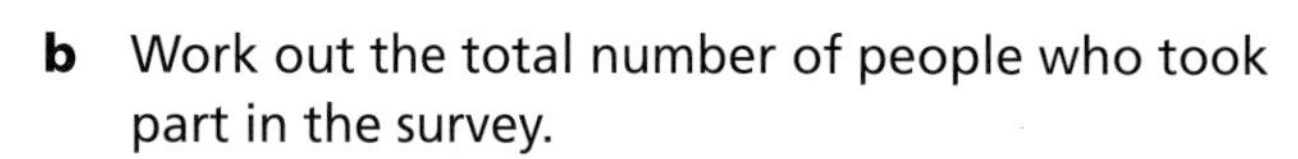

c Two people who like music are picked at random.

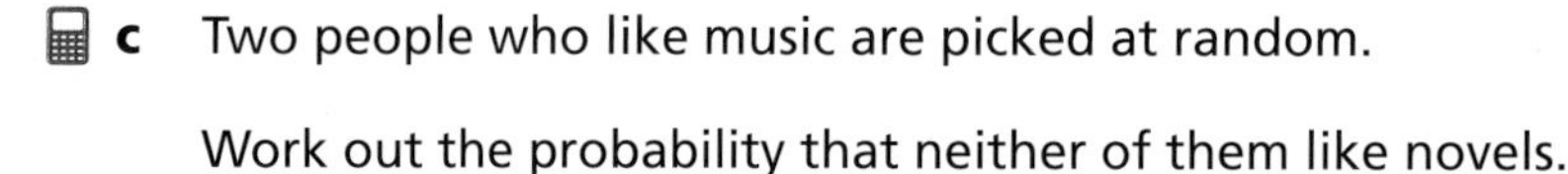

Work out the probability that neither of them like novels.

3 The Venn diagram shows information about animals in a zoo. There are 400 animals in total.

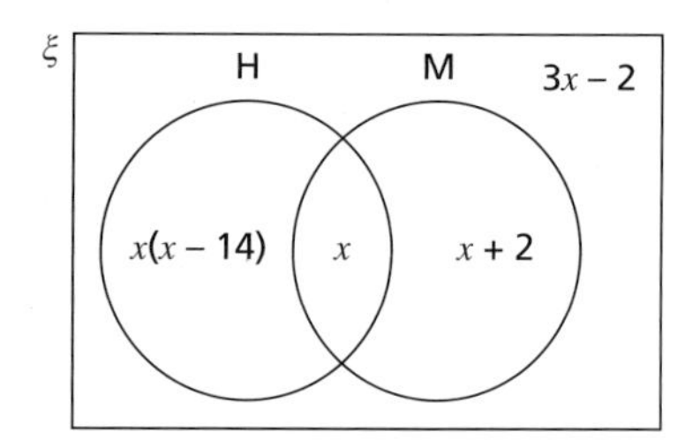

H = animals that are herbivores

M = animals that are mammals

An animal is selected randomly. It is a mammal.

Calculate the probability that it is also a herbivore.

4 The Venn diagram shows information about the number of elements in sets A, B and C.

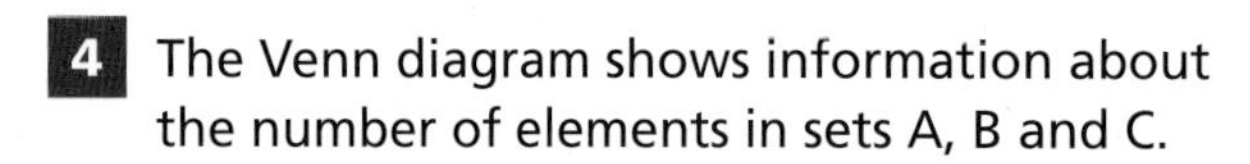

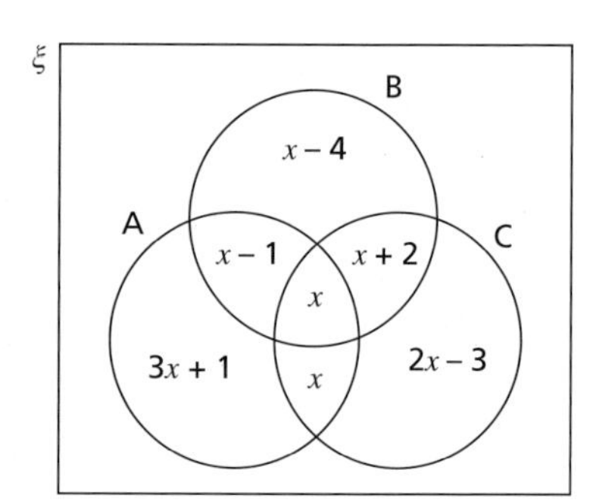

a Write an expression in terms of x for the total number of elements.

b An element is selected at random.

Write an expression in terms of x for P(B′).

c Given that an element is a member of set A, show that the probability it is also in set C is $\frac{1}{3}$

Probability diagrams: exam practice

White Rose MATHS

4–6

1 List all the two-digit numbers that can be made using these three number cards.

Each card can be used once only.

3 6 7

[2 marks]

2 A fair spinner is shown.

The spinner is spun twice and the difference between the numbers shown on the spinner is noted.

Work out the probability that the difference is no more than 1

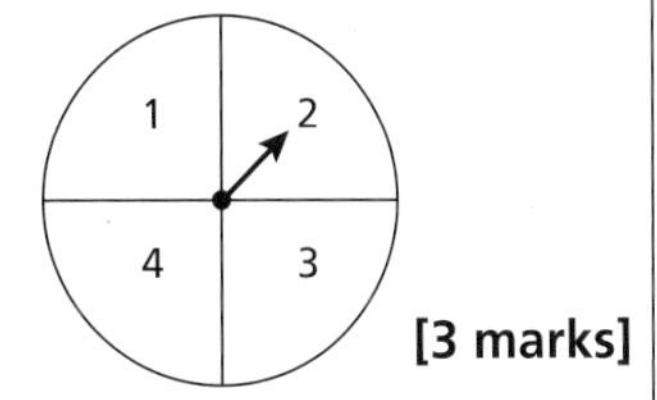

[3 marks]

3 In the Venn diagram,

Set A represents multiples of 5

Set B represents multiples of 3

Set C represents multiples of 10

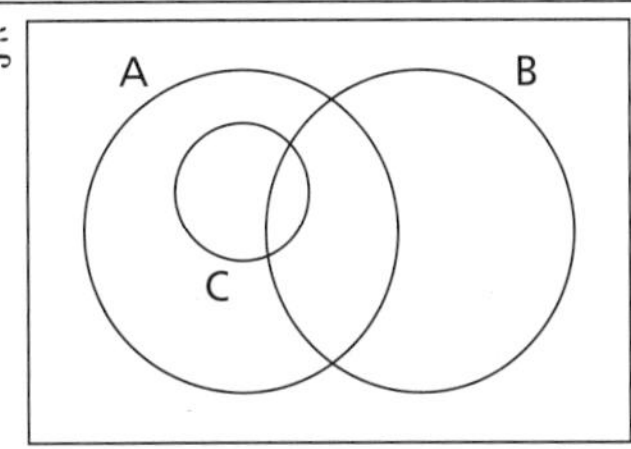

Write the numbers 12, 15, 16, 25, 30 and 40 in the correct regions on a copy of the diagram. **[3 marks]**

4 Emily and Rhys conduct an experiment to estimate the probability that a drawing pin lands point up when dropped.

Emily drops a drawing pin 80 times and it lands point up on 45 occasions.

Rhys drops the same drawing pin 40 times and it lands point up on 18 occasions.

Use this information to work out the best estimate for the probability that the drawing pin lands point up. **[2 marks]**

5 A gym has 200 members.

130 members use the running machines.

110 members use the weights equipment.

94 members use both the running machines and the weights equipment.

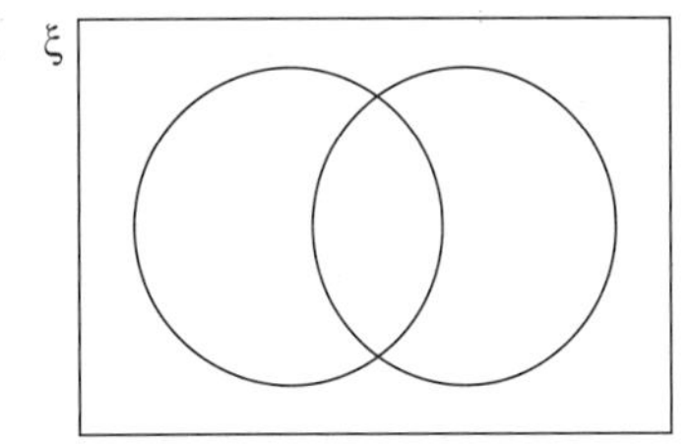

A member of the gym is selected at random.

What is the probability that they use exactly one of either a running machine or the weights equipment? You may use a copy of the Venn diagram to help you. **[3 marks]**

7–9

6 Shade the region that represents $P \cap Q'$ on a copy of the Venn diagram.

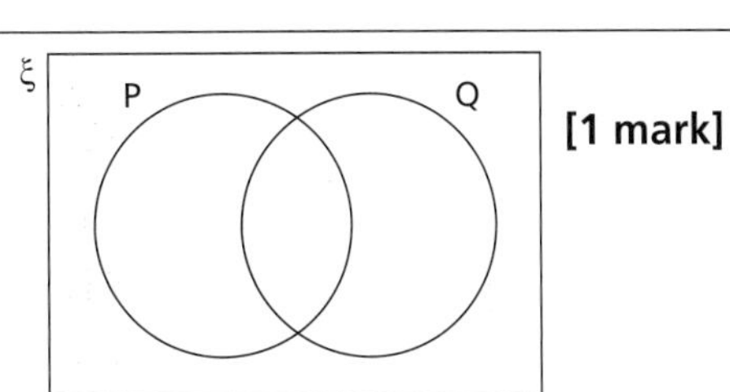

[1 mark]

11 Theory and experiment

In this block, we will cover...

11.1 Relative frequency

Example 1

Huda selects a paragraph from a book at random
vowel (A, E, I, O and U) occurs. She counts 150 vo

a The letter O appears 40 times.

Calculate the relative frequency of the letter O
3 decimal places.

b The relative frequency of the letter U is $\frac{1}{30}$

Work out how many times the letter U appea

c The relative frequency of the letter E is 0.32

Work out how many times the letter E appea

11.2 Expected outcomes

Practice

1. Rob spins a fair spinner 40 times.

   Estimate how many times the spinner wou
   each colour.

2. The table shows the probabilities of a bias
   landing on each possible outcome.

11.1 Relative frequency

Are you ready?

1 Ed scores 21 out of 30 in a science test.

Write his score as a simplified fraction.

2 Work out: **a** $\frac{3}{8}$ of 72 **b** $\frac{11}{20}$ of 120

3 Work out: **a** 0.6×28 **b** $180 \times \frac{1}{5}$

The probability of an event can be estimated by performing an **experiment**, which can be repeated over a number of **trials**.

The **relative frequency** of the event in the experiment tells you how often the event occurs as a proportion of the total number of trials. It is normally expressed as a decimal.

For example, if a coin is flipped 20 times and it lands on tails 12 times then the relative frequency of the coin landing on tails is $\frac{12}{20}$, or 0.6

However, the **theoretical probability** of the event 'the coin lands on tails', based on equally likely outcomes (a '**fair**' coin), would be 0.5

As an experiment is repeated more times (increased number of trials), the relative frequency tends to **converge** (get closer and closer to) to the actual, unknown, probability.

In general, the relative frequency of the event = $\frac{\text{frequency of event}}{\text{total number of trials}}$

Comparing the relative frequency of an event with the theoretical probability can be a useful way of judging, for example, whether a coin is fair or is **biased**.

Example 1

Huda selects a paragraph from a book at random and counts the number of times each vowel (A, E, I, O and U) occurs. She counts 150 vowels in the paragraph.

a The letter O appears 40 times.

Calculate the relative frequency of the letter O. Give your answer as a decimal, correct to 3 decimal places.

b The relative frequency of the letter U is $\frac{1}{30}$

Work out how many times the letter U appears in the paragraph.

c The relative frequency of the letter E is 0.32

Work out how many times the letter E appears in the paragraph.

d The letter A appears three more times than the letter I.

Write the ratio of the number of times the letter A appears to the number of times the letter I appears.

Method

Solution	Commentary
a $\frac{40}{150} = 0.267$	Use $\frac{\text{frequency of letter O}}{\text{total number of vowels}}$ to find the relative frequency.
b $150 \div 30 = 5$	Find $\frac{1}{30}$ of 150
c $0.32 \times 150 = 48$	Multiply the relative frequency by the total number of vowels.
d $40 + 5 + 48 = 93$ $150 - 93 = 57$	Find the sum of the other vowels and subtract from 150 to find the sum of the frequencies of the letters A and I. You could use a comparison bar model labelling the longer bar A, the shorter bar I, and also the total and the difference.
A I 3 57 A I 54	
$54 \div 2 = 27$ I = 27 and A = 30	Subtract 3 from the longer bar to make the bars equal and then divide by 2
30 : 27	Write the ratio in the correct order.

Example 2

Amina rolls a six-sided dice numbered 1 to 6 and keeps count how many times she rolls a 3

She records her results in a frequency table.

Number of rolls	Number of 3s	Relative frequency
10	2	
50	9	
100	17	
500	81	
1000	165	

a If the dice is fair, write the probability of rolling a 3 as a decimal.

b Complete the table, writing the relative frequencies as decimals.

c Do you think the dice is fair? Explain your answer.

Method

Solution	Commentary
a $P(3) = \frac{1}{6} = 0.1\dot{6}$	If the dice is fair then all outcomes are equally likely.
b (see table below)	Divide the number of 3s by the number of trials each time.
c As 0.165 is very close to the theoretical probability, it seems likely that the dice is fair.	Compare the relative frequency with the theoretical probability and state your conclusion.

Number of rolls	Number of 3s	Relative frequency
10	2	2 ÷ 10 = 0.2
50	9	9 ÷ 50 = 0.18
100	17	17 ÷ 100 = 0.17
500	81	81 ÷ 500 = 0.162
1000	165	165 ÷ 1000 = 0.165

Practice

1 A survey is conducted with the residents of a town. They are asked what mode of transport they use most often.

2000 use cars, 1500 use bicycles, and 1000 use public transport.

a How many people take part in the survey?

b Work out the relative frequencies of each mode of transport.

2 Abdullah selects a card from a bag at random.

He records the colour and returns the card to the bag. He repeats this 40 times.

a Copy and complete the table.

Colour	Frequency	Relative frequency
Red	11	
Yellow	15	
Green	14	
Total	40	

b There are 20 cards in the bag.

Estimate how many green cards are in the bag.

c Beca says, "The only colours of cards in the bag are red, yellow and green."

Explain why Beca may **not** be correct.

3 A spinner is split into three equal sections labelled X, Y and Z. It is spun 400 times.

a The spinner lands on Y a total of 85 times.

Work out the relative frequency of the spinner landing on Y.

b The relative frequency of the spinner landing on Z is 0.21

How many times did the spinner land on Z?

4 300 students from Year 10 are asked what their preferred film genre is. Some of the results are recorded in the table.

Genre	Action	Comedy	Horror	Science fiction
Frequency	120	90	60	x

a Work out the relative frequency of students who prefer science fiction films.

b Work out the relative frequency of students who do **not** prefer horror.

5 Mario, Flo and Sven each roll a different six-sided dice numbered 1 to 6

In the table below, they record the number of rolls and how many times their dice lands on 5

	Frequency of 5	**Total number of rolls**
Mario	10	30
Flo	24	120
Sven	15	60

Whose dice is most likely to be fair? Give a reason for your answer.

6 From a survey of 400 students, 240 were found to enjoy playing video games, 180 enjoy reading books, and 100 enjoy both.

Show that the relative frequency of students who enjoy only video games or only books is 0.55

7 A hockey team plays 24 games in a season.

The table shows the relative frequency of the team winning after a certain number of games.

Number of games	6	12	18	24
Relative frequency of a win	0.5	0.25	$0.\dot{4}$	0.5

a How many games did the team win between the 7th and 12th games?

b What proportion of the matches did the team win between game 13 and 18?

What do you think?

1 A spinner is spun 300 times.

Which of the following values are **not** possible relative frequencies for the spinner to land on sector A?

$\frac{2}{7}$	1	0	0.54	71%	$\frac{7}{75}$	75

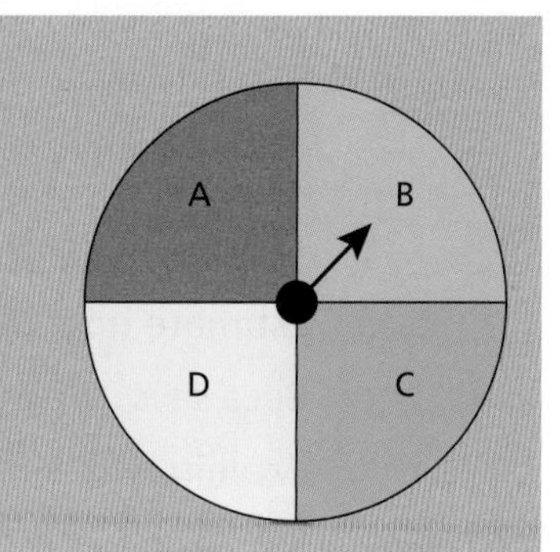

2 A bag contains orange counters, purple counters and grey counters only.

A counter is taken at random, its colour is recorded and it is then returned to the bag.

Benji conducts the experiment 60 times.

a A grey counter is chosen 23 times.

Work out the relative frequency of a grey counter.

b The relative frequency of a purple counter is 0.3

How many times is an orange counter chosen?

The bag contains an equal number of each colour of counter.

c Estimate how many times a grey counter would be picked after 300 trials.

d If 30 000 000 trials were conducted, how many times would you expect a grey counter to be picked?

How exact do you think your answer is?

Consolidate – do you need more?

1 Transport officials conducted a survey with some passengers on how they got to the airport.

149 used their own vehicle, 182 used public transport, and 4 walked.

Work out the relative frequencies of each mode of transport.

2 Emily selects a marble from a bag at random.

She records the pattern and returns the marble to the bag.

She repeats this 15 times.

a Copy and complete the table.

Pattern	Frequency	Relative frequency
Stripes	5	
Spots	2	
Stars	8	
Total	15	

b There are 60 marbles in the bag.

Estimate how many marbles with spots are in the bag.

3 A 20-sided dice numbered 1 to 20 is rolled 120 times.

The dice lands on an even number 42 times.

a Work out the relative frequency of an even number.

b The relative frequency of a prime number is 0.15

How many times did the spinner land on a prime number?

4 280 students from Year 11 were asked what fruit they like the most.

Some of the results are recorded in the table.

Fruit	Apple	Banana	Pear	Grapes
Frequency	75	90	65	g

a Work out the relative frequency of a student liking grapes the most.

b Work out the relative frequency of a student **not** liking banana the most.

Stretch – can you deepen your learning?

1 A coin is flipped x times.

The coin lands on heads three times.

Write an expression for the relative frequency of the coin landing on tails.

2 The number of trials in an experiment is given by the expression $x(11 - x)$.

The number of times the outcome is event A is $3x - 5$

The relative frequency of event A is $\frac{1}{3}$

a Show that $x^2 - 2x - 15 = 0$

b Show that the total number of trials is 30

11.2 Expected outcomes

White Rose MATHS

Are you ready?

1 Work out: **a** $\frac{6}{11}$ of 55 **b** 12% of 50 **c** 0.45×1300

2 Work out: **a** $72 \div 0.6$ **b** $36 \div \frac{2}{5}$ **c** $14 \div 0.35$

You can find an estimate for the number of times an event will occur in an experiment by multiplying the theoretical probability of the event by the total number of trials.

For example, if a fair dice is rolled 120 times then an estimate for the number of sixes that should occur is $\frac{1}{6} \times 120 = 20$. This is called the **expected frequency** of the event. It does not mean that exactly 20 sixes are expected every time the dice is rolled 120 times, but on average after repeating the experiment a large number of times, there should be 20 sixes for every 120 rolls.

So, expected frequency of an event A = P(A) × number of trials

where P(A) is the theoretical probability of A occurring.

Example

A spinner can land on red, blue, yellow or green.

The table shows the probabilities of the spinner landing on red or green.

P(blue) = 2 × P(yellow)

Estimate the number of times the spinner will land on blue if it is spun 600 times.

Colour	Probability
Red	$\frac{5}{12}$
Blue	
Yellow	
Green	$\frac{1}{12}$

Method

Solution	Commentary
P(red) + P(green) = $\frac{5}{12} + \frac{1}{12} = \frac{6}{12} = \frac{1}{2}$	Add together the given probabilities and subtract their total from 1. $1 - \frac{1}{2} = \frac{1}{2}$
P(yellow) = x P(blue) = $2x$ P(yellow or blue): $3x = \frac{1}{2}$ $x = \frac{1}{6}$ P(blue) = $\frac{2}{6}$ or $\frac{1}{3}$	Express the probabilities of yellow and blue algebraically. Divide the remaining probability by 3 to obtain P(yellow) and then double this to find P(blue).
$\frac{1}{3} \times 600 = 200$	Estimate the number of times the spinner lands on blue by multiplying P(blue) by the total number of spins.

Practice

1 Rob spins this fair spinner 40 times.

Estimate how many times the spinner would land on each colour.

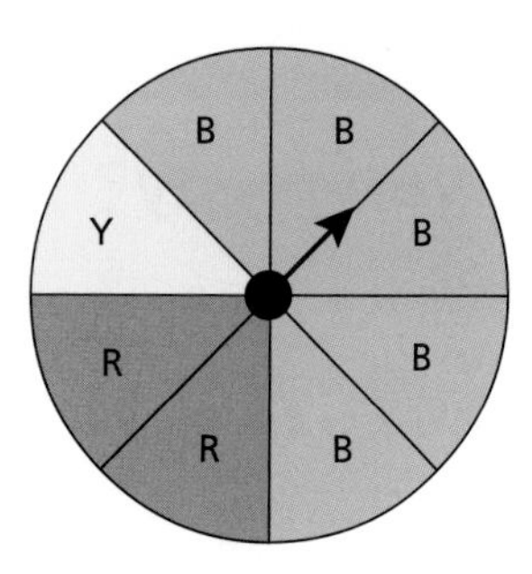

2 The table shows the probabilities of a biased, four-sided spinner landing on each possible outcome.

Outcome	Probability	Expected frequency after 200 spins
1	20%	
2	$\frac{3}{20}$	
3	0.37	
4	$\frac{7}{25}$	

Copy and complete the table.

3 A new medicine is developed to treat an infection. The probability of the new medicine successfully treating the infection is 0.97

Estimate the number of successful treatments if the medicine is used 1000 times.

4 The Wi-Fi connectivity in a shopping centre has an 88% chance of connecting a smartphone to the Internet.

Estimate how many smartphones will make a successful connection if 2525 smartphones attempt to connect to the Wi-Fi.

5 A random number generator selects a number between 1 and 100

30 numbers are selected at random. A number may be chosen more than once.

How many times would you expect to get a number that:

a ends in a 9 **b** is a multiple of 5 **c** is even but **not** a multiple of 10?

6 The table shows the probabilities that a biased, five-sided spinner will land on each of the letters A, B, C, and D.

Letter	A	B	C	D	E
Probability	0.18	0.31	0.15	0.03	

The spinner is spun 300 times.

Work out an estimate for the number of times the spinner will land on A or E.

7. The table shows the probabilities that a biased dice will land on 1, 2, 3, or 6

Number	1	2	3	4	5	6
Probability	0.17	0.08	0.13			0.2

The ratio of the probability of rolling 4 to the probability of rolling 5 is 4 : 3

If the dice is rolled 400 times, estimate how many times it will land on an odd number.

8. The probability of Benji scoring a penalty is 0.75

Benji scored 12 penalties in a season.

a Estimate the total number of penalties Benji took.

b Explain why your answer to part **a** is an estimate.

9. A factory makes 3 000 000 cupcakes a day. There is a 0.002% chance that a cupcake contains a trace of eggshell.

Work out an estimate for the number of cupcakes made on a randomly chosen day that will contain a trace of eggshell.

What do you think?

1. Ed flips a coin 100 times and gets tails 53 times.

He says, "I think the coin is biased because it has not landed on tails 50 times."

a Do you agree with Ed?

b How many tails do you think would need to be recorded before thinking the coin is biased, if the coin is flipped:

i 100 times **ii** 1000 times **iii** 10 000 times?

Consolidate – do you need more?

1. Rhys spins this spinner 30 times.

Estimate how many times the spinner will land on each colour.

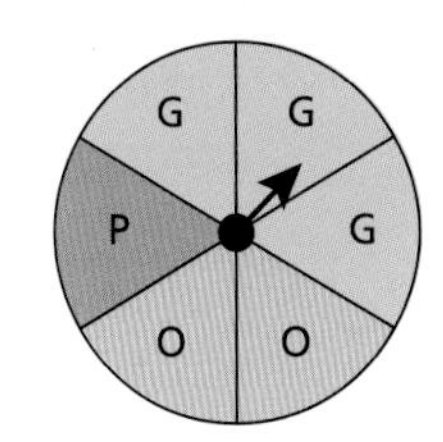

2. The table shows the probabilities of a biased spinner labelled A, B, C, and D landing on each possible outcome.

Outcome	Probability	Expected frequency after 400 spins
A	18%	
B	$\frac{13}{40}$	
C	$\frac{1}{50}$	
D	0.475	

Copy and complete the table.

3 The probability of Marta throwing a basketball into a hoop 3 metres away is 0.63

a Estimate the number of successful shots if Marta throws the basketball:

i 500 times **ii** 1500 times.

b Estimate how many shots Marta will need to make at least 40 successful shots.

4 Two fair, six-sided dice numbered 1 to 6 are rolled and their sum is recorded.

The dice are rolled 72 times.

How many times would you expect to get a score:

a of 3 **b** that is a factor of 12 **c** that is greater than 9?

Stretch – can you deepen your learning?

1 Ali, Junaid and Tiff are playing a game where each person flips a coin at the same time.

Ali wins if exactly one of the three coins lands on tails.

Junaid wins if exactly two of the three coins land on tails.

Tiff wins if all three coins land on tails.

Nobody wins if all three coins land on heads.

They play the game 24 times.

a Estimate the number of times Ali wins the game.

b Estimate the number of times the game ends with a winner.

2 A biased coin is flipped n times. n is a positive integer.

The expected number of heads is twice the expected number of tails.

a Explain why P(heads) = $\frac{2}{3}$

A different biased coin is flipped n times. The expected number of heads is three times the expected number of tails.

b Work out P(heads).

A third biased coin is flipped n times. The expected number of heads is k times the expected number of tails, where k is a positive integer.

c Write an expression for P(heads) in terms of k.

Theory and experiment: exam practice

4–6

1 The probability that a biased coin will land on heads is 0.4

The coin is flipped 300 times.

Work out an estimate for the number of times the coin will land on heads. **[2 marks]**

2 A bag contains only red counters, white counters and blue counters.

A counter is taken at random from the bag.

The table shows the probability of the counter being red or white.

Colour	Red	White	Blue
Probability	0.3	0.5	

A counter is taken from the bag and replaced 200 times.

Work out an estimate for the number of times a blue counter is selected. **[3 marks]**

3 A biased spinner can land on 1, 2, 3 or 4

The probability that the spinner lands on 1 is 0.1

The probability that the spinner lands on 2 is 0.3

The probability that the spinner lands on 3 is three times the probability that the spinner lands on 4

Work out the probability that the spinner lands on an odd number. **[3 marks]**

4 The table shows the results of rolling a dice 60 times.

Result	1	2	3	4	5	6
Frequency	7	9	7	10	17	10

(a) Work out the relative frequency of rolling an even number. **[3 marks]**

(b) Jackson thinks the dice is biased.

Give a reason to support his claim. **[1 mark]**

5 A machine sorts apples and rejects any that have a mass less than 80 grams.

The table shows the number of apples sorted and the number of apples rejected over five 10-minute periods.

Period	1	2	3	4	5
Number of apples sorted	250	300	220	260	220
Number of apples rejected	12	18	12	16	12

Work out the best estimate for the probability that an apple is rejected. **[2 marks]**

12 Combined events

In this block, we will cover...

12.1 The multiplication rule

Example 1

Here are two spinners:

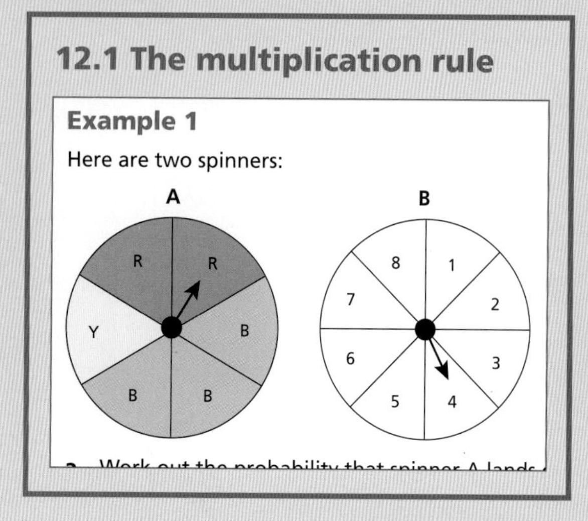

12.2 Tree diagrams

Practice

1. Jakub flips a fair coin and records the resu
   a Draw a tree diagram to represent all th
   b Work out the probability that:
      i the coin lands on heads twice
      ii the coin lands on heads exactly onc
      iii the coin does **not** land on heads.
2. Faith has two fair spinners, A and B. She sp spins spinner B.

12.3 Conditional probability

Consolidate – do you need more

1. Letter tiles are used to spell out the word
   Two letters are randomly selected without
   a What is the probability of choosing th
   b What is the probability of choosing tw
2. Abdullah and Junaid have a box of milk a
   There are 14 milk chocolates and 11 dark
   Abdullah chooses one chocolate at rando

12.1 The multiplication rule

White Rose MATHS

Are you ready?

1 Calculate:

a $\frac{1}{4} \times \frac{1}{6}$ **b** 0.2×0.3 **c** 0.4^2 **d** $\left(\frac{2}{3}\right)^2$

2 Work out:

a $\frac{2}{5} + \frac{1}{6}$ **b** $\frac{7}{10} \times \frac{3}{5} + \frac{3}{10} \times \frac{2}{5}$

Events that do not affect each other are known as **independent events**.

For example, the outcomes of spinning a spinner and randomly selecting a card from a pack are independent of each other. The outcome of the spinner will not affect the choice of card.

A pack contains cards labelled P, Q, R and S in equal numbers.

A spinner has three equal sectors coloured blue, yellow and green.

The sample space diagram shows the outcomes of randomly selecting a particular card and spinning a particular colour.

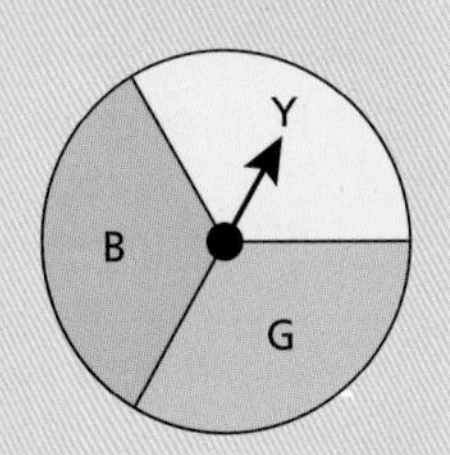

Spinner	Card			
	P	**Q**	**R**	**S**
Yellow	(Y, P)	(Y, Q)	(Y, R)	(Y, S)
Green	(G, P)	(G, Q)	(G, R)	(G, S)
Blue	(B, P)	(B, Q)	(B, R)	(B, S)

The probability of randomly selecting a card labelled Q is $\frac{1}{4}$

The probability of spinning green on the spinner is $\frac{1}{3}$

From the sample space, you can see that the probability of selecting a card labelled Q and spinning green is $\frac{1}{12}$

Notice that P(G and Q) $= \frac{1}{4} \times \frac{1}{3} = \frac{1}{12}$

In general, if events A and B are independent then P(A and B) = P(A) × P(B)

Example 1

Here are two spinners:

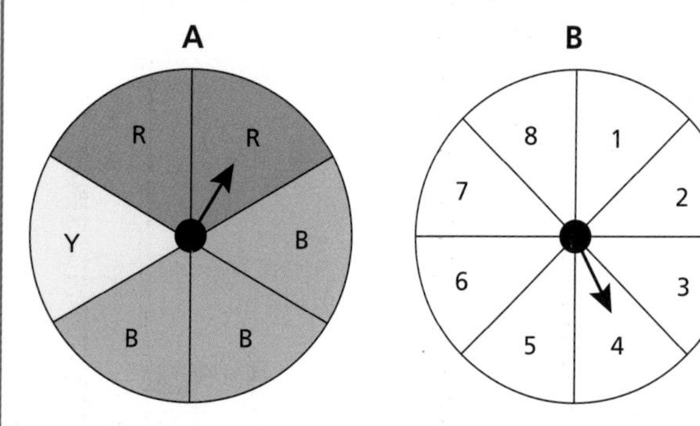

a Work out the probability that spinner A lands on red and spinner B lands on an odd number.

b Work out the probability that spinner A does **not** land on yellow and spinner B lands on a factor of 60

Method

Solution	Commentary
a P(spinner A landing on red) = $\frac{2}{6}$ P(spinner B landing on odd) = $\frac{4}{8}$	Find the probability of each independent event.
$\frac{2}{6} \times \frac{4}{8} = \frac{8}{48} = \frac{1}{6}$	Multiply the probabilities together. In an exam, you don't have to simplify a probability unless you are specifically asked to.
b P(spinner A not landing on yellow) = $\frac{5}{6}$ P(spinner B landing on a factor of 60) = $\frac{6}{8}$	Five of the sectors in spinner A are not yellow. 1, 2, 3, 4, 5 and 6 are factors of 60
$\frac{5}{6} \times \frac{6}{8} = \frac{30}{48} = \frac{5}{8}$	

Example 2

L, M and N are independent events. P(L) = 0.4 P(M) = 0.35 P(M and N) = 0.07

Work out: **a** P(L and M) **b** P(N) **c** P(L′ and N)

Method

Solution	Commentary
a P(L and M) = P(L) × P(M) = 0.4 × 0.35 = 0.14	Find the product of the two probabilities.

b $P(M \text{ and } N) = P(M) \times P(N)$ $0.07 = 0.35 \times P(N)$ $\frac{0.07}{0.35} = P(N)$ $P(N) = 0.2$	Use the probabilities given in the question to set up an equation. Rerrange the equation to find P(N).
c $P(L') = 0.6$ $P(L' \text{ and } N) = P(L') \times P(N)$ $= 0.6 \times 0.2$ $= 0.12$	$P(L') = 1 - P(L)$ Find the product of the two probabilities.

Practice

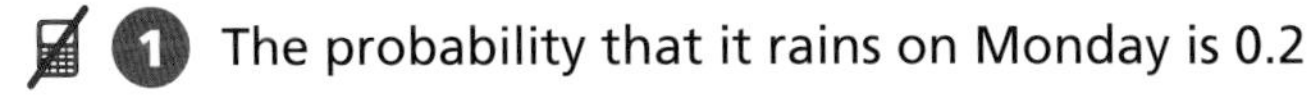

1 The probability that it rains on Monday is 0.2

The probability that Amina is late to school on Monday is 0.1

Assuming that these two events are independent of each other, work out the probability that on Monday it rains and Amina is late to school.

2 The spinners are divided into sectors with equal area.

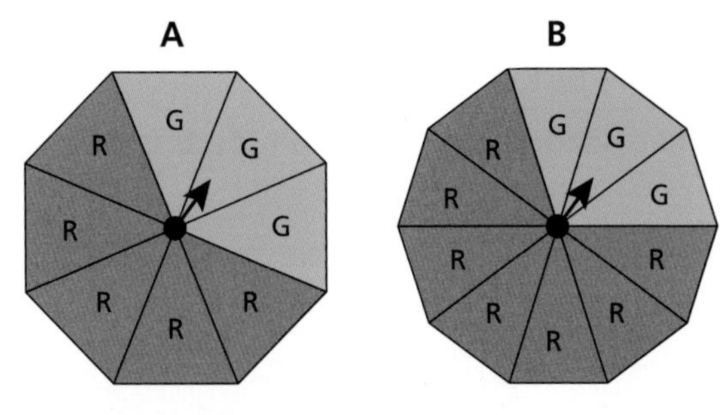

Both spinners are spun.

Work out the probability that:

a both spinners land on red

b both spinners land on green

c both spinners land on the same colour

d the spinners land on different colours.

3 Class 11X has 32 students and Class 11Y has 28 students.

There are 14 blonde-haired students in Class 11X and 10 blonde-haired students in Class 11Y.

A student is picked at random from each class.

a Work out the probability that both students have blonde hair.

b Work out the probability that neither student has blonde hair.

4. The probability of winning a fairground game is $\frac{3}{5}$
Marta plays the fairground game twice.

Assuming that the probability of winning the game stays the same, work out the probability of:

a Marta winning both games **b** Marta losing both games.

5. A and B are independent events. P(A) = 0.45 P(B) = 0.3
Work out:

a P(A and B) **b** P(A′ and B) **c** P(A′ and B′).

6. Two fair, six-sided dice numbered 1 to 6 are rolled.
Work out the probability that both dice show a score that is a factor of 12

7. E and F are independent events. P(E) = $\frac{3}{11}$ P(E and F) = $\frac{4}{15}$
Work out P(F).

8. A fair coin is flipped three times.

a Work out the probability that the coin lands on tails on all three flips.

b Show that P(landing on heads exactly twice) $< \frac{1}{2}$

9. The probability that Kath passes her History test is $\frac{7}{10}$
The probability that Kath passes both her History and Geography tests is 0.63

Work out the probability of Kath passing her Geography test.

What do you think?

1. A security door is unlocked with a four-digit code using digits 0–9
A random four-digit code is set.

a Work out the probability that:

i all four digits are 8s

ii all four digits are even

iii all four digits are prime.

b Which combinations are least likely to be guessed? Discuss.

Consolidate – do you need more?

1 X and Y are independent events. P(X) = 0.64 P(Y) = 0.8

Work out P(X and Y).

2 Bev is travelling by train from Leeds to Cardiff.

She has to change trains at Derby.

The probability that the train from Leeds to Derby is cancelled is 0.25

The probability that the train from Derby to Cardiff is cancelled is 0.1

Work out the probability that neither train is cancelled.

3 Two fair, six-sided dice numbered 1 to 6 are rolled.

a Work out the probability that neither dice shows 5

b Work out the probability that both dice show a square number.

Stretch – can you deepen your learning?

1 Ed has two bags of coins.

One bag contains three 10p coins, four 20p coins and two 50p coins.

The other bag contains two 10p coins, three 20p coins and three 50p coins.

A coin is taken from each bag.

Work out the probability that the two coins:

a are the same **b** are different **c** sum to less than 50p.

2 A and B are independent events.

$P(A) = \frac{1}{x+1}$ $P(A \text{ and } B) = \frac{1}{x+2}$ where $x \geqslant 0$

Work out P(B).

3 X, Y and Z are independent events.

$P(X \text{ and } Y) = \frac{a}{a+b}$ $P(Y \text{ and } Z) = ab$

Work out P(X and Z).

12.2 Tree diagrams

Are you ready?

1 Work out each calculation and give your answers in simplest form.

a $\frac{1}{3} \times \frac{2}{3}$ **b** $\frac{3}{10} \times \frac{4}{5}$ **c** $\frac{8}{9} \times \frac{15}{16}$

2 Work out: **a** 0.4×0.2 **b** 0.15×0.8 **c** 0.48×0.16

3 Work out: **a** $\frac{21}{48} + \frac{13}{48}$ **b** $1 - \frac{12}{17}$ **c** $1 - \left(\frac{8}{23} + \frac{11}{23}\right)$

4 A box contains 3 red pens, 4 blue pens and 8 black pens.

Work out the probability that a pen selected at random is:

a blue **b** purple **c** red or black **d** **not** black.

Tree diagrams are used to represent two or more events and to see all possible outcomes.

The probability of each possible outcome is written **on** the branch of the tree and the outcome is written at the **end** of the branch.

For example, a straw is picked at random from a box with 6 green straws (G) and 5 blue straws (B), and then replaced. Another straw is then taken at random.

This tree diagram represents this situation.

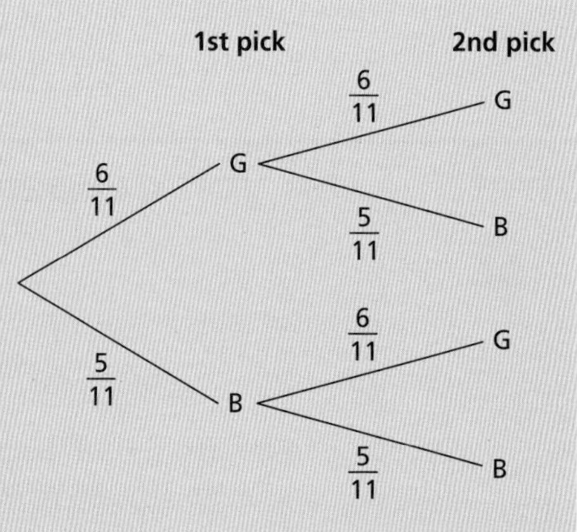

The events are independent (picking the first straw doesn't affect the second choice of straw). You can use the tree diagram to help work out probabilities using the multiplication rule from Chapter 12.1

Example 1

There are 4 pairs of white socks and 3 pairs of grey socks in a drawer.

A pair of socks is chosen at random, then put back in the drawer.

Another pair is then chosen.

a Complete the tree diagram.

b Work out the probability that:

i both pairs are grey

ii one pair of each colour sock is picked

iii at least one pair of socks is white.

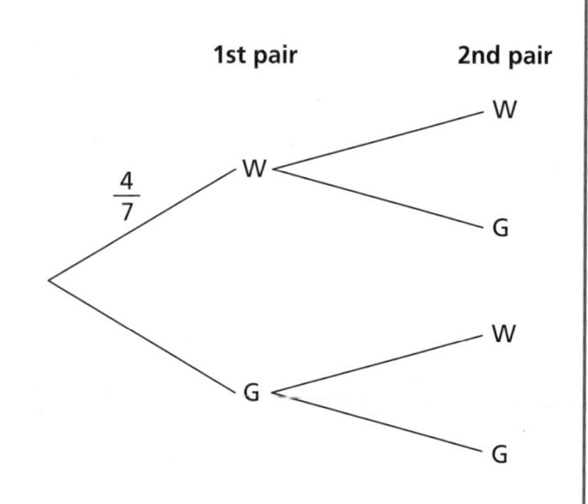

Method

<table>
<tr><th>Solution</th><th>Commentary</th></tr>
<tr><td>

a

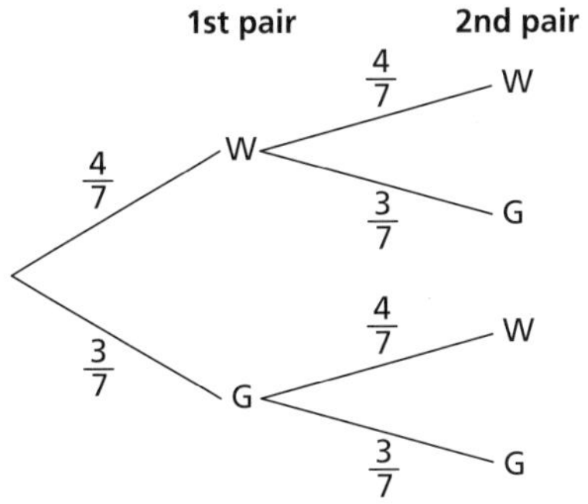

</td><td>There are 3 grey pairs of socks out of a total of 7 pairs of socks.
The probability of choosing a grey pair of socks is $\frac{3}{7}$
This is the same for each branch representing a grey pair of socks.
The probability of choosing a white pair of socks is $\frac{4}{7}$</td></tr>
<tr><td>

b **i** P(both grey) = P(grey) × P(grey)

$= \frac{3}{7} \times \frac{3}{7} = \frac{9}{49}$

</td><td>This corresponds to the bottom pair of branches of the tree diagram. You can use the multiplication rule to multiply along the branches.</td></tr>
<tr><td>

ii P(one pair of socks of each colour)

= P(1st is white, 2nd is grey) + P(1st is grey, 2nd is white)

$= \frac{4}{7} \times \frac{3}{7} + \frac{3}{7} \times \frac{4}{7}$

$= \frac{12}{49} + \frac{12}{49} = \frac{24}{49}$

</td><td>This corresponds to the 'middle two' pairs of branches.
You can use the multiplication rule to work out the probability of each event and then add the two probabilities together.</td></tr>
<tr><td>

iii Method A

P(at least one pair white) = P(both white) + P(1st is white, 2nd is grey) + P(1st is grey, 2nd is white)

$= \frac{4}{7} \times \frac{4}{7} + \frac{4}{7} \times \frac{3}{7} + \frac{3}{7} \times \frac{4}{7} = \frac{16}{49} + \frac{12}{49} + \frac{12}{49} = \frac{40}{49}$

</td><td>This corresponds to the top pair of branches plus the middle two pairs of branches. You can use the multiplication rule to work out the probability of each event and then add the probabilities together.</td></tr>
<tr><td>

Method B

P(at least one pair white) = 1 – P(both grey)

$1 - \left(\frac{3}{7} \times \frac{3}{7}\right) = 1 - \frac{9}{49} = \frac{40}{49}$

</td><td>You can also find the probability of no pairs of white socks being picked and subtract this from 1</td></tr>
</table>

Example 2

The probability of Rhys being on time for school is 0.95

The probability of Huda being on time for school is 0.88

The probabilities are independent of each other.

a Draw a tree diagram to represent this information.

b Work out the probability of both Rhys and Huda being on time.

c Work out the probability of only one of them being on time.

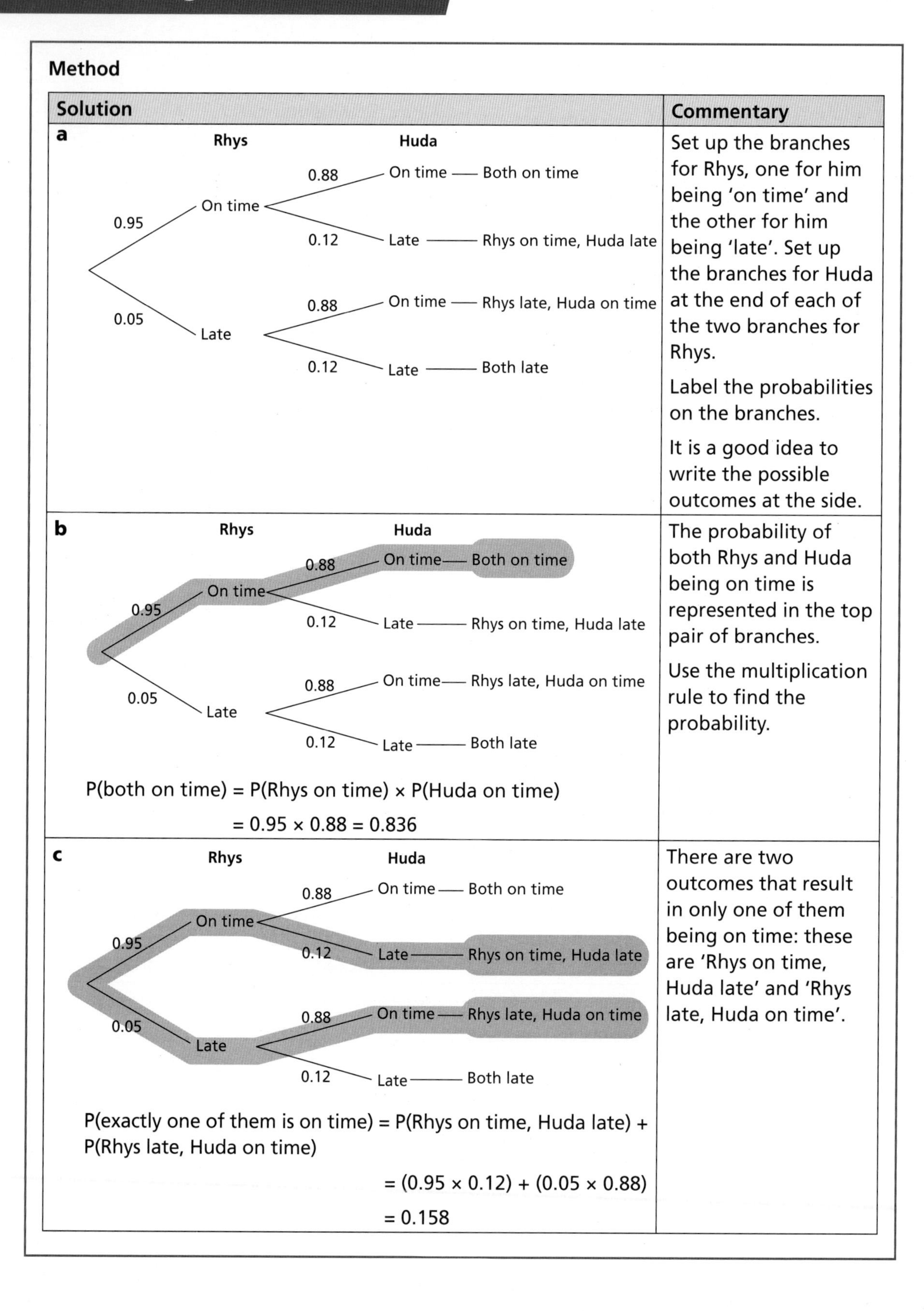

Method

Solution	Commentary
a Tree diagram: Rhys 0.95 On time, 0.05 Late; Huda 0.88 On time, 0.12 Late. Outcomes: Both on time; Rhys on time, Huda late; Rhys late, Huda on time; Both late	Set up the branches for Rhys, one for him being 'on time' and the other for him being 'late'. Set up the branches for Huda at the end of each of the two branches for Rhys. Label the probabilities on the branches. It is a good idea to write the possible outcomes at the side.
b P(both on time) = P(Rhys on time) × P(Huda on time) = 0.95 × 0.88 = 0.836	The probability of both Rhys and Huda being on time is represented in the top pair of branches. Use the multiplication rule to find the probability.
c P(exactly one of them is on time) = P(Rhys on time, Huda late) + P(Rhys late, Huda on time) = (0.95 × 0.12) + (0.05 × 0.88) = 0.158	There are two outcomes that result in only one of them being on time: these are 'Rhys on time, Huda late' and 'Rhys late, Huda on time'.

Practice

1 Jakub flips a fair coin and records the result before flipping the coin again.

a Draw a tree diagram to represent all the possible outcomes of Jakub's experiment.

b Work out the probability that:

i the coin lands on heads twice

ii the coin lands on heads exactly once

iii the coin does **not** land on heads.

2 Faith has two fair spinners, A and B. She spins spinner A, notes the outcome and then spins spinner B.

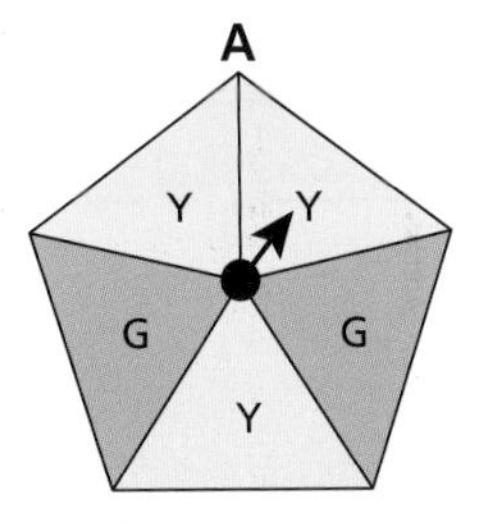

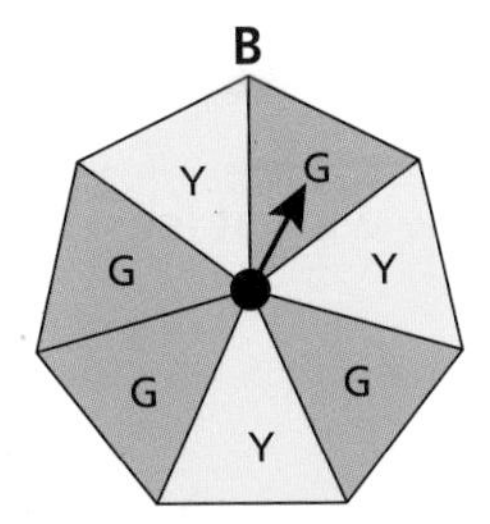

a Draw a tree diagram to represent all the possible outcomes of Faith's experiment.

b Work out the probability that spinner A lands on yellow and spinner B lands on green.

c Faith says, "The probability that both spinners show yellow is equal to the probability that both spinners show green."

Show that Faith is incorrect.

3 A fair, six-sided dice numbered 1 to 6 is rolled.

Jackson notes whether the dice lands on a prime number or not.

He rolls it again and notes whether the dice lands on a factor of 20

a Copy and complete the tree diagram.

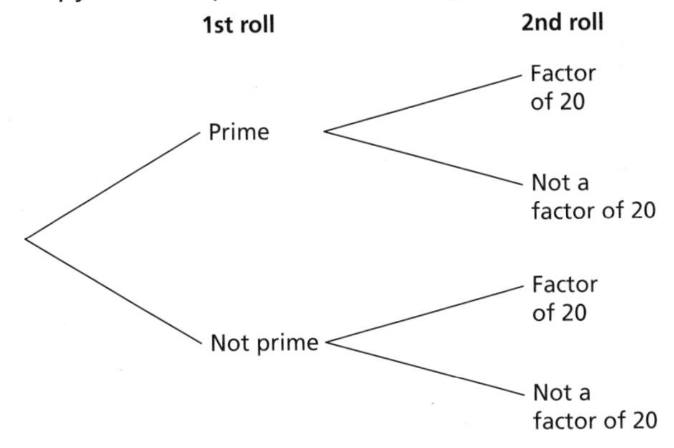

b Work out the probability that:

i the 1st roll is a prime number and the 2nd roll is a factor of 20

ii the 1st roll is not a prime number and the 2nd roll is a factor of 20

iii the 1st roll is a prime number and the 2nd roll is not a factor of 20

iv the 1st roll is not a prime number and the 2nd roll is not a factor of 20

4 The probability that a student picked at random from Class 11P wears glasses is 55%.

The probability that a student picked at random from Class 11Q wears glasses is 40%.

A student is picked at random from each class.

a Draw a tree diagram to represent this information.

b Work out the probability of each of the possible outcomes.

5 A fairground stall offers two games.

Chloe plays both games.

a Draw a tree diagram to represent all the possible outcomes of Chloe playing both games.

b Work out the probability of Chloe winning exactly one game.

c Work out the probability that Chloe wins at least one of the games.

Game A

1 in 5 chance of winning!

Game B

75% chance of winning!

6 Samira spins this spinner twice.

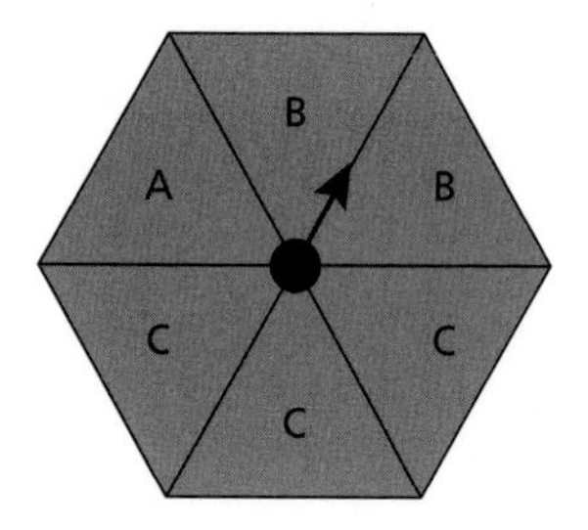

a Copy and complete the tree diagram.

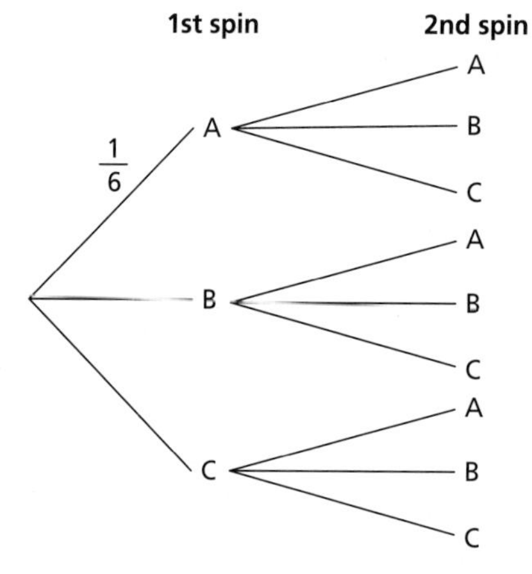

b Work out the probability that the spinner:

i lands on C and then A

ii lands on the same letter each spin

iii lands on A and B in any order.

7 A bag contains 4 red marbles, 3 blue marbles and 5 green marbles.

Lida picks a marble at random from the bag, notes its colour, and then puts it back in the bag.

Her friend Mario then chooses a marble at random from the bag and notes its colour.

a Draw a tree diagram to represent this information.

b Work out the probability that:

i both marbles are green

ii at least one marble is blue.

8 Filipo rolls a fair, six-sided dice numbered 1 to 6 three times. He notes when the dice lands on 6

a Copy and complete the tree diagram.

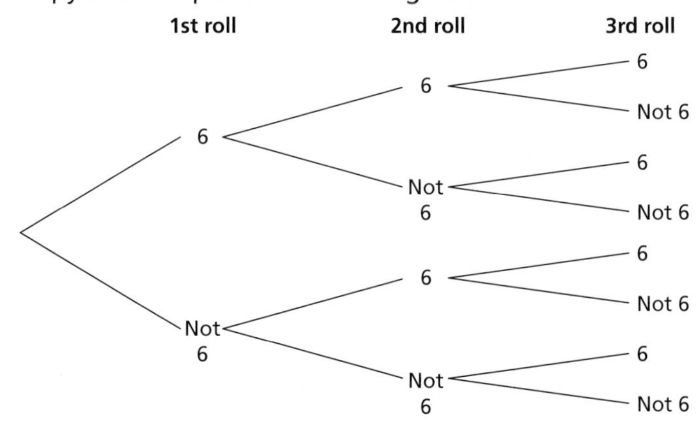

b Work out the probability that Filipo does **not** roll a 6

c Work out the probability that Filipo rolls a 6 exactly once.

What do you think?

1 Chloe and Zach are both sitting an exam.

The probability of Chloe passing the exam is 0.78

The probability of Zach passing the exam is 0.77

By using a probability tree diagram or otherwise, work out:

a P(both pass) **b** P(neither of them pass) **c** P(at least one of them passes).

Consolidate – do you need more?

1 Lida and Junaid both sit a test.

The probability that Lida passes the test is 0.75

The probability that Junaid passes the test is 0.73

a Copy and complete the tree diagram.

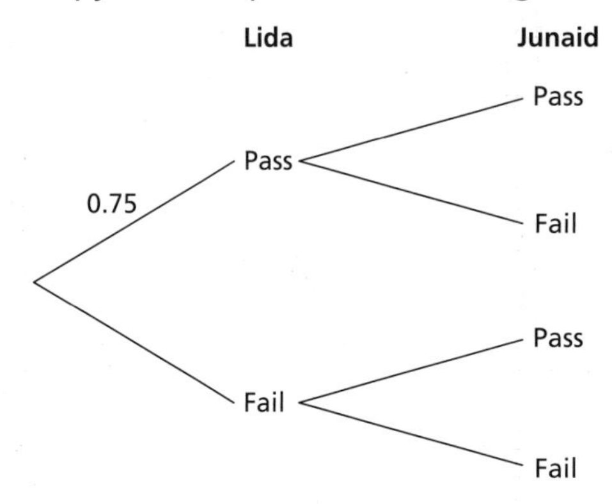

b Work out the probability that exactly one of them passes the test.

2 Sven and Tiff are playing a game that involves rolling a fair, six-sided dice numbered 1 to 6

Sven wins if the dice lands on 1, 2 or 3

Tiff wins if the dice lands on 4, 5, or 6

They play the game twice.

a Draw a tree diagram to represent the possible outcomes of the two games.

b Work out the probability that either Sven or Tiff win both games.

c Work out the probability that Sven does **not** win either game.

3 Samira has two bags that both contain gold ribbons and green ribbons.

Bag A has 12 ribbons, of which 7 are green.

Bag B has 10 ribbons, of which 7 are gold.

Samira chooses a ribbon at random from each bag.

a Draw a tree diagram to represent all the possible outcomes of Samira's choices.

b Work out the probability that she chooses a gold ribbon from both bags.

c Work out the probability that she chooses different colours.

4 A box contains 5 black cards, 3 white cards and 2 grey cards.

Benji picks a card at random from the box and then puts it back. He then picks a second card from the box.

a Draw a tree diagram to represent this information.

b Work out the probability that:

i the two cards are different in colour **ii** exactly one card is grey.

Stretch – can you deepen your learning?

1 Seb picks a fruit at random from a bowl, notes the type of fruit and puts it back. He then picks another fruit.

The probability of Seb picking a banana is 0.3

The probability of Seb picking a banana and then a pear is 0.04

a Work out the probability of Seb picking a pear.

The fruit bowl only has bananas, pears and oranges.

b Work out the probability of Seb choosing an orange and then a banana.

2 Amina and Ali each have a bag of red counters and blue counters. Together they have 30 counters.

Amina has 6 blue counters. She has 3 more red counters than Ali.

Ali has 3 more red counters than blue counters.

a Work out how many of each colour counter Amina and Ali have.

They each pick a counter from their bag at random.

b Work out the probability that they both select the same colour of counter.

3 A biscuit tin contains x custard creams and 3 bourbon biscuits.

A biscuit is selected at random from the tin and then put back.

Another biscuit is then selected at random.

The probability of choosing the same type of biscuit both times is $\frac{73}{121}$

Show that $x = 8$

4 There are k pens in a box.

Four of the pens are blue. The rest of the pens are red.

Bev takes a pen at random from the box and doesn't replace it.

Bev then takes another pen at random from the box.

The probability that Bev takes a pen of each colour is $\frac{8}{15}$

a Show that $k^2 - 16k + 60 = 0$

b Solve $k^2 - 16k + 60 = 0$ to work out two possible values of k.

12.3 Conditional probability

White Rose MATHS

Are you ready?

1 Work out each calculation and give your answers in simplest form.

a $\frac{1}{3} \times \frac{2}{5}$ **b** $\frac{4}{9} \times \frac{3}{8}$ **c** $\frac{14}{15} \times \frac{25}{28}$

2 A box contains 11 gold cards and 14 silver cards.

a What fraction of the cards are:

i gold **ii** silver?

A gold card is removed from the bag.

b Now what fraction of the cards are:

i gold **ii** silver?

3 The Venn diagram shows information about the number of students in a class that enjoy baking (B) or cooking (C).

A student is selected at random.

Given that they enjoy baking, work out the probability that they enjoy cooking.

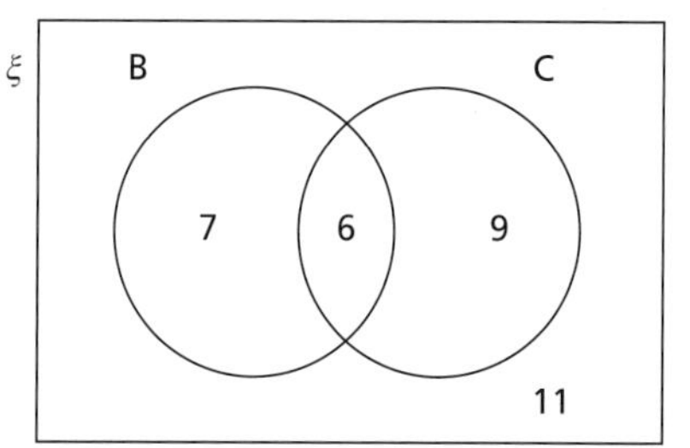

If an event depends on the outcome of another event, then the events are **dependent**. The probability of dependent events is known as **conditional probability**.

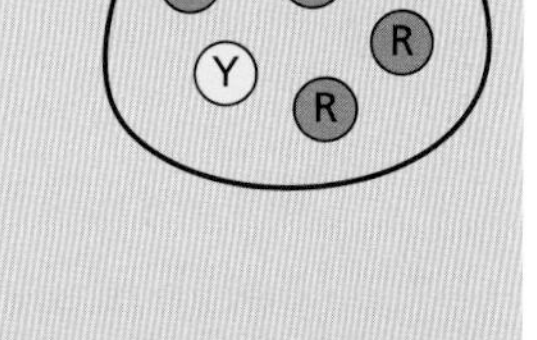

For example, a bag contains 4 red counters and 3 yellow counters.

The probability of randomly choosing a red counter is $\frac{4}{7}$ and the probability of randomly choosing a yellow counter is $\frac{3}{7}$

A counter is selected at random but not replaced.

The total number of counters in the bag has decreased by 1, and there is now one fewer red or yellow counter. The probability of selecting a red counter the second time depends on which counter was selected first:

P(1st counter is red) = $\frac{4}{7}$

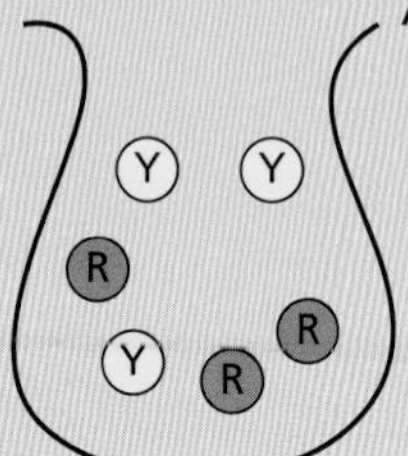

P(2nd counter is red) = $\frac{3}{6}$,

P(2nd counter is yellow) = $\frac{3}{6}$

P(1st counter is yellow) = $\frac{3}{7}$

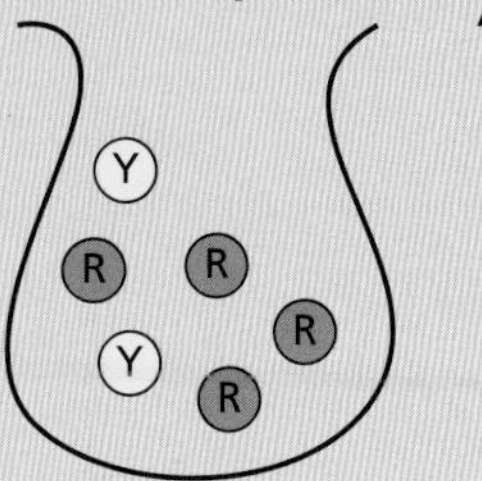

P(2nd counter is red) = $\frac{4}{6}$,

P(2nd counter is yellow) = $\frac{2}{6}$

A tree diagram helps to visualise all the possible outcomes.

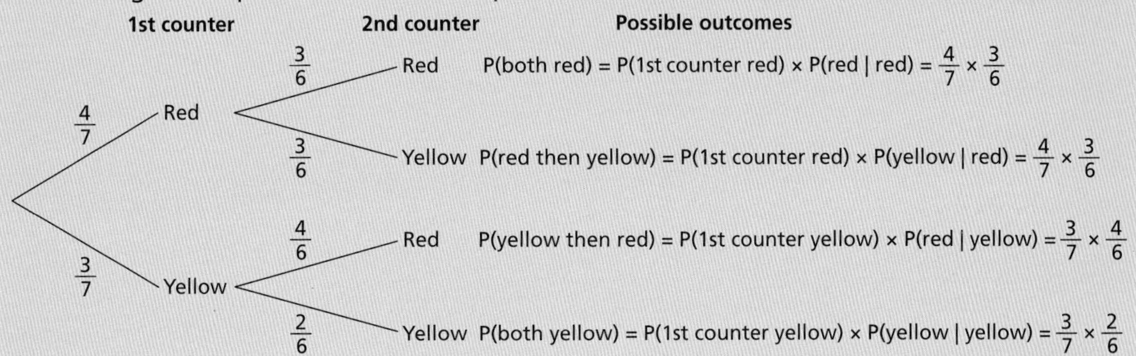

In the diagram above, the notation P(yellow | red) means 'the probability of choosing a yellow counter given that the first counter chosen was red'. Similarly for P(red | yellow), and so on.

To calculate conditional probabilities, you can use the formula:

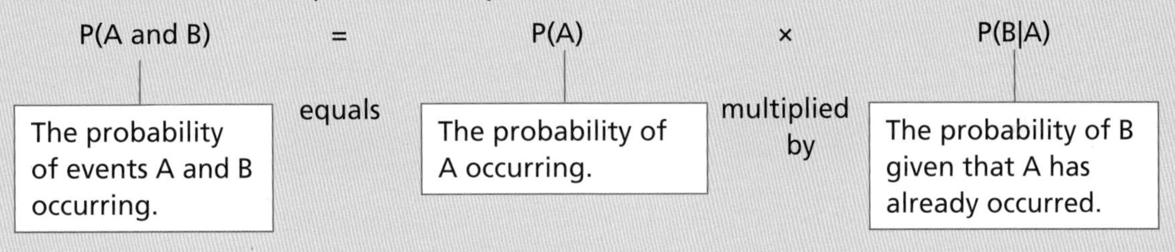

This is similar to the multiplication rule seen in Chapter 12.1

Example 1

Here are nine cards:

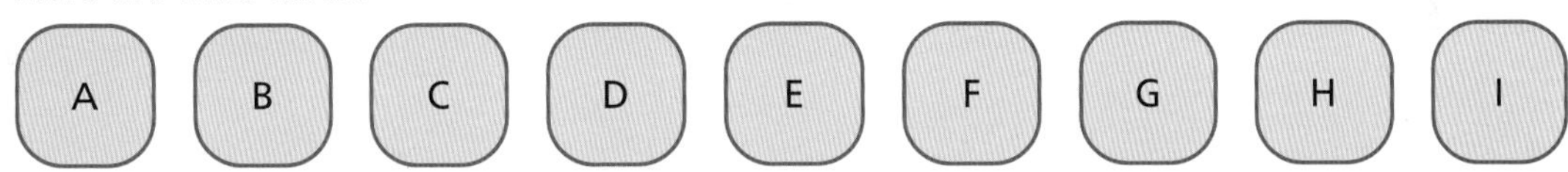

A card is chosen at random and not replaced. A second card is then taken.

a What is the probability that the first card selected is a vowel?

b Given that the first card is a consonant, what is the probability that the second card is a consonant?

Method

Solution	Commentary
a P(vowel) = $\frac{3}{9} = \frac{1}{3}$	There are nine cards, of which three are vowels.
b P(2nd card is a consonant | 1st card is a consonant) = $\frac{5}{8}$	Given that the first card was a consonant means there is one fewer consonant left (5) and one fewer card overall (8).

Example 2

A bag contains 4 black counters and 8 white counters.

Huda takes two counters out of the bag at random, one after the other without replacing the first.

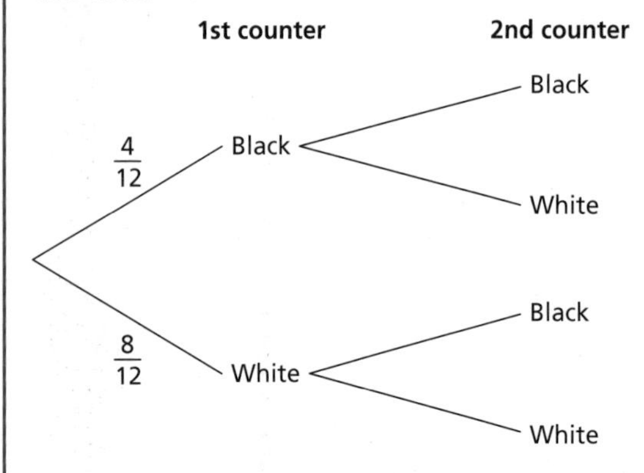

a Complete the tree diagram.

b Work out the probability that Huda takes two counters of the same colour.

c Work out the probability that Huda takes at least one white counter.

Method

Solution	Commentary
a 1st counter / 2nd counter $\frac{4}{12}$ Black: $\frac{3}{11}$ Black, $\frac{8}{11}$ White $\frac{8}{12}$ White: $\frac{4}{11}$ Black, $\frac{7}{11}$ White	Don't simplify the fractions on the branches; you want the denominators on each set of branches to be the same. After the first counter is taken, there will be 11 counters remaining in the bag. Therefore the denominator of each of the probabilities for the second counter will be 11 If Huda takes a black counter, there will only be 3 black counters but still 8 white counters. If she takes a white counter, there will still be 4 black counters but only 7 white counters remaining. Check that the probabilities in each set of branches sum to 1
b P(same colour) = P(BB) + P(WW) $= \frac{4}{12} \times \frac{3}{11} + \frac{8}{12} \times \frac{7}{11}$ $= \frac{68}{132}$	Use the multiplication rule to work out the probability of choosing 2 black counters and 2 white counters. Add together the two probabilities to find the probability of choosing two counters of the same colour.
c P(at least one white counter) = 1 – P(no white counters) $1 - \text{P(BB)} = 1 - \frac{12}{132} = \frac{120}{132}$	Instead of calculating and adding together P(WW) + P(BW) + P(WB), you can use the probability of not choosing a white counter for both choices and subtract this from 1

Example 3

Here are eight number cards:

2

3 5 5 7 7

Two cards are picked without replacement.

Work out:

a P(the sum of the cards is a square number)

b P(the product of the cards is a square number).

Method

Solution	Commentary
a Square numbers: 1, 4, 9, … 1: not possible 4: 2, 2 9: 7, 2 or 2, 7	Listing all the possible pairs of numbers that sum to a square can be a more efficient way to represent the information instead of a tree diagram.
P(4) = P(1st pick 2) × P(2nd pick 2) $= \frac{2}{8} \times \frac{1}{7} = \frac{2}{56}$	4 can only be made one way. There are two 2s out of eight cards in total. The probability of picking a 2 is $\frac{2}{8}$ After 2 is picked, there is one fewer 2 and one fewer card in total. The probability of picking a 2 again is now $\frac{1}{7}$
P(9) = P(1st pick 7) × P(2nd pick 2) + P(1st pick 2) × P(2nd pick 7) $= \frac{2}{8} \times \frac{2}{7} + \frac{2}{8} \times \frac{2}{7} = \frac{4}{56} + \frac{4}{56} = \frac{8}{56}$	9 can be made in two ways. The probability of picking a 7 is $\frac{2}{8}$ The probability of picking a 2 after picking a 7 is $\frac{2}{7}$ If a 2 is picked first, the probability of picking a 7 is $\frac{2}{7}$ Multiply together the probabilities of making 9, and then add them.
P(the sum of the cards is a square number) $= \frac{2}{56} + \frac{8}{56} = \frac{10}{56}$	Finally, add together the probabilities of scoring 4 or 9
b P(same cards) = P(2, 2) + P(3, 3) + P(5, 5) + P(7, 7) $= \frac{2}{8} \times \frac{1}{7} + \frac{2}{8} \times \frac{1}{7} + \frac{2}{8} \times \frac{1}{7} + \frac{2}{8} \times \frac{1}{7}$ $= \frac{2}{56} + \frac{2}{56} + \frac{2}{56} + \frac{2}{56} = \frac{8}{56} = \frac{1}{7}$	Again, instead of drawing a tree diagram with four branches, a listing approach is more efficient. To choose two cards that multiply to make a square number, both numbers must be the same. Multiply together each of the probabilities of choosing the same number twice and then add them together.

Practice

1 A crate has 12 bottles of still water and 10 bottles of sparkling water.

Two bottles are chosen at random and removed from the crate.

a Copy and complete the tree diagram.

b Work out the probability of selecting one still and one sparkling water bottle.

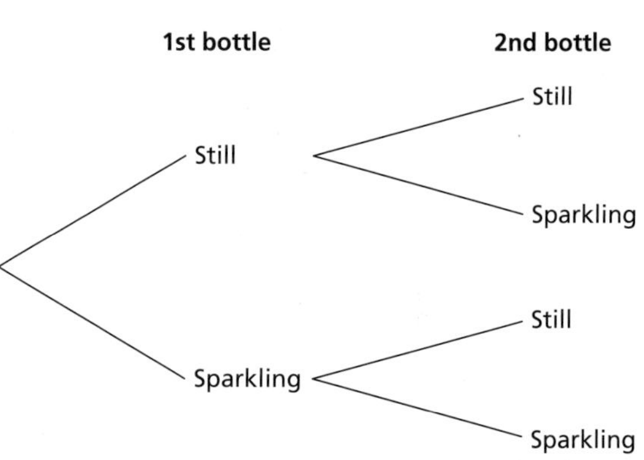

2 Cards numbered 1 to 10 are shuffled. Two are randomly selected.

a What is the probability that the first card shows a square number?

b Given that the first card is a square number, what is the probability that the second card will also be a square number?

c Seb says, "If the first card drawn is a square number, the probability of the second card being a prime number is unchanged."

Explain why Seb is wrong.

3 There are 40 members of a local brass band.

The two-way table shows how many members play brass or percussion, and how long they have been a member of the band.

	Less than 3 years	3 years or more
Brass	10	25
Percussion	3	2

A band member is chosen at random.

a Given that they play percussion, work out the probability that they have been a member for less than 3 years.

b Given that they have been a member for less than 3 years, work out the probability that they play percussion.

4 The tree diagram represents the probability of Mo being stuck in heavy traffic on the way to work and the probability that he is late to work.

Work out:

a the probability that Mo gets stuck in heavy traffic and is late to work

b the probability that he is late to work.

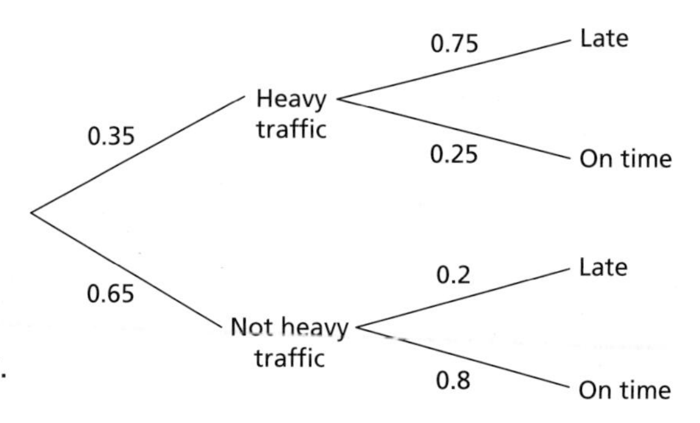

5 Two fair, six-sided dice numbered 1 to 6 are rolled and their scores are added together.

Given that exactly one of the dice rolls a prime number, work out the probability that the sum of the scores is 7

6 Here are five number cards:

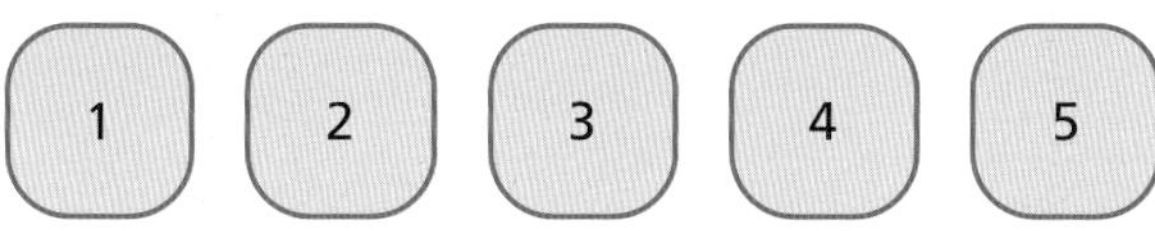

Two cards are selected without replacement.

What is the probability that the difference between the two cards is 3?

7 Bev has 3 white T-shirts, 2 black T-shirts and 4 grey T-shirts.

Bev picks two T-shirts at random to take away on holiday.

Work out the probability that Bev picks two T-shirts of the same colour.

8 Ali and Filipo each have a bag containing red counters and yellow counters.

Ali takes a counter out of his bag at random and puts it in Filipo's bag.

Filipo then takes a counter out of his bag at random.

Work out the probability that Ali and Filipo select different coloured counters.

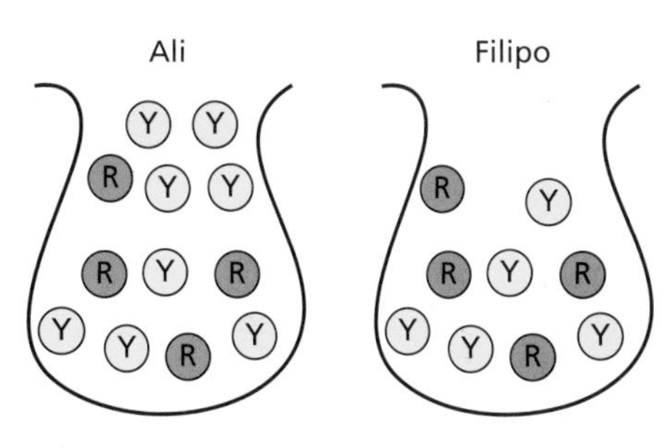

9 Here are eight number cards:

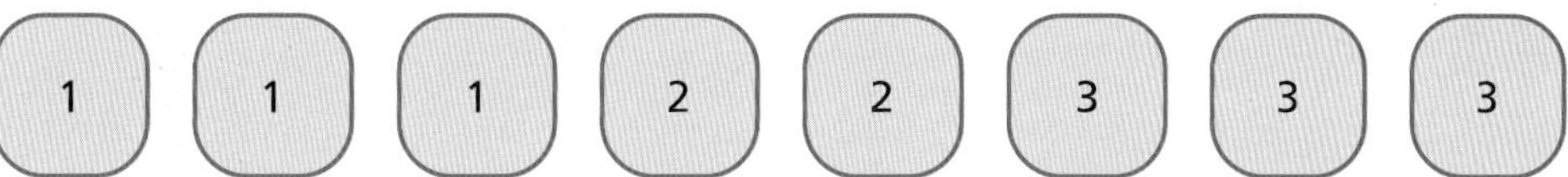

Three cards are selected without replacement.

Work out:

a P(the product of the cards is a cube number) **b** P(the product of the cards is 6).

What do you think?

1 Marta has a bag of 8 blue pencils and 6 green pencils.

Two pencils are taken at random from the bag without replacement.

a Work out the probability that she takes exactly one of each colour of pencil.

b How does your answer to part **a** change if Marta replaced the first pencil after she took it?

Consolidate – do you need more?

1 Letter tiles are used to spell out the word MATHEMATICS.

Two letters are randomly selected without replacement.

a What is the probability of choosing the same letter?

b What is the probability of choosing two letters that are both vowels?

2 Abdullah and Junaid have a box of milk chocolates and dark chocolates.

There are 14 milk chocolates and 11 dark chocolates.

Abdullah chooses one chocolate at random and eats it.

Junaid then chooses one chocolate at random.

a Copy and complete the tree diagram to represent the information.

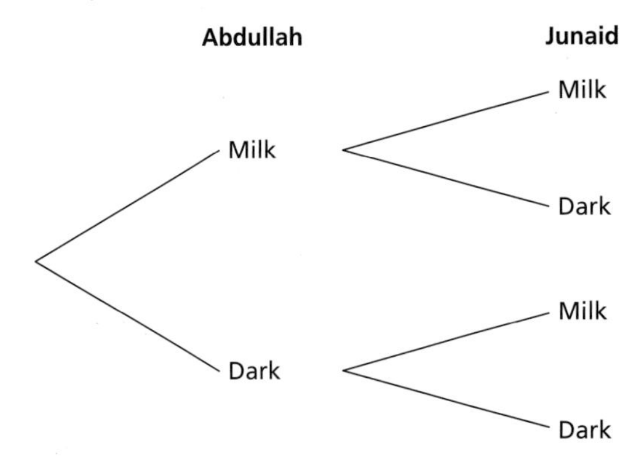

b Work out the probability that Abdullah and Junaid choose different types of chocolate.

c Work out the probability that at least one of them chooses a dark chocolate.

3 A box contains 11 broccoli florets and 13 cauliflower florets. Two florets are taken at random.

a Draw a tree diagram to represent the information.

b Calculate the probability that both florets are cauliflower. Give your answer in its simplest form.

4 There are 4 apples, 3 oranges and 5 bananas in a fruit bowl.

Two fruits are picked at random successively, without replacement.

What is the probability that the second fruit picked is an orange?

5 A computer randomly generates a number between 1 and 20

Once the computer has generated a number, it cannot select it again.

The computer generates two numbers.

a What is the probability that the computer generates two single-digit primes?

b What is the probability that the computer generates two primes that use the digit 3?

The computer generates a third number.

c Work out the probability that the computer generates only odd primes.

Stretch – can you deepen your learning?

1 Tiff and Zach each have a bag of 27 sweets.

The ratio of yellow to orange to red sweets in Tiff's bag is 3 : 4 : 2

The ratio of yellow to orange to red sweets in Zach's bag is 1 : 5 : 3

Zach randomly chooses one of his sweets and gives it to Tiff.

Tiff then randomly chooses a sweet and eats it.

Show that the probability of Tiff eating a yellow sweet is $\frac{41}{126}$

2 A bag contains x red marbles and 5 blue marbles.

Two marbles are selected at random from the bag without replacement.

The probability of choosing a red marble and then a blue marble is $\frac{5}{21}$

Work out the two possible values of x.

3 A box contains y white magnets and 3 black magnets.

Two magnets are successively chosen without replacement.

The probability of choosing two white magnets is $\frac{5}{14}$

a Show that $9y^2 - 39y - 30 = 0$

b Solve $9y^2 - 39y - 30 = 0$ to work out the number of white magnets initially.

c Work out the probability that the two magnets are different colours.

Combined events: exam practice

White Rose MATHS

4–6

1 Amina rolls a fair, six-sided dice numbered 1 to 6 and flips a fair coin.

Work out the probability that the dice lands on an even number and the coin lands on heads. **[2 marks]**

2 Rhys aims for the centre of this board when playing darts.

The probability that he hits the centre is 0.6

Rhys aims two darts at the centre of the board.

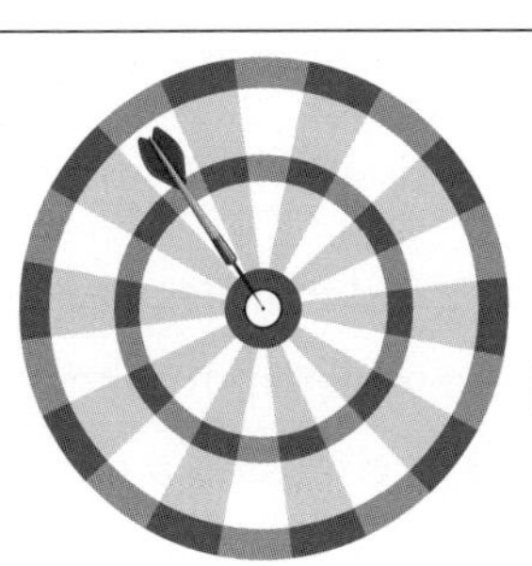

(a) Complete a copy of the tree diagram. **[2 marks]**

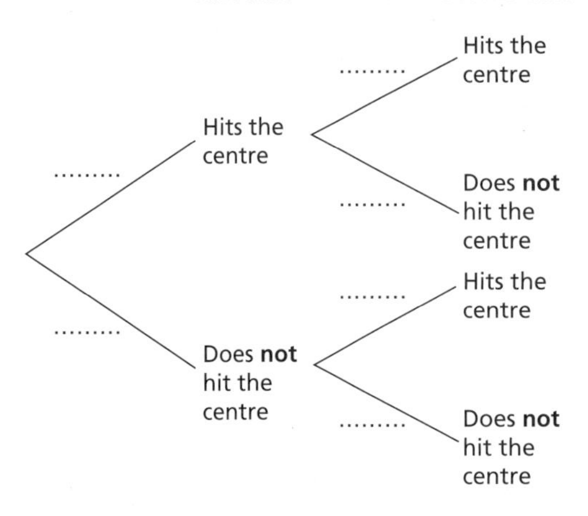

(b) What is the probability that exactly one of Rhys's darts hits the centre? **[2 marks]**

3 2% of items produced by a factory are faulty.

Two items are chosen at random from the factory.

What is the probability that **neither** item is faulty? **[3 marks]**

7–9

4 There are 90 teachers in a school.

12 wear glasses and have blonde hair.

58 wear glasses.

19 have blonde hair.

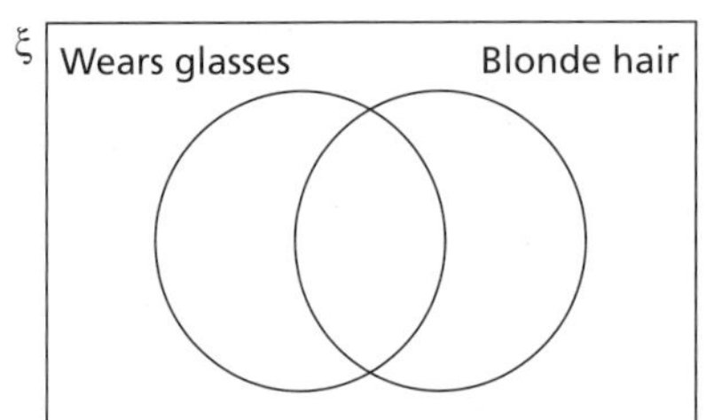

(a) Complete a copy of the Venn diagram. **[2 marks]**

(b) Work out the probability that a teacher chosen at random has blonde hair and also wears glasses. **[2 marks]**

5 A bag contains 6 red counters, 5 blue counters and 4 green counters.

Two counters are taken from the bag at random.

Work out the probability that both counters are the same colour. **[3 marks]**

Probability: exam practice

4–6

1 Two fair, six-sided dice numbered 1 to 6 are rolled.

Calculate the probability that they both land on 5

Give your answer as a fraction. **[2 marks]**

2 Bag A contains only red counters and white counters.

Bag B contains only green counters and blue counters.

The probability of selecting a red counter from bag A is 0.4

The probability of selecting a green counter from bag B is 0.7

Ed selects a counter from each bag at random.

Work out the probability that Ed selects a white counter and a blue counter. **[3 marks]**

3 The probability of winning a fairground game is 0.7

Seb plays the game twice.

Work out the probability that he wins exactly one of the games. **[3 marks]**

4 A biased spinner can land on red, blue, yellow, or green.

The table shows the probabilities that the spinner will land on red, blue or yellow.

Colour	Red	Blue	Yellow	Green
Probability	0.1	0.15	0.3	

The spinner is spun 200 times.

Work out an estimate for the number of times the spinner will land on green. **[3 marks]**

5 In a class of 30 students, each studies History, Geography, or both subjects.

A student is chosen at random.

The probability that they study both subjects is $\frac{1}{5}$

The probability that they study History is 60%

Complete a copy of the Venn diagram. **[4 marks]**

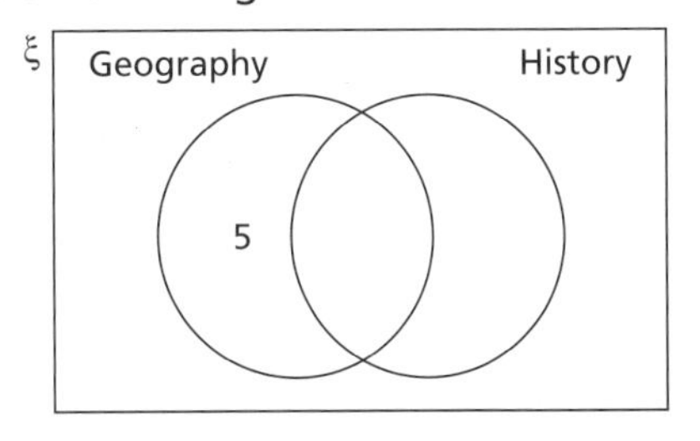

7–9

6 The probability that an event A occurs is 0.8

Event B is independent of event A, and the probability that both A and B occur is 0.48

Calculate the probability that neither event A nor event B occur. **[3 marks]**

7 The first 12 people to finish a race each win a prize.

The winners choose their prizes at random from a box.

The prizes are 1 holiday voucher, 4 gift cards and 7 book tokens.

(a) Work out the probability that the first two winners win the same prize. **[3 marks]**

(b) Work out the probability that after the first two winners win their prizes, the holiday voucher is still in the box. **[2 marks]**

8 A bag contains 4 red counters and 6 green counters.

Three counters are chosen at random without being replaced in the bag.

(a) Work out the probability that the first two counters are red and the third counter is green.

Give your answer as a fraction. **[2 marks]**

(b) Work out the probability that at least one counter is red.

Give your answer as a fraction. **[2 marks]**

13 Statistical measures

In this block, we will cover...

13.1 Basic measures

Example 1

Here are the heights, in metres measured to 1 de

5.6 4.8 9.6 5.6 8.5 7.8 9.1

a Calculate the mean height. b Find th

c State the modal height. d Calcula

Method

Solution
a $\frac{5.6 + 4.8 + 9.6 + 5.6 + 8.5 + 7.8 + 9.1 + 6.1 + \ldots}{10}$

13.2 Grouped continuous data

Practice

1 Twenty students completed a 100 m race. table below.

Time, t (seconds)	Frequency
$12 \leqslant t < 16$	1
$16 \leqslant t < 20$	4
$20 \leqslant t < 22$	7
$22 \leqslant t < 24$	6
$24 \leqslant t < 30$	2
Total	20

13.3 Talking about data

Consolidate – do you need more

1 The times taken, in seconds, to complete a

12.5 13.8 29.4 15 16.35 12.82

a Calculate the mean, median, mode and

b Which time is an outlier?

c Calculate the mean, median, mode and

2 The mean and range of the ages of memb

	Mean	Range

13.1 Basic measures

White Rose MATHS

Are you ready? (A)

1 Write the numbers in order, starting with the smallest.

a 3.2, 3.15, 3.02, 3.52, 2.35

b −7, −8, −5, −6.4, −6.6, −7.2

c 0.8, 0.801, 0.81, 0.832, 0.302

d $\frac{2}{5}, \frac{3}{5}, \frac{3}{7}, \frac{2}{7}, \frac{1}{3}$

2 Work out:

a $\frac{9 + 7 + 8 + 4 + 8}{5}$

b $\frac{6.3 + 9.7 + 2.4 + 6.8}{4}$

c $\frac{(-4) + (-7) + (-2) + (-8) + (-5) + (-7)}{6}$

d $\frac{0.802 + 0.45 + 0.632}{3}$

There are three types of **average** used to represent a set of data:

- The **mean** is found by dividing the total of the data set by the number of items.
- The **median** is the middle item of an ordered set of data. If there is an even number of values in the data set, find the mean of the two values that are in the middle.
- The **mode** is the value in a set of data that appears the most often. A set of data may have one mode, more than one mode, or no mode at all.

You can describe the spread of a set of data by finding the **range**. The range is the difference between the greatest and the smallest values.

When comparing two data sets, the mean and the median are both useful in order to compare the size of a 'typical' member of each data set.

The range can be used to compare the spread of two data sets, showing the 'consistency' of each set of data. The larger the range, the more the data is spread out.

Example 1

Here are the heights, in metres measured to 1 decimal place, of 10 trees in a forest.

5.6 4.8 9.6 5.6 8.5 7.8 9.1 6.1 8.1 6.8

a Calculate the mean height.

b Work out the median height.

c State the modal height.

d Calculate the range of the heights.

Method

Solution	Commentary
a $\frac{5.6 + 4.8 + 9.6 + 5.6 + 8.5 + 7.8 + 9.1 + 6.1 + 8.1 + 6.8}{10}$ $= \frac{72}{10} = 7.2\,\text{m}$	Calculate the mean by finding the total of the data and dividing by how many items there are altogether.

b ~~4.8~~ ~~5.6~~ ~~5.6~~ ~~6.1~~ (6.8 7.8) ~~8.1~~ ~~8.5~~ ~~9.1~~ ~~9.6~~ Median = $\frac{6.8 + 7.8}{2} = 7.3\,\text{m}$	Rewrite the numbers in order. As there is an even number of items, there are two heights 'in the middle'. The median is the mean of these two heights.
c Mode = 5.6 m	'Modal' means the value that is the mode. 5.6 is the only value that appears more than once.
d Range = 9.6 – 4.8 = 4.8 m	Calculate the difference between the greatest and least values in the set of data.

Example 2

The mean height of six students is 1.65 m.

The mean height of another two students is 1.87 m.

Calculate the mean height of all eight students.

Method

Solution	Commentary
Total height of first six students = $6 \times 1.65 = 9.9\,\text{m}$ Total height of the other students = $2 \times 1.87 = 3.74\,\text{m}$	To find the mean of all eight students, you need the overall total. You can find the total height for each group of students by using a rearrangement of the formula for the mean: Mean = total ÷ number of items Total = mean × number of items
Overall total = 9.9 + 3.74 = 13.64 m	Find the total height of all eight students.
Mean of all the students = 13.64 ÷ 8 = 1.705 m	Divide the total height by the total number of students.

Practice (A)

1 Calculate the mean, median, mode and range of each set of data. Round answers where appropriate.

a 18, 14, 21, 19, 18

b 9.8, 8.7, 8.7, 7.25, 9.8, 7.8, 8.1

c $\frac{1}{5}, \frac{2}{5}, \frac{4}{5}, \frac{3}{10}, \frac{7}{10}$

d 0.8, 0.802, 0.28, 0.208, 0.82

e –4, –7, –4, –5, –5.5, –6

f 83 cm, 85 cm, 0.83 m, 0.84 m, 87 cm, 89 cm

g –3, –1, 0, –1, –2, 3, 1

h 78%, 87%, 79%, 82%

2 **a** There are 12 athletes in a team and they have a mean height of 187.2 cm.

There are 16 athletes in another team and they have a mean height of 174.8 cm.

Calculate the mean height of the two teams combined.

b After five rounds of a quiz, a team's average score is 7.5

In the sixth round, the team scores 9 out of 10

Work out the team's mean score after six rounds.

c The mean age of a group of 10 people is 16.8 years old.

Another person, who is 22, joins the group.

What is the new mean age of the group?

d Two teams of cyclists have a mean age of 23

Team A has a total of 18 cyclists and they have an average age of 25

Team B has a total of 15 cyclists.

What is the average age of the cyclists in Team B?

3 By working out the averages and the range, compare each pair of data sets.

Comparison of data sets is covered in detail in Chapter 13.3

a

Group A	15, 19, 17, 15, 16, 18
Group B	18, 19, 17, 16

b

Group C	2.8, 2.7, 2.6, 2.8, 2.5, 2.65
Group D	2.8, 2.9, 2.6, 2.56, 2.7

c

Group E	–8, –7, –8, –6, –6.5, –7.2, –7.1
Group F	–9, –8, –6, –5, –4, –5.4

d

Group G	0.06, 0.08, 0.065, 0.074, 0.082
Group H	0.06, 0.08, 0.072, 0.067, 0.071, 0.068

4 Five numbers have a mean of 13 and a median of 12

Three of the numbers are 9, 10 and 19

What are the other two numbers?

5 The mean of two numbers is 32 and the range is 12

What are the two numbers?

6 Four numbers have a mean of 8, a mode of 9 and a median of 8.5

What is the range of the numbers?

What do you think?

1. What would happen to the mean, median and mode of a set of data containing six values if:

   a all values were doubled

   b three of the values were doubled and three of the values were halved

   c all values were decreased by 3?

2. Investigate the effect of adding or removing an extreme value to or from a data set.

   Which averages change the most? How does the range change?

Are you ready? (B)

1 Calculate the mean, median, mode and range of each set of data.

a 9, 7, 8, 4, 8

b 6.3, 9.7, 2.4, 6.8

c –7, –4, –2, –8, –4, –7

d 0.82, 0.45, 0.632, 0.802, 0.63

When there is a large amount of data, you can present it in a **frequency table** rather than a list.

A frequency table can be used to find the mean, median, mode and range.

If there are n items in a set of data, the median will be in the $\left(\frac{n+1}{2}\right)$th position. The position of the median can be determined using **cumulative frequencies**, which are the running totals of the frequencies.

Example

The table shows data about the ages of a group of dogs.

a Calculate the mean age of the dogs.

b Calculate the median age of the dogs.

c Calculate the modal age of the dogs.

d Calculate the range of the ages of the dogs.

Age	Frequency
8	3
9	8
10	13
11	12
12	19
13	4

Method

<table>
<tr><th>Solution</th><th>Commentary</th></tr>
<tr><td>

a

<table>
<tr><th>Age</th><th>Frequency</th><th>Age × Frequency</th></tr>
<tr><td>8</td><td>3</td><td>24</td></tr>
<tr><td>9</td><td>8</td><td>72</td></tr>
<tr><td>10</td><td>13</td><td>130</td></tr>
<tr><td>11</td><td>12</td><td>132</td></tr>
<tr><td>12</td><td>19</td><td>228</td></tr>
<tr><td>13</td><td>4</td><td>52</td></tr>
<tr><td>Total</td><td>59</td><td>638</td></tr>
</table>

Mean $= \frac{638}{59} = 10.8$ (to 1 d.p.)

</td><td>

Multiply each frequency by the age to give the total age for that row.

Work out the total frequency (the total number of dogs) and the total of the ages (of all the dogs).

To find the mean, divide the total of all ages by the total frequency.

</td></tr>
<tr><td>

b

<table>
<tr><th>Age</th><th>Frequency</th><th>Cumulative frequency</th></tr>
<tr><td>8</td><td>3</td><td>3</td></tr>
<tr><td>9</td><td>8</td><td>3 + 8 = 11</td></tr>
<tr><td>10</td><td>13</td><td>11 + 13 = 24</td></tr>
<tr><td>11</td><td>12</td><td>24 + 12 = 36</td></tr>
<tr><td>12</td><td>19</td><td>36 + 19 = 55</td></tr>
<tr><td>13</td><td>4</td><td>55 + 4 = 59</td></tr>
</table>

Median = 11 years old

</td><td>

The table contains data for 59 dogs.

The median is in the $\left(\frac{59 + 1}{2}\right)$th = 30th position.

The cumulative frequencies show that ages 8 to 10 account for 24 dogs, and ages 8 to 11 account for 36 dogs.

This means that the 30th position is a dog that is 11 years old.

</td></tr>
<tr><td>c The modal age is 12 years old.</td><td>The modal age is the age with the highest frequency.</td></tr>
<tr><td>d Range = 13 – 8 = 5</td><td>The greatest age of the dogs is 13 and the least age is 8</td></tr>
</table>

Practice (B) 🖩

Give answers to 2 decimal places where appropriate.

1 Calculate the mean of each set of data.

a

Age	Frequency	Age × Frequency
11	4	44
12	8	96
13	6	
14	4	
15	3	

b

Number of pets	Frequency	Number × Frequency
0	4	0
1	9	
2	4	
3	2	
4	1	

c

Number of siblings	Frequency	Number × Frequency
0	3	
1	6	
2	8	
3	2	
4	1	

d

Marks	Frequency	Marks × Frequency
3	1	
4	3	
5	4	
6	8	
7	6	
8	3	

2 Calculate the mean, median, mode and range of each set of data.

a

Cost	Frequency	Subtotal
50p	1	
£1	8	
£1.50	6	
£2	7	
£2.50	3	
£3	5	

b

Mass	Frequency	Subtotal
2 kg	4	
3 kg	6	
4 kg	8	
5 kg	2	

c

Height	Frequency	Subtotal
15 cm	8	
16 cm	4	
17 cm	3	
18 cm	5	

d

Mass	Frequency	Subtotal
250 g	3	
500 g	6	
750 g	9	
1 kg	7	

3 Calculate the mean, median, mode and range from each bar chart.

a

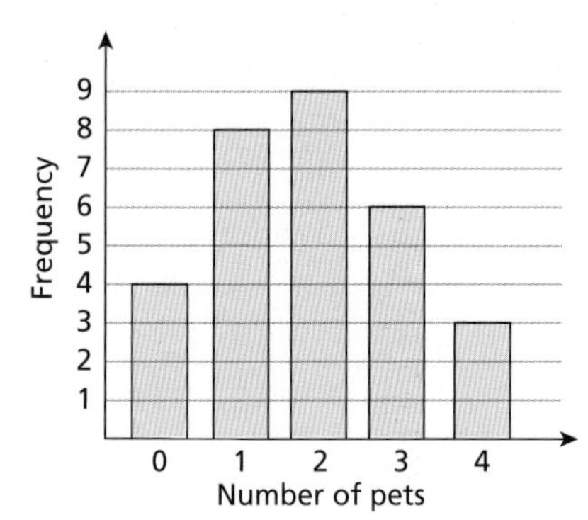

b

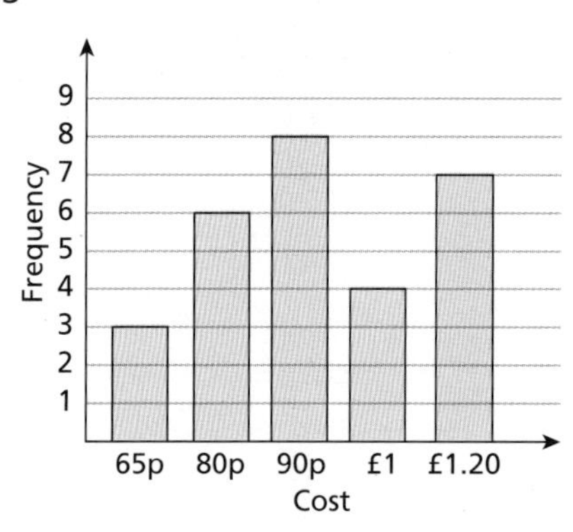

4 The tables show the test results of two classes.

Which class performed better on the test?

Class A

Score	Frequency
13	1
15	4
16	8
18	5
19	6
20	4

Class B

Score	Frequency
15	9
17	5
18	4
19	6
20	2

5 The marks of some students in a test are shown in the table.

The mean mark scored was 3.6

Calculate the value of a and hence state the modal mark.

Marks	Frequency
1	1
2	4
3	9
4	a
5	8

Consolidate – do you need more?

1 Calculate the mean, median, mode and range of each set of data. Round answers where appropriate.

a 19, 17, 21, 20, 17

b £125, £157, £135, £128, £135, £145

c 5.72, 5.72, 5.8, 5.208, 5.208

d 320 g, 0.45 kg, 284 g, 0.32 kg, 0.372 kg, 384 g

e 0.6, 0.06, 0.056, 0.605, 0.62, 0.621

f −4, −4, −3, −5, −7, −5, −2, 0

g $\frac{3}{8}, \frac{5}{8}, \frac{1}{4}, \frac{2}{8}, \frac{3}{4}, \frac{3}{12}$

h $\frac{8}{15}, \frac{2}{3}, \frac{4}{7}, \frac{1}{3}, \frac{7}{14}, \frac{8}{15}$

2 Calculate the mean, median, mode and range of the data in each table.

a

Age	Frequency
7	4
8	6
9	7
10	5
11	5

b

Score	Frequency
1	3
2	7
3	5
4	6
5	8
6	4

c

Price	Frequency
£1500	2
£1750	3
£2000	5
£2250	7
£2500	4
£2750	6
£3000	5

d

Time	Frequency
50 seconds	5
1 minute	12
1 minute 10 seconds	15
1 minute 20 seconds	13
1 minute 30 seconds	9

Stretch – can you deepen your learning?

1 The average mass of four dogs is 19.75 kg.

The masses, in kilograms, are:

$3x + 3$ $x + 15$ 12 $2x + 4$

Calculate the median and the range of the masses of the dogs.

2 The number of days that 50 workers were absent in a month are shown in the table.

The mean number of days absent is 1.12

Calculate the values of a and b.

Number of days	Frequency
0	13
1	25
2	a
3	b
4	0
5	1

3 A library recorded the number of books that were taken out by each member in one day, but missed out some of the information.

The mean number of books taken out was 2.75

Calculate the median and the modal number of books that were taken out.

Number of books	Frequency
1	5
2	
3	3
	1
8	2

13.2 Grouped continuous data

White Rose MATHS

Are you ready?

1 The table shows the score out of 25 that a group of students obtained in a spelling test.

Score	Frequency
17	3
18	5
19	2
20	6
22	3
23	5

Calculate the mean, median, mode and range of the scores.

Data sets are **discrete** when they can only take specific countable values, such as the number of goals scored or the number of people.

Here we consider **continuous** data, which is usually arranged into **class intervals**, as the data is measured rather than counted.

For example, the waiting time, h, for patients in a hospital may be grouped into class intervals of say:

- up to but not including 30 minutes (written $0 < h < 0.5$)
- 30 minutes up to but not including 60 minutes (written $0.5 \leqslant h < 1$), and so on.

Waiting time, h (hours)	Frequency
$0 < h < 0.5$	10
$0.5 \leqslant h < 1$	15
$1 \leqslant h < 2$	8
$2 \leqslant h < 3$	8
$3 \leqslant h < 5$	2

When continuous data is grouped, the value of individual items of data is not known. The **midpoint** of each class interval is used to represent the data and to find an estimate of the mean. The midpoint is the mean of the end points of the class interval.

For example, for $1 \leqslant h < 2$, the midpoint is $\left(\frac{1+2}{2}\right) = 1.5$

Example

The table shows the heights of students in a class.

a Estimate the mean height of the students.

b Identify the class containing the median height.

c Identify the modal class.

Height, h (cm)	Frequency
$150 \leqslant h < 160$	2
$160 \leqslant h < 170$	10
$170 \leqslant h < 175$	7
$175 \leqslant h < 180$	8
$180 \leqslant h < 190$	2

Method

Solution

a

Height, h (cm)	Frequency, f	Midpoint, x	fx
$150 \leqslant h < 160$	2	155	310
$160 \leqslant h < 170$	10	165	1650
$170 \leqslant h < 175$	7	172.5	1207.5
$175 \leqslant h < 180$	8	177.5	1420
$180 \leqslant h < 190$	2	185	370
Total	29		4957.5

Mean $\approx \frac{4957.5}{29} = 170.9$ cm

Commentary

Use the midpoints of the class intervals to find estimates for the total of each class by multiplying them by the frequency of the class.

This is very similar to finding the subtotal for grouped discrete data in Chapter 13.1, section B.

Divide the overall total by the total frequency to give an estimate of the mean.

b

Height, h (cm)	Frequency	Cumulative frequency
$150 \leqslant h < 160$	2	2
$160 \leqslant h < 170$	10	12
$170 \leqslant h < 175$	7	19
$175 \leqslant h < 180$	8	27
$180 \leqslant h < 190$	2	29

Median is in the class $170 \leqslant h < 175$

Commentary

The table contains data for 29 students.

The median is in the $\left(\frac{29 + 1}{2}\right)$th = 15th position.

The cumulative frequencies show that 12 students are of heights $150 \leqslant h < 170$ and 19 students are of heights $150 \leqslant h < 175$.

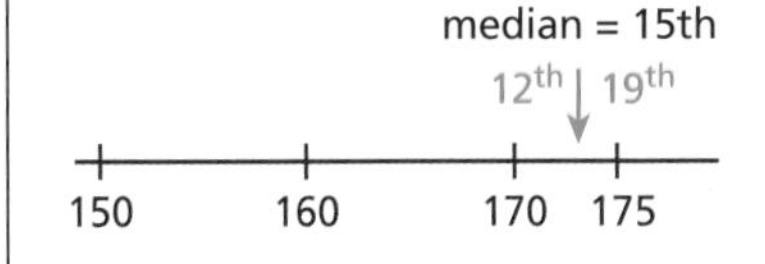

This means that the 15th position is a student who is between 170 cm and 175 cm.

You will further explore estimating the median using cumulative frequency diagrams in Chapter 14.3

c Modal class = $160 \leqslant h < 170$

Commentary

This is the class with the highest frequency.

Practice

1. Twenty students completed a 100 m race. Their times, in seconds, are shown in the table below.

Time, t (seconds)	Frequency	Midpoint	Frequency × Midpoint
$12 \leqslant t < 16$	1	14	14
$16 \leqslant t < 20$	4		
$20 \leqslant t < 22$	7		
$22 \leqslant t < 24$	6		
$24 \leqslant t < 30$	2		
Total	20		

a State the modal class.

b Estimate the mean time taken by the students.

2. The table below shows the ages of visitors to a museum one day.

Age, a years	Frequency
$0 \leqslant a < 18$	12
$18 \leqslant a < 30$	7
$30 \leqslant a < 40$	30
$40 \leqslant a < 50$	22
$50 \leqslant a < 60$	12
$60 \leqslant a < 75$	8
$75 \leqslant a < 90$	4
Total	

a State the modal class.

b Estimate the mean age of the visitors. Give your answer to 2 decimal places.

3. The table shows information about the number of steps taken by 50 people in one day.

Number of steps, x	Frequency
$0 \leqslant x < 2500$	6
$2500 \leqslant x < 5000$	10
$5000 \leqslant x < 7500$	15
$7500 \leqslant x < 10\,000$	17
$10\,000 \leqslant x < 12\,500$	
Total	

a Work out the number of people who took between 10 000 and 12 500 steps.

b Work out an estimate for the mean number of steps taken.

c State the modal class of the number of steps taken.

4 The heights of trees in part of a forest were measured and recorded in the frequency table.

Height, h (m)	Frequency
$5 \leqslant h < 10$	8
$10 \leqslant h < 12$	6
$12 \leqslant h < 15$	4
$15 \leqslant h < 20$	2

a Calculate an estimate for the mean height of the trees. Give your answer to 2 decimal places.

b State the modal class.

c Identify the class that contains the median.

5 The temperatures of places in the United Kingdom on one day were recorded in the cumulative frequency table.

Temperature, t (degrees Celsius)	Cumulative frequency
$0 \leqslant t < 5$	3
$5 \leqslant t < 10$	8
$10 \leqslant t < 15$	12
$15 \leqslant t < 20$	18
$20 \leqslant t < 30$	20

a Work out the class in which the median lies.

b By working out the frequency of each class interval, state the modal class.

c Calculate an estimate for the mean temperature.

6 The times that students spent on their maths homework are recorded in the frequency table.

The estimated mean time taken is 41.2 minutes, to 1 decimal place.

Work out the value of x.

Time, t (minutes)	Frequency
$0 \leqslant t < 20$	2
$20 \leqslant t < 30$	5
$30 \leqslant t < 40$	8
$40 \leqslant t < 50$	x
$50 \leqslant t < 60$	5
$60 \leqslant t < 90$	3

What do you think?

1 Here are two tables showing the ages of the same group of people.

Table A

Age, a (years)	Frequency
$10 \leqslant a < 15$	12
$15 \leqslant a < 20$	17
$20 \leqslant a < 25$	13
$25 \leqslant a < 30$	8

Table B

Age, a (years)	Frequency
$10 \leqslant a < 13$	9
$13 \leqslant a < 16$	6
$16 \leqslant a < 18$	6
$18 \leqslant a < 20$	8
$20 \leqslant a < 23$	9
$23 \leqslant a < 27$	7
$27 \leqslant a < 30$	5

Which table is more likely to give a more accurate estimate for the mean? Give a reason for your answer.

Consolidate – do you need more?

1 Estimate the mean from each table. Give your answers to 2 decimal places where appropriate.

a

Mass, x (kg)	Frequency	Midpoint	Subtotal
$18 \leqslant x < 24$	2		
$24 \leqslant x < 28$	4		
$28 \leqslant x < 30$	3		
$30 \leqslant x < 34$	5		
$34 \leqslant x < 40$	3		
$40 \leqslant x < 50$	3		
Total	20		

b

Height, h (m)	Frequency	Midpoint	Subtotal
$2 \leqslant h < 3$	2		
$3 \leqslant h < 4$	3		
$4 \leqslant h < 4.5$	3		
$4.5 \leqslant h < 5$	4		
$5 \leqslant h < 6$	3		
$6 \leqslant h < 8$	1		
Total	16		

c

Time, t (seconds)	Frequency	Midpoint	Subtotal
$10 \leqslant t < 15$	3		
$15 \leqslant t < 20$	2		
$20 \leqslant t < 22$	3		
$22 \leqslant t < 24$	4		
$24 \leqslant t < 28$	1		
Total			

d

Speed, s (mph)	Frequency
$25 \leqslant s < 30$	1
$30 \leqslant s < 35$	0
$35 \leqslant s < 40$	5
$40 \leqslant s < 50$	8
$50 \leqslant s < 60$	3

2 For each table, state: **i** the modal class **ii** the class that contains the median.

a

Mass, x (kg)	Frequency	Cumulative frequency
$18 \leqslant x < 24$	2	
$24 \leqslant x < 28$	4	
$28 \leqslant x < 30$	3	
$30 \leqslant x < 34$	5	
$34 \leqslant x < 40$	3	
$40 \leqslant x < 50$	3	

b

Height, h (m)	Frequency	Cumulative frequency
$2 \leqslant h < 3$	2	
$3 \leqslant h < 4$	3	
$4 \leqslant h < 4.5$	3	
$4.5 \leqslant h < 5$	4	
$5 \leqslant h < 6$	3	
$6 \leqslant h < 8$	1	

c

Time, t (seconds)	Frequency
$10 \leqslant t < 15$	3
$15 \leqslant t < 20$	2
$20 \leqslant t < 22$	3
$22 \leqslant t < 24$	4
$24 \leqslant t < 28$	1

d

Speed, s (mph)	Frequency
$25 \leqslant s < 30$	1
$30 \leqslant s < 35$	0
$35 \leqslant s < 40$	5
$40 \leqslant s < 50$	8
$50 \leqslant s < 60$	3

Stretch – can you deepen your learning?

1 The scores achieved by 20 gamers are shown in the table below.

Score, s	Frequency
$0 \leqslant s < 4$	3
$4 \leqslant s < 6$	x
$6 \leqslant s < 8$	5
$8 \leqslant s < 12$	y
$12 \leqslant s < 15$	2

The estimated mean score for the group is 6.9 points.

Calculate the values of x and y.

2 The frequency polygon shows the cost of items in a store.

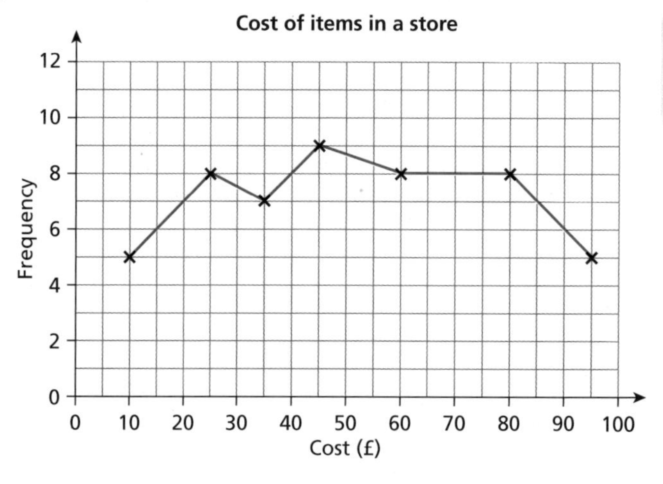

In a frequency polygon, the data points are plotted at the midpoint of each class interval.

Estimate the mean cost of the items.

3 The histogram shows the cost of items in a store.

Histograms are covered in Chapter 14.2

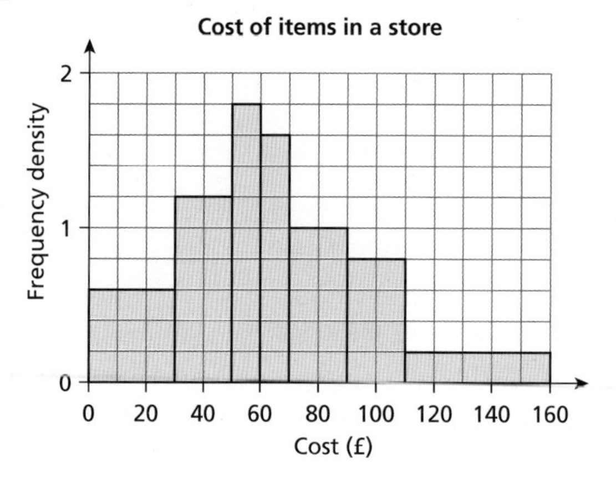

Estimate the mean cost of the items.

13.3 Talking about data

White Rose MATHS

Are you ready?

1 Here are the heights of seven children.

92 cm   87 cm   104 cm   98 cm   101 cm   95 cm   101 cm

a Work out the range of the heights.

b Write down the modal height.

c Work out the median height.

d Calculate the mean height. Give your answer to 1 decimal place.

2 The table shows information about the ages, in completed years, of a group of students.

a Estimate the mean age of the students.

b Work out the median age of the students.

c Identify the modal age of the students in completed years.

d Estimate the range of the ages of the students.

Age	Frequency
13	4
14	7
15	9
16	4

3 The table shows information about the scores of 30 students in a test.

Score, s	Frequency
$0 \leqslant s < 40$	4
$40 \leqslant s < 60$	6
$60 \leqslant s < 70$	5
$70 \leqslant s < 75$	4
$75 \leqslant s < 80$	7
$80 \leqslant s < 100$	4

a Estimate the mean score.

b Identify the interval that contains the median score.

c Identify the modal class interval.

When comparing two sets of data, the mean or median is often used as one comparison and the range is used as the other.

The range shows the spread of a set of data, whilst the mean or median give a value that aims to represent a 'typical' value of the set. Sometimes the mode is useful for categorical data but it is not generally useful for numerical data.

A value in a set of data that lies outside the overall pattern in the distribution is called an **outlier**. For example, if in a set of examination results all the students scored high marks apart from one student who scored a very low mark, this very low mark would be an outlier.

Example 1

Here are the scores obtained by a group of students in a test.

56% 71% 63% 81% 4% 59% 70% 68% 61% 90% 82% 66%

a Calculate the mean and the range of the scores.

b Which data value might be regarded as an outlier?

c How do the mean and the range of the scores change if the outlier is excluded?

Method

Solution	Commentary
a Mean $= \frac{\text{total}}{\text{number of items}} = \frac{771}{12}$ $= 64.3\%$ (to 1 d.p.) Range = 90% − 4% = 86%	To find the range, you subtract the smallest value from the largest value.
b The 4% score is the outlier.	4% is very different from the rest of the scores.
c New mean $= \frac{\text{total}}{\text{number of items}} = \frac{767}{11}$ $= 69.7\%$ (to 1 d.p.) The mean has gone up by more than 5% Range = 90% − 56% = 34% The range is less than half the size it was before.	One outlier can have a significant effect on the mean and the range.

Example 2

The table shows some information about the masses of apples grown on two farms.

	Mean (g)	Range (g)
Farm A	171	22
Farm B	178	59

Compare the masses of the apples grown on the two farms.

Method

Solution	Commentary
On average, the apples from farm B are heavier than the apples from farm A, because the mean mass of the apples is greater.	You can compare the masses using an average; this is usually the mean or the median.
The range of the masses of the apples from farm A is smaller, so they are more consistent in size.	The range tells you how spread out the data is. The smaller the range, the closer together the masses are.

Practice

1 The times taken, in seconds, for eight students to complete a question were recorded.

35 39 8 52 44 35 29 46

a Which measurement is an outlier?

b Calculate the mean, median and mode of the data after the outlier is removed.

2 Here is a set of data: 3 7 37 8 7 5

a Calculate the mean, median and mode of the data.

b Which data item is an outlier?

c Calculate the mean, median and mode of the data after the outlier is removed.

What effect does removing the outlier have on these averages?

3 Class A and class B both take a Maths test.

The table shows the mean and the range of the scores from the test.

	Mean	Range
Class A	68	27
Class B	72	58

Compare the performances in the test, explaining your reasoning.

4 The Internet speed, in megabits per second (Mbps), is recorded for ten households.

50 65 58 57 68 18 54 63 61 52

a Calculate the mean, median and range of the data.

b Which data item is an outlier?

c Calculate the mean, median and range of the data after the outlier is removed.

What effect does removing the outlier have on the mean, median and range?

5 Two classes completed the same quiz.

The scores of the students are recorded in the table below.

Score	Class 1	Class 2
$0 \leqslant s < 30$	2	0
$30 \leqslant s < 50$	1	2
$50 \leqslant s < 60$	4	4
$60 \leqslant s < 75$	6	5
$75 \leqslant s < 80$	4	7
$80 \leqslant s < 90$	7	5
$90 \leqslant s < 100$	4	3

a Calculate an estimate for the mean score of each class.

b Based on the mean, which class performed better in the quiz?

6 The table shows information about the salaries of plumbers in Leeds and Newcastle.

	Leeds	Newcastle
Median	£33 065	£31 205
Maximum	£47 324	£46 280
Minimum	£23 240	£27 974

Make two comparisons about the salaries of plumbers in the two cities.

7 Twenty students complete a Maths test and an English test.

Their scores are recorded in the table.

Maths	15	12	18	19	7	12	11	17	14	15	17	14	14	12	13	6	20	19	17	14
English	12	17	16	15	20	19	9	15	16	14	12	18	11	12	11	14	15	13	17	15

Compare the distribution of scores in the tests.

Consolidate – do you need more?

1 The times taken, in seconds, to complete a puzzle are shown below.

12.5 13.8 29.4 15 16.35 12.82 14.07

a Calculate the mean, median, mode and range of the times.

b Which time is an outlier?

c Calculate the mean, median, mode and range once the outlier is removed.

2 The mean and range of the ages of members at two golf clubs are shown in the table.

	Mean	Range
West Cliffe	48.72	49
East Lake	47.34	37

Use this information to make two comparisons about the two golf clubs.

3 Ten people participated in a quiz.

Their scores were: 8, 9, 7, 8, 9, 10, 10, 8, 5, 7

a Work out the mean score in the quiz.

b Work out the range of scores.

Another 15 people completed the same quiz the next day.

The table shows the mean and the range for their scores.

Mean	8.4
Range	7

c Use your answers to parts **a** and **b** to compare the scores in the quiz each day.

4 The table shows the median and the range of the number of children living in houses on two streets.

	Median	Range
Top Lane	2.5	4
Upper Street	2	2

Make two comparisons about the number of children on the two streets.

5 The table shows information about the Maths and English scores of 15 students.

Maths (%)	82	96	54	58	62	84	52	73	78	81	59	67	94	68	82
English (%)	85	72	97	34	67	28	48	61	94	51	42	83	87	92	74

Make two comparisons about the scores in each subject.

Stretch – can you deepen your learning?

1 Look at sets of data from other subjects, such as Science, that include outliers.

Discuss whether the outliers are due to errors in recording or genuine extreme values, and whether it is always possible to tell.

2 Using data sets of your choice, compare the effect of outliers on the mean, median, mode and range.

Discuss which measures are most and least reliable.

Statistical measures: exam practice

White Rose MATHS

4–6

1 Emily and Junaid sat five weekly mental arithmetic tests.

Emily's marks were 5, 6, 7, 3, 8

Junaid had a higher median mark than Emily.

Junaid had a lower mean mark than Emily.

Junaid had the same range in marks as Emily.

List a possible set of marks that Junaid could have had. **[3 marks]**

2 The mean of five numbers is 28

The mean of three of the numbers is 31

Work out the mean of the other two numbers. **[3 marks]**

3 The chart shows the number of pets owned by a group of students.

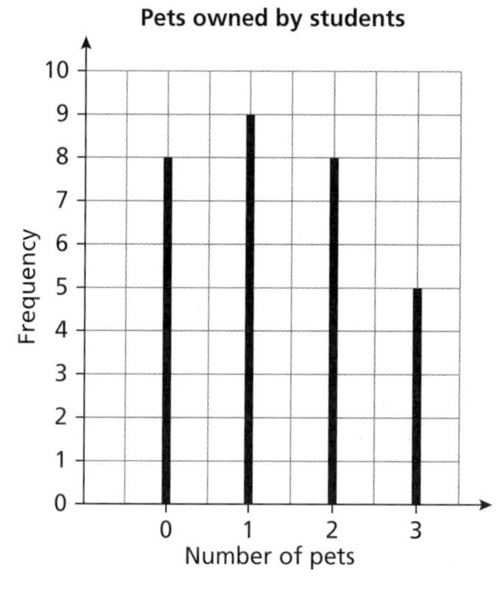

Work out the mean number of pets owned by the students in the group. **[3 marks]**

4 The table shows the times taken by some people to complete a puzzle.

Time, t (minutes)	$0 < t \leqslant 2$	$2 < t \leqslant 5$	$5 < t \leqslant 10$	$10 < t \leqslant 15$	$15 < t \leqslant 25$
Number of people	3	7	13	18	9

(a) In which class does the median time lie? **[1 mark]**

(b) Calculate an estimate of the mean time. **[4 marks]**

14 Charts and diagrams

In this block, we will cover...

14.1 Time series

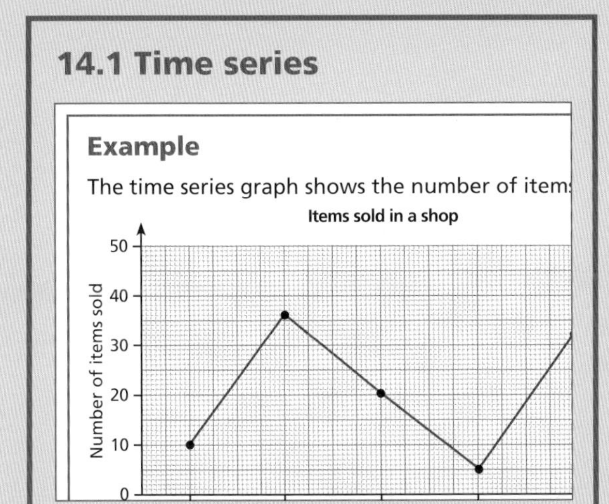

14.2 Histograms

Practice

1. Draw a histogram for each set of data.

a

Mass, x (kg)	Frequency
$18 \leqslant x < 24$	3
$24 \leqslant x < 28$	4
$28 \leqslant x < 30$	3
$30 \leqslant x < 34$	5
$34 \leqslant x < 40$	3
$40 \leqslant x < 50$	3

c

14.3 Cumulative frequency diagrams

Consolidate – do you need more

1. Draw a cumulative frequency diagram for

a

Weight, x (kg)	Frequency	Cumulative frequency
$18 < x \leqslant 24$	2	
$24 < x \leqslant 28$	4	
$28 < x \leqslant 30$	3	
$30 < x \leqslant 34$	5	
$34 < x \leqslant 40$	3	

14.4 Box plots

Stretch – can you deepen your le

1. The box plot shows the masses of 400 dog

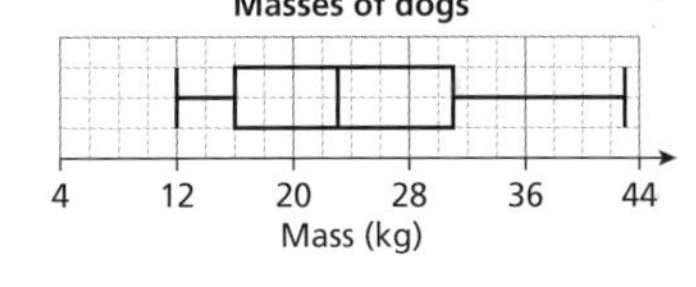

a Estimate how many dogs have a mass

b Estimate how many dogs have a mass

2. Some information about the cost of some

14.5 Two-way tables

Example

In a school, there are 300 students in total in Yea

Students are asked if they travel to school by bus

160 students are in Year 11.

40 students cycle.

The same number of students get the bus as w

65 Year 11 students walk to school.

17 Year 10 students cycle to school.

Represent this information as a two-way table, a

14.1 Time series

Are you ready?

1 The graph shows the number of drinks sold in a café over a five-day period.

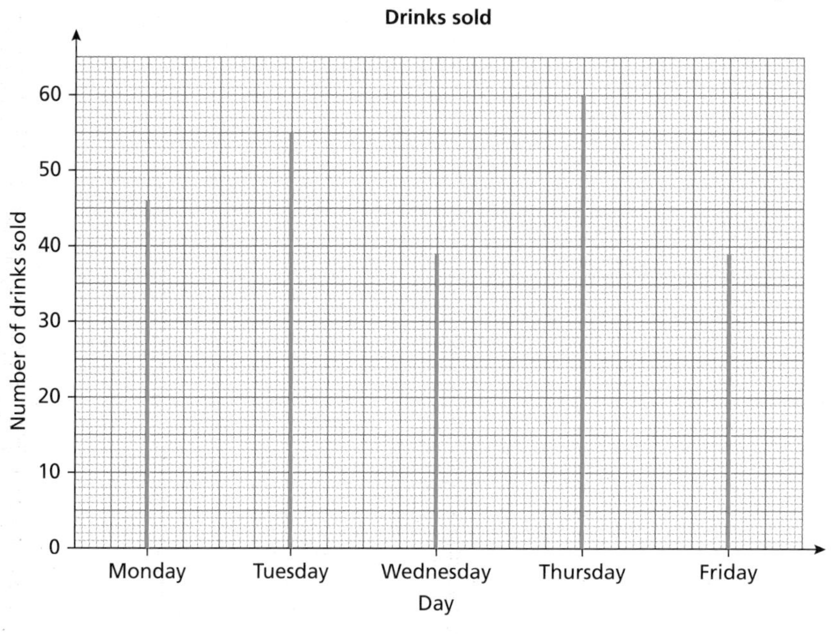

a How many drinks were sold on Tuesday?

b On which day was the greatest number of drinks sold?

c On which two days were the same number of drinks sold?

2 Calculate the mean of these numbers.

1300 1150 1200 900 600 400

A **time series** graph is a line graph that shows how data changes over time.

This time series graph shows information about the amount of profit made in a shop on a given day during the week.

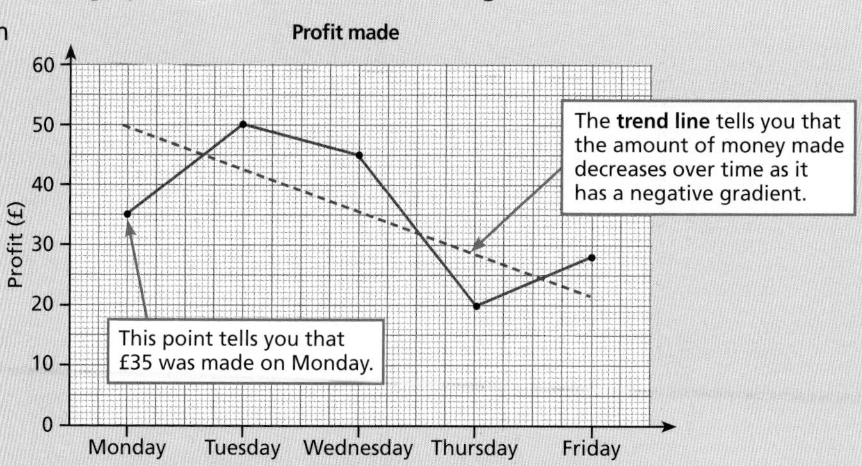

You can read off time series and estimate the value of points between those given. It is not reliable to estimate values outside of the given range as you do not know if the trend continues in the same way or not; this is called **extrapolation**.

Example

The time series graph shows the number of items sold in a shop from Monday to Friday.

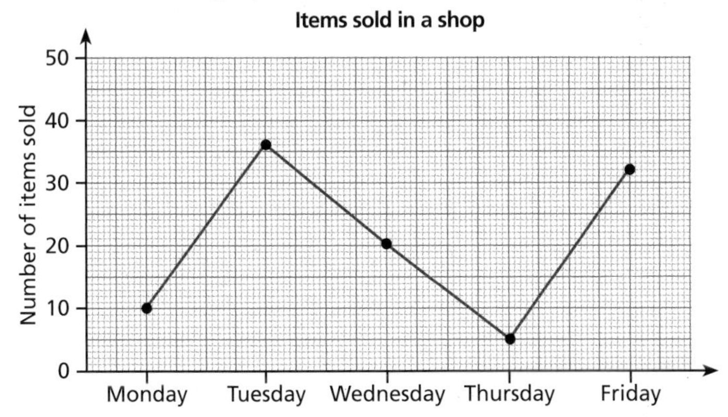

a On which day was the fewest number of items sold?

b How many items were sold in total on Tuesday and Wednesday?

Method

Solution	Commentary
a Thursday	This is the lowest point of the graph.
b 36 + 20 = 56	Add the sales for the two days.

Practice

1 The time series graph shows the number of customers that visit a shop in a week.

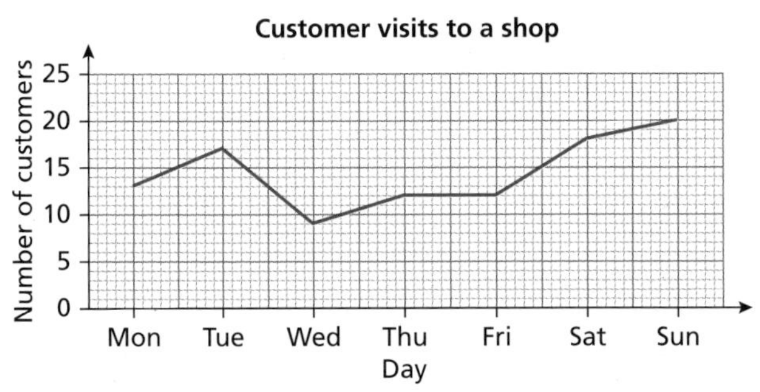

a How many customers visited the shop on Tuesday?

b On which day did the greatest number of customers visit the shop?

c Which two days had the same number of customers?

2 The time series graph shows the quarterly sales of a company for 2023 and 2024.

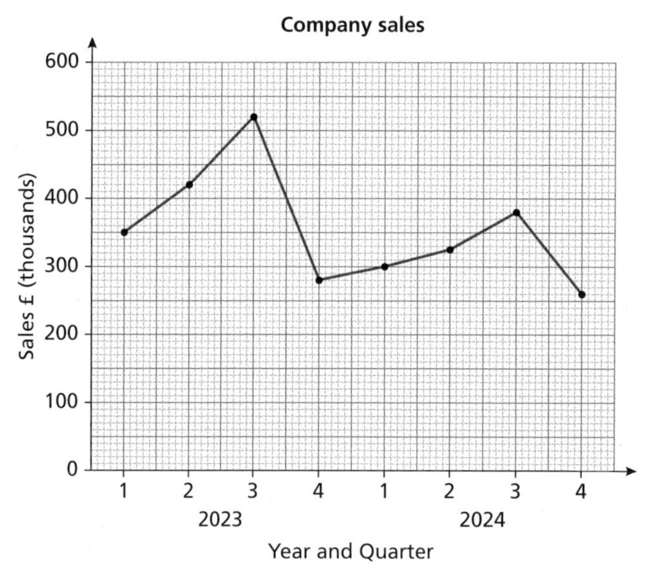

a In 2023, which quarter had the highest sales?

b How much money was made in sales in quarter 2 in 2024?

c In which year and quarter did the company record its highest sales?

d Calculate the mean sales made per quarter.

e Work out the mean sales made per month.

3 The table shows the profit made by a shop over Monday to Friday one week.

Day	Monday	Tuesday	Wednesday	Thursday	Friday
Profit (£)	45	50	30	55	40

a Copy the axes and plot a time series graph.

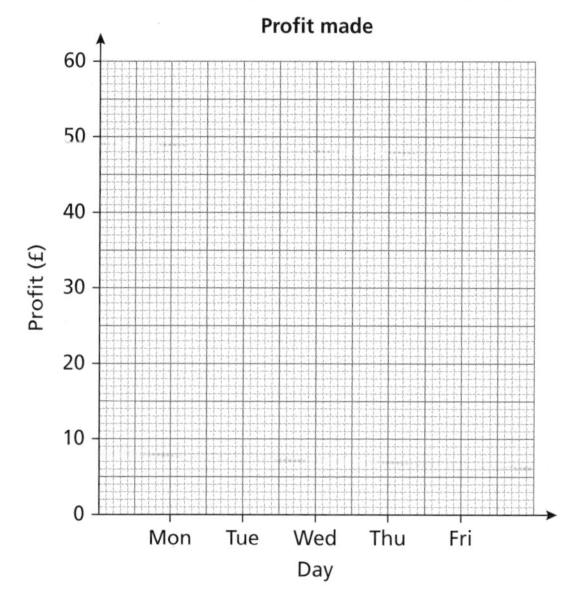

b On which day was the most profit made?

c What is the range of profit made?

d What is the average profit made per day?

4 The time series graph shows information about visitors to a museum.

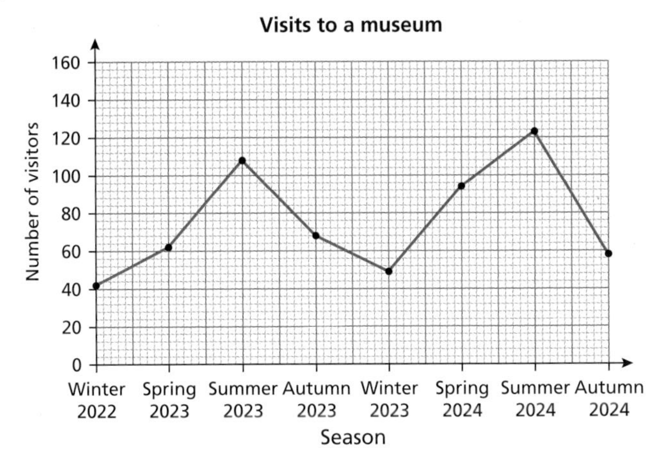

Make two comments about the trend of the number of visitors to the museum from the graph.

What do you think?

1 The time series graph shows information about the population of tigers in an area.

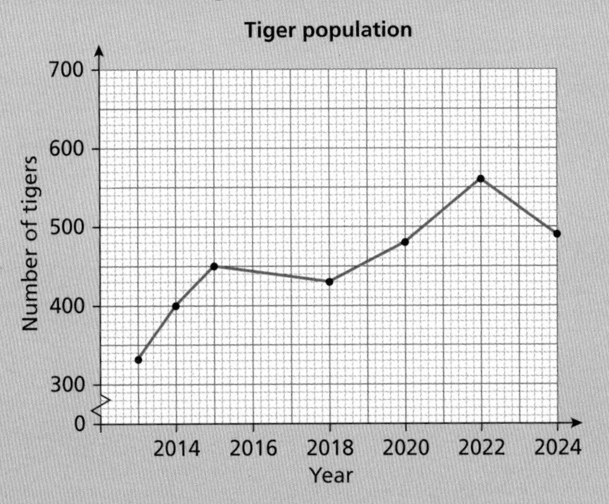

If a trend line was drawn, would this be a reliable way of making an estimate for the population in 2028?

Consolidate – do you need more?

1 The time that students spent on their smartphone was recorded.

The time series graph shows the average daily time spent per student for each year from 2017 to 2023.

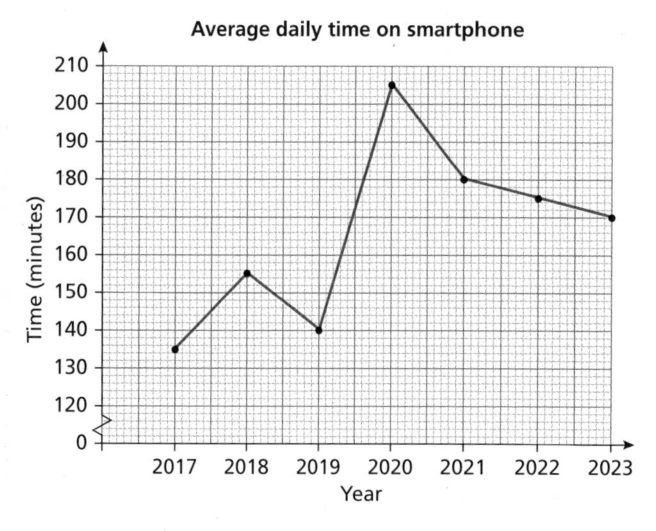

a Write down the average daily time spent on a smartphone per student in 2018.

b In which year was the greatest average daily time spent on a smartphone?

c In which year was the least average daily time spent on a smartphone?

d How many more average daily minutes were spent on smartphones in 2022 compared to 2019?

2 Copy the axes and plot a time series graph for this data.

Month	Jun	Jul	Aug	Sep	Oct	Nov
Average amount of rainfall (mm)	48	42	36	50	42	54

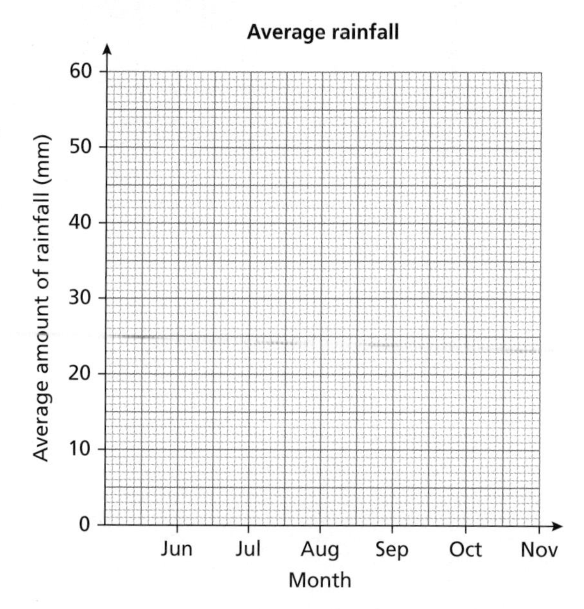

Stretch – can you deepen your learning?

1 Energy is measured in kWh (kilowatt hours).

The graph shows the amount of energy used each month in a particular year in one household.

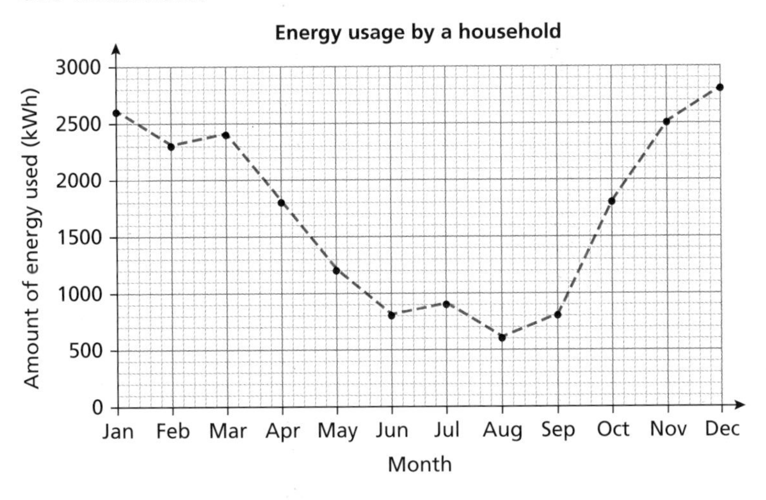

a In which month did the household use the least amount of energy?

b What is the difference in energy consumption in July compared to the energy consumption in November?

c Why is adding a trend line **not** suitable for this graph?

d Zach says, "The average energy consumption for the first six months of the year is greater than for the second six months of the year."

Show that Zach is correct.

2 **a** Investigate the meaning of 'moving averages'.

b The table shows information about the number of burgers sold by a snack van over a six-week period.

Week	1	2	3	4	5	6
Number of burgers sold	130	167	156	160	191	183

i Draw a time series graph to represent the data.

ii Calculate the three-point moving averages and copy and complete the table.

Three-point period	Week 1 to 3	Week 2 to 4	Week _ to _	Week _ to _
Three-point moving average				

iii Plot the moving averages onto your time series graph and draw a trend line.

14.2 Histograms

Are you ready?

1 Work out the midpoint of each class interval.

a $0 \leqslant x < 4$ **b** $4 \leqslant x < 9$ **c** $9 \leqslant x < 13$ **d** $13 \leqslant x < 25$

2 John receives 300 emails over a two-hour period.

On average, how many emails does John receive: **a** per hour **b** per minute?

3 A small business receives an average of 35 items of post per day.

Estimate the total number of items the business receives: **a** in a week **b** in a month.

In a bar chart, the bar widths are always the same and the height of each bar represents the frequency of each item.

In a **histogram**, the widths can be different and it is the **area** of the bar that represents the frequency.

Histograms represent continuous data organised into groups, known as **classes**.

The height of each bar in a histogram corresponds to the **frequency density** of the class, which is found using the formula:

Frequency density = frequency ÷ class width

Example 1

Thirty people were asked to complete a puzzle.

The time that each of them took is shown in the table.

Draw a histogram to represent this data.

Time, s (seconds)	Frequency
$0 \leqslant s < 20$	1
$20 \leqslant s < 30$	4
$30 \leqslant s < 40$	8
$40 \leqslant s < 50$	9
$50 \leqslant s < 60$	6
$60 \leqslant s < 80$	2
Total	30

Method

Solution

Time, s (seconds)	Frequency	Class width	Frequency density
$0 \leqslant s < 20$	1	20	0.05
$20 \leqslant s < 30$	4	10	0.4
$30 \leqslant s < 40$	8	10	0.8
$40 \leqslant s < 50$	9	10	0.9
$50 \leqslant s < 60$	6	10	0.6
$60 \leqslant s < 80$	2	20	0.1
Total	30		

Commentary

$$\text{Frequency density} = \frac{\text{frequency}}{\text{class width}}$$

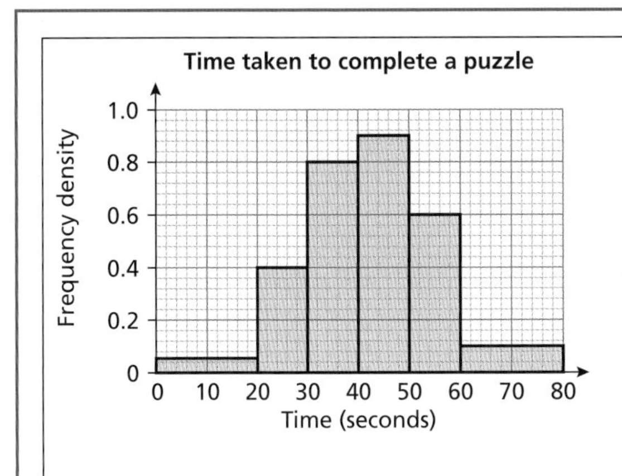

When drawing a histogram, the vertical axis is always frequency density.

In this example, the horizontal axis needs to go from 0 to 80

The vertical axis can go to 1

Draw the rectangular bars with bases that correspond to the class intervals and heights that correspond to the frequency densities.

Example 2

The histogram shows the ages of a group of people.

a Work out the total number of people in the group.

b Estimate the number of people who are under 13 years old.

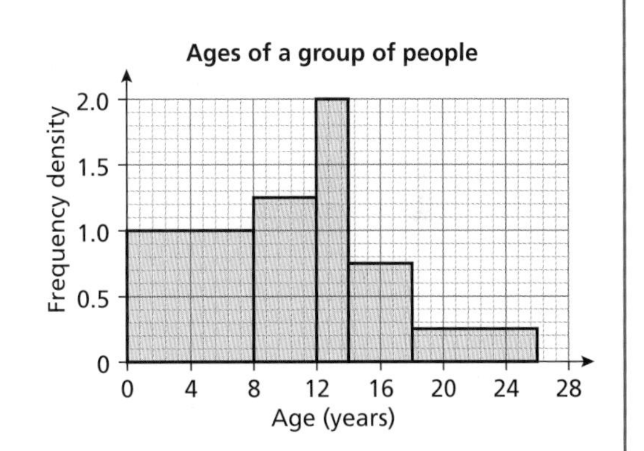

Method

Solution	Commentary
a	The area of each bar in the histogram represents the frequency of the class.

Age, y (years)	Frequency density	Class width	Area (frequency)
$0 \leqslant y < 8$	1	8	8
$8 \leqslant y < 12$	1.25	4	5
$12 \leqslant y < 14$	2	2	4
$14 \leqslant y < 18$	0.75	4	3
$18 \leqslant y < 26$	0.25	8	2
Total			22

b

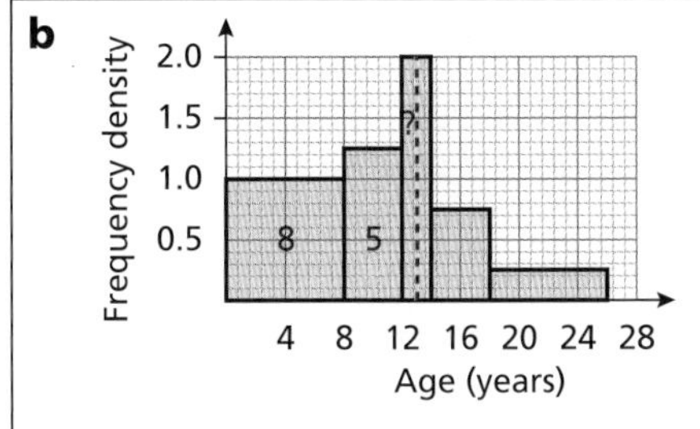

Bar 1 = 8   Bar 2 = 5   Bar 3 = 4 ÷ 2 = 2

Number of people aged under 13 = 8 + 5 + 2 = 15

Since the area represents the frequency, the frequency below 13 years is left of the dotted line.

Bar 1 has a frequency of 8

Bar 2 has a frequency of 5

13 is halfway between 12 and 14, which means between 12 and 13 is half of the bar.

Practice

1 Draw a histogram for each set of data.

a

Mass, x (kg)	Frequency
$18 \leqslant x < 24$	3
$24 \leqslant x < 28$	4
$28 \leqslant x < 30$	3
$30 \leqslant x < 34$	5
$34 \leqslant x < 40$	3
$40 \leqslant x < 50$	3

b

Height, h (m)	Frequency
$2 \leqslant h < 3$	3
$3 \leqslant h < 4$	3
$4 \leqslant h < 4.5$	2
$4.5 \leqslant h < 5$	1
$5 \leqslant h < 6$	3
$6 \leqslant h < 8$	1

c

Time, t (seconds)	Frequency
$10 \leqslant t < 15$	3
$15 \leqslant t < 20$	2
$20 \leqslant t < 22$	3
$22 \leqslant t < 24$	4
$24 \leqslant t < 28$	1

d

Speed, s (mph)	Frequency
$25 \leqslant s < 30$	1
$30 \leqslant s < 35$	3
$35 \leqslant s < 40$	5
$40 \leqslant s < 50$	8
$50 \leqslant s < 60$	2

2 Copy and complete the tables and histograms from the given information.

a

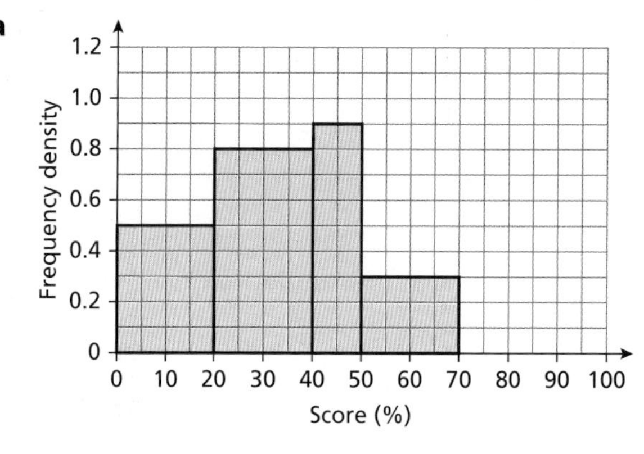

Score, s (%)	Frequency
$0 \leqslant s < 20$	
$20 \leqslant s < 40$	
$40 \leqslant s < 50$	
$50 \leqslant s < 70$	
$70 \leqslant s < 80$	5
$80 \leqslant s < 90$	4
$90 \leqslant s < 100$	1

b

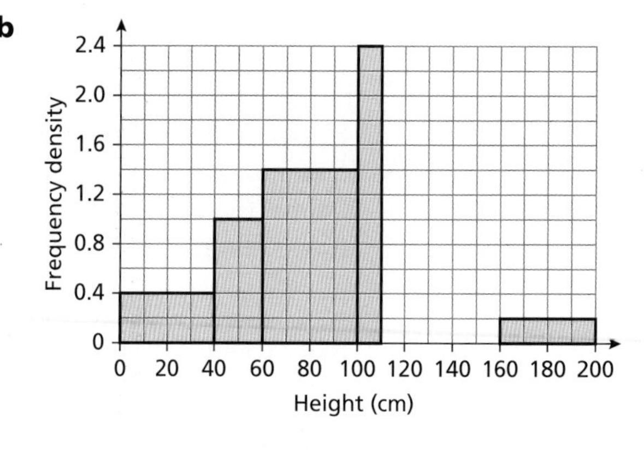

Height, h (cm)	Frequency
$0 \leqslant h < 40$	
$40 \leqslant h < 60$	
$60 \leqslant h < 100$	
$100 \leqslant h < 110$	
$110 \leqslant h < 120$	16
$120 \leqslant h < 140$	40
$140 \leqslant h < 160$	20
$160 \leqslant h < 200$	

c

Length, l (m)	Frequency
$0 \leqslant l < 40$	
$40 \leqslant l < 80$	
$80 \leqslant l < 160$	40
$160 \leqslant l < 240$	
$240 \leqslant l < 260$	
$260 \leqslant l < 340$	20
$340 \leqslant l < 400$	

d

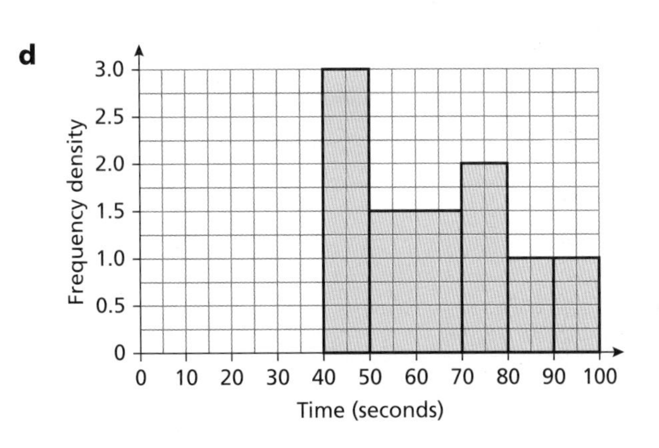

Time, t (seconds)	Frequency
$0 \leqslant t < 20$	5
$20 \leqslant t < 25$	5
$25 \leqslant t < 40$	30
$40 \leqslant t < 50$	
$50 \leqslant t < 70$	
$70 \leqslant t < 80$	
$80 \leqslant t < 90$	
$90 \leqslant t < 100$	

3 **a** Estimate the frequency for a mass below 40 kg.

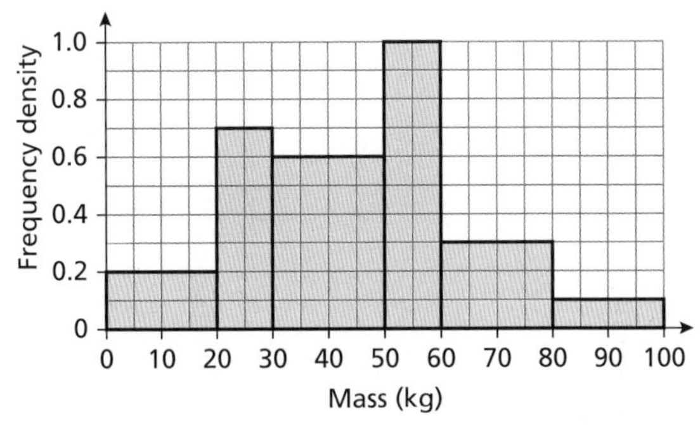

b Estimate the frequency for a score above 40

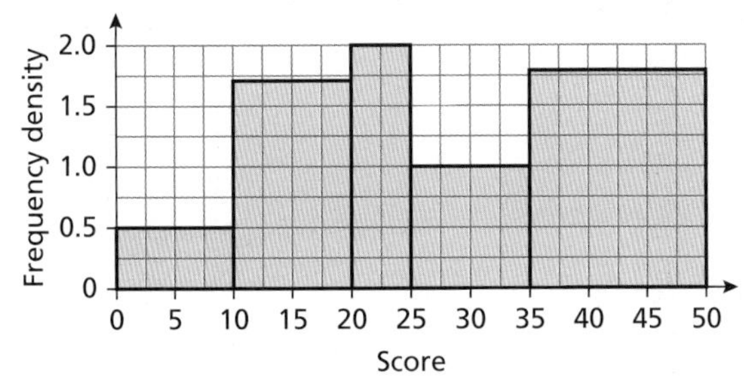

c Estimate the frequency for a time over 42 seconds.

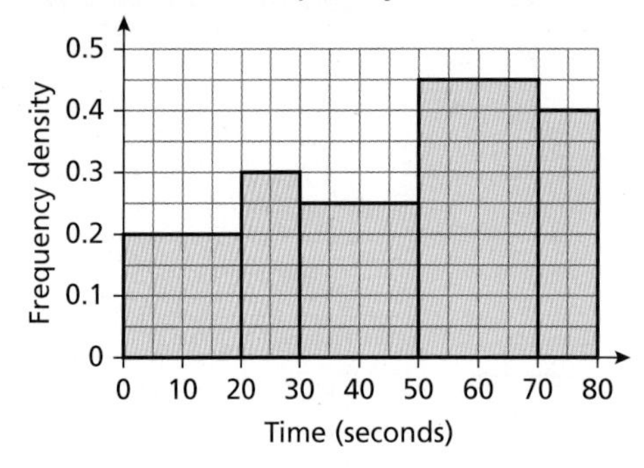

d Estimate the frequency for a distance between 2.5 m and 6 m.

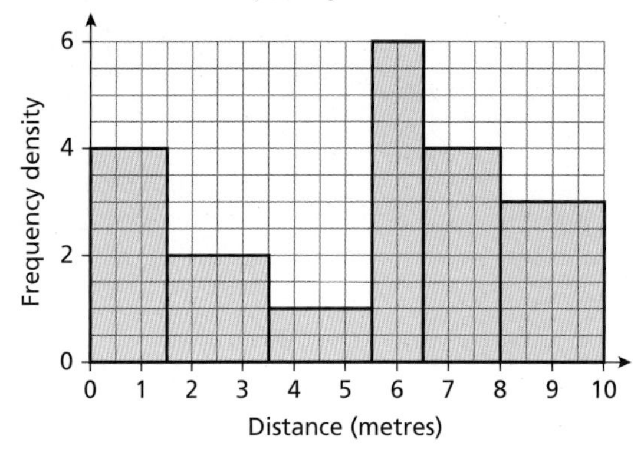

4 The histogram shows the time taken by some people to complete a puzzle.

Eight people completed the puzzle in less than 20 seconds.

How many people does the histogram represent altogether?

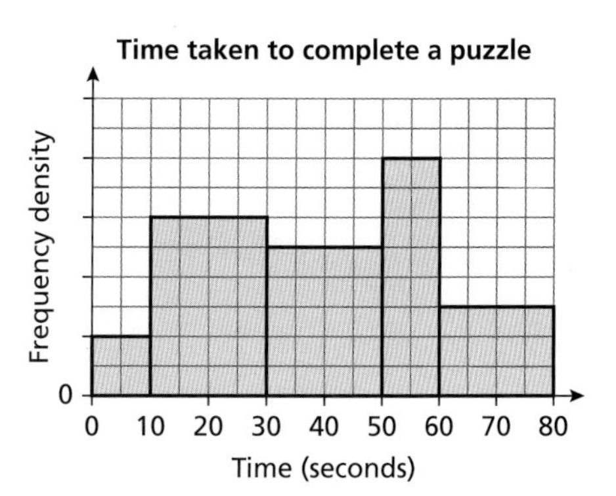

Consolidate – do you need more?

1 Draw a histogram for each frequency table.

a

Age, a (years)	Frequency	Class width	Frequency density
$0 \leqslant a < 10$	7	10	$7 \div 10 = 0.7$
$10 \leqslant a < 18$	8		
$18 \leqslant a < 25$	14		
$25 \leqslant a < 30$	10		
$30 \leqslant a < 40$	17		
$40 \leqslant a < 60$	26		
$60 \leqslant a < 100$	20		

b

Time, t (minutes)	Frequency
$0 \leqslant t < 5$	12
$5 \leqslant t < 10$	7
$10 \leqslant t < 20$	15
$20 \leqslant t < 30$	19
$30 \leqslant t < 50$	28
$50 \leqslant t < 60$	12

c

Height, h (cm)	Frequency
$130 \leqslant h < 150$	14
$150 \leqslant h < 170$	20
$170 \leqslant h < 180$	15
$180 \leqslant h < 190$	13
$190 \leqslant h < 200$	17

d

Mass, m (kg)	Frequency
$30 \leqslant m < 40$	9
$40 \leqslant m < 50$	12
$50 \leqslant m < 70$	16
$70 \leqslant m < 80$	15
$80 \leqslant m < 90$	8

2 Draw a frequency table for each histogram.

a

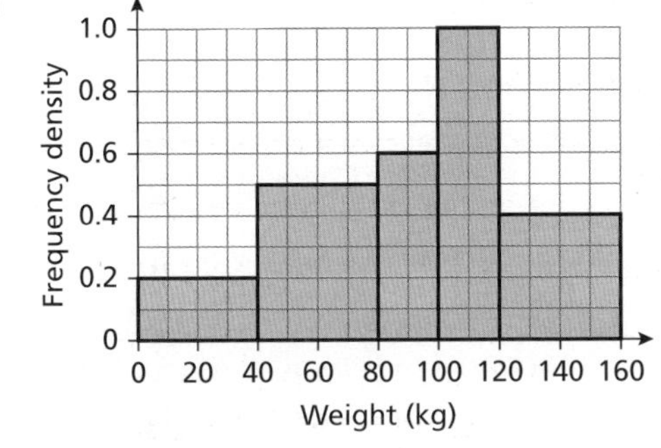

b

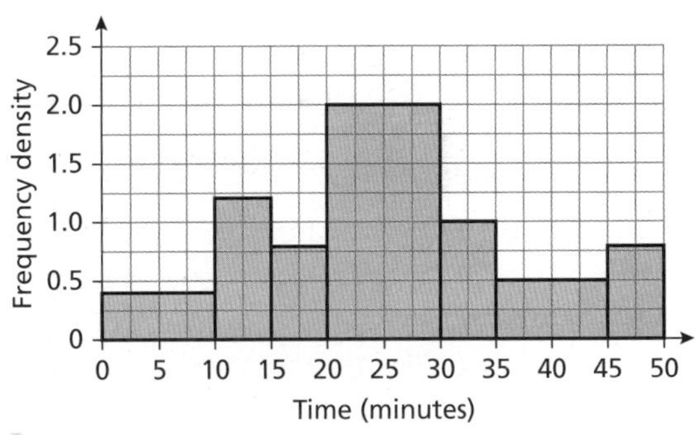

c

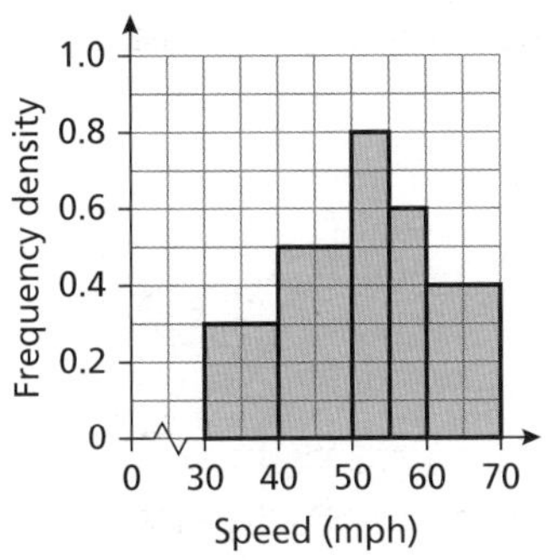

d

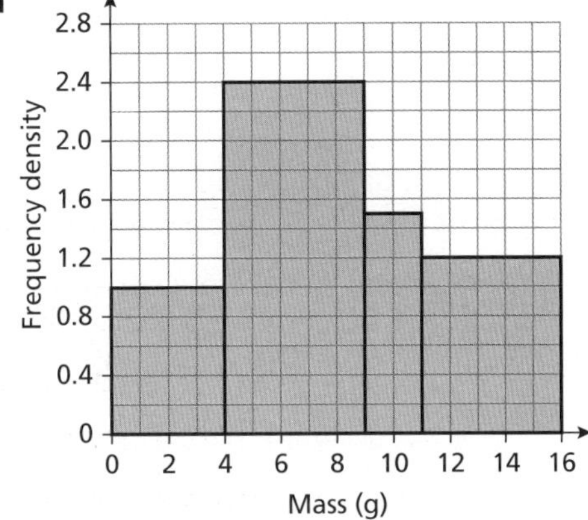

Stretch – can you deepen your learning?

1 The diagrams show the number of steps taken by two groups of people one day.

Estimate the mean number of steps taken by each group.

Group A

Frequency
50
40
30
20
10
0
0 4 5 6 7 8 9 10 11 12 13 14
Steps (thousands)

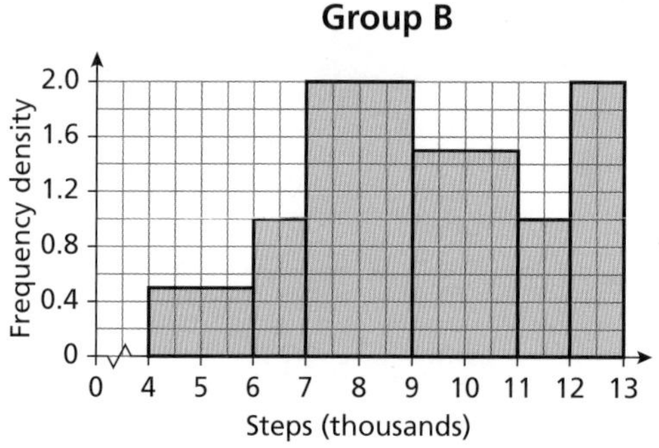
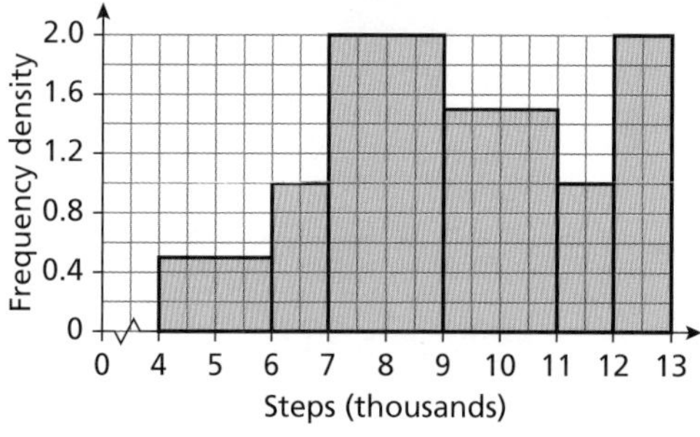

2 The histogram shows the test results of 200 people.

a 35% of people scored no more than a marks.

Work out the value of a.

b 27.5% of people scored more than b marks.

Work out the value of b.

c 65% of people scored no more than c marks.

Work out the value of c.

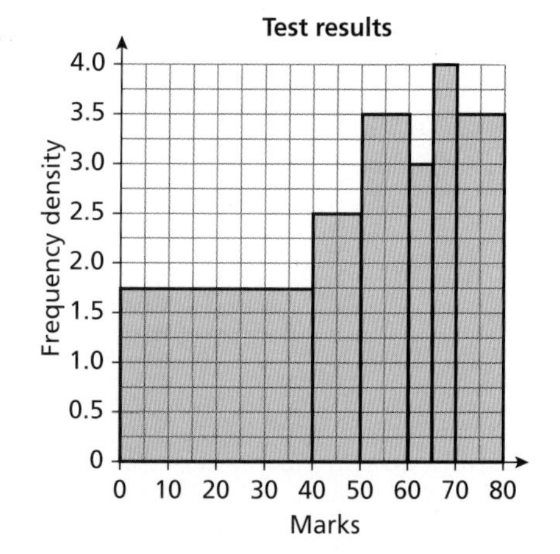

3 The histogram shows the ages of members of a pool team.

There are 11 people over the age of 60.

Work out an estimate for the mean age of the members.

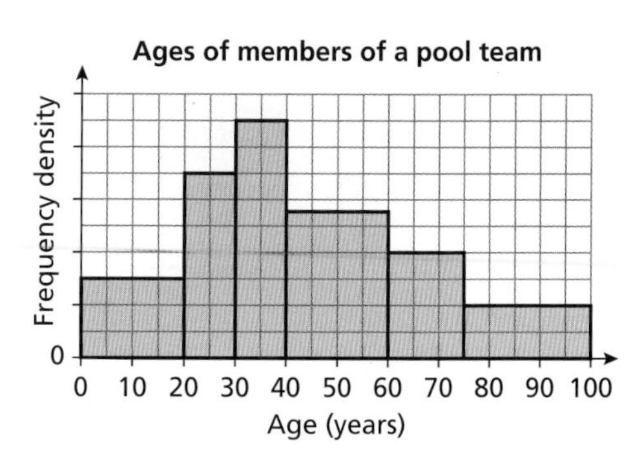

14.3 Cumulative frequency diagrams

Are you ready? (A)

1 The table shows information about the ages in years of some people.

Age, a (years)	$0 < a \leqslant 10$	$10 < a \leqslant 20$	$20 < a \leqslant 30$	$30 < a \leqslant 40$
Frequency	15	22	17	6

a How many people are there altogether?

b How many people are aged 20 or under?

c Write down the modal class.

d How many people are over the age of 10?

e In which interval does the median lie?

Cumulative frequency shows the total number of items of data up to a given point in a frequency distribution.

Mass, m (kg)	Frequency	Cumulative frequency
$0 < m \leqslant 10$	8	8
$10 < m \leqslant 20$	14	$8 + 14 = 22$
$20 < m \leqslant 30$	17	$22 + 17 = 39$
$30 < m \leqslant 40$	3	$39 + 3 = 42$

There are 22 items with mass up to and including 20 kg. There are 39 items with mass up to and including 30 kg, and 42 items with mass up to and including 40 kg.

A **cumulative frequency diagram** can be drawn to represent the amount of data up to given points in a distribution.

You can use a cumulative frequency diagram to estimate:

- the **median**; the middle value of the distribution
- the **lower quartile**; the value below which 25% of the distribution lies
- the **upper quartile**; the value below which 75% of the distribution lies
- any other value you are interested in; for example, a pass mark so that 40% of people pass a test would correspond to 60% of the data below the pass mark.

The **interquartile range** is found by subtracting the lower quartile from the upper quartile.

Example

The table shows information about the ages of 50 people.

a Draw a cumulative frequency diagram to represent the data.

b Use the cumulative frequency diagram to work out an estimate for the median age.

Age, a (years)	Frequency
$0 < a \leqslant 10$	6
$10 < a \leqslant 20$	15
$20 < a \leqslant 30$	21
$30 < a \leqslant 40$	8

Method

Solution	Commentary
a	Cumulative frequency is the running total of the frequencies. The cumulative frequency points are plotted at the upper end points of the intervals.

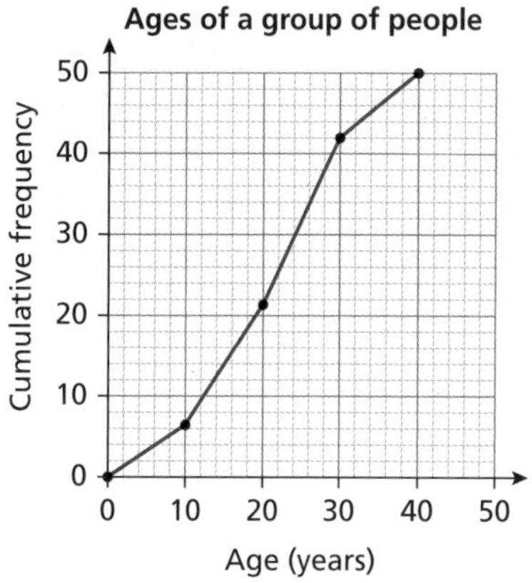

Age, a (years)	Frequency	Cumulative frequency	Point to plot
$0 < a \leqslant 10$	6	6	(10, 6)
$10 < a \leqslant 20$	15	21	(20, 21)
$20 < a \leqslant 30$	21	42	(30, 42)
$30 < a \leqslant 40$	8	50	(40, 50)

Ages of a group of people

Cumulative frequency

0 10 20 30 40 50

Age (years)

You can also plot the point (0, 0) as the first interval starts at age 0

The points are then joined together using straight line segments.

b

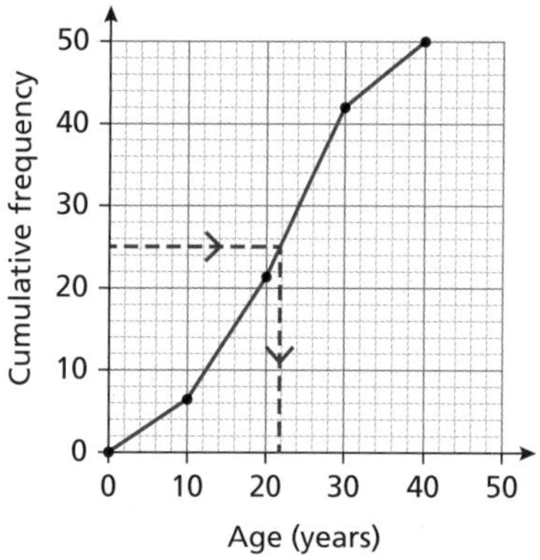

Cumulative frequency

0 10 20 30 40 50

Age (years)

An estimate for the median age is 22

The median is at 50% of the cumulative frequency.

For this example, this is at 25 (50% of 50).

Draw a line from the vertical axis to the graph then down to the horizontal axis to find an estimate for the median age.

Practice (A)

1 Draw a cumulative frequency diagram for each set of data.

a

Weight, x (kg)	Frequency	Cumulative frequency	Point to plot
$18 < x \leqslant 24$	2	2	(24, 2)
$24 < x \leqslant 28$	4	6	
$28 < x \leqslant 30$	3		
$30 < x \leqslant 34$	5		
$34 < x \leqslant 40$	3		
$40 < x \leqslant 50$	3		

b

Height, h (m)	Frequency	Cumulative frequency	Point to plot
$2 < h \leqslant 3$	2		
$3 < h \leqslant 4$	3		
$4 < h \leqslant 4.5$	3		
$4.5 < h \leqslant 5$	4		
$5 < h \leqslant 6$	3		
$6 < h \leqslant 8$	1		

c

Time, t (s)	Frequency
$10 < t \leqslant 15$	3
$15 < t \leqslant 20$	2
$20 < t \leqslant 22$	4
$22 < t \leqslant 24$	5
$24 < t \leqslant 28$	1

d

Speed, s (mph)	Frequency
$25 < s \leqslant 30$	12
$30 < s \leqslant 35$	13
$35 < s \leqslant 40$	25
$40 < s \leqslant 50$	8
$50 < s \leqslant 60$	2

2 Estimate the median from each cumulative frequency diagram.

a

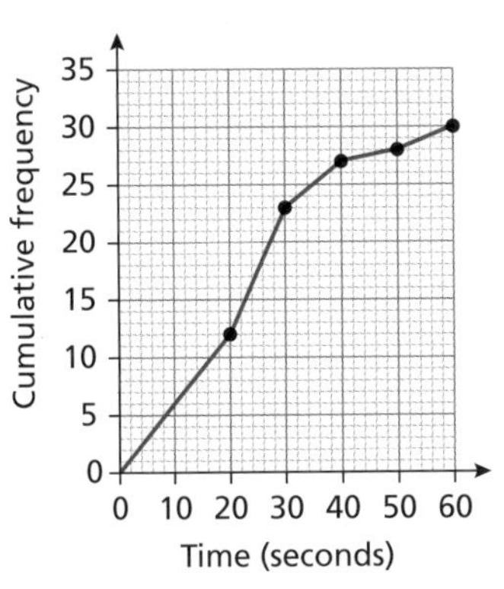

b

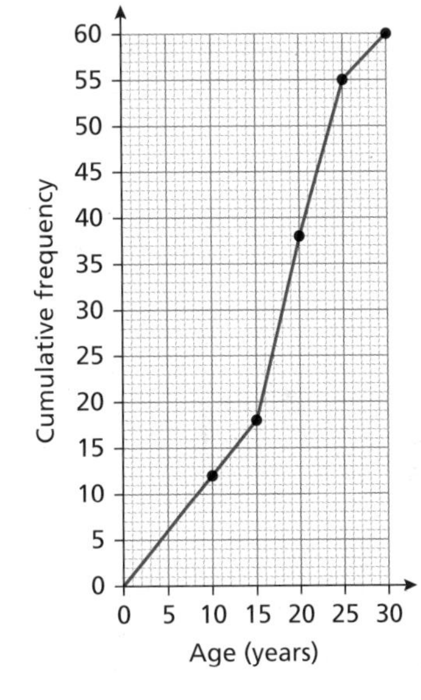

c

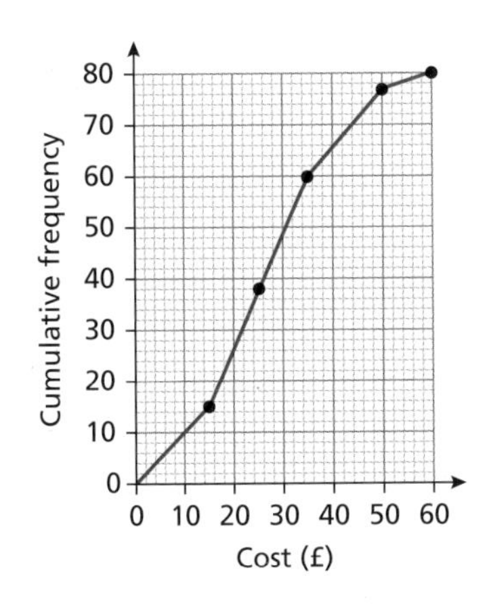

d

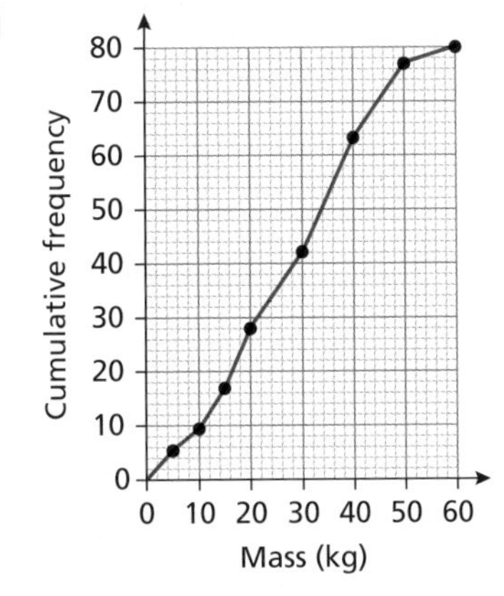

3 The heights of 60 people are shown in the table.

a Draw a cumulative frequency diagram for the heights.

b Estimate the median height.

c Estimate the number of people who are taller than 170 cm.

Use a vertical line from 170 on the height axis to your cumulative frequency graph to help.

Height, h (cm)	Frequency
$120 < h \leqslant 140$	6
$140 < h \leqslant 150$	7
$150 < h \leqslant 160$	11
$160 < h \leqslant 175$	15
$175 < h \leqslant 190$	12
$190 < h \leqslant 200$	9

4 120 people run a race.

Their times, in seconds, are shown in the table.

a Draw a cumulative frequency diagram for the times.

b Estimate the median time taken.

c Estimate the number of people who finished the race in less than 40 seconds.

Time, t (s)	Frequency
$0 < t \leqslant 20$	8
$20 < t \leqslant 30$	21
$30 < t \leqslant 45$	27
$45 < t \leqslant 60$	39
$60 < t \leqslant 90$	21
$90 < t \leqslant 120$	4

Are you ready? (B)

1 Work out:

a 50% of 90 **b** 25% of 80 **c** 75% of 120

2 Work out the range of each set of numbers.

a 7, 9, 4, 8, 2, 5 **b** 17, 19, 31, 28, 21, 19 **c** 1.08, 2.06, 3.08, 1.51, 3.51, 2.8

Example 1

The cumulative frequency diagram shows the battery life of some laptop batteries.

Use the cumulative frequency diagram to estimate:

a the median battery life

b the lower quartile of battery life

c the upper quartile of battery life

d the interquartile range of battery life.

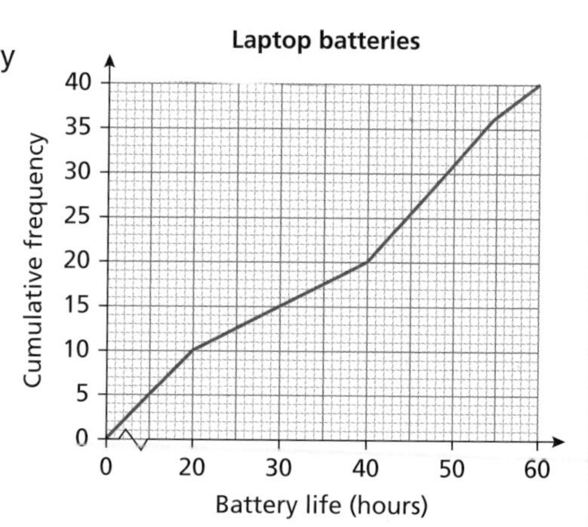

Method

Solution	Commentary
a 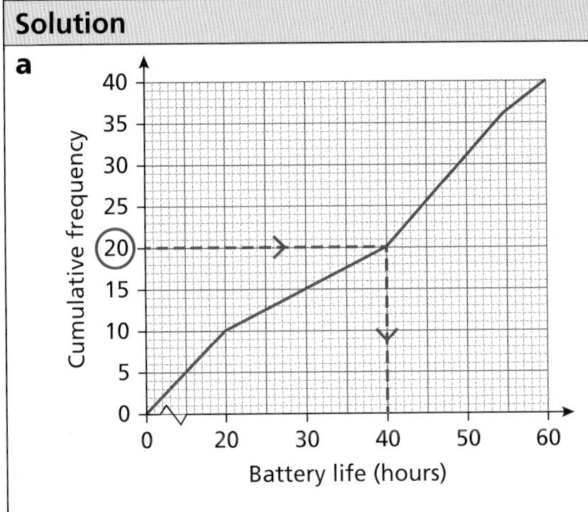 Median = 40 hours	There are 40 batteries, so the median is at the $\left(\frac{40}{2}\right)$th = 20th position On the cumulative frequency diagram, draw a line across from 20 on the cumulative frequency axis and then down from the line to find the median battery life.
b 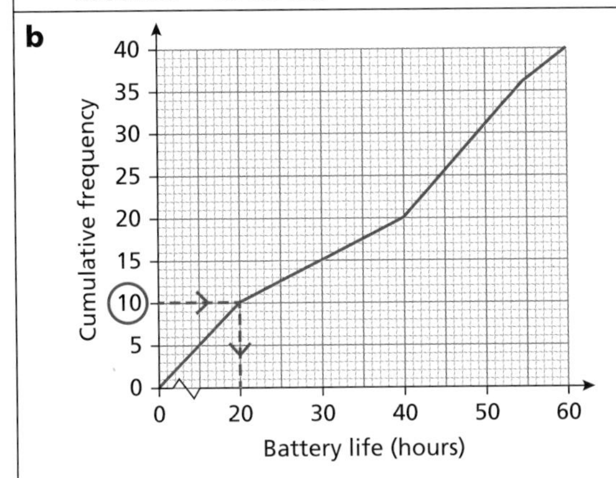 Lower quartile = 20 hours	The lower quartile is one-quarter of the way along the distribution. $\frac{40}{4} = 10$ On the cumulative frequency diagram, draw a line across from 10 on the cumulative frequency axis, and then down from the line to find the lower quartile.
c 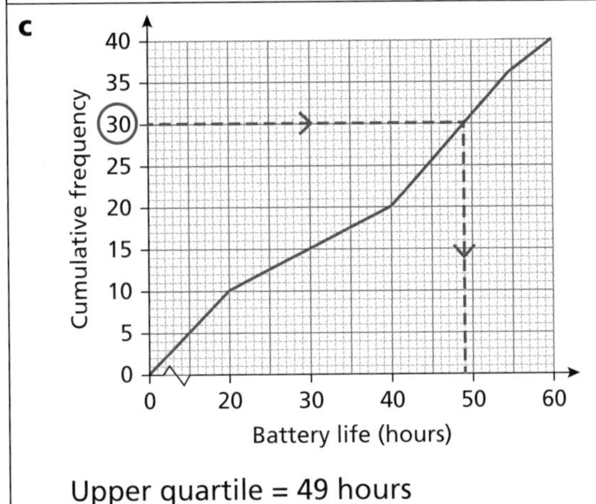 Upper quartile = 49 hours	The upper quartile is three-quarters of the way along the distribution. $\frac{40}{4} \times 3 = 30$ On the cumulative frequency diagram, draw a line across from 30 on the cumulative frequency axis, and then down from the line to find the upper quartile.
d Interquartile range = 49 – 20 = 29 hours	The interquartile range is found by working out the difference between the upper and lower quartiles.

Example 2

The table shows information about the speeds of 100 vehicles.

a Draw a cumulative frequency graph to show the information.

b Estimate the median speed of the vehicles.

c 20% of the vehicles are exceeding speed x km/h. Estimate the value of x.

Speed, s (km/h)	Cumulative frequency
$0 < s \leqslant 20$	10
$0 < s \leqslant 40$	25
$0 < s \leqslant 60$	47
$0 < s \leqslant 80$	73
$0 < s \leqslant 100$	88
$0 < s \leqslant 120$	100

Method

Solution	Commentary
a 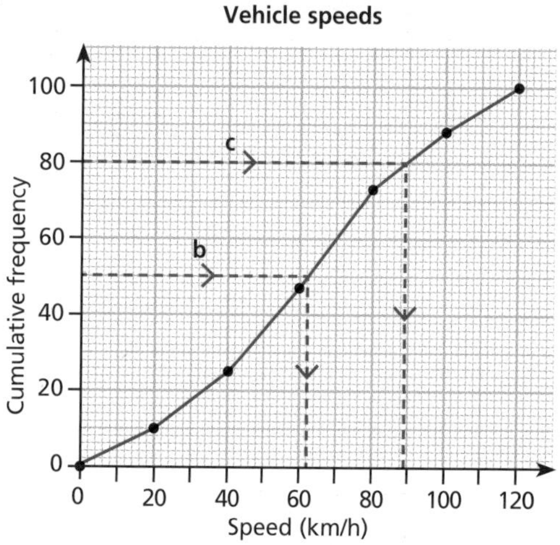	The first point to plot is (0, 0). Then plot the upper bound of each class and the cumulative frequency: (20, 10), (40, 25), (60, 47), (80, 73), (100, 88) and (120, 100) Join each point to the previous point with a line segment.
b Median is approximately 62 km/h	There are 100 vehicles, so the median is at the $\frac{100}{2}$th = 50th position. Use the line labelled **b**.
c x is approximately 89 km/h	As 20% of vehicles are exceeding speed x, this means 100% – 20% = 80% are going slower than x. Use the line labelled **c**.

Practice (B)

1 Use a cumulative frequency diagram to estimate the median for each set of data.

a

Weight, x (kg)	Frequency	Cumulative frequency
$10 < x \leqslant 20$	8	8
$20 < x \leqslant 30$	14	22
$30 < x \leqslant 45$	12	
$45 < x \leqslant 60$	19	
$60 < x \leqslant 70$	12	
$70 < x \leqslant 80$	5	

b

Height, h (m)	Frequency	Cumulative frequency
$2 < h \leqslant 3$	10	
$3 < h \leqslant 4$	14	
$4 < h \leqslant 4.5$	16	
$4.5 < h \leqslant 5$	18	
$5 < h \leqslant 6$	15	
$6 < h \leqslant 8$	7	

c

Time, t (s)	Frequency
$10 < t \leqslant 15$	19
$15 < t \leqslant 20$	27
$20 < t \leqslant 22$	24
$22 < t \leqslant 24$	18
$24 < t \leqslant 28$	12

d

Speed, s (mph)	Frequency
$25 \leqslant s < 30$	8
$30 \leqslant s < 35$	13
$35 \leqslant s < 40$	21
$40 \leqslant s < 50$	12
$50 \leqslant s < 60$	6

2 From each cumulative frequency diagram, estimate the:

i median **ii** lower quartile **iii** upper quartile **iv** interquartile range.

a

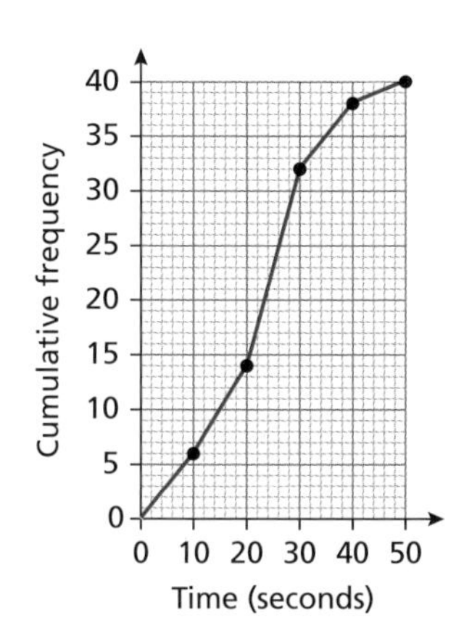

b

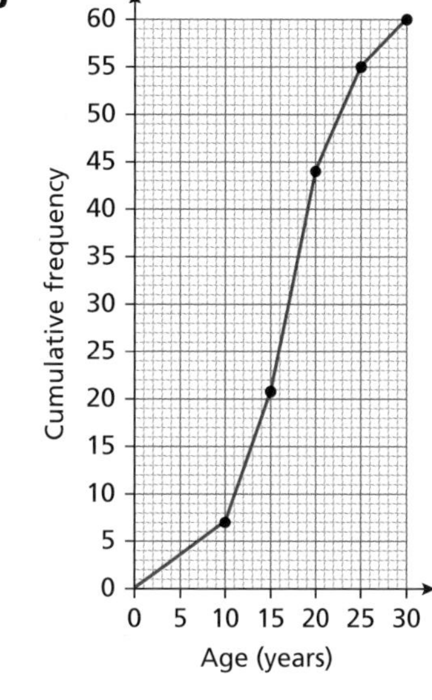

c

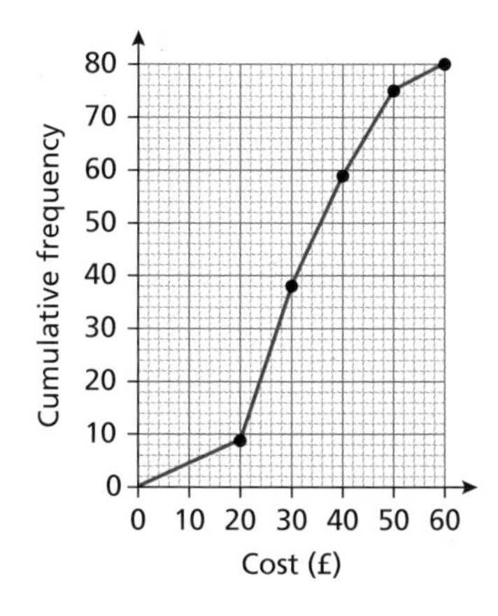

d

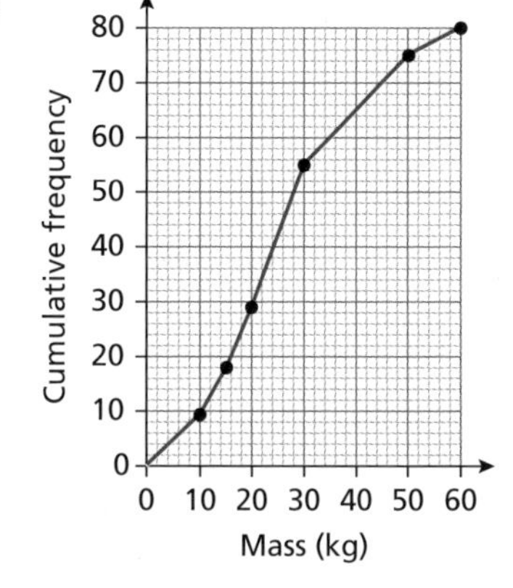

3 The heights of 80 people are shown in the table.

a Draw a cumulative frequency diagram for the heights.

b Estimate the median height.

c Estimate the interquartile range of the heights.

Height, h (cm)	Frequency
$120 < h \leqslant 140$	6
$140 < h \leqslant 150$	14
$150 < h \leqslant 160$	17
$160 < h \leqslant 175$	19
$175 < h \leqslant 190$	15
$190 < h \leqslant 200$	9

4 120 people run a race.

Their times, in seconds, are shown in the table.

a Draw a cumulative frequency diagram for the times.

b Estimate the median time taken.

c Estimate how many people finished the race in more than 40 seconds.

d Estimate the interquartile range for the times.

Time, t (s)	Frequency
$0 < t \leqslant 20$	9
$20 < t \leqslant 30$	23
$30 < t \leqslant 45$	26
$45 < t \leqslant 60$	37
$60 < t \leqslant 90$	19
$90 < t \leqslant 120$	6

Consolidate – do you need more?

1 Draw a cumulative frequency diagram for each table.

a

Weight, x (kg)	Frequency	Cumulative frequency
$18 < x \leqslant 24$	2	
$24 < x \leqslant 28$	4	
$28 < x \leqslant 30$	3	
$30 < x \leqslant 34$	5	
$34 < x \leqslant 40$	3	
$40 < x \leqslant 50$	3	

b

Height, h (m)	Frequency	Cumulative frequency
$2 < h \leqslant 3$	7	
$3 < h \leqslant 4$	13	
$4 < h \leqslant 4.5$	18	
$4.5 < h \leqslant 5$	24	
$5 < h \leqslant 6$	13	
$6 < h \leqslant 8$	5	

c

Time, t (s)	Frequency
$10 < t \leqslant 15$	13
$15 < t \leqslant 20$	22
$20 < t \leqslant 22$	33
$22 < t \leqslant 24$	24
$24 < t \leqslant 28$	8

d

Speed, s (mph)	Frequency
$25 < s \leqslant 30$	15
$30 < s \leqslant 35$	4
$35 < s \leqslant 40$	21
$40 < s \leqslant 50$	17
$50 < s \leqslant 60$	3

2 For each set of data:

i draw a cumulative frequency diagram

ii estimate the median

iii estimate the interquartile range.

a

Length, l (cm)	Frequency	Cumulative frequency
$0 < l \leqslant 10$	2	
$10 < l \leqslant 15$	11	
$15 < l \leqslant 20$	19	
$20 < l \leqslant 30$	7	
$30 < l \leqslant 40$	1	

b

Height, h (m)	Frequency	Cumulative frequency
$0 < h \leqslant 0.8$	6	
$0.8 < h \leqslant 1.5$	14	
$1.5 < h \leqslant 2$	18	
$2 < h \leqslant 2.5$	27	
$2.5 < h \leqslant 3$	12	
$3 < h \leqslant 4$	3	

c

Age, a (years)	Frequency
$0 < a \leqslant 18$	8
$18 < a \leqslant 25$	22
$25 < a \leqslant 30$	27
$30 < a \leqslant 40$	19
$40 < a \leqslant 60$	4

d

Mass, m (kg)	Frequency
$10 < m \leqslant 14$	8
$14 < m \leqslant 18$	42
$18 < m \leqslant 22$	68
$22 < m \leqslant 30$	64
$30 < m \leqslant 40$	18

Stretch – can you deepen your learning?

1 The histogram shows the times taken by 48 people to complete a puzzle.

By drawing a cumulative frequency diagram, estimate the interquartile range for the time taken to complete the puzzle.

Discuss the accuracy of your estimates.

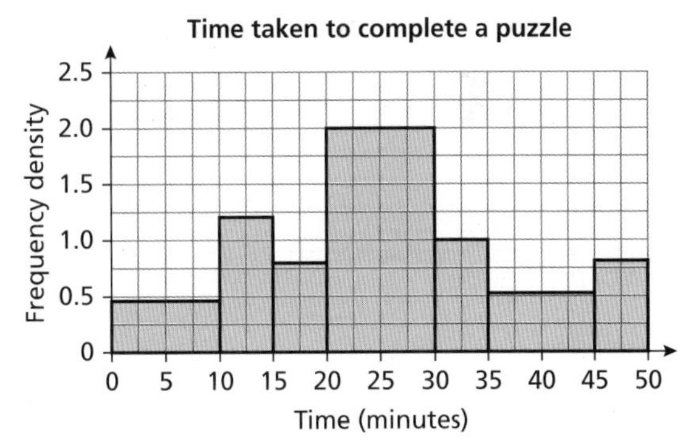

2 Calculate an estimate for the mean cost of the 80 items shown on the cumulative frequency diagram.

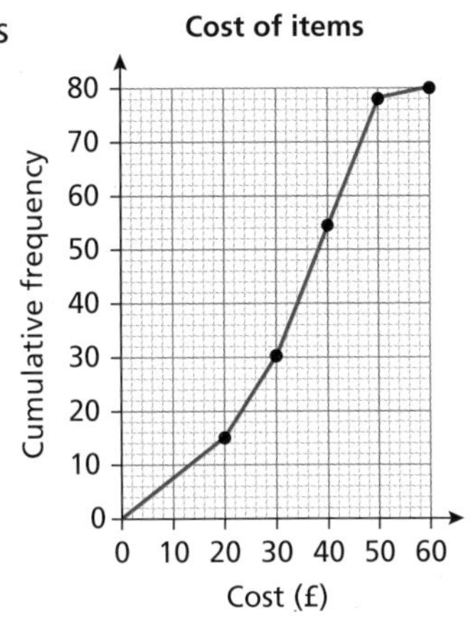

14.4 Box plots

Are you ready?

1 For each set of data, work out: **i** the median **ii** the range.

a 32, 34, 38, 41, 42, 48, 54, 55, 55, 57, 58

b 2.4, 3.8, 3.4, 2.8, 1.8, 3.8, 4.1, 3.4, 3.6, 4.2, 2.8

c 145, 134, 184, 148, 152, 168, 157

d 15.7, 16, 18, 14.5, 15.9, 16.2, 14.7, 16.9, 17.1, 15.1, 16.8

A **box plot** shows key information about a set of data.

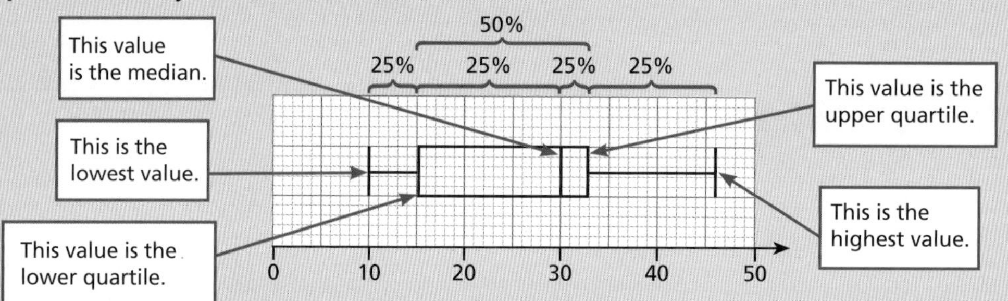

The box plot clearly shows the four quarters of a distribution with the box representing the 'middle 50%' of the data.

When comparing box plots from two sets of data, you can see the shapes of the distributions and use the medians to compare average values. You can also use the ranges or interquartile ranges to compare the spread of the data.

Example 1

Draw a box plot to represent the data in the table.

Lowest value	8
Highest value	25
Median	17
Lower quartile	13
Upper quartile	20

Method

Solution	Commentary
4 8 12 16 20 24 28	Once the scale of the axis has been decided or given, draw each of the five values and join them together to create the correct box plot.

Example 2

Draw a box plot to represent the data.

10, 10, 14, 16, 17, 21, 21, 24, 25, 26, 33

Method

Solution	Commentary
10 10 14 16 17 (21) 21 24 25 26 33 14 ↑ LQ; 21 ↑ (circled); 25 ↑ UQ	
Lowest value = 10 Highest value = 33 Median = 21	The median value is the $\frac{(n+1)}{2}$th For 11 items, this is the sixth item.
Lower quartile = 14 Upper quartile = 25	The lower and upper quartiles are the middle values of the two sets either side of the median.
8 16 24 32 40	Once all the values have been found, the box plot can be drawn.

Example 3

The box plots show the distribution of the masses of dogs examined on two days in a vet's surgery.

Compare the distributions.

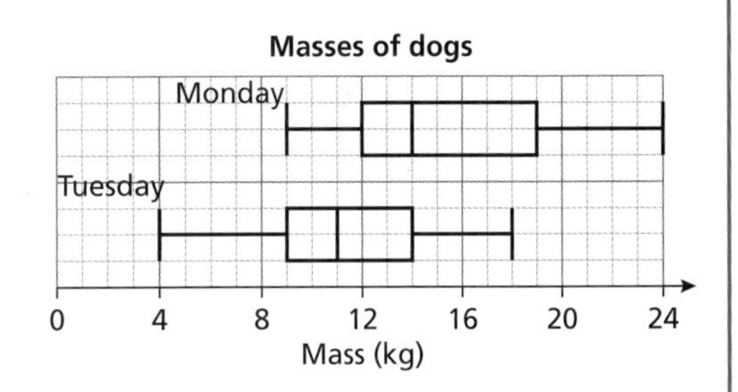

Method

Solution	Commentary
Monday Tuesday 0 4 8 12 16 20 24 Mass (kg)	Comparing the medians gives an average mass for the dogs each day.
The median mass on Monday is 14 kg. The median mass on Tuesday is 11 kg.	Quote the value of the median for each day.
On average, the dogs that visited the vets on Monday had a greater mass.	Make the comparison.

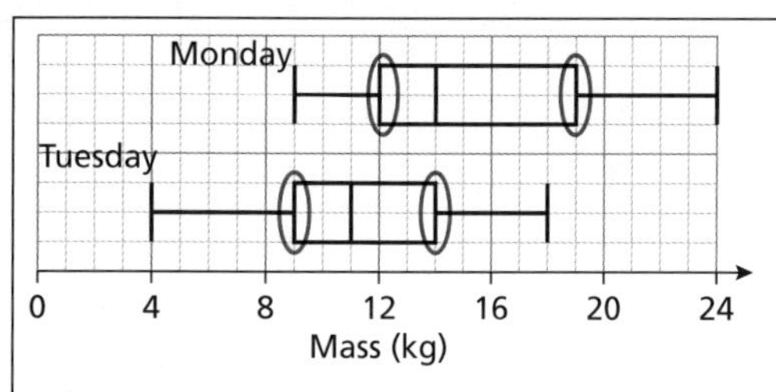

Comparing the ranges or the interquartile ranges gives the spread of the distribution.

The lower the spread, the more consistent the data is.

The interquartile range for Monday is 7 kg (19 – 12).

The interquartile range for Tuesday is 5 kg (14 – 9).

The masses of the dogs that visited the vets on Tuesday were more consistent.

To find the interquartile range, subtract the lower quartile from the upper quartile.

Make the comparison.

The interquartile range is a better measure of consistency than the range because it is less likely to be affected by outliers.

Practice

1 Draw a box plot for each set of data.

a

Lowest value	15
Highest value	38
Median	27
Lower quartile	21
Upper quartile	30

b

Lowest value	8
Highest value	19
Median	12
Lower quartile	11
Upper quartile	17

c

Lowest value	19
Highest value	84
Median	64
Lower quartile	48
Upper quartile	72

d

Lowest value	2.8
Highest value	5.4
Median	4.1
Lower quartile	3.7
Upper quartile	4.6

2 For each box plot, write down or work out:

i the median **ii** the range **iii** the interquartile range.

a

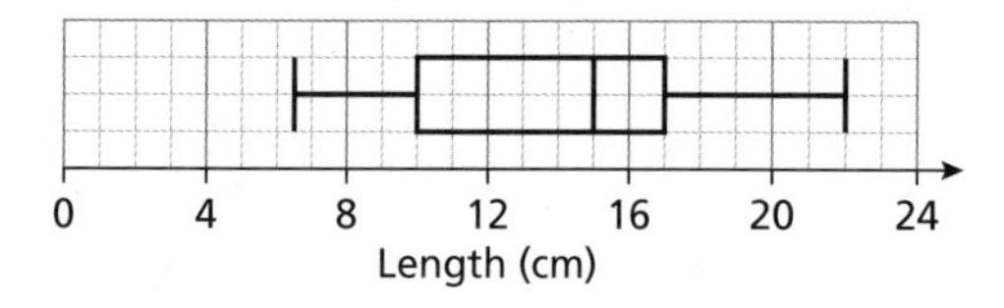

b

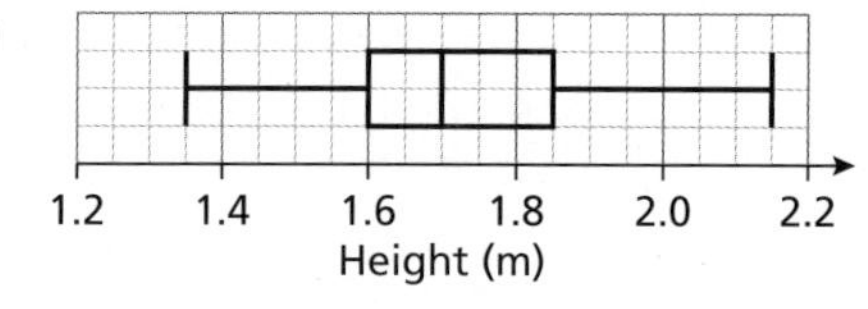

c

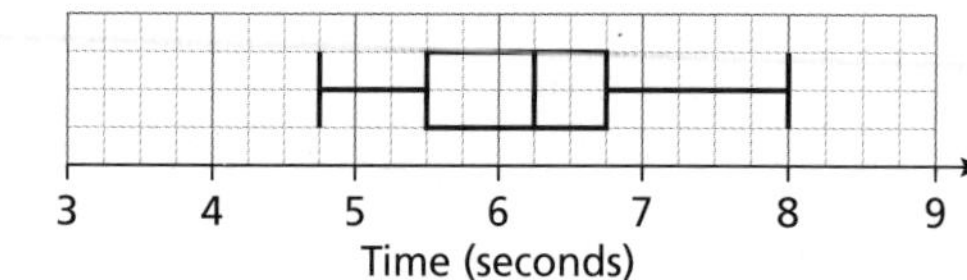

d

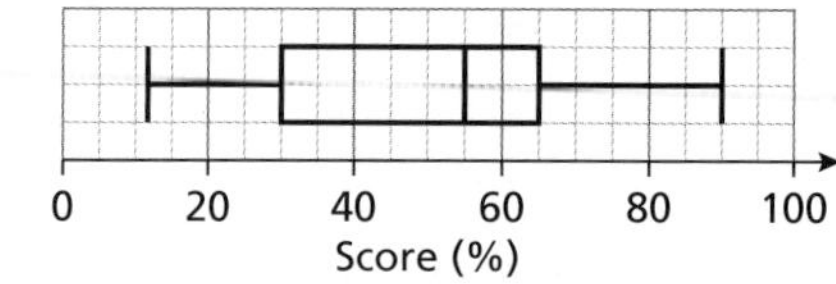

3 Draw a box plot for each set of data.

a 32, 34, 38, 41, 42, 48, 54, 55, 55, 57, 58

b 2.4, 3.8, 3.4, 2.8, 1.8, 3.8, 4.1, 3.4, 3.6, 4.2, 2.8

c 145, 134, 174, 148, 152, 168, 157

d 15.7, 16, 18, 14.5, 15.9, 16.2, 14.7, 16.9, 17.1, 15.1, 16.8

4 Compare the distributions of each pair of box plots.

a

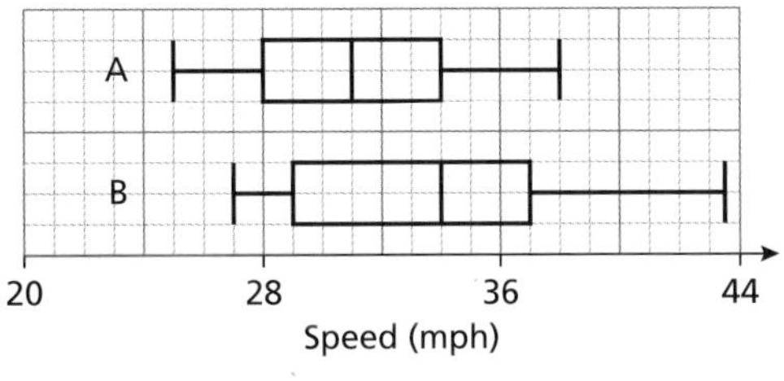

b

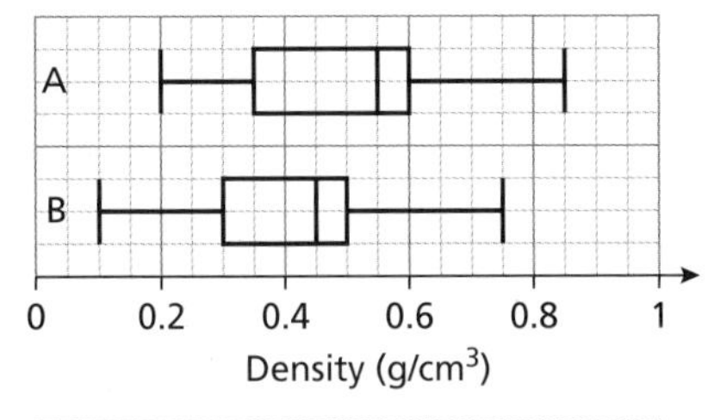

c

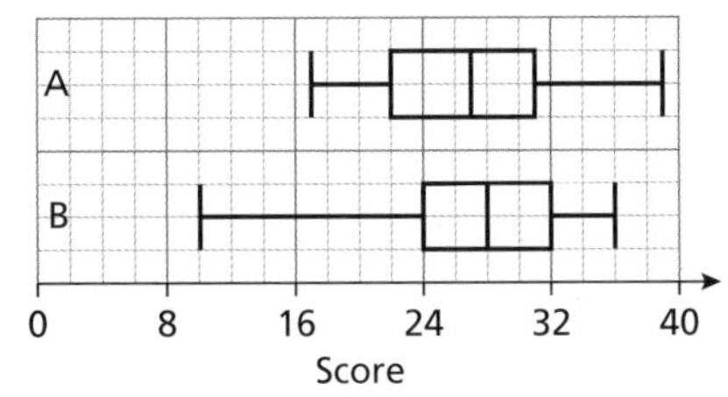

d

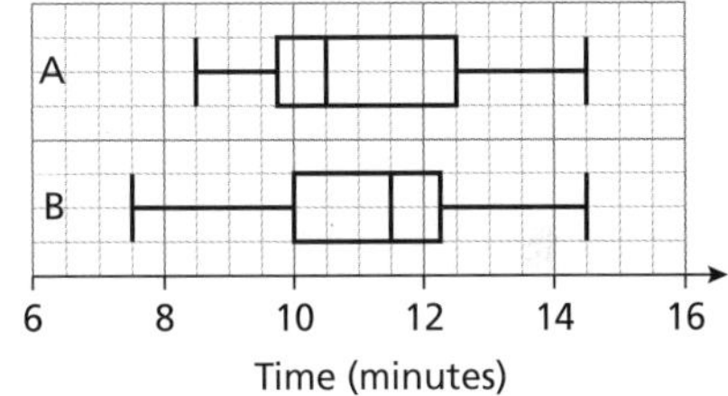

5 Use each cumulative frequency diagram to draw a box plot.

a Lowest value = 5 Greatest value = 49

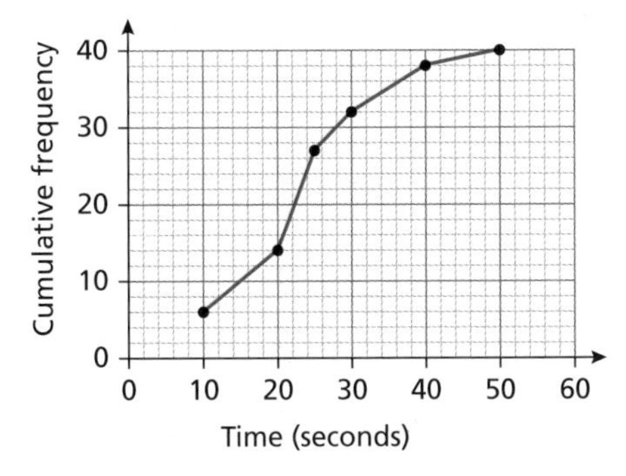

b Lowest value = 7 Greatest value = 30

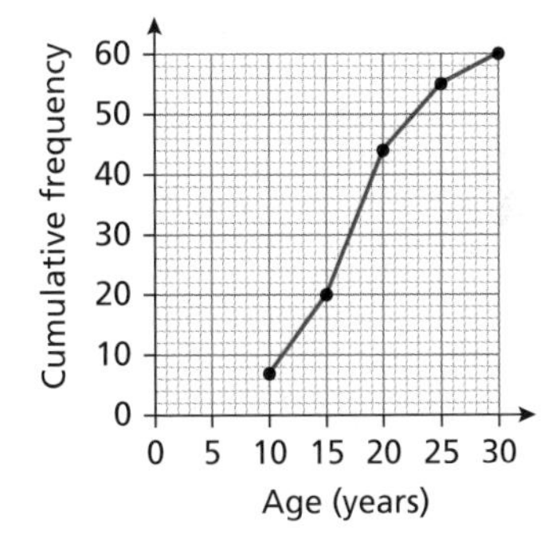

c Lowest value = 12 Greatest value = 58

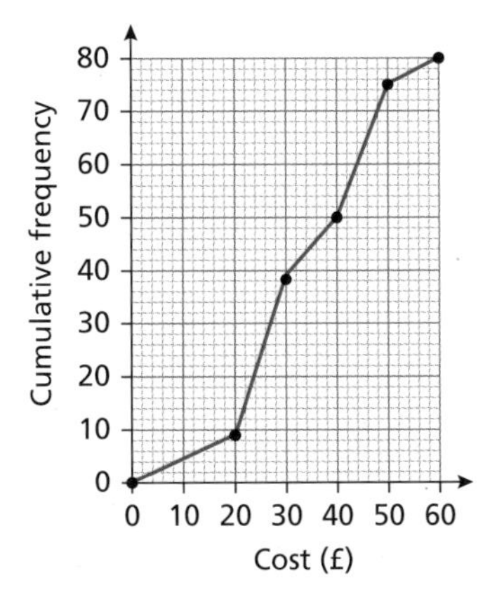

d Lowest value = 2 Greatest value = 57

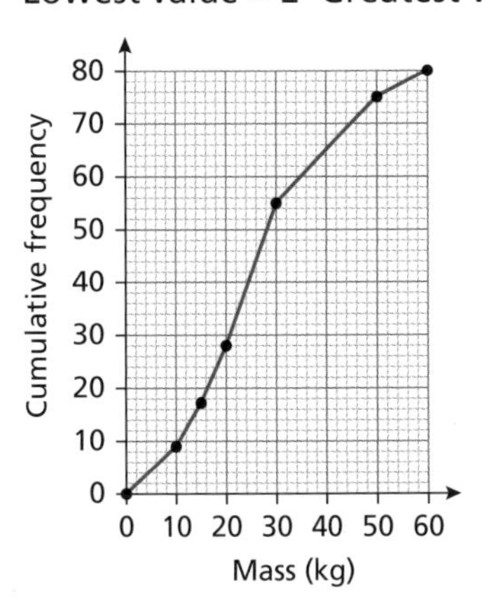

6 **a** Draw a cumulative frequency diagram for the information in the table.

Age, a (years)	Frequency
$0 < a \leqslant 10$	6
$10 < a \leqslant 15$	14
$15 < a \leqslant 20$	28
$20 < a \leqslant 25$	9
$25 < a \leqslant 30$	3

b The lowest age was 8 years old.

The range of ages was 18 years.

Use this and the information from the cumulative frequency diagram to draw a box plot.

What do you think?

1 Huda thinks that the distance between the lower quartile and the median is always equal to the distance between the median and the upper quartile, as they both represent 25% of the data set.

Do you agree with Huda? Explain your answer.

Consolidate – do you need more?

1 Draw a box plot for each table.

a

Lowest value	15
Highest value	48
Median	32
Lower quartile	24
Upper quartile	36

b

Lowest value	2.8
Highest value	8.4
Median	6.5
Lower quartile	5.4
Upper quartile	7.2

c

Lowest value	89
Highest value	120
Median	98
Lower quartile	96
Upper quartile	114

d

Lowest value	0.51
Highest value	1.19
Median	0.89
Lower quartile	0.74
Upper quartile	1.1

2 For each box plot, work out:

i the median **ii** the range **iii** the interquartile range.

a

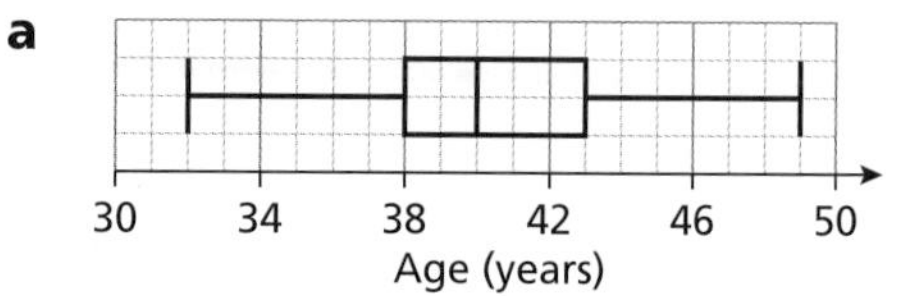

b

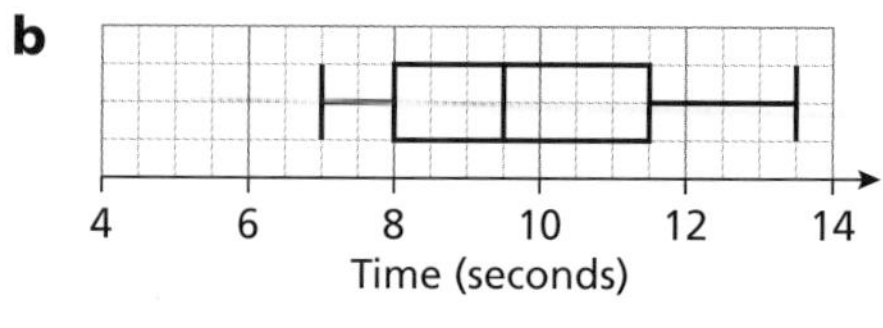

c

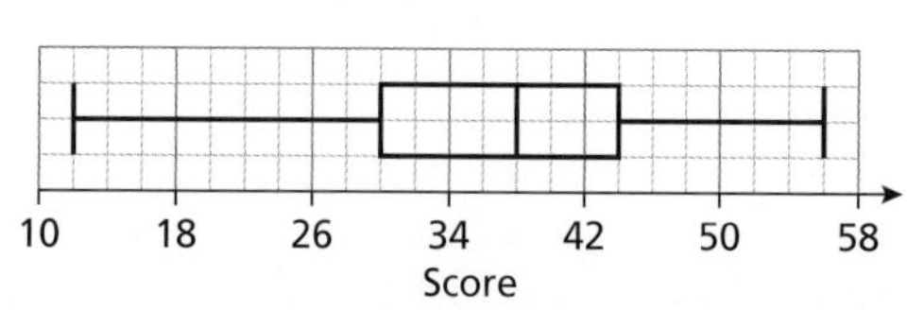

d

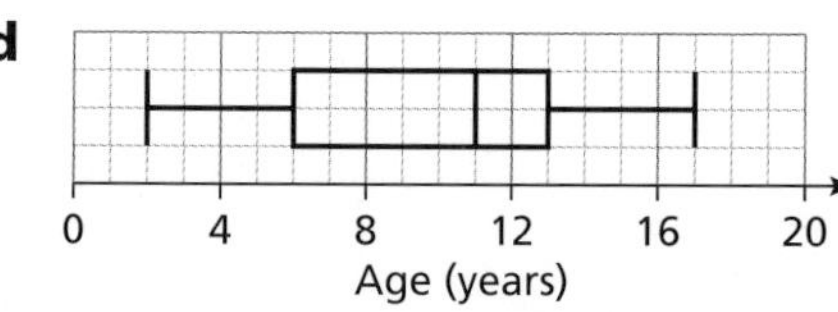

3 Draw a box plot for each set of data.

a 9, 14, 18, 12, 16, 8, 17, 13, 14, 11, 15

b 48, 42, 38, 31, 58, 51, 41

c 8.4, 8.9, 6.7, 6.1, 5, 6.5, 7.8, 6.9, 5.7, 5.2, 8.1, 7, 7.8, 8.2, 7.9

d 0.5, 0.25, 0.4, 0.86, 0.35, 0.74, 0.49, 0.58, 0.84, 0.67, 0.55

4 Compare the distributions of each pair of box plots.

a

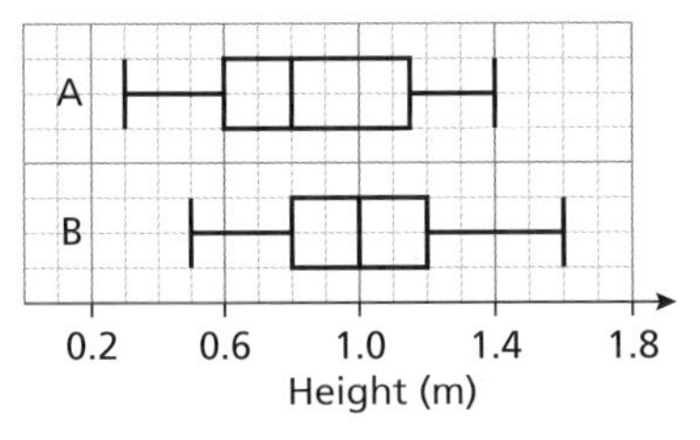

b

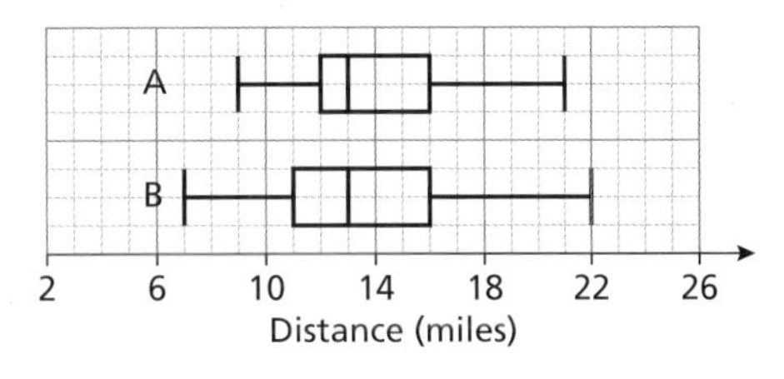

c

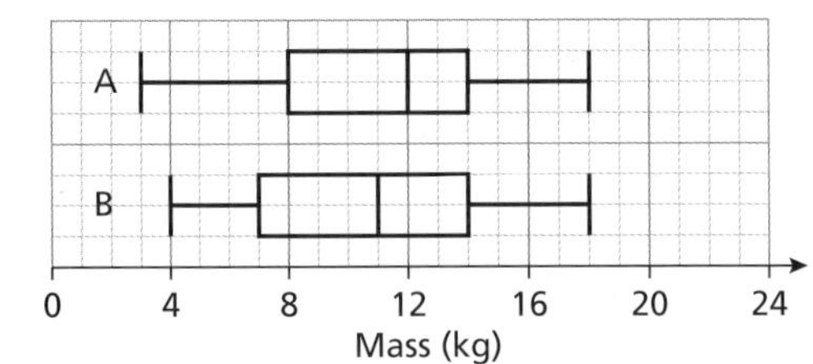

d

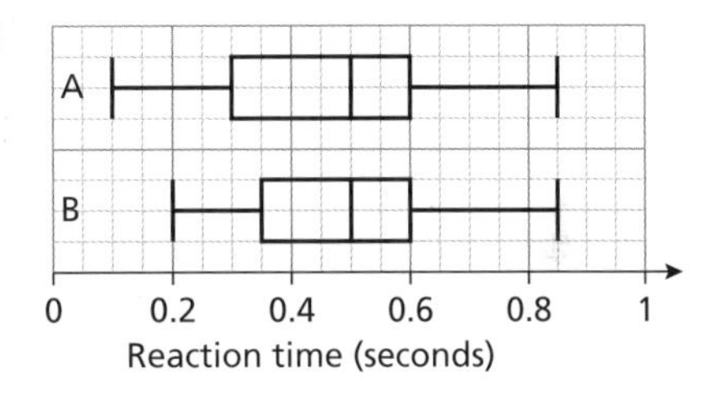

5 For each set of data:

i draw a cumulative frequency diagram **ii** draw a box plot.

a Lowest value = 8 Greatest value = 38

Length, l (cm)	Frequency
$0 < l \leqslant 10$	2
$10 < l \leqslant 15$	11
$15 < l \leqslant 20$	19
$20 < l \leqslant 30$	7
$30 < l \leqslant 40$	1

b Lowest value = 0.2 Greatest value = 3.7

Height, h (m)	Frequency
$0 < h \leqslant 0.8$	6
$0.8 < h \leqslant 1.5$	14
$1.5 < h \leqslant 2$	18
$2 < h \leqslant 2.5$	27
$2.5 < h \leqslant 3$	12
$3 < h \leqslant 4$	3

c Lowest value = 7 Greatest value = 55

Age, a (years)	Frequency
$0 < a \leqslant 18$	8
$18 < a \leqslant 25$	22
$25 < a \leqslant 30$	27
$30 < a \leqslant 40$	19
$40 < a \leqslant 60$	4

d Lowest value = 10 Greatest value = 40

Mass, m (kg)	Frequency
$10 < m \leqslant 14$	8
$14 < m \leqslant 18$	42
$18 < m \leqslant 22$	68
$22 < m \leqslant 30$	64
$30 < m \leqslant 40$	18

Stretch – can you deepen your learning?

1 The box plot shows the distribution of the masses of 400 dogs.

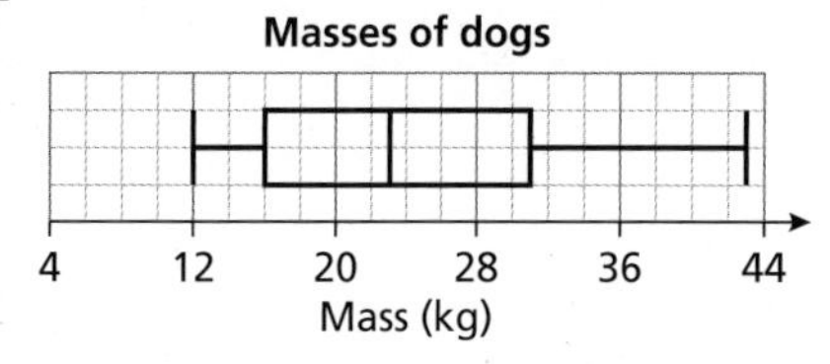

a Estimate how many dogs have a mass of more than 23 kg.

b Estimate how many dogs have a mass between 12 kg and 31 kg.

2 Information about the cost of some phone cases is given below.

The lowest cost is £1.50

The range of costs is £15.50

$\frac{1}{2}$ of the cases cost less than £8

75% of the cases cost less than £14

The interquartile range is £8

Draw a box plot to represent this data.

14.5 Two-way tables

White Rose MATHS

Are you ready?

1 The two-way table shows information about the language subject taken by a group of students in Year 10 and Year 11.

a Copy and complete the two-way table.

b One of the students is chosen at random.

What is the probability that they are a Year 10 student who studies Spanish?

	Year 10	Year 11	Total
French			100
Spanish	65		
Total	135		250

A **two-way table** displays two sets of data in rows and columns.

	Under 16	16 or over	Total
Male	22	25	47
Female	19	34	53
Total	41	59	100

Example

In a school, there are 300 students in total in Year 10 and Year 11.

Students are asked if they travel to school by bus, walking or cycling.

160 students are in Year 11.

40 students cycle.

The same number of students get the bus as walk to school.

65 Year 11 students walk to school.

17 Year 10 students cycle to school.

Represent this information as a two-way table, and hence work out how many Year 10 students walk to school.

Method

Solution

	Bus	Walk	Cycle	Total
Year 10			17	
Year 11		65		160
Total			40	300

Commentary

Once the two-way table is set up, write in as much as possible of the information given in the question.

Number who use bus/walk = $\frac{300 - 40}{2} = 130$

	Bus	Walk	Cycle	Total
Year 10			17	
Year 11		65		160
Total	130	130	40	300

The question says, 'the same number of students get the bus as walk to school'. So both of these can be worked out using the calculation shown.

	Bus	Walk	Cycle	Total
Year 10	58	65	17	140
Year 11	72	65	23	160
Total	130	130	40	300

65 Year 10 students walk to school.

The remainder of the values can be found by subtraction. Start with columns or rows with only one unknown value.

Practice

1 50 people take part in a show. The show is made up of singers, dancers and actors.

27 people in the show are male.

20 of the males are dancers.

There are 13 female singers.

Of the 12 actors, 7 are male.

a Draw a two-way table to show this information.

b How many dancers are in the show?

2 150 students in Year 11 either study French, German or Spanish.

64 of the students are boys and the rest are girls.

18 boys study German.

18 boys and 31 girls study French.

A total of 38 students study Spanish.

Work out the number of girls who study German.

3 A football team played 38 games in a season. They played the same number of home and away games.

They won half of their games, including 7 at home.

They drew 6 of their away games.

They lost a total of 4 games during the season.

Work out the number of games that they lost at home.

4. 100 people were asked if they preferred tea or coffee.

   47 of the people asked were male, and the rest were female.

   27 of the 52 people who preferred coffee were male.

   Work out the number of females who preferred tea.

5. In a library, books are either fiction or non-fiction, and hardback or paperback.

   There are 70 hardback fiction books.

   There are a total of 500 paperback books.

   The ratio of hardback books to paperback books is 1 : 4

   $\frac{3}{5}$ of the paperback books are non-fiction.

   Work out the number of paperback fiction books in the library.

6. Deliveries from a factory are either classified as on time or late, and accepted or rejected. There were 600 deliveries in one particular week.

   The number of deliveries that were on time and late were in the ratio 4 : 1

   125 deliveries were rejected. The remainder were accepted.

   $\frac{3}{4}$ of the deliveries were on time and accepted.

   Work out the number of late deliveries that were accepted.

What do you think?

1. The hair colour and eye colour of 77 people are identified and recorded in the two-way table.

		Eye colour			
		Blue	Green	Brown	Total
Hair colour	Blonde	15			
	Brunette				
	Red				
	Total	25		31	77

   10 people with red hair are included.

   One more person has blonde hair than brunette hair.

   $\frac{3}{11}$ of the people have green eyes.

   The people with green eyes have blonde hair to brunette hair to red hair in the ratio 4 : 2 : 1

   60% of the people with red hair have blue eyes.

   24% of the people with blue eyes have red hair.

   A person is chosen at random. What is the probability that they have brunette hair and blue eyes?

Consolidate – do you need more?

1 200 students in a school are asked about their favourite sport.

88 students are in Year 10.

24 of the 38 students whose favourite sport is netball are in Year 11.

63 students say their favourite sport is rugby.

5 more students like football than cricket.

41 Year 10 students say rugby is their favourite sport.

32 Year 11 students say cricket is their favourite sport.

How many Year 11 students say their favourite sport is rugby?

2 150 students attend either Maths, English or Science revision sessions over a weekend.

35 of the 82 students who attend the revision sessions on Saturday do Maths.

63 students attend a Maths revision session over the weekend.

One more student attends English on Sunday than on Saturday.

27 students attend Science on Saturday.

Work out the total number of students who attend a Science revision session over the weekend.

3 350 teachers, students and parents go to watch either football, rugby or athletics.

14 of the 45 teachers go to watch athletics.

135 people go to watch football.

$\frac{1}{9}$ of the people who watch football are teachers.

41 of the 104 people who watch athletics are parents.

57 out of 195 students watch rugby.

Work out how many students watch football.

4 233 students in Year 10 and Year 11 study either French, German or Spanish.

A total of 99 Year 10 students study a language.

For Year 10 students, the ratio who study French to German to Spanish is 4 : 2 : 5

The ratio of Year 10 students who study French to Year 11 students who study French is 2 : 3

The number of students who study German is $\frac{4}{9}$ of the number of students who study French.

Work out the number of Year 11 students who study Spanish.

Stretch – can you deepen your learning?

1 The two-way table shows the cost of 2 litres of milk in shops in the UK and in France. The costs are given in pounds and pence for easy comparison.

Cost, c	UK	France
$0p \leqslant c < 50p$	x	$2x$
$50p \leqslant c <$ £1	$5x$	$4x$
£1 $\leqslant c <$ £1.50	$4x$	$5x$
£1.50 $\leqslant c <$ £2	$2x$	$2x$

The price is recorded for 150 shops altogether.

In how many of the surveyed shops in the UK do 2 litres of milk cost less than £1?

2 Some students in Year 10 and Year 11 are going on a school trip of their choice.

The two-way table shows some information about their choices.

	Bowling	Theme park	Cinema	Total
Year 10	y	$y - 4$	$y - 5$	
Year 11	$x + 2$	$2x - 1$		$4x$
Total	73	$3x + 8$		

Work out the total number of students on the trips.

Charts and diagrams: exam practice

White Rose MATHS

4–6

1 A netball team played 18 games in a season.

Nine games were played at home and the rest were played away.

The team won a total of 10 games.

They drew two games away.

One of the five games they lost was at home.

Complete a copy of the two-way table. **[3 marks]**

	Won	Drew	Lost	Total
Home				
Away				
Total				

2 The population of a new village was recorded at the start of each year.

The table shows the population for each of the first eight years.

Year	1	2	3	4	5	6	7	8
Population	189	175	171	160	151	178	184	181

(a) Draw a time series graph to represent the data using a copy of the axes. **[3 marks]**

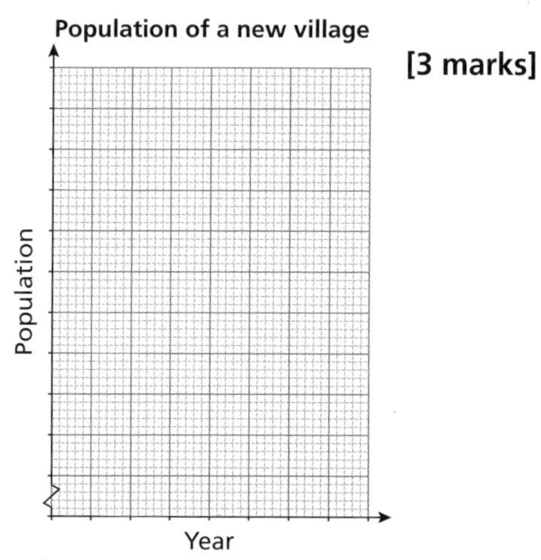

(b) Could the graph be used to give an estimate of the population in year 9? Give a reason for your answer. **[1 mark]**

3 The frequency table shows the time taken by 120 people to travel to a festival by train.

(a) Draw a cumulative frequency graph for this information. **[3 marks]**

(b) Estimate the median time it took for people to travel to the festival by train. **[2 marks]**

(c) Calculate an estimate for the number of people who took longer than $\frac{3}{4}$ of an hour to travel to the festival by train. **[2 marks]**

Time taken, x (minutes)	Frequency
$20 < x \leqslant 30$	10
$30 < x \leqslant 40$	13
$40 < x \leqslant 50$	14
$50 < x \leqslant 60$	40
$60 < x \leqslant 70$	38
$70 < x \leqslant 100$	5

7–9

4 Last Saturday, 80 teachers and 80 teenagers answered a survey about how many minutes they spent using social media.

This box plot shows the results for the **teenagers** who answered the survey.

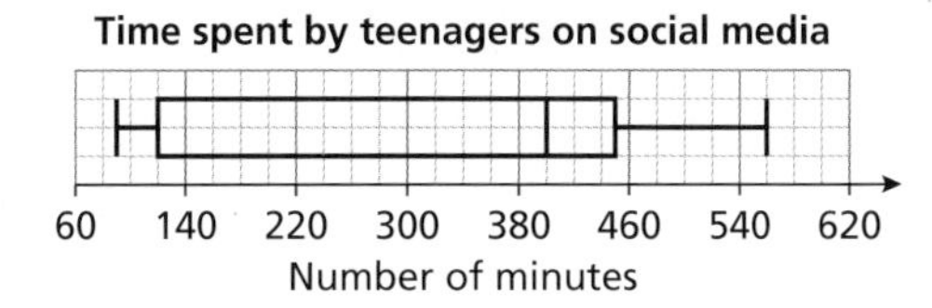

The results from the **teachers** who answered the survey were as follows:

- The median time was 120 minutes.
- The lower quartile was 80 minutes.
- The upper quartile was 300 minutes.
- The shortest time was exactly half an hour.
- The longest time was exactly 6 hours.

(a) Draw a box plot to represent this data from the teachers. **[3 marks]**

(b) Compare the length of time spent by teachers and teenagers on social media. **[3 marks]**

5 The table shows information about the time, in seconds, taken for some people to run a 100-metre race.

Time, x (seconds)	Frequency
$10 < x \leqslant 12$	2
$12 < x \leqslant 13$	3
$13 < x \leqslant 14$	6
$14 < x \leqslant 16$	
$16 < x \leqslant 20$	

(a) Use the information in the table to complete a copy of the histogram. **[3 marks]**

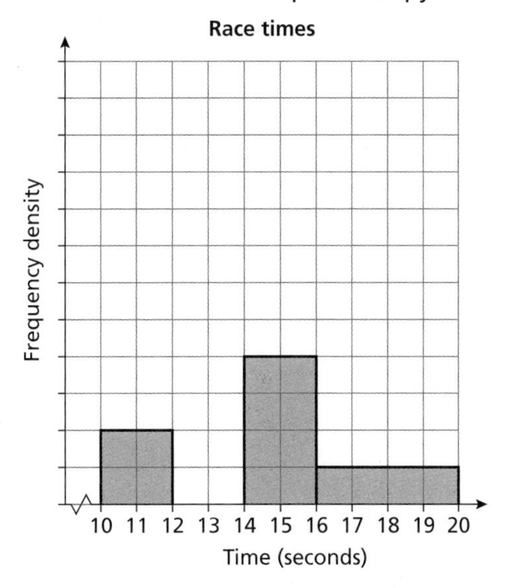

(b) Use the histogram to complete a copy of the table. **[2 marks]**

15 Applying statistics

White Rose MATHS

In this block, we will cover...

15.1 Scatter graphs

Example

Here is a scatter graph comparing temperature and the number of ice creams sold in a park.

a Show this additional information on the grap

Temperature (°C)	22	25
Number of ice creams sold	78	77

b Describe the relationship between the temperature and the number of ice creams sold.

15.2 Making predictions

Practice

1. Sven, Beca and Flo have all drawn a line of
   Which lines of best fit are correctly drawn

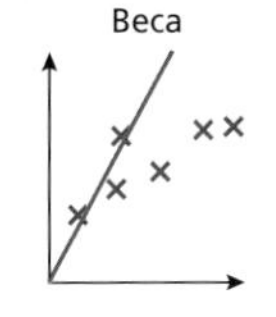

2. The scatter graph shows the number of ice for 10 days.

15.3 Samples and populations

Consolidate – do you need more

1. Jakub is carrying out a survey to find out i the number of after-school clubs offered.
   He selects seven of his friends to take part
   a What is the population Jakub is studyi
   b Give a reason why his sample may be b
   c Suggest a way Jakub's sample could be
2. Decide if the following statements are **tru**
   a A sample uses all the population.

15.1 Scatter graphs

White Rose MATHS

Are you ready?

1 Write the coordinates of each point marked on the grid.

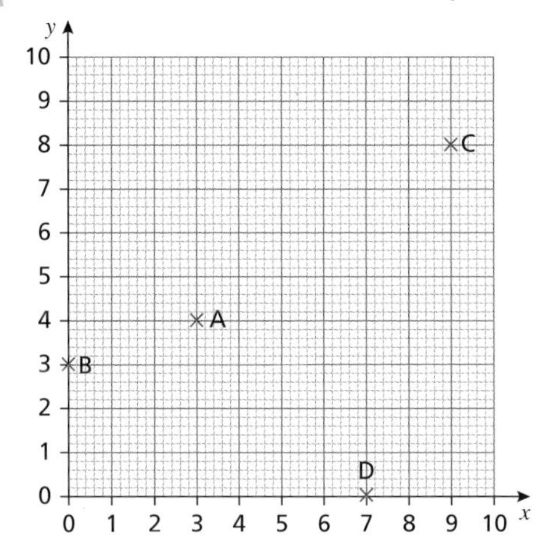

2 Draw a grid with x and y values from 0 to 10. Plot these points on your grid.

A (4, 6) B (0, 7) C (6, 0) D (4.5, 7.5)

Scatter graphs display two sets of data to show if there is a correlation or connection.

IQ

Height

Sales of ice creams

Temperature

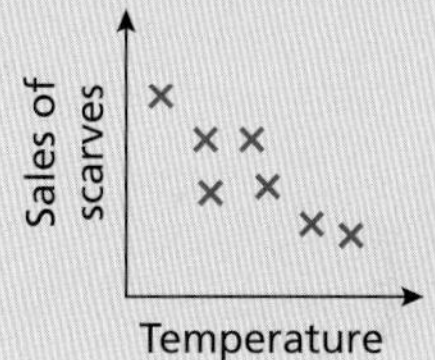

This graph shows there is **no correlation** between the height of a person and their IQ.

This graph shows a **positive correlation**. The warmer it is, the more ice creams are sold.

This graph shows a **negative correlation**. The warmer it is, the fewer scarves are sold.

Scatter graphs don't always have a continuous scale from the **origin**.

When values close to 0 are not required, a **broken scale** may be used.

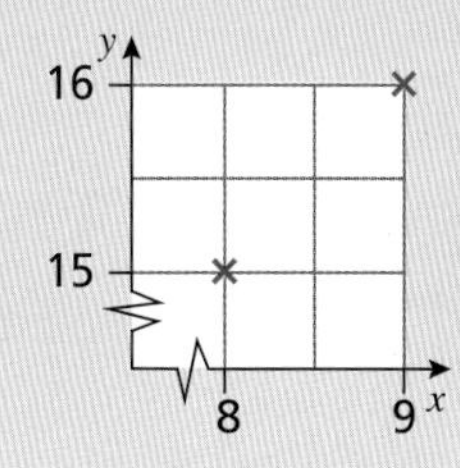

Example

Here is a scatter graph comparing temperature and the number of ice creams sold in a park.

a Show this additional information on the graph.

Temperature (°C)	22	25
Number of ice creams sold	78	77

b Describe the relationship between the temperature and the number of ice creams sold.

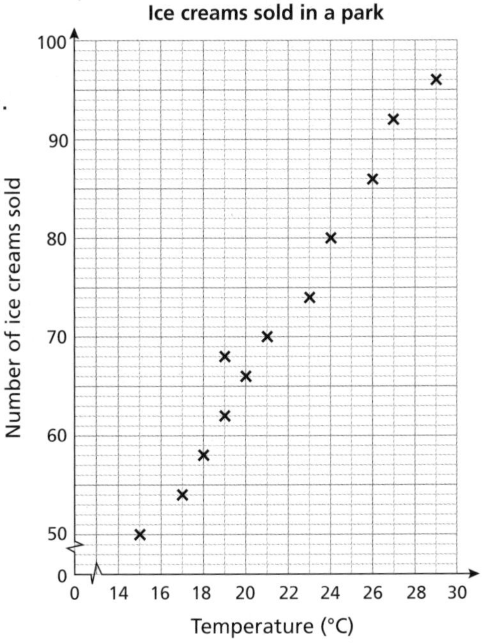

Method

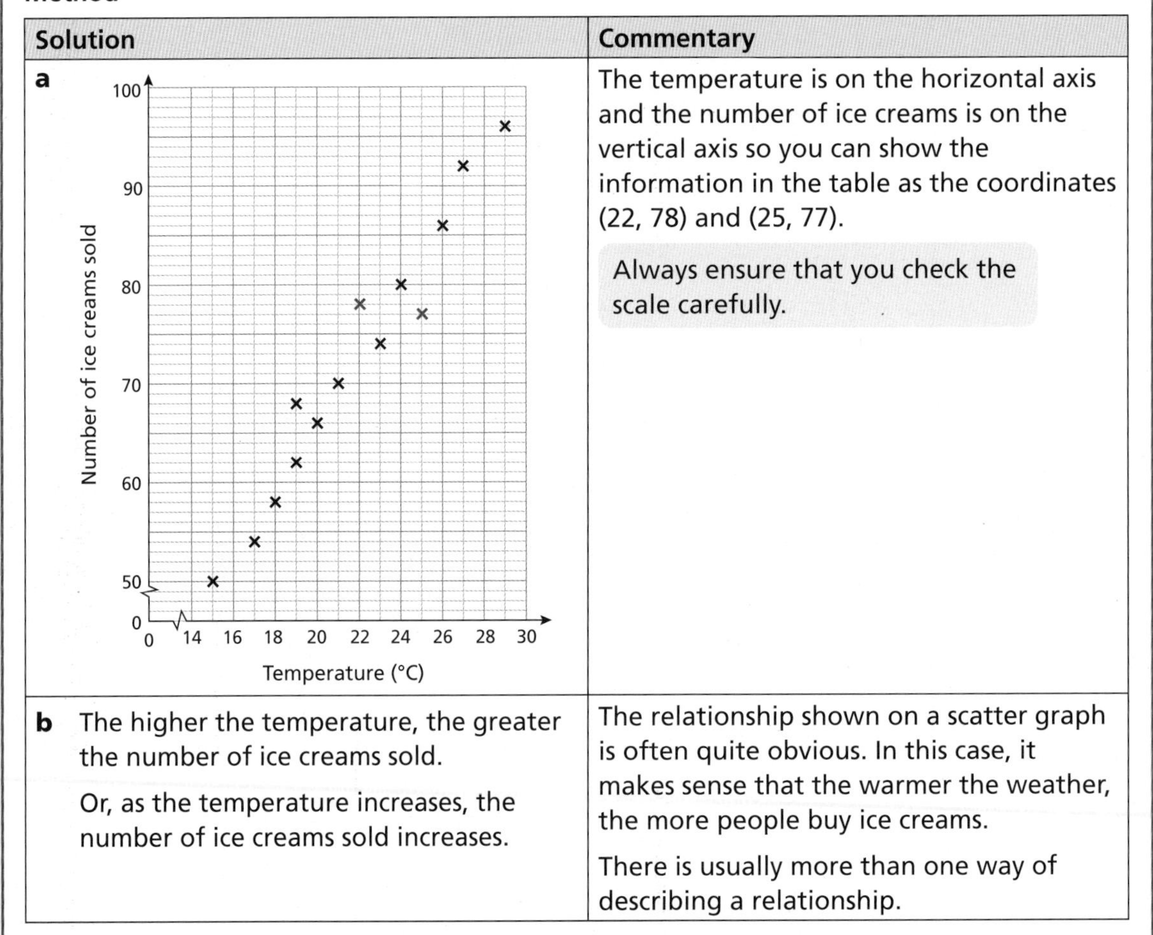

Solution	Commentary
a (graph)	The temperature is on the horizontal axis and the number of ice creams is on the vertical axis so you can show the information in the table as the coordinates (22, 78) and (25, 77). Always ensure that you check the scale carefully.
b The higher the temperature, the greater the number of ice creams sold. Or, as the temperature increases, the number of ice creams sold increases.	The relationship shown on a scatter graph is often quite obvious. In this case, it makes sense that the warmer the weather, the more people buy ice creams. There is usually more than one way of describing a relationship.

Practice

1 Match each graph to the correct type of correlation.

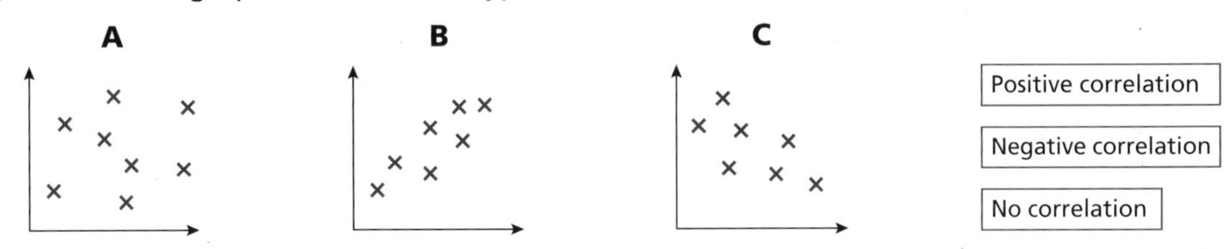

2 The table shows information about test scores and the number of hours of study for a group of students.

Number of hours of study	16	15	10	11	18	12	10	13	14	15
Test score	18	19	15	14	20	12	11	15	16	17

a On a copy of the grid, plot the information from the table, with number of hours of study on the horizontal axis and test score on the vertical axis.

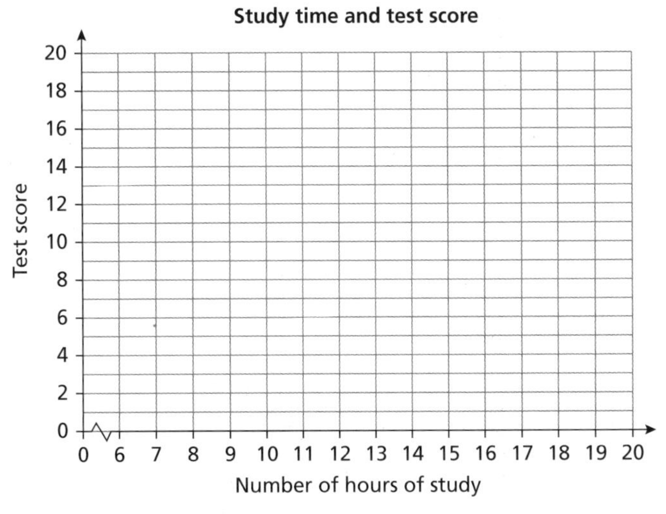

b Describe the relationship between the test score and the number of hours of study.

3 The scatter graph shows information about the temperature and the cost of gas.

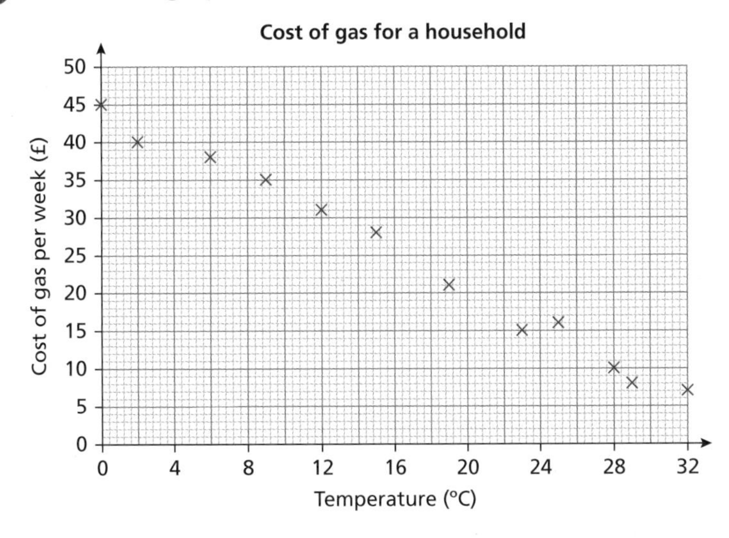

a Estimate the cost of gas per week when the temperature is 0°C.

b Estimate the cost of gas per week when the temperature is 25°C.

c Estimate the temperature when the cost of gas is £28 per week.

d Describe the relationship between the temperature and the cost of gas per week.

4 The scatter graph shows the heights and weights of some students.

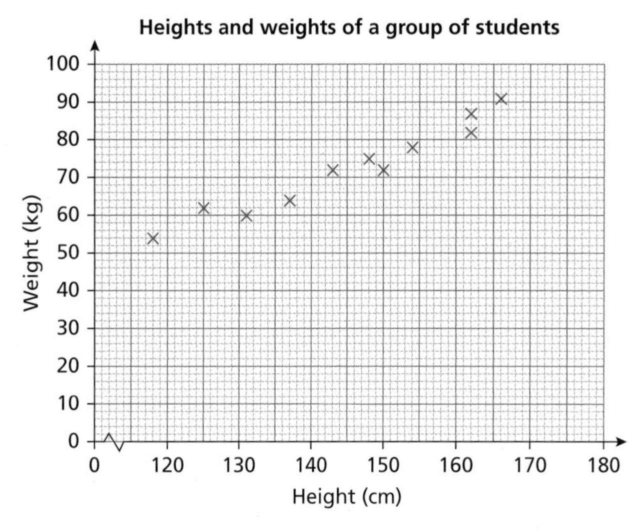

a Estimate the weight of a student who has a height of 125 cm.

b Estimate the weight of a student who has a height of 148 cm.

c Two students have a weight of 72 kg. Estimate their heights.

The table shows the heights and weights of four more students.

Height (cm)	126	157	169	170
Weight (kg)	58	80	88	95

d Copy the scatter graph and plot the information from the table.

e What type of correlation does this scatter graph show?

5 **a** Draw a scatter graph for each set of data.

i

Number of siblings	2	1	3	4	1	2	3	5	2	0
Favourite number	10	4	8	3	2	5	1	6	8	3

ii

Customer wait time (minutes)	3	10	15	2	5	9	7	12
Customer satisfaction (%)	90	60	20	95	75	65	70	55

iii

Age (years)	25	70	35	60	45	40	55	50	30	65
Blood pressure (mm Hg)	110	145	118	135	122	120	130	125	115	140

b Describe the relationship shown by each scatter graph in part **a**.

Consolidate – do you need more?

1 The scatter graph shows information about the age and the number of hours of exercise per week taken by a group of people.

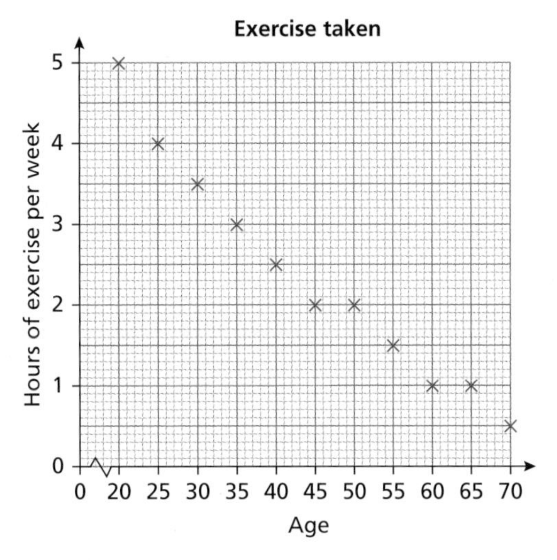

a Estimate how many hours of exercise a 25-year-old takes per week.

b Estimate the age of a person who exercises 1.5 hours per week.

c Describe the relationship between age and the number of hours of exercise each week.

2 The scatter graph shows the distance that students live from school and their test result.

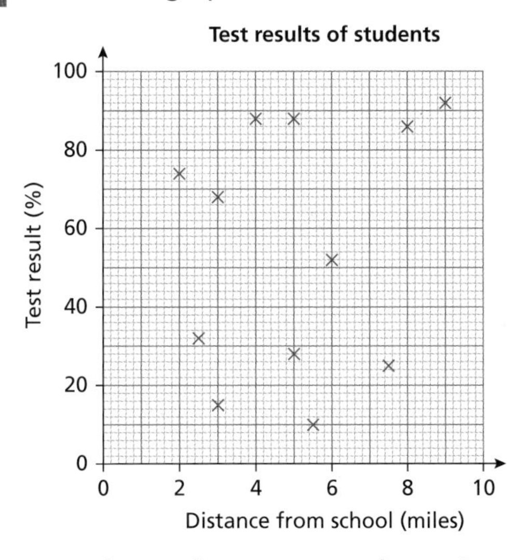

a Estimate the test score of a student who lives 2 miles from school.

b Estimate how far from school a student lives who got a test score of 52%.

c Describe the relationship between the distance a student lives from school and their test result.

3 **a** Draw a scatter graph for each set of data.

i

Commuting time per week (hours)	20	24	10	6	36	16	30	2	40	10	14	5
Job satisfaction (1–10)	5	4	7	8	2	6	3	9	1	8	6	8

ii

Test 1	13	15	19	19	18	14	11	18	20	10
Test 2	12	13	18	20	19	13	12	17	19	10

b Describe the relationship shown by each scatter graph in part **a**.

Stretch – can you deepen your learning?

1 Here are some graphs.

1 2  3

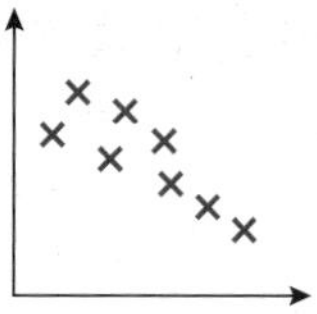

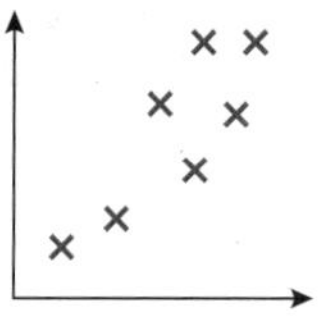

4 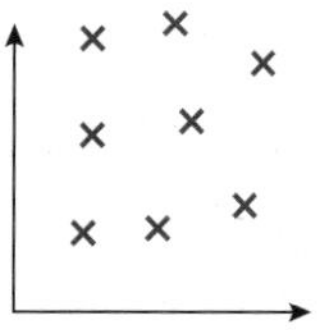 5 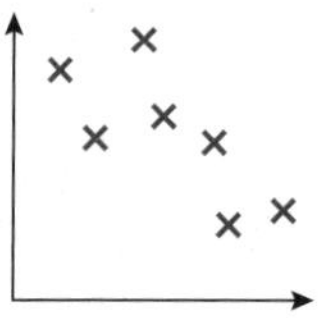 6

a Write the numbers of all the graphs that show:

i a positive correlation **ii** a negative correlation **iii** no correlation.

b Which graph shows the strongest positive correlation?

c Which graph shows the strongest negative correlation?

d Suggest a context for each type of correlation.

2 The table shows some information about 12 cars. For each car it shows the engine size (in litres) and the distance (in miles) that the car travels on 1 litre of petrol.

Engine size (litres)	1	1.4	2	1.8	3	1.5	3	3.5	4	1.6	2.8	3.2
Distance (miles)	15	14.2	11.8	13.6	10.2	13.8	9.8	8.6	7.8	13.6	16.1	9.2

a Plot these points on a scatter graph.

b Which car do you think is an outlier? Explain your reason.

c Another car has an engine size of 2.5 litres.

Estimate the distance that it is likely to travel on 1 litre of petrol, giving a reason for your answer.

15.2 Making predictions

Are you ready?

1 Draw a grid with x and y values from 0 to 10. Plot these points on your grid.

A (6, 4) B (8, 0) C (0, 3.5) D (2.5, 0.5)

2 For each scatter graph, write whether it shows a **positive** correlation, a **negative** correlation or **no** correlation.

a

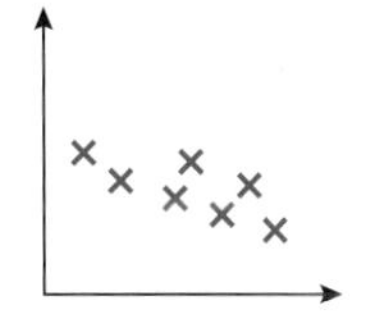

b

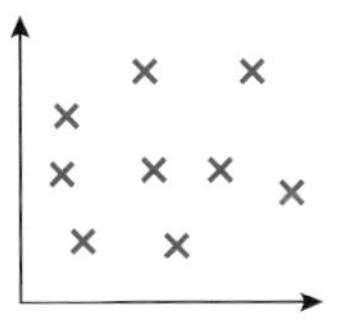

c

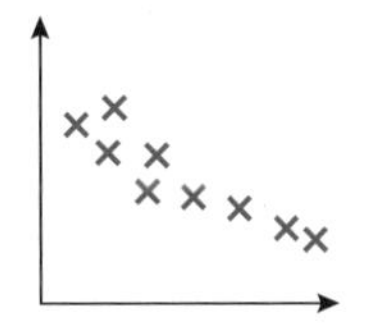

d

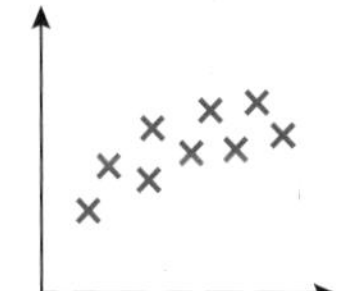

Remember that scatter graphs can be used to represent **bivariate data** (data on each of two variables, where each value of one of the variables is paired with a value of the other variable) to identify whether there is a correlation.

Positive correlation

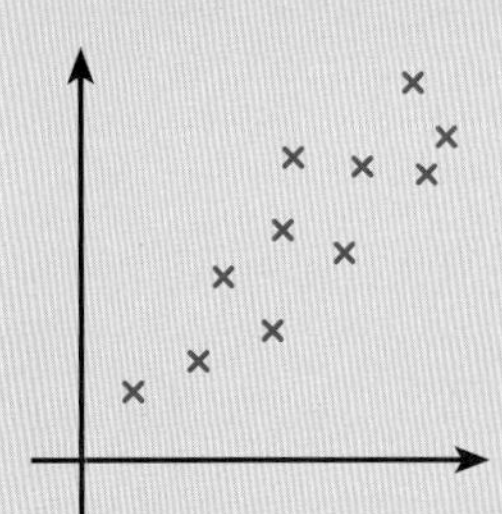

Negative correlation

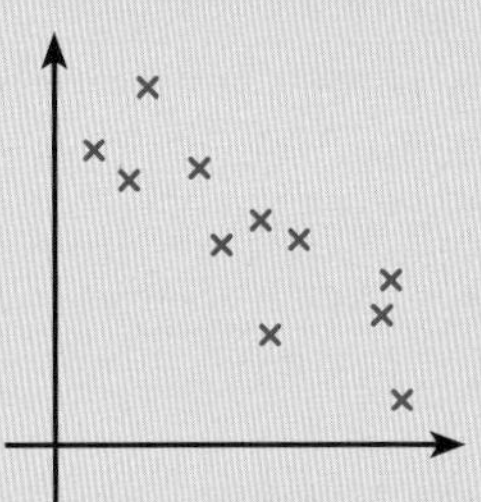

No correlation

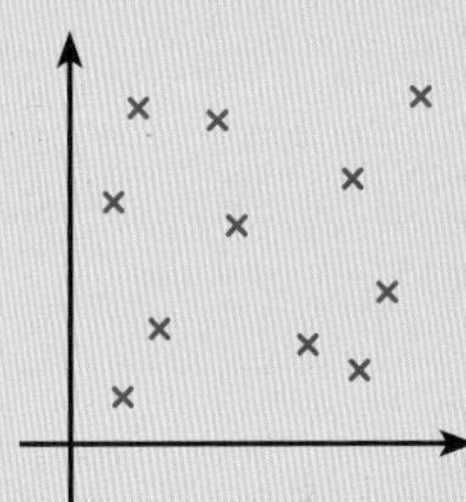

Lines of best fit can be added to scatter graphs. These can then be used to make estimates from the data.

Here is an example of a line of best fit on a scatter graph showing a positive correlation. It does not necessarily need to pass through the origin or any of the points in the data set. Correlation can be described as **strong** if the points lie very close to the line of best fit or **weak** if they lie further away from it.

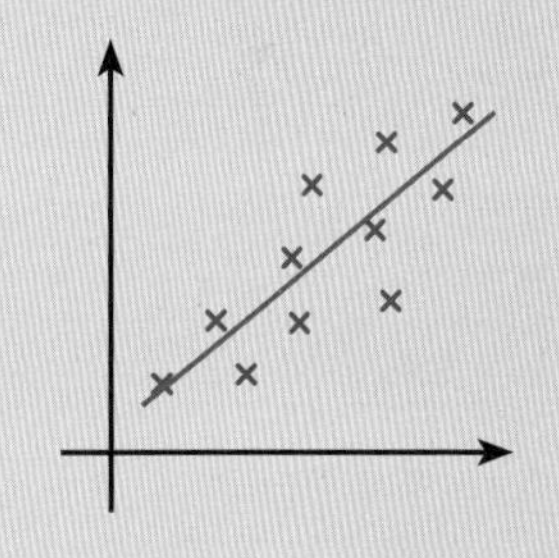

Using a line of best fit to estimate a value is called **interpolation**.

Using a line of best fit outside of the range of data is called **extrapolation**. This is less reliable as the trend may not continue outside the range of the given data.

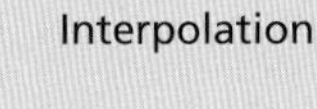

Interpolation

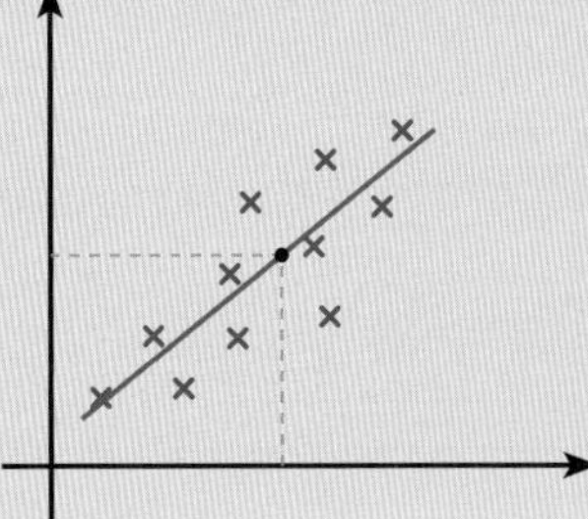

Extrapolation

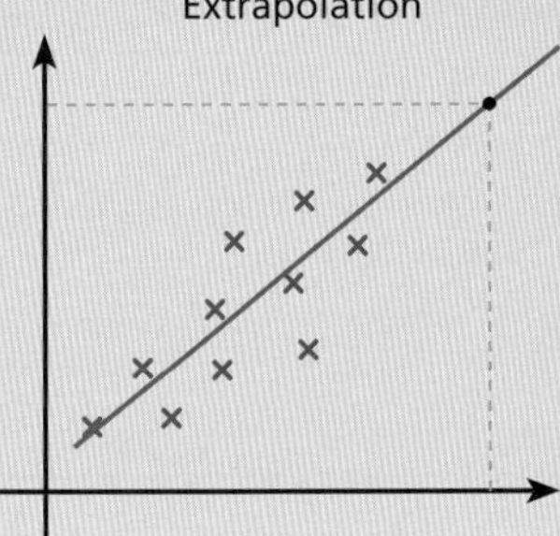

Example

The scatter graph shows infomation about the heights and arm spans of 10 people.

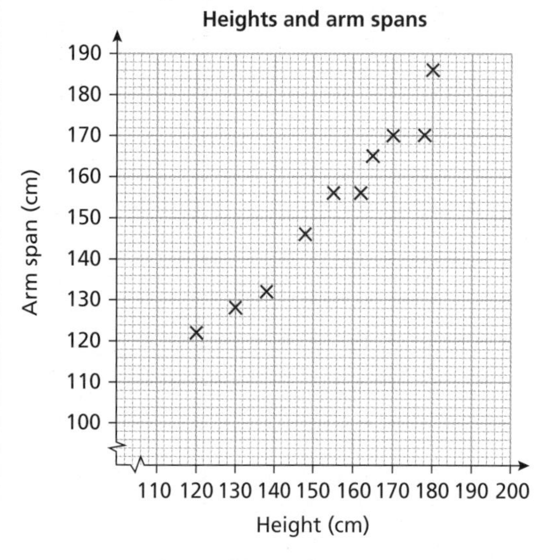

a Draw a line of best fit.

b Describe the correlation.

c Estimate the arm span of a person who is 145 cm tall.

d Seb wants to use the scatter graph to estimate the arm span of someone who is 110 cm tall.

Explain why this might **not** be accurate.

Method

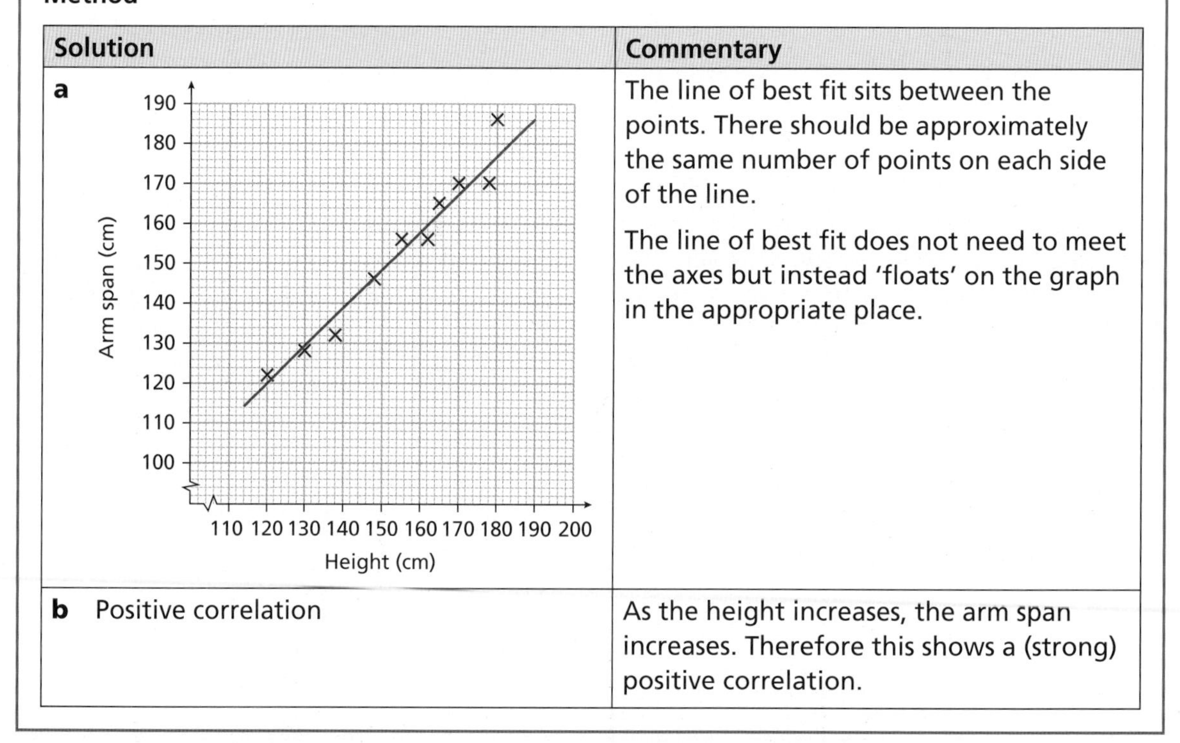

Solution	Commentary
a (graph with line of best fit)	The line of best fit sits between the points. There should be approximately the same number of points on each side of the line. The line of best fit does not need to meet the axes but instead 'floats' on the graph in the appropriate place.
b Positive correlation	As the height increases, the arm span increases. Therefore this shows a (strong) positive correlation.

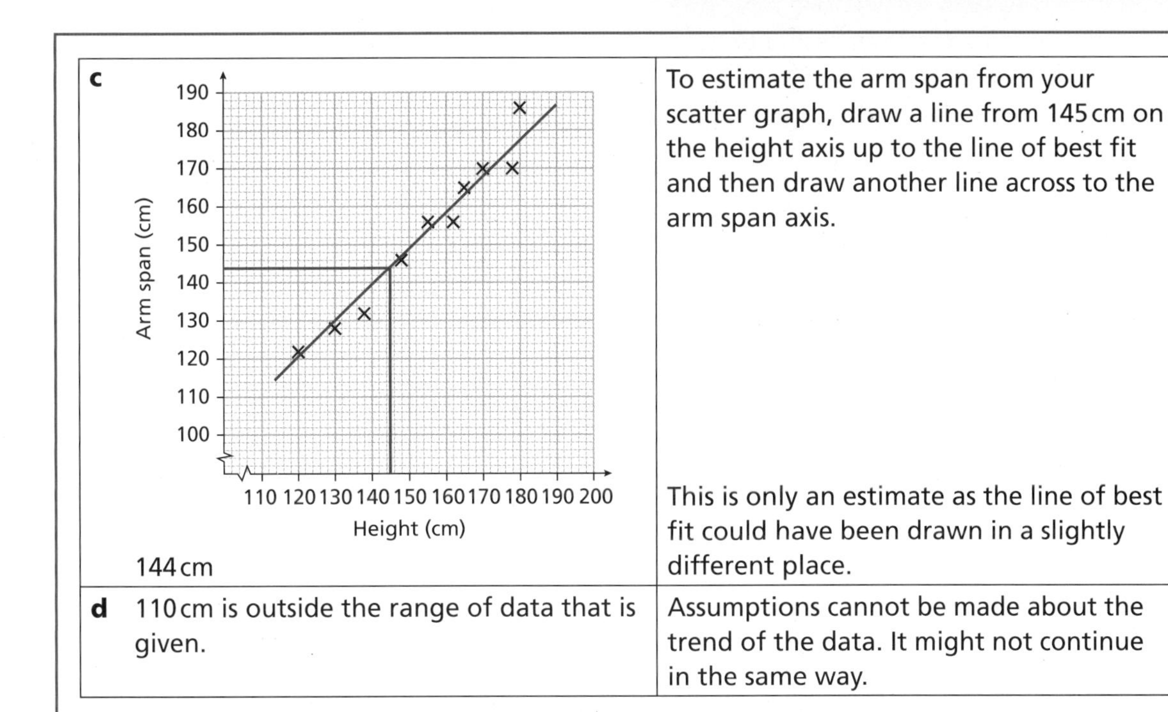

c 144 cm	To estimate the arm span from your scatter graph, draw a line from 145 cm on the height axis up to the line of best fit and then draw another line across to the arm span axis. This is only an estimate as the line of best fit could have been drawn in a slightly different place.
d 110 cm is outside the range of data that is given.	Assumptions cannot be made about the trend of the data. It might not continue in the same way.

Practice

1 Sven, Beca and Flo have all drawn a line of best fit.

Which lines of best fit are correctly drawn and which are not? Explain your answers.

Sven

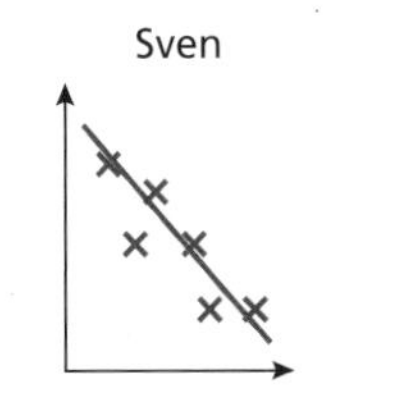

Beca

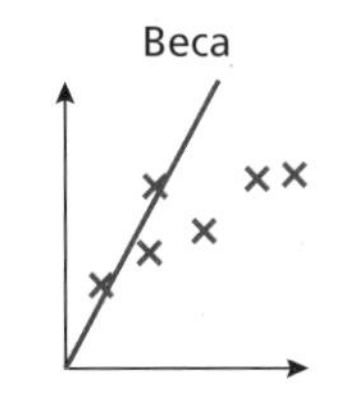

Flo

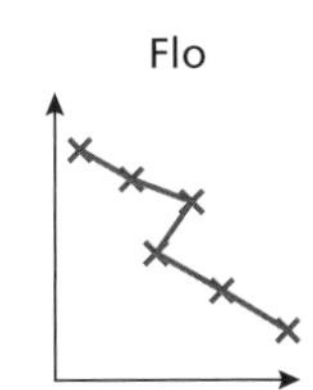

2 The scatter graph shows the number of ice creams sold and the maximum temperature for 10 days.

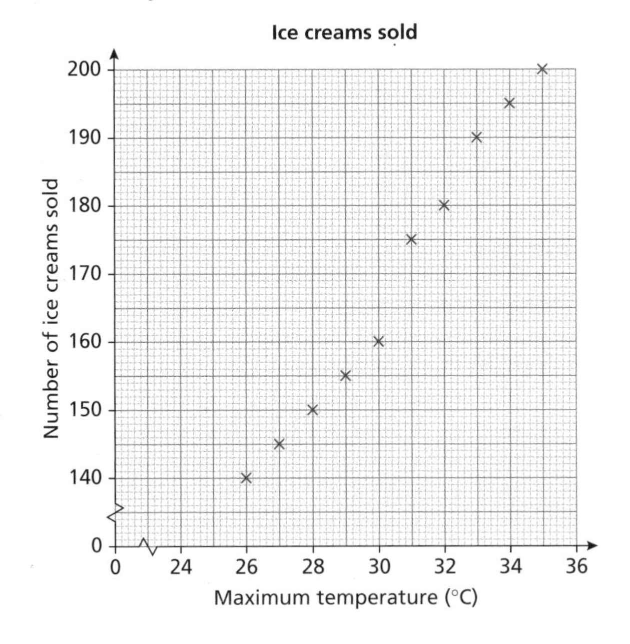

a Describe the relationship between the sales of ice creams and the maximum temperature.

b **i** On a copy of a graph, draw a line of best fit.

ii Use your line of best fit to estimate the maximum temperature on a day that 160 ice creams were sold.

3 The table shows the amount of rainfall, in millimetres, and the number of visitors to an outdoor event.

Rainfall (mm)	5	8	20	25	30	10	15	18	22	28
Number of visitors	200	150	100	60	30	180	120	80	50	20

a Plot this information on a scatter graph.

b Draw a line of best fit and use it to estimate the number of visitors that would attend the event if there was 12 mm of rainfall.

c Explain why it may not be appropriate to use your scatter graph to predict the number of visitors that would attend the event if there was 35 mm of rainfall.

4 The scatter graph shows information about the values of 10 used cars.

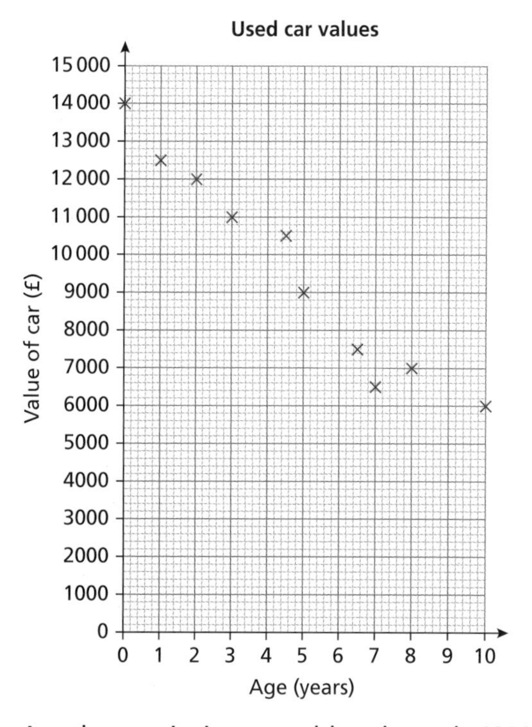

Another car is six years old and worth £8000

a On a copy of the scatter graph, show this information.

b What type of correlation does the graph show between the age of a car and the value of the car?

c A different car is four years old. Estimate the value of this car.

d If the graph was extended to include cars up to 50 years old, do you think the trend shown by the line of best fit would continue? Explain your answer.

Consolidate – do you need more?

1 Filipo records the ages of 10 people and their reaction times, in seconds.

Person	1	2	3	4	5	6	7	8	9	10
Age (years)	18	30	40	50	60	25	35	45	55	65
Reaction time (s)	3	2.5	2.1	1.7	1.3	2.7	2.3	1.9	1.5	1.1

The points for the first eight people have been plotted on the scatter graph below.

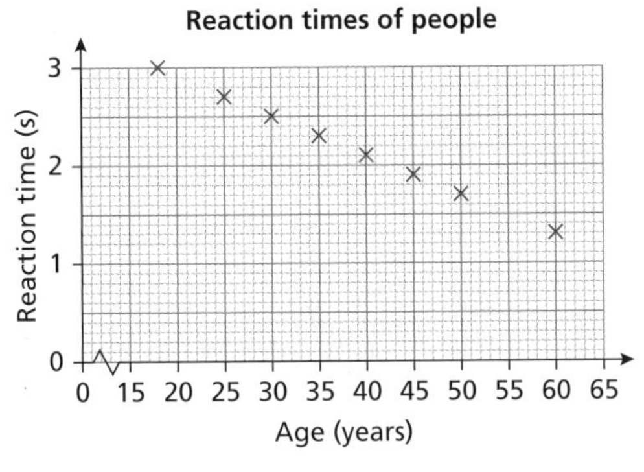

a On a copy of the scatter graph, plot the points for the remaining two people.

b A 20-year-old person claims their reaction time is 2.9 seconds.

Does the scatter graph support his statement? Explain your answer.

c Estimate the age of a person who has a reaction time of 2 seconds.

2 The table shows the number of hours worked by nine electricians and the cost of their work.

Number of hours worked	5	10	15	20	8	12	18	22	25
Cost (£)	150	300	400	500	250	350	450	550	600

a Plot this information on a scatter graph.

b Draw a line of best fit and use it to estimate the cost of a 6-hour job.

c Explain why it may not be appropriate to use your line of best fit to estimate the cost of a job lasting 30 hours.

Stretch – can you deepen your learning?

1 The scatter graph shows the relationship between the number of people wearing sunglasses and the number of cups of coffee sold per day in a town, following a survey conducted over 10 days.

Ali says, "The more sunglasses worn per day, the less coffee is sold. This means that people who wear sunglasses buy less coffee."

Explain why Ali's statement may be incorrect.

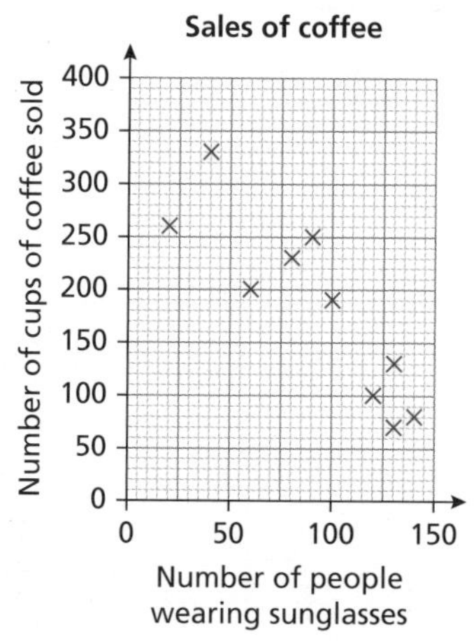

2 The scatter graph shows information about the number of people at a cinema and the maximum temperature on 10 different days.

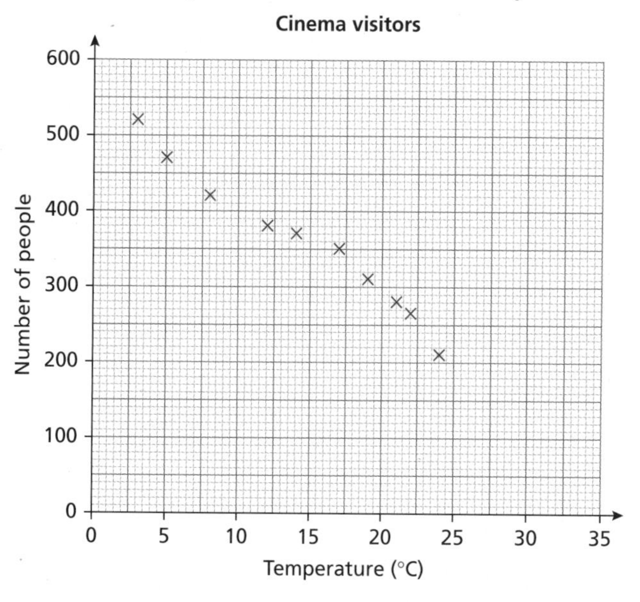

a Estimate the number of people at the cinema when the maximum temperature is 15°C.

b Marta draws a line of best fit on the graph and says, "On a day with a temperature of 35°C, there will be approximately 95 people."

Explain why this estimate may be unreliable.

c On a day when the maximum temperature was 22°C, $\frac{3}{5}$ of the people who went to the cinema were adults.

How many adults were at the cinema that day?

d The ratio of the number of adults to the number of children was 9 : 5 on the day it was 21°C.

How many adults were at the cinema that day?

3 For the lines of best fit you have drawn in this chapter, work out their equations in the form $y = mx + c$

Compare your answers with a partner's.

See Block 9 in *Collins White Rose Maths AQA GCSE 9–1 Higher Student Book 1* for help if needed.

15.3 Samples and populations

White Rose MATHS

Are you ready? (A)

1 Work out 30% of 1500

2 Kath scores 24 out of 50 on a test.

Write her score as a percentage.

3 Work out: **a** $\frac{1}{15} \times 75$ **b** $\frac{7}{29} \times 145$

When collecting statistical data, the **population** is the whole group being studied.

As it might be impractical or expensive to study every member of a population, it is more common to select a **sample**. A sample is a smaller group from the population.

A sample is **random** if every member of the population has an equal chance of being selected.

A **biased** sample is one that does not represent the population fairly; for example, leaving out or having too many people from certain groups, such as gender or age. A sample that is too small may also be biased.

The size of each group in a **stratified sample** is proportional to the size of the same group in the population.

Example 1

Rob is investigating whether people in his town would support the construction of a new library. He selects 20 of his friends to give a questionnaire to.

a What is the population that Rob is studying?

b Give a reason why his sample might be biased.

c Suggest a way that Rob's sample could be improved.

Method

Solution	Commentary
a The population is the people who live in the town.	The population is the whole group being studied.
b His sample of 20 of his friends might all be his age, so won't necessarily represent the whole population.	The sample is the people involved in answering the questionnaire. Another answer could be that 20 is a very small sample for the population size. There are various other valid responses to this question.
c He could take a random sample of people from different age groups within the town.	Another suggestion is that a larger sample is needed.

Example 2

A survey was carried out to see how many investors own cryptocurrency.

18 investors said they did own cryptocurrency, and 7 said they did not.

From a population of 4500 investors, estimate how many do not own cryptocurrency.

Method

Solution	Commentary
$18 + 7 = 25$	Find the total sample size.
$\frac{7}{25}$ said they don't own cryptocurrency.	Find the fraction of the sample who said they don't own cryptocurrency.
$\frac{7}{25} \times 4500 = 1260$ investors	Work out this fraction of the whole population.

Example 3

A group of 200 students went on four different trips.

80 of the 200 students went to a theme park.

A stratified sample of 40 students will be taken to survey their thoughts on the trips.

How many of the students in the sample went to the theme park?

Method

Solution	Commentary
$\frac{x}{40} = \frac{80}{200}$	The proportion of students in the sample $\left(\frac{x}{40}\right)$ will be the same as the proportion of students in the population $\left(\frac{80}{200}\right)$.
$x = 16$	You can solve the equation or use equivalent fractions to work out the value of x.

Practice (A)

1. Ed wants to know if students who attend his school are interested in joining a debating club.

   He asks 80 people.

   a What is the population? **b** What is the sample?

2. A sports club has 530 members.

   The owner of the club wants to know if the members are happy with the facilities.

   The owner asks 10 members who go to the sports club on a Monday morning.

   Give two reasons why this may not be a good sample to use.

3. Abdullah is doing a survey to find out how often people go to the theatre and how much they spend. He stands outside a theatre and asks people as they go in.

   Explain why the sample is biased.

4 A supermarket has 200 employees.

The manager takes a sample of 76 employees.

What percentage of the supermarket employees does this sample represent?

5 A sample of 45 employees is taken. This is 15% of the total number of employees.

How many employees are there in total?

6 A sample of 30 people at a leisure centre are asked which facility they use most often.

Facility	Number of people
Gym	12
Swimming pool	8
Spa	3
Sports hall	7

The leisure centre has 4500 members.

Estimate the number of members that use the swimming pool most often.

7 The table shows the number of students in each year at a school.

Year	7	8	9	10	11
Students	240	219	198	237	216

A sample of 60 students is going to be taken.

Estimate the number of Year 7 students that are going to be included in the sample.

Are you ready? (B)

1 Find the missing numbers in the equivalent fractions.

a $\frac{2}{9} = \frac{\square}{45}$ **b** $\frac{8}{15} = \frac{32}{\square}$ **c** $\frac{\square}{6} = \frac{18}{27}$ **d** $\frac{80}{\square} = \frac{16}{350}$

Capture/recapture is a sampling method used to estimate the size of a population.
It involves a sample of objects (often animals) that are captured, marked, released, and then another sample is taken.

The proportion of marked objects in the second sample should be the same as the proportion in the population as a whole.

This can be expressed as a formula: $\frac{M}{P} = \frac{R}{S}$

where M = the number marked the first time

P = the total population

R = the number marked in recapture

S = the number captured the second time.

Example

A scientist wants to track the movement of rats in an area of a town.

The scientist captures 90 rats and puts a microchip in them, then releases them. A week later, the scientist captures 60 rats. 18 of these have a microchip.

Estimate the population of rats in this area of the town.

Method

Solution	Commentary
$\frac{90}{P} = \frac{18}{60}$	Equate the proportions.
$P = \frac{90 \times 60}{18}$	Rearrange to solve the equation for P.
$P = 300$	In this case, P is an integer. Sometimes you may need to round your answer.

Practice (B)

1. Bev is researching the number of fish in a pond.

   She catches 20 fish, marks them and puts them back in the pond.

   Two weeks later, she catches 70 fish, 10 of which are marked.

   Estimate the total number of fish in the pond.

2. Jackson wants to know the number of bees in a beehive.

   He catches 40 bees, marks them and puts them back in the beehive.

   The next day, he catches 240 bees, 15 of which are marked.

   Estimate the total number of bees in the beehive.

Consolidate – do you need more?

1. Jakub is carrying out a survey to find out if the students in his school are happy with the number of after-school clubs offered.

   He selects seven of his friends to take part in the survey.

   a What is the population Jakub is studying?

   b Give a reason why his sample may be biased.

   c Suggest a way of improving Jakub's sample.

2. Decide if the following statements are **true** or **false**. Give reasons for your answers.

   a A sample uses all the population.

   b A larger sample gives better results than a smaller sample.

   c Every member of the population has an equal chance of being selected if a sample is random.

   d A biased sample represents the population fairly.

3 Ten people are given a number from 1 to 10

A fair, 10-sided dice is then rolled to choose a person.

Explain why this method selects one of the people at random.

4 There are 1500 students in a school.

A survey was carried out to find out how many students like pizza.

Six students said they did not like pizza and 39 students said they did like pizza.

Estimate the number of students in the school who like pizza.

5 A chocolate maker asked a sample of 50 people for their preferred type of chocolate:

- 12 said dark chocolate
- 29 said milk chocolate
- 9 said white chocolate.

The chocolate maker makes 3000 bars of chocolate per day.

How many of each chocolate bar should they make?

6 Samira is estimating the number of giraffes in an area.

On Tuesday, she tagged 12 giraffes.

On Friday, she went back to the same area and spotted 18 giraffes, four of which were tagged.

Estimate the number of giraffes in the area.

Stretch – can you deepen your learning?

1 A sample of 60 is taken from a population.

The sample represents 15% of the population.

What is the size of the population?

2 A sample of 20 students from Year 11 are asked which flavour of crisps they prefer.

The results are shown in the table.

There are 240 students in Year 11.

Flavour	Frequency
Ready salted	8
Salt and vinegar	5
Prawn cocktail	3
Cheese and onion	4

Estimate how many students in Year 11 prefer each flavour of crisp.

3 On Monday, Tiff tags 25 fish from a pond.

On Tuesday, she tags another 12 fish from the same pond.

On Friday, she captures 70 fish, 15 of which are tagged.

Estimate the total number of fish in the pond.

Applying statistics: exam practice

4–6

1 Use a different card to label each scatter graph. **[3 marks]**

No correlation | Strong negative correlation | Weak negative correlation

Strong positive correlation | Weak positive correlation

(a)

(b)

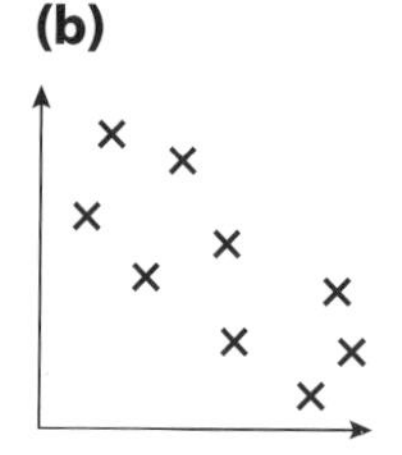

(c)

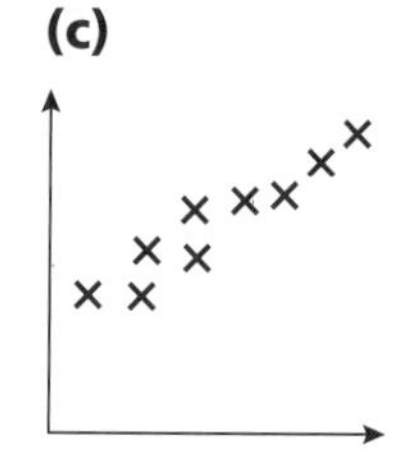

2 The age and price of eight silver spoons in an antique shop are recorded in the table.

Age (years)	14	25	38	28	8	40	80	68
Price (£)	52	36	28	48	10	8	40	20

(a) Draw a scatter graph to display these ages and prices. **[2 marks]**

(b) Is it appropriate to draw a line of best fit? Explain your answer. **[1 mark]**

3 A factory makes 1500 laptops. A random sample of 200 laptops is required so that the quality can be checked.

(a) Describe a method that could be used to select the sample. **[1 mark]**

Eight of the laptops are found to be faulty.

(b) Work out an estimate for how many of the 1500 laptops are faulty. **[2 marks]**

4 A newspaper reports that as smartphone sales have increased since 2015, the sales of chewing gum have decreased.

Amina says, "This means that if people spend more on smartphones, they buy less chewing gum."

Do you agree with Amina? Give a reason for your answer. **[1 mark]**

7–9

5 A survey is used to assess the wellbeing of hospital staff.

The table shows the total number of staff in each job type in the hospital.

Job type	Nurse	Doctor	Other
Number of staff	560	220	340

The survey is going to be given to 100 members of staff.

Calculate the number of staff from each job type that should be asked to complete the survey. **[4 marks]**

Statistics: exam practice

White Rose MATHS

4–6

1 Rhys has these three number cards.

Samira also has three number cards, but you can only see the value of one of them.

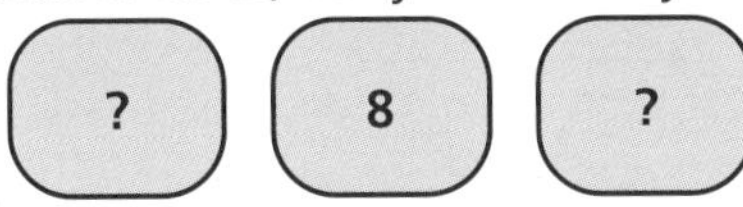

The range of the numbers on Samira's cards is twice the range of the numbers on Rhys's cards.

The mean of the numbers on Samira's cards is the same as the mean of the numbers on Rhys's cards.

Given that the cards can only take integer values, work out the numbers that must be on Samira's two unknown cards. **[2 marks]**

2 The table shows some information about the number of sweets in some packets.

The mean number of sweets in the packets is 47.8

Work out the value of x.

Number of sweets in a packet	Frequency
45	1
46	2
47	4
48	x
49	3
50	2

[3 marks]

3 Company A and Company B both sell second-hand cars.

The graphs show how many cars they each sell per week over a 12-week period.

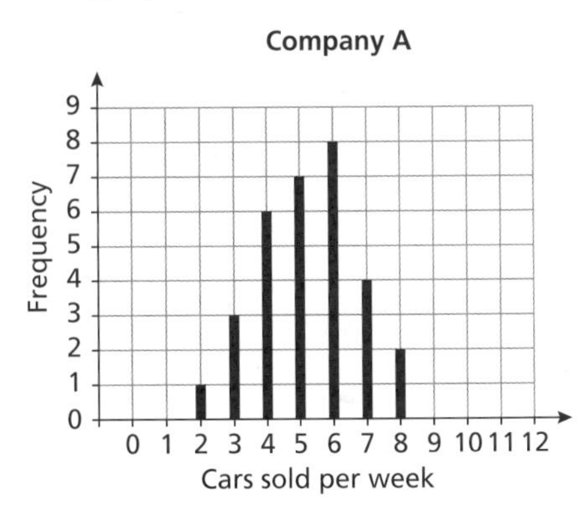

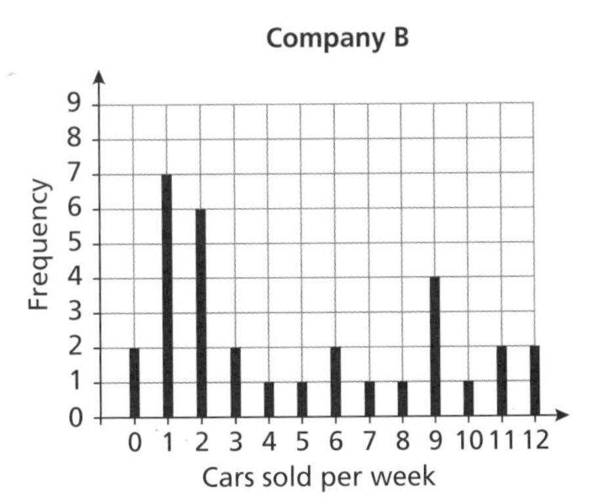

Use statistical measures to compare the sales of the two companies. **[6 marks]**

4–6

4 The cumulative frequency graph shows some information about the heights of a group of children.

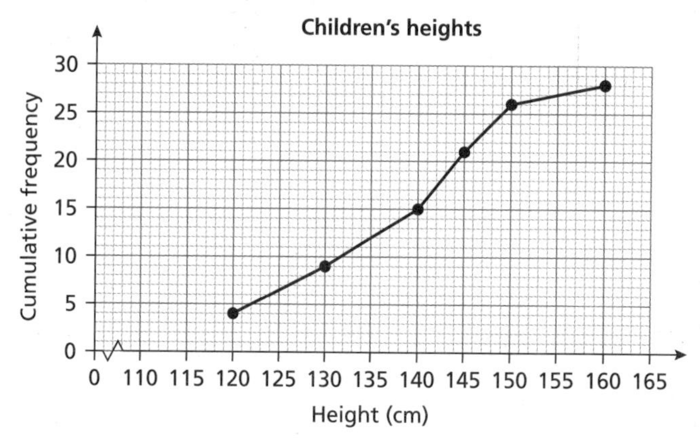

The shortest child had a height of 113 cm and the tallest child had a height of 158 cm.

Use this information and the cumulative frequency graph to draw a box plot to represent the heights of the children. **[4 marks]**

5 In a survey in the village of Amberley, 6 out of 40 adults say they exercise for more than 30 minutes, four times a week.

There are 2500 adults in Amberley.

Estimate how many of these adults exercise for more than 30 minutes, four times a week. **[2 marks]**

7–9

6 The table shows information about the ages of 100 bus drivers.

Age (years)	Frequency
$20 < x \leqslant 30$	20
$30 < x \leqslant 40$	25
$40 < x \leqslant 45$	25
$45 < x \leqslant 50$	15
$50 < x \leqslant 65$	15

Draw a histogram to represent this information. **[3 marks]**

7 Each of 180 Year 11 students at a school study one foreign language.

65 study French, 45 study German and the rest study Spanish.

A representative sample of 36 students is to be taken.

How many of the sample should be students who study Spanish? **[3 marks]**

Glossary

2D shape – a flat shape with two dimensions such as length and width

3D shape – a shape with three dimensions: length, width and height

Acute angle – an angle less than 90°

Adjacent – next to each other

Alternate angles – a pair of angles between a pair of lines on opposite sides of a transversal

Alternate segment – in a circle divided by a chord, the alternate segment lies on the other side of the chord

Angle – a measure of turn between two lines around their common point

Anticlockwise – in the opposite direction to the way the hands of an analogue clock move

Arc – a part of the circumference of a circle

Area – the space inside a 2D shape

Ascending – increasing in size

Associative law for addition – when you add numbers it does not matter how they are grouped

Average – a number representing the typical value of a set of data

Average speed – the total distance travelled divided by the total time taken

Axis – a line on a graph that you can read values from

Bar chart – a chart that uses horizontal or vertical rectangles to show frequencies

Base – a side of a shape that is used as the foundation of the shape

Bearing – the angle measured clockwise from a North line at a fixed point

Biased – all possible outcomes are not equally likely

Bisect – cut in half

Bisector – a line that divides something into two equal parts

Box plot – a graphical representation of the distribution of a set of data, showing the median, quartiles and the extremes of the data set

Capacity – how much space a container holds

Centre – the point at the middle of a shape

Centre of enlargement – the point from which the enlargement is made

Centre of rotation – the point around which a shape can rotate

Chord – a line joining two points on the circumference of a circle

Circumference – the distance around the edge of a circle

Class interval – the range of data in a group

Clockwise – in the same direction as the hands of an analogue clock move

Co-interior angles – a pair of angles between a pair of lines on the same side of a transversal

Collinear – passing through, or lying on, the same straight line

Column vector – a description of a vector showing its horizontal and vertical components

Complement – the elements of the universal set that do not belong to a set

Compound shape – also known as a composite shape, this is a shape made up of two or more other shapes

Common factor – a factor that is shared by two or more numbers

Conditional probability – the probability that an event will occur given that another event has already occurred

Cone – a 3D shape with a plane circular face, a curved surface and one vertex

Congruent – exactly the same size and shape, but possibly in a different orientation

Construct – draw accurately using a ruler and compasses

Continuous data – data that is measured and so does not take exact values

Correlation – the relationship between the values of two variables

Corresponding angles – a pair of angles in matching positions compared with a transversal

Cosine – the cosine of an angle is the ratio of the length of the adjacent side to the length of the hypotenuse in a right-angled triangle

Cross-section – the shape of a slice through a solid

Cube – a 3D shape with six square faces

Cuboid – a 3D shape with six rectangular faces

Cumulative frequency – the running total of frequencies calculated from a frequency table

Cyclic quadrilateral – a quadrilateral with all four vertices on the circumference of a circle

Cylinder – a 3D shape with a constant circular cross-section

Data – a collection of numbers or information

Degree – a unit of measurement of temperature; a degree is also a unit of measurement of angles

Derive – find or discover something from existing knowledge

Describe – say what you see, or what is happening

Diagonal – a line segment that joins two opposite vertices of a polygon

Diagram – a simplified drawing showing the appearance, structure, or workings of something

Diameter – a straight line across a circle, from circumference to circumference and passing through the centre

Dimensions – measurements such as the length, width and height of an object

Discrete data – data that can only take certain values

Edge – a line segment joining two vertices of a 3D shape; it is where two faces of a 3D shape meet

Element – a member of a set

Enlargement – a transformation of a shape that makes it bigger or smaller

Equally likely – having the same probability of happening

Equidistant – at the same distance from

Equilateral – all sides of equal length

Event – a set of one or more outcomes of an experiment

Expected outcome – an estimate of how many times a possible outcome will occur based on relative frequency (experimental probability) or theoretical probability

Experiment – a test or a trial

Exterior angle – an angle between the side of a shape and a line extended from the adjacent side

Face – the flat surface or side of a 3D shape

Fair – an item or event that isn't biased

Frequency – the number of times something happens

Frequency density – the frequency per unit for the data in each class, found by diving the frequency of class interval by the width of class interval

Frequency tree – a diagram showing a number of people/objects grouped into categories

Front elevation – the 2D view of a 3D shape or object as seen from the front

Give a reason – state the mathematical rule(s) you have used, not just the calculations you have done

Gradient – the measure of the steepness of a line or a curve

Grouped data – data that is organised into groups

Hemisphere – half of a sphere

Histogram – a chart that is used to show continuous data where frequency is proportional to the area of each bar

Hypothesis – an idea to investigate that might be true or false

Image – the result of a transformation of an object

Independent events – two events are independent if the outcome of one event isn't affected by the outcome of the other event

Interior angle – an angle on the inside of a shape

Interquartile range – the difference between the lower quartile and the upper quartile of a set of data

Intersection of sets A and B – the set containing all the elements of A that also belong to B

Invariant points – points on a line or shape which do not move when a transformation is applied

Irregular – not regular; a shape where sides and/or angles are not all equal

Isosceles – having two sides the same length

Key – used to identify the categories present in a graph or chart

Kite – a quadrilateral with two pairs of adjacent sides that are equal in length

Line graph – this has connected points and shows how a value changes over time

Line of symmetry – a line that cuts a shape exactly in half

Line segment – a part of a line that connects two points

Locus (plural: **loci**) – a set of points that describe a property

Lower quartile – the value below which one quarter of a set of data lies

Mean – the result of sharing the total of a set of data equally between them

Measure of spread – shows how similar or different a set of values are

Median – the middle number in an ordered list

Misleading graph – a graph that suggests an incorrect conclusion or assumption

Modal class – the class in a grouped frequency distribution that has the greatest frequency

Mode – the item which appears most often in a set of data

Multiple bar chart – a way to represent several, related sets of data

Mutually exclusive events – two or more events that cannot happen at the same time

Net – a 2D shape that can be folded to make a 3D shape

Obtuse angle – an angle more than 90° but less than 180°

Opposite sides/angles – sides or angles that are not next to each other in a quadrilateral

Order of rotational symmetry – the number of positions where the shape looks the same when rotated through 360°

Orientation – the position of an object based on the direction it is facing

Outcome – the possible result of an experiment

Outlier – a value that differs significantly from the others in a data set

Parallel – always the same distance apart and never meeting

Parallelogram – a quadrilateral with two pairs of parallel sides

Perimeter – the total distance around a 2D shape

Perpendicular – at right angles to

Perpendicular bisector – a line that is drawn at right angles to the midpoint of a line segment

Perpendicular height – the height of a shape measured at a right angle to the base

Pi – pronounced 'pie' and written using the symbol π. It is the ratio of the circumference of a circle to its diameter

Pie chart – a graph in which a circle is divided into sectors that each represent a proportion of the whole

Plan view – the 2D view of a 3D shape or object as seen from above

Polygon – a closed 2D shape with straight sides

Population – the whole group that is being investigated

Predict – make a statement as to what will happen based on information you already know

Primary data – data you collect yourself

Prime factor decomposition – writing numbers as a product of their prime factors

Prism – a 3D shape with a constant cross-section

Probability – how likely an event is to occur

Proof – an argument that shows that a statement is true

Properties – features of something that are always true

Pyramid – a 3D shape with triangular faces meeting at a vertex

Qualitative – data that describes characteristics

Quantitative – numerical data

Questionnaire – a list of questions to gather information

Radius – the distance from the centre of a circle to a point on the circle

Random – each outcome is equally likely to occur

Random sample – each item of the population has an equal probability of being chosen

Range – the difference between the greatest value and the smallest value in a set of data

Reflection – a type of geometrical transformation, where an object is flipped to create a mirror image

Reflex angle – an angle more than 180° but less than 360°

Regular – a shape that has all equal sides and all equal angles

Regular polygon – a polygon whose sides are all equal in length and whose angles are all equal in size

Relative frequency – the number of times an event occurs divided by the number of trials

Replacement – putting back – when an item is replaced, the probabilities do not change; when an item is not replaced, the probabilities do change

Represent – draw or show

Right angle – an angle of exactly 90°

Rotation – a geometrical transformation in which every point on a shape is turned through the same angle about a given point

Sample – a selection taken from a population

Sample space – the set of all possible outcomes or results of an experiment

Scale – the ratio of the length in a drawing or a model to the length on the actual object

Scale drawing – a diagram that represents a real object with accurate sizes reduced or enlarged by a ratio

Scale factor – how much a shape has been enlarged by

Scatter diagram – a statistical graph that compares two variables

Secondary data – data already collected by someone else

Sector – a part of a circle formed by two radii and a fraction of the circumference

Segment – a chord splits a circle into two segments

Set – a collection of objects or numbers

Side – a line segment that joins two vertices in a 2D shape

Side elevation – the 2D view of a 3D shape or object as seen from the side

Similar – two shapes are similar if their corresponding sides are in the same ratio, so one shape is an exact enlargement of the other shape

Sine – the sine of an angle is the ratio of the length of the opposite side to the length of the hypotenuse in a right-angled triangle

Sketch – a rough drawing

Sphere – a 3D shape in which all points of its surface are equidistant from its centre.

Stratified sample – a sample that has the same proportion of items in each of a set of groups as the whole population

Surface area – the sum of the areas of all the faces of a 3D shape

Tabular – organised into a table

Tangent (to a circle) – a straight line that touches the circumference of a circle at one point only

Tangent (trigonometry) – the tangent of an angle is the ratio of the length of the opposite side to the length of the adjacent side in a right-angled triangle

Timetable – a table showing times

Translation – a type of geometrical transformation, where an object is moved left or right and/or up or down

Transversal – a line that crosses at least two other lines

Trapezium – a quadrilateral with one pair of parallel sides

Tree diagram – a way of recording possible outcomes that can be used to find probabilities

Trigonometry – the study of lengths and angles in triangles

Two-step – when a calculation involves two processes rather than one

Two-way table – a table that displays two sets of data in rows and columns

Union of sets A and B – the set containing all the elements of A or B or both A and B

Universal set – the set containing all relevant elements

Upper quartile – the value below which three quarters of a set of data lies

Vector – a quantity with both magnitude and direction, e.g. velocity, force, displacement

Venn diagram – a diagram used for sorting data

Vertex (plural: **vertices**) – a point where two line segments meet; a corner of a shape

Vertically opposite angles – angles opposite each other when two lines cross

Volume – the amount of solid space in a 3D shape

Answers

White Rose MATHS

Block 1 Constructions

Chapter 1.1

Are you ready? (A)

1

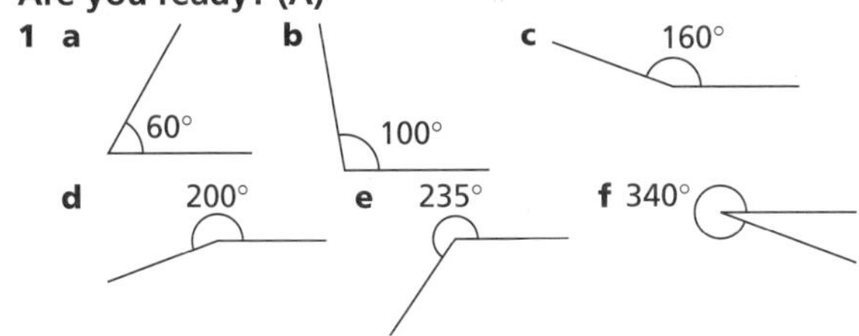

2 **a** 50° **b** 60° **c** 120°
d 80° **e** 150° **f** 250°

3 Clockwise from top: North; North East; East; South East; South; South West; West; North West.

Practice (A)

1 **a** 060° **b** 120° **c** 210° **d** 330°

2 **a** 060° **b** 110° **c** 150° **d** 230°

3 **a** N N 70° A B **b** N N 100° C D

c N N P 200° Q **d** N N M L 300°

4 Any suitable estimations, for example:
a 070° **b** 100° **c** 170° **d** 280°

What do you think?

1 A and B are true

2 250°

Are you ready? (B)

1 Check accuracy with a partner.

2

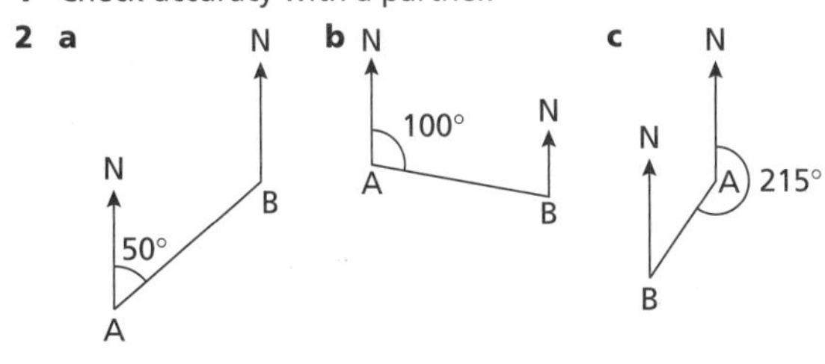

3 **a** 10 km **b** 8 cm

Practice (B)

1

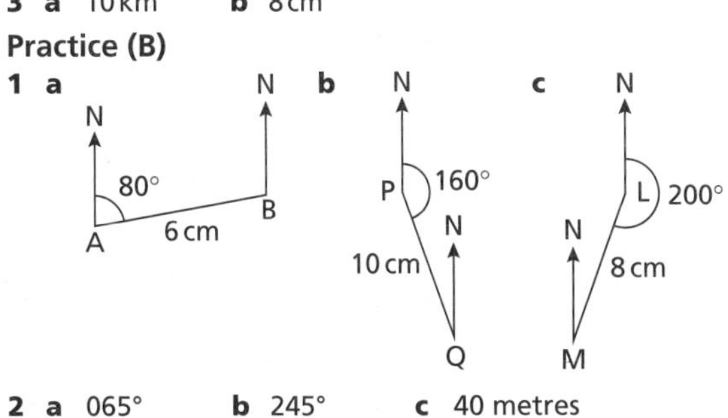

2 **a** 065° **b** 245° **c** 40 metres

3 N N 70° 8 cm House

4

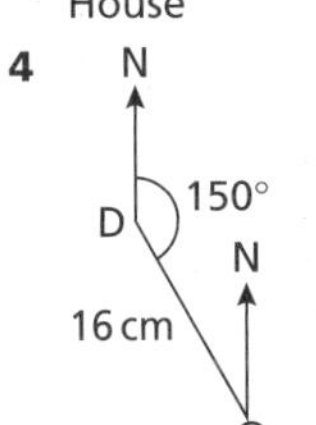

5

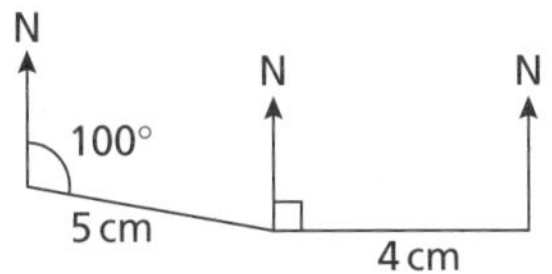

6

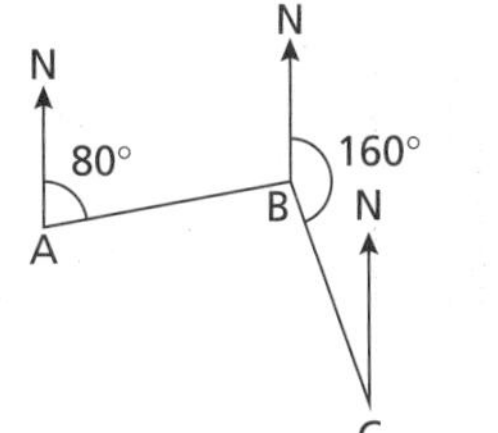

7 **a**

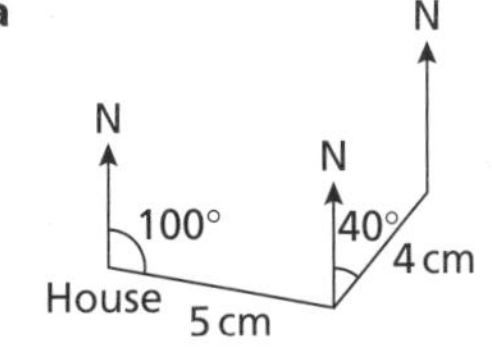

b 074° (accept answers from 071° to 077° inclusive)

Consolidate

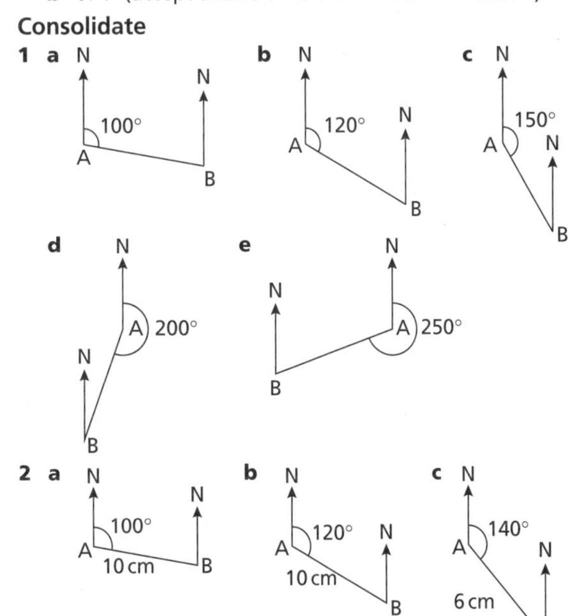

d **e**

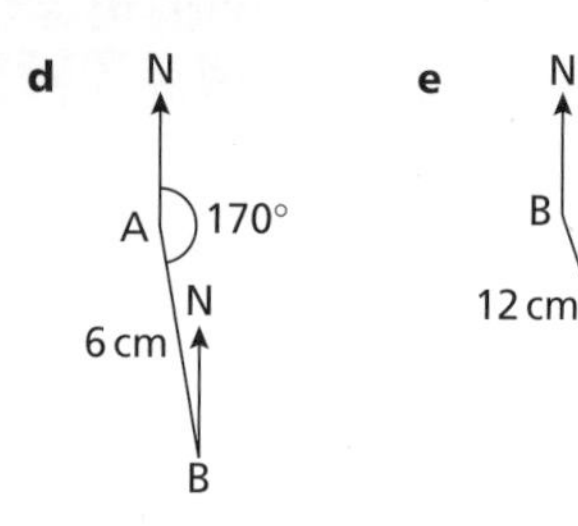

Stretch

1 Sometimes. If the bearing of B from A is less than or equal to 180°, this is true. If the bearing of B from A is greater than 180°, this is not true.
2 No, you need to consider the distances between A, B and C as well.

Chapter 1.2

Are you ready? (A)

1 Compare accuracy of diagrams as a class.
2 Compare accuracy of diagrams as a class.

Practice (A)

1 **a** Check the two angles produced are both 35°
  b Check the two angles produced are both 50°
  c Check the two angles produced are both 75°
2 **a** Check the two angles produced are both 20°
  b Check the two angles produced are both 25°
  c Check the two angles produced are both 55°
3

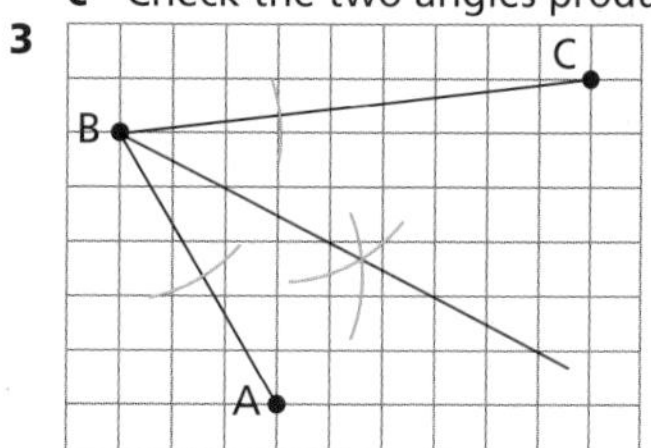

4 **a**

B C

A D

b

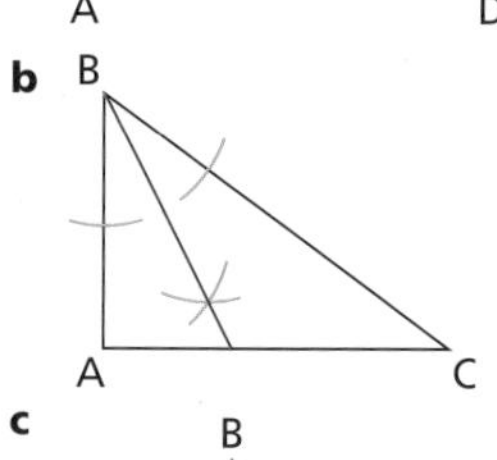

c

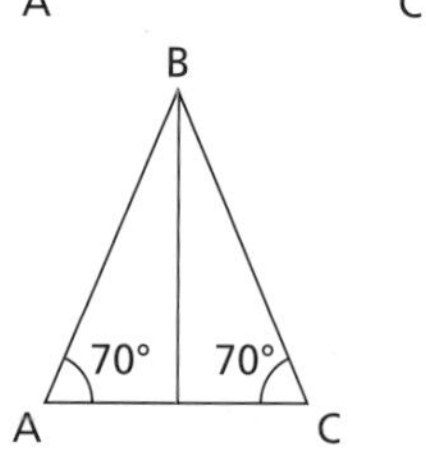

5 **a** Diagram drawn to dimensions shown.
  b

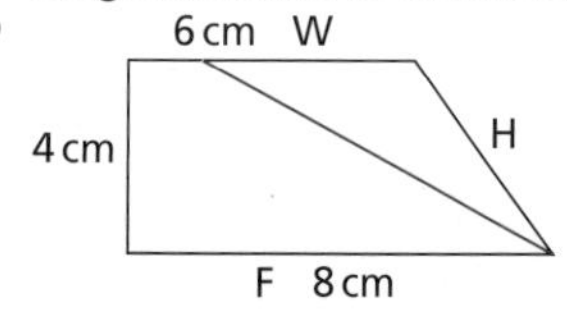

What do you think? (A)

1 No
2 Compare accuracy with a partner.

Are you ready? (B)

1

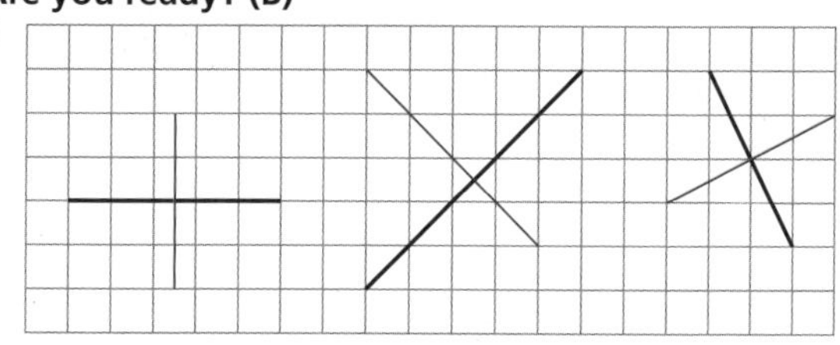

2 Compare accuracy of diagrams as a class.

Practice (B)

1 **a**

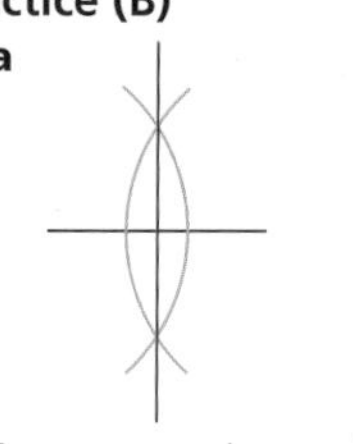

b

c

d

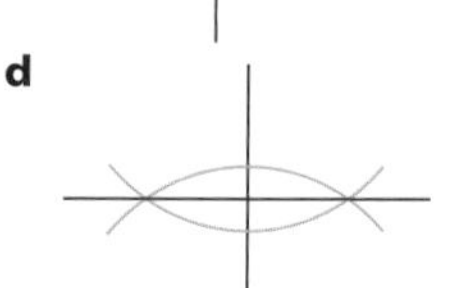

2 **a**

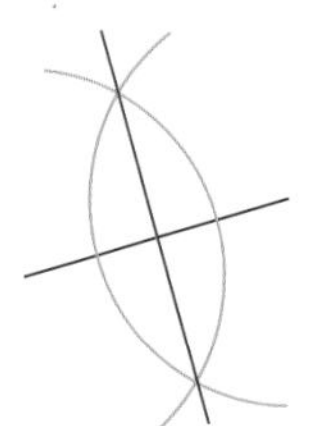

b

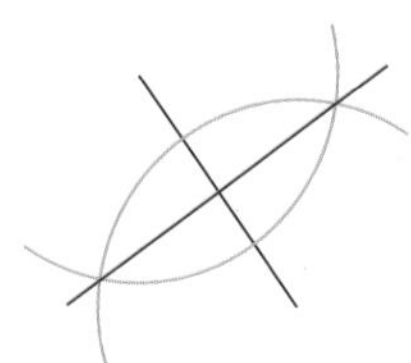

c

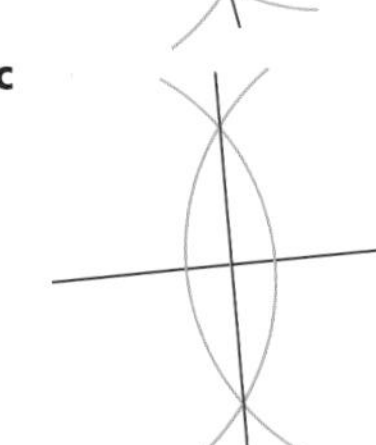

d

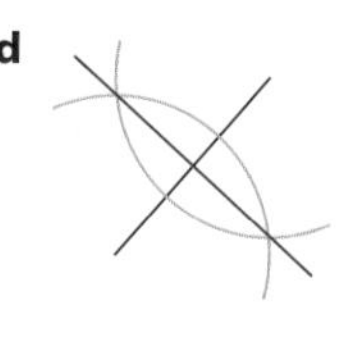

3

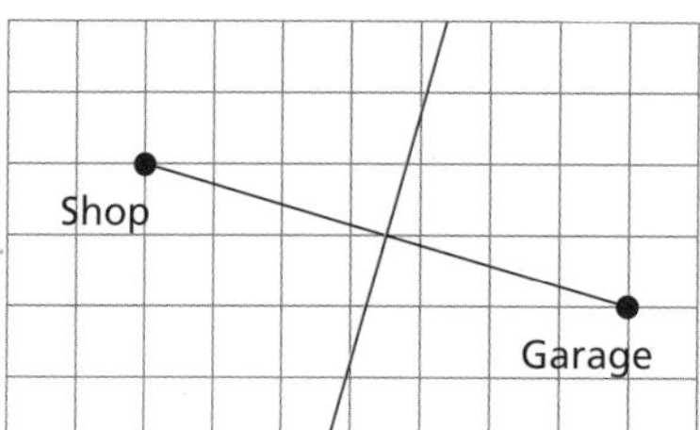

4

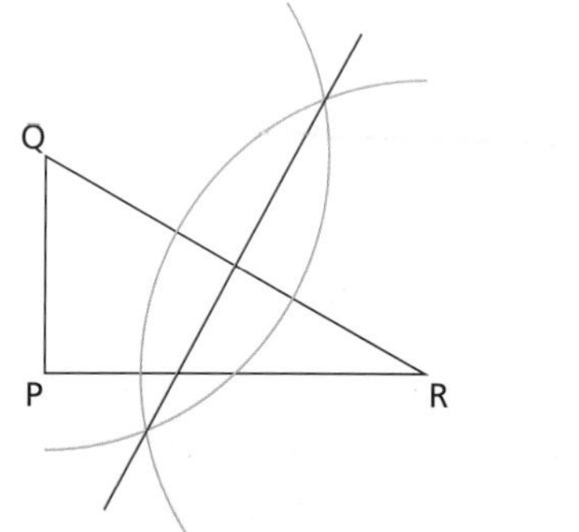

What do you think? (B)

1

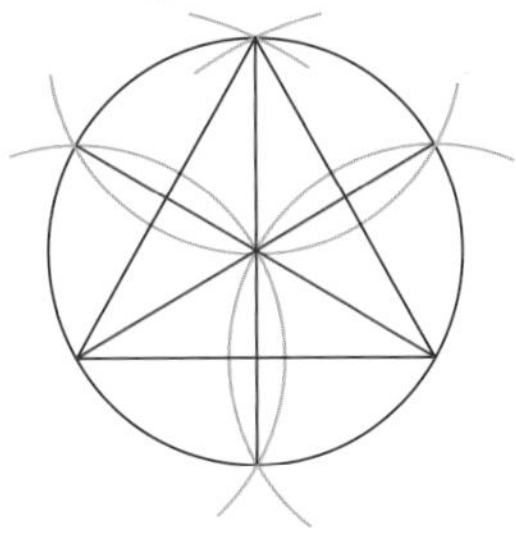

2 Junaid is correct. The result will not be true for rectangles and trapezia.

3 a b Compare answers as a class.

Consolidate

1 a

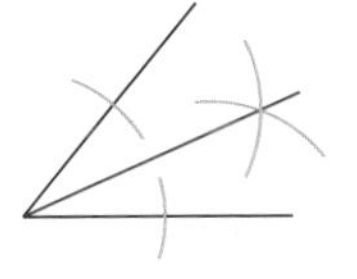

b

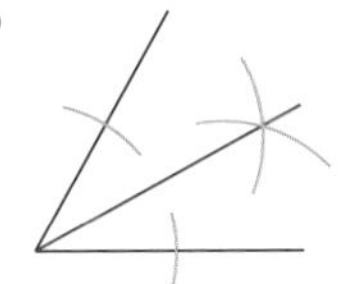

c

d

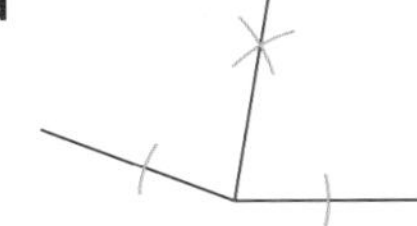

2 a

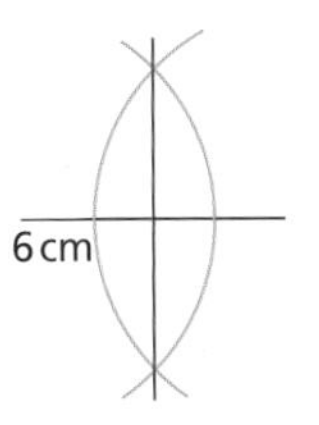

b

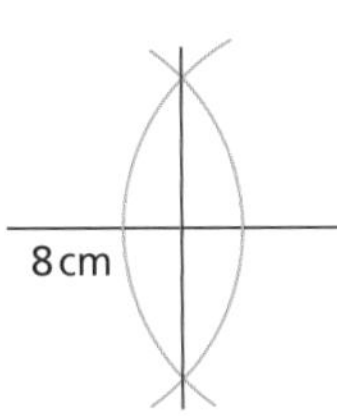

c

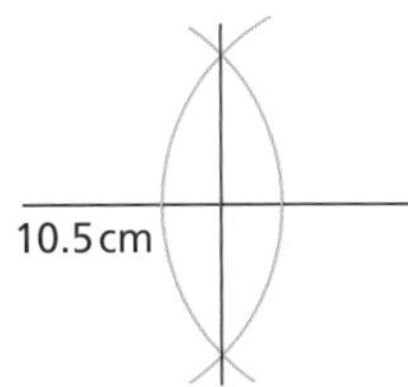

d

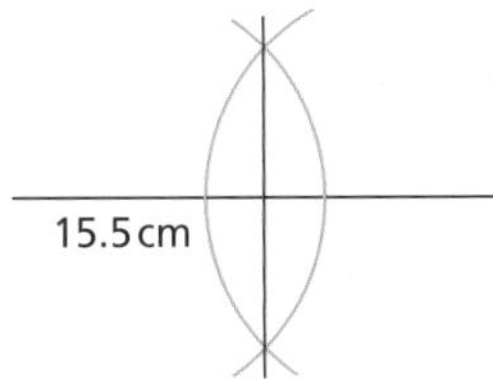

3 Compare answers as a class.

Stretch

1

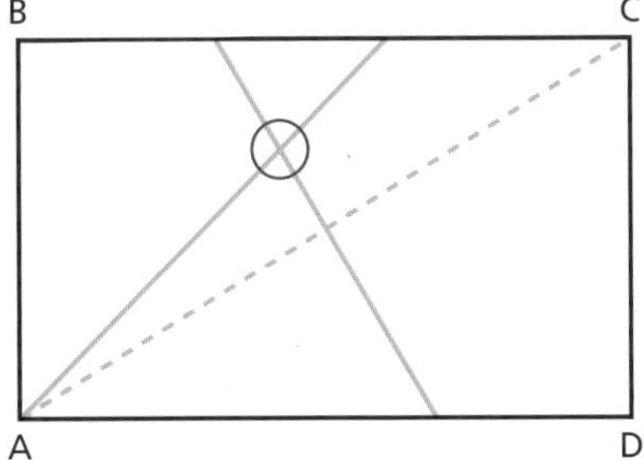

2 The theorem is true.

3 a Check construction using a protractor.

b Various answers, including 60°

4 a Check construction using a protractor.

b Compare answers as a class.

Chapter 1.3

Are you ready? (A)

1

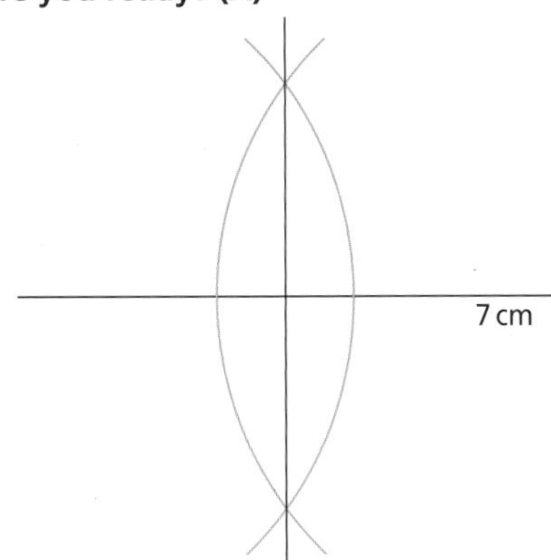

2

A

B

Practice (A)

1 a

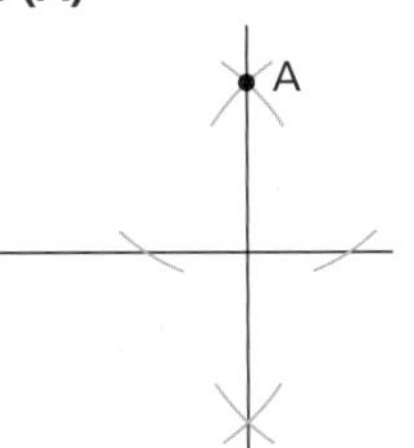

b

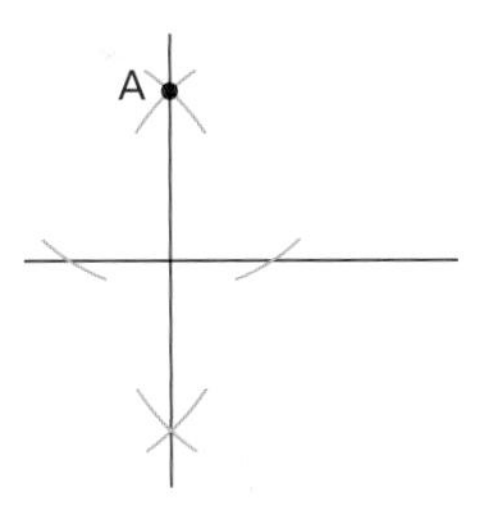

c

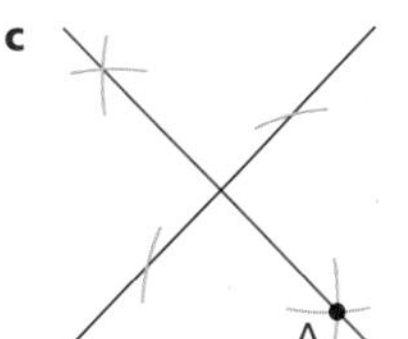

d

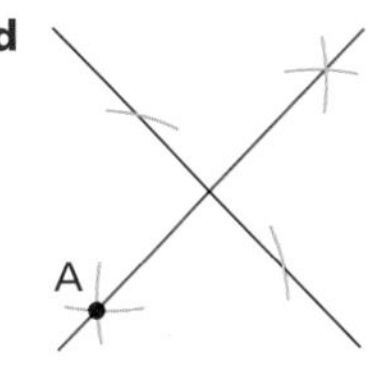

2 a

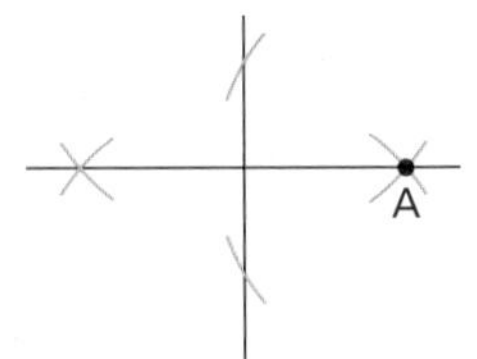

b

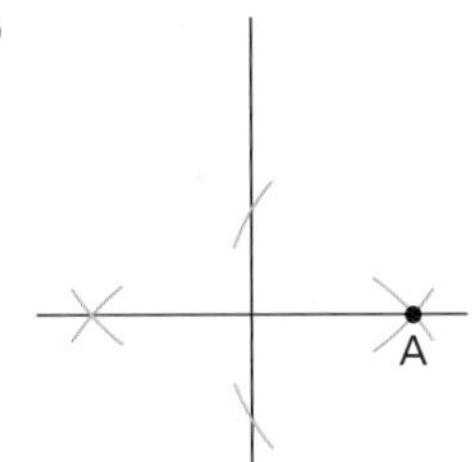

c

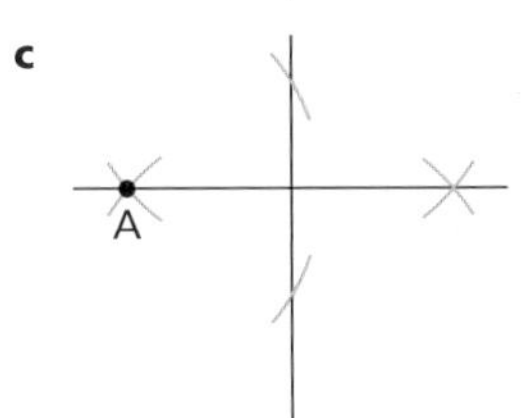

d

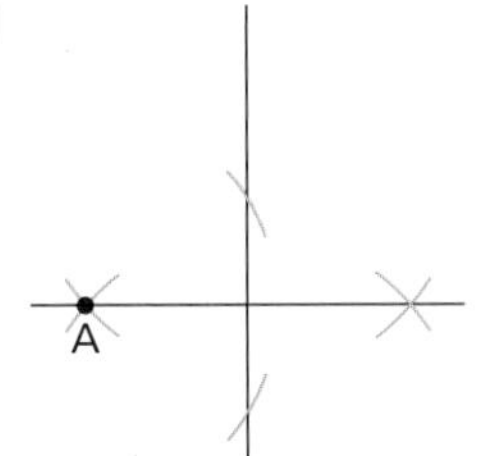

3 a

b

c

4

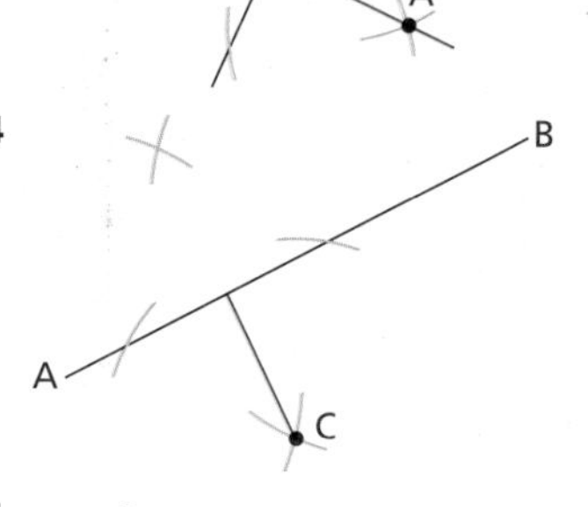

5

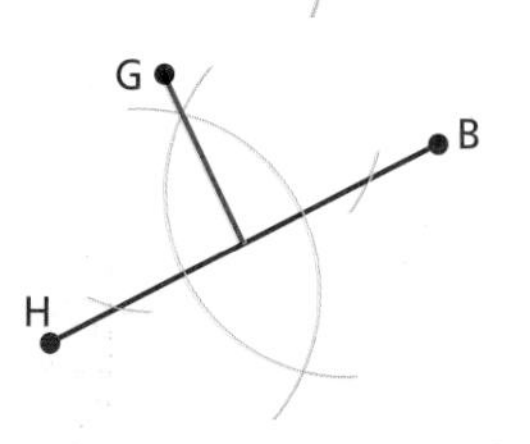

What do you think? (A)

1

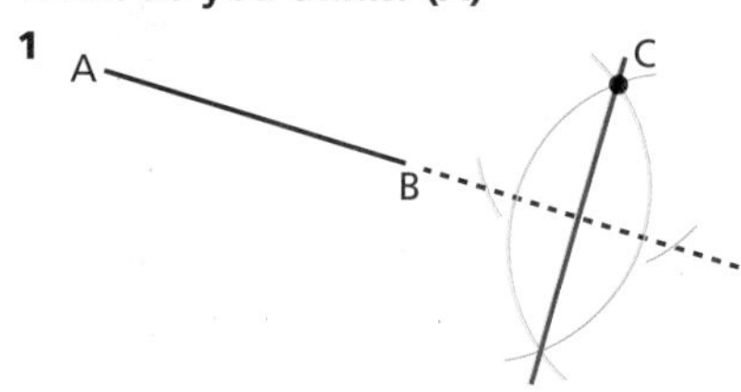

Are you ready? (B)

1

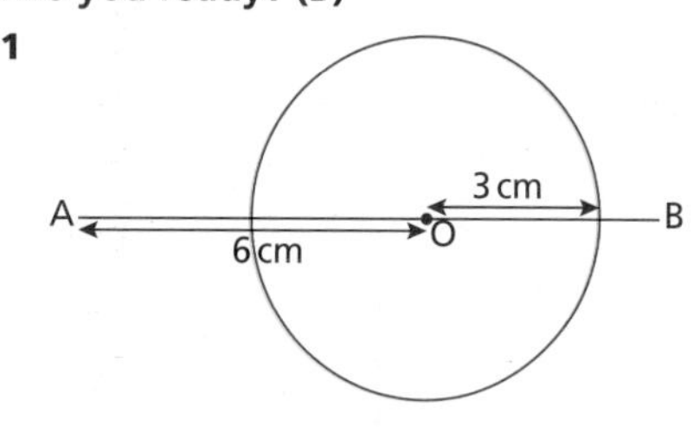

Practice (B)

1 a

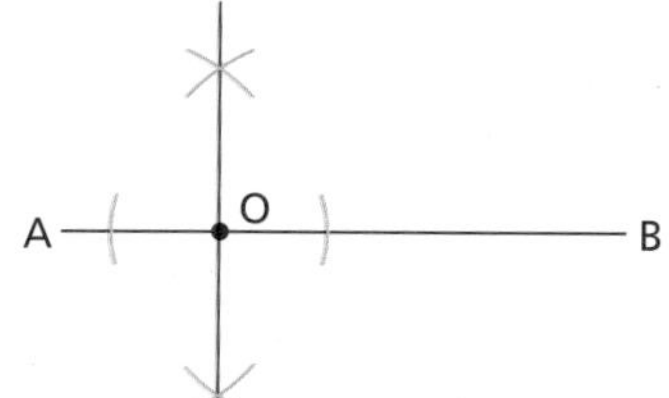

b

c

2

3 a

b

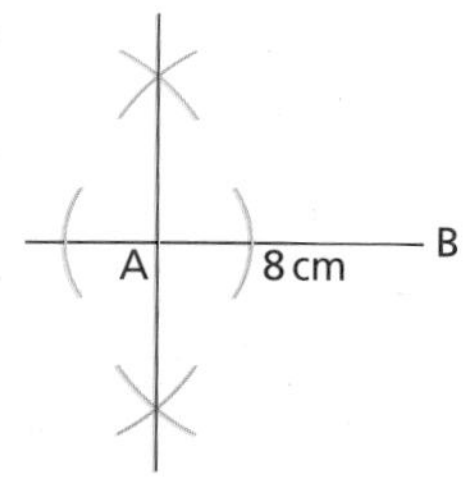

What do you think? (B)

1 There are many correct solutions, e.g. as shown. Triangle DBE will always be similar to triangle ABC as DE is parallel to AC and the angles are the same in each.

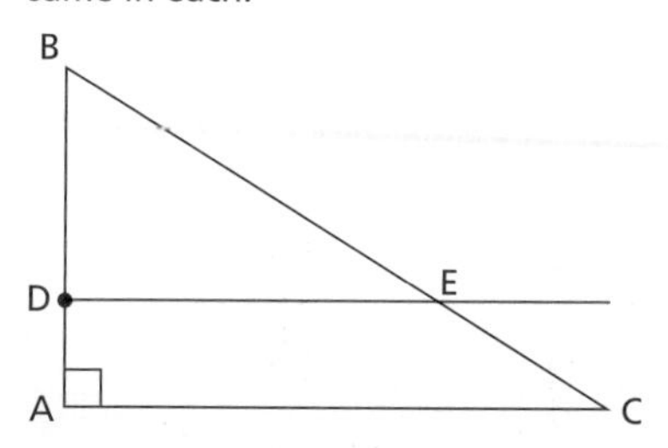

Consolidate

1 a

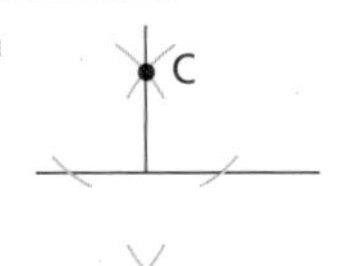

b

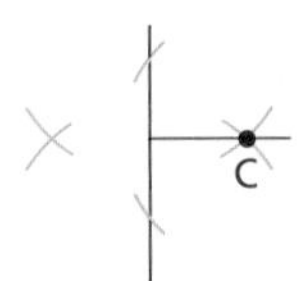

c

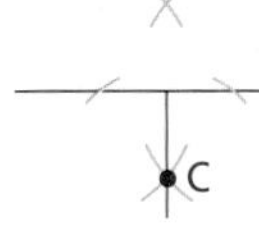

d

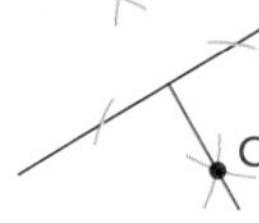

e

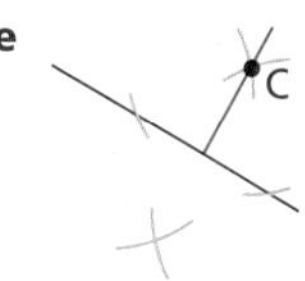

f

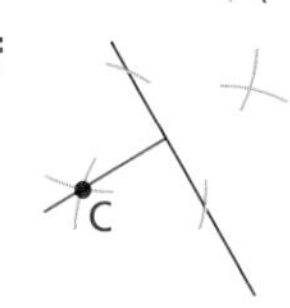

2 a–f Check all lines are set correctly at 90° to line AB and pass through O.

Stretch

1 a

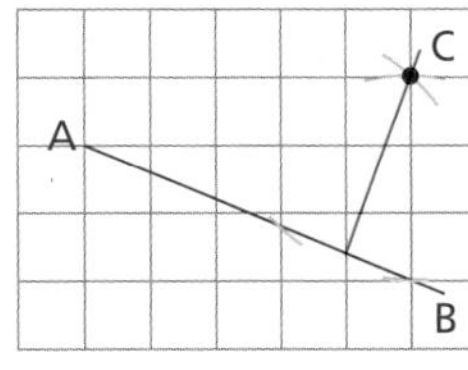

b The gradients are $-\frac{2}{5}$ and $\frac{5}{2}$
They are negative reciprocals of each other.
This will always happen with two perpendicular lines.

2 a Check accuracy with a partner.
b $y = 2x - 1$
c (0, –1)
d $y = -\frac{1}{2}x - 1$
e The product of the gradients is –1

Chapter 1.4

Are you ready? (A)

1

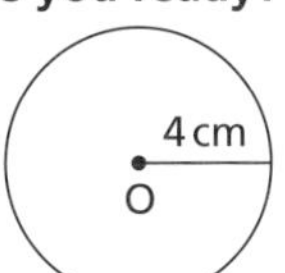

2

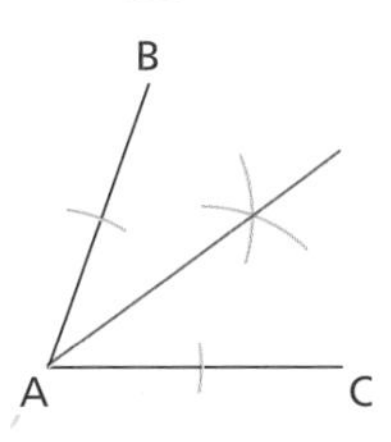

3

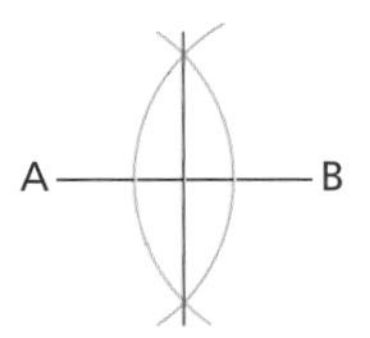

Practice (A)

1

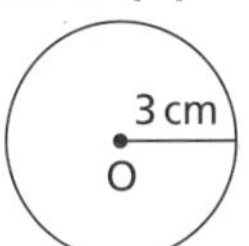

2 a

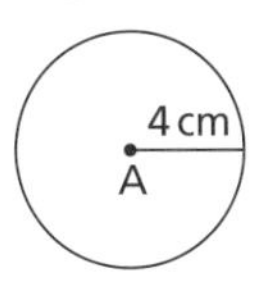

b

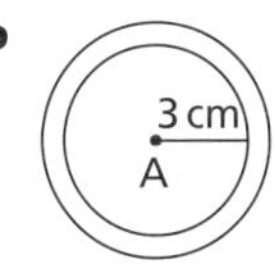

c

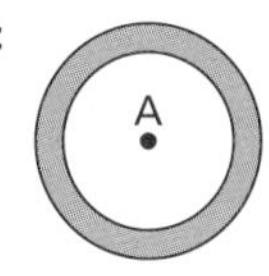

3 a A circle drawn with centre A, radius 7 cm.
b A circle drawn with centre B, radius 5 cm.
c

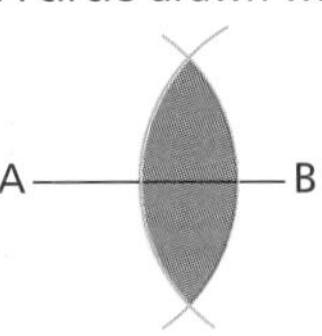

4

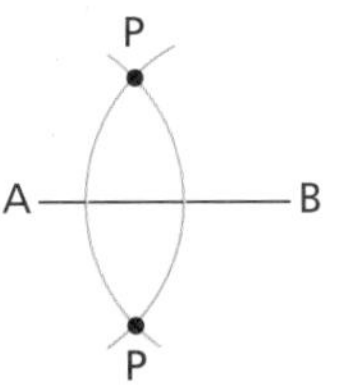

5 a i An arc of radius 4 cm, centre A, drawn within the rectangle.
ii An arc of radius 5 cm, centre C, drawn within the rectangle.

b

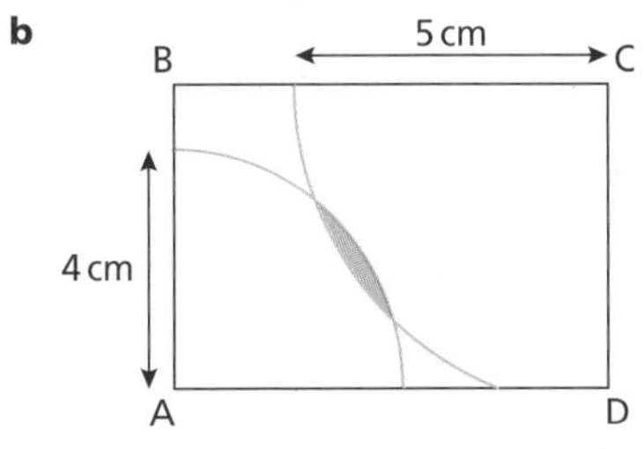

What do you think? (A)

1 No. The circles radius 4 cm from A and radius 3 cm from B do not intersect.

2 a ii or v **b** v or iv **c** v

Are you ready? (B)

1 a

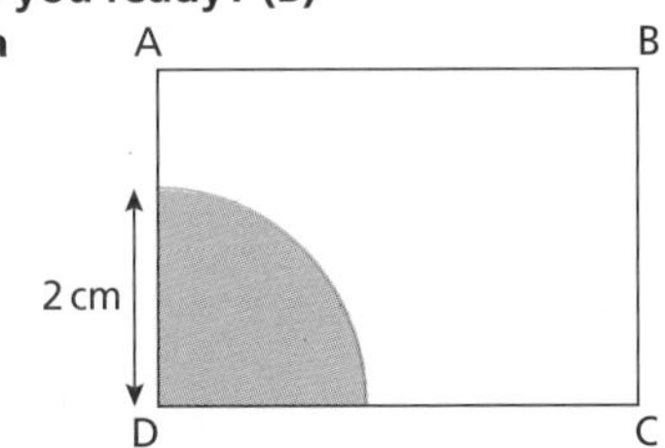

b

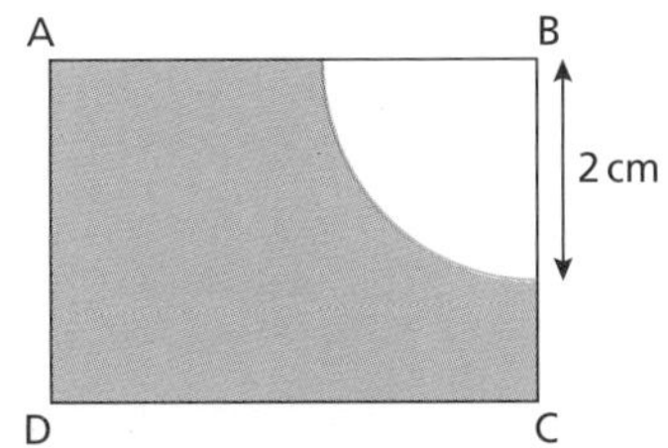

Practice (B)

1

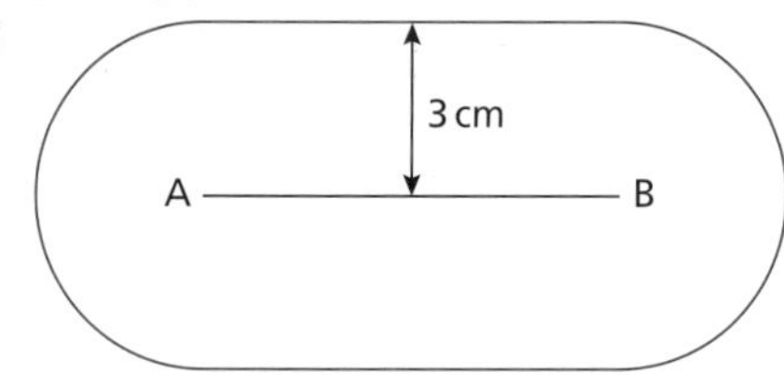

2

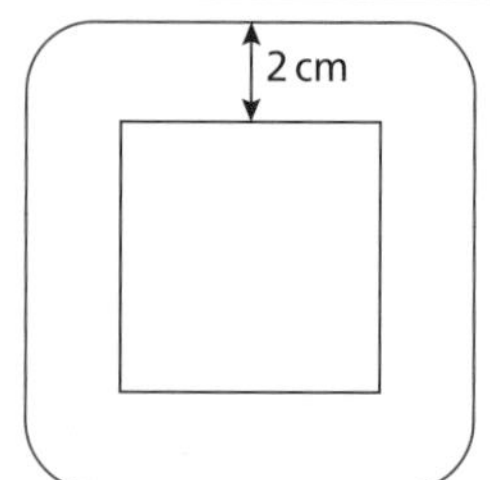

3 **a** **b**

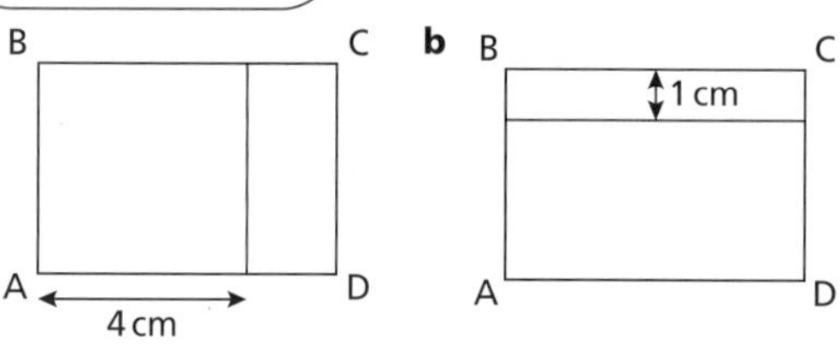

4 **a** Rectangle drawn to the dimensions shown in part **b**.

b **i** **ii**

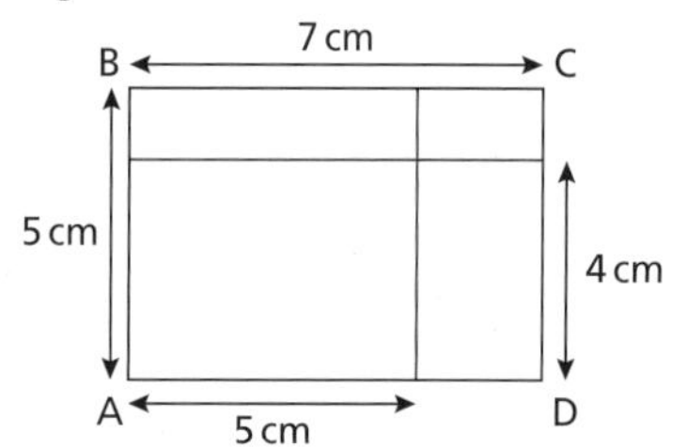

c

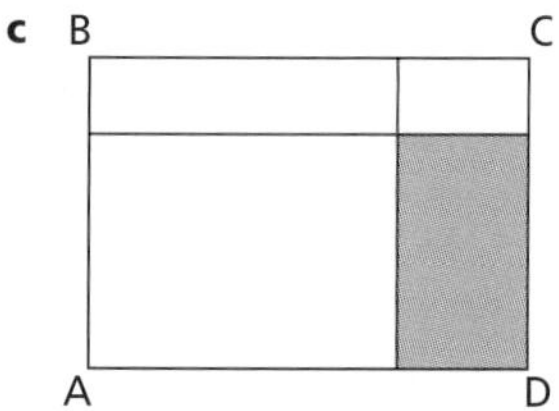

5 **a** Correct lines drawn as shown in part **b**.

b

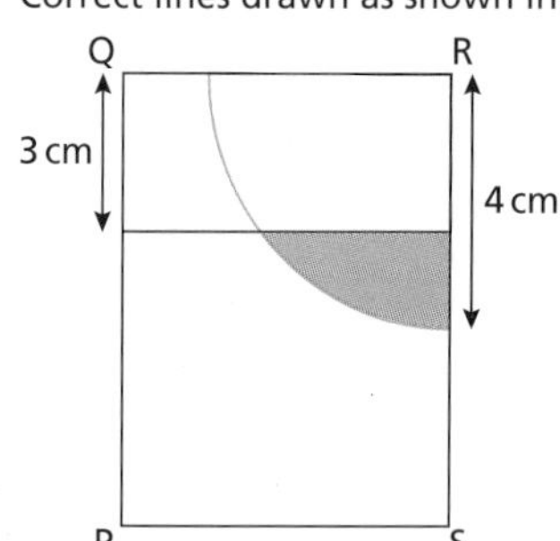

6

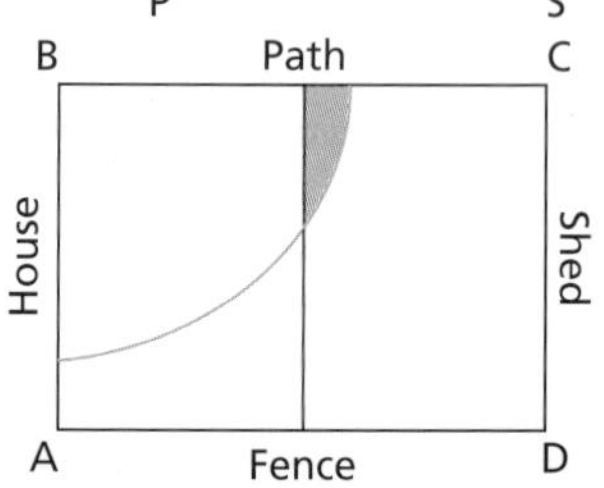

7 The points less than 4 cm from A and more than 2 cm from AB (or less than 6 cm from CD)

8

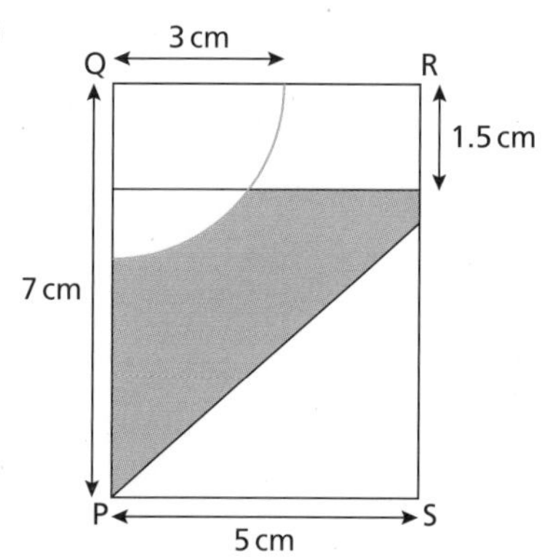

9

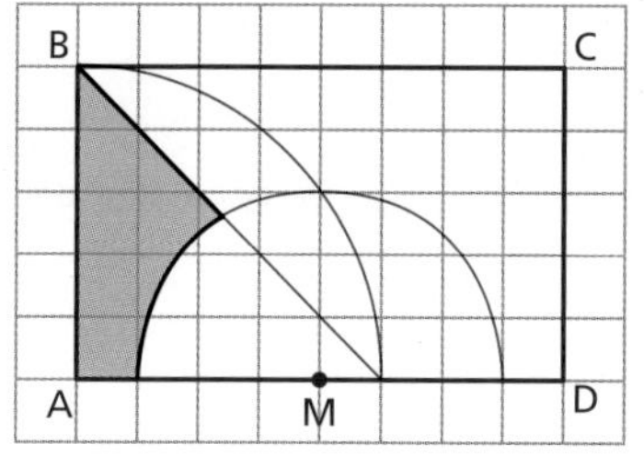

10

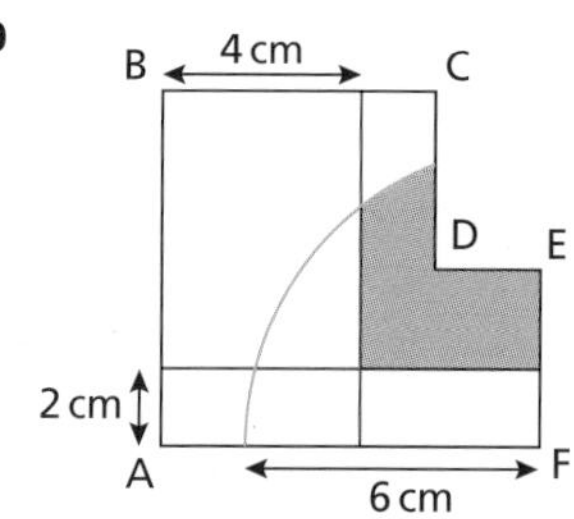

11

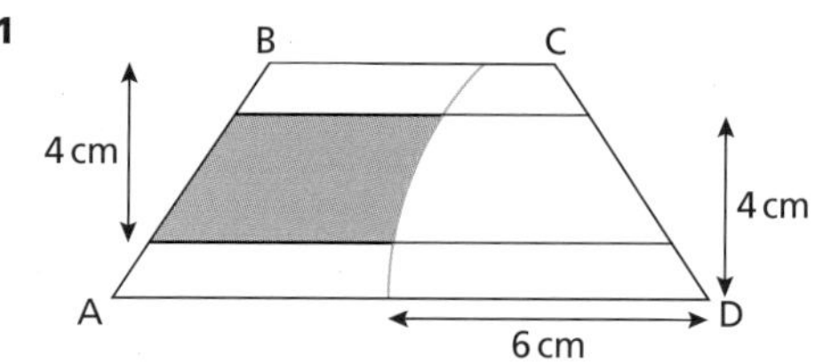

12

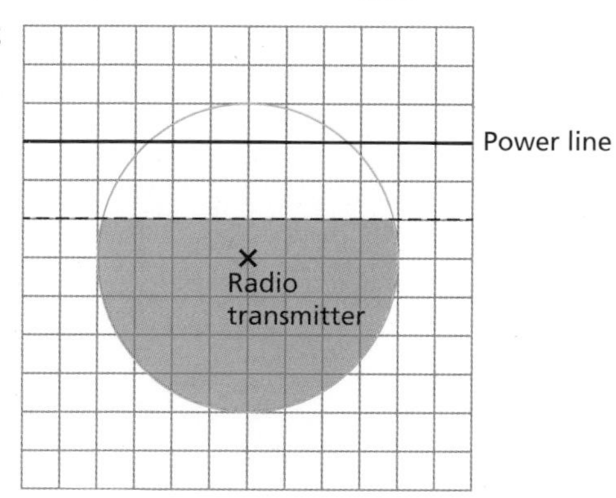

13 **a** Diagonal line drawn

b Midpoint of diagonal line marked

c

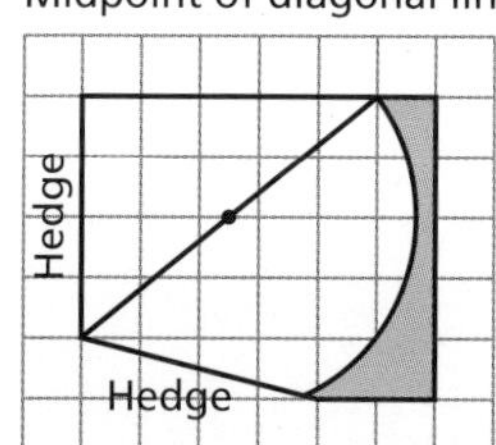

What do you think? (B)

1 Discuss solutions as a class.

Consolidate

1 a

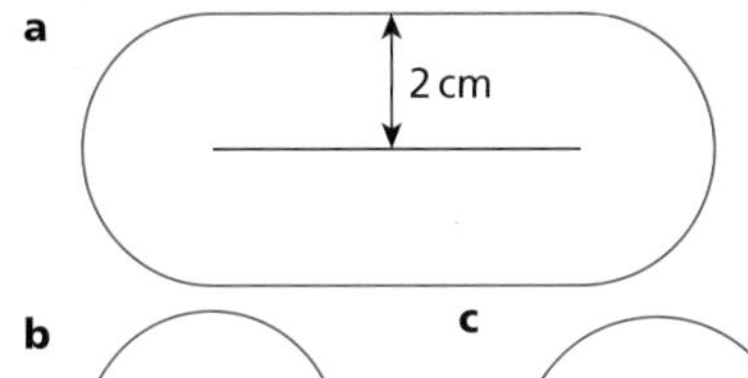

b **c**

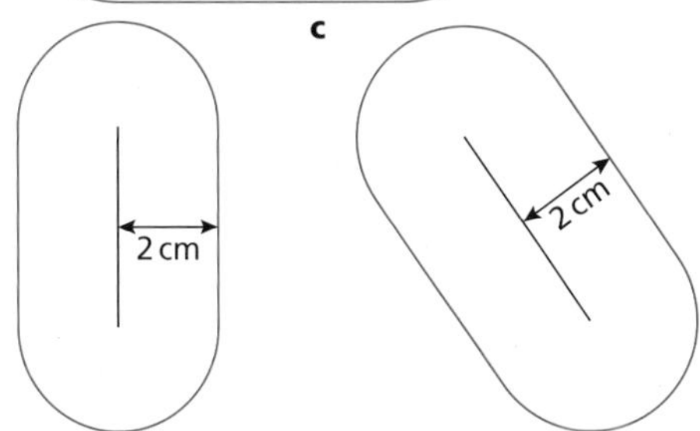

2 a

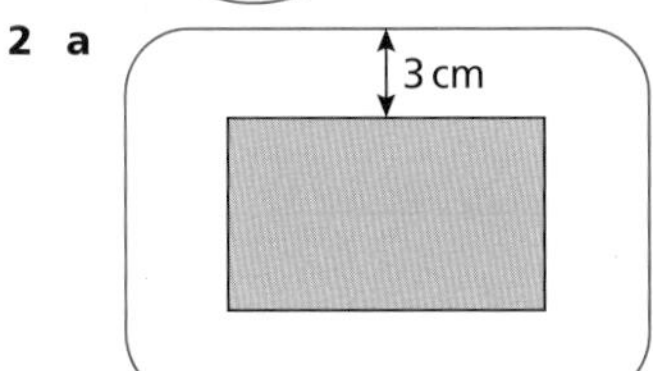

b **c**

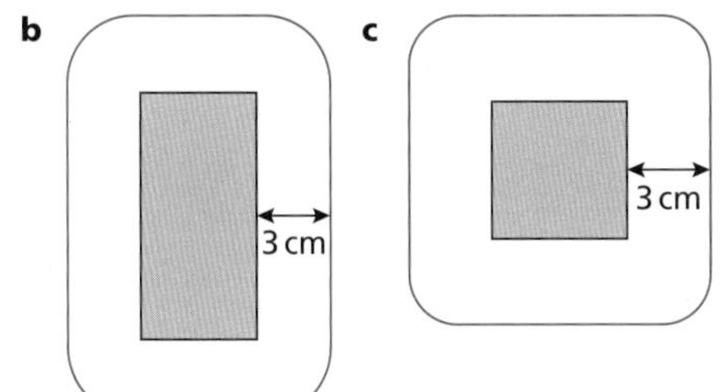

3 a

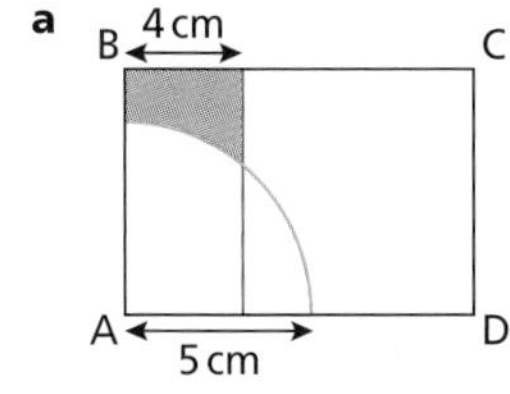

b

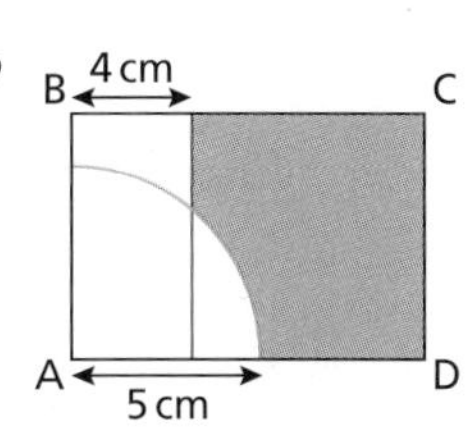

c **d**

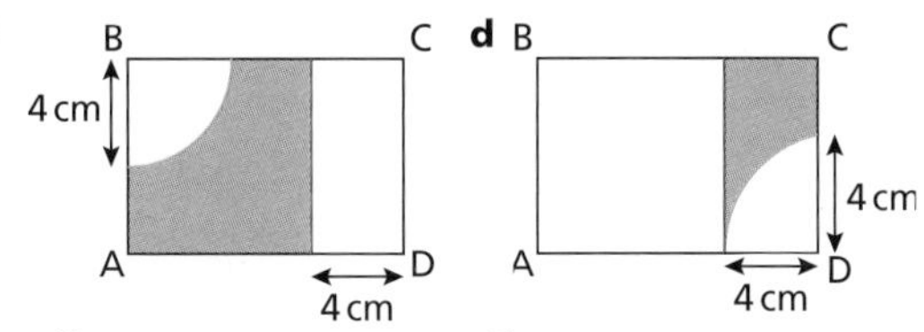

4

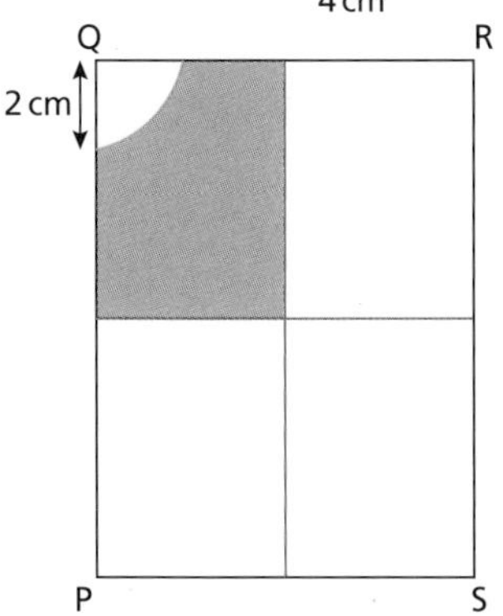

Stretch

1 64.3 cm^2

2 69.1 m^2

3

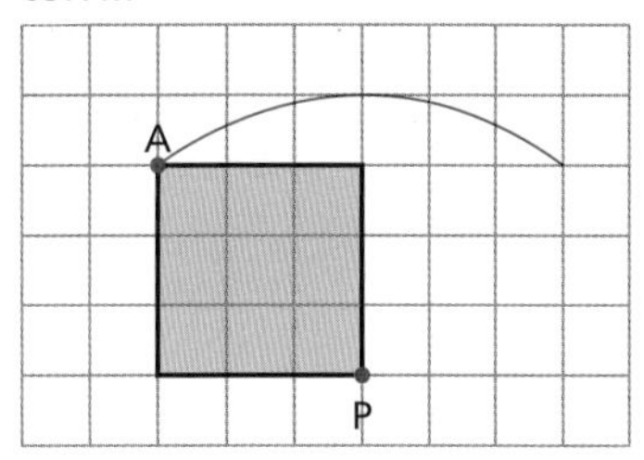

Constructions: exam practice

1 Check accuracy with a partner.

2 Check accuracy with a partner.

3 Check accuracy with a partner.

4 Check accuracy with a partner.

5 360° – (180° – 110°) = 290° shown on a diagram.

6 Check accuracy with a partner.
Should be parallel lines and arcs from the vertices as here:

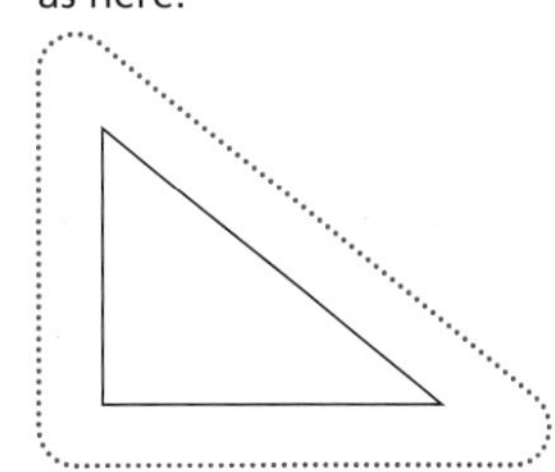

Block 2 Congruence and similarity

Chapter 2.1

Are you ready?

1 a 5 : 6 **b** 7 : 20 **c** 5 : 2 **d** 2 : 5

2 a 3 : 4 **b** 3 : 4 **c** 1 : 3 **d** 1 : 3

Practice

1 a 21 cm **b** 2 cm **c** 7 mm **d** 6.5 cm

2 a a = 7 cm and b = 3.25 cm
b c = 72° and d = 123°

3 a E, H **b** C, G

4 Compare examples as a class.

5 B, D, F

6 Compare examples as a class.

7 x = 27.625 cm

8 x = 20 cm, y = 13.5 cm

9 a a = 1.75 cm, b = 6 cm
b c = 55°

What do you think?

1 Yes she is correct (if the figures are the same size then they are also congruent).

2 a Similar shapes can have different corresponding side lengths and therefore not necessarily congruent.
b The ratio of corresponding sides of two congruent shapes is the same (1 : 1), also the angles, so therefore congruent shapes are similar.

3

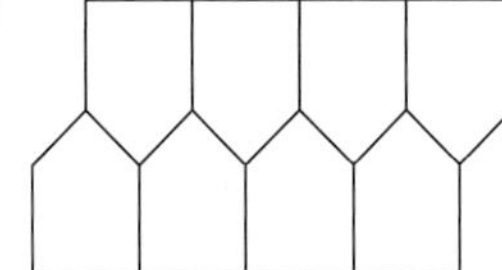

Examples of shapes which do not tessellate are circles or regular hexagons.

Consolidate

1 a $x = 18$ cm and $y = 4$ cm
  b $x = 18$ cm and $y = 10$ cm
  c $x = 14$ cm and $y = 1.75$ cm
  d $x = 19.25$ cm and $y = 4$ cm
2 a 12.75 cm b 50°
3 a C b D c B, D and E

Stretch

1 $x = 24$
2 60 cm^2
3 26.4 cm
4 a $10\sqrt{3} + 6$ b $\frac{5\sqrt{3} - 1}{2}$

Chapter 2.2

Are you ready?

1 a Isosceles b Isosceles c Scalene
  d Isosceles e Equilateral

Practice

1 a Congruent b Not enough information
  c Not congruent
2 a AB b CB c 11 cm
3 A and C. SAS
4 B and C. ASA (or AAS)
5 a RHS b SSS c SAS
6 a Example answer: DEF and DGF are congruent using the condition SSS with two equal sides and a shared side.
  b Example answer: PSQ and QRS are congruent using the condition RHS with a right angle, equal sides and hypotenuse in common.
  c Example answer: WYZ and XWZ are congruent using the condition SSS with three equal sides.
  d Example answer: JKM and KLM are congruent using the condition SAS with two equal sides and the angle between them.
7 $x = 3$, $y = 5$
8 ∠BCD = ∠AEB because alternate angles are equal.
  ∠ABE = ∠CBD because vertically opposite angles are equal.
  BC = BE (given)
  The triangles are congruent using the condition ASA (or AAS).
9 a ∠BAC = ∠AED because alternate angles are equal.
  ∠ABD = ∠BDE because alternate angles are equal.
  ∠ACB = ∠DCE because vertically opposite angles are equal.
  All three angles are equal, therefore the triangles are similar.
  b A corresponding length on both triangles.

What do you think?

1 SAS or AAS
2 Squares, rectangles, parallelograms, rhombuses, isosceles trapezia and kites can be split into two congruent triangles.
3 AC is a shared side and all sides are equal, therefore they are congruent using the condition SSS.
  ∠ADC = ∠ABC, and all sides are equal therefore they are congruent using the condition SAS.

Consolidate

1 a SSS b SAS c ASA (or AAS)
  d RHS e AAS f SAS
  g SSS h SAS i AAS
2 A and C. SAS
3 Example answer: AB = DC = 7.5 cm, AD = BC = 3.9 cm. BD is a common side. The triangles are congruent using the condition SSS.
  Discuss other proofs as a class.

Stretch

1 If AE = a and AH = b, it follows that the side length of ABCD is $a + b$.
  The triangles are all congruent using the conditions SSS or RHS
2 ∠BCG = ∠DCE (140°)
  BC = CE
  CG = CD
  The triangles are congruent using the condition SAS.
3 $k = 3$

Chapter 2.3

Are you ready?

1 a 25 cm b 5.6 cm
2 a YX b QR c ∠YZX (or ∠XZY)

Practice

1 a $a = 10$ cm, $b = 7.5$ cm
  b $a = 6$ cm, $b = 16$ cm
  c $a = 24$ cm, $b = 61°$
2 a 9 m b 8 m
3 a 15 mm b 6 mm
4 a 11 cm b 19.5 cm
5 a 37.5 m b 16 m
6 a Example answer: ∠BAC is common, ABC = ADE and ACB = AED as corresponding angles are equal. So the triangles are similar as they have the same angles.
  b 12 cm b 4 cm
7 9 cm
8 a 14 cm b 11.428… cm
9 Example answer: $\frac{4.5}{2.5} = \frac{3.6}{2} = \frac{2.7}{1.5} = 1.8$
  The sides are all in the same ratio, so the triangles are similar.
10 a 6 cm b $(12 + 12\sqrt{3})$ cm
11 $x = 6.5$
12 a $\frac{3}{5}x$ b $26.4 - \frac{3}{5}x$
  c PQ = 17.65, EF = 10.59 and FG = 15.81

What do you think?

1

5.9 cm 4.8 cm 8 cm

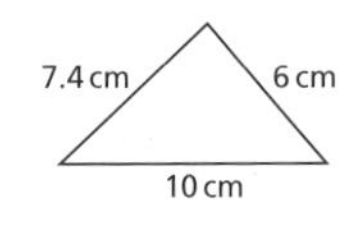

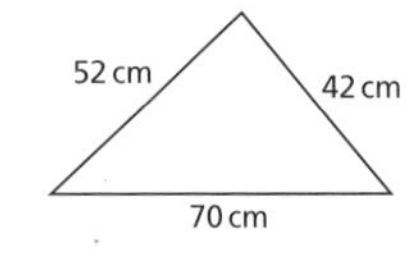

2 a $\frac{4}{3}, \frac{8}{15}$ or $\frac{4}{5}$
  b Triangles sketched with the following side lengths:
  8 cm, 13.3 cm and 20 cm
  3.2 cm, 5.3 cm and 8 cm
  4.8 cm, 8 cm and 12 cm
3 If two pairs of angles are equal, the third pair must be equal as they will both be the difference between the sum of the first two and 180°. As all three pairs of angles are equal, the triangles will be similar.

Consolidate

1 a $a = 18$ cm $b = 7$ cm
  b $a = 13.5$ cm $b = 8$ cm
  c $a = 14$ cm $b = 21$ cm
2 a $a = 9$ cm $b = 12$ cm
  b $a = 15$ cm $b = 9$ cm
  c $a = 2.5$ cm $b = 8.5$ cm

Stretch

1 $(60 + 36\sqrt{5})$ cm
2 $x = \frac{7}{5}$ or $x = \frac{221}{7}$

Congruence and similarity: exam practice

1. Any three from: ABC and ADC; ABE and CDE; ABD and BCD; ADE and BCE
2. Any rectangle with sides in the ratio 1 : 4 and shorter side greater than 1 cm.
3. 3.36 cm
4. Corresponding sides are in the same ratio. Congruent when $a = b$.
5. 3.92 cm
6. Any valid proof, e.g. using SAS, RHS or SSS and stating the properties given.

Block 3 Angles and circles

Chapter 3.1

Are you ready? (A)

1 **a** $a = 146°$ **b** $b = 57°$
c $c = 42°$, $d = 138°$ and $e = 138°$
2 **a** $a = 63°$ and $b = 117°$
b $c = 71°$ and $d = 109°$
c $e = 62°$ and $f = 56°$
3 **a** $a = 78°$ **b** $b = 115°$ **c** $c = 67°$

Practice (A)

1 Example reasons are given.
a $a = 157°$, alternate angles are equal.
b $b = 72°$, corresponding angles are equal.
c $c = 104°$, corresponding angles are equal and angles on a straight line sum to 180°.
d $d = 101°$, co-interior angles sum to 180°.
e $e = 109°$, corresponding angles are equal.
f $f = 94°$, co-interior angles sum to 180°.
g $g = 72°$, alternate angles are equal.
2 Example reasons are given.
a 71°, alternate angles are equal.
b 38°, base angles in an isosceles triangle are equal and angles in a triangle sum to 180°.
3 Example reasons are given.
a 43°, alternate angles are equal.
b 68.5°, base angles in an isosceles triangle are equal and angles in a triangle sum to 180°.
4 Example reasons are given.
a $a = 65°$, alternate angles are equal.
$b = 107°$, co-interior angles have a sum of 180°.
b $c = 98°$, co-interior angles have a sum of 180°.
$d = 66°$, co-interior angles have a sum of 180°.
c $e = 85°$, co-interior angles have a sum of 180°.
$f = 118°$, alternate angles are equal.
5 Example reasons are given.
59°. CAE = CBD because corresponding angles are equal.
BCD = 59° because angles in a triangle sum to 180°.
6 **a** $x = 54$ **b** $y = 30$ **c** $z = 15$

What do you think? (A)

1 **a** AB and CD are not parallel.
b EF is parallel to CD.
Discuss justifications as a class.
2 **a** Compare answers as a class.
b For example: ∠BDE and ∠CED
Compare answers as a class.
3 **a** $x = 80°$ **b** $y = 235°$
Compare answers as a class.
4 ∠AFE = 57° and ∠ABE = 66°. ∠ACB = ∠ECF.
The triangles have three pairs of equal angles.

Are you ready? (B)

1 **a** $a = 45°$ **b** $b = 105°$ **c** $c = 51°$ and $d = 129°$
2 **a** Pentagon **b** Hexagon **c** Octagon

Practice (B)

1 **a** **i** 540° **ii** 108°
b **i** 720° **ii** 120°
c **i** 1080° **ii** 135°
2 **a** 540° **b** $a = 122°$
3 $b = 232°$
4 **a** $a = 65°$ **b** $b = 60°$ and $c = 120°$
5 $f = 134°$ and $g = 63°$
6 $x = 120°$
7 $y = 36°$
8 **a** $y = 162°$ **b** $y = 60°$ **c** $y = 27°$
9 $x = 54°$
10 12

What do you think? (B)

1 A regular polygon with an interior angle of 170° (36 sides)
A regular polygon with an exterior angle of 18° (20 sides)
2 Equilateral triangles, squares and regular hexagons. Internal angles must be a factor of 360

Consolidate

1 Example reasons are given.
a $a = 62°$, alternate angles are equal.
b $b = 125°$, corresponding angles are equal.
c $c = 37°$, co-interior angles have a sum of 180°.
d $d = 43°$, co-interior angles have a sum of 180°.
e $e = 121°$, corresponding angles are equal.
f $f = 129°$, alternate angles are equal.
2 **a** $x = 67°$ **b** $y = 67°$
3 **a** 63° **b** 135° **c** 128°
d 115° **e** 121° **f** 106°

Stretch

1 $x = 30$ and $y = 25$
2 $x = 10$
3 $\angle BDC = y$

$$\angle DBC = \angle DCB = \frac{180 - y}{2}$$

$$x + \frac{180 - y}{2} = 90$$

$$x - \frac{y}{2} = 0$$

$$y = 2x$$

4 $x = 2$. The shape has 15 sides.

Chapter 3.2

Are you ready? (A)

1 Diameter is C; Radius is A; Circumference is B; Chord is D; Segment is E
2 **a** $a = 39°$ **b** $b = 69°$ **c** $c = 34°$

Practice (A)

1 **a** $a = 108°$ **b** $b = 66°$ **c** $c = 67.5°$
d $d = 73°$ **e** $e = 87.5°$ **f** $f = 41°$
g $g = 103°$ **h** $h = 222°$ **i** $i = 90°$
2 **a** ∠ABC **b** 60°
3 **a** $a = 72°$ **b** $b = 21°$ **c** $c = 37.5°$
4 **a** $x = 240°$ **b** $x = 73.5°$
5 55°. Base angles in an isosceles triangle are equal. Angles in a triangle have a sum of 180°. The angle at the centre is twice the angle at the circumference.
6 **a** $a = 29°$ **b** $b = 61°$ **c** $c = 11°$ **d** $d = 45°$
7 **a** $a = 36°$ **b** $b = 54°$ **c** $c = 9°$
8 62°
9 **a** 11.4 cm **b** 28°

What do you think? (A)

1 ∠ABC = 111°, ∠OCB = 82°
2 Seb and Flo are both correct.
3 **a** Example answer:
$\angle OXY = \angle OYX = x$
$\angle XOY = 180 - 2x$
$\angle OZY = \angle OYZ = y$
$\angle ZOY = 180 - 2y$
$(180 - 2x) + (180 - 2y) + \angle XOZ = 360$
$\angle XOZ = 2x + 2y$
$2x + 2y = 2(x + y)$

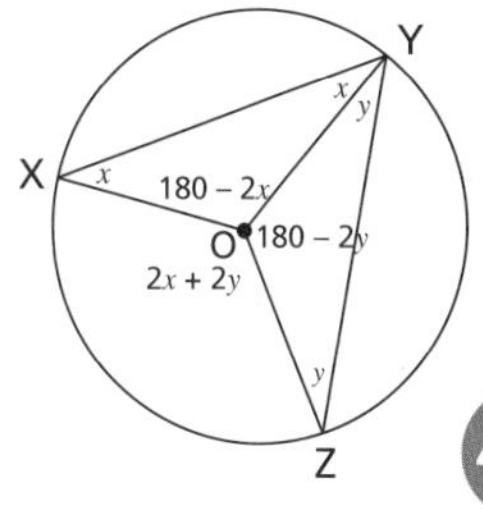

b Example answer:
$\angle OPQ = \angle OQP = x$
$\angle POQ = 180 - 2x$
$\angle ORQ = \angle OQR = y$
$\angle ROQ = 180 - 2y$
$(180 - 2x) + (180 - 2y)$
$= 180$
$180 = 2x + 2y$
$90 = x + y$

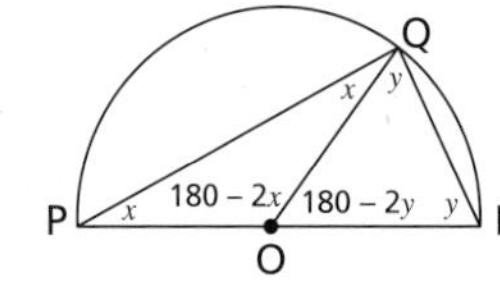

Practice (B)

1 a $a = 39°$ **b** $b = 28°$ **c** $c = 15°$ **d** $d = 124°$
2 a $a = 45°, b = 31°$ **b** $c = 42°, d = 25°$
c $e = 18°, f = 105°$ **d** $g = 35°, h = 35°$
3 65°. Angles in the same segment are equal. Vertically opposite angles are equal. Angles in a triangle have a sum of 180°.
4 a 62.5°. The angle at the centre is twice the angle at the circumference.
b 62.5°. Angles in the same segment are equal.
5 a $a = 55°$
b $b = 119°$
c $c = 92°, d = 106°$
d $e = 135°, f = 85°$
e $g = 75°, h = 75°, i = 105°$
f $j = 123°, k = 123°, l = 57°$
6 a $a = 108°, b = 72°$
b $c = 126°, d = 125°$
c $e = 103°, f = 46°$
d $g = 67°, h = 67°, i = 113°$
7 a $a = 92°$ **b** $b = 12°$
8 $a = 90°$. Angles in a triangle have a sum of 180°.
$b = 90°$. Opposite angles in a cyclic quadrilateral have a sum of 180°.
$c = 45°, d = 45°$. Base angles in an isosceles triangle are equal and angles in a triangle have a sum of 180°.

What do you think? (B)

1 False, there are many different cyclic quadrilaterals with a right angle.
2 They could be congruent or neither.
3 a Example answer: If $\angle WXY = x$, then the reflex angle $\angle WOY = 2x$
If $\angle WZY = y$, then the obtuse angle $\angle WOY = 2y$
$2x + 2y = 360$
$x + y = 180$
b Example answer: If $\angle WXZ = x$, then the obtuse angle $\angle WOZ = 2x$
If $\angle WYZ = y$, then the obtuse angle $\angle WOZ = 2y$
$\angle WOZ = 2x = 2y$
$2x = 2y$
$x = y$

Consolidate

1 a $a = 96°$ **b** $b = 54°$ **c** $c = 78°$
d $d = 70°$ **e** $e = 67.5°$ **f** $f = 142°$
2 a $a = 33°$ **b** $b = 51°$ **c** $c = 13°$
3 a $a = 52°$ **b** $b = 31°$ **c** $c = 17°$ **d** $d = 112°$
4 a $a = 56°$ **b** $b = 149°$ **c** $c = 89°, d = 96°$

Stretch

1 Example answer: AC is a diameter, thus triangle ADC is a right-angled triangle.
Since $\angle DCA = 28°$ (given) then $\angle DAC = 62°$, since the angle sum in a triangle is 180°.
$\angle DAC$ and $\angle DBC$ are angles in the same segment DC, and thus they are equal.
Therefore, $\angle DBC = \angle DAC = 62°$
2 Example answer: $\angle ACD = 58°$. The angle in a semicircle is 90° and angles in a triangle have a sum of 180°.
$\angle ABD = 58°$. Angles in the same segment are equal.
$\angle BEA = 70°$. Angles on a straight line have a sum of 180°.
$\angle BAC = 52°$. Angles in a triangle have a sum of 180°.

3 Example answer: $2x + 2y = 180°$
$x + y = 90°$
$x = 90° - y$
4 107 cm²
5 $x = 12.5, y = 77.5$

Chapter 3.3

Are you ready? (A)

1 a Radius **b** Diameter **c** Tangent **d** Chord
2 a $a = 28°$ **b** $b = 62.5°$ **c** $c = 42°$ **d** $d = 77°$

Practice (A)

1 a 90°. The angle between a chord midpoint and a radius is 90°.
b 58°
2 62°. The angle between a chord midpoint and a radius is 90° and angles in a triangle sum to 180°.
3 a 7.5 cm **b** 6.4 cm **c** 12.8 cm
4 129°. The angle between a chord midpoint and a radius is 90°, angles in a triangle sum to 180° and angles on a straight line sum to 180°.
5 a $a = 67°$ **b** $b = 18°$ **c** $c = 45°$
6 a 33°. The angle between a radius and a tangent is 90°.
b 33°. The angle in a semicircle is 90° and angles in a triangle sum to 180°.
7 26°. The angle between a radius and a tangent is 90° and angles in a quadrilateral sum to 360°.
8 $x = 12.5$
9 54°. The angle between a radius and a tangent is 90° and angles in the same segment are equal.
10 a 12 cm. Tangents which meet at a point are equal in length.
b $x = 66°$
11 a 10 cm **b** 48 cm²
12 48.75 cm

Are you ready? (B)

1 A, C and D

Practice (B)

1 a

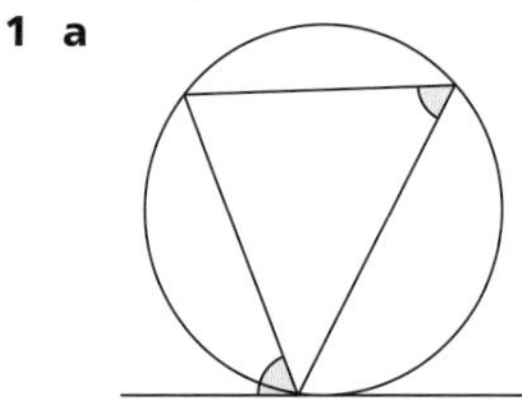

b

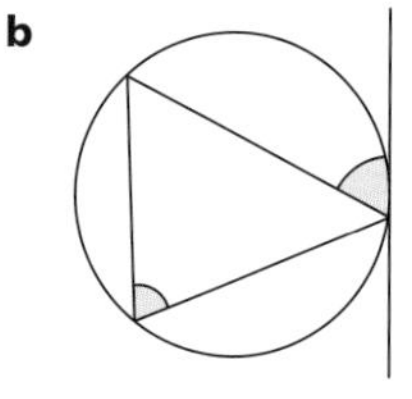

2 a $a = 65°$ **b** $b = 72°$ **c** $c = 32°$
3 a $x = 64°, y = 48°$ **b** $x = 81°, y = 24°$
4 a 48° **b** 57° **c** 75°
5 a $x = 52°$. Angles in alternate segments are equal and the base angles in an isosceles triangle are equal.
b $y = 52°$. Angles in alternate segments are equal.
6 64°. The angle at the centre is twice the angle at the circumference and angles in alternate segments are equal.
7 37°. Angles in alternate segments are equal, base angles in an isosceles triangle are equal and the angle between a radius and a tangent is 90°.
8 a 58°. Angles in alternate segments are equal, the angle in a semicircle is 90° and angles in a triangle sum to 180°.
b 58°. Angles in alternate segments are equal.

What do you think?

1 BCD could be isosceles or equilateral.

Consolidate

1 50°. The angle between a chord midpoint and a radius is 90° and angles in a triangle sum to 180°.
2 55°. The angle between a chord midpoint and a radius is 90°, and angles in a triangle sum to 180°.

3 **a** $a = 49°$ **b** $b = 22°$ **c** $c = 72°$

4 18 m. Tangents that meet at a point are equal in length.

5 5 cm

6 **a** $a = 35°$ **b** $b = 84°$
c $c = 28°$ **d** $d = 60°$, $e = 50°$
e $f = 23°$, $g = 42°$ **f** $h = 74°$, $i = 38°$

Stretch

1 59°. Opposite angles in a cyclic quadrilateral sum to 180°, angles in a triangle sum to 180°, alternate angles are equal and angles in alternate segments are equal.

2 $x = 11$, $y = 10$

3 $x = \frac{180}{17}$

4 $2x$

5 Let $\angle STQ = x$
$\angle SOQ = 2x$
$\angle OSQ = \angle OQS = 90° - x$
$\angle PQS = 90° - (90° - x) = x$

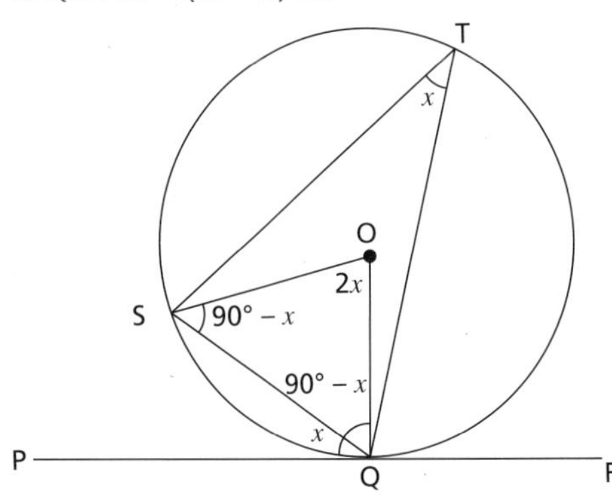

Chapter 3.4

Are you ready? (A)

1 **a** 14.8 m **b** 79.8 mm

2 **a** 11π cm **b** 8π m

Practice (A)

1 **a** 7.9 cm **b** 8.4 m **c** 20.9 mm **d** 31.4 km

2 **a** 3.14 cm **b** 43.7 m **c** 16.3 km **d** 160 cm

3 **a** 2π cm **b** 6π m **c** 6π km **d** 5π mm

4 **a** 13.4 cm **b** 29.9 cm **c** 65.1 cm

5 **a** 2.98 cm **b** 20.98 cm

6 60.0 m

7 $(6\pi + 36)$ cm

8 9.5 cm

9 **a** 7.6 cm **b** 27.5 m **c** 3.6 km

10 45 cm

11 84°

12 199°

13 11.4 cm

Are you ready? (B)

1 **a** 113 cm² **b** 272 m² **c** 804 km²

2 **a** 16π cm² **b** 225π cm² **c** 64π cm²

Practice (B)

1 25.1 cm²

2 **a** 50.3 cm² **b** 37.7 cm² **c** 16.4 cm²

3 **a** 35.9 cm² **b** 123 cm² **c** 521 cm²

4 **a** 16π m² **b** 3π cm² **c** 96π cm²

5 75π cm²

6 9.8 cm

7 **a** 21.4 cm **b** 8.29 cm **c** 14.0 mm

8 140°

9 **a** 64° **b** 43° **c** 279°

What do you think?

1 For a sector with a radius of 3 cm, the area is $\frac{9\theta\pi}{360}$
Doubling the radius would give an area of $\frac{36\theta\pi}{360}$
$\frac{36\theta\pi}{360}$ does not equal $2 \times \frac{9\theta\pi}{360}$

2 17.4 cm. Changes to 24.6 cm if the area is doubled. Discuss as a class.

Consolidate

1 **a** 4.71 cm **b** 23.0 cm **c** 5.01 m **d** 17.7 cm
e 4.52 cm **f** 45.4 cm **g** 37.3 cm **h** 354 m

2 **a** 33.5 cm² **b** 31.8 m² **c** 1010 cm²
d 68.4 cm² **e** 665 cm² **f** 607 m²

Stretch

1 123°

2 **a** 19.3 cm **b** 5.08 cm²

3 $\frac{\pi r^2}{5} = 9$
$\pi r^2 = 45$
$r = \sqrt{\frac{45}{\pi}} = \frac{3\sqrt{5}}{\sqrt{\pi}}$ cm

4 8π cm²

5 $(100\pi - 200\sqrt{2})$ cm²

Angles and circles: exam practice

1 144°

2 120°

3 40

4 80

5 Example answer:
ACB = 35° (alternate segment theorem)
So ACD = 145° (angles on a straight line add up to 180°)

Block 4 Transformations

Chapter 4.1

Are you ready? (A)

1 **a**

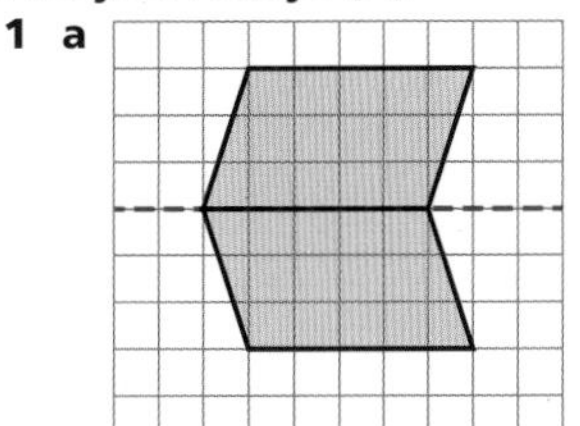

b

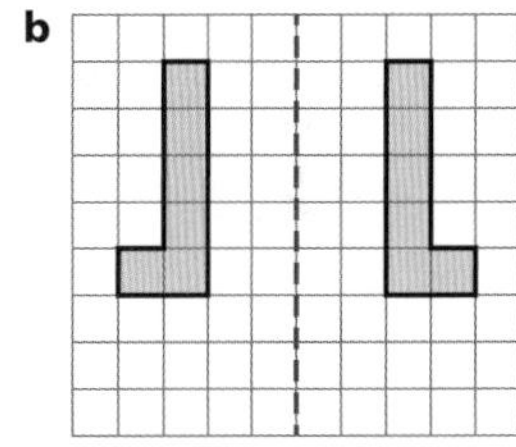

2 A $y = 5$ B $x = -6$ C $x = 6$ D $y = -3$ E $y = x$

Practice (A)

1

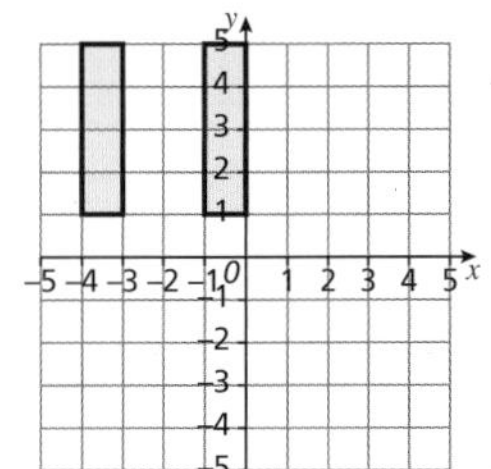

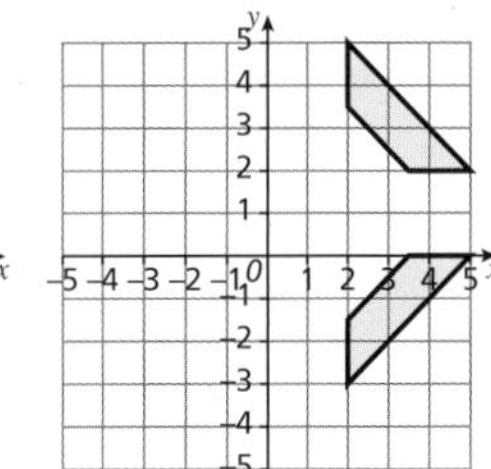

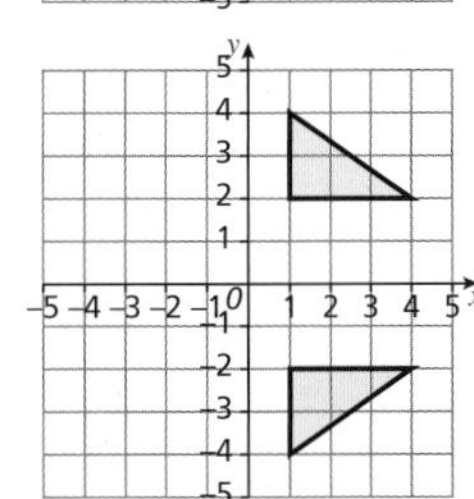

2 a $x = 1$ **b** $x = 0$ **c** $y = 0.5$

3 a

b

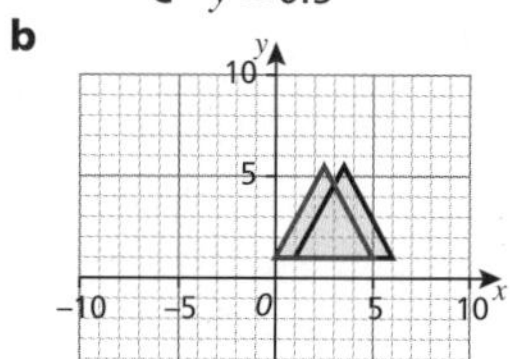

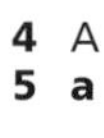

4 A

5 a

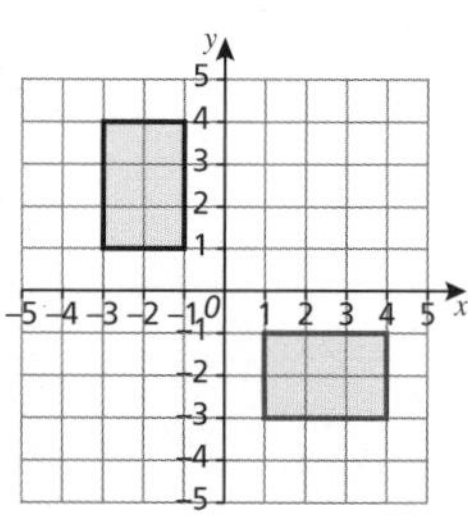

b

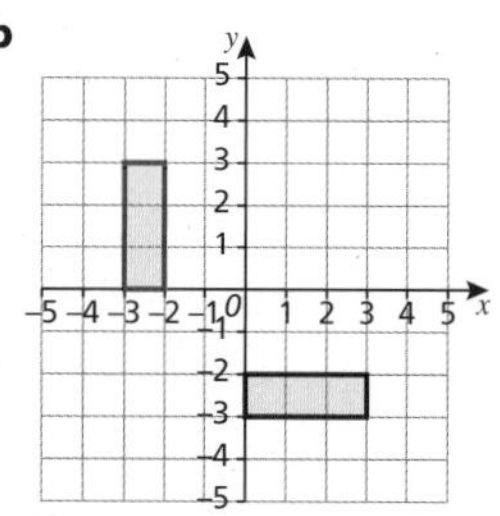

c

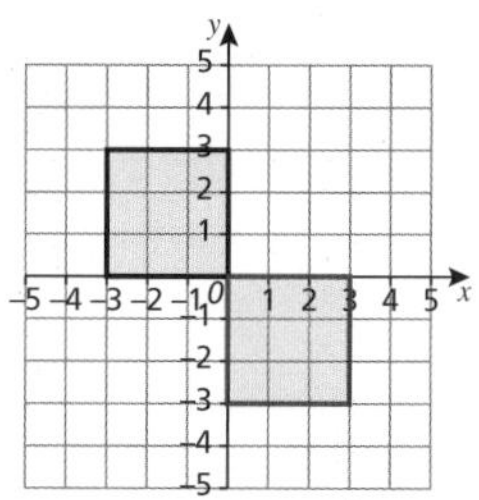

d

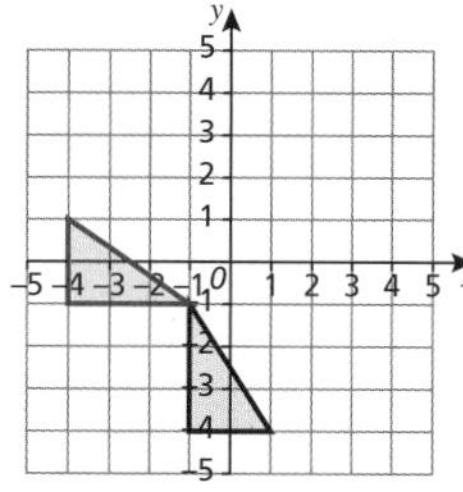

6 a

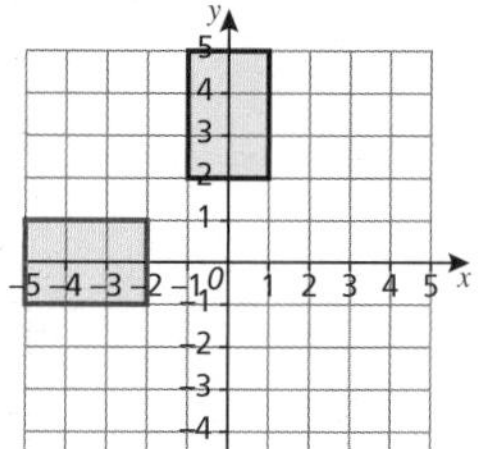

b

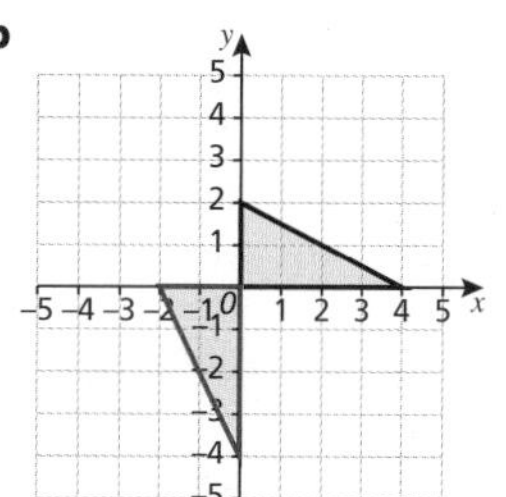

c

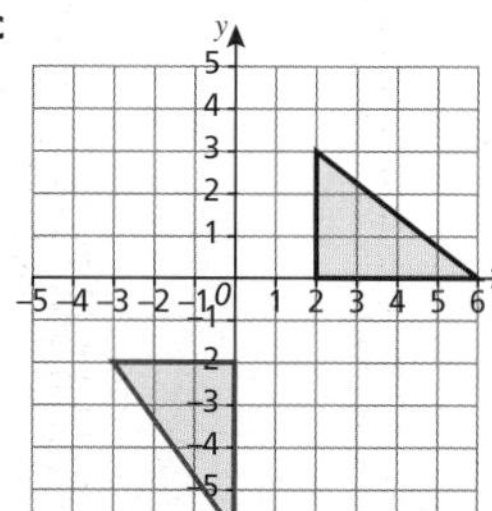

d

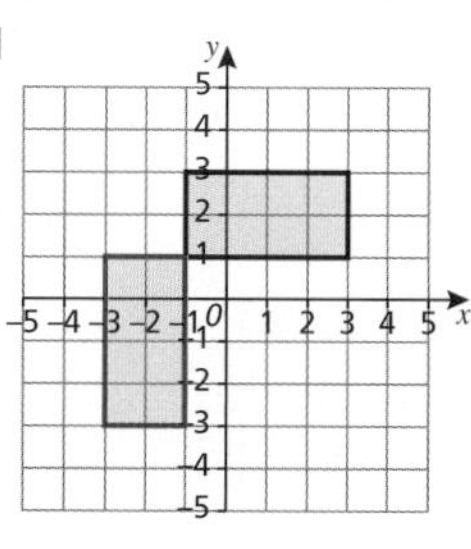

7

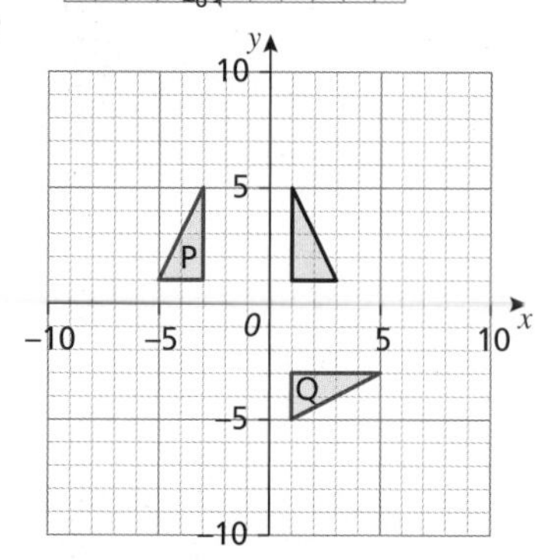

What do you think? (A)

1 $x = 0$, $y = 0$, $y = x$, $y = -x$
A circle with centre (0, 0) or a regular octagon would achieve the same results.

2 The x and y coordinates are always the same, e.g. (1, 1), (3, 3), as the midpoint will always lie on the line $x = y$.

Are you ready? (B)

1 a 1 **b** 4 **c** 2

2 a i

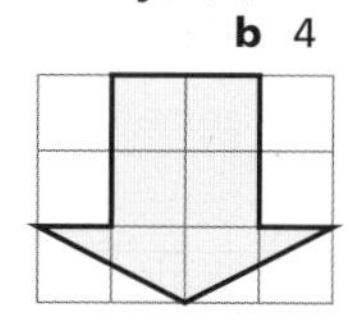

b i

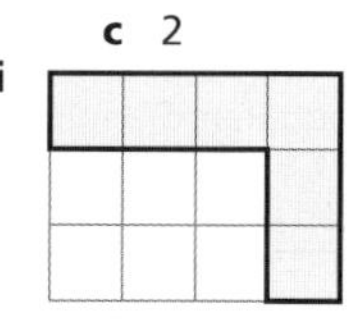

ii

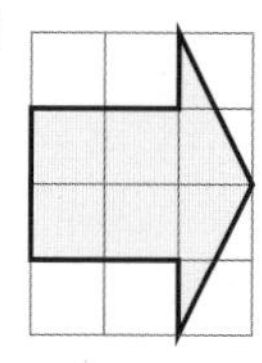

ii

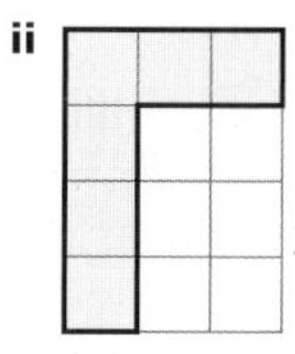

c i

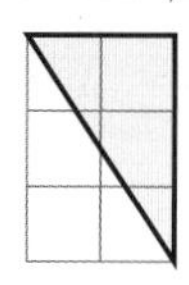

d i

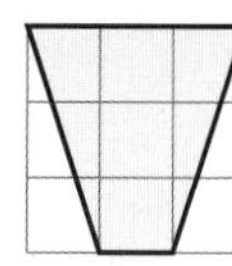

ii

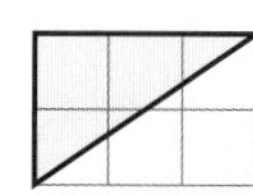

ii

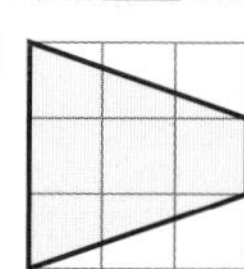

Practice (B)

1 a

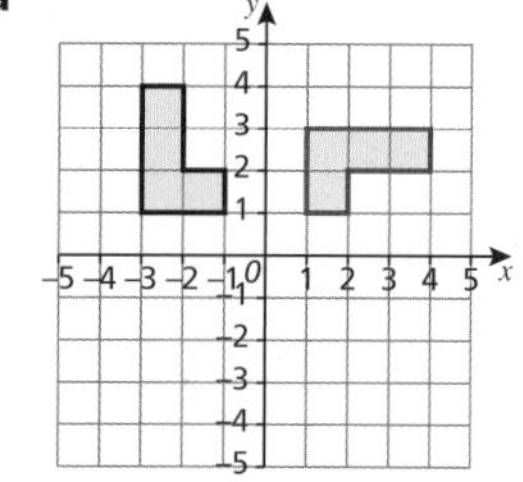

b

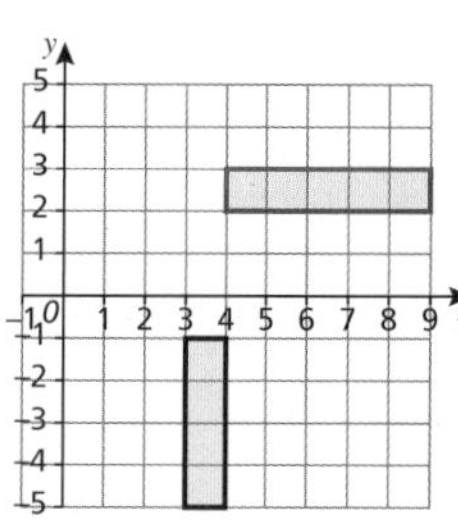

c

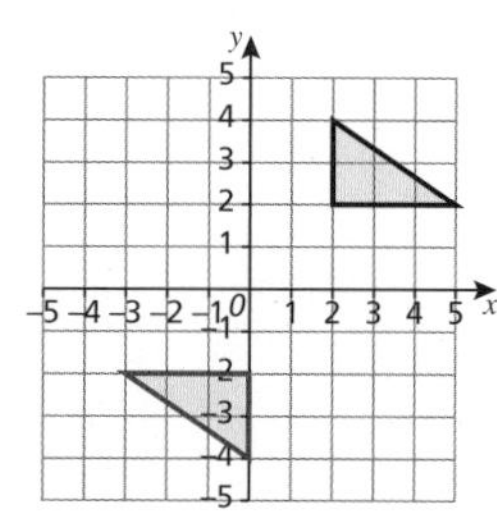

2

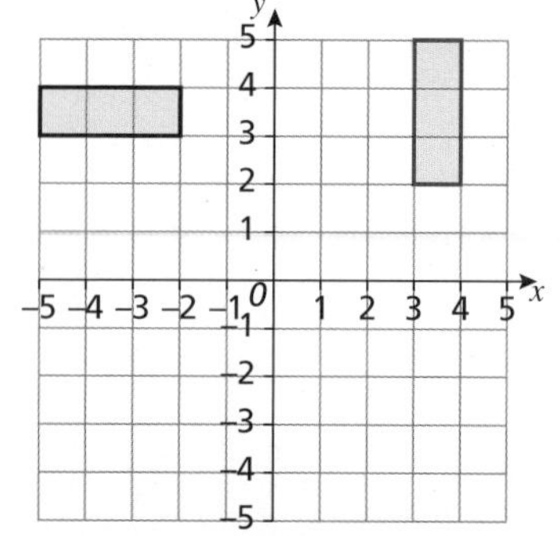

3 a

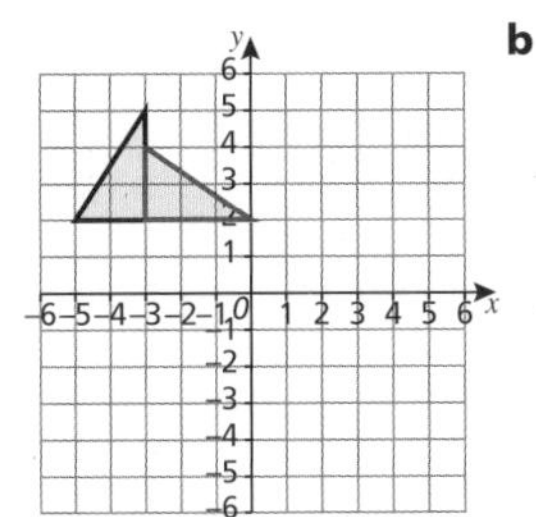

b

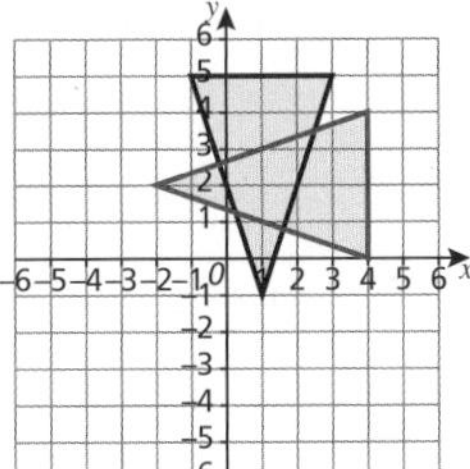

c

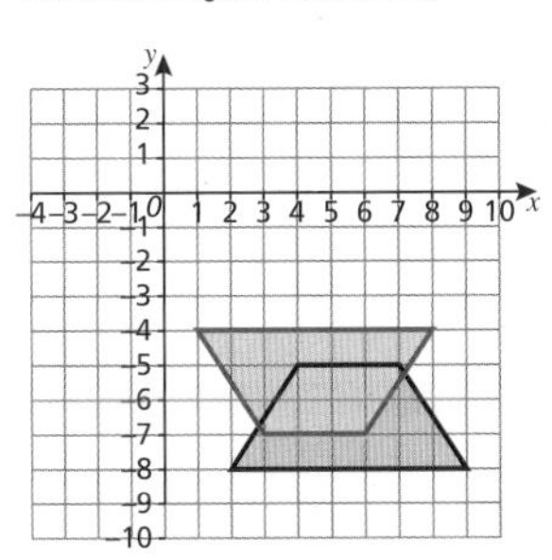

4 B

5 a Rotation 90° anticlockwise about (0, –2) or 90° clockwise about (7.5, –0.5)

b Rotation 180° about (–6, –4)

6 Rotation 90° anticlockwise about (–1, 2.5) or 90° clockwise about (–1.5, 3).

What do you think? (B)

1 Sometimes. This is true if the rotation is 180°.

2 Discuss as a class.

3 Example answers: 180°, 900°

4 Example answers: when points are on the mirror line; when the centre of rotation is a point on the shape

Consolidate

1 a

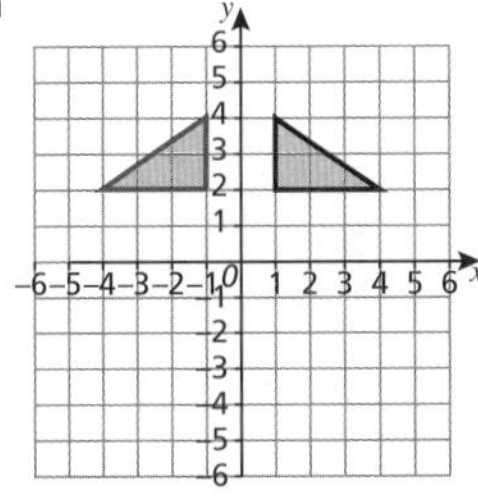

b

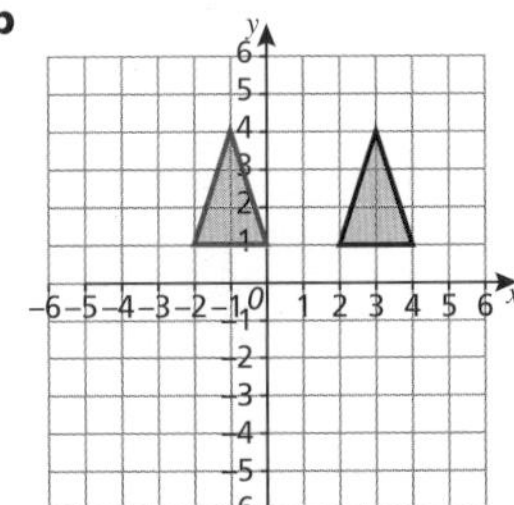

c

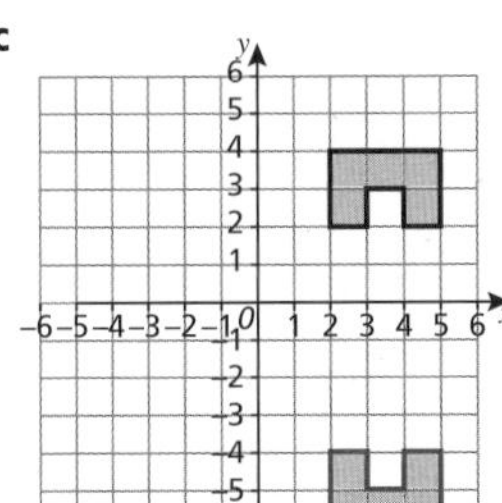

d

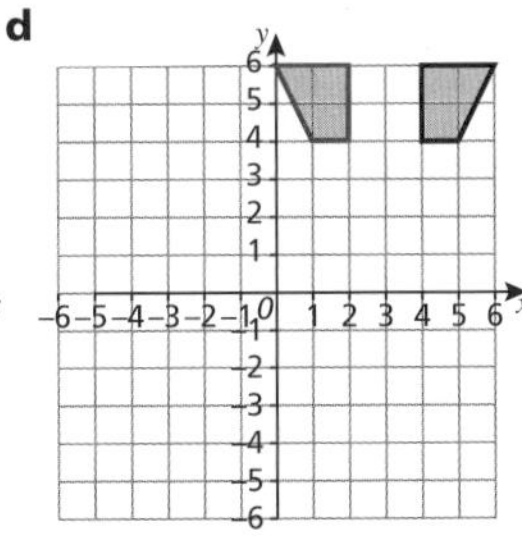

2 a

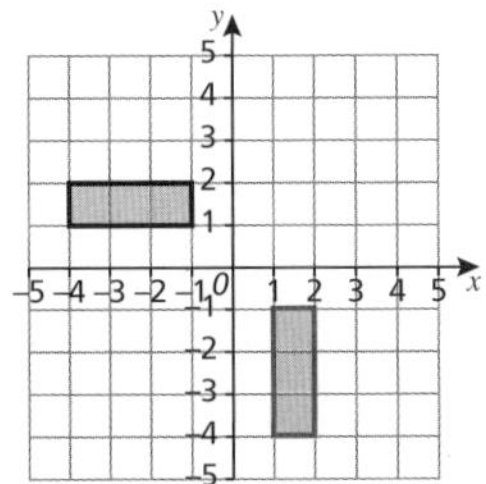

b

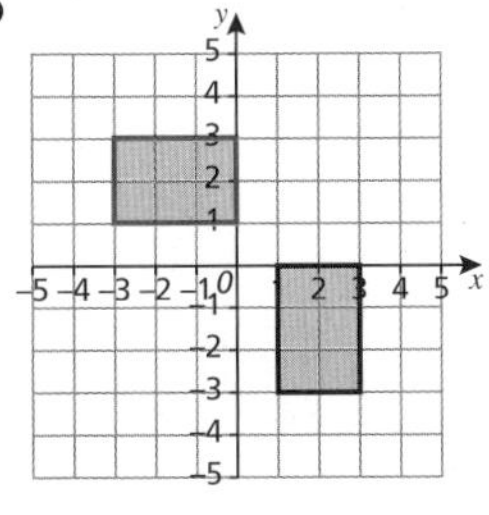

c

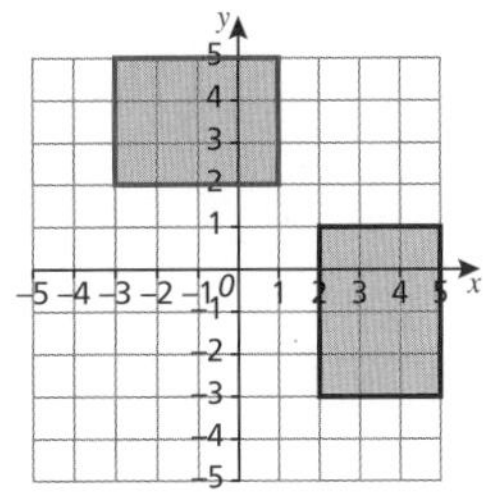

3 a Reflection in the line $y = 1$

b Reflection in the line $x = 2$

c Reflection in the line $y = x$

4 a **b**

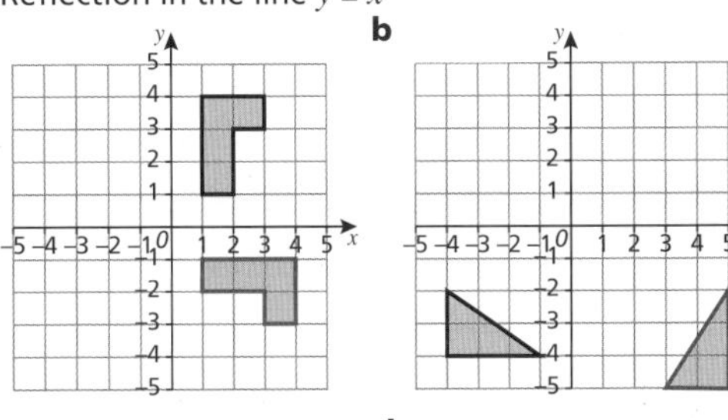

c

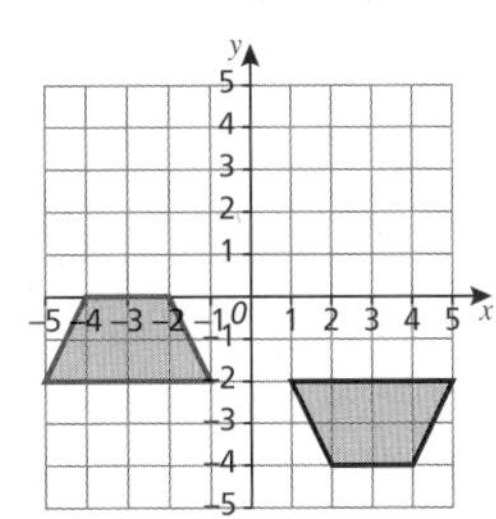

d

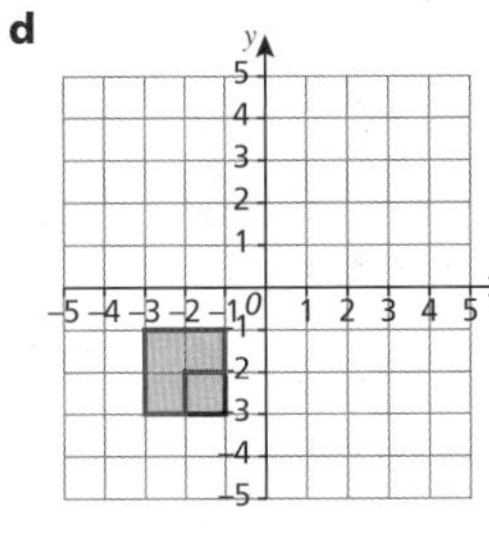

Stretch

1 (–1, 3), (–1, 1), (2, 1) and (3, 3)

2 a Square **b** 16 square units

3 Discuss different solutions as a class, e.g. $m = 1$ and $n = 4$

4 $y = x + 1$

Chapter 4.2

Are you ready?

$\begin{pmatrix} -6 \\ 1 \end{pmatrix}$ joined to 6 left and 1 up

$\begin{pmatrix} 6 \\ 1 \end{pmatrix}$ joined to 6 right and 1 up

$\begin{pmatrix} -6 \\ -1 \end{pmatrix}$ joined to 6 left and 1 down

$\begin{pmatrix} 6 \\ -1 \end{pmatrix}$ joined to 6 right and 1 down

$\begin{pmatrix} -1 \\ 6 \end{pmatrix}$ joined to 1 left and 6 up

$\begin{pmatrix} 1 \\ 6 \end{pmatrix}$ joined to 1 right and 6 up

Practice

1 a

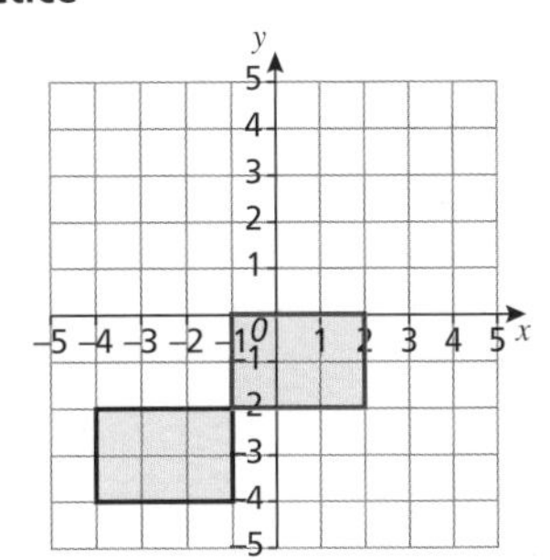

b

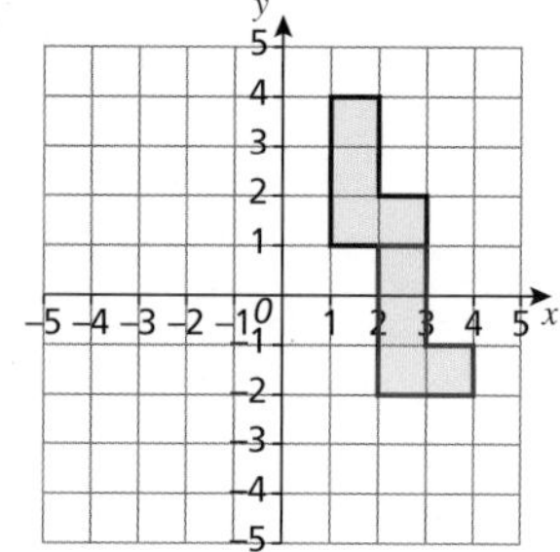

c

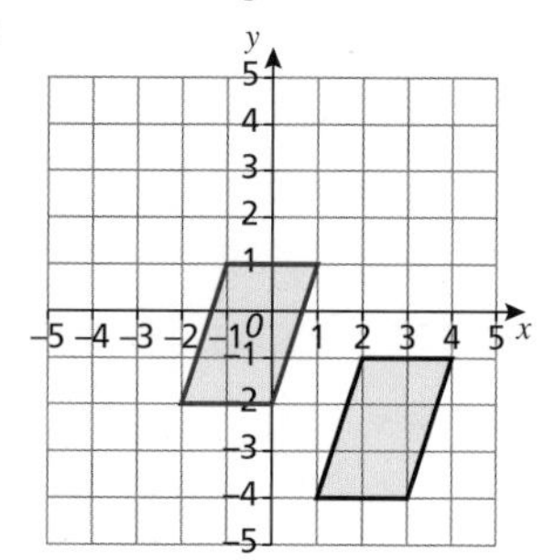

2 a Translation by vector $\begin{pmatrix} 5 \\ 3 \end{pmatrix}$

b Translation by vector $\begin{pmatrix} 6 \\ 0 \end{pmatrix}$

c Translation by vector $\begin{pmatrix} -5 \\ 3 \end{pmatrix}$

3 Translation by vector $\begin{pmatrix} 2 \\ -2 \end{pmatrix}$

4 Junaid has translated the shape up 5 units.
Tiff has written the translation of B to A.

5 B and F
A, E and G

What do you think?

1 Translation by vector $\begin{pmatrix} 9 \\ 0 \end{pmatrix}$

Reflection in the line $x = 1$
Rotation of 180° about (1, 3.5)
Enlargement, scale factor –1, centre (1, 3.5)

Consolidate

1 a

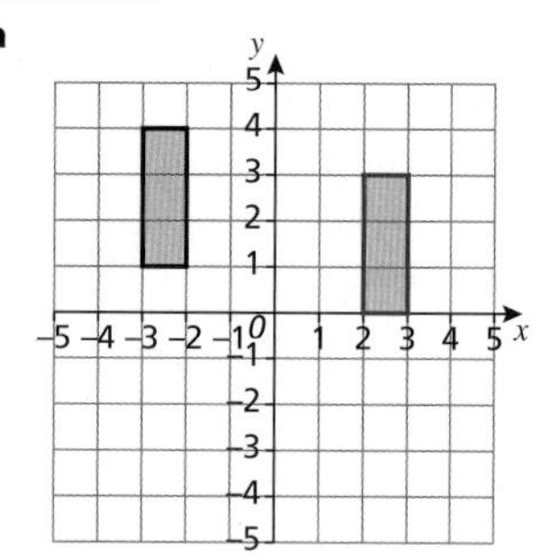

b

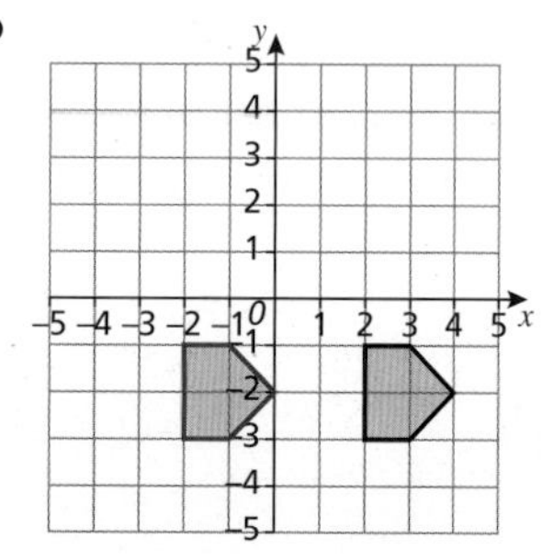

c

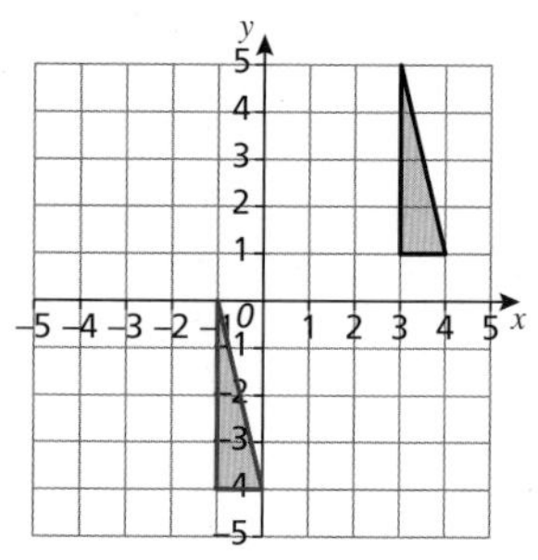

2 a Translation by vector $\begin{pmatrix} -6 \\ 1 \end{pmatrix}$

b Translation by vector $\begin{pmatrix} 6 \\ 6 \end{pmatrix}$

c Translation by vector $\begin{pmatrix} 0 \\ -6 \end{pmatrix}$

Stretch

1 (8, –2), (19, 3) and (30, –2)

2 Ed is correct. $\begin{pmatrix} 3 \\ 7 \end{pmatrix} + \begin{pmatrix} -6 \\ 12 \end{pmatrix} + \begin{pmatrix} 2 \\ 19 \end{pmatrix} = \begin{pmatrix} -1 \\ 0 \end{pmatrix}$

3 $p = +\sqrt{128}$ and $p = -\sqrt{128}$

4 $a = -7$ and $b = -4$

5 Compare answers as a class, discussing whether you think $\begin{pmatrix} 0 \\ 0 \end{pmatrix}$ counts as a translation or not.

Chapter 4.3

Are you ready?

1 $a = 6$ cm $\quad b = 10.5$ cm $\quad c = 10$ cm

2 a

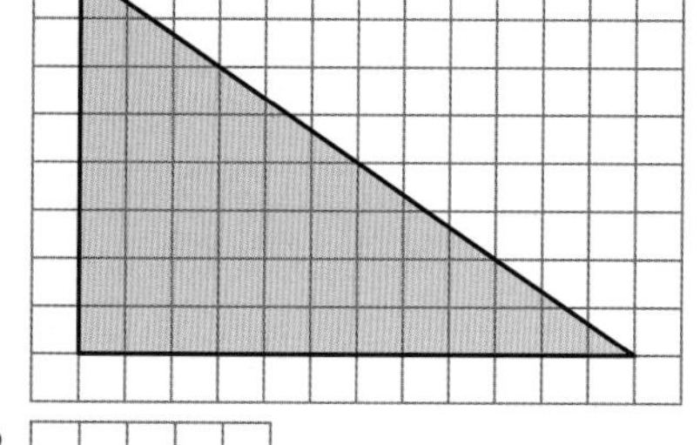

b

Practice

1 a

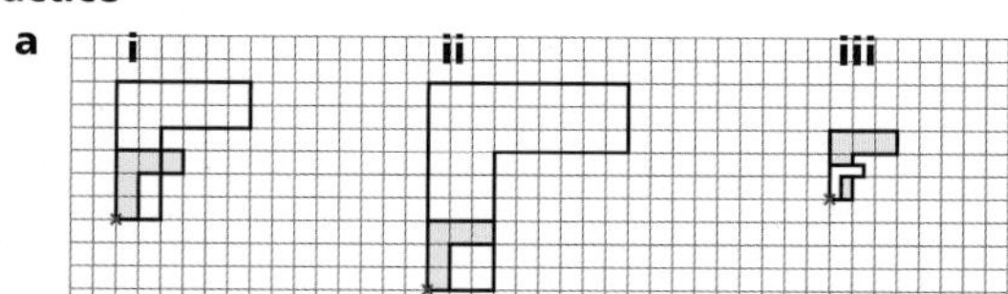

b

c

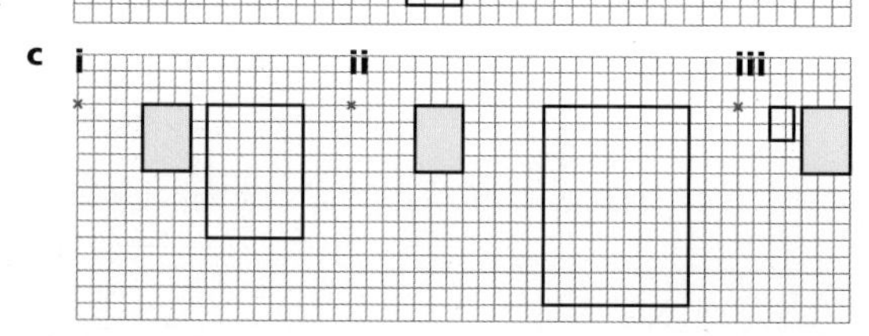

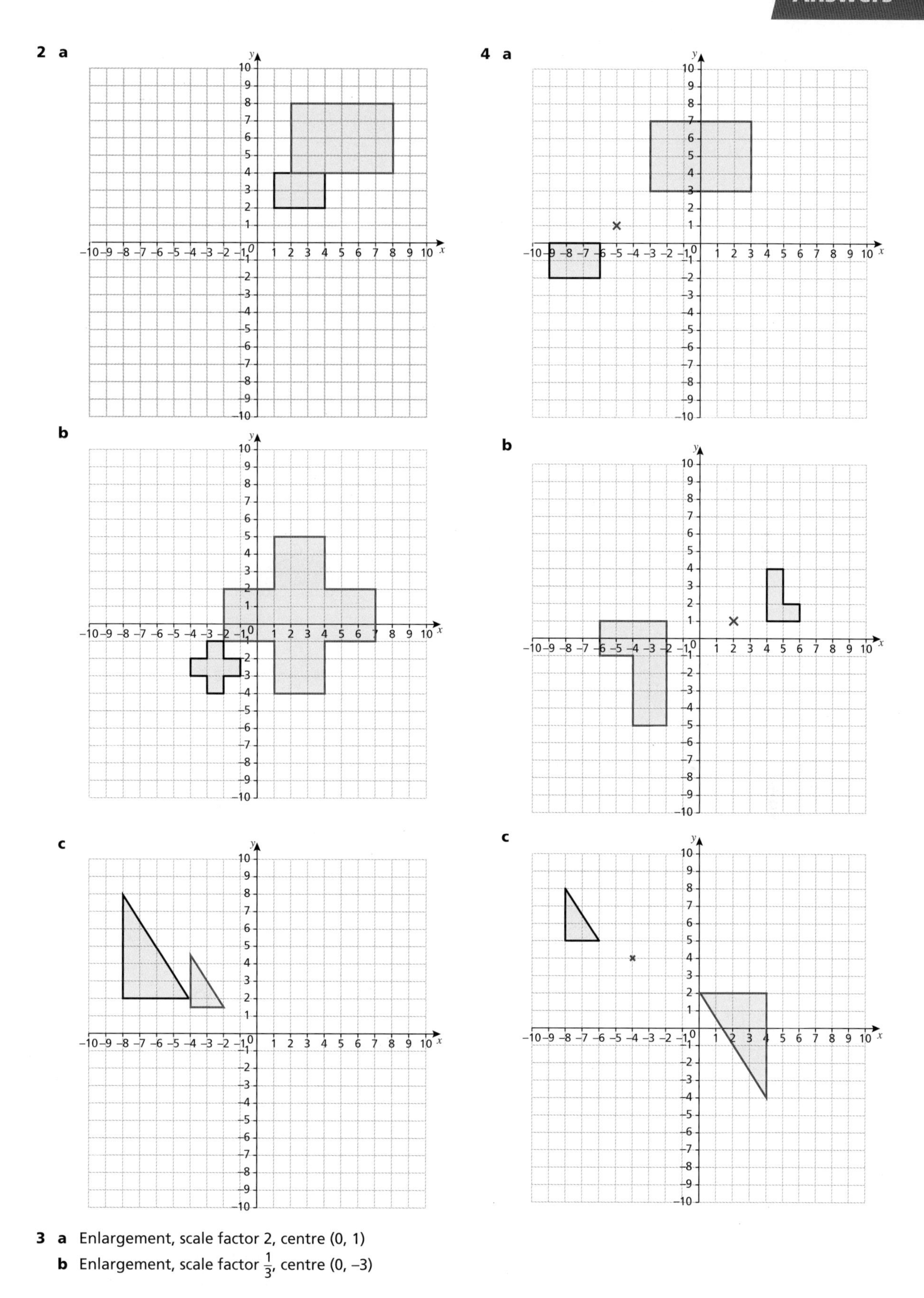

3 a Enlargement, scale factor 2, centre (0, 1)

b Enlargement, scale factor $\frac{1}{3}$, centre (0, –3)

d

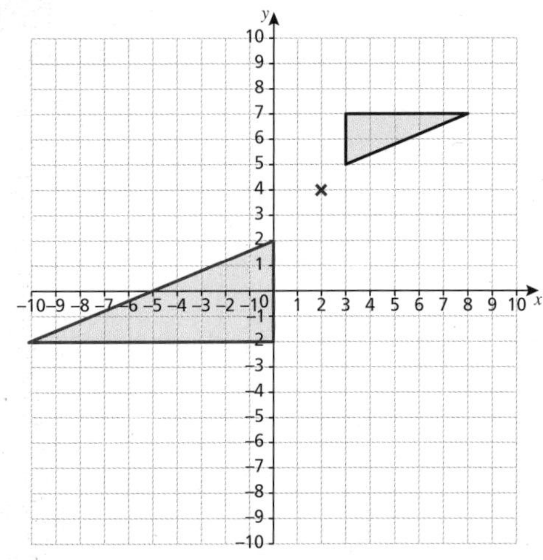

c

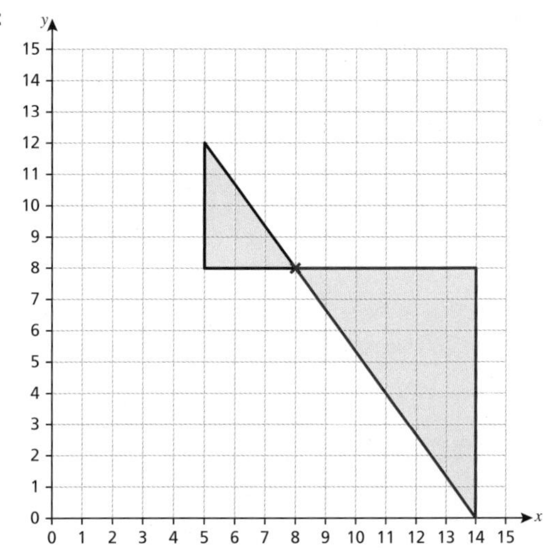

5 a

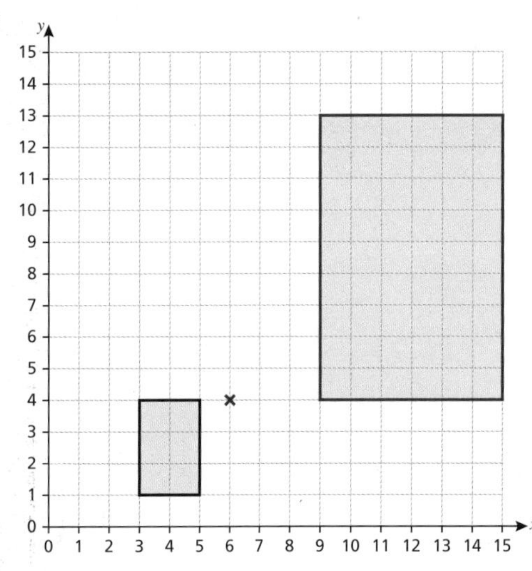

d

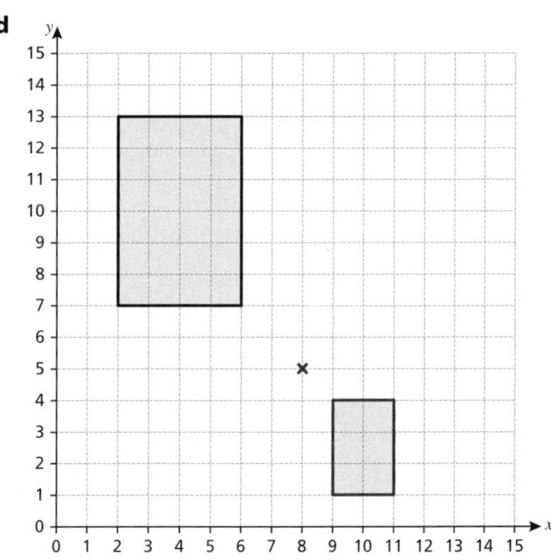

b

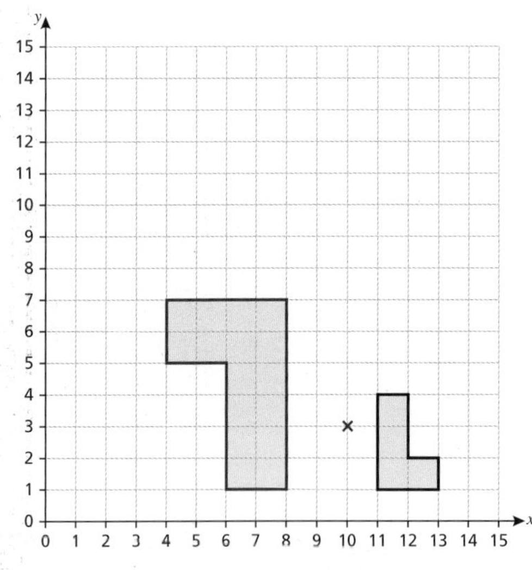

6 a

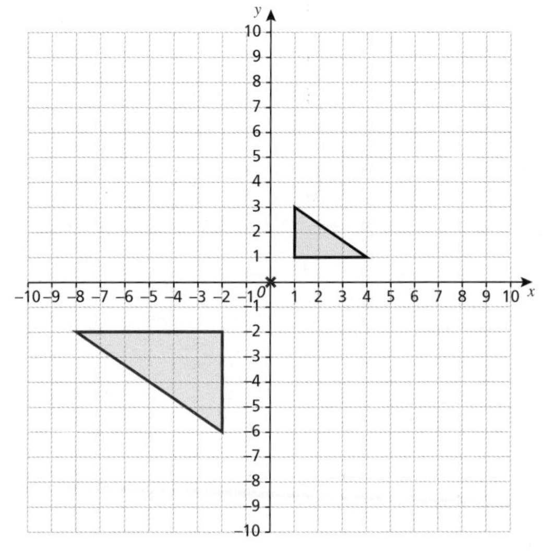

b

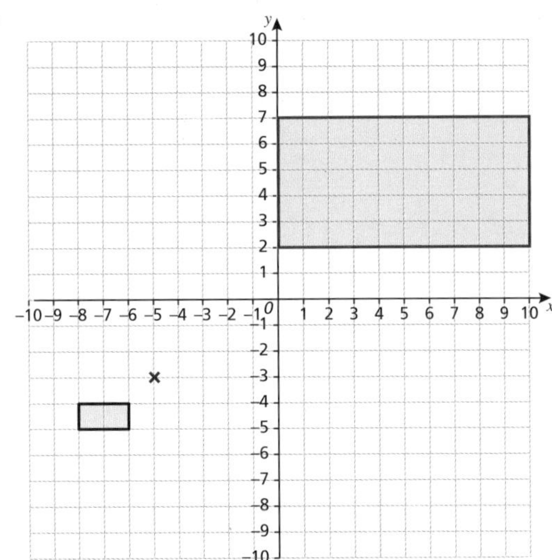

c

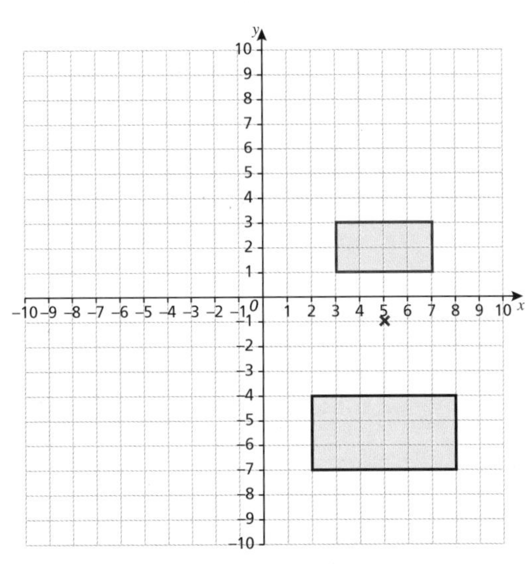

d

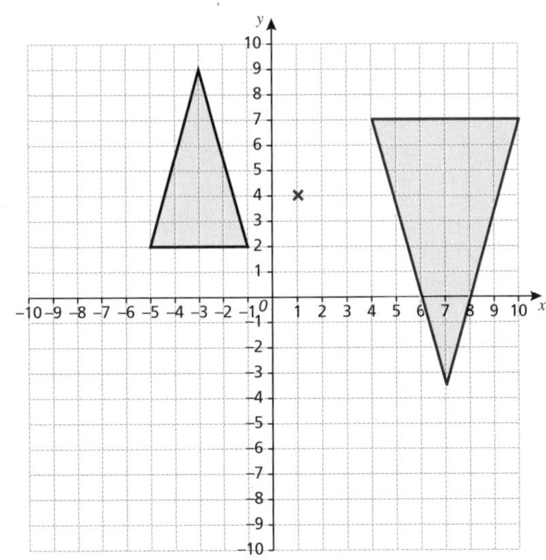

What do you think?

1 Ed is correct.
Discuss explanations as a class.

Consolidate

1 a

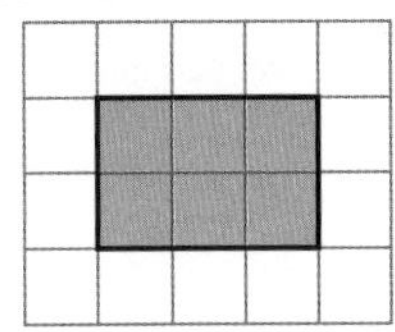

b

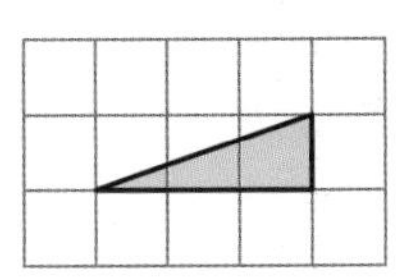

c

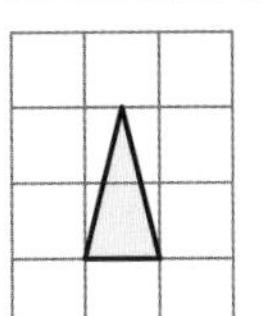

d

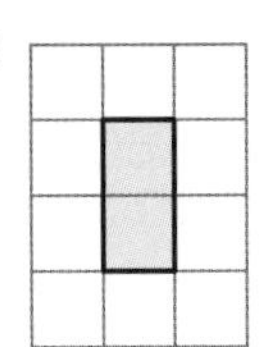

2 a Enlargement, scale factor 2

b Enlargement, scale factor $\frac{1}{2}$

3 a

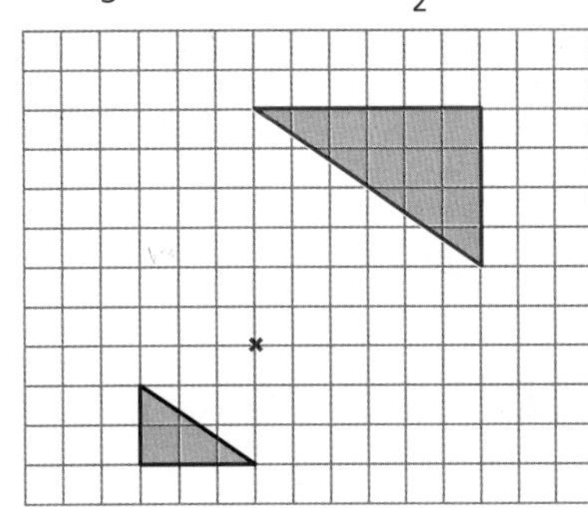

b

Stretch

1 Enlargement, scale factor –4, centre (0, 0)

2 The area of the new shape will be 180 cm^2.

3 Eva is incorrect, as the enlargement of Q would be a different orientation to P. Enlarging Q by a scale factor of $-\frac{2}{3}$ would achieve this result.

4 The statement is true.

Chapter 4.4

Are you ready?

1 a

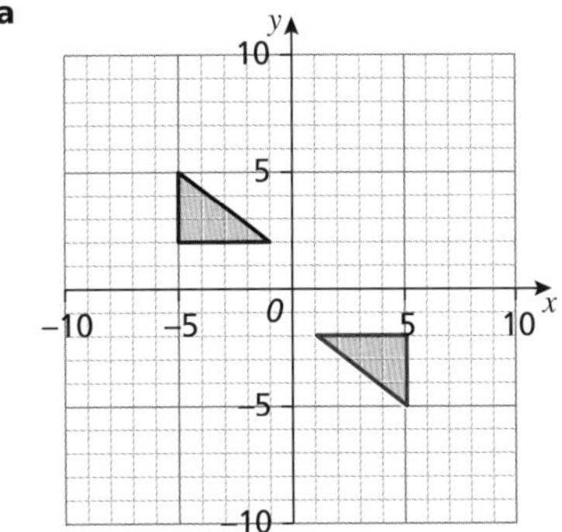

b

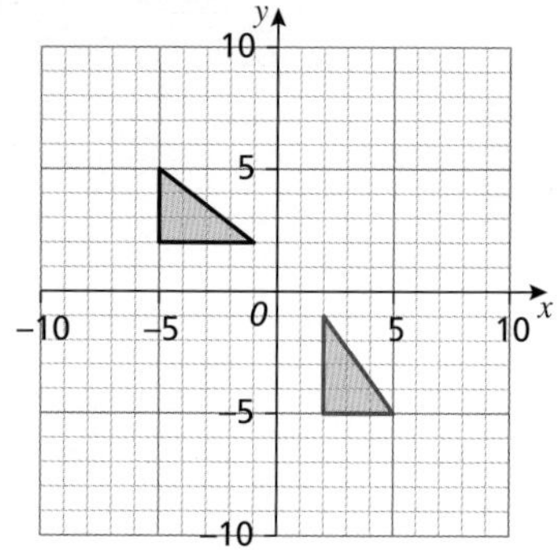

c

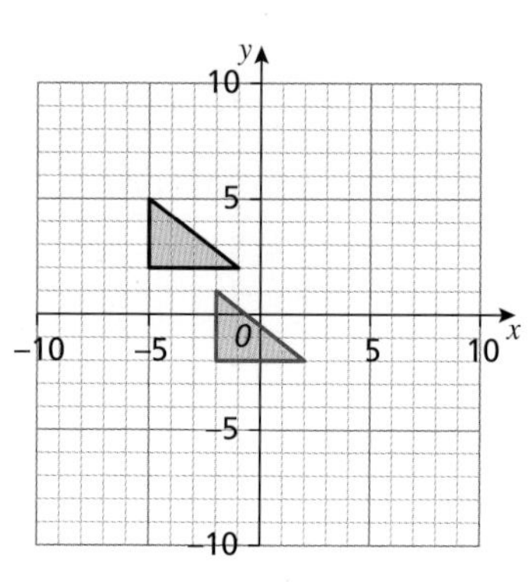

d

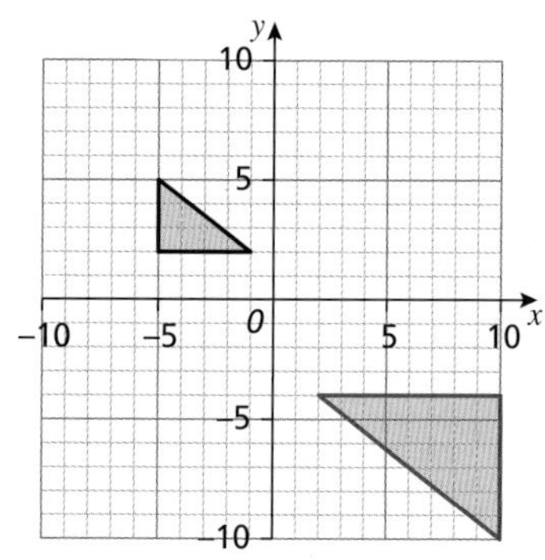

Practice

1 a b

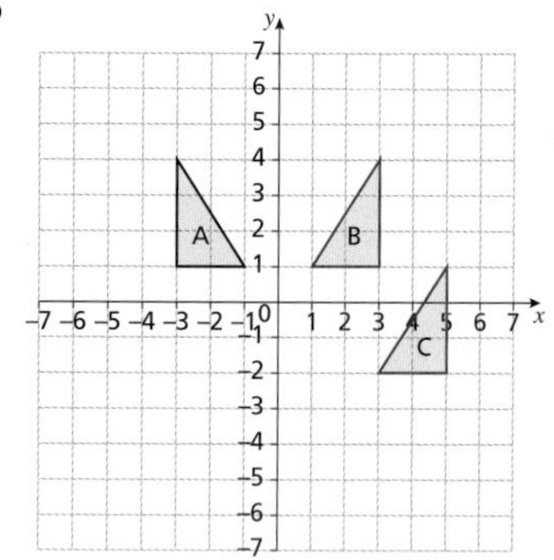

2 a b

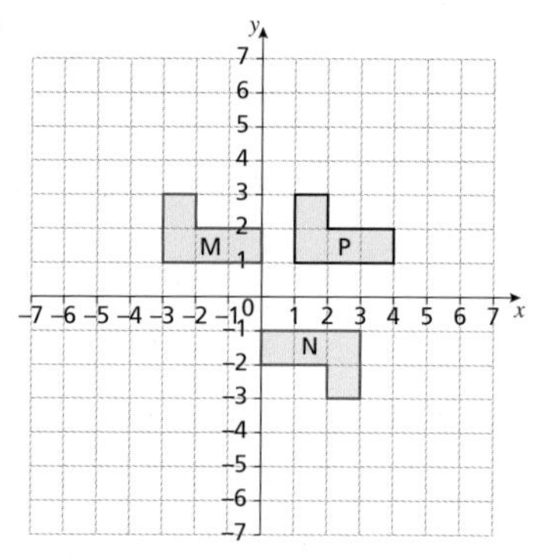

3 a b

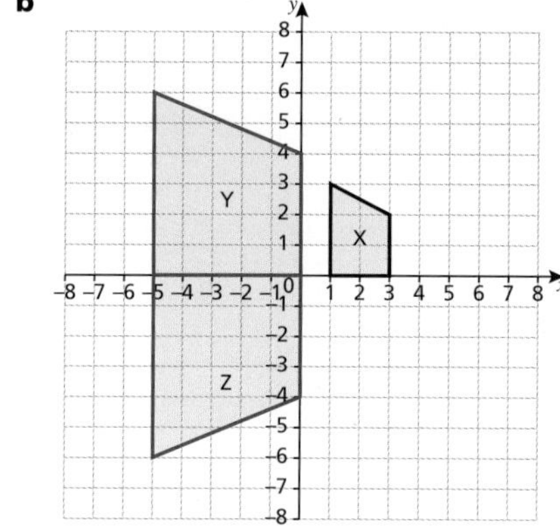

4 a b

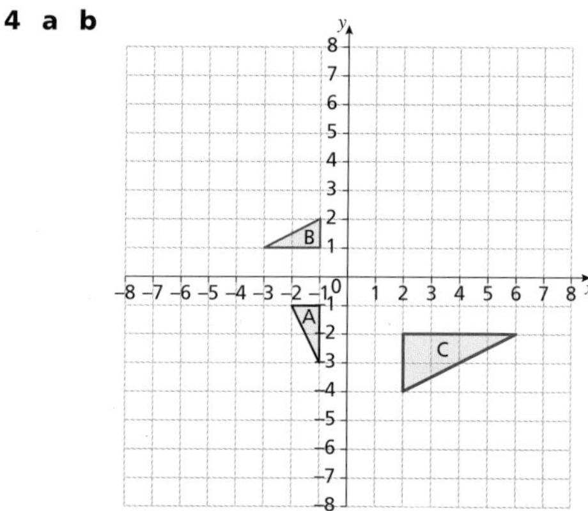

5 Rotation, 180°, centre (0, 0)
6 Rotation, 180°, centre (0, –2)
7 Rotation, 180°, centre (3, –1)
8 Rotation, 90° clockwise, centre (–2, 0)
9 $p = 4$ and $q = 2$

What do you think?

1 The combination of transformations maps the original shape onto itself.
2 Huda is incorrect. Discuss answers as a class.

3 A translation by the vector $\begin{pmatrix}-2\\0\end{pmatrix}$. $\begin{pmatrix}2\\-5\end{pmatrix} + \begin{pmatrix}-4\\5\end{pmatrix}$

$= \begin{pmatrix}-2\\0\end{pmatrix}$. In general, a translation by the vector $\begin{pmatrix}a\\b\end{pmatrix}$

followed by a translation by the vector $\begin{pmatrix}c\\d\end{pmatrix}$ is

equivalent to a translation by the vector $\begin{pmatrix}a+c\\b+d\end{pmatrix}$.

Consolidate

1 a b

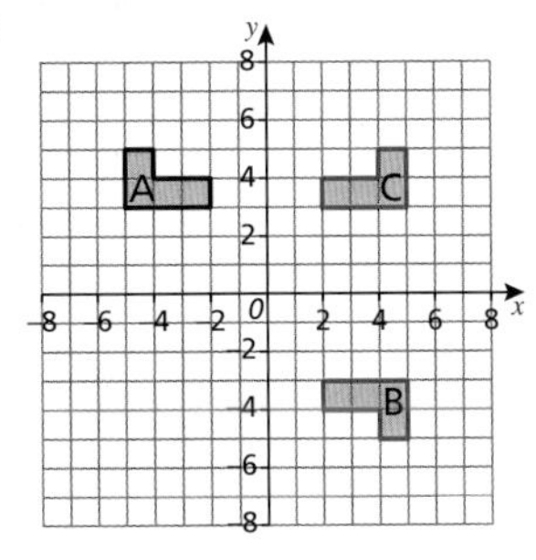

2 a b

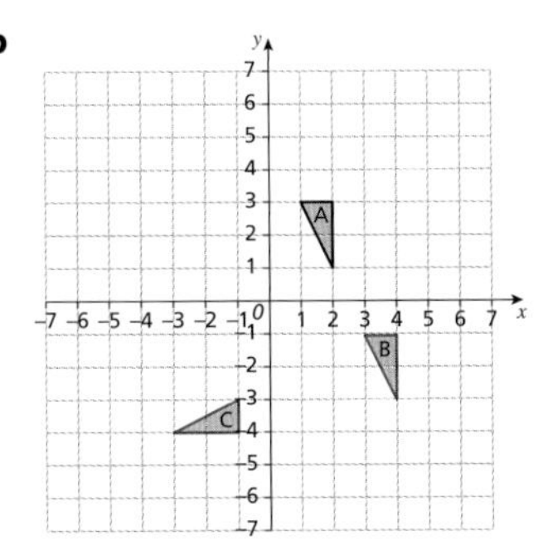

3 Reflection in the line $x = -2$
4 Enlargement, scale factor –2 about (4, 6)
5 Rotation, 90° clockwise about (0, 0)

Stretch

1 a Translation by vector $\begin{pmatrix}-5\\1\end{pmatrix}$

b Reflection in the line $x = 2$, then rotation 180° about $(-\frac{1}{2}, \frac{1}{2})$.
Discuss other solutions as a class.

2

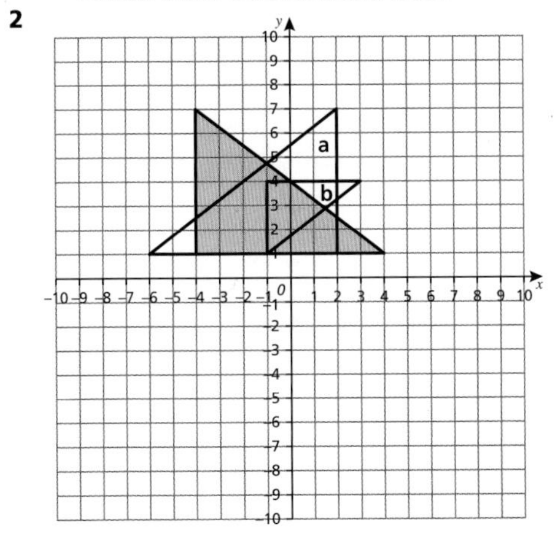

Transformations: exam practice

1

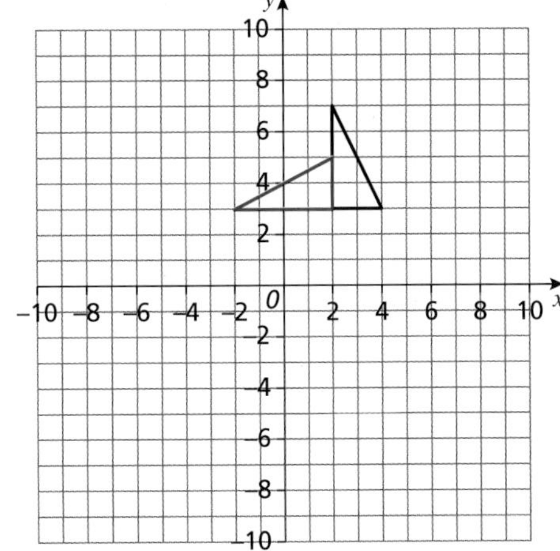

2

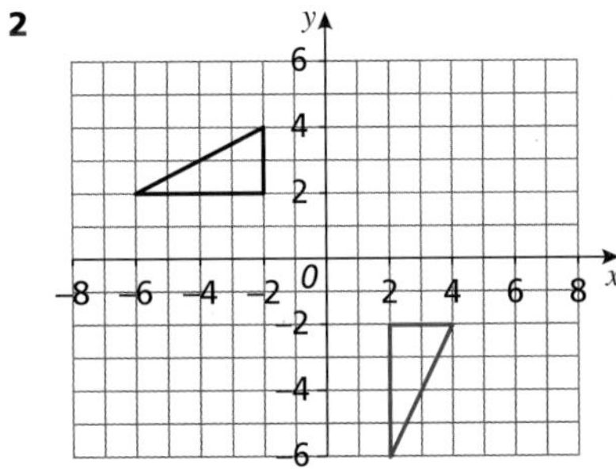

3 Enlargement, scale factor –2 and centre (4, 3)

4

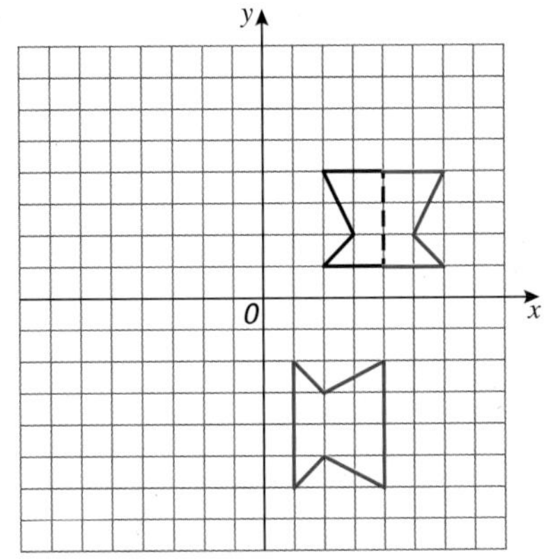

5 Enlargement
6 (2, 0)

Block 5 2D and 3D shapes

Chapter 5.1

Are you ready?

1 a i 6 **ii** 12 **iii** 8
b i 7 **ii** 15 **iii** 10
c i 5 **ii** 9 **iii** 6

2 a **b** **c**

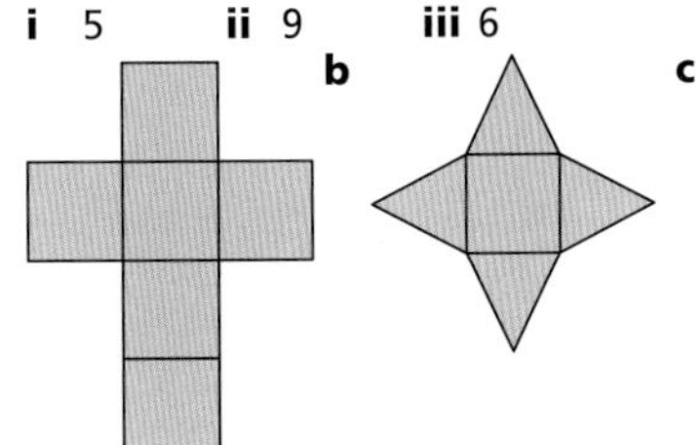

Practice

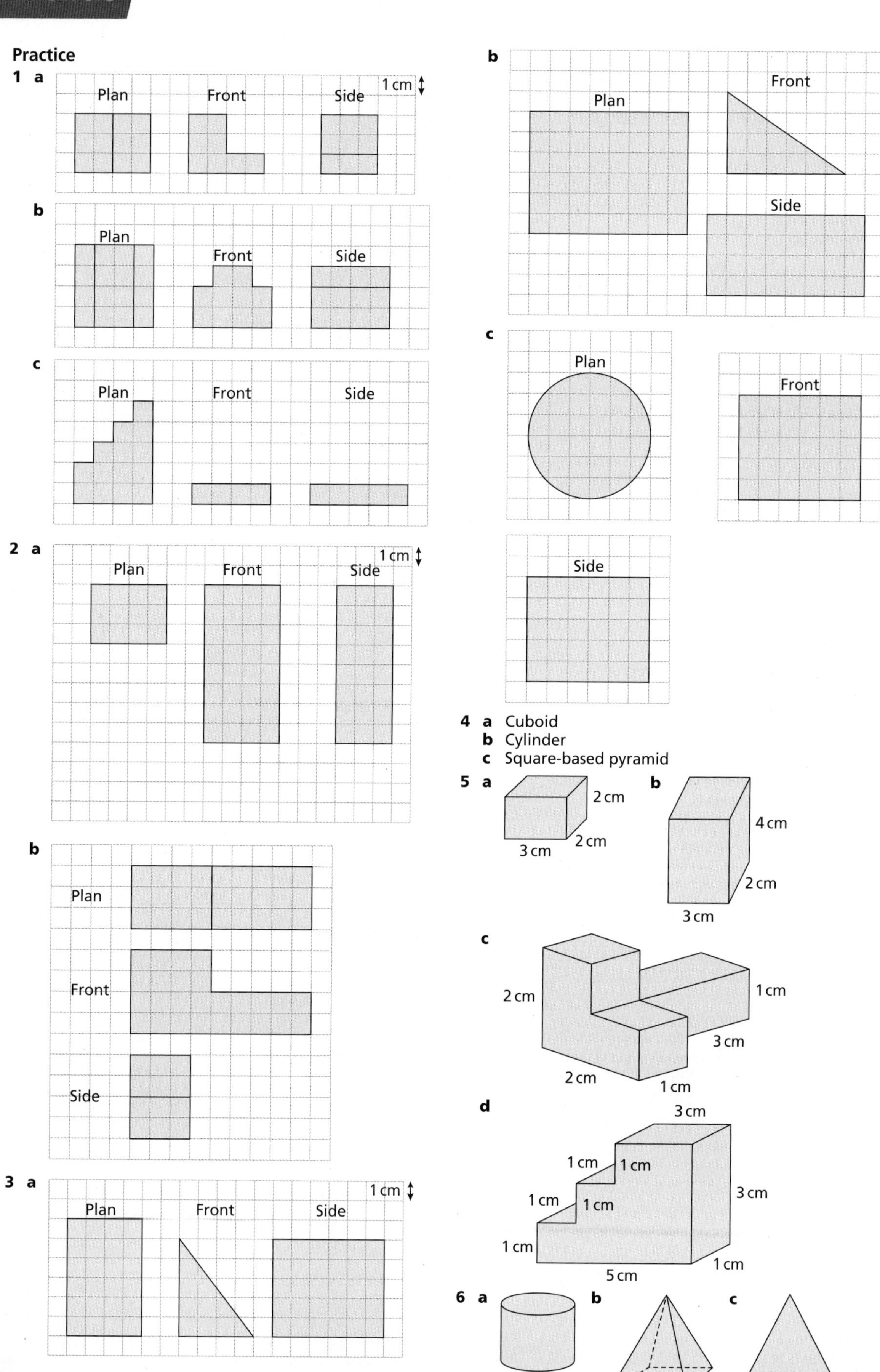
1 a
1 cm
Plan
Front
Side
b
Plan
Front
Side
c
Plan
Front
Side
2 a
1 cm
Plan
Front
Side
b
Plan
Front
Side
3 a
1 cm
Plan
Front
Side
b
Front
Plan
Side
c
Plan
Front
Side
4 a Cuboid
b Cylinder
c Square-based pyramid
5 a
2 cm
2 cm
3 cm
b
4 cm
2 cm
3 cm
c
2 cm
1 cm
3 cm
2 cm
1 cm
d
3 cm
1 cm
1 cm
1 cm
1 cm
3 cm
1 cm
5 cm
1 cm
6 a
b
c

What do you think?

1 Example answer:

Plan Front Side

Discuss other possibilities as a class.

2

1 cm

Plan Front Side

Consolidate

1

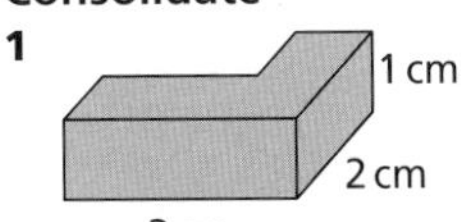

2

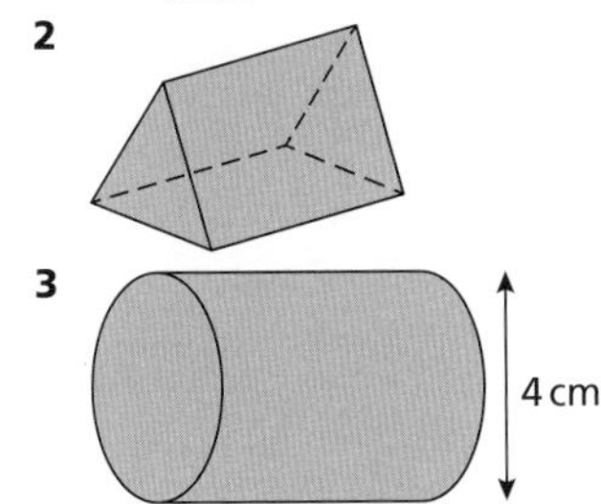

3

4 cm

4 cm

4 a

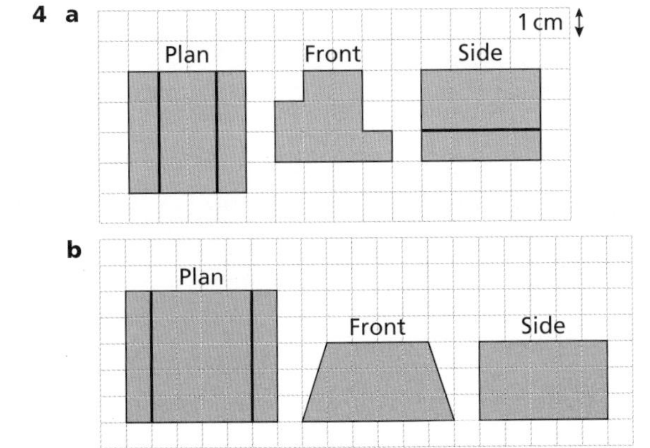

b

Plan Front Side

Stretch

1 3.6 m

2 a 18 cm³ b 12π cm³

Chapter 5.2

Are you ready? (A)

1 a Area = 19.5 cm², Perimeter = 19 cm
  b Area = 35 mm², Perimeter = 27 mm

2 a Area = 113.1 m², Circumference = 37.7 m
  b Area = 124.7 cm², Circumference = 39.6 cm

Practice (A)

1 a Area = 42 cm², Perimeter = 34 cm
  b Area = 71 cm², Perimeter = 40 cm
  c Area = 34 m², Perimeter = 30 m

2 145 cm²

3 52 cm

4 a 140 cm² b 202 m² c 622 m²

5 a 57 cm² b 136.5 cm² c 18 cm²

6 39.8 cm

7 a 86 m² b 12.02 cm²

8 a 119.3 cm² b 41.7 cm

9 a $\left(\frac{9}{4}\pi + 6\right)$ cm² b $\left(\frac{3}{2}\pi + 12\right)$ cm

What do you think? (A)

1 Example answer: A right-angled triangle with sides 5, 12 and 13

2 Huda is wrong. For example, a square with side length 0.5 cm, or a right-angled triangle with base 1.2 m, height 0.5 m and hypotenuse 1.3 m. Discuss examples as a class, and whether it makes any difference if you convert to a different unit of measure.

3 Investigate and discuss as a class.

Practice (B)

1 294 cm²

2 a 216 cm² b 118 m² c 249.94 m²

3 132 cm²

4 640 cm²

5 203 cm²

6 $x = 11$ m

7 1.32 m² so £2.40

8 $x = 4$ m

What do you think? (B)

1 Ed is wrong as two of the faces of the cubes are now hidden.

2 Compare findings as a class.

Consolidate

1 a Area = 34 cm², Perimeter = 34 cm
  b Area = 92 cm², Perimeter = 48 cm
  c Area = 47 cm², Perimeter = 34 cm

2 a 121 cm² b 49 cm² c 56 cm²

3 a 126 cm² b 270 cm² c 1006 cm²

Stretch

1 75 m²

2 $\frac{27\sqrt{3}}{2}$ cm²

3 $2x^2 + 3x = 20$, $x = \frac{5}{2}$

Chapter 5.3

Are you ready?

1 a 78 cm² b 30 cm² c 49 m²

2 212 cm²

Practice

1 a 160 cm³ b 270 cm³ c 8000 m³ d 291.6 cm³

2 a 64 cm³ b 125 mm³ c 704.969 m³

3 a 12 cm b 8 m

4 a 168 cm³ b 75 mm³ c 96 m³

5 140 cm³

6 a 520 mm³ b 2640 m³

7 9 cm

8 a $a = 10$ m b $b = 2$ cm

9 500

10 10 minutes

What do you think?

1 Example answer: Sketch of a cube with a side of 6 units
  Discuss other answers as a class.

2 Example answer: Sketch of a cuboid 5 × 5 × 8 units or a cuboid 10 × 10 × 2 units
  Discuss other answers as a class.

Consolidate

1 a 70 cm³ b 300 cm³ c 1367.3 m³

2 a 27 cm³ b 512 ft³ c 19.683 m³

3 a 120 cm³ b 288 mm³ c 240 cm³

Stretch

1 $1331\,cm^3$
2 4.3 cm
3 $\frac{(210\sqrt{3} + 63)}{2}\,cm^3$
4 $9860\,mm^3$

Chapter 5.4

Are you ready? (A)

1 a i $113.1\,cm^2$ ii 37.7 cm
b i $196.1\,m^2$ ii 49.6 m
c i $248.8\,mm^2$ ii 55.9 mm
2 a i $126\,cm^3$ ii $162\,cm^2$
b i $576\,cm^3$ ii $480\,cm^2$

Practice (A)

1 a $502.7\,cm^3$ b $471.2\,m^3$
c $628.3\,mm^3$ d $2889.3\,ft^3$
2 a $40\pi\,cm^3$ b $48\pi\,m^3$
c $200\pi\,mm^3$ d $1125\pi\,in^3$
3 2.5 cm
4 3.5 m
5 9.3 mm
6 6.5 cm
7 a $282.7\,cm^2$ b $301.6\,cm^2$
c $226.2\,cm^2$ d $1570.8\,cm^2$
8 a $48\pi\,cm^2$ b $120\pi\,m^2$ c $170\pi\,mm^2$
9 39.4 cm
10 a $(2\pi r^2 + 14\pi r)\,cm^2$ b 5 cm

What do you think? (A)

1 Huda is incorrect. You can fit 8 large tubes (total area of ends 32π) or 24 small tubes (area 24π). Smaller tubes have more wasted space between them.
2 a i $251.3\,cm^3$ ii $255.9\,cm^2$
b i $212\,cm^3$ ii $259.6\,cm^2$
3 $401.6\,cm^2$

Are you ready? (B)

1 a i $226.2\,cm^3$ ii $207.3\,cm^2$
b i $603.2\,cm^3$ ii $402.1\,cm^2$

Practice (B)

1 a $75.4\,cm^3$ b $167.6\,m^3$ c $392.7\,mm^3$
d $394.5\,cm^3$ e $523.6\,m^3$ f $742.2\,mm^3$
2 a $30\pi\,cm^3$ b $50\pi\,cm^3$ c $400\pi\,cm^3$ d $\frac{28}{3}\pi\,cm^3$
3 29.8 cm
4 12.6 cm
5 $811.6\,cm^3$
6 a $175.9\,cm^2$ b $361.3\,cm^2$ c $245.0\,cm^2$ d $479.9\,m^2$
7 a $75\pi\,cm^2$ b $135\pi\,cm^2$ c $189\pi\,m^2$
8 7.7 cm
9 a $(\pi r^2 + 12\pi r)\,cm^2$ b 3 cm
10 4.6 cm

What do you think? (B)

1 The volume of the original cone is $V = \frac{1}{3}\pi r^2 h$
The volume of the cone when the radius and perpendicular height are both doubled is
$V = \frac{1}{3}\pi(2r)^2 \times 2h = \frac{8}{3}\pi r^2 h$
Thus the volume of the larger cone is 8 times that of the original cone.
2 $4618\,cm^3$

Are you ready? (C)

1 a i $28.3\,cm^2$ ii $9\pi\,cm^2$
b i $113.1\,m^2$ ii $36\pi\,m^2$
c i $19.6\,mm^2$ ii $\frac{25}{4}\pi\,mm^2$

Practice (C)

1 a $33.5\,cm^3$ b $113.1\,m^3$ c $7238.2\,mm^3$
d $5276.7\,cm^3$ e $523.6\,m^3$ f $2572.4\,mm^3$
2 a $36\pi\,cm^3$ b $288\pi\,m^3$ c $\frac{256}{3}\pi\,mm^3$
3 $261.8\,cm^3$
4 $763.4\,cm^3$
5 5.9 cm
6 $\sqrt[3]{\frac{2250}{\pi}}$ cm
7 3.4 cm
8 a $804\,cm^2$ b $1810\,m^2$
c $283\,000\,mm^2$ d $17\,700\,ft^2$
9 a $16\pi\,cm^2$ b $400\pi\,m^2$ c $144\pi\,mm^2$
10 5.1 cm
11 15 cm
12 4.8 ft

What do you think? (C)

1 Yes. $\frac{\text{vol. sphere}}{\text{vol. cube}} = \frac{\frac{4}{3}\pi r^3}{(2r)^3} = \frac{\frac{4}{3}\pi r^3}{8r^3} = \frac{\pi}{2}$ so ratio is independent of radius.
2 Huda is correct.

Consolidate

1 a i $283\,cm^3$ ii $245\,cm^2$
b i $628\,m^3$ ii $408\,m^2$
c i $1360\,mm^3$ ii $679\,mm^2$
d i $2220\,cm^3$ ii $977\,cm^2$
2 a $36\pi\,cm^3$ b $377\,m^3$
c $360\pi\,mm^3$ d $312\,ft^3$
3 a $267.0\,cm^2$ b $182.4\,m^2$
c $247.4\,mm^2$ d $573.2\,ft^2$
4 a i $268.1\,cm^3$ ii $201.1\,cm^2$
b i $904.8\,m^3$ ii $452.4\,m^2$
c i $248.5\,mm^3$ ii $191.1\,mm^2$
d i $3591.4\,in^3$ ii $1134.1\,in^2$

Stretch

1 $100.2\,cm^2$
2 a $\frac{1275\pi}{4}\,m^3$ b $\frac{901\pi}{2}\,m^2$
3 $x = \sqrt[3]{18}\,y$

Chapter 5.5

Are you ready? (A)

1 a $51.84\,cm^2$ b $15.99\,cm^2$ c $75.6\,m^2$
2 a $a = 21$ cm, $b = 7$ cm
b $c = 6$ m, $d = 3.4$ m

Practice (A)

1 a 3 b 9
2 a 5 b $12\,cm^2$ c $300\,cm^2$ d 25
3 a i 5 ii 25
b i 3 ii 9
4 a 4 b 16 c $240\,cm^2$
5 a 36 b $1944\,cm^2$
6 $93.75\,cm^2$
7 a $3\,cm^2$ b $12.5\,cm^2$
8 $64\,m^2$
9 $13.44\,cm^2$
10 $x = 3.5$

What do you think? (A)

1 Huda is correct.
Scale factor length = $\frac{3}{4}$
Scale factor area = $\left(\frac{3}{4}\right)^2$

Are you ready? (B)

1 a $84\,cm^3$ b $412.5\,cm^3$ c $1002\,cm^3$ d $19\,060\,cm^3$

Practice (B)

1 a 4 b 64
2 a 3 b 27
3 a 64 b $2240\,cm^3$
4 $7560\,cm^3$

5 a 1080 cm³ **b** 421.9 m³ **c** 8918 mm³ **d** 113.2 cm³
6 a 4 **b** 2 **c** 2520 m³
7 8192 cm³

What do you think? (B)

1 The area scale factor is 2, therefore the length scale factor is $\sqrt{2}$ and the volume scale factor is $2\sqrt{2}$. So the volume of cuboid B is $2\sqrt{2}$ times greater than the volume of cuboid A.
2 Compare answers as a class.

Consolidate

1 a 288 cm² **b** 320 mm² **c** 84 cm²
2 a 675 cm³ **b** 6650 m³ **c** 1170 mm³ **d** 19.2 cm³

Stretch

1 $25 : 2\sqrt{10}$
2 1 : 3.95
3 14.444… cm³
4 a $y = \frac{x+2}{2}$ **b** $\frac{x(x+2)^2}{8}$

2D and 3D shapes: exam practice

1 36π cm²
2 20 cm
3 262 cm³
4

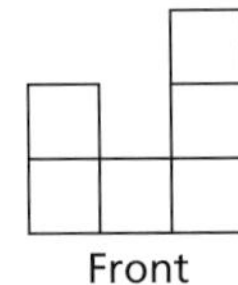

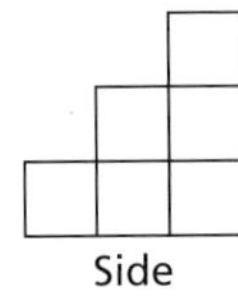

5 a 6
b Yes, e.g. from volume, area is 24 cm² so base is 6 cm. By Pythagoras' theorem, in right-hand small triangle, base = 3 cm so left-hand side triangle base must be 6 – 3 = 3 cm, so will also be a 3, 4, 5 triangle.
6 35 cm
7 13 cm

Block 6 Right-angled triangles

Chapter 6.1

Are you ready? (A)

1 a 25 **b** 121 **c** 400 **d** 72.25
2 a 6 **b** 6.08 **c** 6.86 **d** 2.17
3 a i 3.2 **ii** 3.25 **iii** 3.25
b i 12.7 **ii** 12.68 **iii** 12.7
c i 19.1 **ii** 19.06 **iii** 19.1
d i 20.4 **ii** 20.37 **iii** 20.4

Practice (A)

1 10 cm
2 a 4.5 cm **b** 7.6 cm **c** 7.1 cm
d 7.5 cm **e** 7.1 cm **f** 8.5 cm
3 a 6.40 m **b** 9.57 km **c** 26.2 mm
4 7.1 cm
5 a 8 cm **b** 48 cm²
6 a 17.2 cm **b** 137.9 cm²
7 5.7 m
8 21.19 m
9 $\sqrt{41}$ m
10 a 13 units **b** 10.8 units
11 $\sqrt{100} = 10$ but $7^2 + 7^2 = 98$

What do you think? (A)

1 a A and C
Discuss justifications as a class.
b No. If B was right angled, the hypotenuse would be 6.4. The longest side is 6 cm, meaning that the opposite angle is less than 90°.
2 CD = $\sqrt{61}$ = 7.8
If AB = x, AD = y and BC = z then CD = $\sqrt{x^2 + y^2 + z^2}$
3 Compare answers as a class.

Are you ready? (B)

1 a BC, FG and EH
b HG, DC and AB
c Example answer: BE
d Example answer: AH
2 Example answers: ADC, CDG or AHD

Practice (B)

1 a 13.89 cm **b** 14.76 cm
2 a 7.07 cm **b** 8.66 cm
3 23.4 cm
4 a 12.7 cm **b** 13.3 cm
5 18.6 cm
6 62.9 m
7 $4\sqrt{3}$ cm
8 $6\sqrt{2}$ cm

What do you think? (B)

1 No. The longest diagonal is 10.7 cm (1 d.p.)
2 a b Proofs using Pythagoras in 3D

Consolidate

1 a 7.8 cm **b** 9.8 cm **c** 12.1 cm
d 6.7 m **e** 18.6 km **f** 7.1 cm
2 a 7.9 cm **b** 12.7 m **c** 16.2 km
d 12.8 mm **e** 10.5 m **f** 27.5 cm
3 14.2 km
4 4.5 units

Stretch

1 $a = 4, b = 3$
2 13.687 ⩽ PR < 13.803
3 $\frac{1}{2} \times \pi \times \left(\frac{a}{2}\right)^2 = \frac{a^2}{8}\pi$

$\frac{1}{2} \times \pi \times \left(\frac{b}{2}\right)^2 = \frac{b^2}{8}\pi$

$\frac{1}{2} \times \pi \times \left(\frac{c}{2}\right)^2 = \frac{c^2}{8}\pi$

$a^2 + b^2 = c^2$

$\frac{a^2\pi}{8} + \frac{b^2\pi}{8} = \frac{\pi}{8}(a^2 + b^2) = \frac{c^2\pi}{8}$

4 $k = 17$ or $k = -7$
5 120 cm³
6 $\sqrt{(a-c)^2 + (b-d)^2}$

Chapter 6.2

Are you ready? (A)

1 a $m = 14$ **b** $t = 14$ **c** $y = \frac{2}{7}$
2 a $p = 20$ **b** $k = 22.75$ **c** $r = \frac{4}{7}$
3 a i 6.7 **ii** 6.74 **iii** 6.74
b i 12.3 **ii** 12.25 **iii** 12.3
c i 10.1 **ii** 10.07 **iii** 10.1
d i 201.0 **ii** 200.97 **iii** 201
4 a b c

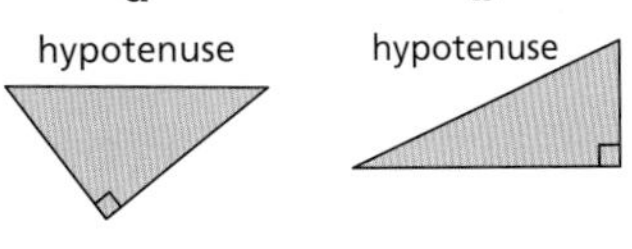

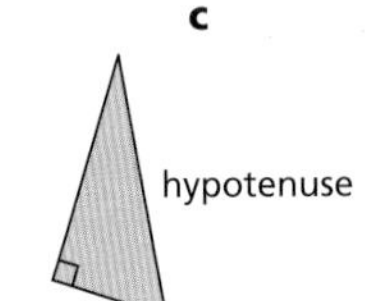

Practice (A)

1 a

b

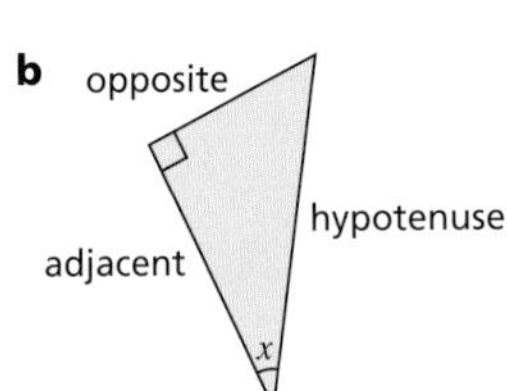

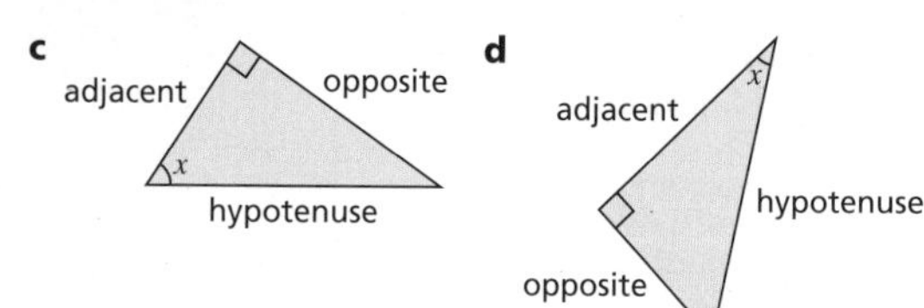

2 **a** 4.0 cm **b** 3.8 mm **c** 7.5 m **d** 8.2 km
3 **a** 5.2 cm **b** 9.8 m **c** 6.8 cm **d** 9.0 mm
4 **a** 4.2 mm **b** 5.3 m **c** 3.5 cm **d** 12.3 km
5 **a** cos **b** tan **c** sin **d** cos
6 **a** 4.2 cm **b** 5.2 mm **c** 6.2 m
d 4.0 cm **e** 9.0 mm **f** 8.8 km
7 **a** 8.3 cm **b** 7.4 mm **c** 17.3 m
8 **a** Marta should have written $\tan 28° = \frac{PQ}{11}$ **b** 5.8 cm
9 **a** 3.29 cm **b** 14.91 cm **c** 12.42 mm
d 6.09 mm **e** 6.68 km **f** 21.57 km
10 3.01 m
11 83.2 cm^2

What do you think? (A)

1 x = 13.21 cm
Compare methods as a class.
2 Any values of x and y where $x + y = 90$
One of the angles in a right-angled triangle is 90°, therefore the sum of the remaining two angles must be 90°.

Are you ready? (B)

1 $\sin x = \frac{\text{opposite}}{\text{hypotenuse}}$ $\cos x = \frac{\text{adjacent}}{\text{hypotenuse}}$ $\tan x = \frac{\text{opposite}}{\text{adjacent}}$
2 **a** 10.9 cm **b** 5.07 cm **c** 5.60 cm **d** 25.7 cm

Practice (B)

1 **a** 11.5° **b** 69.5° **c** 32.0°
d 72.5° **e** 56.4° **f** 63.0°
2 **a** 30° **b** 39.5° **c** 56.1° **d** 27.9°
3 **a** 53.1° **b** 41.4° **c** 34.7° **d** 31.9°
4 **a** 40.6° **b** 60.9° **c** 49.3° **d** 22.0°
5 **a** 36.9° **b** 51.1° **c** 32.0°
d 16.1° **e** 24.2° **f** 55.0°
6 28.9°
7 54.0°

What do you think? (B)

1 36.254…° (to 3 d.p.)
2 The new triangle will be similar to the original triangle, so the size of the angle will not change.
3 Ed should have halved the length of the base to form a right-angled triangle.
The two equal angles are 71° to the nearest degree.

Are you ready? (C)

1 **a** 9.80 cm **b** 7.36 cm **c** 5.85 cm **d** 10.7 cm

Practice (C)

1 **a** 9.58 cm **b** 11.83 cm
2 **a** 8.67 mm **b** 30.0°
3 13.8 cm
4 116°
5 67.4°
6 38.0°
7 37.6°
8 **a** 3.32 m **b** 33.6°
9 **a** 145 cm **b** 26.6°
10 122°

11 **a**

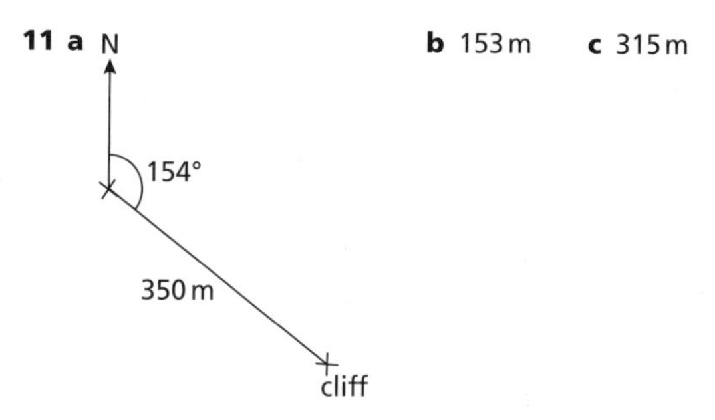

b 153 m **c** 315 m

Consolidate

1 **a** 5.2 cm **b** 8.2 mm **c** 15.5 km
d 18.0 cm **e** 17.2 km **f** 12.2 mm
2 **a** 51.3° **b** 36.0° **c** 69.1°
d 45.8° **e** 30.7° **f** 37.4°
3 51.1°
4 77.3 mm
5 **a** 9.60 cm **b** 22.63 cm

Stretch

1 AC = 8.60 cm, BD = 13.5 cm
2 True. $\frac{\sin\theta}{\cos\theta} = \tan\theta$
3 $\sin^2 x + \cos^2 x = 1$
4 **a** $(\sin x + 3)(\sin x + 2)$
b **i** $(\sin x + 2)(\sin x + 6)$
ii $(\cos x + 5)(\cos x - 3)$
iii $(2\tan x + 5)(\tan x + 2)$
5 $\sin 30° = \frac{1}{2}$
You can also work out the values of $\sin\theta$, $\cos\theta$ and $\tan\theta$, when θ = 30°, 45° or 60°

Chapter 6.3

Are you ready?

1 **a** 12.0 cm **b** 7.7 mm **c** 17.6 cm
2 **a** 3.7 cm **b** 10.4 mm **c** 65.6 m
3 **a** 55.8° **b** 49.6° **c** 39.6°

Practice

1 36.9°
2 **a** 7.81 cm **b** 21.0°
3 **a** 9.85 cm **b** 26.9°
4 **a** 8.49 m **b** 35.2°
5 23.7°
6 9.22°
7 **a** 6.36 cm **b** 13.7 cm **c** 21.5°

What do you think?

1 True. All cubes are mathematically similar.

Consolidate

1 **a** 6.40 cm **b** 25.1°
2 **a** 11.4 cm **b** 23.7°
3 **a** 5.83 cm **b** 45.8°
4 **a** 7.07 m **b** 35.3°

Stretch

1 15 600 cm^3
2 9.88 cm
3 14.9°

Chapter 6.4

Are you ready?

1 **a** $a = \sqrt{2}$ cm **b** $b = \sqrt{3}$ cm
2 **a** $a = 3\sqrt{3}$ cm **b** $b = 3\sqrt{2}$ cm
3 **a** $\sqrt{3}$ **b** $4\sqrt{3}$ **c** $5\sqrt{2}$
4 **a** $a = 50$ **b** $b = 1.5$ **c** $c = \frac{5\sqrt{3}}{2}$ **d** $d = 10\sqrt{2}$

Practice

1 a $\frac{\sqrt{3}}{2}$ b $\frac{1}{2}$ c $\frac{1}{\sqrt{3}}$ or $\frac{\sqrt{3}}{3}$
d $\frac{1}{2}$ e $\frac{\sqrt{3}}{2}$ f $\sqrt{3}$

2 1

3 a $\frac{\sqrt{2}}{2}$ b $\frac{\sqrt{2}}{2}$ c 1

4 a 0 b 1 c 1
d 0 e 0 f undefined

5 a 1 b $\sqrt{3}$ c $\frac{2\sqrt{3}}{3}$
d $\sqrt{3}$ e 8 f $4\sqrt{2}$
g $5\sqrt{3}$ h 10 i 0

6 a $\frac{\sqrt{2}}{2}$ b 3 c 3

7 a $p = 5$ b $q = 3$ c $r = \sqrt{3}$

8 a $a = 4$ b $b = 6\sqrt{3}$ c $c = 12\sqrt{3}$
d $d = 3\sqrt{2}$ e $e = \frac{5\sqrt{3}}{3}$

9 $\frac{49\sqrt{3}}{2}$ cm²

10 a $a = 60°$ b $b = 60°$ c $c = 30°$

11 $\frac{5}{2}$

12 $4\sqrt{3}$ cm

What do you think?

1 True. Discuss as a class.

2 sin 30° cos 45° cos 30° tan 60°

Consolidate

1 a $\frac{1}{2}$ b $\frac{\sqrt{3}}{2}$ c $\frac{\sqrt{2}}{2}$ d $\frac{\sqrt{3}}{2}$ e $\frac{1}{2}$
f $\frac{\sqrt{2}}{2}$ g $\frac{1}{\sqrt{3}}$ *or* $\frac{\sqrt{3}}{3}$ h $\sqrt{3}$ i 1

2 a 10 cm b 7.5 mm c 6 m
d $4\sqrt{2}$ km e 10 mm

Stretch

1 $\frac{8\sqrt{6}}{3}$ cm

2 $\frac{7\sqrt{3}}{6}$

3 $\frac{3}{2}$

4 30°

5 $\sqrt{6} : 2\sqrt{6} : 2$

6 $a = 3$ and $b = 2$

Right-angled triangles: exam practice

1 25.4°

2 Yes, $36^2 + 77^2 = 85^2$

3 4.5 cm

4 21.2 cm

5 40 cm

6 a 1 b $\frac{1}{2}$

7 34.7°

8 $8\sqrt{3}$ cm

Block 7 Non-right-angled triangles

Chapter 7.1

Are you ready?

1 a 20 b 10 c 0.57 d 5.74

2 a $a = 20.3$ b $\theta = 38.7°$

Practice

1 a 19.4 cm² b 29.1 m² c 25.9 km²

2 160 cm²

3 a 15.7 cm² b 43.3 cm²

4 32.7 m²

5 a 27.0 cm b 6.0 cm c 6.1 cm

6 $x = 11.3$

7 57.2°

What do you think?

1 Triangle is right-angled so both equations are valid (because sin 90° = 1)

2 Calculation gives $\sin\theta = \frac{6}{7}$ which has two solutions, 59.0° and 121.0°

Consolidate

1 a 12.0 cm² b 15.1 cm² c 15.0 m²
d 18.7 km² e 23.4 m² f 63.6 km²

2 a $a = 72.2°$ b $b = 56.4°$ c $c = 65.1°$

Stretch

1 a 12.5 cm² b $\frac{35\sqrt{3}}{2}$ ft² c $\frac{117\sqrt{2}}{4}$ cm²

2 a 11.0 m² (to 3 s.f.)
b A triangle with angle 120° has the same area but the sector of the circle will be twice as big.

3 $\frac{1}{2} \times 2x \times (x+3) \times \sin 60°$

$$= \frac{1}{2} \times 2x \times (x+3) \times \frac{\sqrt{3}}{2}$$
$$= \frac{1}{2} \times (2x^2 + 6x) \times \frac{\sqrt{3}}{2}$$
$$= \frac{\sqrt{3}(2x^2 + 6x)}{4}$$
$$= \frac{\sqrt{3}(x^2 + 3x)}{2}$$

Chapter 7.2

Are you ready? (A)

1 a $a = 0.9$ b $b = 28.0$ c $c = 4.9$ d $d = 9.0$

Practice (A)

1 C

2 M

3 a Triangle labelled correctly.
b 4.57 cm

4 a 4.5 cm b 6.4 cm c 8.0 mm
d 33.9 cm e 4.8 m f 502.2 mm

5 8.1 cm

6 5.98 cm

7 23.9 cm

8 a 14.3 cm b 7.89 mm

9 a 23° b 12.88 km

10 a 12.4 cm b 8.1 cm c 11.1 cm

11 42.9 cm

12 13.4 cm

13 4.07 cm

14 21.6 cm

What do you think? (A)

1 C (Both). Discuss explanations as a class.

2 In all cases $\frac{a}{\sin A} = \frac{b}{\sin B} = \frac{c}{\sin C} = 2R$, where R is the radius of the circumscribed circle

Are you ready? (B)

1 a $x = 36.9°$ b $x = 53.1°$ c $x = 31.1°$

2 a 45.6° b 49.1°

Practice (B)

1 a $x = 24.2°$ b $x = 33.8°$ c $x = 42.4°$

2 a B
b 30.7°

3 a 34.8° b 27.0° c 43.3°
d 38.6° e 37.4° f 47.1°

4 31.4°

5 59.1°

6 a 43.1° **b** 156.8°
7 a 41.5° **b** 14.5°
8 33.04°
9 a 53° **b** 127°
10 42.7° and 137.3°

What do you think? (B)

1 Mo is correct. It is possible to use right-angled trigonometry.

2

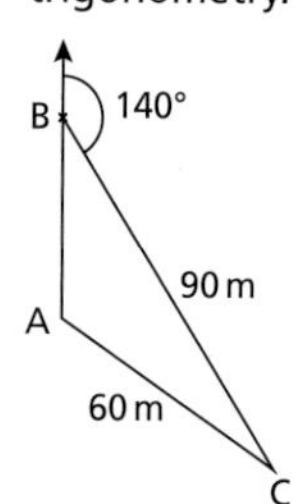

105°

3 a Example answer:

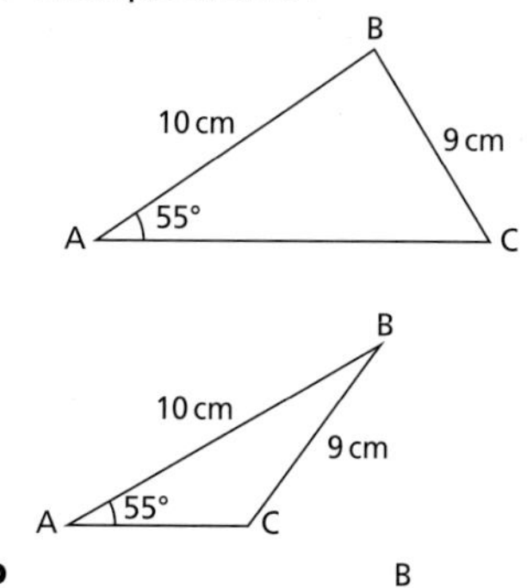

b

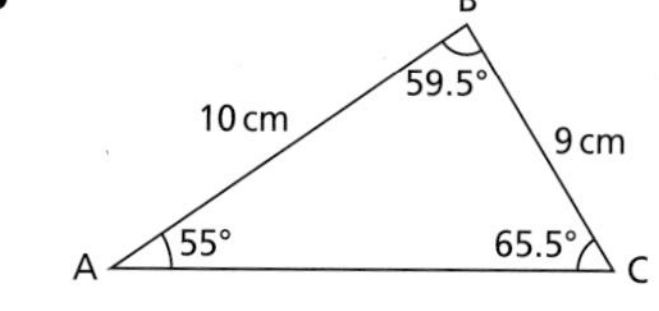

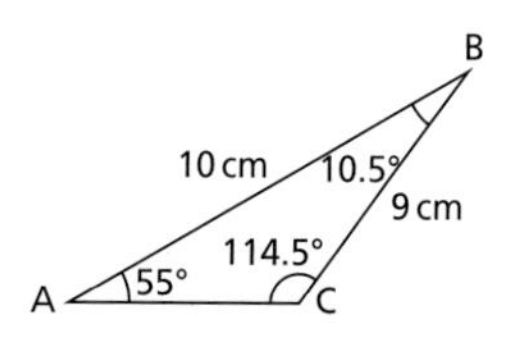

4 a Two solutions **b** One solution **c** No solutions
Discuss explanations as a class.

Consolidate

1 a 12.0 cm **b** 7.3 mm **c** 4.6 m **d** 15.8 km
2 a 29.7° **b** 38.4° **c** 36.5° **d** 26.0°

Stretch

1 10.7 cm
2 $y = \frac{2x \sin 32°}{\sin 48°}$
3 $8\sqrt{2}$
4 a 51.55115° **b** 43.04329°

Chapter 7.3

Are you ready? (A)

1 a 25 **b** 9 **c** 30 **d** 17.2 **e** 16.8
2 a **b** **c**

B, c, a, A, b, C A, c, b, B, a, C C, a, B, c, b, A

Practice (A)

1 a

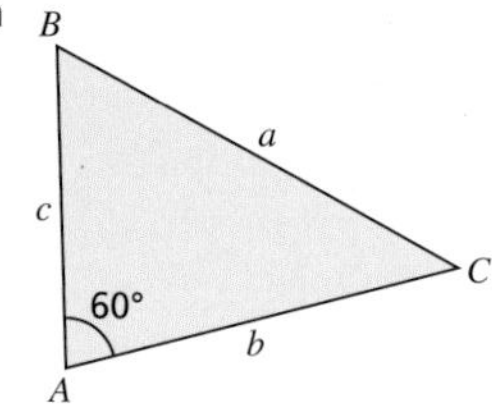

b C **c** 9.2 cm
2 a 4.2 cm **b** 6.8 mm **c** 13.8 m **d** 12.9 km
3 8.0 m
4 9.10
5 7.3 cm
6 335 m
7 $2\sqrt{13}$
8 $3\sqrt{7}$
9 7.41 cm
10 a

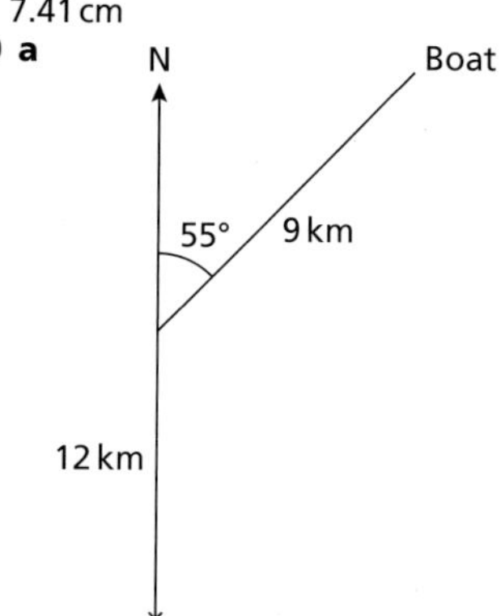

b 18.7 km

What do you think? (A)

1 When $A = 90°$, $\cos A = 0$
So $a^2 = b^2 + c^2 - 2bc \cos A$ becomes $a^2 = b^2 + c^2$

Are you ready? (B)

1 a 45.6° **b** 30.1° **c** 72.5° **d** 60°

Practice (B)

1 a

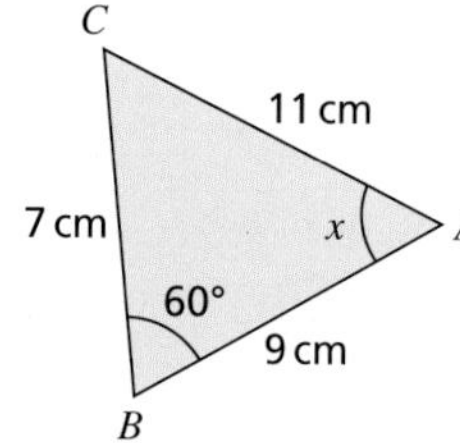

b C **c** 39.4°
2 a 83.3° **b** 69.1° **c** 52.0° **d** 45.7°
3 70.0°
4 55.6°
5 24.8°
6 73.1°
7 $\cos A = \frac{5^2 + 7^2 - 3^2}{2 \times 5 \times 7} = \frac{65}{70} = \frac{13}{14}$

What do you think? (B)

1 85.9°
2 $\cos^{-1}\left(\frac{21^2 + 20^2 - 29^2}{2 \times 21 \times 20}\right) = 90°$
3 a cosine **b** cosine **c** sine **d** cosine

Consolidate

1 **a** 3.7 cm **b** 5.9 mm **c** 28.0 m
d 12.5 km **e** 7.5 m **f** 328.3 m

2 **a** 54.7° **b** 78.6° **c** 75.7°
d 75.9° **e** 55.5° **f** 84.6°

3 $\sqrt{21}$

4 $8^2 = 5^2 + 11^2 - 2 \times 5 \times 11 \cos A$ gives

$\cos A = \frac{5^2 + 11^2 - 8^2}{2 \times 5 \times 11} = \frac{41}{55}$

Stretch

1 $2\sqrt{21}$ cm

2 $y^2 = 7x^2 - 2x + 4$

Non-right-angled triangles: exam practice

1 126 cm²

2 101°

3 10.5 cm

4 38.6 cm²

5 30.7°

6 52.4 cm²

Block 8 Vectors

Chapter 8.1

Are you ready?

1

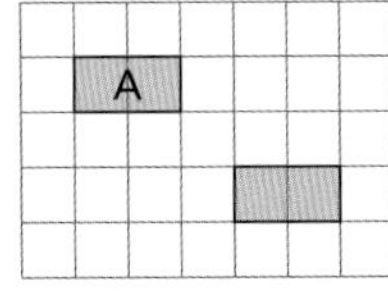

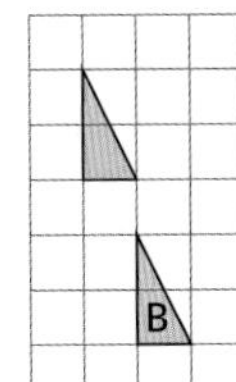

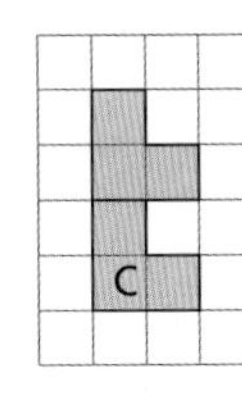

2 **a** Translation by $\begin{pmatrix} 4 \\ 1 \end{pmatrix}$ **b** Translation by $\begin{pmatrix} -1 \\ -3 \end{pmatrix}$

3 **a** –4 **b** –10 **c** 10 **d** 4
e –21 **f** –21 **g** 21 **h** 21

Practice

1 a = $\begin{pmatrix} 3 \\ 2 \end{pmatrix}$ b = $\begin{pmatrix} -3 \\ -2 \end{pmatrix}$ c = $\begin{pmatrix} 3 \\ -2 \end{pmatrix}$ d = $\begin{pmatrix} 2 \\ -3 \end{pmatrix}$

e = $\begin{pmatrix} 2 \\ -5 \end{pmatrix}$ f = $\begin{pmatrix} 5 \\ 2 \end{pmatrix}$ g = $\begin{pmatrix} -5 \\ -2 \end{pmatrix}$ h = $\begin{pmatrix} 0 \\ -5 \end{pmatrix}$

2

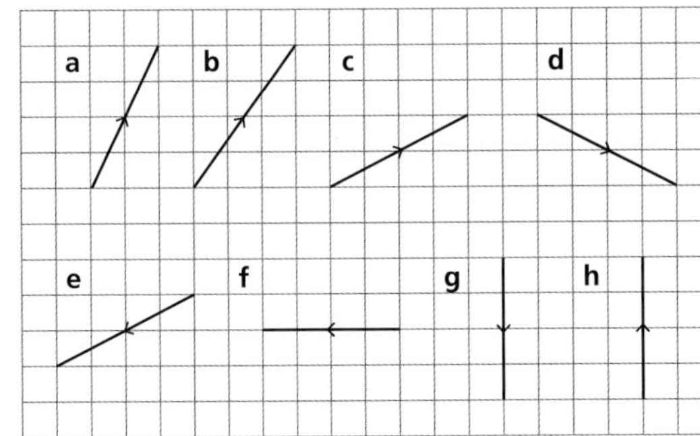

3

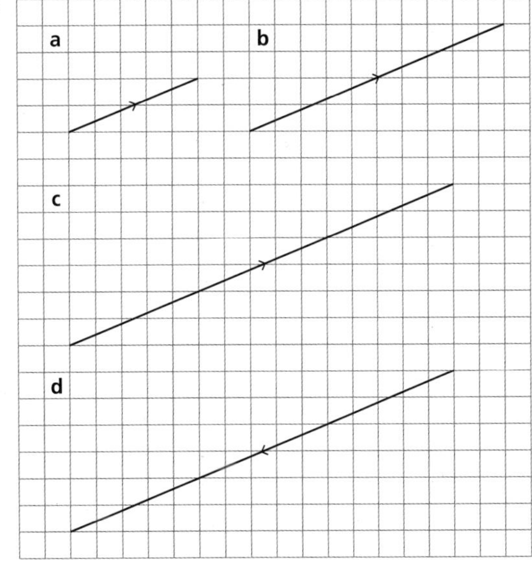

4 **a** $\begin{pmatrix} 8 \\ -6 \end{pmatrix}$ **b** $\begin{pmatrix} 20 \\ -15 \end{pmatrix}$ **c** $\begin{pmatrix} -16 \\ 12 \end{pmatrix}$ **d** $\begin{pmatrix} 2 \\ -1.5 \end{pmatrix}$

5 **a** $\begin{pmatrix} 2 \\ 4 \end{pmatrix}$ **b** $\begin{pmatrix} 5 \\ -2 \end{pmatrix}$ **c** $\begin{pmatrix} 7 \\ 2 \end{pmatrix}$

6 **a** $\begin{pmatrix} 3 \\ -1 \end{pmatrix}$ **b** $\begin{pmatrix} -1 \\ 7 \end{pmatrix}$ **c** $\begin{pmatrix} 2 \\ -1 \end{pmatrix}$ **d** $\begin{pmatrix} 0 \\ 7 \end{pmatrix}$

e $\begin{pmatrix} 1 \\ 0 \end{pmatrix}$ **f** $\begin{pmatrix} 3 \\ -8 \end{pmatrix}$ **g** $\begin{pmatrix} -2 \\ 1 \end{pmatrix}$ **h** $\begin{pmatrix} -3 \\ 8 \end{pmatrix}$

7 **a** $\begin{pmatrix} -1 \\ 9 \end{pmatrix}$ **b** $\begin{pmatrix} 1 \\ 13 \end{pmatrix}$ **c** $\begin{pmatrix} -4 \\ 14 \end{pmatrix}$ **d** $\begin{pmatrix} 0 \\ 22 \end{pmatrix}$

e $\begin{pmatrix} -12 \\ -2 \end{pmatrix}$ **f** $\begin{pmatrix} -12 \\ -2 \end{pmatrix}$ **g** $\begin{pmatrix} -21 \\ 13 \end{pmatrix}$ **h** $\begin{pmatrix} 9 \\ -37 \end{pmatrix}$

8 **a**

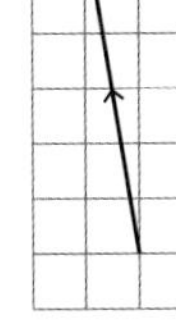

b $\begin{pmatrix} 13 \\ 2 \end{pmatrix}$

9 $k = 4$

10 **a** $\begin{pmatrix} 3 \\ 5 \end{pmatrix}$ **b** $\begin{pmatrix} -3 \\ -5 \end{pmatrix}$ **c** $\begin{pmatrix} 9 \\ 15 \end{pmatrix}$

11 **a** $\begin{pmatrix} 2 \\ -9 \end{pmatrix}$ **b** $\begin{pmatrix} -2 \\ 9 \end{pmatrix}$ **c** $\begin{pmatrix} -5 \\ 5 \end{pmatrix}$ **d** $\begin{pmatrix} -10 \\ 10 \end{pmatrix}$

e $\begin{pmatrix} -3 \\ -4 \end{pmatrix}$ **f** $\begin{pmatrix} -7 \\ 14 \end{pmatrix}$ **g** $\begin{pmatrix} -12 \\ 19 \end{pmatrix}$

12 $p = 0$ and $q = 15$

What do you think?

1 $\begin{pmatrix} 5 \\ -3 \end{pmatrix}$, $\begin{pmatrix} -10 \\ 6 \end{pmatrix}$ and $\begin{pmatrix} 20 \\ -12 \end{pmatrix}$

Check other vectors are a multiple of $\begin{pmatrix} -5 \\ 3 \end{pmatrix}$.

2 Scalene

3 Ed has drawn $\begin{pmatrix} -5 \\ -3 \end{pmatrix}$

Vector correctly drawn 5 units to the right and 3 units down.

4 True: A, B False: C, D, E

Consolidate

1 **a** $\begin{pmatrix}4\\2\end{pmatrix}$ **b** $\begin{pmatrix}-4\\-2\end{pmatrix}$ **c** $\begin{pmatrix}4\\-2\end{pmatrix}$
d $\begin{pmatrix}-4\\2\end{pmatrix}$ **e** $\begin{pmatrix}2\\4\end{pmatrix}$ **f** $\begin{pmatrix}-2\\-4\end{pmatrix}$
g $\begin{pmatrix}-2\\4\end{pmatrix}$ **h** $\begin{pmatrix}2\\-4\end{pmatrix}$ **i** $\begin{pmatrix}2\\-5\end{pmatrix}$
j $\begin{pmatrix}-2\\-5\end{pmatrix}$ **k** $\begin{pmatrix}-5\\-5\end{pmatrix}$ **l** $\begin{pmatrix}5\\0\end{pmatrix}$ **m** $\begin{pmatrix}0\\5\end{pmatrix}$

2

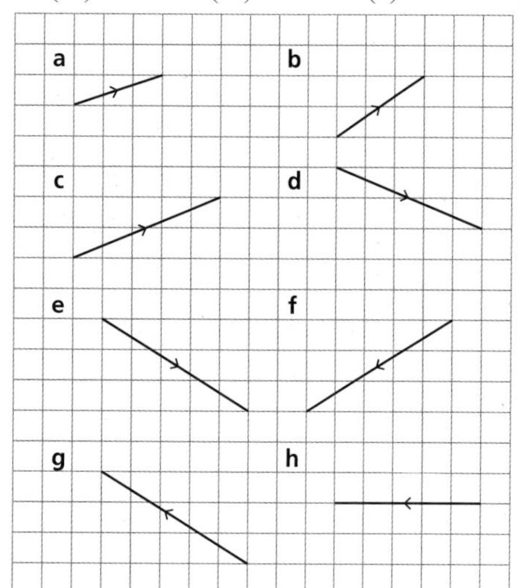

3

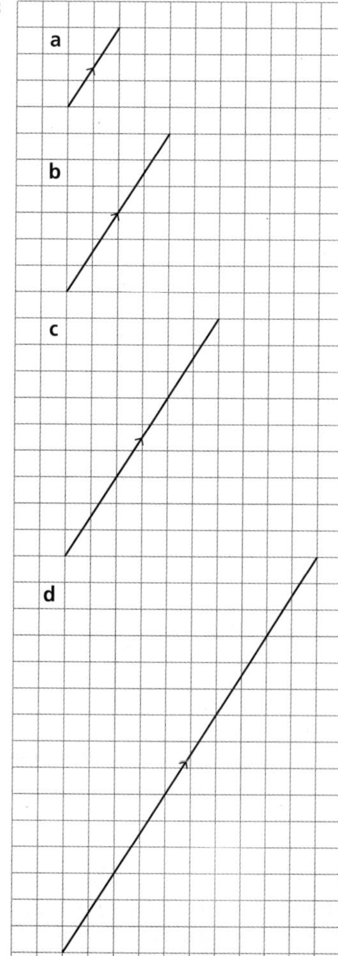

4 **a** $\begin{pmatrix}-4\\6\end{pmatrix}$ **b** $\begin{pmatrix}-6\\9\end{pmatrix}$ **c** $\begin{pmatrix}-8\\12\end{pmatrix}$ **d** $\begin{pmatrix}-10\\15\end{pmatrix}$

5 **a** $\begin{pmatrix}4\\8\end{pmatrix}$ **b** $\begin{pmatrix}1\\13\end{pmatrix}$ **c** $\begin{pmatrix}-2\\18\end{pmatrix}$ **d** $\begin{pmatrix}-5\\23\end{pmatrix}$
e $\begin{pmatrix}13\\-7\end{pmatrix}$ **f** $\begin{pmatrix}-13\\7\end{pmatrix}$ **g** $\begin{pmatrix}-11\\11\end{pmatrix}$ **h** $\begin{pmatrix}11\\-11\end{pmatrix}$

Stretch

1 $m = 2$ and $n = 6$
2 $2\sqrt{17}$
3 $m = 6$ or $m = -6$

Chapter 8.2

Are you ready? (A)

1 **a** $4a - b$ **b** $2a - b$
c $-4a - b$ **d** $-3a - 2b$
2 **a** $-2a - 2b$ **b** $-6a - 2b$
c $-6a + 2b$ **d** $-6a + 10b$
3 **a** $-a - b$ **b** $-a + b$
c $-2a + b$ **d** $2a + b$
4 $\frac{4}{7}$

Practice (A)

1 $\overrightarrow{AB} = \overrightarrow{DC}$ $\overrightarrow{AD} = \overrightarrow{BC}$
2 **a** $\mathbf{a}$ **b** $\mathbf{b}$ **c** $2\mathbf{b}$ **d** $3\mathbf{a}$
e $\mathbf{a} + \mathbf{b}$ **f** $\mathbf{a} + 2\mathbf{b}$ **g** $2\mathbf{a} - \mathbf{b}$ **h** $-2\mathbf{a} - \mathbf{b}$
3 **a** $\mathbf{n} - \mathbf{o}$ **b** $\mathbf{m} - \mathbf{p}$ **c** $-\mathbf{m} + \mathbf{n}$ and $-\mathbf{p} + \mathbf{o}$
4 **a** $2\mathbf{a} + \mathbf{b}$ **b** $-2\mathbf{a} - \mathbf{b}$
5 **a** $3\mathbf{a} - 2\mathbf{b}$ **b** $-3\mathbf{a} + 2\mathbf{b}$
6 **a** $\mathbf{a} + \mathbf{b}$ **b** $\mathbf{a} - \mathbf{b}$ **c** $-\mathbf{a} - \mathbf{b}$
7 **a** $-3\mathbf{a}$ **b** $-2\mathbf{a} + \mathbf{b}$
8 $\mathbf{a} + 2\mathbf{b}$
9 **a** $\mathbf{a} + \mathbf{b}$ **b** $\frac{1}{2}\mathbf{a} + \frac{1}{2}\mathbf{b}$ **c** $-\frac{1}{2}\mathbf{a} + \frac{1}{2}\mathbf{b}$
10 **a** $-\mathbf{b}$ **b** $\mathbf{a} + 3\mathbf{b}$
11 **a** $\frac{3}{2}\mathbf{a} - 2\mathbf{b}$ **b** $-\frac{1}{2}\mathbf{a} - \mathbf{b}$
12 **a** $3\mathbf{a} + 3\mathbf{b}$ **b** $2\mathbf{a} + 5\mathbf{b}$ **c** $\mathbf{a} + 4\mathbf{b}$
13 $\frac{3}{4}\mathbf{x} + \frac{1}{4}\mathbf{y}$

Are you ready? (B)

1 **a** $3(a + b)$ **b** $2(5a - 8b)$
c $3(a - 9b)$ **d** $-4(5a - 3b)$
2 **a** $2a + 2b$ **b** $6a - 3b$
c $-4a - 8b$ **d** $-36a + 45b$

Practice (B)

1 $4\mathbf{a} + 6\mathbf{b} = 2(2\mathbf{a} + 3\mathbf{b})$ and $10\mathbf{a} + 15\mathbf{b} = 5(2\mathbf{a} + 3\mathbf{b})$
2 $4(\mathbf{a} + 3\mathbf{b}) = 4\mathbf{a} + 12\mathbf{b}$
3 **a** $4\mathbf{a} - 2\mathbf{b}$ **b** $4\mathbf{a} - 2\mathbf{b} = 2(2\mathbf{a} - \mathbf{b})$
4 $4\mathbf{a} - 4\mathbf{b} = 4(\mathbf{a} - \mathbf{b})$
5 $\overrightarrow{AB} = -2\mathbf{a} + 2\mathbf{b}$
$\overrightarrow{MN} = -\mathbf{a} + \mathbf{b}$
$\overrightarrow{AB} = -2\mathbf{a} + 2\mathbf{b} = 2(-\mathbf{a} + \mathbf{b})$

What do you think?

1 Yes, they are parallel.
$\overrightarrow{PR} = -\mathbf{a} + \mathbf{b}$ and $\overrightarrow{MN} = -\frac{1}{2}\mathbf{a} + \frac{1}{2}\mathbf{b}$
$-\frac{1}{2}\mathbf{a} + \frac{1}{2}\mathbf{b} = \frac{1}{2}(-\mathbf{a} + \mathbf{b})$

Are you ready? (C)

1 **a** $4\mathbf{a} - 6\mathbf{b}$ **b** $10\mathbf{a} - 15\mathbf{b}$ **c** $-6\mathbf{a} + 9\mathbf{b}$ **d** $\mathbf{a} - \frac{3}{2}\mathbf{b}$

Practice (C)

1 $8\mathbf{a} - 12\mathbf{b} = 4(2\mathbf{a} - 3\mathbf{b})$. C is a common point.
2 **a**

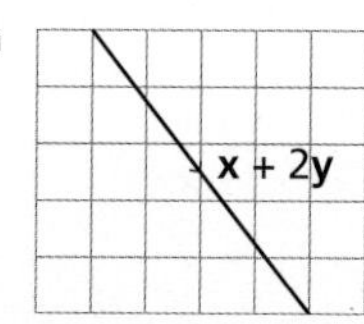

b $\begin{pmatrix} 4 \\ -5 \end{pmatrix}$

c

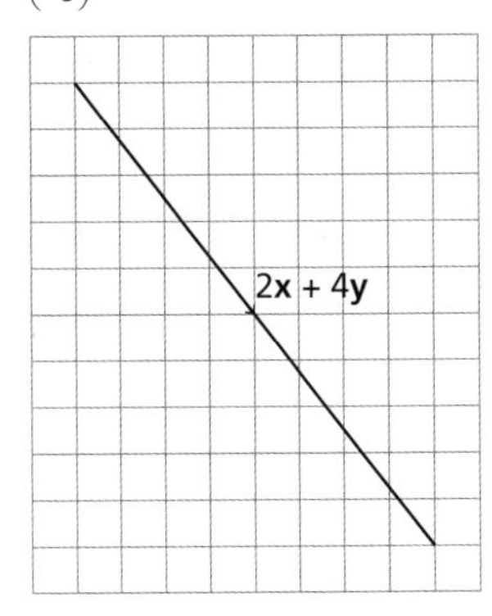

d $\begin{pmatrix} 8 \\ -10 \end{pmatrix}$

e $\begin{pmatrix} 4 \\ -5 \end{pmatrix} \times 2 = \begin{pmatrix} 8 \\ -10 \end{pmatrix}$. A is a common point.

3 a $-\mathbf{a} + 4\mathbf{b}$ **b** $-3\mathbf{a} + 12\mathbf{b}$ **c** $-3\mathbf{a} + 8\mathbf{b}$

4 a $-\frac{1}{4}\mathbf{a} + \frac{3}{2}\mathbf{b}$

b $\overrightarrow{MN} = -\frac{1}{4}\mathbf{a} + \frac{3}{2}\mathbf{b} = -\frac{1}{4}(\mathbf{a} - 6\mathbf{b})$

$\overrightarrow{MD} = -\frac{1}{2}\mathbf{a} + 3\mathbf{b} = -\frac{1}{2}(\mathbf{a} - 6\mathbf{b})$

$\overrightarrow{MN}$ and $\overrightarrow{MD}$ are parallel and share a common point M.

5 a $-3\mathbf{y} + \mathbf{x}$ **b** $-\frac{1}{2}\mathbf{y} + \frac{1}{2}\mathbf{x}$

c $\overrightarrow{NS} = -2\mathbf{y} + 2\mathbf{x}$

$-2\mathbf{y} + 2\mathbf{x} = -2(\mathbf{y} - \mathbf{x})$

$\overrightarrow{NM} = -\frac{1}{2}\mathbf{y} + \frac{1}{2}\mathbf{x} = -\frac{1}{2}(\mathbf{y} - \mathbf{x})$

$\overrightarrow{NM}$ and $\overrightarrow{NS}$ are parallel and share a common point N.

6 $\overrightarrow{MN} = \frac{1}{3}(2\mathbf{x} + \mathbf{y}) - \mathbf{x} = -\frac{1}{3}(\mathbf{x} - \mathbf{y})$

$\overrightarrow{MC} = -\frac{4}{3}\mathbf{x} + \frac{4}{3}\mathbf{y} = -\frac{4}{3}(\mathbf{x} - \mathbf{y})$

$\overrightarrow{MN}$ and $\overrightarrow{MC}$ are parallel and share a common point M.

Consolidate

1 a x **b** y **c** 2y **d** 2x **e** y **f** 2y **g** x + y **h** x + 2y

2 a x + y **b** x − y **c** x − 2y

3 $4\mathbf{a} + 12\mathbf{b} = 4(\mathbf{a} + 3\mathbf{b})$

4 $\begin{pmatrix} 1 \\ 2 \end{pmatrix} \times 3 = \begin{pmatrix} 3 \\ 6 \end{pmatrix}$

$\overrightarrow{AC}$ and $\overrightarrow{BC}$ are parallel and share a common point C.

5 $\begin{pmatrix} -3 \\ 4 \end{pmatrix} \times 2 = \begin{pmatrix} -6 \\ 8 \end{pmatrix}$

$\overrightarrow{PQ}$ and $\overrightarrow{PR}$ are parallel and share a common point P.

6 $3\mathbf{a} - 6\mathbf{b} = 3(\mathbf{a} - 2\mathbf{b})$

$\overrightarrow{AC}$ and $\overrightarrow{BC}$ are parallel and share a common point C.

Stretch

1 a $\overrightarrow{DB} = -3\mathbf{b} + \mathbf{a}$

b $\overrightarrow{NM} = \frac{1}{2}(-\mathbf{b} + \mathbf{a})$ $\overrightarrow{NC} = 2(-\mathbf{b} + \mathbf{a})$

Therefore, parallel and share a common point N.

2 $k = \frac{3}{7}$

Example method:

$\overrightarrow{QM} = \frac{1}{2}(\mathbf{x} - \mathbf{y})$

$\overrightarrow{PM} = \mathbf{y} + \frac{1}{2}(\mathbf{x} - \mathbf{y}) = \frac{1}{2}(\mathbf{x} + \mathbf{y})$

$\overrightarrow{PO} = \frac{3}{5}\overrightarrow{PM} = \frac{3}{10}(\mathbf{x} + \mathbf{y})$

$\overrightarrow{QN} = -\mathbf{y} + k\mathbf{x}$

$\overrightarrow{ON} = \lambda(-\mathbf{y} + k\mathbf{x})$ where $\lambda = \frac{ON}{QN}$

$\overrightarrow{PN} = k\mathbf{x} = \overrightarrow{PO} + \overrightarrow{ON}$

$k\mathbf{x} = \frac{3}{10}(\mathbf{x} + \mathbf{y}) + \lambda(-\mathbf{y} + k\mathbf{x})$

y component $= 0 \therefore \frac{3}{10}\mathbf{y} - \lambda\mathbf{y} = 0$

$\mathbf{y} = \frac{3}{10}$

$k\mathbf{x} = \frac{3}{10}\mathbf{x} + \frac{3}{10}k\mathbf{x}$

$\Rightarrow k = \frac{3}{10} + \frac{3}{10}k$

$\Rightarrow \frac{7}{10}k = \frac{3}{10}$

$\Rightarrow k = \frac{3}{7}$

Vectors: exam practice

1 $\begin{pmatrix} -2 \\ 3 \end{pmatrix}$

2 Answer should show $3\mathbf{p} + \mathbf{q} = \begin{pmatrix} -1 \\ 3 \end{pmatrix} = \frac{1}{4}\mathbf{r}$

3 Trapezium

4 $\begin{pmatrix} 9 \\ -6 \end{pmatrix}$

5 $\overrightarrow{XY}$ is parallel to $\overrightarrow{PQ}$ and three times as long.

6 $\overrightarrow{AC} = -\mathbf{a} + \mathbf{c}$, $\overrightarrow{XY} = \frac{1}{2}\mathbf{c} - \frac{1}{3}\mathbf{a}$

So $\overrightarrow{AC}$ is not a multiple of $\overrightarrow{XY}$.

7 $\overrightarrow{AX} = \frac{3}{4}\overrightarrow{AB} = \frac{3}{4}(3\mathbf{b} - 2\mathbf{a})$ so $\overrightarrow{OX} = 2\mathbf{a} + \frac{3}{4}(3\mathbf{b} - 2\mathbf{a})$

$= \frac{9}{4}\mathbf{b} + \frac{1}{2}\mathbf{a} = \frac{1}{4}(9\mathbf{b} + 2\mathbf{a})$

So $k = \frac{1}{4}$

8 Answer should show $\overrightarrow{AY} = -\mathbf{a} + \frac{1}{2}\mathbf{b}$ and $\overrightarrow{YC} = -2\mathbf{a} + \mathbf{b}$

$\overrightarrow{YC}$ is a multiple of $\overrightarrow{AY}$ and they have a common point.

Geometry and measures: exam practice

1 Check accuracy with a partner.

2 $x = 6$ cm and $y = 7.2$ cm

3 $\begin{pmatrix} -1 \\ 3 \end{pmatrix}$

4

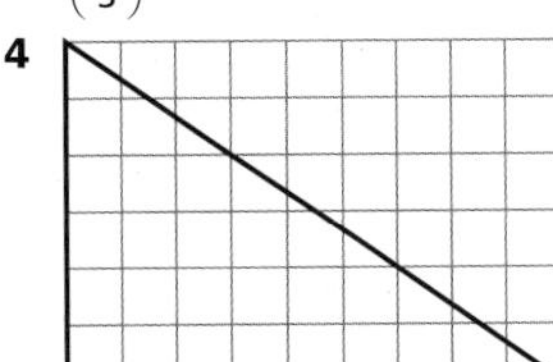

5 75°

6 42.5 cm²

7 143°

8 Example answer could show ABC = 33° and DCE = 86°, and states condition AAS or ASA.

9 Use $\frac{1}{2}ab\sin C = 35$ to show OA ≈ 4 cm and then use area of circle $= \pi r^2$.

10 58.0°

Block 9 Basic probability

Chapter 9.1

Are you ready?

1 $\frac{7}{20}$

2 a $\frac{15}{41}$ **b** $\frac{24}{41}$ **c** 1

3 1, $\frac{3}{7}$, 55% and 0.35

Practice

1 a $\frac{1}{13}$ b $\frac{2}{13}$ c 0 d $\frac{6}{13}$ e $\frac{7}{13}$

2 a $\frac{3}{7}$ b $\frac{1}{7}$ c $\frac{2}{7}$ d $\frac{4}{7}$

3 a 6 blue, 10 red and 14 black
b $\frac{14}{30}$ c $\frac{16}{30}$

4 0.97

5 $\frac{4}{15}$

6 0.6

7 Example answer: The probability of winning a football match is not necessarily $\frac{1}{3}$. Other factors can affect the outcome.

8 Table completed for Stripey: $\frac{5}{8}$

9 P(yellow) = $\frac{2}{7}$ P(purple) = $\frac{4}{7}$ P(orange) = $\frac{1}{7}$

10 75

What do you think?

1 a No. If there are 8 counters in the bag, it is not possible to have these probabilities.
b 40

2 Because the probability of landing on red and the probability of landing on 1 are not mutually exclusive events, their probabilities cannot be added together.

3 Rob is correct. $\frac{1}{6} \times \frac{1}{6} = 36$

Consolidate

1 a $\frac{2}{7}$ b $\frac{3}{7}$ c $\frac{2}{7}$ d $\frac{5}{7}$ e 0

2 a 3 green, 5 yellow and 12 pink
b $\frac{12}{20}$ c $\frac{17}{20}$

3 0.29

4 $\frac{16}{80}$

Stretch

1 Not clear if 50% greater means add 0.5 or multiply by 1.5

2 $\frac{ab - a - b}{ab}$

3 $\frac{x + 2}{4x + 2}$

Basic probability: exam practice

1 Sections completed with 1, 2, 3, 4, 4, 4 in any order.

2 Example answers: $\frac{150}{360}$, $\frac{5}{12}$, $0.41\dot{6}$, $41.\dot{6}\%$

3 0.3

4 Example answers: $\frac{8}{20}$, $\frac{2}{5}$, 0.4, 40%

5 0.25, 0.4, 0.35

6 $\frac{n - 5}{n}$

Block 10 Probability diagrams

Chapter 10.1

Are you ready?

1 Garlic bread with Spaghetti; Garlic bread with BBQ burger; Garlic bread with Tofu stir-fry; Mozzarella sticks with Spaghetti; Mozzarella sticks with BBQ burger; Mozzarella sticks with Tofu stir-fry; Loaded nachos with Spaghetti; Loaded nachos with BBQ burger; Loaded nachos with Tofu stir-fry

2 a Heads, Heads Heads, Tails
Tails, Heads Tails, Tails
b i $\frac{1}{4}$ ii $\frac{3}{4}$

3 $\frac{3}{8}$

Practice

1 15

2 24

3 $\frac{1}{12}$

4 378

5 Kath is wrong because each card can only be used once.

6 $38 \times 37 \times 36 = 50616$

7 4

8 a 216 b $\frac{6}{216} = \frac{1}{36}$

9 a 6 b 120

10 24

11 864

What do you think?

1 a 190 connections
b 19900 connections
c $\frac{n(n-1)}{2}$ connections

Consolidate

1 35

2 $\frac{1}{60}$

3 a 81 b $\frac{3}{81}$

4 a 105 b $\frac{1}{105}$ c $\frac{14}{105}$ d $\frac{1}{91}$

Stretch

1 25

2 $\frac{1}{10}$

3 Let Class Y have y students.
Class X has $y + 1$ students and Class Z has $y - 1$ students.
The total number of ways of selecting one student from Class X and one student from Class Z is $(y + 1)(y - 1) \equiv y^2 - 1$, which is 1 less than the square of y.

4 a $10 \times 10 \times 10 \times 10 = 10000$
b 5040 c 625 d 81 e 35

5 a 12 b 24

Chapter 10.2

Are you ready?

1 a

	Left-hand dominant	Right-hand dominant	Total
Left-foot dominant	18	**32**	50
Right-foot dominant	**4**	46	**50**
Total	22	**78**	**100**

b $\frac{18}{100}$

2 a $\frac{25}{225}$ b $\frac{45}{225}$

3 2, 3, 4, 5, 6, 7, 8, 9, 10, 11, 12

Practice

1 a

		Spinner A				
		1	2	3	4	5
Spinner B	1	2	3	4	5	6
	2	3	4	5	6	7
	3	4	5	6	7	8
	4	5	6	7	8	9

b $\frac{10}{20}$ c $\frac{4}{20}$

2 a

	1	2	3	4	5	6
H	(H, 1)	(H, 2)	(H, 3)	(H, 4)	(H, 5)	(H, 6)
T	(T, 1)	(T, 2)	(T, 3)	(T, 4)	(T, 5)	(T, 6)

b $\frac{10}{12}$

3 a

	Banana	Orange	Pear	Grapefruit
Orange	(O, B)	(O, O)	(O, P)	(O, G)
Apple	(A, B)	(A, O)	(A, P)	(A, G)
Mango	(M, B)	(M, O)	(M, P)	(M, G)

b i $\frac{1}{12}$ **ii** $\frac{5}{12}$

4 a

		Red					
		1	2	3	4	5	6
Blue	1	(1, 1) 2	(1, 2) 3	(1, 3) 4	(1, 4) 5	(1, 5) 6	(1, 6) 7
	2	(2, 1) 3	(2, 2) 4	(2, 3) 5	(2, 4) 6	(2, 5) 7	(2, 6) 8
	3	(3, 1) 4	(3, 2) 5	(3, 3) 6	(3, 4) 7	(3, 5) 8	(3, 6) 9
	4	(4, 1) 5	(4, 2) 6	(4, 3) 7	(4, 4) 8	(4, 5) 9	(4, 6) 10
	5	(5, 1) 6	(5, 2) 7	(5, 3) 8	(5, 4) 9	(5, 5) 10	(5, 6) 11
	6	(6, 1) 7	(6, 2) 8	(6, 3) 9	(6, 4) 10	(6, 5) 11	(6, 6) 12

b i $\frac{5}{36}$ **ii** $\frac{15}{36}$ **iii** $\frac{9}{36}$ **iv** $\frac{20}{36}$

Consolidate

1 a

		Mario's spinner				
		1	3	5	7	9
Jackson's spinner	2	(2, 1)	(2, 3)	(2, 5)	(2, 7)	(2, 9)
	4	(4, 1)	(4, 3)	(4, 5)	(4, 7)	(4, 9)
	6	(6, 1)	(6, 3)	(6, 5)	(6, 7)	(6, 9)
	8	(8, 1)	(8, 3)	(8, 5)	(8, 7)	(8, 9)

b $\frac{2}{20}$ **c** $\frac{14}{20}$

2 a

		Chloe		
		Milk (M)	White (W)	Dark (D)
Filipo	Milk (M)	(M, M)	(M, W)	(M, D)
	White (W)	(W, M)	(W, W)	(W, D)
	Dark (D)	(D, M)	(D, W)	(D, D)

b i $\frac{1}{9}$ **ii** $\frac{4}{9}$ **iii** $\frac{4}{9}$

3 a

	1	2	3	4	5	6
A	(A, 1)	(A, 2)	(A, 3)	(A, 4)	(A, 5)	(A, 6)
B	(B, 1)	(B, 2)	(B, 3)	(B, 4)	(B, 5)	(B, 6)
C	(C, 1)	(C, 2)	(C, 3)	(C, 4)	(C, 5)	(C, 6)
D	(D, 1)	(D, 2)	(D, 3)	(D, 4)	(D, 5)	(D, 6)

b $\frac{4}{24}$ **c** $\frac{23}{24}$

Stretch

1 a

		Sum of first two tokens														
	+	2	3	3	4	4	4	5	5	5	5	6	6	6	7	7	8
Third token	1	3	4	4	5	5	5	6	6	6	6	7	7	7	8	8	9
	2	4	5	5	6	6	6	7	7	7	7	8	8	8	9	9	10
	3	5	6	6	7	7	7	8	8	8	8	9	9	9	10	10	11
	4	6	7	7	8	8	8	9	9	9	9	10	10	10	11	11	12

b $\frac{4}{8}$ **c** Discuss as a class.

2 a $\frac{67}{1296}$ **b** $\frac{56}{1296}$ **c** $\frac{430}{1296}$ **d** $\frac{297}{1296}$

3 a $\frac{10}{32}$ **b** $\frac{1}{2^n}$

Chapter 10.3

Are you ready? (A)

1 41, 43, 47

2 a 4 and 8 **b** 1, 3, 5, 7 and 9

3 a 58 **b** 11 **c** 29 **d** 23

Practice (A)

1 a

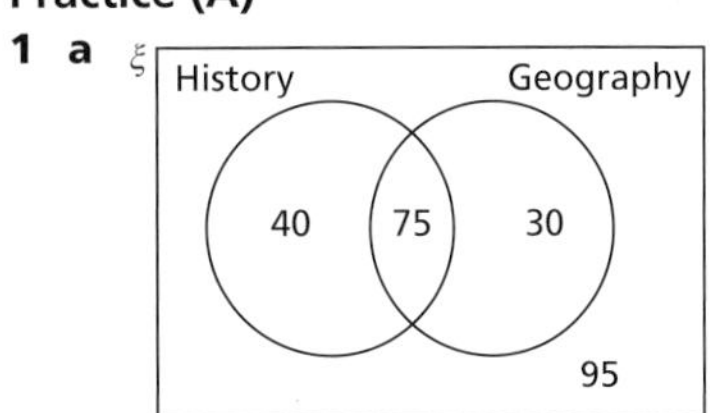

b $\frac{145}{240}$

2 a

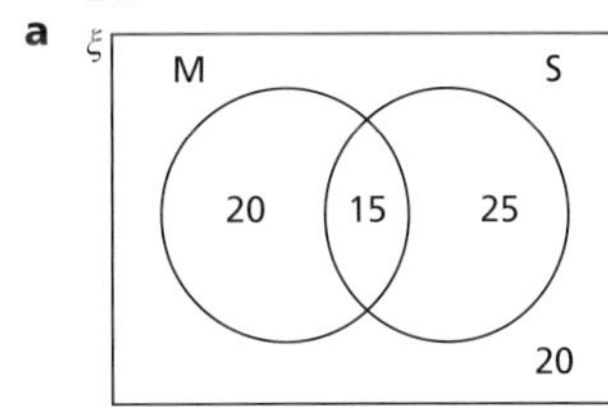

b 20 **c** $\frac{60}{80}$

3 a

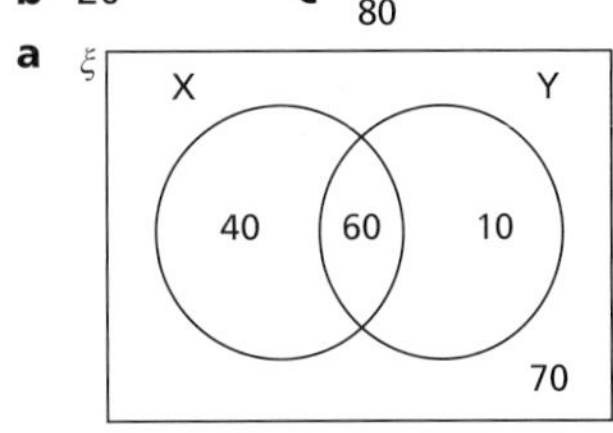

b 70 **c** $\frac{10}{180}$

4 $\frac{130}{200}$

5 $\frac{50}{60}$

6

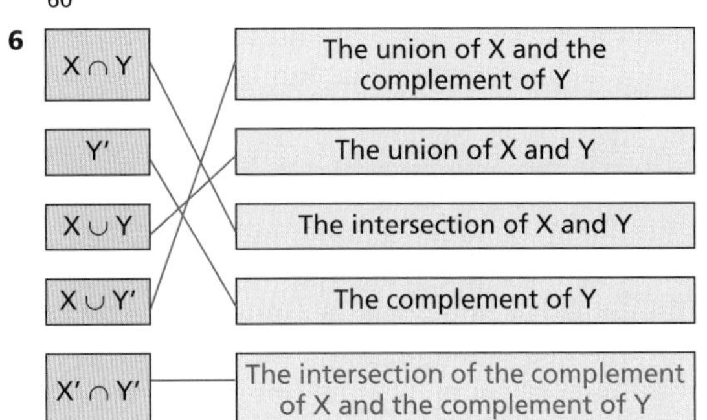

7 a $\frac{6}{12}$ **b** $\frac{3}{12}$ **c** $\frac{2}{12}$ **d** $\frac{6}{12}$ **e** $\frac{9}{12}$

f $\frac{7}{12}$ **g** $\frac{10}{12}$ **h** $\frac{4}{12}$ **i** $\frac{5}{12}$

What do you think? (A)

1 a $\frac{35}{43}$ **b** $\frac{14}{17}$

2 Example answer:

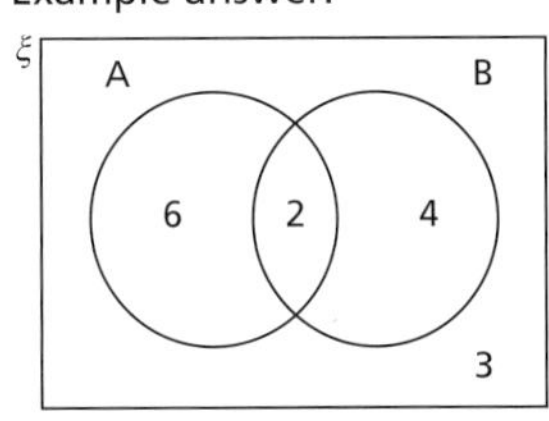

3 a Q **b** $P \cup Q$ **c** $P \cap Q$ **d** Q'

4 a All multiples of 4 are multiples of 2
b Multiples of 6
c Multiples of 12

Are you ready? (B)

1 a 20 **b** 200 **c** 60 **d** 10 **e** 60

Practice (B)

1 a 100 **b i** $\frac{60}{100}$ **ii** $\frac{25}{100}$ **iii** $\frac{97}{100}$
c A student who does not have a smartphone but has a tablet.

2 a 769 **b** 66.1% **c** $\frac{404}{1500}$ **d** $\frac{104}{650}$

3 a

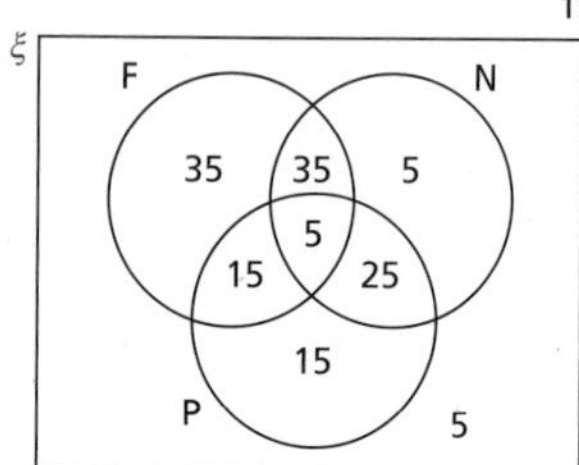

b 5 **c** $\frac{100}{140}$

4 a

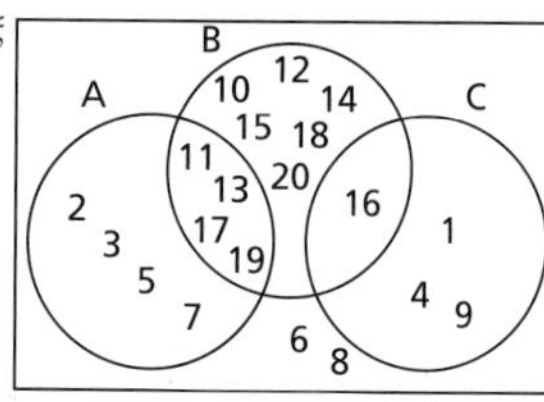

b i $\frac{1}{20}$ **ii** $\frac{18}{20}$

5 a

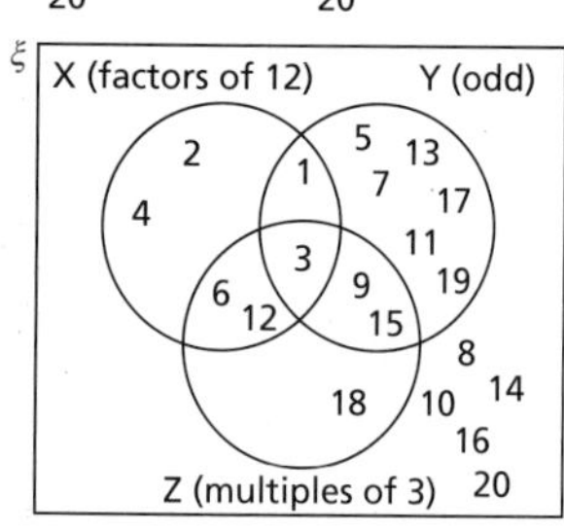

b i $\frac{3}{20}$ **ii** $\frac{1}{20}$ **iii** $\frac{15}{20}$ **iv** $\frac{17}{20}$

6 a

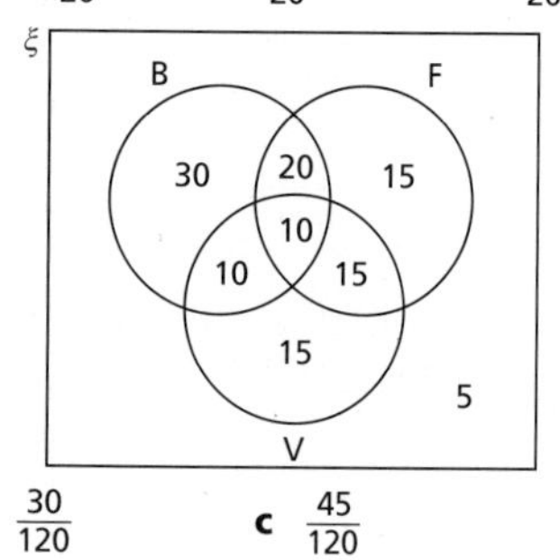

b $\frac{30}{120}$ **c** $\frac{45}{120}$ **d** $\frac{10}{60}$

7 40

8 $\frac{60}{260}$

What do you think? (B)

1 a $F \cup G \cup H$ **b** $F \cap G \cap H$ **c** $(G \cup H)'$

Consolidate

1 a

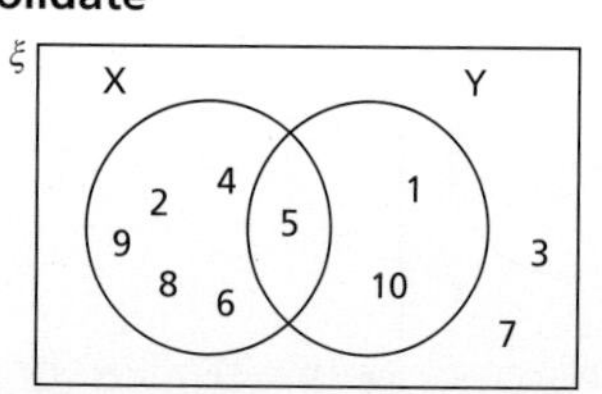

b $\frac{1}{10}$ **c** $\frac{7}{10}$

2 a

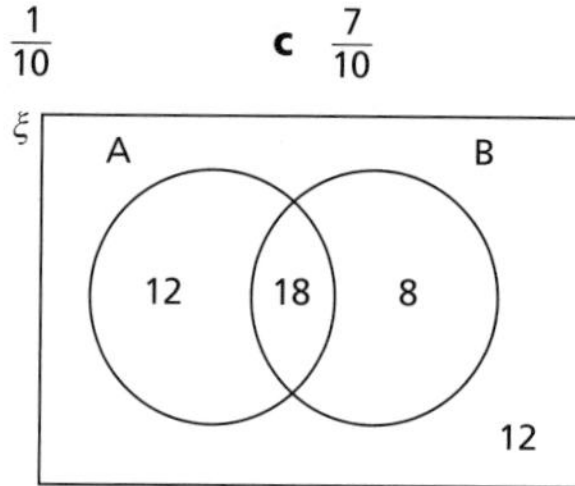

b $\frac{8}{50}$ **c** $\frac{20}{50}$

3 a

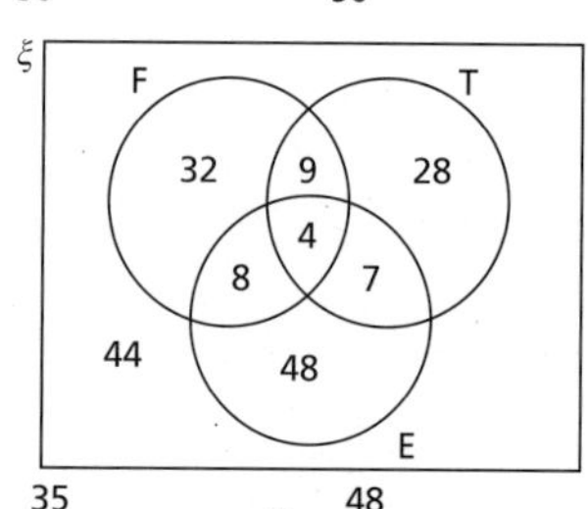

b $\frac{35}{180}$ **c** $\frac{48}{180}$ **d** $\frac{4}{53}$

4 a

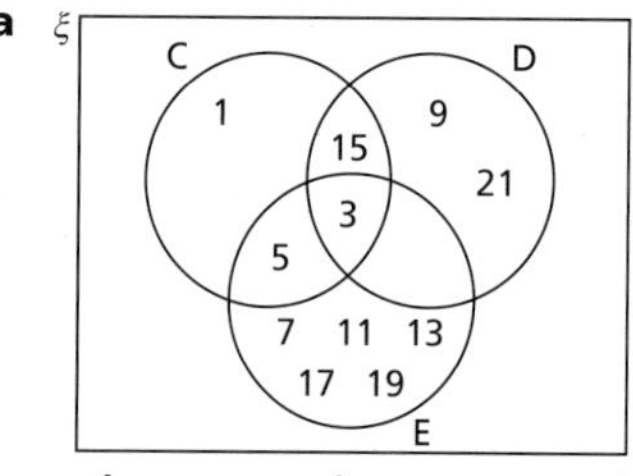

b i $\frac{4}{11}$ **ii** $\frac{1}{11}$ **iii** $\frac{10}{11}$

Stretch

1 a 50 people **b** $\frac{11}{50}$

2 a $x = 33$ **b** 173 people **c** $\frac{1711}{4186}$

3 $\frac{25}{52}$

4 a $5(2x - 1)$ **b** $\frac{2(3x - 1)}{5(2x - 1)}$ **c** $\frac{2x}{6x} = \frac{1}{3}$

Probability diagrams: exam practice

1 36, 37, 63, 67, 73, 76

2 Example answers: $\frac{10}{16}$, $\frac{5}{8}$, 0.625, 62.5%

3

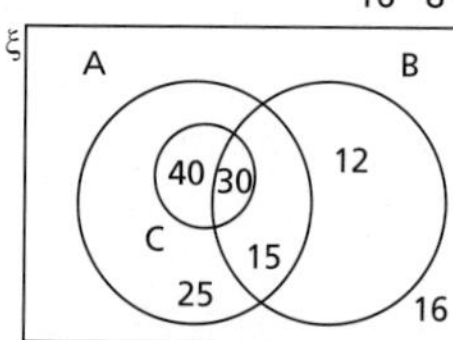

4 Example answers: $\frac{63}{120}$, $\frac{21}{40}$, 0.525, 52.5%

5 $\frac{52}{200}$, $\frac{13}{50}$, 0.26, 26%

6

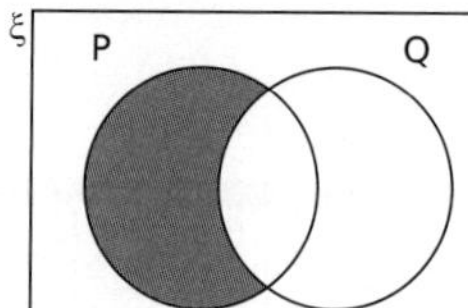

Block 11 Theory and experiment

Chapter 11.1

Are you ready?

1 $\frac{7}{10}$

2 a 27 **b** 66

3 a 16.8 **b** 36

Practice

1 a 4500
  b Car: $0.\dot{4}$ Bicycle: $0.\dot{3}$ Public transport: $0.\dot{2}$

2 a Relative frequencies: Red $\frac{11}{40}$, Yellow $\frac{15}{40}$, Green $\frac{14}{40}$
  b 7
  c There may be other colours in the bag, which haven't been selected.

3 a 0.2125 b 84

4 a $\frac{30}{300}$ b $\frac{240}{300}$

5 Flo because she completed the most trials and the relative frequency of 5s on her dice is closest to what you would expect.

6 220 students enjoy only video games or only books.
  $\frac{220}{400} = 0.55$

7 a The team did not win a game.
  b $\frac{5}{6}$

What do you think?

1 $\frac{2}{7}$ and 75

2 a $\frac{23}{60}$ b 19 c 100
  d 10 000 000. Discuss answers as a class.

Consolidate

1 Own vehicle: $\frac{149}{335}$ Public transport: $\frac{182}{335}$
  Walked: $\frac{4}{335}$

2 a Relative frequencies: Stripes $\frac{5}{15}$, Spots $\frac{2}{15}$, Stars $\frac{8}{15}$
  b 8

3 a $\frac{42}{120}$ b 18

4 a $\frac{50}{280}$ b $\frac{190}{280}$

Stretch

1 $\frac{x-3}{x}$

2 a Relative frequency of event A: $\frac{3x-5}{x(11-x)} = \frac{1}{3}$
  $3(3x-5) = x(11-x)$
  $9x - 15 = 11x - x^2$
  $x^2 - 2x - 15 = 0$
  b $x^2 - 2x - 15 = 0$
  $(x-5)(x+3) = 0$
  $x = 5$ or $x = -3$
  Since the number of trials is $x(11-x)$, x must be > 0 so $x = 5$ only.
  $5(11-5) = 5 \times 6 = 30$

Chapter 11.2

Are you ready?

1 a 30 b 6 c 585
2 a 120 b 90 c 40

Practice

1 Blue: 25 Red: 10 Yellow: 5

2 Expected frequencies after 200 spins: 1 = 40, 2 = 30, 3 = 74, 4 = 56

3 970

4 2222

5 a 3 b 6 c 12

6 153

7 192

8 a 16
  b It is not guaranteed that Benji will always score 3 out of every 4 penalties.

9 60

What do you think?

1 a Example answer: As probability is random, it is likely that a fair coin yields these results.
  b i Discuss as a class.
  ii Discuss as a class.
  iii Discuss as a class.

Consolidate

1 Green: 15 Orange: 10 Purple: 5

2 Expected frequencies after 400 spins: A = 72, B = 130, C = 8, D = 190

3 a i 315 ii 945 b 64

4 a 4 b 24 c 12

Stretch

1 a 9 b 21

2 a Expected number of tails = $n \times$ P(tails)
  Expected number of heads = $n \times$ P(heads)
  So $n \times$ P(heads) = $2 \times n \times$ P(tails)
  P(heads) = $2 \times$ P(tails)
  P(heads) + P(tails) = 1
  $2 \times$ P(tails) + P(tails) = 1
  $3 \times$ P(tails) = 1
  P(tails) = $\frac{1}{3}$
  P(heads) = $\frac{2}{3}$
  b P(heads) = $\frac{3}{4}$
  c P(heads) = $\frac{k}{k+1}$

Theory and experiment: exam practice

1 120

2 40

3 0.55

4 a $\frac{29}{60}$
  b Example answer: the relative frequency of 5 is much greater than $\frac{1}{6}$

5 0.056

Block 12 Combined events

Chapter 12.1

Are you ready?

1 a $\frac{1}{24}$ b 0.06 c 0.16 d $\frac{4}{9}$

2 a $\frac{17}{30}$ b $\frac{27}{50}$

Practice

1 0.02

2 a $\frac{35}{80}$ b $\frac{9}{80}$ c $\frac{44}{80}$ d $\frac{36}{80}$

3 a $\frac{5}{32}$ b $\frac{81}{224}$

4 a $\frac{9}{25}$ b $\frac{4}{25}$

5 a 0.135 b 0.165 c 0.385

6 $\frac{25}{36}$

7 $\frac{44}{45}$

8 a $\frac{1}{8}$
  b Example answer: HHH, **HHT**, **HTH**, HTT, **THH**, THT, TTH, TTT

9 $\frac{9}{10}$

What do you think?

1 a i $\frac{1}{10000}$ ii $\frac{625}{10000}$ iii $\frac{256}{10000}$
  b Discuss as a class.

Consolidate

1 0.512

2 0.675

3 a $\frac{25}{36}$ b $\frac{4}{36}$

Stretch

1 a $\frac{24}{72}$ **b** $\frac{48}{72}$ **c** $\frac{35}{72}$

2 $\frac{x+1}{x+2}$

3 $\frac{b}{a+b}$

Chapter 12.2

Are you ready?

1 a $\frac{2}{9}$ **b** $\frac{6}{25}$ **c** $\frac{5}{6}$

2 a 0.08 **b** 0.12 **c** 0.0768

3 a $\frac{17}{24}$ **b** $\frac{5}{17}$ **c** $\frac{4}{23}$

4 a $\frac{4}{15}$ **b** 0 **c** $\frac{11}{15}$ **d** $\frac{7}{15}$

Practice

1 a **1st flip** **2nd flip**

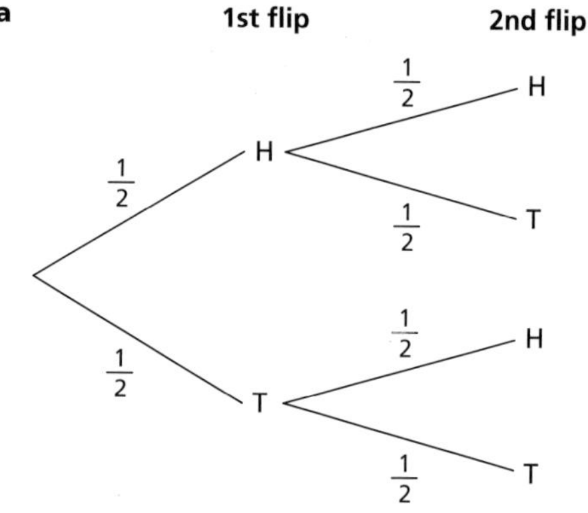

b i $\frac{1}{4}$ **ii** $\frac{1}{2}$ **iii** $\frac{1}{4}$

2 a **Spinner A** **Spinner B**

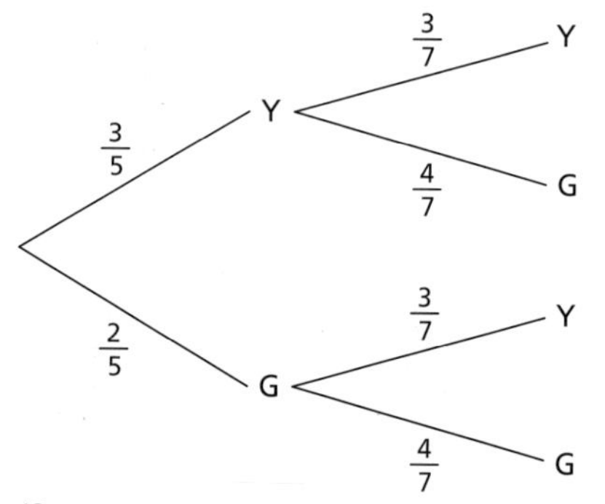

b $\frac{12}{35}$

c P(YY) = $\frac{9}{35}$ P(GG) = $\frac{8}{35}$ $\frac{9}{35} \neq \frac{8}{35}$

3 a **1st roll** **2nd roll**

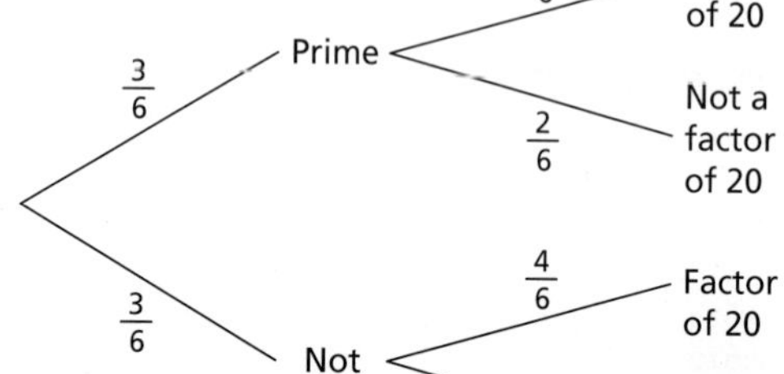

b i $\frac{12}{36}$ **ii** $\frac{12}{36}$ **iii** $\frac{6}{36}$ **iv** $\frac{6}{36}$

4 a **11P** **11Q**

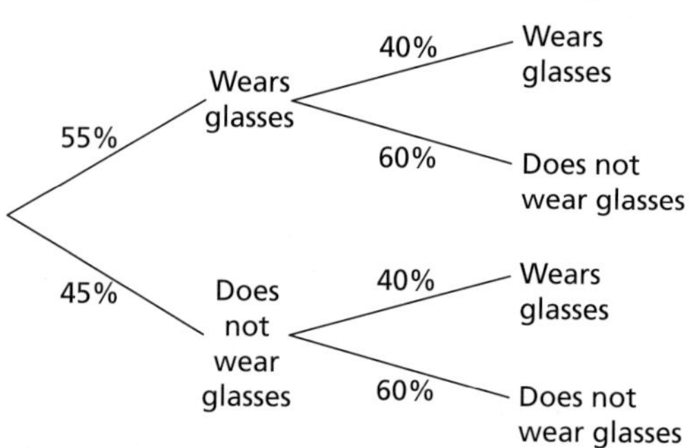

b P(11P – glasses, 11Q – glasses) = 22%
P(11P – glasses, 11Q – no glasses) = 33%
P(11P – no glasses, 11Q – glasses) = 18%
P(11P – no glasses, 11Q – no glasses) = 27%

5 a **A** **B**

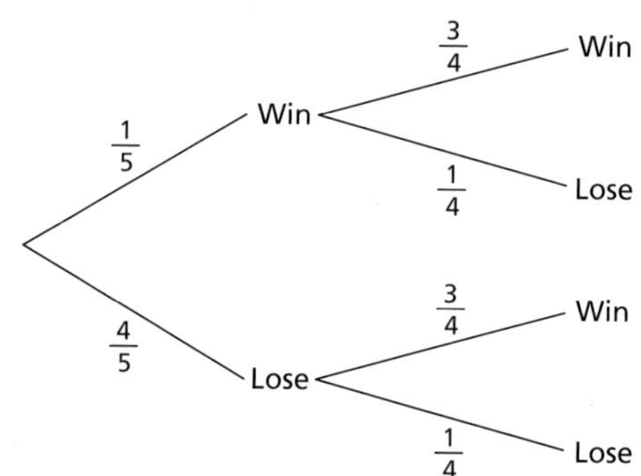

b $\frac{13}{20}$ **c** $\frac{16}{20}$

6 a **1st spin** **2nd spin**

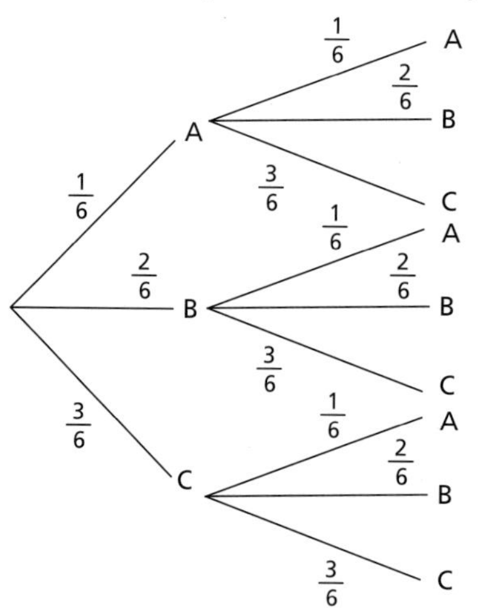

b i $\frac{3}{36}$ **ii** $\frac{14}{36}$ **iii** $\frac{4}{36}$

7 a Lida Mario

R: $\frac{4}{12}$ → R $\frac{4}{12}$, B $\frac{3}{12}$, G $\frac{5}{12}$

B: $\frac{3}{12}$ → R $\frac{4}{12}$, B $\frac{3}{12}$, G $\frac{5}{12}$

G: $\frac{5}{12}$ → R $\frac{4}{12}$, B $\frac{3}{12}$, G $\frac{5}{12}$

b i $\frac{25}{144}$ **ii** $\frac{63}{144}$

8 a

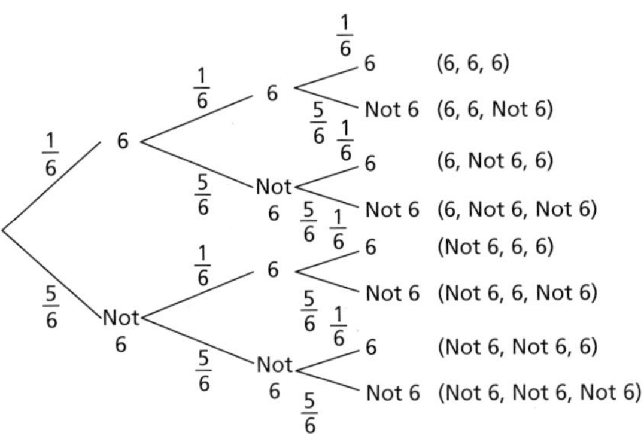

b $\frac{125}{216}$ **c** $\frac{75}{216}$

What do you think?

1 a 0.6006 **b** 0.0506 **c** 0.9494

Consolidate

1 a

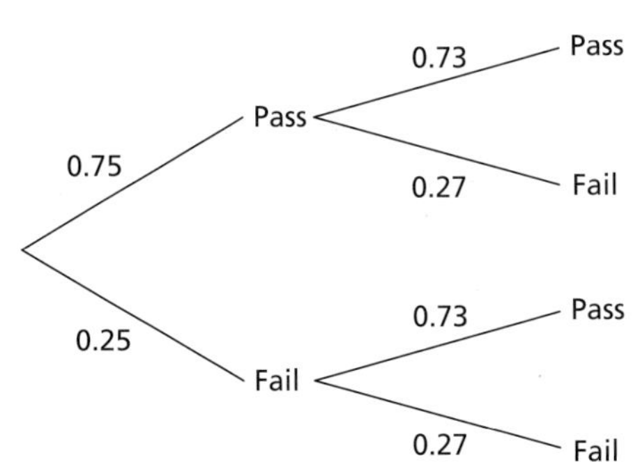

b 0.385

2 a

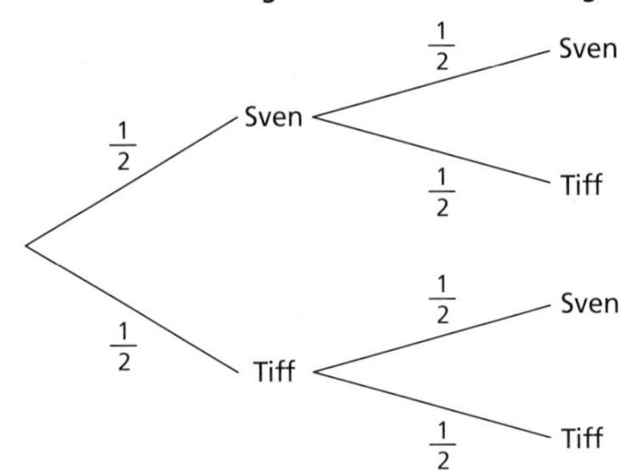

b $\frac{1}{2}$ **c** $\frac{1}{4}$

3 a

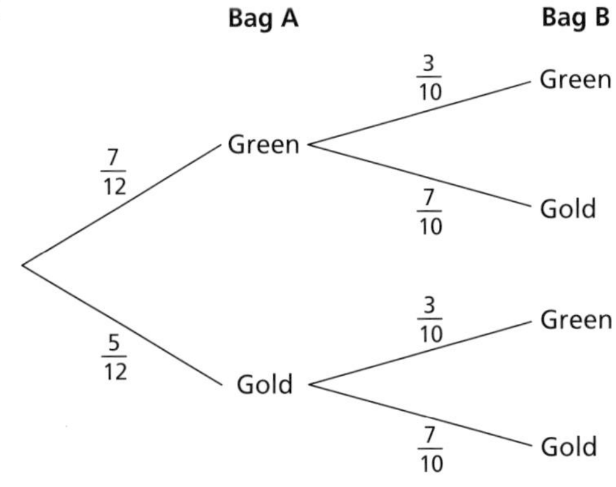

b $\frac{35}{120}$ **c** $\frac{64}{120}$

4 a

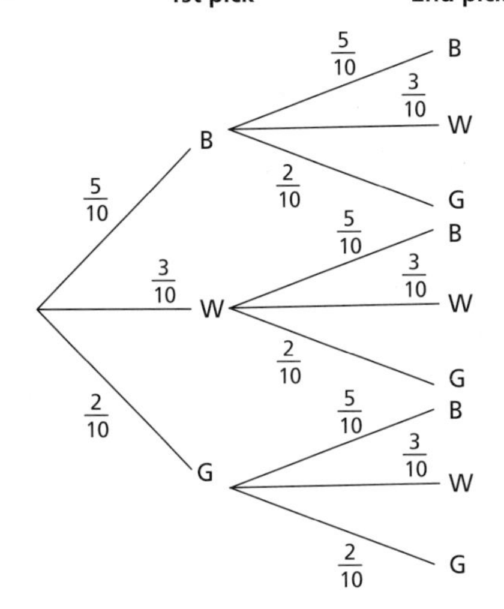

b i $\frac{62}{100}$ **ii** $\frac{32}{100}$

Stretch

1 a $\frac{2}{15}$ **b** $\frac{17}{100}$

2 a Amina has 11 red counters and 6 blue counters; Ali has 8 red counters and 5 blue counters.

b $\frac{118}{221}$

3 Testing $x = 8$

$\left(\frac{8}{11} \times \frac{8}{11}\right) + \left(\frac{3}{11} \times \frac{3}{11}\right)$

$= \frac{64}{121} + \frac{9}{121}$

$= \frac{73}{121}$

4 a P(blue) $= \frac{4}{k}$ P(red) $= \frac{k-4}{k}$

If blue chosen first, P(red) $= \frac{k-4}{k-1}$

If red chosen first, P(blue) $= \frac{4}{k-1}$

P(one of each colour) = P(1st blue, 2nd red) +

P(1st red, 2nd blue) $= \frac{8}{15}$

$\left(\frac{4}{k} \times \frac{k-4}{k-1}\right) + \left(\frac{k-4}{k} \times \frac{4}{k-1}\right) = \frac{8}{15}$

$\frac{4(k-4)}{k(k-1)} + \frac{4(k-4)}{k(k-1)} = \frac{8}{15}$

$\frac{8(k-4)}{k(k-1)} = \frac{8}{15}$

$120(k-4) = 8k(k-1)$

$120k - 480 = 8k^2 - 8k$

$8k^2 - 128k + 480 = 0$

$k^2 - 16k + 60 = 0$

b $k = 6, k = 10$

Chapter 12.3

Are you ready?

1 a $\frac{2}{15}$ **b** $\frac{1}{6}$ **c** $\frac{5}{6}$

2 a i $\frac{11}{25}$ **ii** $\frac{14}{25}$

b i $\frac{10}{24}$ **ii** $\frac{14}{24}$

3 $\frac{6}{13}$

Practice

1 a

1st bottle — 2nd bottle

$\frac{12}{22}$ Still: $\frac{11}{21}$ Still, $\frac{10}{21}$ Sparkling

$\frac{10}{22}$ Sparkling: $\frac{12}{21}$ Still, $\frac{9}{21}$ Sparkling

b $\frac{240}{462}$

2 a $\frac{3}{10}$ **b** $\frac{2}{9}$

c The overall number of cards is fewer so the probability of choosing a prime does not stay the same.

3 a $\frac{3}{5}$ **b** $\frac{3}{13}$

4 a 0.2625 **b** 0.3925

5 $\frac{1}{9}$

6 $\frac{4}{20}$

7 $\frac{5}{18}$

8 $\frac{13}{30}$

9 a $\frac{2}{56}$ **b** $\frac{18}{56}$

What do you think?

1 a $\frac{48}{91}$

b Example answer: with replacement, it is not as likely $\frac{48}{91} > \frac{24}{49}$

Consolidate

1 a $\frac{6}{110}$ **b** $\frac{12}{110}$

2 a

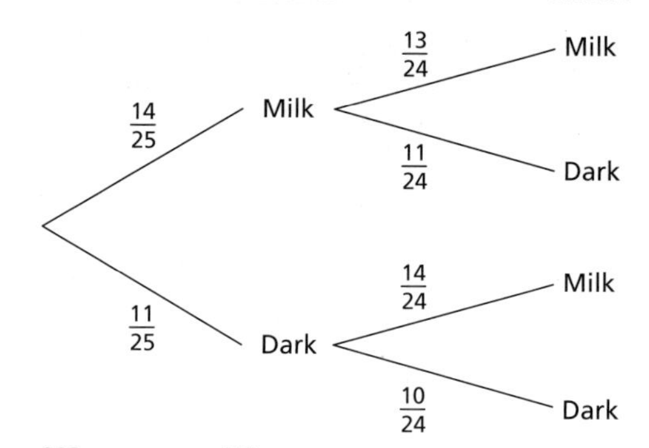

b $\frac{308}{600}$ **c** $\frac{418}{600}$

3 a

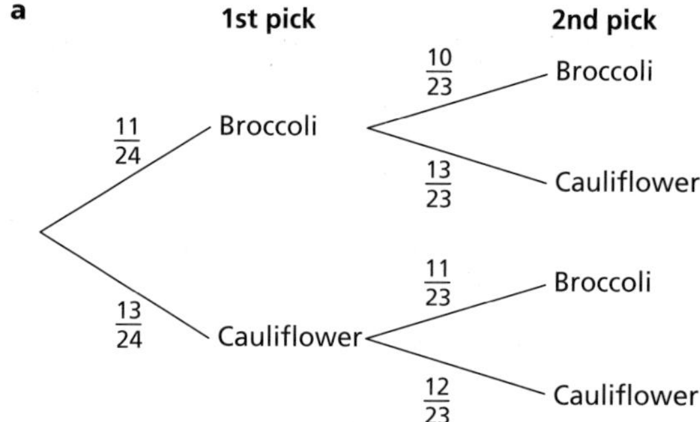

b $\frac{13}{46}$

4 $\frac{33}{132}$

5 a $\frac{3}{95}$ **b** $\frac{1}{190}$ **c** $\frac{7}{228}$

Stretch

1 P(Tiff eats a yellow sweet) $= \left(\frac{3}{27} \times \frac{10}{28}\right) + \left(\frac{15}{27} \times \frac{9}{28}\right) + \left(\frac{9}{27} \times \frac{9}{28}\right) = \frac{246}{756} = \frac{41}{126}$

2 $x = 2, x = 10$

3 a P(WW) $= \frac{y}{y+3} \times \frac{y-1}{y+2} = \frac{y(y-1)}{(y+3)(y+2)} = \frac{5}{14}$

$14y(y-1) = 5(y+3)(y+2)$

$14y^2 - 14y = 5y^2 + 25y + 30$

$9y^2 - 39y - 30 = 0$

b $y = 5$ or $y = -\frac{2}{3}$ so there were 5 white magnets initially.

c $\frac{15}{28}$

Combined events: exam practice

1 $\frac{1}{4}$
2 a 0.6 and 0.4 on each pair of branches
  b 0.48
3 0.9604
4 a

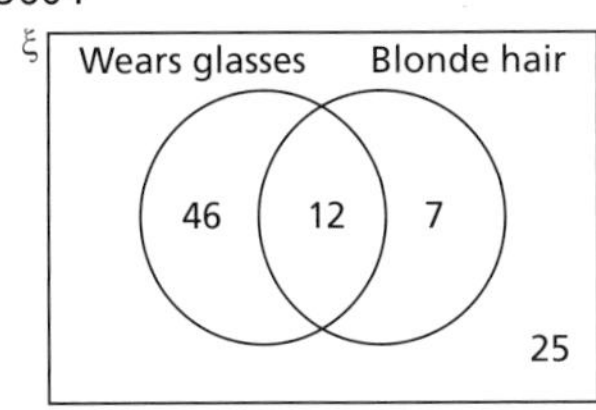

  b $\frac{12}{90}$
5 $\frac{31}{105}$

Probability: exam practice

1 $\frac{1}{36}$
2 0.18
3 0.42
4 90
5

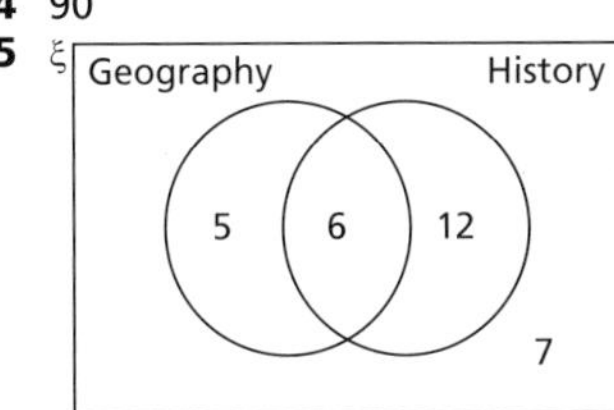

6 0.08
7 a $\frac{9}{22}$ b $\frac{5}{6}$
8 a $\frac{1}{10}$ b $\frac{5}{6}$

Block 13 Statistical measures

Chapter 13.1

Are you ready? (A)

1 a 2.35, 3.02, 3.15, 3.2, 3.52
  b −8, −7.2, −7, −6.6, −6.4, −5
  c 0.302, 0.8, 0.801, 0.81, 0.832
  d $\frac{2}{7}, \frac{1}{3}, \frac{2}{5}, \frac{3}{7}, \frac{3}{5}$
2 a 7.2 b 6.3 c −5.5 d 0.628

Practice (A)

1 a Mean = 18; Median = 18; Mode = 18; Range = 7
  b Mean = 8.59; Median = 8.7; Mode = 8.7 and 9.8; Range = 2.55
  c Mean = $\frac{12}{25}$; Median = $\frac{2}{5}$; Mode = no mode; Range = $\frac{3}{5}$
  d Mean = 0.582; Median = 0.8; Mode = no mode; Range = 0.612
  e Mean = −5.25; Median = −5.25; Mode = −4; Range = 3
  f Mean = 85.17 cm; Median = 84.5 cm; Mode = 83 cm; Range = 6 cm
  g Mean = −0.43; Median = −1; Mode = −1; Range = 6
  h Mean = 81.5%; Median = 80.5%; Mode = no mode; Range = 9%
2 a 180.1 cm (to 1 d.p.) b 7.75
  c 17.3 years old (to 1 d.p.) d 20.6 years old
3 a

	Mean	Median	Mode	Range
Group A	16.67	16.5	15	4
Group B	17.5	17.5	No mode	3

Group B has a higher mean and median, meaning the data is higher on average.
Group B has more consistent data as the range is smaller.

  b

	Mean	Median	Mode	Range
Group C	2.675	2.675	2.8	0.3
Group D	2.712	2.7	No mode	0.34

Group D has a higher mean and median, meaning the data is higher on average.
Group C has more consistent data as the range is smaller.

  c

	Mean	Median	Mode	Range
Group E	−7.11	−7.1	−8	2
Group F	−6.23	−5.7	No mode	5

Group F has a higher mean and median, meaning the data is higher on average.
Group E has more consistent data as the range is smaller.

  d

	Mean	Median	Mode	Range
Group G	0.0722	0.074	No mode	0.022
Group H	0.0697	0.0695	No mode	0.02

Group G has a higher mean and median, meaning the data is higher on average.
Group H has more consistent data as the range is smaller.

4 12 and 15
5 26 and 38
6 6, 8, 9, 9
  Range = 3

What do you think?

1 a The mean, median and mode would all double.
  b The mean, median and mode would change depending on which numbers were halved and which were doubled.
  c The mean, median and mode would all decrease by 3
2 The mean will change the most.
  The range would become greater if an extreme value is added, and would decrease if an extreme value is removed.

Are you ready? (B)

1 a Mean = 7.2; Median = 8; Mode = 8; Range = 5
  b Mean = 6.3; Median = 6.55; Mode = no mode; Range = 7.3
  c Mean = −5.3; Median = −5.5; Mode = −4 and −7; Range = 6
  d Mean = 0.6668; Median = 0.632; Mode = no mode; Range = 0.37

Practice (B)

1 a 12.76 b 1.35 c 1.6 d 5.96
2 a Mean = £1.80; Median = £1.75; Mode = £1; Range = £2.50
  b Mean = 3.4 kg; Median = 3.5 kg; Mode = 4 kg; Range = 3 kg
  c Mean = 16.25 cm; Median = 16 cm; Mode = 15 cm; Range = 3 cm
  d Mean = 700 g; Median = 750 g; Mode = 750 g; Range = 750 g
3 a Mean = 1.87 pets; Median = 2 pets; Mode = 2 pets; Range = 4 pets
  b Mean = 94.1p; Median = 90p; Mode = 90p; Range = 55p
4 Class A: Mean = 17.321; Median = 18; Range = 7
  Class B: Mean = 17.154; Median = 17; Range = 5
  Class A had the higher mean and median, so they had better scores on average.
  Class B had more consistent scores as the range is smaller.
5 a = 8, modal mark = 3

Consolidate

1 a Mean = 18.8; Median = 19; Mode = 17; Range = 4
  b Mean = £137.50; Median = £135; Mode = £135; Range = £32

c Mean = 5.53; Median = 5.72; Mode = 5.72 and 5.208; Range = 0.592
d Mean = 355 g; Median = 346 g; Mode = 320 g; Range = 166 g
e Mean = 0.427; Median = 0.6025; Mode = no mode; Range = 0.565
f Mean = –3.75; Median = –4; Mode = –4 and –5; Range = 7
g Mean = $\frac{5}{12}$; Median = $\frac{15}{48}$; Mode = $\frac{1}{4}$; Range = $\frac{1}{2}$
h Mean = $\frac{659}{1260}$; Median = $\frac{8}{15}$; Mode = $\frac{8}{15}$; Range = $\frac{1}{3}$

2 a Mean = 9.04; Median = 9; Mode = 9; Range = 4
b Mean = 3.63; Median = 4; Mode = 5; Range = 5
c Mean = £2359.38; Median = £2250; Mode = £2250; Range = £1500
d Mean = 71.7 s; Median = 1 min 10 s; Mode = 1 min 10 s; Range = 40 s

Stretch

1 $x = 7.5$
Masses = 25.5 kg, 22.5 kg, 12 kg, 19 kg
Median = 20.75 kg; Range = 13.5 kg
2 $a = 7$, $b = 4$
3 Mode = 1 and 2 books
Median = 2 books

Chapter 13.2

Are you ready?

1 Mean = 20; Median = 20; Mode = 20; Range = 6

Practice

1 a $20 \leqslant t < 22$ **b** 21.25 seconds
2 a $30 \leqslant a < 40$ **b** 40.48 years old
3 a 2 **b** 6200 steps **c** $7500 \leqslant x < 10000$
4 a 10.75 m **b** $5 \leqslant h < 10$ **c** $10 \leqslant h < 12$
5 a $10 \leqslant t < 15$ **b** $15 \leqslant t < 20$ **c** 12.5 degrees Celsius
6 $x = 6$

What do you think?

1 Table B as the class intervals are narrower.

Consolidate

1 a 31.95 kg **b** 4.42 m (to 2 d.p.)
c 19.5 seconds **d** 43.53 mph (to 2 d.p.)
2 a i $30 \leqslant x < 34$ **ii** $30 \leqslant x < 34$
b i $4.5 \leqslant h < 5$ **ii** $4.5 \leqslant h < 5$
c i $22 \leqslant t < 24$ **ii** $20 \leqslant t < 22$
d i $40 \leqslant s < 50$ **ii** $40 \leqslant s < 50$

Stretch

1 $x = 6$, $y = 4$
2 £49.90
3 £64.02

Chapter 13.3

Are you ready?

1 a 17 cm **b** 101 cm **c** 98 cm **d** 96.9 cm
2 a 14.54 (to 2 d.p.) **b** 15 **c** 15 **d** 3
3 a 63 **b** $70 \leqslant s < 75$ **c** $75 \leqslant s < 80$

Practice

1 a The outlier is 8
b Mean = 40 s; Median = 39 s; Mode = 35 s
2 a Mean = 11.17 ; Median = 7; Mode = 7
b 37
c Mean = 6; Median = 7; Mode = 7
The median and mode have stayed the same but the mean has decreased.
3 Class B has a higher mean and a greater range, indicating that, on average, Class B performed better but also had a greater variability in scores.
4 a Mean = 54.6 Mbps; Median = 57.5 Mbps; Range = 50 Mbps
b The outlier is 18
c Mean = 58.67; Median = 58; Range = 18
The median and mean have increased. The range has decreased.
5 a Class 1 = 70.71 Class 2 = 72.69
b Class 2 performed better in the quiz, as their mean was higher.
6 Comparison 1: On average, plumbers in Leeds earn more.
Comparison 2: The range in Leeds is £24 084 and in Newcastle is £18 306, which means the salaries in Newcastle are more consistent.
7

	Mean	Median	Range
Maths	14.3	14	14
English	14.55	15	11

Looking at both the mean and the median, students performed better in the English test. The range of the scores in the English test are smaller, which means they are more consistent.

Consolidate

1 a Mean = 16.28 s (to 2 d.p.); Median = 14.07 s; Mode = no mode; Range = 16.9 s
b 29.4 s
c Mean = 14.09 s; Median = 13.935 s; Mode = no mode; Range = 3.85 s
2 Comparison 1: The mean age at West Cliffe is higher than at East Lake.
Comparison 2: The range of ages at East Lake is smaller, meaning the ages are more consistent.
3 a Mean = 8.1
b Range = 5
c Comparison 1: The mean for the group of 15 is higher, so their score is higher on average.
Comparison 2: The range for the group of 10 is lower, so their scores are more consistent.
4 Comparison 1: The median for Top Lane is higher, meaning there are more children in each house on average.
Comparison 2: The range for Upper Street is lower, meaning the numbers of children in each house are closer together (more consistent).
5

	Mean	Median	Range
Maths	72.7%	73%	44%
English	67.7%	72%	69%

Comparison 1: The mean and median for Maths are higher, meaning on average the test scores were higher in Maths.
Comparison 2: The range for Maths is lower, meaning the scores were more consistent.

Stretch

1 Discuss as a class.
2 Discuss as a class.

Statistical measures: exam practice

1 Any five numbers with total less than 29, median greater than 6, and a range of 5
2 23.5
3 1.3 (to 1 d.p.)
4 a $10 < t \leqslant 15$ **b** 10.6 minutes

Block 14 Charts and diagrams

Chapter 14.1

Are you ready?

1 a 55 **b** Thursday **c** Wednesday and Friday
2 925

Practice

1 a 17 **b** Sunday **c** Thursday and Friday

2 a Quarter 3 **b** £325 000 **c** In 2023 in quarter 3
d £354 375 **e** £118 125

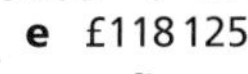

3 a

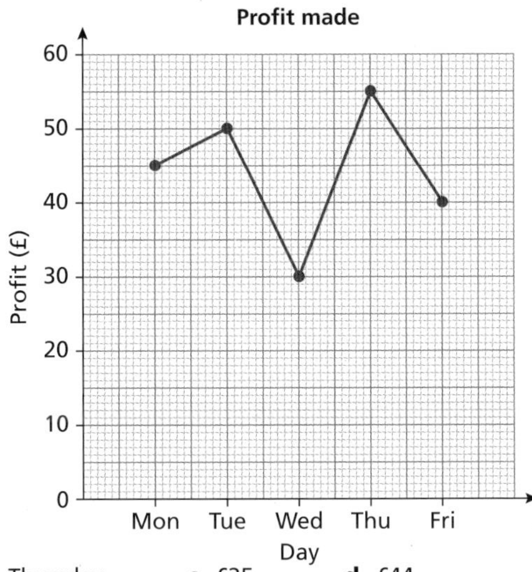

b Thursday **c** £25 **d** £44

4 Example answers: The lowest number of visitors for each year was in winter. The highest number of visitors for each year was in summer.

What do you think?

1 A trend line shows the general trend of the information but it won't necessarily be useful for estimating the data that doesn't fit with it, e.g. in 2018 and 2024 the population has dropped.

Consolidate

1 a 155 minutes **b** 2020
c 2017 **d** 35 minutes

2

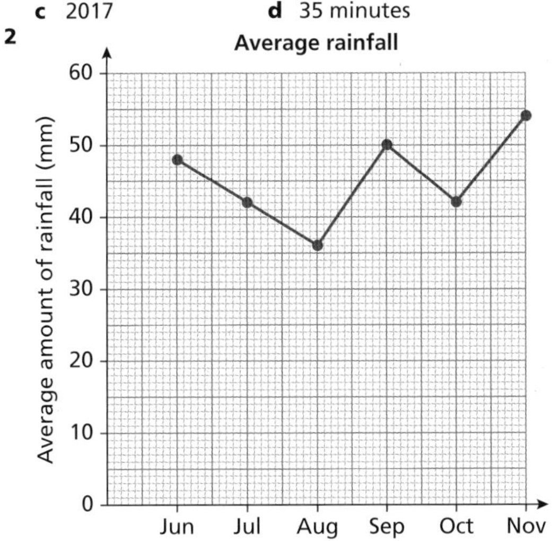

Stretch

1 a August **b** 1600 kWh

c A trend line would not be suitable as the first points of the graph have similar y values as the end points on the graph.

d Mean energy consumption for the first six months

$= \frac{2600 + 2300 + 2400 + 1800 + 1200 + 800}{6} = 1850\text{ kWh}$

Mean energy consumption for the second six months $= \frac{900 + 600 + 800 + 1800 + 2500 + 2800}{6}$

$= 1566.7\text{ kWh}$

$1850 > 1566.7$ so Zach is correct.

2 a Discuss as a class.

b i

Burgers sold

Number of burgers sold: 200, 190, 180, 170, 160, 150, 140, 130, 0

Week: 0, 1, 2, 3, 4, 5, 6

ii

Three-point period	Week 1 to 3	Week 2 to 4	Week 3 to 5	Week 4 to 6
Three-point moving average	151	161	169	178

iii

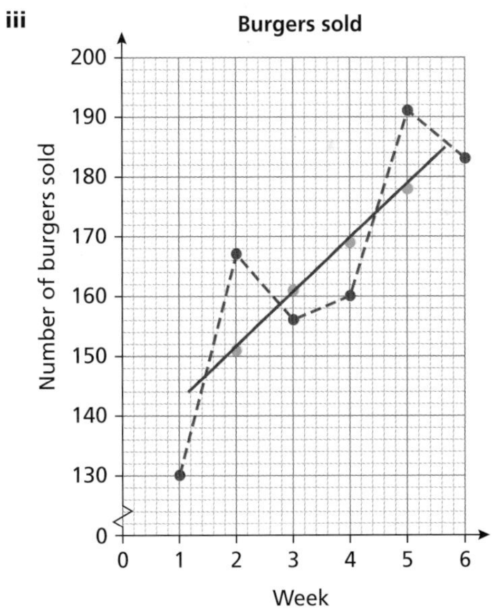

Chapter 14.2

Are you ready?

1 a 2 **b** 6.5 **c** 11 **d** 19

2 a 150 emails per hour **b** 2.5 emails per minute

3 a 245 items per week **b** 1050 items per month (30 days)

Practice

1 a

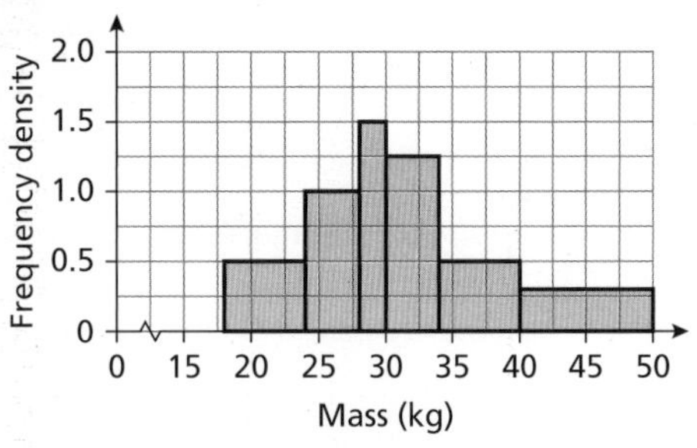

b

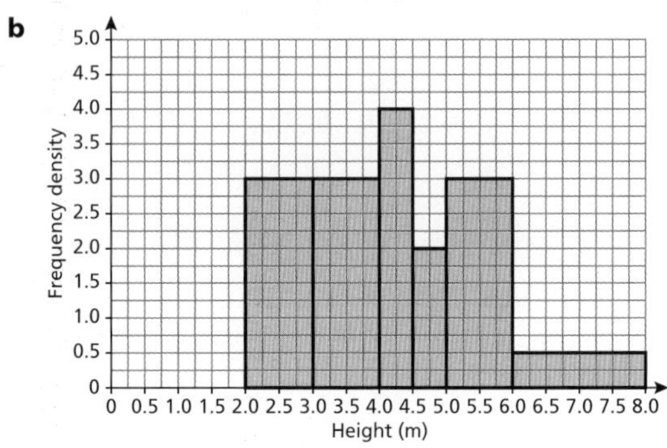

c

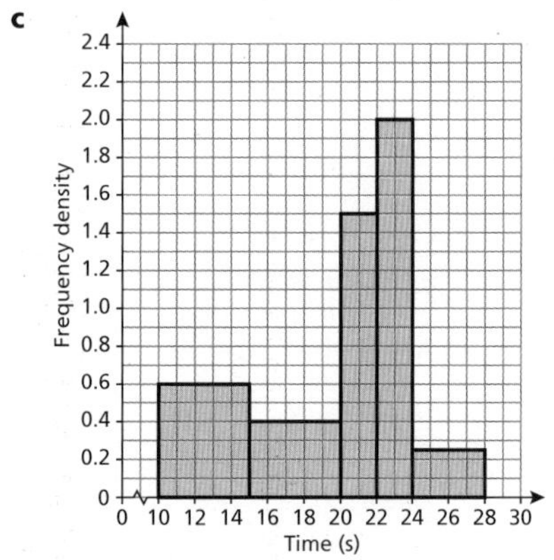

d

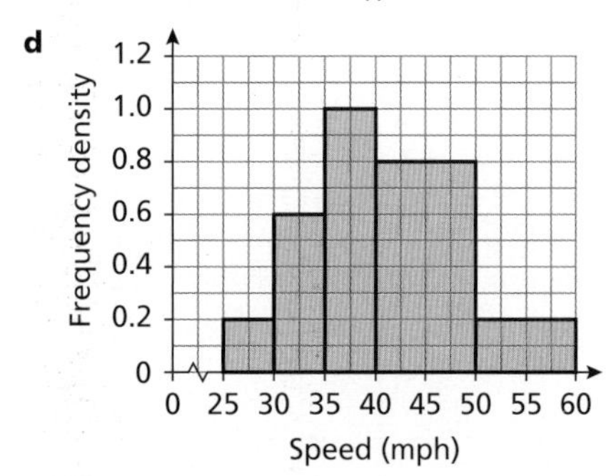

2 a

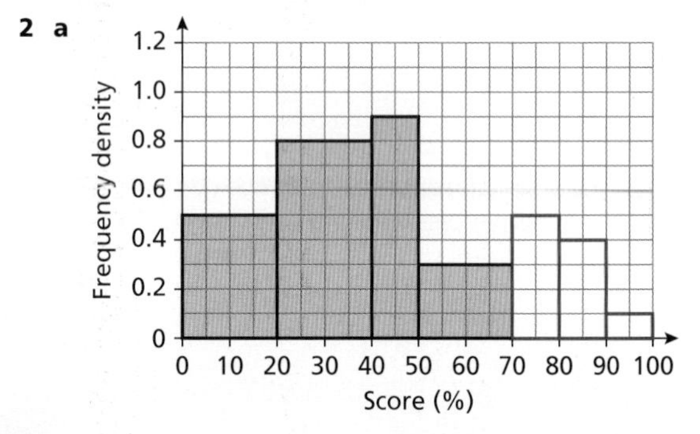

Score, s (%)	Frequency
$0 \leqslant s < 20$	**10**
$20 \leqslant s < 40$	**16**
$40 \leqslant s < 50$	**9**
$50 \leqslant s < 70$	**6**
$70 \leqslant s < 80$	5
$80 \leqslant s < 90$	4
$90 \leqslant s < 100$	1

b

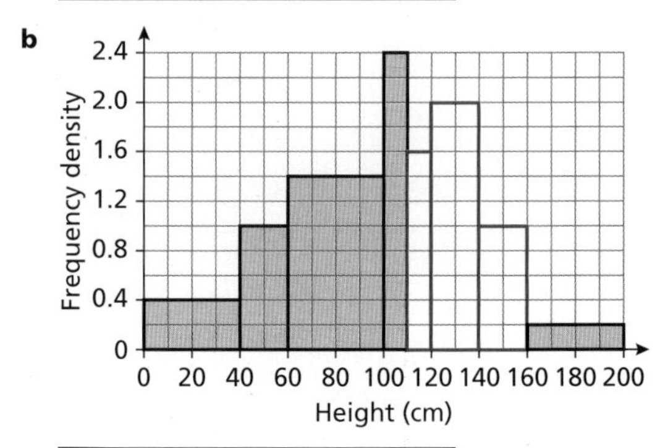

Height, h (cm)	Frequency
$0 \leqslant h < 40$	**16**
$40 \leqslant h < 60$	**20**
$60 \leqslant h < 100$	**56**
$100 \leqslant h < 110$	**24**
$110 \leqslant h < 120$	16
$120 \leqslant h < 140$	40
$140 \leqslant h < 160$	20
$160 \leqslant h < 200$	**8**

c

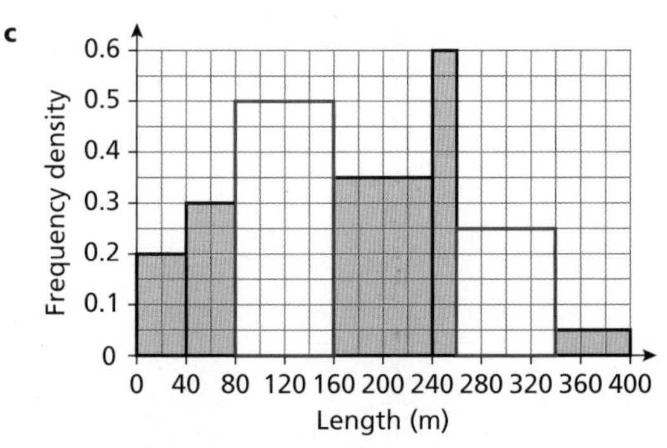

Length, l (m)	Frequency
$0 \leqslant l < 40$	**8**
$40 \leqslant l < 80$	**12**
$80 \leqslant l < 160$	40
$160 \leqslant l < 240$	**28**
$240 \leqslant l < 260$	**12**
$260 \leqslant l < 340$	20
$340 \leqslant l < 400$	**3**

d

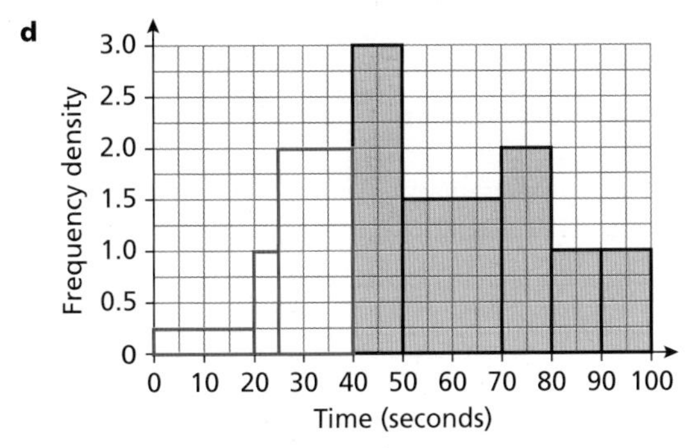

Time, t (seconds)	Frequency
$0 \leqslant t < 20$	5
$20 \leqslant t < 25$	5
$25 \leqslant t < 40$	30
$40 \leqslant t < 50$	**30**
$50 \leqslant t < 70$	**30**
$70 \leqslant t < 80$	**20**
$80 \leqslant t < 90$	**10**
$90 \leqslant t < 100$	**10**

3 **a** 17 **b** 18 **c** 15 **d** 7

4 38

Consolidate

1 **a**

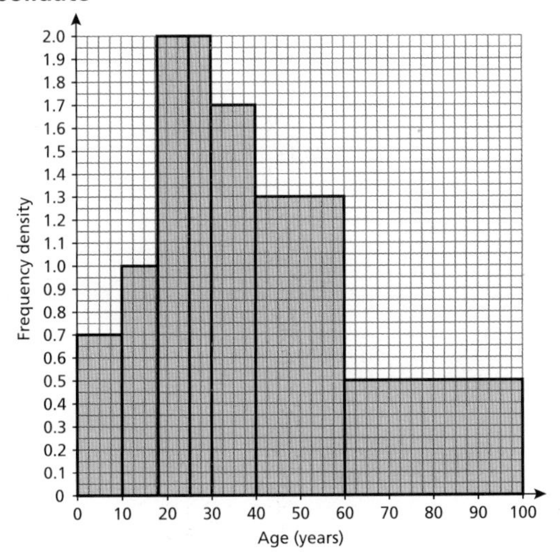

b

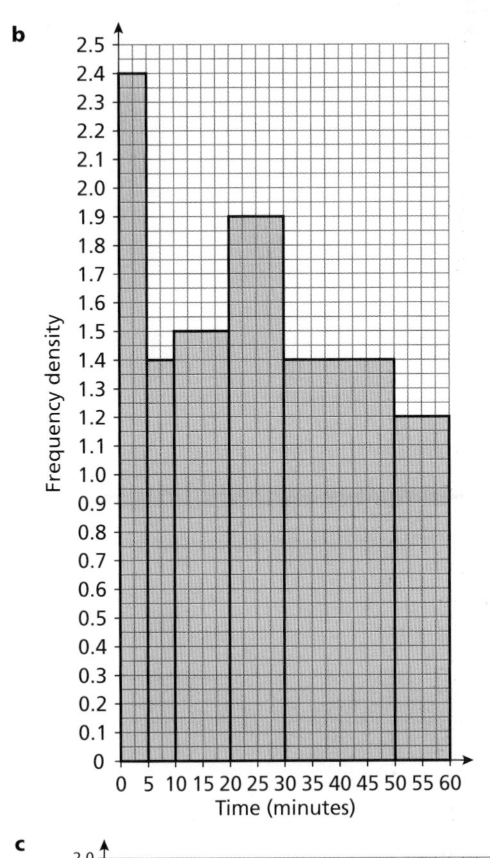

c

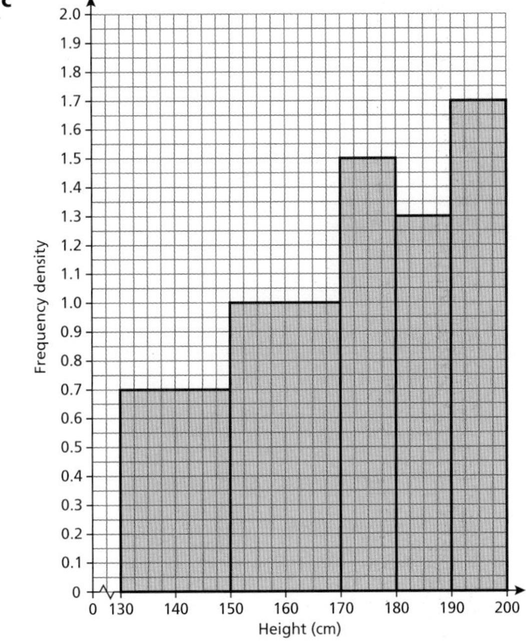

d

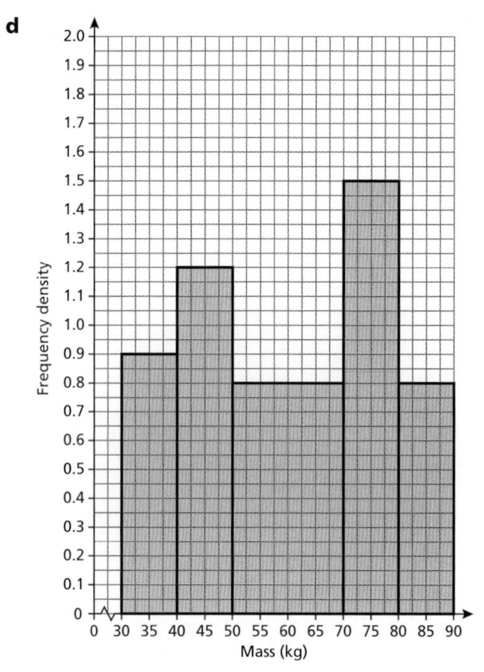

2 a

Weight, w (kg)	Frequency
$0 \leqslant w < 40$	8
$40 \leqslant w < 80$	20
$80 \leqslant w < 100$	12
$100 \leqslant w < 120$	20
$120 \leqslant w < 160$	16

b

Time, t (minutes)	Frequency
$0 \leqslant t < 10$	4
$10 \leqslant t < 15$	6
$15 \leqslant t < 20$	4
$20 \leqslant t < 30$	20
$30 \leqslant t < 35$	5
$35 \leqslant t < 45$	5
$45 \leqslant t < 50$	4

c

Speed, s (mph)	Frequency
$30 \leqslant s < 40$	3
$40 \leqslant s < 50$	5
$50 \leqslant s < 55$	4
$55 \leqslant s < 60$	3
$60 \leqslant s < 70$	4

d

Mass, m (grams)	Frequency
$0 \leqslant m < 4$	4
$4 \leqslant m < 9$	12
$9 \leqslant m < 11$	3
$11 \leqslant m < 16$	6

Stretch

1 Group A

Steps, s (thousands)	Frequency
$4 \leqslant s < 7$	25
$7 \leqslant s < 10$	25
$10 \leqslant s < 12$	40
$12 \leqslant s < 14$	35

Mean = 9.96 thousand (or 9960)

Group B

Steps, s (thousands)	Frequency
$4 \leqslant s < 6$	1
$6 \leqslant s < 7$	1
$7 \leqslant s < 9$	4
$9 \leqslant s < 11$	3
$11 \leqslant s < 12$	1
$12 \leqslant s < 13$	2

Mean = 9.17 thousand (or 9170)

2 **a** $a = 40$ **b** $b = 65$ **c** $c = 60$

3 44.15

Chapter 14.3

Are you ready? (A)

1 **a** 60 **b** 37 **c** $10 < a \leqslant 20$
d 45 **e** $10 < a \leqslant 20$

Practice (A)

1 a

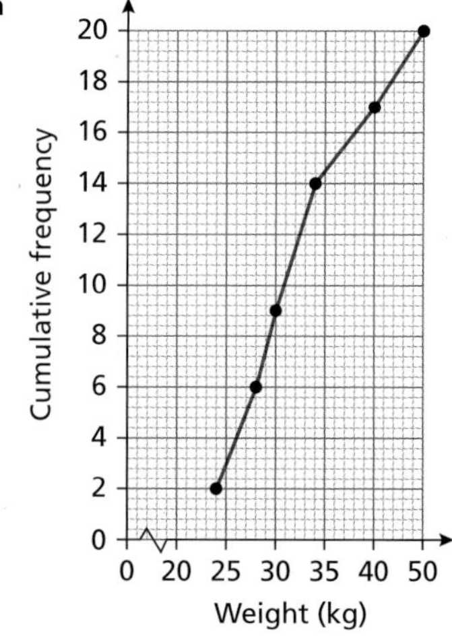

b

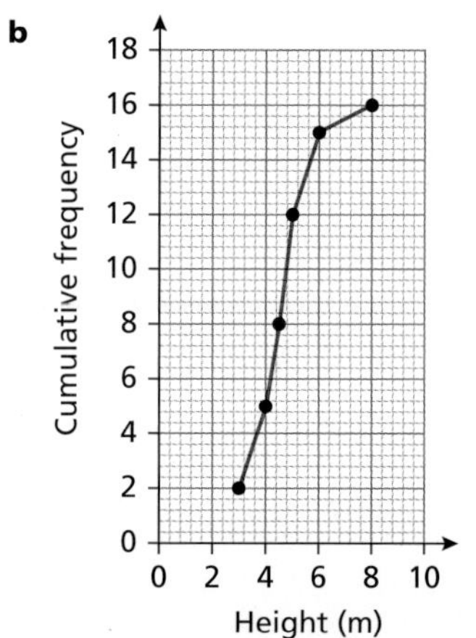

c

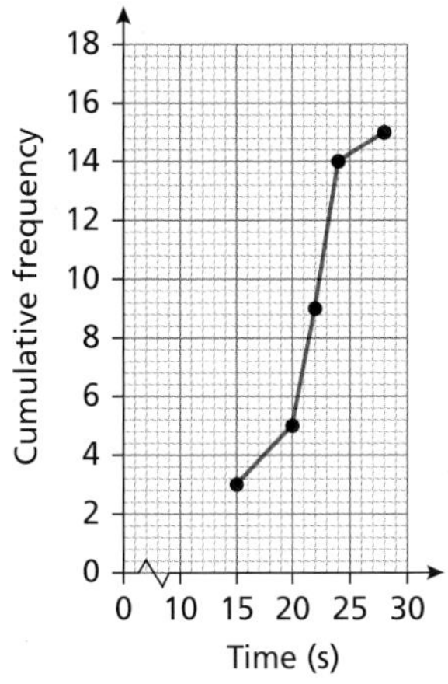

d

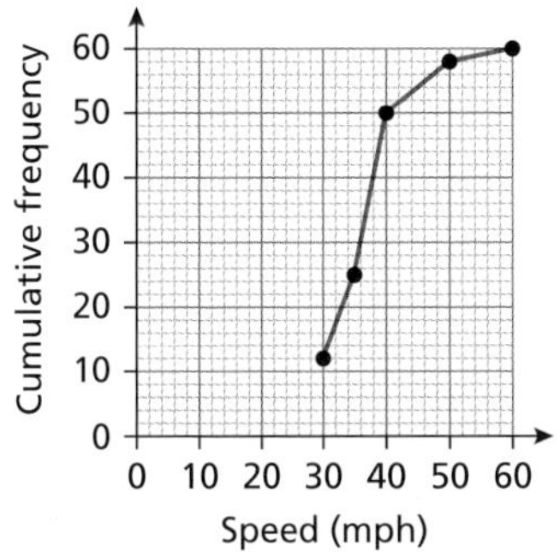

2 Approximate answers:
a 22 seconds **b** 18 years **c** £26 **d** 29 kg

3 a

Heights of people

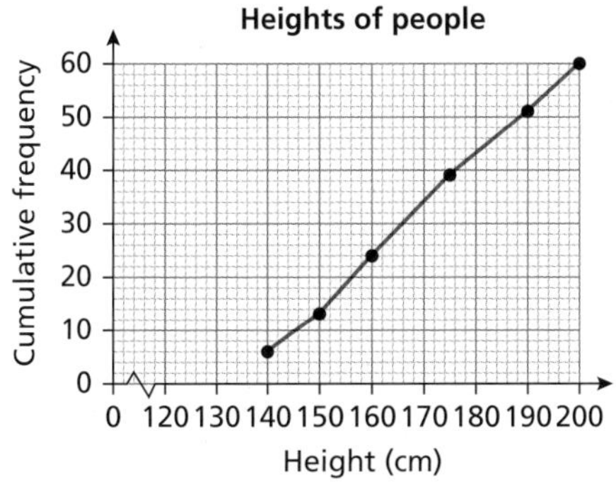

b Approx. 166 cm **c** Approx. 26

4 a

Race times

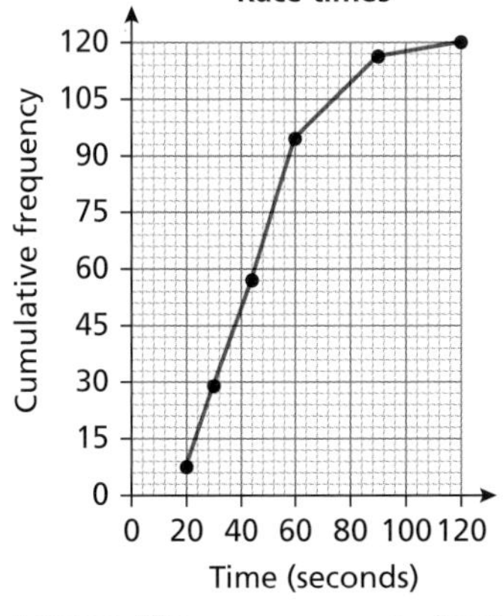

b Approx. 47 s **c** Approx. 47

Are you ready? (B)

1 a 45 **b** 20 **c** 90
2 a 7 **b** 14 **c** 2.43

Practice (B)

1 Approximate answers:
a 45 kg **b** 4.5 m **c** 20 s **d** 37 mph

2 Approximate answers:

a	**i** 22 s	**ii** 14 s	**iii** 28 s	**iv** 14 s
b	**i** 17 years	**ii** 13 years	**iii** 21 years	**iv** 8 years
c	**i** £31	**ii** £23	**iii** £40	**iv** £17
d	**i** 24 kg	**ii** 16 kg	**iii** 35 kg	**iv** 19 kg

3 a

Heights of people

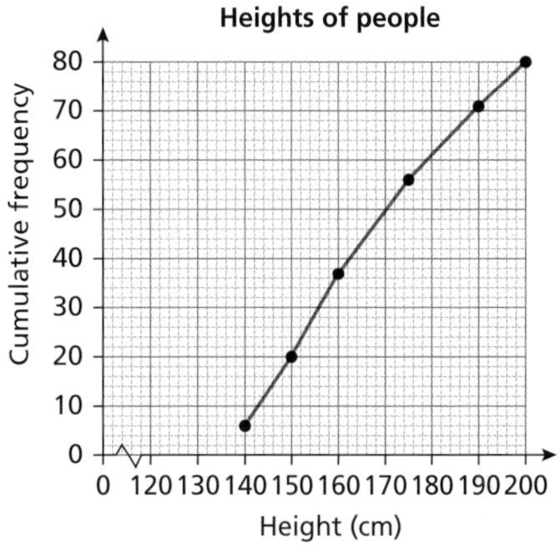

b Approx. 162 cm **c** Approx. 29 cm

4 a

Race times

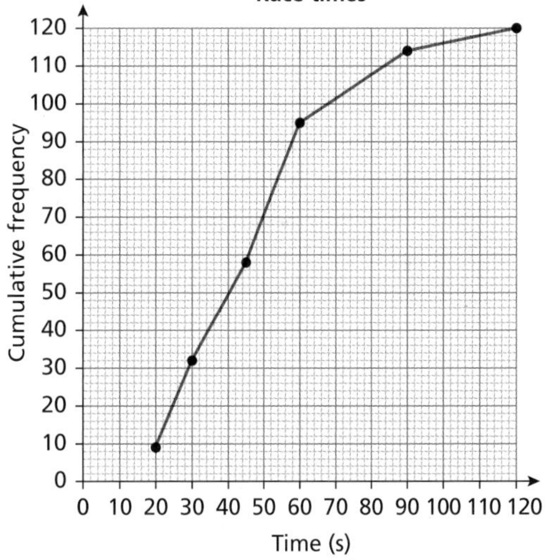

b Approx. 46 s **c** Approx. 71 **d** Approx. 29 s

Consolidate

1 a

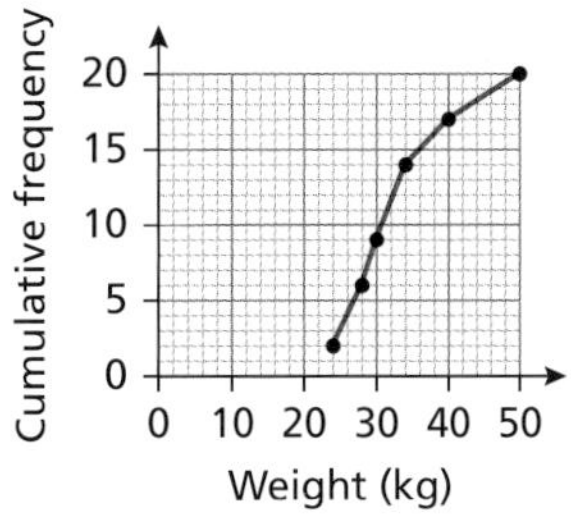

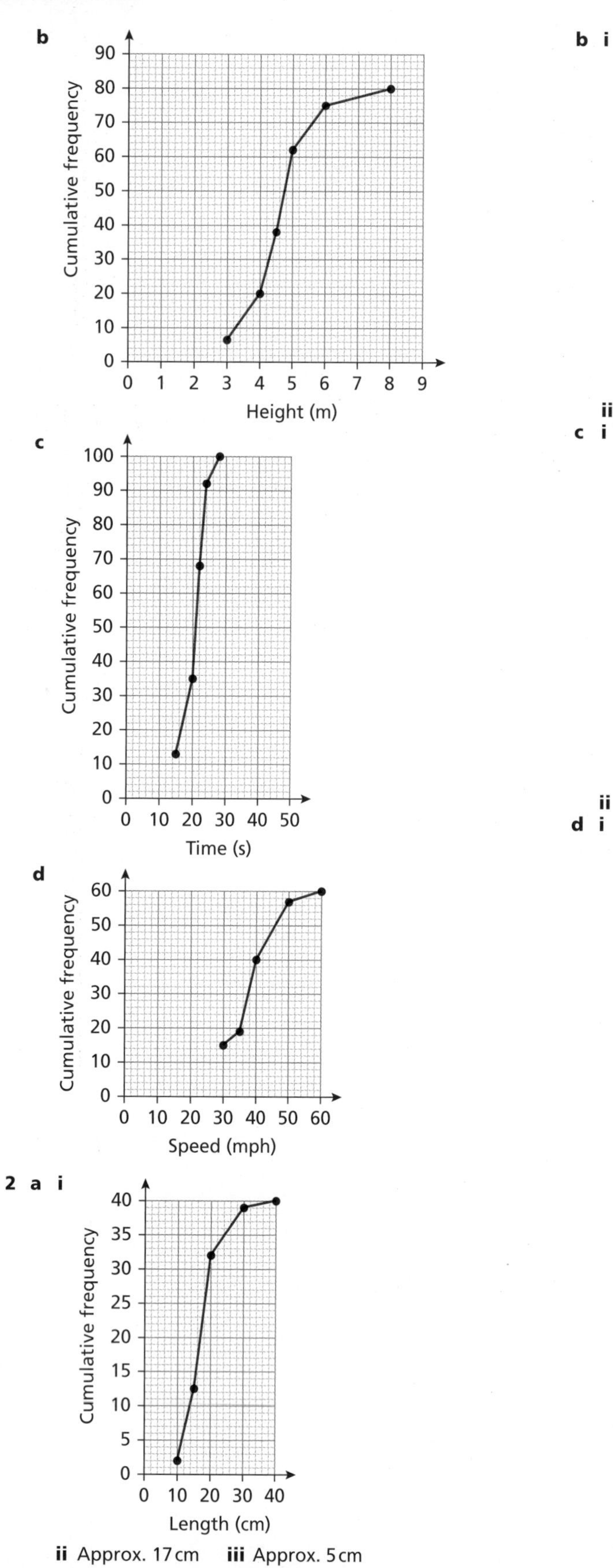

ii Approx. 17 cm **iii** Approx. 5 cm

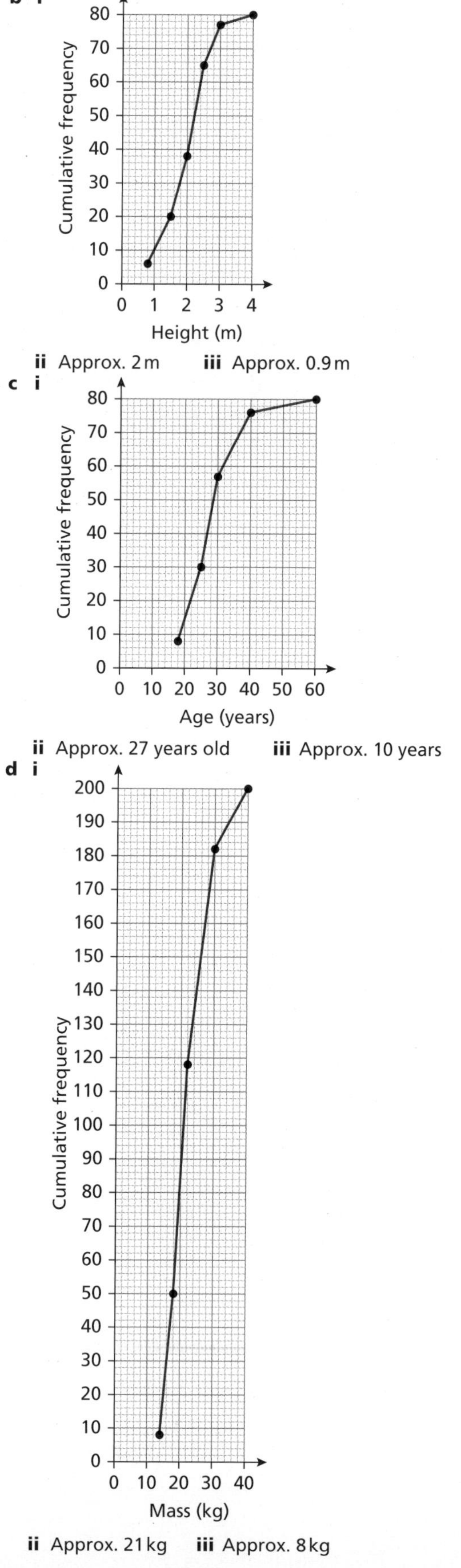

ii Approx. 2 m **iii** Approx. 0.9 m

ii Approx. 27 years old **iii** Approx. 10 years

ii Approx. 21 kg **iii** Approx. 8 kg

Stretch

1 15 minutes
Discuss accuracy as a class.
2 Answer in range £31.75 to £31.95

Chapter 14.4

Are you ready?

1 a i 48 **ii** 26 **b i** 3.4 **ii** 2.4
c i 152 **ii** 50 **d i** 16 **ii** 3.5

Practice

1 a

b

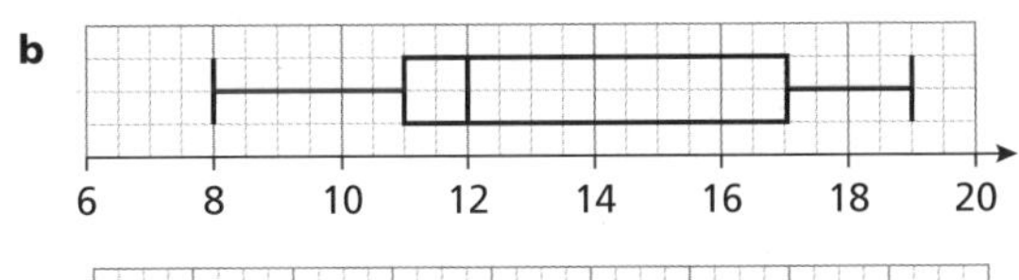

c

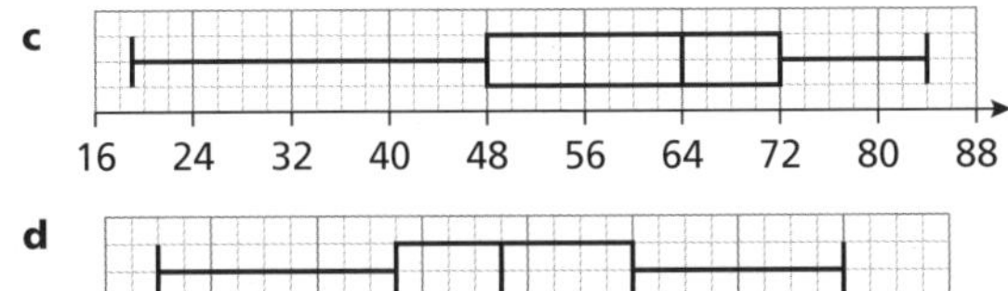

d

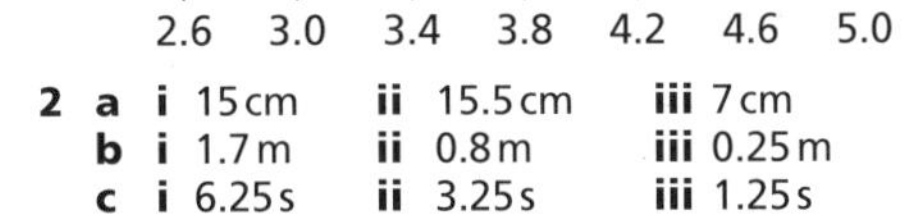

2 a i 15 cm **ii** 15.5 cm **iii** 7 cm
b i 1.7 m **ii** 0.8 m **iii** 0.25 m
c i 6.25 s **ii** 3.25 s **iii** 1.25 s
d i 55% **ii** 78% **iii** 35%

3 a

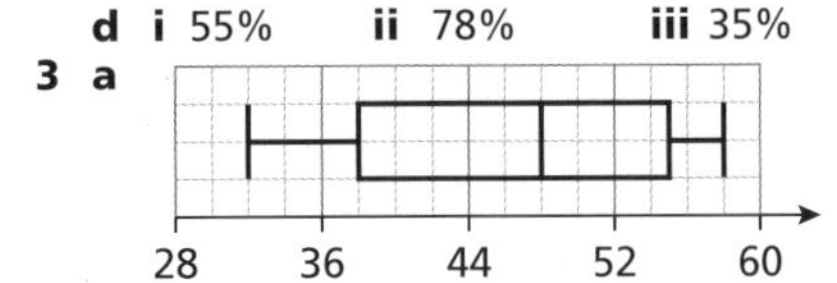

b

c

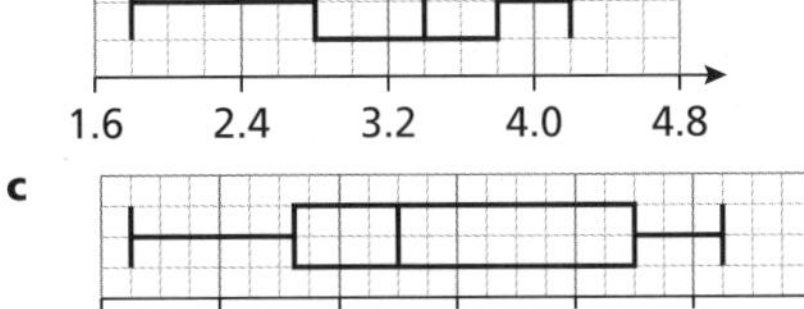

d

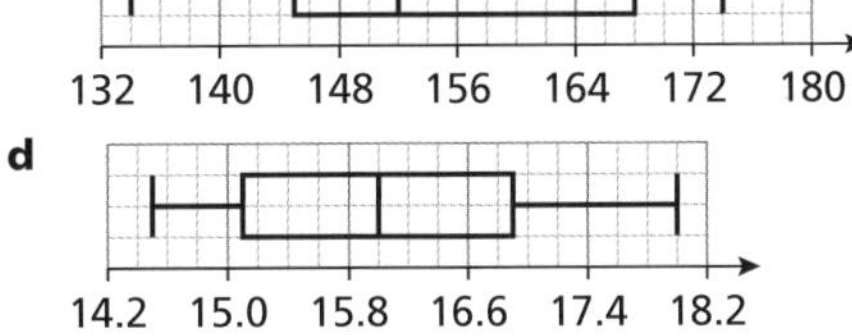

4 a The median speed for group B is higher than for A (34 mph compared to 31 mph).
The range for group B is higher, meaning the speeds are less consistent (A = 13 mph, B = 16.5 mph).
The interquartile range for group B is higher, meaning the speeds without outliers are less consistent (A = 6 mph, B = 8 mph).

b The median of A is higher than B (0.55 g/cm^3 compared to 0.45 g/cm^3).
The range for both A and B is 0.65 g/cm^3 meaning that the spread of the whole data set is the same.
The interquartile range for B is lower, meaning the density is more consistent across the middle half of the data (A = 0.25 g/cm^3, B = 0.2 g/cm^3).

c The median score of B is higher than A (28 compared to 27).
The range for group A is lower than for B, meaning the scores are more consistent (A = 22, B = 26).
The interquartile range for group B is lower than for A, meaning the scores without outliers are more consistent (A = 9, B = 8).

d The median time for group B is higher than for A, meaning the task took longer on average (A = 10.5 minutes, B = 11.5 minutes).
The range for group A is smaller than for B, meaning the time taken is more consistent for group A. (A = 6, B = 7).
The interquartile range for group B is lower than for A, meaning the time taken without outliers is more consistent (A = 2.75, B = 2.25).

5 a

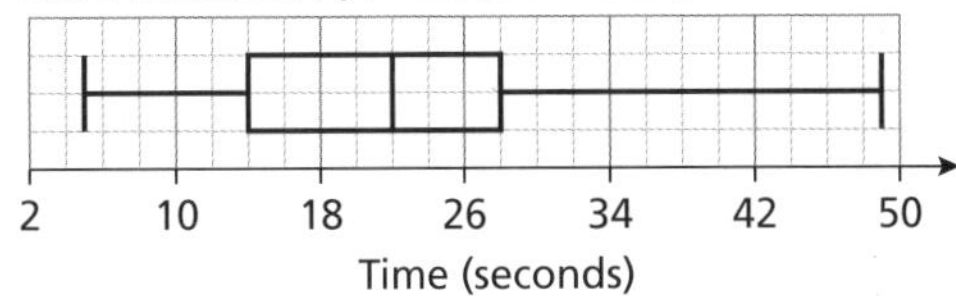

b

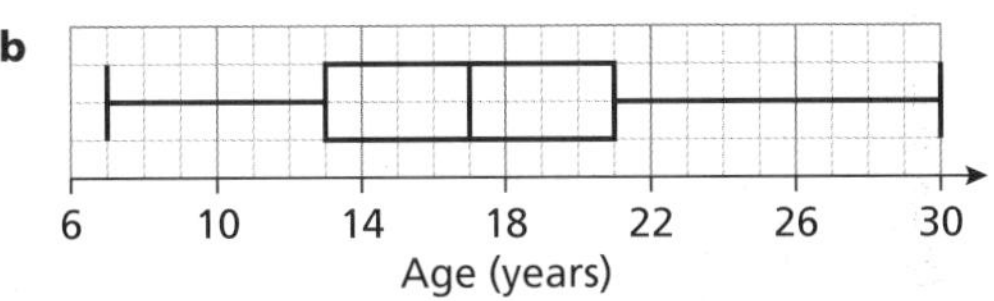

c

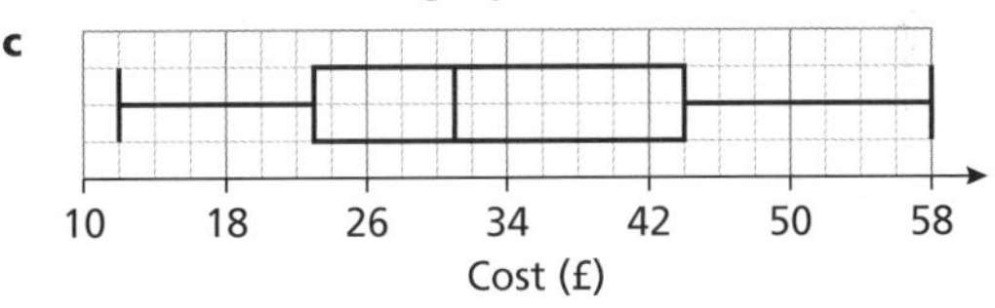

d

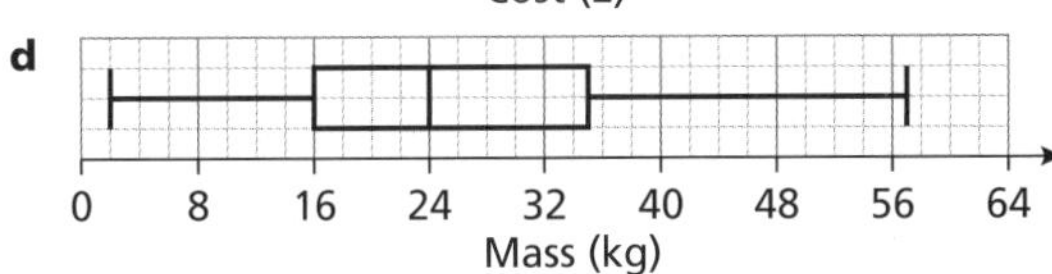

6 a

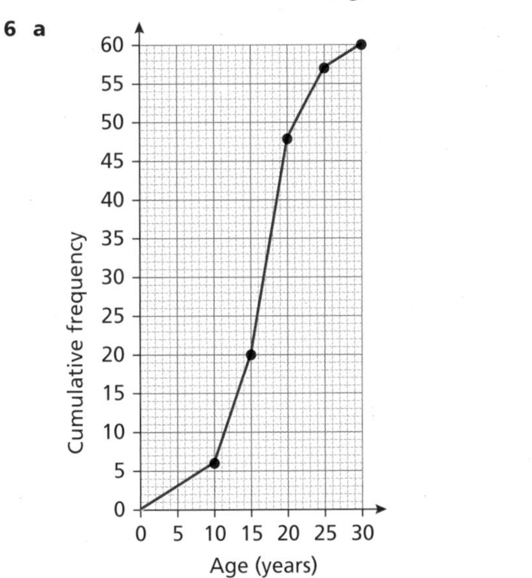

b

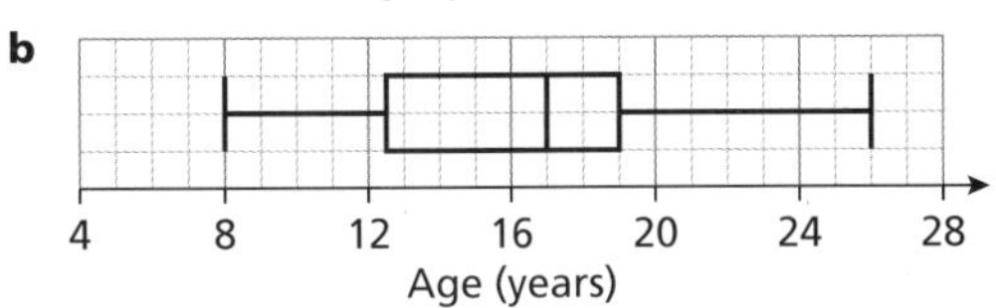

What do you think?

1 No, 25% of the data isn't always equal on either side of the median.

Consolidate

1 a

12 16 20 24 28 32 36 40 44 48

b

2.4 3.2 4.0 4.8 5.6 6.4 7.2 8.0 8.8

c

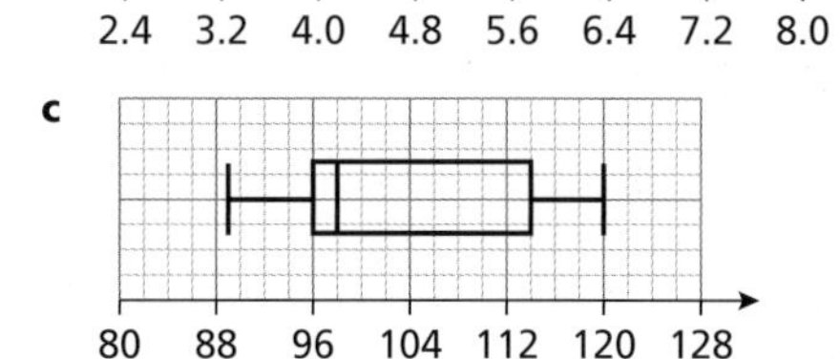

d

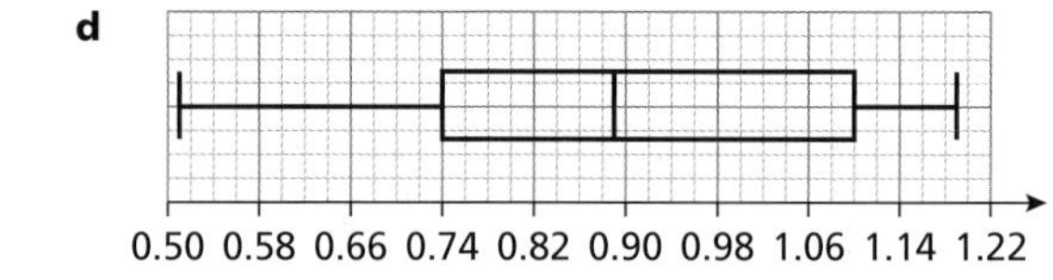

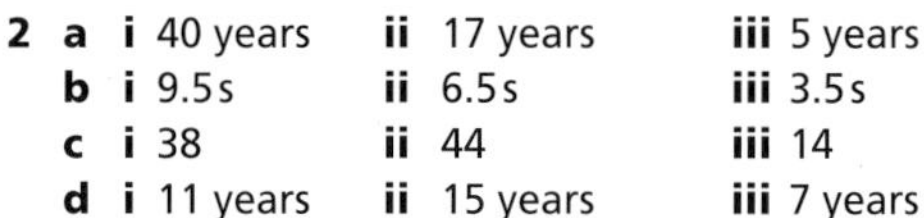

2 a i 40 years ii 17 years iii 5 years
b i 9.5 s ii 6.5 s iii 3.5 s
c i 38 ii 44 iii 14
d i 11 years ii 15 years iii 7 years

3 a

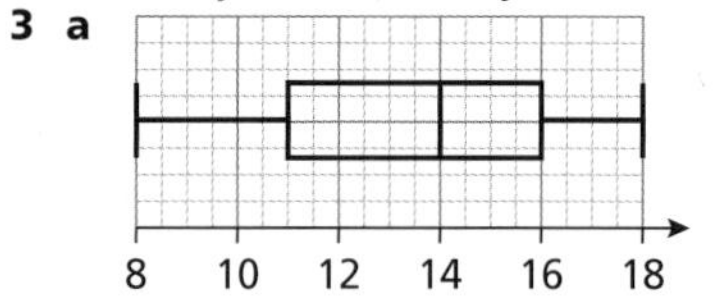

b

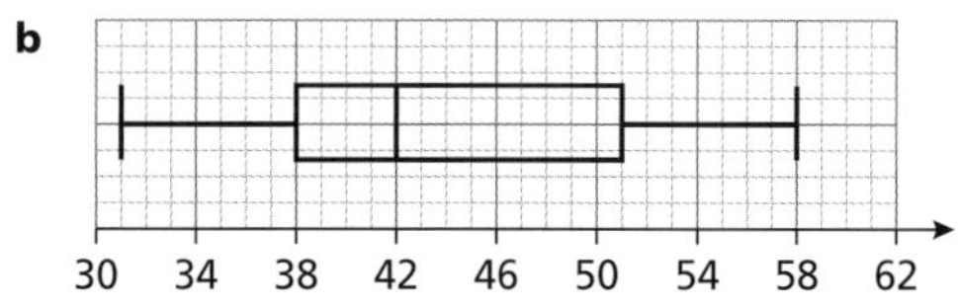

c

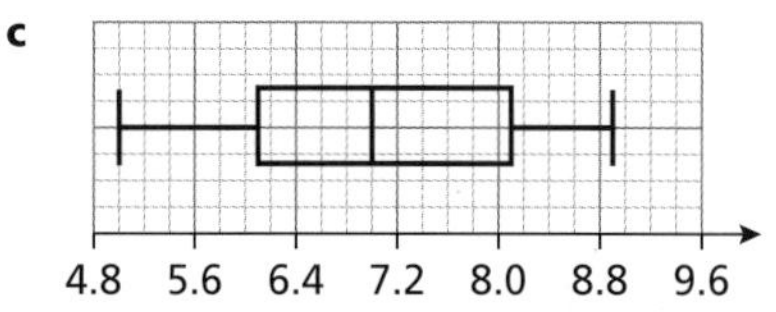

d

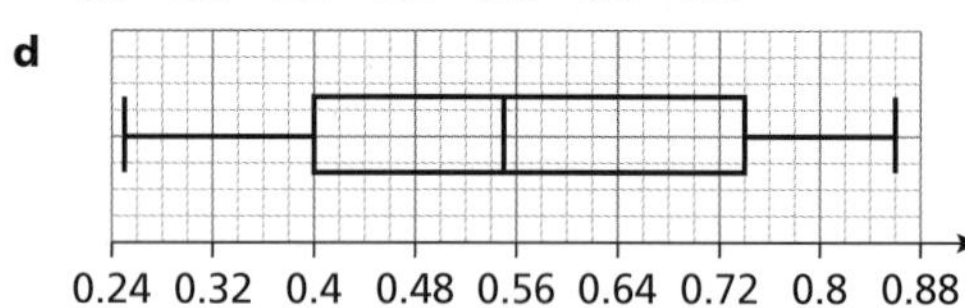

4 a The median height of B is higher than A meaning they are taller on average. (A = 0.8 m, B = 1 m).
The range of each group is the same (1.1 m).
The interquartile range for group B is lower than for A, meaning the heights are more consistent (A = 0.55 m, B = 0.4 m).

b The median distance of both groups is 13 miles meaning on average the distance is the same in both groups.
The range of A is lower than the range of B, meaning the distances are more consistent (A = 12 miles, B = 15 miles).
The interquartile range for A is lower than for B, meaning the distances are more consistent (A = 4 miles, B = 5 miles).

c The median mass of group A is higher than B (12 kg compared to 11 kg).
The range of group A is higher than the range of B, meaning the masses are less consistent (A = 15 kg, B = 14 kg).
The interquartile range for group A is smaller than for B, meaning that the masses are more consistent when the outliers are ignored (A = 6 kg, B = 7 kg).

d The median reaction times are the same for both groups (both 0.5 seconds).
The range of reaction times for B is lower than for A (A = 0.75 seconds, B = 0.65 seconds). This means that the reaction times are more consistent for group B.
The interquartile range for group B is lower than for A (A = 0.3 seconds, B = 0.25 seconds). This means that the reaction times are more consistent when the outliers are ignored.

5 a i

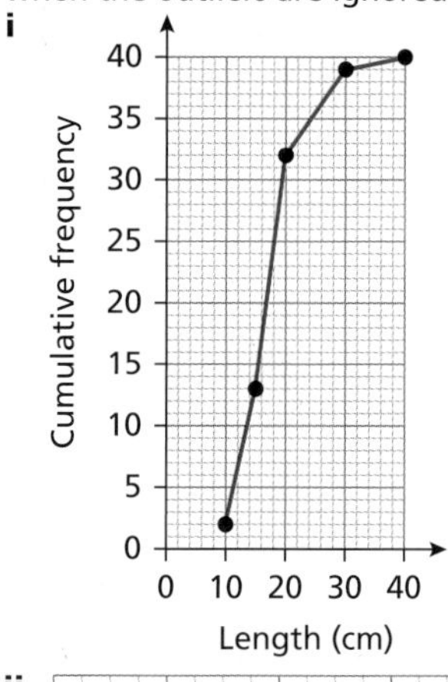

ii

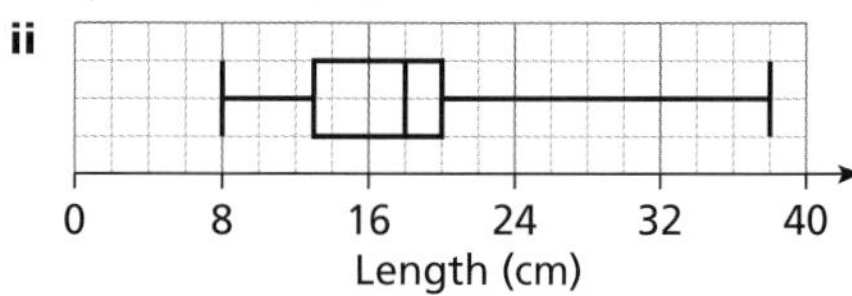

b i

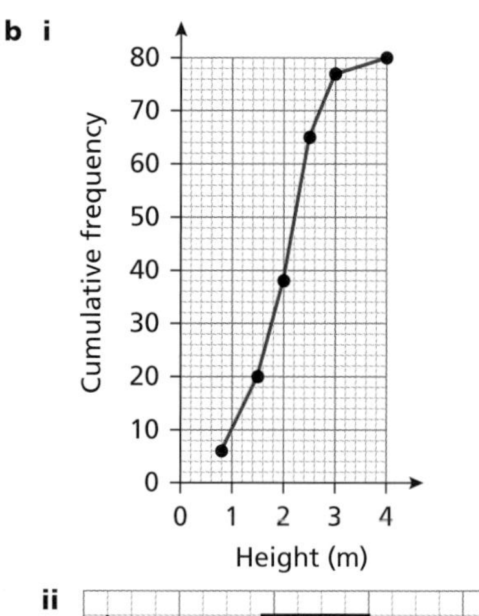

ii

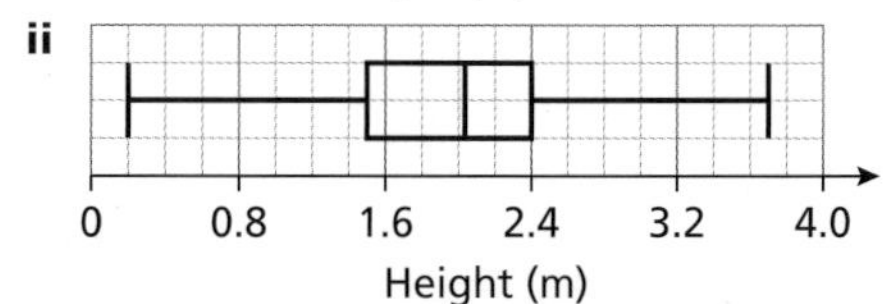

c i

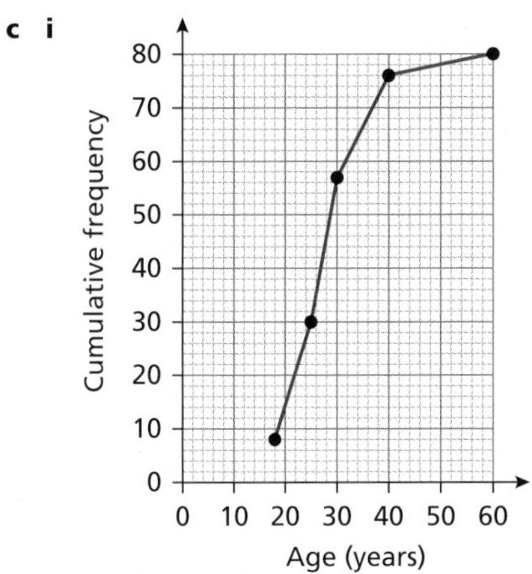

ii

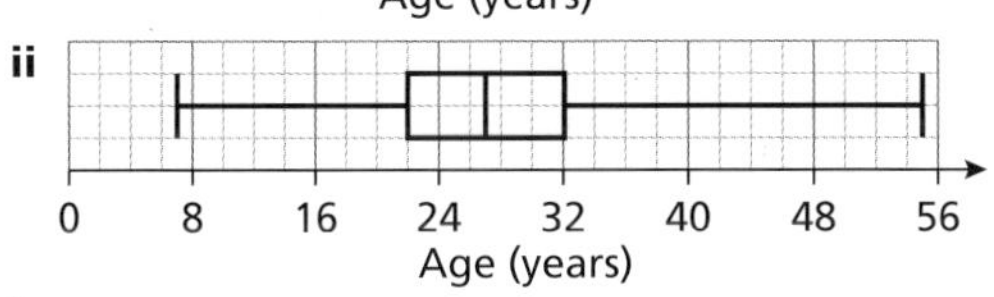

d i

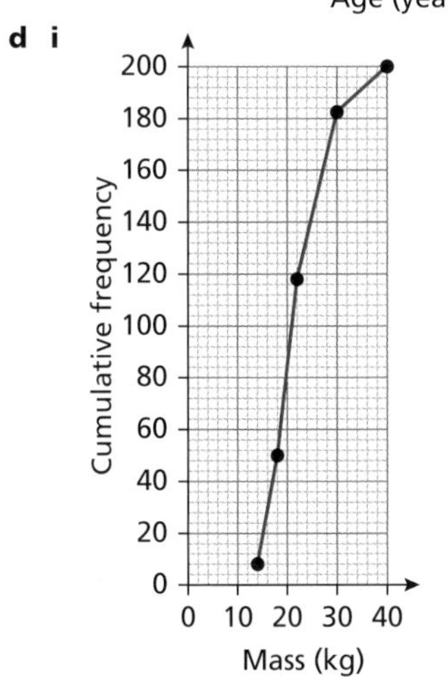

ii

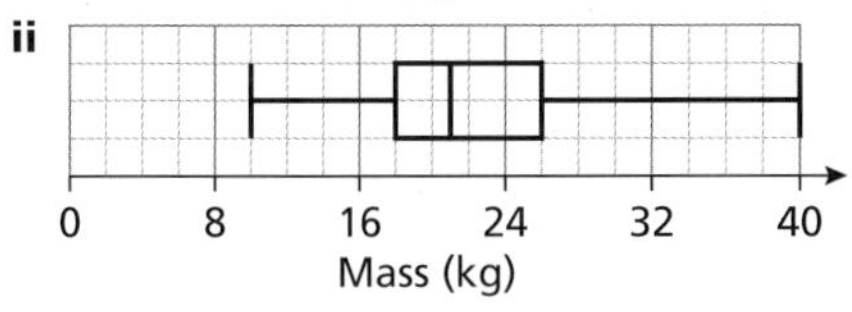

Stretch

1 a 200 **b** 300

2

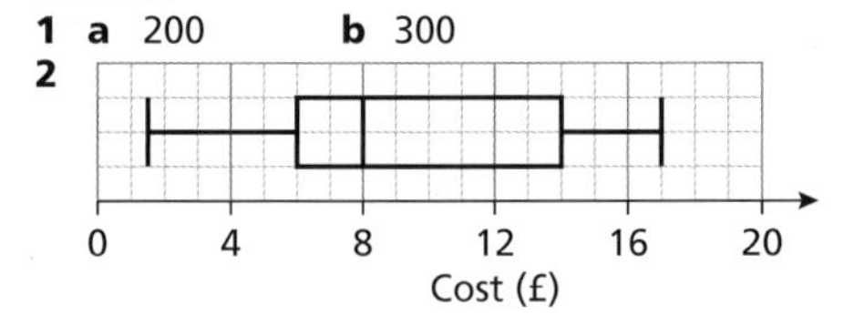

Chapter 14.5

Are you ready?

1 a

	Year 10	Year 11	Total
French	**70**	**30**	100
Spanish	65	**85**	**150**
Total	135	**115**	250

b $\frac{65}{250}$

Practice

1 a

	Singers	Dancers	Actors	Total
Male	0	20	7	27
Female	13	5	5	23
Total	13	25	12	50

b 25 dancers

2

	French	German	Spanish	Total
Boys	18	18	28	64
Girls	31	45	10	86
Total	49	63	38	150

45 girls study German.

3

	Win	Draw	Lose	Total
Home	7	9	3	19
Away	12	6	1	19
Total	19	15	4	38

The team lost 3 games at home.

4

	Tea	Coffee	Total
Male	20	27	47
Female	28	25	53
Total	48	52	100

28 females preferred tea.

5

	Fiction	Non-fiction	Total
Paperback	200	300	500
Hardback	70	55	125
Total	270	355	625

There are 200 paperback fiction books.

6

	On time	Late	Total
Accepted	450	25	475
Rejected	30	95	125
Total	480	120	600

25 orders were late and accepted.

What do you think?

1

	Blue	Green	Brown	Total
Blonde	15	12	7	34
Brunette	4	6	23	33
Red	6	3	1	10
Total	25	21	31	77

P(blue, brunette) = $\frac{4}{77}$

Consolidate

1

	Football	Rugby	Cricket	Netball	Total
Year 10	18	41	15	14	88
Year 11	34	22	32	24	112
Total	52	63	47	38	200

22 Year 11 students say their favourite sport is rugby.

2

	Maths	English	Science	Total
Saturday	35	20	27	82
Sunday	28	21	19	68
Total	63	41	46	150

46 students attend a Science revision session.

3

	Football	Rugby	Athletics	Total
Teachers	15	16	14	45
Students	89	57	49	195
Parents	31	38	41	110
Total	135	111	104	350

89 students watch football.

4

	French	German	Spanish	Total
Year 10	36	18	45	99
Year 11	54	22	58	134
Total	90	40	103	233

58 Year 11 students study Spanish.

Stretch

1

Cost, c	UK	France
$0p \leqslant c < 50p$	6	12
$50p \leqslant c <$ £1	30	24
£1 $\leqslant c <$ £1.50	24	30
£1.50 $\leqslant c <$ £2	12	12

In 36 shops in the UK, 2 litres of milk cost less than £1

2

	Bowling	Theme park	Cinema	Total
Year 10	42	38	37	117
Year 11	31	57	28	116
Total	73	95	65	233

$x = 29$ $y = 42$
233 students in total.

Charts and diagrams: exam practice

1

	Won	Drew	Lost	Total
Home	7	1	1	9
Away	3	2	4	9
Total	10	3	5	18

2 a

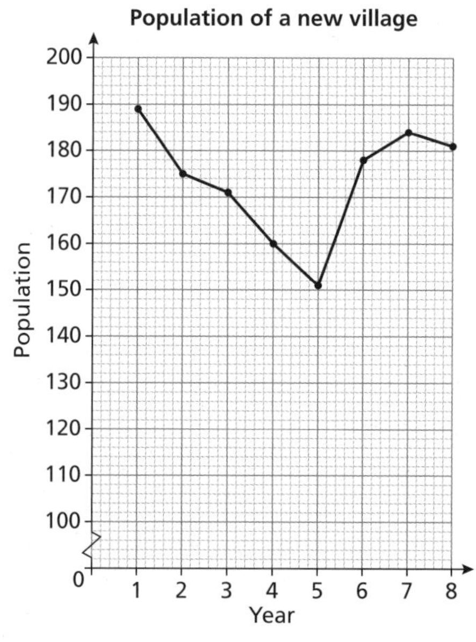

b No. Year 9 is outside the data range.

3 a

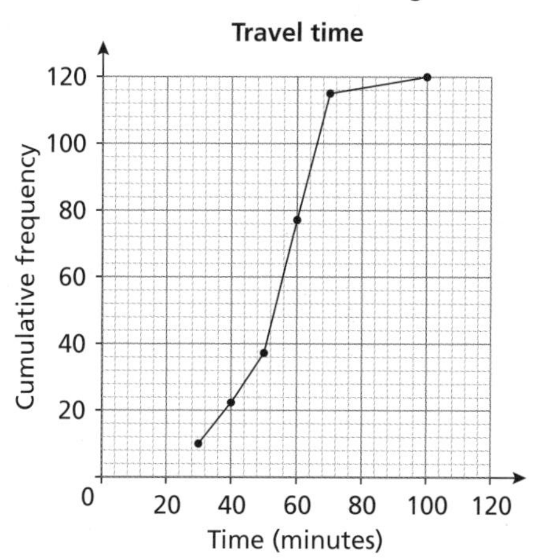

b 56 minutes **c** 90 people

4 a

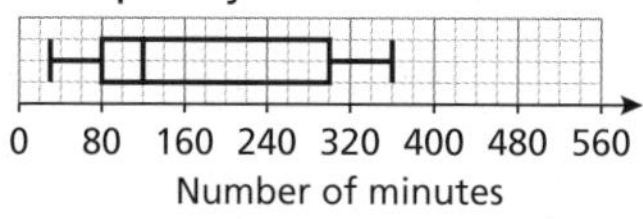

b Example answers:

- The shortest amount of time spent on social media was by one of the teachers, whereas the longest time spent on social media was by one of the teenagers.
- On average, the teenagers spent longer than the teachers on social media.
- The teenagers varied more in the time that they spent on social media when compared to the teachers (as shown by the larger overall range, and the larger interquartile range).
- The teachers had a smaller interquartile range than the teenagers, showing that they were less varied in the time that they spent on social media.

5 a

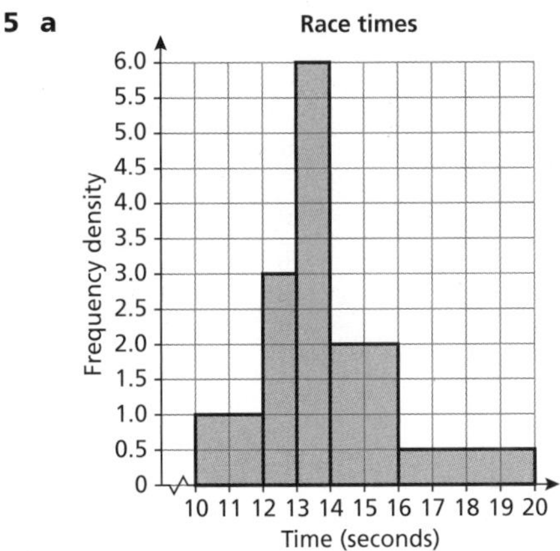

b

Time, x (seconds)	Frequency
$10 < x \leqslant 12$	2
$12 < x \leqslant 13$	3
$13 < x \leqslant 14$	6
$14 < x \leqslant 16$	**4**
$16 < x \leqslant 20$	**2**

Block 15 Applying statistics

Chapter 15.1

Are you ready?

1 A (3, 4) B (0, 3) C (9, 8) D (7, 0)

2

y 10 9 8 7 6 5 4 3 2 1 0 ×B ×D ×A ×C 0 1 2 3 4 5 6 7 8 9 10 x

Practice

1 Graph A: No correlation
Graph B: Positive correlation
Graph C: Negative correlation

2 a

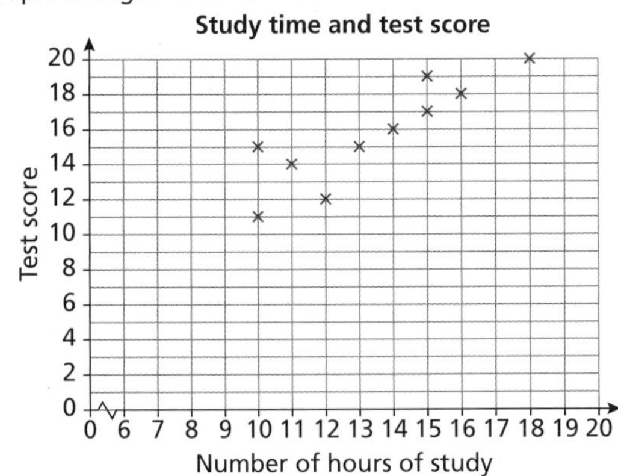

b There is a positive correlation because as the number of hours of study increases, the test score increases.

3 a £45 **b** £16 **c** 15°C

d There is a negative correlation because as the temperature increases, the gas cost per week decreases.

4 a 62 kg **b** 75 kg

c 143 cm and 150 cm

d

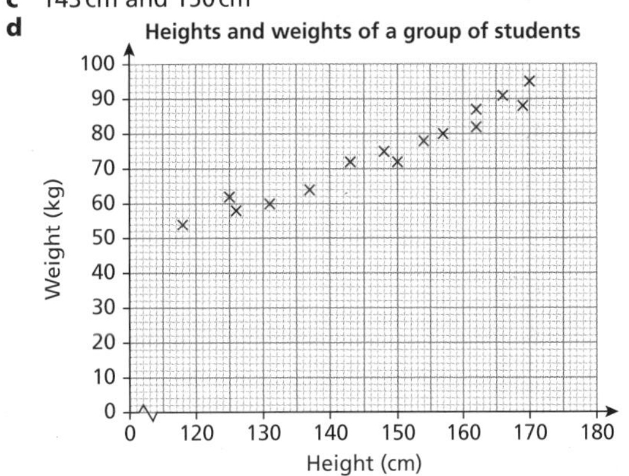

e Positive correlation

5 a i

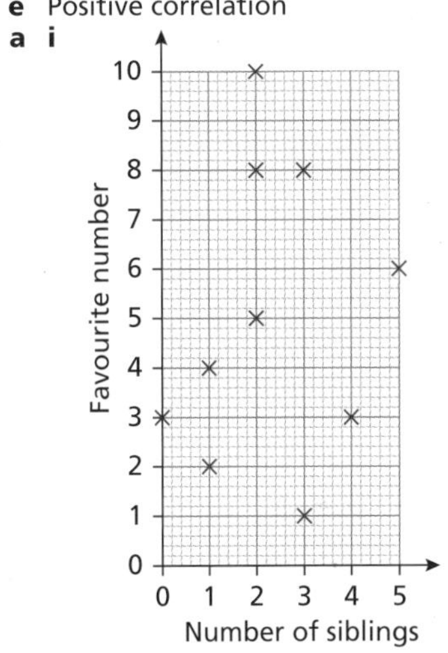

ii

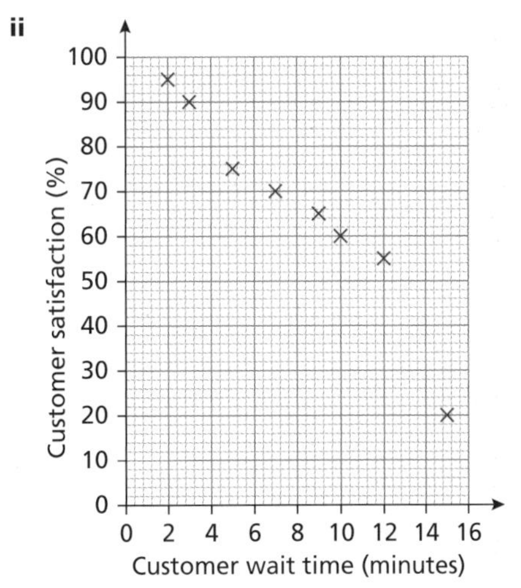

iii

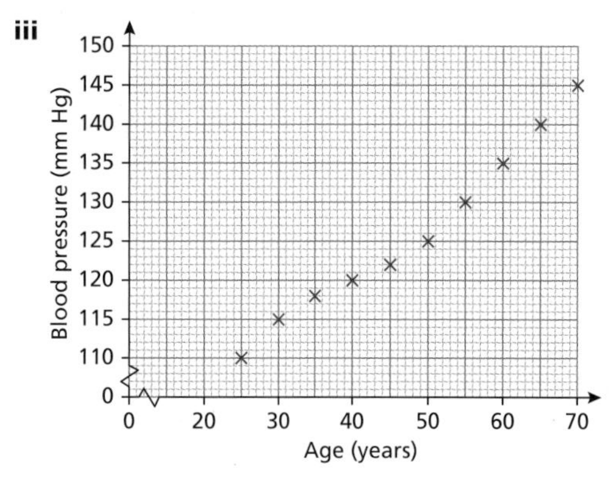

b **i** There is no correlation because the number of siblings has no relationship with a person's favourite number.

ii There is a negative correlation because as the customer wait time increases, customer satisfaction decreases.

iii There is a positive correlation because as age increases, blood pressure also tends to increase.

Consolidate

1 **a** 4 hours **b** 55 years

c There is a negative correlation because as the age increases, the number of hours of exercise each week decreases.

2 **a** 74% **b** 6 miles

c There is no correlation because the distance a student lives from school has no connection with their test result.

3 **a** **i**

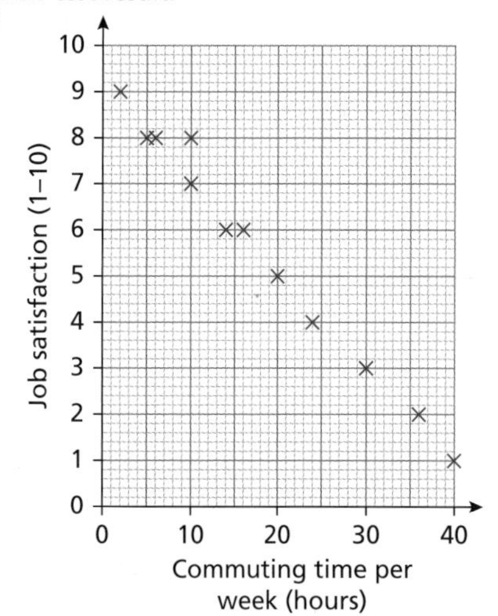

ii

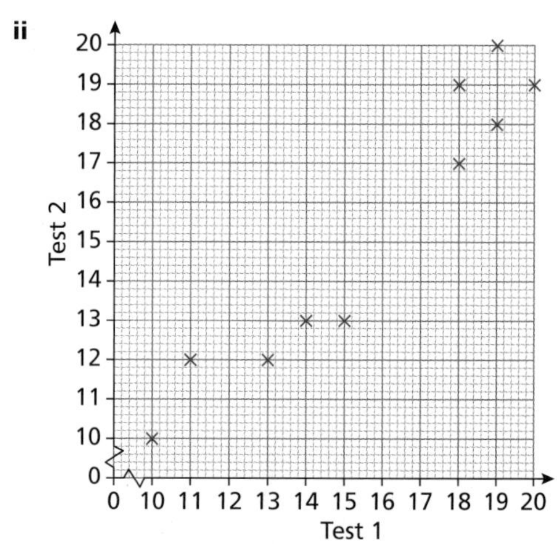

b **i** There is a negative correlation because as the commute increases, the job satisfaction decreases.

ii There is a positive correlation because the higher the score in test 1, the higher the score in test 2.

Stretch

1 **a** **i** 2, 3 and 6 **ii** 1 and 5 **iii** 4

b 2 **c** 1

d Various possible answers.

2 **a**

Car engines

Distance (miles)

Engine size (litres)

b 2.8 litres and 16.1 miles, as these values do not fit with the rest of the data.

c Various possible answers between 10 and 12 miles. Reason: fits with the trend of the data.

Chapter 15.2

Are you ready?

1

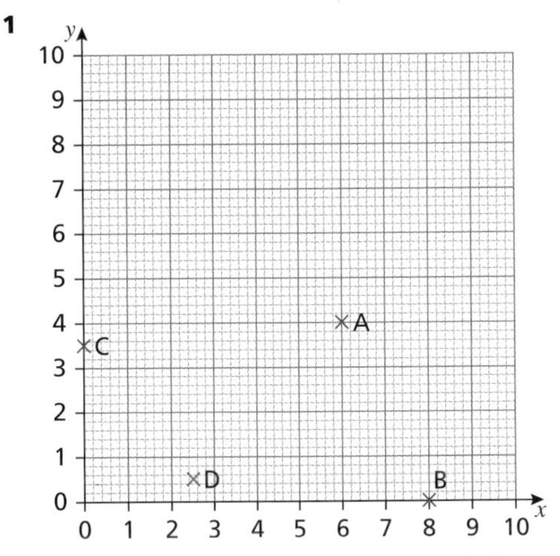

2 **a** Negative correlation **b** No correlation
c Negative correlation **d** Positive correlation

Practice

1 Sven's line of best fit is drawn correctly, as his line shows the general trend of the data. Beca's line of best fit does not go with the trend of the data and does not need to start at origin. Flo's line of best fit isn't a straight line, it is a collection of lines joining the points like a dot-to-dot picture.

2 **a** The graph shows a positive correlation because as the temperature increases, the number of ice creams sold increases.

b **i** Ice creams sold

Number of ice creams sold

Maximum temperature (°C)

ii 29°C (estimate can be between 28°C and 30°C)

3 **a** Attendance at an event

Number of visitors

Rainfall (mm)

b Any suitable line of best fit drawn on the graph (example shown above)
142 visitors (estimate can be between 140 and 144)

c It would not be appropriate as 35 mm is outside of the data that has been provided.

4 **a** Point plotted at (6, 8000)
b Negative correlation
c £10 500 (estimate can be between £10 300 and £10 700)
d No. The extended graph would suggest that the value of a 50-year-old car would be a negative amount.

Consolidate

1 **a** Points plotted at (55, 1.5) and (65, 1.1)
b Yes, as it follows the trend of the data.
c 43 years old (estimate can be between 41 and 45 years)

2 **a**

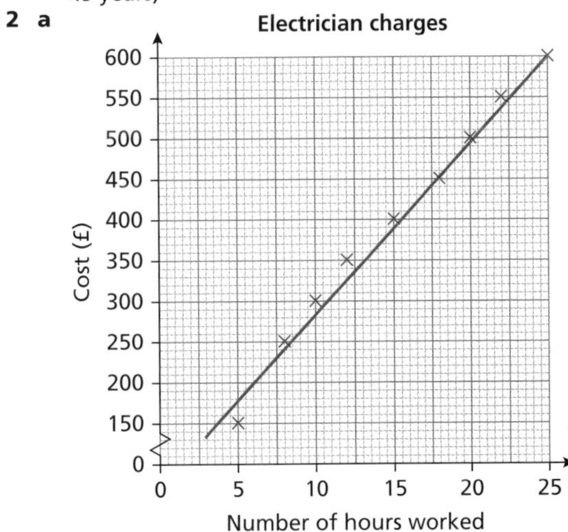

b Any suitable line of best fit drawn on the graph (example shown above)
£200 (estimate can be between £195 and £205)
c 30 hours is outside the data range and you can't assume that the trend will continue beyond the given data.

Stretch

1 There is a correlation between the variables but this does not mean that one directly affects the other. The correlation is likely to be caused by both variables being affected by the temperature.

2 **a** 350 (estimate can be between 345 and 355)
b 35°C is outside the data range and you can't assume that the trend will continue beyond the given data.
c 159 **d** 180

3 Compare answers with a partner.

Chapter 15.3

Are you ready? (A)

1 450
2 48%
3 **a** 5 **b** 35

Practice (A)

1 **a** Students in Ed's school
b 80 students

2 Example answer: The sample size is too small. It only takes the views of people who attend the club on a Monday morning.

3 If someone is not at the theatre, then they can't be included in the sample and the sample is not representative of the whole population.

4 38%

5 300
6 1200
7 13

Are you ready? (B)

1 a 10 b 60 c 4 d 1750

Practice (B)

1 140
2 640

Consolidate

1 a The students who attend his school
 b Example answer: His friends may not represent the school as a whole. The sample size is too small.
 c Example answer: He could take a random sample of students in his school. He could use a larger sample.
2 a False. A sample is a part of the population, not the entire population.
 b True. A larger sample size provides more accurate estimates of the population.
 c True. Random sampling helps ensure that the sample is representative of the entire population and each member of the population has an equal opportunity to be included in the sample.
 d False. A biased sample is one that does not represent the population fairly.
3 The number obtained when throwing a dice is random, therefore the choice of person with the corresponding number is also random.
4 1300
5 720 dark chocolate, 1740 milk chocolate, 540 white chocolate
6 54

Stretch

1 400
2 96 ready salted, 60 salt and vinegar, 36 prawn cocktail, 48 cheese and onion.
3 173 fish

Applying statistics: exam practice

1 a Strong negative correlation
 b Weak negative correlation
 c Strong positive correlation
2 a

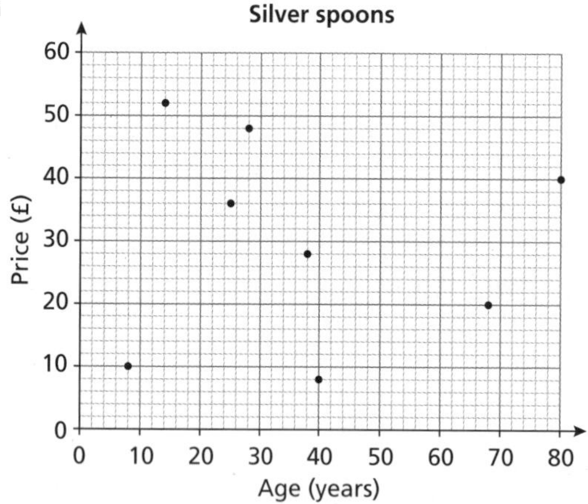

 b It is not appropriate to draw a line of best fit as there is no correlation between the age of the spoon and the price of the spoon.

3 a Example answer: The laptops could be numbered from 1 to 1500. A random number generator could be used to generate 200 numbers between 1 and 1500 (ignoring any repeats). The numbers generated would indicate which laptops should be selected for the random sample.
 b 60
4 No, even if there appears to be a correlation between purchases of smartphones and purchases of chewing gum, it doesn't mean that greater expenditure on smartphones causes less expenditure on chewing gum.
5 50 nurses, 20 doctors, 30 other members of staff

Statistics: exam practice

1 1 and 15
2 8
3 A statement that compares the median values, e.g. on average, Company A sells more cars than Company B as their median sales are 5 cars per week compared to Company B's median sales of 3 cars per week.
A statement that compares the ranges, e.g. Company A's range value of 6 is smaller than Company B's range value of 12, showing that they are more consistent when selling cars. They regularly sell around 5 cars per week whereas Company B can sometimes sell many cars per week, or none at all.
4

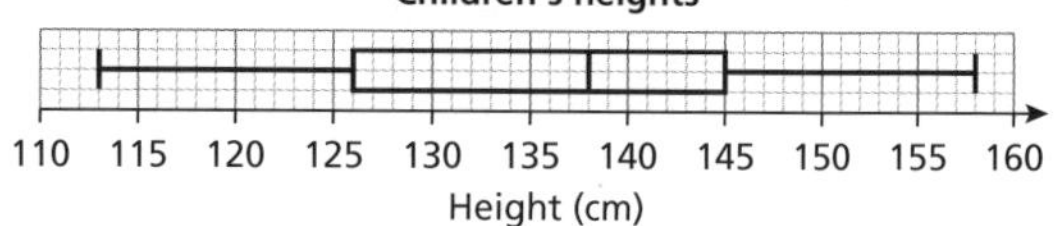

5 375
6

7 14

This page has deliberately been left blank

This page has deliberately been left blank